T0189180

Lecture Notes in Artificial Intelligence 10842

Subseries of Lecture Notes in Computer Science

More information about this series at http://www.springer.com/series/1244

Leszek Rutkowski · Rafał Scherer
Marcin Korytkowski · Witold Pedrycz
Ryszard Tadeusiewicz · Jacek M. Zurada (Eds.)

Artificial Intelligence and Soft Computing

17th International Conference, ICAISC 2018
Zakopane, Poland, June 3–7, 2018
Proceedings, Part II

 Springer

Editors
Leszek Rutkowski
Częstochowa University of Technology
Częstochowa
Poland

and

University of Social Sciences
Lodz
Poland

Rafał Scherer
Częstochowa University of Technology
Częstochowa
Poland

Marcin Korytkowski
Częstochowa University of Technology
Częstochowa
Poland

Witold Pedrycz
University of Alberta
Edmonton, AB
Canada

Ryszard Tadeusiewicz
AGH University of Science and Technology
Kraków
Poland

Jacek M. Zurada
University of Louisville
Louisville, KY
USA

ISSN 0302-9743   ISSN 1611-3349 (electronic)
Lecture Notes in Artificial Intelligence
ISBN 978-3-319-91261-5   ISBN 978-3-319-91262-2 (eBook)
https://doi.org/10.1007/978-3-319-91262-2

Library of Congress Control Number: 2018942345

LNCS Sublibrary: SL7 – Artificial Intelligence

Printed on acid-free paper

This Springer imprint is published by the registered company Springer Nature Switzerland AG
The registered company address is: Gewerbestrasse 11, 6330 Cham, Switzerland

Preface

This volume constitutes the proceedings of 17th International Conference on Artificial Intelligence and Soft Computing ICAISC 2018, held in Zakopane, Poland, during June 3–7, 2018. The conference was organized by the Polish Neural Network Society in cooperation with the University of Social Sciences in Łódź, the Institute of Computational Intelligence at the Częstochowa University of Technology, and the IEEE Computational Intelligence Society, Poland Chapter. Previous conferences took place in Kule (1994), Szczyrk (1996), Kule (1997) and Zakopane (1999, 2000, 2002, 2004, 2006, 2008, 2010, 2012, 2013, 2014, 2015, 2016, and 2017) and attracted a large number of papers and internationally recognized speakers: Lotfi A. Zadeh, Hojjat Adeli, Rafal Angryk, Igor Aizenberg, Cesare Alippi, Shun-ichi Amari, Daniel Amit, Albert Bifet, Piero P. Bonissone, Jim Bezdek, Zdzisław Bubnicki, Andrzej Cichocki, Swagatam Das, Ewa Dudek-Dyduch, Włodzisław Duch, Pablo A. Estévez, João Gama, Erol Gelenbe, Jerzy Grzymala-Busse, Martin Hagan, Yoichi Hayashi, Akira Hirose, Kaoru Hirota, Adrian Horzyk, Eyke Hüllermeier, Hisao Ishibuchi, Er Meng Joo, Janusz Kacprzyk, Jim Keller, Laszlo T. Koczy, Tomasz Kopacz, Zdzislaw Kowalczuk, Adam Krzyzak, Rudolf Kruse, James Tin-Yau Kwok, Soo-Young Lee, Derong Liu, Robert Marks, Evangelia Micheli-Tzanakou, Kaisa Miettinen, Krystian Mikołajczyk, Henning Müller, Ngoc Thanh Nguyen, Andrzej Obuchowicz, Erkki Oja, Witold Pedrycz, Marios M. Polycarpou, José C. Príncipe, Jagath C. Rajapakse, Šarunas Raudys, Enrique Ruspini, Jörg Siekmann, Roman Słowiński, Igor Spiridonov, Boris Stilman, Ponnuthurai Nagaratnam Suganthan, Ryszard Tadeusiewicz, Ah-Hwee Tan, Shiro Usui, Thomas Villmann, Fei-Yue Wang, Jun Wang, Bogdan M. Wilamowski, Ronald Y. Yager, Xin Yao, Syozo Yasui, Gary Yen, Ivan Zelinka, and Jacek Zurada. The aim of this conference is to build a bridge between traditional artificial intelligence techniques and so-called soft computing techniques. It was pointed out by Lotfi A. Zadeh that "soft computing (SC) is a coalition of methodologies which are oriented toward the conception and design of information/intelligent systems. The principal members of the coalition are: fuzzy logic (FL), neurocomputing (NC), evolutionary computing (EC), probabilistic computing (PC), chaotic computing (CC), and machine learning (ML). The constituent methodologies of SC are, for the most part, complementary and synergistic rather than competitive." These proceedings present both traditional artificial intelligence methods and soft computing techniques. Our goal is to bring together scientists representing both areas of research. This volume is divided into five parts:

- Computer Vision, Image and Speech Analysis
- Bioinformatics, Biometrics and Medical Applications
- Data Mining
- Artificial Intelligence in Modeling, Simulation and Control
- Various Problems of Artificial Intelligence

The conference attracted a total of 242 submissions from 48 countries and after the review process, 140 papers were accepted for publication.

I would like to thank our participants, invited speakers, and reviewers of the papers for their scientific and personal contribution to the conference. The Program Committee and additional reviewers were very helpful in reviewing the papers.

Finally, I thank my co-workers Łukasz Bartczuk, Piotr Dziwiński, Marcin Gabryel, Marcin Korytkowski and the conference secretary, Rafał Scherer, for their enormous efforts to make the conference a very successful event. Moreover, I appreciate the work of Marcin Korytkowski, who was responsible for the Internet submission system.

June 2018   Leszek Rutkowski

Organization

ICAISC 2018 was organized by the Polish Neural Network Society in cooperation with the University of Social Sciences in Łódź and the Institute of Computational Intelligence at Częstochowa University of Technology.

ICAISC Chairs

Honorary Chairmen

Hojjat Adeli	Ohio State University, USA
Witold Pedrycz	University of Alberta, Edmonton, Canada
Jacek Żurada	University of Louisville, USA

General Chairman

Leszek Rutkowski — Częstochowa University of Technology, Poland and University of Social Sciences, Łodz, Poland

Co-chairmen

Wlodzislaw Duch	Nicolaus Copernicus University, Torun, Poland
Janusz Kacprzyk	Systems Research Institute, Polish Academy of Sciences, Poland
Józef Korbicz	University of Zielona Góra, Poland
Ryszard Tadeusiewicz	AGH University of Science and Technology, Poland

ICAISC Program Committee

Rafał Adamczak, Poland
Cesare Alippi, Italy
Shun-ichi Amari, Japan
Rafal A. Angryk, USA
Jarosław Arabas, Poland
Robert Babuska, The Netherlands
Ildar Z. Batyrshin, Russia
James C. Bezdek, Australia
Marco Block-Berlitz, Germany
Leon Bobrowski, Poland
Piero P. Bonissone, USA
Bernadette Bouchon-Meunier, France
Tadeusz Burczynski, Poland
Andrzej Cader, Poland
Juan Luis Castro, Spain

Yen-Wei Chen, Japan
Wojciech Cholewa, Poland
Kazimierz Choroś, Poland
Fahmida N. Chowdhury, USA
Andrzej Cichocki, Japan
Paweł Cichosz, Poland
Krzysztof Cios, USA
Ian Cloete, Germany
Oscar Cordón, Spain
Bernard De Baets, Belgium
Nabil Derbel, Tunisia
Ewa Dudek-Dyduch, Poland
Ludmiła Dymowa, Poland
Andrzej Dzieliński, Poland
David Elizondo, UK

Olga Rebrova, Russia
Vladimir Red'ko, Russia
Raúl Rojas, Germany
Imre J. Rudas, Hungary
Enrique H. Ruspini, USA
Khalid Saeed, Poland
Dominik Sankowski, Poland
Norihide Sano, Japan
Robert Schaefer, Poland
Rudy Setiono, Singapore
Paweł Sewastianow, Poland
Jennie Si, USA
Peter Sincak, Slovakia
Andrzej Skowron, Poland
Ewa Skubalska-Rafajłowicz, Poland
Roman Słowiński, Poland
Tomasz G. Smolinski, USA
Czesław Smutnicki, Poland
Pilar Sobrevilla, Spain
Janusz Starzyk, USA
Jerzy Stefanowski, Poland
Vitomir Štruc, Slovenia
Pawel Strumillo, Poland
Ron Sun, USA
Johan Suykens, Belgium
Piotr Szczepaniak, Poland
Eulalia J. Szmidt, Poland

Przemysław Śliwiński, Poland
Adam Słowik, Poland
Jerzy Świątek, Poland
Hideyuki Takagi, Japan
Yury Tiumentsev, Russia
Vicenç Torra, Spain
Burhan Turksen, Canada
Shiro Usui, Japan
Michael Wagenknecht, Germany
Tomasz Walkowiak, Poland
Deliang Wang, USA
Jun Wang, Hong Kong, SAR China
Lipo Wang, Singapore
Paul Werbos, USA
Slawo Wesolkowski, Canada
Sławomir Wiak, Poland
Bernard Widrow, USA
Kay C. Wiese, Canada
Bogdan M. Wilamowski, USA
Donald C. Wunsch, USA
Maciej Wygralak, Poland
Roman Wyrzykowski, Poland
Ronald R. Yager, USA
Xin-She Yang, UK
Gary Yen, USA
Sławomir Zadrożny, Poland
Ali M. S. Zalzala, United Arab Emirates

ICAISC Organizing Committee

Rafał Scherer, Secretary
Łukasz Bartczuk
Piotr Dziwiński
Marcin Gabryel, Finance Chair
Rafał Grycuk
Marcin Korytkowski, Databases and Internet Submissions
Patryk Najgebauer

Additional Reviewers

J. Arabas
T. Babczyński
M. Baczyński
Ł. Bartczuk
P. Boguś
B. Boskovic
J. Botzheim
J. Brest
T. Burczyński
R. Burduk
L. Chmielewski
W. Cholewa
K. Choros
P. Cichosz
P. Ciskowski
B. Cyganek
J. Cytowski
I. Czarnowski
K. Dembczynski
J. Dembski
N. Derbel
L. Diosan
G. Dobrowolski
A. Dockhorn
A. Dzieliński
P. Dziwiński
B. Filipic
M. Gabryel
E. Gelenbe
M. Giergiel
P. Głomb
F. Gomide
Z. Gomółka
M. Gorzałczany
D. Grabowski
M. Grzenda
J. Grzymala-Busse
L. Guo
H. Haberdar
C. Han
Y. Hayashi
T. Hendtlass
Z. Hendzel

F. Hermann
H. Hikawa
K. Hirota
A. Horzyk
E. Hrynkiewicz
J. Ishikawa
D. Jakóbczak
E. Jamro
A. Janczak
W. Kamiński
E. Kerre
J. Kluska
L. Koczy
Z. Kokosinski
A. Kołakowska
J. Konopacki
J. Korbicz
P. Korohoda
J. Koronacki
M. Korytkowski
M. Korzeń
J. Kościelny
L. Kotulski
Z. Kowalczuk
J. Kozlak
M. Kretowska
D. Krol
R. Kruse
B. Kryzhanovsky
A. Kubiak
E. Kucharska
P. Kudová
J. Kulikowski
O. Kurasova
V. Kurkova
M. Kurzyński
J. Kusiak
H. Lenz
Y. Li
A. Ligęza
J. Łęski
B. Macukow
W. Malina

J. Mańdziuk
M. Marques
F. Masulli
A. Materka
R. Matuk Herrera
J. Mazurkiewicz
V. Medvedev
M. Mernik
J. Michalkiewicz
Z. Mikrut
S. Misina
W. Mitkowski
W. Moczulski
F. Mokom
W. Mokrzycki
O. Mosalov
W. Muszyński
H. Nakamoto
G. Nalepa
M. Nashed
S. Nemati
F. Neri
M. Nieniewski
R. Nowicki
A. Obuchowicz
S. Osowski
E. Ozcan
M. Pacholczyk
W. Palacz
G. Paragliola
A. Paszyńska
K. Patan
A. Pieczyński
A. Piegat
Z. Pietrzykowski
P. Prokopowicz
A. Przybył
R. Ptak
E. Rafajłowicz
E. Rakus-Andersson
A. Rataj
Ł. Rauch
L. Rolka

Contents – Part II

Bioinformatics, Biometrics and Medical Applications

Data Mining

Artificial Intelligence in Modeling, Simulation and Control

Various Problems of Artificial Intelligence

Contents – Part I

Evolutionary Algorithms and Their Applications

Pattern Classification

Computer Vision, Image and Speech Analysis

Moving Object Detection and Tracking Based on Three-Frame Difference and Background Subtraction with Laplace Filter

Beibei Cui[✉] and Jean-Charles Créput

Le2i FRE2005, CNRS, Arts et Métiers, Univ. Bourgogne Franche-Comté,
90010 Belfort Cedex, France
beibei.cui@utbm.fr

Abstract. Moving object detection and tracking is an important research field. Currently, ones of the core algorithms used for tracking include frame difference method (FD), background subtraction method (BS), and optical flow method. Here, authors are looking at the first two approaches since very adequate for very fast real-time treatments whereas optical flow has higher computation cost since related to a dense estimation. Combination of FD and BS with filters and edge detectors is a way to achieve sparse detection fast. This paper presents a tracking algorithm based on a new combination of FD and BS, using Canny edge detector and Laplace filter. Laplace filter occupies a leading role to sharpen the outlines and details. Canny edge detector identifies and extracts edge information. Morphology processing is used to eliminate interfering items finally. Experimental results show that 3FDBD-LC method has higher detection accuracy and better noise suppression than current combination methods on sequence images from standard datasets.

Keywords: Frame difference · Background subtraction
Laplace filter · Canny detector

1 Introduction

Object detection and tracking [1,2] from video sequences with a static camera can be used in many applications such as intelligent surveillance, moving target detection, monitoring and vehicle tracking. Currently, ones of the main tracking algorithms can be based on frame difference method, background subtraction method and optical flow method. Optical flow [3] spends more time than other methods so it is the more complex since it computes a dense optical flow field, which makes its application to video rate image processing difficult. Background subtraction method can extract objects completely with a relatively simple algorithm, but it is sensitive to the changes of the light. Frame difference method is

© Springer International Publishing AG, part of Springer Nature 2018
L. Rutkowski et al. (Eds.): ICAISC 2018, LNAI 10842, pp. 3–13, 2018.
https://doi.org/10.1007/978-3-319-91262-2_1

relatively easy to implement, it can adapt to the changing of the environment, but it is sensitive to the noise, so its results are not accurate enough. Although there are problems on frame difference (FD) and background subtraction (BS), these problems are under addressing gradually by some literature on the field.

Many of the state-of-the-art approaches are combination of methods that include filters, edge detection with BS and FD. In 2007, Zhan [4] proposed an object detection algorithm based on two frame difference and Canny edge detection (2FD-C). In 2010, Zhang [5] presented a motion human detection method based on standard background subtraction. In 2012, Zhang [6] demonstrated a three-frame difference (3FD) algorithm research based on mathematical morphology. In 2013, Gang [7] presented an improved moving objects detection algorithm which combined three-frame difference method with Canny edge detection algorithm (3FD-C). After that, some of the object tracking algorithms that combined BS and FD [8,9] were put forward but without edge detector (FDBS). Although there are numerous works on BS and FD, no systematic method today seems to be came up. Considering all of this, authors present a new combined tracking algorithm mainly based on three-frame difference method, background subtraction method, Laplace filter and Canny edge detector (called 3FDBS-LC).

Section 2 presents the basic operations as filters and edge detector most often used in BS and FD methods. Section 3 presents BS and FD and the new combined proposed approach. Section 4 presents experiments, whereas last section concludes.

2   Basic Processing Operations

Basic processing operations such as color conversion, image binarization, filter processing, edge detection and morphology processing, are current basic operations in BS/FD tracking. We detail the standard treatments that are to be combined in the proposed tracking method.

(a)   (b)   (c)

Fig. 1. (a) Original image, (b) Gray scale image and (c) Binary image.

First, color scale images are converted into grayscale images to improve computational efficiency. Grayscale image has gray values ranging from 0 to 255, where 0 represents the color of black, and 255 represents white. A binary image is usually quantized with two possible intensity values, 0 and 1, representing black and white respectively. The purpose of image binarization is to accelerate the logical decision process when merging information. Figure 1 shows an original RGB image and its corresponding grayscale image and binary image.

Among many filters used in image processing, some of the most commonly used in BS/FD tracking are Mean filter, Median filter and Gaussian filter. Mean filter and Median filter are common linear smoothing algorithm for reducing noise but blur the picture. Gaussian filter outputs a weighted average of each pixel's neighborhood, with the average weighted more towards the center pixel's value but contrast to mean filter's uniformly weighted average. Whereas Laplacian is the second-order derivative of the Gaussian equation. It has stronger edge localization capability and better sharpening effect than Gaussian filter. The effect of image sharpening is enhancing the contrast of the grays and making the blurred image clearer. The basic method of Laplacian sharpening can be expressed as:

$$\nabla^2 G_\sigma(x,y) = \frac{\partial^2 G_\sigma(x,y)}{\partial x^2} + \frac{\partial^2 G_\sigma(x,y)}{\partial y^2} = \frac{x^2 + y^2 - 2\sigma^2}{\sigma^4} e^{-\frac{x^2+y^2}{2\sigma^2}}, \qquad (1)$$

where x, y are the pixel coordinates, and σ is the standard deviation of the Gaussian distribution. It is a proposal of this paper to integrate and experiment the Laplace filter in a combined BS/FD tracking method. So we will adopt Laplace filter which can not only produce the sharpening effect, but also preserve the background information.

Edge detection is an image processing technique used to find the boundaries of objects within an image. There are many different kinds of edge detections, common edge detection algorithms include Roberts, Sobel, Prewitt, and Canny. The Robert operator can point out the target precisely, but it is less sensitive to noise because it is not smoothed. Prewitt operator and Sobel operator belong to first-order differential operators, while the former is an average filter, the latter is the weighted average filter. Both of them have better detection effect on grayscale low-noise images, but they do not work very well for mixing images with many complicated noises. It looks commonly admitted that Canny edge detector is more accurate compared to Sobel, Prewitt and Roberts operators, at least for BS/FD tracking methods. From the Fig. 2, it can estimate that the Canny edge detector can discover the edge information of the object more completely. In this work, Canny edge detector was chosen.

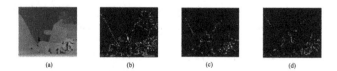

Fig. 2. (a) Original image, and its corresponding processed image: (b) Canny operator, (c) Prewitt operator, (d) Roberts operator.

The function of morphology processing is to eliminate interferences, fill small apertures and smooth boundary. The most fundamental morphological operators are erosion and dilation. In this algorithm, the closing operation was adopted which is the dilation operation followed by erosion operation. The formula is

$$close\,(A,B) = erode\,(dilate\,(A,B)\,,B) = (A \oplus B) \ominus B, \qquad (2)$$

where $\oplus$ is dilation operator, $\ominus$ is erosion operator, A is the image, B is a structural element which is specified as 3×3 matrix. Therefore, after this morphology processing, the apertures should be filled and small clearances should be connected by closing operation.

3 Proposed Approach

We now present the most central part that are frame difference method and background subtraction method and their combination in tracking methods. We present standard approaches and our proposed new combination.

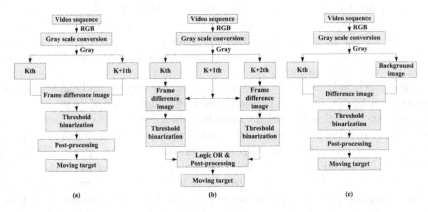

Fig. 3. From the left to the right: (a) Two-frame difference method, (b) Three-frame difference method and (c) Background subtraction method.

Frame difference method is achieved by a series of continuous images. The basic principle of this method is calculating the difference between two adjacent pictures by comparing point-by-point gray value to determine the information of moving objects. The formula of difference between frames can be written as

$$D_k\,(x,y) = |f_k\,(x,y) - f_{k-1}\,(x,y)\,|, \qquad (3)$$

where the current frame image gray value is f_k, the adjacent frame image gray value is f_{k-1}, and D_k is image after difference between f_k and f_{k-1}. We define R_k as the binary translation of the difference image. If $D_k\,(x,y) > T$, $R_k\,(x,y)$ belongs to foreground and is set to 1, on the contrary, it belongs to the background and will be set to 0, where T is a threshold that is fixed empirically. The disadvantage of two frame difference is to produce foreground aperture and ghosting problem. In contrast, the three-frame difference method can weaken this issue better. This is achieved by subtraction operation of current frame image with the previous frame and the latter frame separately and then performing a logic OR operation on the results, just as Zhang [6] did and based on mathematical morphology at post-processing part. The flow chart of standard two-frame difference method and three-frame difference method are shown in Fig. 3(a) and (b).

The principle of background subtraction is to use the difference method to subtract the background image from the current frame. The steps of the BS algorithm can be divided into the following steps: firstly, from the video sequence the current frame image K_{th} was obtained, and through the background modeling method get the background image. Secondly, the current frame image and the background image were used to get frame difference image. Zhang [5] used this BS method but with morphological filtering and contour projection analysis as post-processing. The specific flow chart is shown in Fig. 3(c).

As explained before, the frame difference method and background subtraction method have some disadvantages which are sensitivity to noise or brightness. Moreover, when the image is processed in a complex scene, the edge information of the moving target is easily influenced by the background scene and can not be extracted completely. Then, it is especially important to propose a way to avoid the effects of interferences and to extract the edge information of moving objects at the same time. Here, Canny edge detector will be inserted into three-frame-difference and background subtraction method respectively that will be combined together.

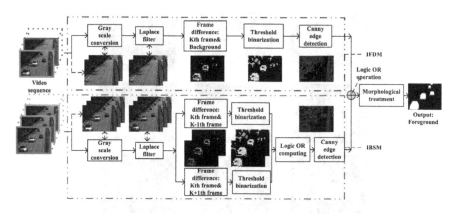

Fig. 4. Framework of improved 3FDBS-LC method.

An outline of the whole steps and treatments of this proposed method is summarized in Fig. 4. As we can see, firstly, after color images converted into grayscale image, Laplace filter which occupies a leading role will sharpen the outlines and details of target on grayscale. Secondly, three-frame difference and background subtraction operation will be performed respectively, both accompanied by threshold binarization and Canny edge detection to identify and extract edge information. Lastly, a combination of these two main methods goes through a logical OR operation followed by a morphological operation for getting the final moving object detection.

4   Experiments

In this section, three standard benchmarks, SABS dataset [10], Wallflower dataset [11] and Multivision dataset [12], are used. They are used for visual evaluation, whereas the second and third sets for numerical evaluation and comparison with other standard methods.

4.1   Visual Presentation of Results

The sequence SABS-Bootstrap with 352×288 images, encoded frame-by-frame as PNG is first used to illustrate the results at different steps of proposed 3FDBS-LC method. The visual presentation of results are shown in Fig. 5. From the images in the figure, we can appreciate quality from standard BS method, FD method and their combination with or without Laplace filter as shown in (d)–(h) in comparison with the proposed method in (i). We think it more clearly finds the moving objects: the running cars, walking pedestrians and the swinging tree blown by the wind.

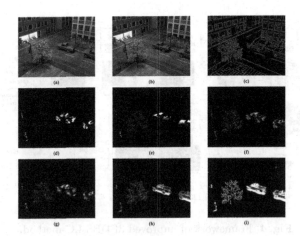

Fig. 5. From the left to the right: (a) Original color scale image, (b) Gray scale image, (c) Image processed by the Canny edge detector, (d) Image extracted by standard three-frame difference, (e) Image extracted by standard background difference, (f) The logic OR operation between (d) and (e), (g) The improved three-frame difference method after Laplace filter, (h) The improved background subtraction method after Laplace filter and (i) The improved 3FDBS-LC method.

The second set of experiments, with comparison evaluation, is based on the Wallflower and Multivision datasets. We adopt ten image sequences: Camouflage, Foreground Aperture, Ground Truth Sequences, Chair Box, Hallway, Lab Door, LCD Screen, Wall, Crossing and Suitcase. Comparative visual results are shown in Fig. 6. The first row is background images, the second row displays

Fig. 6. From the left to the right: (a) Background image, (b) Frame image, (c) Ground truth image, (d) Background subtration method, (e) Frame difference method and (f) The proposed 3FDBS-LC method.

only one sample frame per sequences, the third row shows ground truth images, the fourth and fifth rows demonstrate the detected foreground under standard background subtraction and frame difference method respectively, whereas, the last row presents the proposed 3FDBS-LC method.

4.2   Evaluation Criteria

Here, authors define the evaluation criteria to evaluate and compare the accuracy of the different tracking methods based on ground truth evaluation. In information retrieval and pattern recognition, precision is a measure of result relevancy, while recall is a measure of how many truly relevant results are returned. We use the metrics of precision, recall, and also accuracy and F-measure.

More precisely, accuracy is defined as the number of true positives (TP) plus the number of true negatives (TN) over all of the samples. Recall is defined as the number of true positives (TP) over the number of true positives plus the number of false negatives (FN). Precision is defined as the number of true positives (TP) over the number of true positives plus the number of false positives (FP). F-measure is defined as the harmonic mean of Precision and Recall. High scores for F-measure show that the classifier is returning accurate results (high precision), as well as returning a majority of positive results (high recall). The evaluation criteria are formally defined as follows:

$$Accuracy = (TP + TN) \div (TP + FP + TN + FN) \tag{4}$$

$$Recall = TP \div (TP + FN) \tag{5}$$

$$Precision = TP \div (TP + FP) \tag{6}$$

$$F - measure = \frac{2Recall * Precision}{(Recall + Precision)} \tag{7}$$

where TP is the number of foreground pixels that are correctly defined as foreground, TN is the number of background pixels that are correctly defined as background, FP is the number of background pixels that are mistakenly defined as foreground and FN is the number of foreground pixels that are mistakenly defined as background.

4.3   Comparison Evaluation

Experimental results are displayed in Table 1. It shows a comparative evaluation on accuracy, precision, recall and F-measure for three different tracking methods, include proposed method and two standard methods, under ten image sequences. Figure 7 shows the corresponding histograms. From these results, it can be found that this proposed 3FDBS-LC algorithm can achieve good detection results that outperform standard BS and FD methods.

Table 1. Different types of metrics of three different kinds of methods.

Datasets	Accuracy			Precision		
	Our proposed method	Traditional background subtraction	Traditional frame difference	Our proposed method	Traditional background subtraction	Traditional frame difference
Camouflage	0.9153	0.8279	0.4522	0.9090	0.9693	0.5882
F-A	0.9319	0.9022	0.7525	0.8230	0.9087	0.5746
GT-S	0.8889	0.9221	0.9144	0.5207	0.9681	0.8702
Chair Box	0.9244	0.9145	0.8534	0.8980	0.9976	0.6264
Hallway	0.9119	0.8625	0.8022	0.8552	0.9952	0.6764
Lab Door	0.9547	0.9469	0.8996	0.8369	0.8936	0.5137
LCD Screen	0.9601	0.9415	0.9146	0.8484	0.9403	0.6844
Wall	0.9625	0.9541	0.9405	0.6904	0.7759	0.4795
Crossing	0.9579	0.8417	0.8311	0.8399	0.5977	0.4760
Suitcase	0.9822	0.8997	0.9318	0.9438	0.2978	0.6573
Datasets	Recall			F-measure		
Camouflage	0.9380	0.7130	0.0115	0.9232	0.8216	0.0226
F-A	0.9405	0.7007	0.1871	0.8778	0.7913	0.2823
GT-S	0.6922	0.3494	0.3195	0.5943	0.5135	0.4674
Chair Box	0.6363	0.5215	0.3736	0.7449	0.6850	0.4680
Hallway	0.6842	0.3412	0.0767	0.7602	0.5082	0.1379
Lab Door	0.6858	0.5412	0.1307	0.7539	0.6741	0.2084
LCD Screen	0.6864	0.3928	0.1417	0.7589	0.5591	0.2348
Wall	0.6313	0.3032	0.0908	0.6595	0.4360	0.1527
Crossing	0.9104	0.1773	0.0251	0.8737	0.2735	0.0477
Suitcase	0.7897	0.3194	0.0657	0.8599	0.3082	0.1195

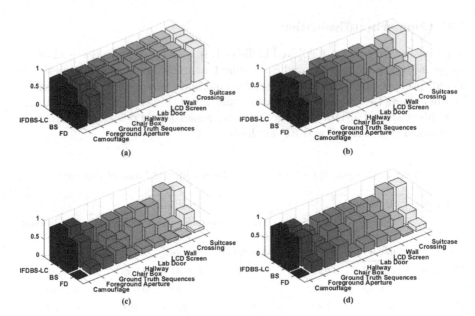

Fig. 7. The comparison histograms of three different kinds of methods in (a) Accuracy, (b) Precision, (c) Recall and (d) F-measure.

5   Conclusion

In this paper, an improved object tracking algorithm is proposed which mainly uses combination of Laplace filter, frame difference method, background subtraction method and Canny edge detector. We adopt the Laplace filter to strengthen image information and improve the detection effect. Canny detector is highly correlated with the edge contour, and as usual, it is more helpful for dividing the pixels into foreground and background. The proposed algorithm was tested on standard datasets with the following evaluation criteria: accuracy, recall, precision and F-measure. The experiments show that the proposed method has competitive performances with state-of-the-art methods to get the desired result. In our future research, authors will focus on the parallelization strategies to realize this optimization algorithm in multi-core and Graphics Processing Unit systems in combination with the use of more elaborated tracking structures.

References

1. Rout, R.K.: A Survey on Object Detection and Tracking Algorithms (2013)
2. Prasad, P., Gupta, A.: Moving object tracking and detection based on kalman filter and saliency mapping. In: Satapathy, S.C., Bhateja, V., Raju, K.S., Janakiramaiah, B. (eds.) Data Engineering and Intelligent Computing. AISC, vol. 542, pp. 639–646. Springer, Singapore (2018). https://doi.org/10.1007/978-981-10-3223-3_61

3. Guo, Z., Wu, F., Chen, H., Yuan, J., Cai, C.: Pedestrian violence detection based on optical flow energy characteristics. In: 4th International Conference on In Systems and Informatics (ICSAI), pp. 1261–1265. IEEE (2017)
4. Zhan, C., Duan, X., Xu, S., Song, Z., Luo, M.: An improved moving object detection algorithm based on frame difference and edge detection. In: Fourth International Conference on Image and Graphics (ICIG), pp. 519–523. IEEE (2007)
5. Zhang, L., Liang, Y.: Motion human detection based on background subtraction. In: 2010 Second International Workshop on Education Technology and Computer Science (ETCS), vol. 1, pp. 284–287. IEEE (2010)
6. Zhang, Y., Wang, X., Qu, B.: Three-frame difference algorithm research based on mathematical morphology. Procedia Eng. **29**, 2705–2709 (2012)
7. Gang, L., Shangkun, N., Yugan, Y., Guanglei, W., Siguo, Z.: An improved moving objects detection algorithm. In: International Conference on Wavelet Analysis and Pattern Recognition (ICWAPR), pp. 96–102. IEEE (2013)
8. Lavanya, M.P.: Real time motion detection using background subtraction method and frame difference. Int. J. Sci. Res. (IJSR) **3**(6), 1857–1861 (2014)
9. Liu, H., Dai, J., Wang, R., Zheng, H., Zheng, B.: Combining background subtraction and three-frame difference to detect moving object from underwater video. In: OCEANS, Shanghai, pp. 1–5. IEEE (2016)
10. Brutzer, S., Höferlin, B., Heidemann, G.: Evaluation of background subtraction techniques for video surveillance. In: 2011 IEEE Conference on Computer Vision and Pattern Recognition (CVPR), pp. 1937–1944. IEEE (2011)
11. Toyama, K., Krumm, J., Brumitt, B., Meyers, B.: Wallflower: principles and practice of background maintenance. In: The Proceedings of the Seventh IEEE International Conference on Computer Vision, vol. 1, pp. 255–261. IEEE (1999)
12. Fernandez-Sanchez, E.J., Rubio, L., Diaz, J., Ros, E.: Background subtraction model based on color and depth cues. Mach. Vis. Appl. **25**(5), 1211–1225 (2014)

Robust Lane Extraction Using Two-Dimension Declivity

Mohamed Fakhfakh[1,2(✉)], Nizar Fakhfakh[2], and Lotfi Chaari[1,3]

[1] University of Sfax, Sfax, Tunisia
mohamed.fakhfakh.research@gmail.com
[2] NAVYA, Paris, France
nizar.fakhfakh@navya.tech
[3] IRIT-ENSEEIHT, University of Toulouse, Toulouse, France
lotfi.chaari@enseeiht.fr

Abstract. A new robust lane marking extraction algorithm for monocular vision is proposed based on Two-Dimension Declivity. It is designed for the urban roads with difficult conditions (shadow, high brightness, etc.). In this paper, we propose a locating system which, from an embedded camera, allows lateral positioning of a vehicle by detecting road markings. The primary contribution of the paper is that it supplies a robust method made up of six steps: *(i)* Image Pre-processing, *(ii)* Enhanced Declivity Operator (DE), *(iii)* Mathematical Morphology, *(iv)* Labeling, *(v)* Hough Transform and *(vi)* Line Segment Clustering. The experimental results have shown the high performance of our algorithm in various road scenes. This validation stage has been done with a sequence of simulated images. Results are very promising: more than 90% of marking lines are extracted for less than 12% of false alarm.

Keywords: Curve lane detection · Declivity operator
Road marking · Clustering · Hough transform

1 Introduction

Since the last decade, autonomous driving has become a reality due to algorithmic and computational advancements. According to the Society of Automotive Engineers, different levels of driving automation are defined. The highest level consists of a "full-time performance by an automated driving system of all aspects of the dynamic driving task under all roadway and environmental conditions that can be managed by a human driver". This high level of automation requires well-designed algorithms with high performances to deal with all of challenging use cases usually encountered in real-world. Algorithms must be real-time, accurate, reliable and robust to achieve a dynamic secure driving.

Such a vehicle is equipped with a set of active and passive sensors to perform functions, inter alia, obstacles detection and tracking [1,2] objects recognition [3], free space detection [4], global and local vehicle guidance [5]. During the

© Springer International Publishing AG, part of Springer Nature 2018
L. Rutkowski et al. (Eds.): ICAISC 2018, LNAI 10842, pp. 14–24, 2018.
https://doi.org/10.1007/978-3-319-91262-2_2

last decade, many researchers have proposed solutions for lateral positioning by embedded cameras.

In order to perform global vehicle positioning, several algorithms were designed by using advanced technologies, such as lidars [6] and differential GPS [7]. These solutions cannot respond to all of contexts and are mainly dependent on the environment. However, a lidar-based perception of loosely structured scenes could not give accurate global localization and can potentially fail in certain cases. In contrast, by using GPS technology, the localization is inaccurate because of the multipath problem in dense and highly structured scenes. In recent years, Visual SLAM algorithms [8] have been proposed as an alternative to improve the localization task and to deal with drawbacks of lidar and GPS technologies. Nevertheless, Visual SLAM-based algorithms remain insufficient and suffer from limits related to the processing time and the difficulty to accurately detect landmarks in challenging environments.

In this context, the present work is dedicated to Road Marking Extraction by mono camera in complex environments. Such an algorithm is usually used to perform path planning, lane departure warning or lane changing functions.

This paper is organized as follows: after an introduction covering the context of this research and the problem to be handled, Sect. 2 is dedicated for a state of the art from which our research is based. In Sect. 3, an overview of the proposed algorithm is presented and all of steps are detailed for road marking extraction in challenging environment by embedded camera for lateral positioning. Evaluation and experimental results are detailed in Sect. 4, and we finish by a discussion about the different axes to go further.

2   State of the Art of Road Marking Approaches

By referring to the literature, one of the first approaches for road marking detection is to apply a simple Hough Transform on the image in order to detect straight lines. This step is usually performed either on original images obtained with the projective model [9] or after applying an inverse perspective transform [10] which reflects the real geometry of lanes, but may introduce a loss of data because of the transformation step.

The most common way to address this task is to apply a thresholding step on the input image to extract pixels having intensities higher than a given threshold which typically correspond to road markings. Clearly, this is only true in the special case of a dark road and bright markings, which is not the case in the presence of partial shadows or high local variation of intensities.

There exist different methods of thresholding. It is commonly applied by using the histogram of gray levels and allows correctly separating the foreground from the background if the histogram clearly contains distinct peaks. Either global and local thresholding procedures can be considered [11,12].

In [13], the Otsu method requires the calculation of a gray level histogram before estimating an overall threshold. The idea of the OTSU algorithm is to divide the image histogram into two classes of pixels, the foreground and the background. The optimal threshold is calculated to separate the two classes so that their intra-class variances are minimal. However, the main limitation is the High brightness or shadows in the image, where the result of the road features extraction is not satisfactory.

The standard Otsu algorithm is affected by the presence of shadows or high brightness. Scholars X. Jia from Jilin University developed the optimized Otsu method based on the maximal variance threshold in order to solve the problems posed by the standard Otsu method [13]. The general idea of "optimized Otsu" consists of cutting the region of the road into several rectangular regions and treating each sub-region respectively according to the Otsu method. Then apply the maximum variance threshold segmentation method to process the contrast sequence. This family of approaches present the disadvantage of having a fairly large processing time.

Tang [14] proposed a method based on progressive threshold segmentation. The objective of considering each line of the image as a unit of image is to considerably reduce the impact of the contrast, light or shadow changes. The distribution of the gray levels of each image line follows a certain distribution and generally presents two peaks: a large peak corresponding to the background or the road, and a second peak corresponding to the markings. The threshold taken is the one that separates these two distributions. To reduce the computational time, the upper region of the image defined as being beyond the horizon line is set to zero.

In [15,16], the contrast in the image is automatically adjusted for each image. Adaptive thresholding was used for each pixel to improve the primitive extraction and binarization step.

Contrary to the first approach, a classification of the primitives was carried out on a bird's eye view. This view is obtained by a geometric transformation of the basic image to another projective plane, using the intrinsic parameters of the camera and the Homography matrix, this is an Inverse Reverse Transformation (IRT). Thereafter, an image segmentation process is applied whose purpose is to have a robust marking extraction. The procedure consists mainly of a filtering step with the Gaussian filter, and an adaptive thresholding method.

Based on a thorough reading of the various methods of binarization, we noticed some limitations due to the low or high brightness in the image:

- Images with presence of shadows: no method is able to extract the line marking in these types of complex images. Some markings are not extracted because the gray level values associated with the pixels are often below the threshold.
- Bright images: most methods of extracting the road markings give poor results with this type of images. Usually, there is a difficulty in identifying and extracting markings. In this case, the distinguishing between the regions corresponds to the line markings remains a difficult operation since the gray level values of the pixels corresponding to the markings and those of the road are almost the same.

3   Strategy for Lane Marking Extraction

We present here our approach which has been designed to cope with the limitations of previous methods of the literature. Figure 1 shows the flow diagram of lane extraction.

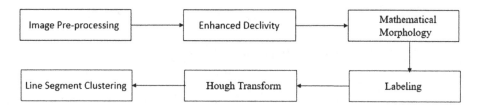

Fig. 1. The structure of our algorithm.

3.1   Image Preprocessing

This step consists of preparing the images by filtering them in order to reduce the noise caused by the acquisition process or by the impact of the illumination conditions of the environment. We chose to apply a Gaussian filter with a 1D kernel. This choice is justified by the fact that we are estimating a threshold for each image line by applying an improved gradient approach described in the following paragraph.

3.2   Enhanced Declivity

The declivity [17,18] has been proposed for contour detection. It is of the same family as the Canny and Deriche methods [19] which propose an optimal filter for contour detection. The response of these filters largely depends on their setting. Indeed, it is necessary to fix in advance the size of the filter for the Canny algorithm and a Deriche scale setting parameter. In order to produce satisfactory results on all types of images, the estimation of optimal values is still a challenging problem.

A declivity is applied on an image line whose gray levels are represented according to their position. The declivity makes it possible to identify the increasing and decreasing peaks, and is characterized by the following parameters:

- The bounds: x_i and x_j.
- The direction of the declivity: increasing "e" if $I(x_i) < I(x_j)$ or decreasing "d" otherwise.
- The width "l": $x_j - x_i$.

Figure 2 summarizes the concept of declivity. From the short gray levels per line, the objective is to identify ascendants and descents that have a high probability to correspond to the marking. We propose a new criterion which consists

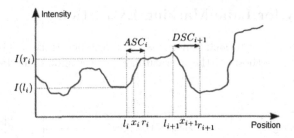

Fig. 2. Principle of the declivity.

of matching the ascending slopes with the descending slopes. We also propose
a new thresholding by line which is defined as the average of the height of the
ascendant declivity. The width of a slope has been defined differently from the
basic version: the width "l" corresponds to the width of the cut of a ascending
slope and a descending one at height α h. Note D the set of declivities and N
the total number of declivity per image line. The threshold of the line j, denoted
S_j is given by the following criterion:

$$S_j = \sum_{i=1}^{N} D_i(h, c),\tag{1}$$

where $D_i(h, c)$ corresponds to the height of the rising i.

To merge declivities, we apply the following algorithm: for each declivity
upward slope, we seek for the most appropriate descending slope in an interval
defined for each line. This interval is related to the possible width of a marking
in the image. A correspondence table is created from the width in pixel of the
nearest marking (the lowest in the image). This procedure will assign an ascendant peak to a descendant one. It is thus possible to identify a potential marking
segment. Figure 3 shows the gray levels of a real image line, the ascendant and
descendant peaks.

A declivity is valid if each ascending slope (in red) corresponds to a valid
descending slop (in blue) in a given interval, and if the height of the declivity is

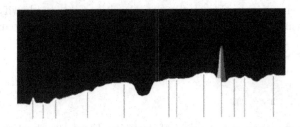

Fig. 3. Example of a valid declivity on an image line. (Color figure online)

greater than a threshold S_j. The algorithm of the improved declivity is summarized in Algorithm 1.

Algorithm 1. *E*xtraction algorithm.

---

1.   Preprocessing: Application of a 1D Gaussian filter
2.   For each line j do
    2.1   Compute the declivity on the appearance of gray levels
    2.2   Association of ascending and descending declivities
    2.3   Compute a threshold per line: $S_j = \sum\limits_{i=1}^{N} D_i(h, c)$
    2.4   For each declivity D_i
      2.4.1   If $D_i(x_j) < S_j$ and $l < l_s$; l_s : is interval limit
          Retain the pair of declivities.
          Consider the width of the valid associated declivities as the width of the marking.

The Association step allows regrouping each main ascending declivity MAD with the next main descending declivity MDD. A MAD could correspond to the left side of a road marking and a MDD probably correspond to the right side of the same road marking. It could exist some local variation between a MAD and a MDD which might correspond to a shadow or other noise. These artifacts are represented as small, negligible, and insignificant declivities and consequently removed to consider only relevant declivities.

The previous algorithm is designed to highlight all of possibles road marking segments per image line from which are based the next steps of extraction.

The obtained result is a binary image with segments which the left side for each corresponds to the beginning of a road marking, and the right side is the end.

3.3   Morphology Filtering

This step consists of applying a mathematical morphology operation on theenerated binary image. This image contains the center of each segment detected with the Enhanced Declivity algorithm. We found that the result of this first phase of detection highlights the true marking which has the particularity of being compact, dense and continuous while the distribution of noise, or false detections, is rather random. We chose to apply a dilatation in order to connect the segments and keep this continuity between segments.

3.4   Labeling

The labeling consists of defining primitives by grouping the related pixels. A pixel is added to a component if there are neighboring pixels in the previous row and/or column belonging to the same component. Once the image is labeled, a primitive is defined for each object in order to detect and characterize all the objects in the image: the surface, the bounding box (*Umin, Umax, Vmin, Vmax*).

Primitives with a surface area below a certain threshold are not taken into account in the detection phase. Only the most significant primitives are detected. This filter allows us to avoid accounting for small objects and large objects that do not correspond to the channels, as well as reducing the background noise. Figure 4 below shows the result of the morphological filtering on the obtained binary image and the image after thresholding.

Fig. 4. Left: Dilatation. Right: After thresholding.

The morphological operation allows connecting neighboring pixels within small local regions in order to highlight the true road markings. The filtering step allows removing all small and isolated clusters of pixels.

3.5   Line Detection and Segments Grouping (Clustering)

On the obtained image, we apply the Hough transform [20,21] for the detection of straight lines. We get a set of segments that will be grouped into different

(a)   (b)

(c)   (d)

Fig. 5. Some results on a complicated image set.

classes from a clustering algorithm [22] that we have developed. Two segments are classified in the same class if they respect the alignment and orientation constraints. Figure 5 shows the result of our different algorithms on certain images considered very complicated in the state of the art.

4 Experimental Results

The proposed method is tested on different conditions of road marking (simple and complex real conditions). The results of our experiment indicate the good performance of our algorithm for lane detection, especially under some challenging scenarios.

4.1 Quantitative Evaluation

The lane marking detection algorithm described in this paper has been implemented on a computer that had an Intel i7, 2.40 GHz CPU. The algorithm was executed in Visual C++ with OpenCV. The processing time was approximately 25 ms. per frame in good and complex road condition. As part of the evaluation of our approach, we have images from the "ROMA" database. It comprises more than 100 Heterogeneous images of various road scenes. Each image is delivered with a manually constructed ground truth, which indicates the position of the visible road markings.

Table 1. Quantitative performance of our algorithm.

Progressive threshold			Optimized Otsu			Our approach		
TDR	TPR	FPR	TDR	TPR	FPR	TDR	TPR	FPR
67.42%	32.55%	20.22%	80.63%	20.11%	51.39%	91.72%	7.26%	11.98%

Table 1 shows the performance of our method in terms of the algorithm. Figure 6 show the final results of our algorithm. TDR denotes the successful points marking detection rate while, TPR is the True Positive Rate, and FPR is the False Positive Rate.

$$TDR = \frac{\text{the number of successful lane markings}}{\text{the number of white points in ground truth}} \qquad (2)$$

$$FPR = \frac{\text{the number of points markings detected but not exist in ground truth}}{\text{the number of white points in ground truth}}$$
$$(3)$$

$$TPR = \frac{\text{the number of points markings exist in ground truth but not detected}}{\text{the number of white points in ground truth}}$$
$$(4)$$

Our extraction approach has been evaluated and compared with other state-of-the-art algorithms. We selected and implemented two approaches: static thresholding and the OTSU method detailed in Sect. 2. The images on which the evaluation was made are chosen for their complexity. The algorithms selected have given very poor results and only select pixels with high luminosity. Until today, no approach is able to extract the markings well under different conditions. From Table 1 and the images in Fig. 6, we can easily conclude that our algorithm is robust and accurate especially under severe experimental conditions.

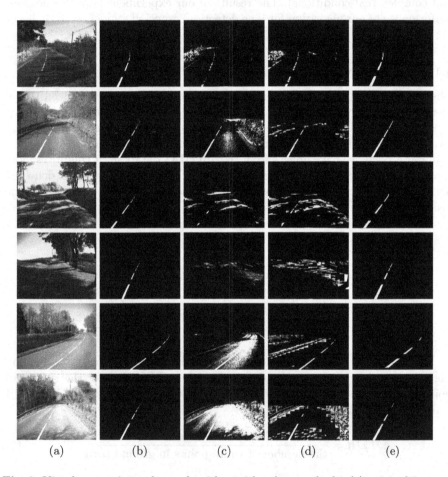

   (a)   (b)   (c)   (d)   (e)

Fig. 6. Visual comparison of our algorithm with other methods. (a) original images (b) ground truth (c) progressive threshold method (d) optimized otsu method (e) our approach.

5 Conclusion

In this paper, we represented a new algorithm of moving and stationary object extraction based on improving declivity. Different algorithms have been implemented to improve the quality of markings extraction, each of them bring some improvements in one sense.

The experimentation showed that the method is applicable in real-world and complex scenes. The foreground extraction method is based on two-dimensional declivity. It has already been evaluated in terms of precision on a set of images from the "ROMA" database. Real-world datasets have been shot at four different environments, including a hundred scenario per place under different illumination and weather conditions.

Future improvements will consider machine learning methods in order to learn a model of the different types, forms, appearances, etc.

References

1. Hwang, S., Kim, N., Choi, Y., Lee, S., Kweon, I.S.: Fast multiple objects detection and tracking fusing color camera and 3D LIDAR for intelligent vehicles. In: International Conference on Ubiquitous Robots and Ambient Intelligence (URAI), Sotitel Xian on Renmin Square, Xian, China, 19–22 August 2016
2. Yousif, T.M., Alharam, A.K., Elmedany, W., AlKhalaf, A.A., Fardan, Z.: GPRS-based robotic tracking system with real time video streaming. In: 4th International Conference on Future Internet of Things and Cloud Workshops, Vienna, Austria. IEEE (2016)
3. Xiaozhu, X., Cheng, H.: Object detection of armored vehicles based on deep learning in battlefield environment. In: 4th International Conference on Information Science and Control Engineering (ICISCE), Changsha, China, 21–23 July. IEEE (2017)
4. Saleem, N.H., Klette, R.: Accuracy of free-space detection: monocular versus binocular vision. In: International Conference on Image and Vision Computing New Zealand (IVCNZ), Palmerston North, New Zealand, 21–22 November. IEEE (2016)
5. Wedde, H.F., Senge, S.: BeeJamA: a distributed, self-adaptive vehicle routing guidance approach. IEEE Trans. Intell. Transp. Syst. $14(4)$, 1882–1895 (2013)
6. Magnier, V., Gruyer, D., Godelle, J.: Automotive LIDAR objects detection and classification algorithm using the belief theory. In: 2017 IEEE Intelligent Vehicles Symposium (IV), Redondo Beach, CA, USA, 11–14 June 2017
7. Nimvari, Z.E., Mosavi, M.R.: Accurate prediction of differential GPS corrections using fuzzy cognitive map. In: 3rd Iranian Conference on Signal Processing and Intelligent Systems (ICSPIS), 20–21 December 2017
8. Xu, F., Wang, Z.: An embedded visual SLAM algorithm based on Kinect and ORB features. In: Proceedings of the 34th Chinese Control Conference, Hangzhou, China, 28–30 July 2015
9. Chen, Y., He, M.: Sharp curve lane boundaries projective model and detection. In: 10th International Conference on Industrial Informatics (INDIN), Beijing, China, 25–27 July. IEEE (2012)
10. Basri, R., Rivlin, E., Shimshoni, I.: Image-based robot navigation under the perspective model. In: Proceedings of International Conference on Robotics and Automation, Detroit, MI, USA, 10–15 May 1999

11. Bali, A., Singh, S.N.: A review on the strategies and techniques of image segmentation. In: Fifth International Conference on Advanced Computing and Communication Technologies, Haryana, India, 21–22 February. IEEE (2015)
12. Kwon, D.: An image segmentation method based on improved watershed algorithm. In: International Conference on Computational and Information Sciences (ICCIS), Chengdu, China, 17–19 December. IEEE (2010)
13. Otsu, N.: A threshold selection method from gray-level histograms. IEEE Trans. Syst. Man Cybern. 9(1), 62–66 (1979)
14. Tang, G.: Road recognition and obstacle detection based on machine vision research, pp. 14–19 (2005)
15. Huang, J., Liang, H., Wang, Z., Mei, T., Song, Y.: Robust lane marking detection under different road conditions. In: International Conference on Robotics and Biomimetics (ROBIO), China, 12–14 December 2013. IEEE (2013)
16. Huang, J., Liang, H., Wang, Z., Song, Y., Deng, Y.: Lane marking detection based on adaptive threshold segmentation and road classification. In: Proceedings of the International Conference on Robotics and Biomimetics, 5–10 December 2014. IEEE (2014)
17. Michi, P., Debrie, R.: Fast and self-adaptive image segmentation using extended declivity. Ann. Télécommun. 50(3–4), 401 (1995)
18. Elhassouni, F., Ezzine, A., Alami, S.: Modelisation of raindrops based on declivity principle. In: 13th International Conference Computer Graphics, Imaging and Visualization (CGiV), 29 March–1 April 2016. IEEE (2016)
19. Yan, X., Li, Y.: A method of lane edge detection based on canny algorithm. In: Chinese Automation Congress (CAC), 20–22 October 2017
20. Jung, C.R., Kelber, C.R.: A robust linear-parabolic model for lane following. In: Proceedings of 17th Brazilian Symposium on Computer Graphics and Image Processing, pp. 72–77. IEEE (2004)
21. Takahashi, A., Ninomiya, Y., Ohta, M., Nishida, M., Yoshikawa, N.: Image processing technology for rear view camera (1): development of lane detection system. R&D Rev. Toyota CRDL 38(2), 31–36 (2003)
22. Niu, J., Lu, J., Xu, M., Lv, P., Zhao, X.: Robust lane detection using two-stage feature extraction with curve fitting. Pattern Recogn. 59, 225–233 (2016)

Segmentation of the Proximal Femur by the Analysis of X-ray Imaging Using Statistical Models of Shape and Appearance

Joel Oswaldo Gallegos Guillen[1]📧, Laura Jovani Estacio Cerquin[1]📧,
Javier Delgado Obando[2], and Eveling Castro-Gutierrez[1(📧)]📧

[1] San Agustín National University of Arequipa, Arequipa, Peru
{jgallegosg,lestacio,ecastro}@unsa.edu.pe
[2] Austral University of Chile, Valdivia, Chile
jdelgado@uach.cl
http://www.unsa.edu.pe/, http://www.uach.cl/

Abstract. Using image processing to assist in the diagnostic of diseases is a growing challenge. Segmentation is one of the relevant stages in image processing. We present a strategy of complete segmentation of the proximal femur (right and left) in anterior-posterior pelvic radiographs using statistical models of shape and appearance for assistance in the diagnostics of diseases associated with femurs. Quantitative results are provided using the DICE coefficient and the processing time, on a set of clinical data that indicate the validity of our proposal.

Keywords: Segmentation · AP X-ray
Statistical shape models (SSM)
Statistical appearance models (SAM) · Gold standard
DICE coefficient

1 Introduction

Currently, the routine clinical process requires radiological imaging for the purpose of diagnosing diseases. There are different radiological modalities such as computed tomography (CT), magnetic resonance imaging (MRI) and X-ray (Rx); X-ray images are the most recommended in any clinical process because CT scans can be harmful to health due to their high radiation in the acquisition of the image; and, MRI are not usually recommended for patients who have implants [1]. Lesions located in the pelvic area can be detected and analyzed radiologically by means of an anteroposterior (AP) view of X-ray images [2], due to significant information that they provide. The AP X-ray view allows specialists to make decisions on clinical diagnosis; as well as obtaining information for the planning of the treatment to be followed [3,4].

© Springer International Publishing AG, part of Springer Nature 2018
L. Rutkowski et al. (Eds.): ICAISC 2018, LNAI 10842, pp. 25–35, 2018.
https://doi.org/10.1007/978-3-319-91262-2_3

The analysis of digital medical images requires the use of computational vision techniques that facilitate their evaluation taking into account explicit and implicit characteristics present in the image and that will be obtained from the processing of it. In this context, the segmentation of the proximal femur in X-ray images is fundamental for the development of automated tools for the clinical evaluation process. However, the process of analyzing X-ray images is affected by the low resolution of the images due to factors such as: (a) variations in illumination, (b) voluntary or involuntary movements of the patient, (c) overlapping of bones due to variations in bone shape, (d) presence of gas in the colon and factors that are not adequately controlled by radiology laboratories [5–9].

To achieve accurate segmentation, imaging improvements are still required. Finally, the improvements made in this proposal will serve as a solid basis for the construction of a 3D model of the femur [10–13].

2   Related Work

The role played by medical images in the clinical decision process has aroused the interest of various researchers, due to the growing need for automated methods. In this sense, the segmentation stage occupies an essential place for the analysis of the medical images. It is thus that in the step of automatically segregating images, achieving highly accurate results is usually a difficult problem due to the characteristics of the device or the characteristics of the anatomical structure evaluated [14,15].

A large number of researchers have tried to increase the accuracy of the segmentation of images, in the last decade, segmentation has been widely applied to medical imaging to help the physician diagnose a variety of diseases, such as traumatic injuries. In this context, X-ray images are the most used for diagnosis and clinical evaluations of bone structures; however, they are affected by factors related to the contrast between tissues and bones (overlapping bones and tissue). Due to these difficulties, most of the segmentation methods are unable to segment the bones with the desired precision [9,16].

In the year 2013 [12], the use of Sparse Shape Composition and Random Forest Regression was proposed. The combination of both methods allowed the detection of reference points on the femur and the pelvis through the placement of patches based on the construction of statistical models of shape and appearance. This work reached a 98% accuracy. However, its computational cost in time was around 5 min per X-ray images. Based on the approach of statistical models, in 2014 [9] the construction of statistical models of shape and appearance was proposed, followed by a non-rigid B-Spline deformation or registration for a precise segmentation of the pelvis; reaching a satisfactory precision, with the advantage that the processing time results being less than one minute per X-ray image. This same year [17], a method based on Data Driven Joint Estimation was proposed for a set of patches randomly sampled on the pelvis, based on statistical models similar to what was done by [12]; with the difference that a vector of characteristics of each patch generated was obtained by means of the Multi-level Oriented Gradient Histogram method.

In the year 2015 [18], a method was proposed based on the combination of the methods proposed in [9,17] for the detection of osteoarthritis; achieving satisfactory results in the segmentation of the pelvic structure. Another method proposed in this year is given by [19]; which consists of the use of a model female pelvis to separate the bone from the air present in a patient's organism by implementing deformable B-spline record. Other methods that have received special attention in 2016 consist of a modification of the conventional statistical models; these methods are known as articulated statistical models. In this sense, [20] the use of these methods was proposed for the segmentation of adipose tissue, for which the proposed method is based on a statistical model of the entire surface of the body, which is learned from geometric scans. The body model is factored into deformations or register of position and shape, which allows a compact parametrization of large variations in the human form. The experiments carried out show that the proposed model can be used to effectively segment the geometry of subcutaneous fat in subjects with different body mass indexes. Finally, [16] the segmentation-registration stage was proposed with clustering methods based on entropy. According to the experiments carried out, an accuracy of 80% for X-ray images was found of complex bone structures such as the femur.

In this way, the proposed article focuses the use of statistical models of shape and appearance; since they have proven to be efficient against complex bone structures; reaching recognition within the bone research community according to that cited in [21].

3   Materials and Methods

In this section we will deal with the generation of the Gold Standard (GS) for the purpose of validating results. Subsequently, the approach of the segmentation methods proposed with the results obtained will be shown; having as main points of reference the authors Tim Cootes et al. [22] and Xie et al. [9]. Finally, the evaluation metric used for the evaluation of the results will be shown. Figure 1 shows the steps used for the femur segmentation.

3.1   Gold Standard

Digital images of X-rays in anteroposterior view were obtained from a radiology laboratory, so both the right and left femur were segmented by a specialist in the area in order to validate the results obtained with the proposed segmentation method.

3.2   Statistical Shape Model (SSM)

For the construction of the statistical shape model, we considered training images delimited from the proximal femur; taking into account the placement of reference points along the femur as shown in Fig. 2.

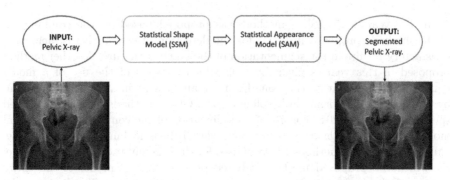

Fig. 1. Drawing showing each task involved in the process of segmentation of the proximal femur. To this end, the construction of statistical models of shape and appearance for each femur has been considered to obtain the final segmentation of the right and left femur. Source: All the images presented in this paper were self-elaborated.

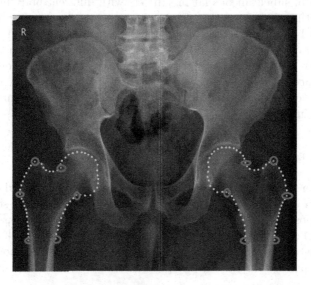

Fig. 2. The reference points were placed along the proximal femur; considering as key points of reference those that are in places with high curvature (points enclosed in red color). (Color figure online)

Each femur was delimited with 62 reference points making a total of 124 points of reference. The alignment of the training set was carried out with the use of an iterative method proposed by [22]; which consists in the application of the Analysis of Procrustes approach. The iterative method consists of the following steps:

1. Move each example of the training set to the origin.
2. Select an example as the initial estimate of the average shape.
3. Record the first estimate to define the frame of references by effect.

4. Align all shapes with the current estimate of the average shape.
5. Re-estimate the average shape based on the aligned shapes.
6. Apply rigid registration restrictions.
7. If it does not converge, repeat from step 4.

Through the implementation of this iterative method, the average shape of each femur was estimated, as can be seen in Fig. 3

a)   b)

Fig. 3. (a) Average model of shape denoted in red color of the left femur. (b) Average model of shape denoted in red color of the right femur. (Color figure online)

The modeling of the shape variation consists of generating new examples based on the model of average shape found. For this, the following steps were carried out:

– Calculate the covariance of the data.

$$\mathbf{S} = \frac{1}{s-1} \sum_{i=1}^{s} (x_i - \overline{x})(x_i - \overline{x})^T \tag{1}$$

– Calculate the eigenvectors, ϕ_i and their corresponding eigenvalues λ_i of the training set (ordered in such a manner that $\lambda_i \geq \lambda_{i+1}$).

If Φ contains the eigenvectors corresponding to the largest eigenvalues, we can then approximate any of the x training sets, using:

$$x \approx \overline{x} + \Phi b \tag{2}$$

Where $\Phi = (\phi_1, \phi_2, ..., \phi_t)$ and $\mathbf{b}$ is a t dimensional vector given by:

$$b = \Phi^T (x - \overline{x}) \tag{3}$$

Vector $\mathbf{b}$ defines a set of parameters of a deformable model. Through the variation of the elements of $\mathbf{b}$ we can vary the form, x using Eq. 2. The variance of the i^{mo} parameter of b, b_i, through the set of training is given by λ_i. By applying limits of $\pm 3\sqrt{\lambda_i}$ to the b_i parameter we make sure that the generated shape is similar to the shapes that comprise the training set.

The number of modes to retain, t, can be chosen in many ways. The simplest way to choose t is to represent some proportion of the variance displayed in the training set.

Given λ_i as the eigenvalues of the covariance matrix of the training data. The total variance in the training data is the sum of all the eigenvalues, $V_T = \sum \lambda_i$.

Then, the election of the t largest eigenvalues is represented by:

$$\sum i = 1^t \lambda_i \geq f_v V_T \tag{4}$$

Where f_v defines the total proportion of the variation of the data.

The total variation selected in our research was 98%; in this way, it is sought to have satisfactory results in the construction of the model in a statistical way. In this way, 17 modes of shape were generated with their respective variation on each side of the femur by applying the limits $\pm 3\sqrt{\lambda_i}$.

Figure 4 shows some examples of the selected modes of the right and left femur respectively:

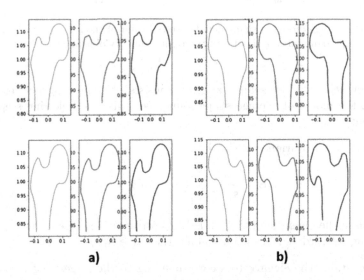

Fig. 4. Examples of 2 main modes of shape of the right and left femur. The selected shape modes of the right (**a**) and left (**b**) femur occupy the central place (red) while their respective shape variation occupies the extremes; **b** being modified on the side left (green) by $-3\sqrt{\lambda_i}$ and on the right side (blue) by $+3\sqrt{\lambda_i}$. (Color figure online)

3.3 Statistical Appearence Model (SAM)

The construction of statistical appearence model follows the same steps as the construction of the statistical shape model; with the difference that in this case, we do not work with a single reference point (pixel); but with a set of pixels around the reference point; thus forming a patch. Bearing in mind that the shape model is made up of 124 reference points (62 reference points for each femur), we will have 124 patches.

A. Generation of Patches

The generation of patches was performed for each reference point that makes up the delimitation of the proximal femur. The patches were drawn in a rectangular shape; having as its center the reference point. Each patch has a dimension of 20 pixels wide by 10 pixels high.

B. Alignment of the Patch Assembly

The alignment of the set of patches for each femur is made based on the training images and the alignment process performed in the statistical model followed; in such a way, that each patch is aligned with respect to the patches that occupy its same position; in order to obtain the average appearance model. Figure 5 represents the average appearance model of the right and left femur respectively:

C. Variation of Appearance and Choice of the Number of Modes

For the case of variation of appearance and choice of the number of modes, as in the shape model, 98% of variation was taken into account. In this way, the variation modes for the appearance model were also obtained.

Figures 6 and 7 show some examples of the selected modes of the right and left femur respectively:

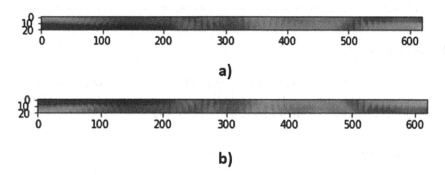

Fig. 5. (a) Average appearance model of the right femur obtained from the alignment of the 62 patches that make up the right femur. (b) Average appearance model of the left femur obtained from the alignment of the 62 patches that make up the left femur.

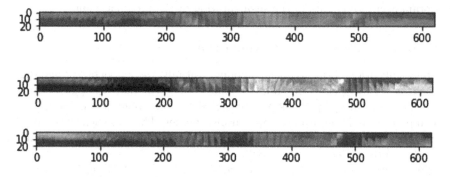

Fig. 6. An example of the appearance mode of the right femur (central position) and its respective variation in the upper part by $+3\sqrt{\lambda_i}$ and in the lower part by $-3\sqrt{\lambda_i}$.

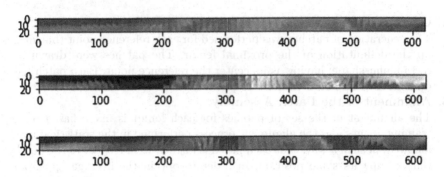

Fig. 7. An example of the appearance mode of the left femur (central position) and its respective variation in the upper part by $+3\sqrt{\lambda_i}$ and in the lower part by $-3\sqrt{\lambda_i}$.

3.4   Evaluation Metric

Considering that there is a Gold Segmentation Standard, different evaluation methods can be used. In this research the use of the DICE coefficient has been considered to evaluate the overlap between the segmentation of the proposed method and the manual segmentation performed by the specialist (GS).

4   Results

The present work focused on the use of statistical models of shape and appearance for the segmentation of the proximal femur; having as a tool of validation of results a GS generated by a specialist in the medical field.

4.1   Data Set

The data set used was of 60 images of anteroposterior X-rays (AP) of the pelvic structure. Using 50 images for training purposes of the statistical models of shape and appearance and 10 images for purposes of validation of the performance of the proposed method. The digital X-ray images were taken from healthy patients to obtain accurate results in the segmentation of the proximal femur.

4.2   Segmentation of the Proximal Femur

For the application of the statistical models, initially, the model of average shape of each femur was placed on the new image. Once the model is placed on top of the new image, the patches are extracted for each reference point in order to compare it with the average appearance model. That is, each patch of the average appearance model is compared with respect to the patches obtained from the overlap of the shape model.

This comparison is also made considering horizontal and vertical displacement; in such a way that it proceeds to move from left to right in a limit of horizontal displacement d_h; where $-20 \leq d_h \leq 20$. Likewise, displacement from top to bottom in a limit of vertical displacement d_v; where $-40 \leq d_v \leq 40$. For each displacement made, the patches are compared obtained from the displacement against the patches of the 17 appearance modes in order to segment the contour correctly.

After performing the previous steps, the same comparison is made, but now with the modes obtained from the shape model, to verify which of them is better adapted to the new input image; considering that this does not guarantee a correct segmentation. The obtained shape is then passed to a deformation step to reach a greater precision.

The image shown below in Fig. 8 refers to the results obtained from the segmentation process using the statistical models of shape and appearance:

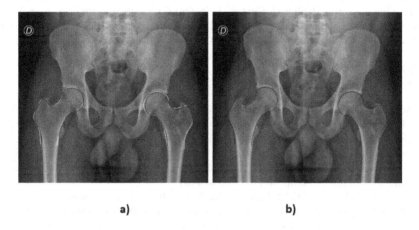

a)   b)

Fig. 8. Segmentation example. (Color figure online)

For the previous segmentation examples, statistical models of shape and appearance were used. A comparison is made between the Segmentation GS (denoted with red lines) and the segmentation of the proposed method (denoted with green lines). Likewise, the coefficient DICE is estimated according to the overlap between both segmentations.

The coefficient of DICE is 80% and average response time obtained is 1 min and 50 s in the segmentation process of the proximal femur using statistical models of shape and appearance.

5   Conclusions

In the present work, a strategy of complete segmentation of the femur was presented using statistical models of shape (SSM) and appearance (SAM), reaching an 80% accuracy in the evaluation stage. The GS was used as a reference against automatic segmentation with the constructed models, thus demonstrating the robustness of the models constructed in relation to complex anatomical structures and to medical imaging with lower resolution. It was also shown that the processing time of the statistical models SSM and SAM is 1 min and 50 s per X-ray image compared to 5 min found in the literature.

6   Future Works

The construction of a 3D model of the proximal femur for the diagnosis of osteoporosis has been proposed as future work. In this way we are going to seek to analyze the trabecular patterns present in the femur.

Acknowledgements. This research project was subsidized by the San Agustín National University. RDE No. 121-2016-FONDECYT-DE, RV. No. 004-2016-VR.INV-UNSA. Thanks to the "Research Center, Transfer of Technologies and Software Development R + D + i" – CiTeSoft-UNSA for their collaboration in the use of their equipment and facilities, for the development of this research work.

References

1. Weidman, E.K., Dean, K.E., Rivera, W., Loftus, M.L., Stokes, T.W., Min, R.J.: MRI safety: a report of current practice and advancements in patient preparation and screening. Clin. Imaging **39**(6), 935–937 (2015)
2. Kandasamy, M.S., Duraisamy, M., Ganeshsankar, K., Kurup, V.G.K., Radhakrishnan, S.: Acetabular fractures: an analysis on clinical outcomes of surgical treatment. Int. J. Res. Orthop. **3**(1), 122–126 (2016)
3. Wu, J., Davuluri, P., Ward, K.R., Cockrell, C., Hobson, R., Najarian, K.: Fracture detection in traumatic pelvic CT images. J. Biomed. Imaging **2012**, 1 (2012)
4. Jeuthe, J.: Automatic Tissue Segmentation of Volumetric CT Data of the Pelvic Region (2017)
5. Edeh, V.I., Olowoyeye, O.A., Irurhe, N.K., Abonyi, L.C., Arogundade, R.A., Awosanya, G.O., Eze, C.U., Omiyi, O.D.: Common factors affecting radiographic diagnostic quality in X-ray facilities in lagos. J. Med. Imaging Radiat. Sci. **43**, 108–111 (2012)
6. Alginahi, Y.: Preprocessing techniques in character recognition. In: Character Recognition, Minoru Mori (2010)
7. Pandey, M., Bhatia, M., Bansal, A.: An anatomization of noise removal techniques on medical image. In: 2016 21st International Conference on Innovation and Challenges in Cyber Security (ICICCS-INBUSH), pp. 224–229 (2016)
8. Ramamurthy, P.: Factors controlling the quality of radiography and the quality assurance. National Tuberculosis Institute (NTI), Bangalore, vol. 31, pp. 37–41 (1995)
9. Xie, W., Franke, J., Chen, C., Gruetzner, P., Schumann, S., Nolte, L.P., Zheng, G.: A complete pelvis segmentation framework for image-free total hip arthroplasty (THA): methodology and clinical study. Int. J. Med. Robot. Comput. Assist. Surg. **11**, 166–180 (2014)
10. Schumann, S., Sato, Y., Nakanishi, Y., Yokota, F., Takao, M., Sugano, N., Zheng, G.: Cup implant planning based on 2-D/3-D radiographic pelvis reconstruction – first clinical results. IEEE Trans. Biomed. **62**, 2665–2673 (2015)

11. Yu, W., Zheng, G.: 2D-3D regularized deformable B-spline registration: application to the proximal femur. In: Proceedings of International Symposium on Biomedical Imaging, vol. 1, pp. 829–832 (2015)

12. Chen, C., Zheng, G.: Fully automatic segmentation of AP pelvis X-rays via random forest regression and hierarchical sparse shape composition. In: Wilson, R., Hancock, E., Bors, A., Smith, W. (eds.) CAIP 2013. LNCS, vol. 8047, pp. 335–343. Springer, Heidelberg (2013). https://doi.org/10.1007/978-3-642-40261-6_40

13. Xie, W., Franke, J., Chen, C., Grützner, P.A., Schumann, S., Nolte, L.P., Zheng, G.: Statistical model-based segmentation of the proximal femur in digital anteroposterior (AP) pelvic radiographs. Int. J. Comput. Assist. Radiol. Surg. **9**, 165–176 (2014)

14. Akkus, Z., Carvalho, D.D., van den Oord, S.C., Schinkel, A.F., Niessen, W.J., de Jong, N., van der Steen, A.F., Klein, S., Bosch, J.G.: Fully automated carotid plaque segmentation in combined contrast-enhanced and B-mode ultrasound. Ultrasound Med. Biol. **41**(2), 517–531 (2015)

15. Viergever, M.A., Maintz, J.A., Klein, S., Murphy, K., Staring, M., Pluim, J.P.: A survey of medical image registration-under review. Med. Image Anal. **33**, 140–144 (2016)

16. Tamouk, J., Acan, A.: Entropy guided clustering improvements and statistical rule-based refinements for bone segmentation of X-ray images. J. Comput. Sci. **4**(1), 39–66 (2016)

17. Chen, C., Xie, W., Franke, J., Grutzner, P., Nolte, L.P., Zheng, G.: Automatic X-ray landmark detection and shape segmentation via data-driven joint estimation of image displacements. Med. Image Anal. **18**, 487–499 (2014)

18. Krishnakumari, P.K.: Supervised learning for measuring hip joint distance in digital X-ray images. Master thesis, Faculty of Electrical Engineering, Mathematics and Computer Science, Department of Computer Graphics and Visualization. Delft University of Technology, August 2015

19. Liu, L., Cao, Y., Fessler, J.A., Jolly, S., Balter, J.M.: A female pelvic bone shape model for air/bone separation in support of synthetic CT generation for radiation therapy. Phys. Med. Biol. **61**(1), 169 (2015)

20. Yeo, S., Romero, J., Loper, M., Machann, J., Black, M.: Shape estimation of subcutaneous adipose tissue using an articulated statistical shape model. Comput. Methods Biomech. Biomed. Eng.: Imaging Vis. **6**, 1–8 (2016)

21. Raudaschl, P., Fritscher, K.: Statistical shape and appearance models for bone quality assessment. In: Statistical Shape and Deformation Analysis: Methods, Implementation and Applications, p. 409 (2017)

22. Cootes, T.F., Edwards, G.J., Taylor, C.J.: Active appearance models. IEEE Trans. Pattern Anal. Mach. Intell. **23**(6), 681–685 (2001)

Architecture of Database Index for Content-Based Image Retrieval Systems

Rafał Grycuk[1], Patryk Najgebauer[1], Rafał Scherer[1(✉)],
and Agnieszka Siwocha[2,3]

[1] Computer Vision and Data Mining Lab, Institute of Computational Intelligence,
Częstochowa University of Technology,
Al. Armii Krajowej 36, 42-200 Częstochowa, Poland
{rafal.grycuk,patryk.najgebauer,rafal.scherer}@iisi.pcz.pl
[2] Information Technology Institute, University of Social Sciences,
90-113 Lodz, Poland
[3] Clark University, Worcester, MA 01610, USA
http://iisi.pcz.pl

Abstract. In this paper, we present a novel database index architecture for retrieving images. Effective storing, browsing and searching collections of images is one of the most important challenges of computer science. The design of architecture for storing such data requires a set of tools and frameworks such as relational database management systems. We create a database index as a DLL library and deploy it on the MS SQL Server. The CEDD algorithm is used for image description. The index is composed of new user-defined types and a user-defined function. The presented index is tested on an image dataset and its effectiveness is proved. The proposed solution can be also be ported to other database management systems.

Keywords: Content-based image retrieval · Image indexing

1 Introduction

The emergence of content-based image retrieval (CBIR) in the 1990s enabled automatic retrieval of images and allowed to depart from searching collections of images by keywords and meta tags or just by manual browsing them. Content-based image retrieval (CBIR) is a group of technologies which general purpose is to organize digital images by their visual content. Many methods, algorithms or technologies can be aggregated into this definition. CBIR takes unique place within the scientific society. This challenging field of study involves scholars from various fields, such as [4] computer vision (CV), machine learning, information retrieval, human-computer interaction, databases, web mining, data mining, information theory, statistics. Bridging of this fields proved to be very effective and provided interesting results and practical implementations thus, it creates new fields of research [32]. The current CBIR state of the art allows

© Springer International Publishing AG, part of Springer Nature 2018
L. Rutkowski et al. (Eds.): ICAISC 2018, LNAI 10842, pp. 36–47, 2018.
https://doi.org/10.1007/978-3-319-91262-2_4

using its methods in real-world applications used by millions of people globally (e.g. Google image search, Microsoft image search, Yahoo, Facebook, Instagram, Flickr and many others). The databases of these applications contain millions of images thus, the effective storing and retrieving images is extremely challenging. Images are created every day in tremendous amount and there is ongoing research to make it possible to efficiently search these vast collections by their content. Recognizing images and objects on images relies on suitable feature extraction which can be basically divided into several groups, i.e. based on color representation [19], textures [29], shape [17], edge detectors [14] or local invariant features [7,9,18,21], e.g. SURF [1], SIFT [25], neural networks [16] bag of features [6,23,30] or image segmentation [10,12].

A process associated with retrieving images in the databases is query formulation (similar to the 'select' statement in the SQL language). In the literature, it is possible to find many algorithms which operate on one of the three levels: [24]

1. Level 1: Retrieval based on primary features like color, texture and shape. A typical query is "search for a similar image".
2. Level 2: Retrieval of a certain object which is identified by extracted features, e.g. "search for a flower image".
3. Level 3: Retrieval of abstract attributes, including a vast number of determiners about the presented objects and scenes. Here, it is possible to find names of events and emotions. An example query is: "search for satisfied people".

Such methods require the use of algorithms from many different areas such as computational intelligence, mathematics and image processing. There are many content-based image processing systems developed so far, e.g. [8,11,13]. A good review of such systems is provided in [31]. To the best of our knowledge, no other system uses a similar set of tools to the system proposed in the paper.

2   Color and Edge Directivity Descriptor

In this section we briefly describe the Color and Edge Directivity Descriptor (CEDD) [3,20,22]. CEDD is a global feature descriptor in the form of a histogram obtained by so-called fuzzy-linking. The algorithm uses a two-stage fuzzy [2,27, 28] system in order to generate the histogram. A term fuzzy-linking defines that the output histogram is composed of more than one histogram. In the first stage, image blocks in the HSV colour space channels are used to compute a ten-bin histogram. The input channels are described by fuzzy sets as follows [20]:

- the hue (H) channel is divided in 8 fuzzy areas,
- the saturation (S) channel is divided in 2 fuzzy regions,
- the value (V) channel is divided in 3 fuzzy areas.

The membership functions are presented in Fig. 1. The output of the fuzzy system is obtained by a set of twenty rules and provides a crisp value [0:1] in order to produce the ten-bin histogram. The histogram bins represent ten preset colours:

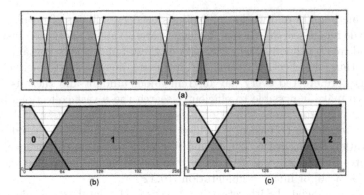

Fig. 1. Representations of fuzzy membership functions for the channels in the HSV color space, respectively: H (a), S (b), V (c) [20].

black, grey, white, red, etc. In the second stage of the fuzzy-linking system, a brightness value of seven colours is computed (without black, grey, white). Similar to the previous step, S and V channels and image blocks are inputs of the fuzzy system. The output of the second-stage is a three-bin histogram of crisp values, which describes the brightness of the colour (light, dark, normal and dark). Both histogram outputs (the first and the second stage) are combined, which allows producing the final 24-bin histogram. Each bin corresponds with color [20]: (0) Black, (1) Grey, (2) White, (3) Dark Red, (4) Red, (5) Light Red, (6) Dark Orange, (7) Orange, (8) Light Orange, (9) Dark Yellow, (10) Yellow, (11) Light Yellow, (12) Dark Green, (13) Green, (14) Light Green, (15) Dark Cyan, (16) Cyan, (17) Light Cyan, (18) Dark Blue, (19) Blue, (20) Light Blue, (21) Dark Magenta, (22) Magenta, (23) Light Magenta. In parallel to the Colour Unit, a Texture Unit of the Image-Block is computed, which general schema is presented in Fig. 2.

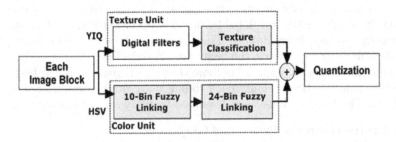

Fig. 2. A general schema of computing the CEDD descriptor [20].

In the first step of the Texture Unit, an image block is converted to the YIQ colour space. In order to extract texture information, MPEG-7 digital filters are used. One of these filters is the Edge Histogram Descriptor, which represents five edge types: vertical, horizontal, 45 diagonal, 135 diagonal, and isotropic (Fig. 3).

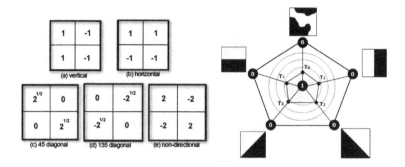

Fig. 3. Edge filters used to compute the texture descriptor [20].

The output of the Texture Unit is a six-bin histogram. When both histograms are computed, we obtain a 144-bin vector for every image block. Then, the vector is normalized and quantized into 8 predefined levels. This is the final step of computing the CEDD descriptor and now it can be used as a representation of the visual content of the image.

3   Database Index for Content-Based Image Retrieval System

In this section, we present a novel database architecture used to image indexing. The presented approach has several advantages over the existed ones:

- It is embedded into Database Management System (DBMS),
- Uses all the benefits of SQL and object-relational database management systems (ORDBMSs),
- It does not require any external program in order to manipulate data. A user of our index operate on T-SQL only, by using Data Modification Language (DML) by INSERT, UPDATE, and DELETE,
- Provides a new type for the database, which allows storing images along with the CEDD descriptor,
- It operates on binary data (vectors are converted to binary) thus, data processing is much faster as there is no JOIN clause used.

Our image database index is designed for Microsoft SQL Server, but it can be also ported to other platforms. A schema of the proposed system is presented in Fig. 4. It is embedded in the CLR (Common Language Runtime), which is a part of the database engine. After compilation, our solution is a .NET library, which is executed on CLR in the SQL Server. The complex calculations

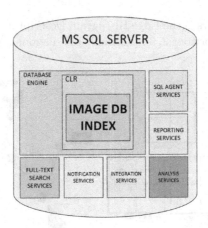

Fig. 4. The location of the presented image database index in Microsoft SQL Server.

of the CEDD descriptor cannot be easily implemented in T-SQL thus, we decided to use the CLR C#, which allows implementing many complex mathematical transformations.

In our solution we use two tools:

- SQL C# User-Defined Types - it is a project for creating a user defined types, which can be deployed on the SQL Server and used as the new type,
- SQL C# Function - it allows to create SQL Function in the form of C# code, it can also be deployed on the SQL Server and used as a regular T-SQL function. It should be noted that we use table-valued functions instead of scalar-valued functions.

At first we need to create a new user-defined type for storing binary data along with the CEDD descriptor. During this stage we encountered many issues which were resolved eventually. The most important ones are described below:

- The *Parse* method cannot take the *SqlBinary* type as a parameter, only *SqlString* is allowed. This method is used during INSERT clause. Thus, we resolve it by encoding binary to string and by passing it to the *Parse* method. In the body of the method we decode the string to binary and use it to obtain the descriptor,
- Another interesting problem is registration of external libraries. By default the library *System.Drawing* is not included. In order to include it we need to execute an SQL script.
- We cannot use reference types as fields or properties and we resolve this issue by implementing the *IBinarySerialize* interface.

We designed three classes: *CeddDescriptor, UserDefinedFunctions, QueryResult* and one static class *Extensions* (Fig. 5). The *CeddDescriptor* class implements two interfaces *INullable* and *IBinarySerialize*. It also contains one field *_null* of type *bool*. The class also contains three properties and five methods. A *IsNull*

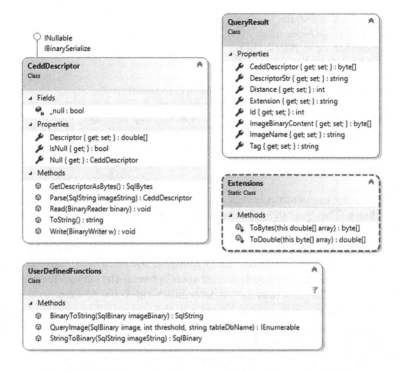

Fig. 5. Class diagram of the proposed database visual index.

and *Null* properties are required by user defined types and they are mostly generated. The *Descriptor* property allows to set or get the CEDD descriptor value in the form of a double array. A method *GetDescriptorAsBytes* provides a descriptor in the form of a byte array. Another very important method is *Parse*. It is invoked automatically when the T-SQL *Cast* method is called (Listing 1.2). Due to the restrictions implemented in UDT, we cannot pass parameter of type *SqlBinary* as it must be *SqlString*. In order to resolve the nuisance we encode byte array to string by using the *BinaryToString* method from the *UserDefinedFunctions* class. In the body of the *Parse* method we decode the string to byte array, then we create a bitmap based on the previously obtained byte array. Next, the Cedd descriptor value is computed. Afterwards, the obtained descriptor is set as a property. The pseudo-code of this method is presented in Algorithm 1 The *Read* and *Write* method are implemented in order to use reference types as fields and properties. They are responsible for writing and reading to or from a stream of data. The last method (*ToString*) represents the *CeddDescriptor* as *string*. Each element of the descriptor is displayed as a string with a separator, this method allows to display the descriptor value by the SELECT clause.

INPUT: *EncodedString*
OUTPUT: *CeddDescriptor*
if *EncodedString = NULL* **then**
  | RETURN NULL;
end
ImageBinary := DecodeStringToBinary(EncodedString);
ImageBitmap := CreateBitmap(ImageBinary);
CeddDescriptor := CalculateCeddDescriptor(ImageBitmap);
SetAsPropertyDescriptor(CeddDescriptor)

Algorithm 1. Steps of the *Parse* method.

Another very important class is *UserDefinedFunctions*, it is composed of three methods. The *QueryImage* method performs the image query on the previously inserted images and retrieves the most similar images with respect to the *threshold* parameter. The method has three parameters: *image, threshold, tableDbName*. The first one is the query image in the form of a binary array, the second one determines the threshold distance between the image query and the retrieved images. The last parameter determines the table to execute the query on (it possible that many image tables exist in the system). The method takes the *image* parameter and calculates the *CeddDescriptor*. Then, it compares it with those existed in the database. In the next step the similar images are retrieved. The method allows filtering the retrieved images by the distance with the threshold. The two remaining methods *BinaryToString* and *StringToBinary* allow to encode and decode images as string or binary. The *QueryResult* class is used for presenting the query results to the user. All the properties are self-describing (see Fig. 5). The static *Extension* class contains two methods which extend double array and byte array, what allows to convert a byte array to a double array and vice versa.

4   Simulation Environment

The presented visual index was built and deployed on Microsoft SQL Server as a CLR DLL library written in C#. Thus, we needed to enable CLR integration on the server. Afterwards, we also needed to add *System.Drawing* and index assemblies as trusted. Then, we published the index and created a table with our new *CeddDescriptor* type. The table creation is presented on Listing 1.1. As can be seen, we created the *CeddDescriptor* column and other columns for the image meta-data (such as *ImageName, Extension* and *Tag*). The binary form of the image is stored in the *ImageBinaryContent* column.

Listing 1.1. Creating a table with the CeddDescriptor column.

```
CREATE TABLE CbirBow.dbo.CeddCorelImages
(
        Id int primary key identity(1,1),
        CeddDescriptor CeddDescriptor not null,
```

```
         ImageName  varchar (max)  not  null ,
         Extension  varchar (10)  not  null ,
         Tag       varchar (max)  not  null ,
         ImageBinaryContent  varbinary (max)  not  null
) ;
```

Now we can insert data into the table what requires a binary data that will be loaded into a variable and passed as a parameter. This process is presented in Listing 1.2

Listing 1.2. Inserting data to a table with the CeddDescriptor.

```
DECLARE @filedata AS varbinary (max) ;
SET @filedata = (SELECT *
FROM OPENROWSET(BULK N'{ path_to_file }' ,
SINGLE_BLOB) as BinaryData)
INSERT INTO dbo . CeddCorelImages
( CeddDescriptor , ImageName , Extension , Tag, ImageBinaryContent)
VALUES (
CONVERT( CeddDescriptor , dbo . BinaryToString ( @filedata )) ,
'644010.jpg' , '.jpg' , 'art_dino ' , @filedata );
```

Such prepared table can be used to insert images from any visual dataset, e.g. Corel, Pascal, ImageNet, etc. Afterwards, we can execute queries by the *QueryImage* method and retrieve images. For the experimental purposes, we used the PASCAL Visual Object Classes (VOC) dataset [5]. We split the image sets of each class into a training set of images for image description and indexing (90%) and evaluation, i.e. query images for testing (10%). In Table 1 we presented the retrieved factors of multi-query. As can be seen, the results are satisfying which allows us to conclude that our method is effective and proves to be useful in CBIR techniques. For the purposes of the performance evaluation we used two well-known measures: *precision* and *recall* [26]. These measures are widely used in CBIR for evaluation. The representation of measures is presented in Fig. 6.

- *AI* - appropriate images which should be returned,
- *RI* - returned images by the system,
- *Rai* - properly returned images (intersection of *AI* and *RI*),
- *Iri* - improperly returned images,
- *anr* - proper not returned images,
- *inr* - improper not returned images.

These measures allows to define *precision* and *recall* by the following formulas [26]

$$precision = \frac{|rai|}{|rai + iri|},\tag{1}$$

$$recall = \frac{|rai|}{|rai + anr|}.\tag{2}$$

Table 1. Simulation results (Multi-Query). Due to limited space only a small part of the query results is presented.

Table 2. Example query results. The image with the border is the query image.

Image Id	RI	AI	rai	iri	anr	Precision	Recall
598(pyramid)	50	47	33	17	14	66	70
599(pyramid)	51	47	31	20	16	61	66
600(revolver)	73	67	43	30	24	59	64
601(revolver)	72	67	41	31	26	57	61
602(revolver)	73	67	40	33	27	55	60
603(revolver)	73	67	42	31	25	58	63
604(revolver)	73	67	44	29	23	60	66
605(revolver)	71	67	40	31	27	56	60
606(revolver)	73	67	40	33	27	55	60
607(rhino)	53	49	39	14	10	74	80
608(rhino)	53	49	42	11	7	79	86
609(rhino)	53	49	42	11	7	79	86
610(rhino)	52	49	38	14	11	73	78
611(rhino)	52	49	39	13	10	75	80
612(rooster)	43	41	36	7	5	84	88
613(rooster)	43	41	33	10	8	77	80
614(rooster)	43	41	34	9	7	79	83
615(rooster)	44	41	35	9	6	80	85
616(saxophone)	36	33	26	10	7	72	79
617(saxophone)	36	33	26	10	7	72	79
618(saxophone)	35	33	26	9	7	74	79
619(schooner)	56	52	37	19	15	66	71
620(schooner)	56	52	37	19	15	66	71
621(schooner)	56	52	39	17	13	70	75
622(schooner)	55	52	37	18	15	67	71
623(schooner)	56	52	35	21	17	62	67
624(scissors)	35	33	22	13	11	63	67
625(scissors)	36	33	22	14	11	61	67
626(scissors)	36	33	20	16	13	56	61
627(scorpion)	75	69	59	16	10	79	86
628(scorpion)	73	69	57	16	12	78	83
629(scorpion)	73	69	58	15	11	79	84
630(scorpion)	73	69	59	14	10	81	86

Table 2 shows the visualization of experimental results from a single image query. As can be seen, most images were correctly retrieved. Some of them are improperly recognized because they have similar features such as

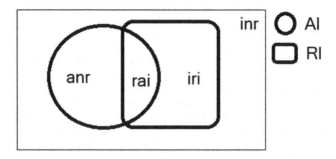

Fig. 6. Performance measures diagram [15].

shape or colour background. The image with the red border is the query image. The *AveragePrecision* value for the entire dataset equals 71 and for *AverageRecall* 76.

5 Conclusion

The presented system is a novel architecture of a database index for content-based image retrieval. We used Microsoft SQL Server as the core of our architecture. The approach has several advantages: it is embedded into RDBMS, it benefits from the SQL commands, thus it does not require external applications to manipulate data, and finally, it provides a new type for DBMSs. The proposed architecture can be ported to other DBMSs (or ORDBMSs). It is dedicated to being used as a database with CBIR feature. The performed experiments proved the effectiveness of our architecture. The proposed solution uses the CEDD descriptor but it is open to modifications and can be relatively easily extended to other types of visual feature descriptors.

References

1. Bay, H., Ess, A., Tuytelaars, T., Van Gool, L.: Speeded-up robust features (SURF). Comput. Vis. Image Underst. **110**(3), 346–359 (2008)
2. Beg, I., Rashid, T.: Modelling uncertainties in multi-criteria decision making using distance measure and topsis for hesitant fuzzy sets. J. Artif. Intell. Soft Comput. Res. **7**(2), 103–109 (2017)
3. Chatzichristofis, S.A., Boutalis, Y.S.: CEDD: color and edge directivity descriptor: a compact descriptor for image indexing and retrieval. In: Gasteratos, A., Vincze, M., Tsotsos, J.K. (eds.) ICVS 2008. LNCS, vol. 5008, pp. 312–322. Springer, Heidelberg (2008). https://doi.org/10.1007/978-3-540-79547-6_30
4. Datta, R., Joshi, D., Li, J., Wang, J.Z.: Image retrieval: ideas, influences, and trends of the new age. ACM Comput. Surv. (CSUR) **40**(2), 5 (2008)
5. Everingham, M., Van Gool, L., Williams, C.K.I., Winn, J., Zisserman, A.: The pascal visual object classes (VOC) challenge. Int. J. Comput. Vis. **88**(2), 303–338 (2010)

6. Gabryel, M.: The bag-of-words methods with pareto-fronts for similar image retrieval. In: Damaševičius, R., Mikašytė, V. (eds.) ICIST 2017. CCIS, vol. 756, pp. 374–384. Springer, Cham (2017). https://doi.org/10.1007/978-3-319-67642-5_31
7. Gabryel, M., Damaševičius, R.: The image classification with different types of image features. In: Rutkowski, L., Korytkowski, M., Scherer, R., Tadeusiewicz, R., Zadeh, L.A., Zurada, J.M. (eds.) ICAISC 2017. LNCS (LNAI), vol. 10245, pp. 497–506. Springer, Cham (2017). https://doi.org/10.1007/978-3-319-59063-9_44
8. Gabryel, M., Grycuk, R., Korytkowski, M., Holotyak, T.: Image indexing and retrieval using GSOM algorithm. In: Rutkowski, L., Korytkowski, M., Scherer, R., Tadeusiewicz, R., Zadeh, L.A., Zurada, J.M. (eds.) ICAISC 2015. LNCS (LNAI), vol. 9119, pp. 706–714. Springer, Cham (2015). https://doi.org/10.1007/978-3-319-19324-3_63
9. Grycuk, R.: Novel visual object descriptor using surf and clustering algorithms. J. Appl. Math. Comput. Mech. **15**(3), 37–46 (2016)
10. Grycuk, R., Gabryel, M., Korytkowski, M., Romanowski, J., Scherer, R.: Improved digital image segmentation based on stereo vision and mean shift algorithm. In: Wyrzykowski, R., Dongarra, J., Karczewski, K., Waśniewski, J. (eds.) PPAM 2013. LNCS, vol. 8384, pp. 433–443. Springer, Heidelberg (2014). https://doi.org/10.1007/978-3-642-55224-3_41
11. Grycuk, R., Gabryel, M., Korytkowski, M., Scherer, R.: Content-based image indexing by data clustering and inverse document frequency. In: Kozielski, S., Mrozek, D., Kasprowski, P., Małysiak-Mrozek, B., Kostrzewa, D. (eds.) BDAS 2014. CCIS, vol. 424, pp. 374–383. Springer, Cham (2014). https://doi.org/10.1007/978-3-319-06932-6_36
12. Grycuk, R., Gabryel, M., Korytkowski, M., Scherer, R., Voloshynovskiy, S.: From single image to list of objects based on edge and blob detection. In: Rutkowski, L., Korytkowski, M., Scherer, R., Tadeusiewicz, R., Zadeh, L.A., Zurada, J.M. (eds.) ICAISC 2014. LNCS (LNAI), vol. 8468, pp. 605–615. Springer, Cham (2014). https://doi.org/10.1007/978-3-319-07176-3_53
13. Grycuk, R., Gabryel, M., Nowicki, R., Scherer, R.: Content-based image retrieval optimization by differential evolution. In: 2016 IEEE Congress on Evolutionary Computation (CEC), pp. 86–93. IEEE (2016)
14. Grycuk, R., Gabryel, M., Scherer, M., Voloshynovskiy, S.: Image descriptor based on edge detection and crawler algorithm. In: Rutkowski, L., Korytkowski, M., Scherer, R., Tadeusiewicz, R., Zadeh, L.A., Zurada, J.M. (eds.) ICAISC 2016. LNCS (LNAI), vol. 9693, pp. 647–659. Springer, Cham (2016). https://doi.org/10.1007/978-3-319-39384-1_57
15. Grycuk, R., Gabryel, M., Scherer, R., Voloshynovskiy, S.: Multi-layer architecture for storing visual data based on WCF and microsoft SQL server database. In: Rutkowski, L., Korytkowski, M., Scherer, R., Tadeusiewicz, R., Zadeh, L.A., Zurada, J.M. (eds.) ICAISC 2015. LNCS (LNAI), vol. 9119, pp. 715–726. Springer, Cham (2015). https://doi.org/10.1007/978-3-319-19324-3_64
16. Grycuk, R., Knop, M.: Neural video compression based on SURF scene change detection algorithm. In: Choraś, R.S. (ed.) Image Processing and Communications Challenges 7. AISC, vol. 389, pp. 105–112. Springer, Cham (2016). https://doi.org/10.1007/978-3-319-23814-2_13
17. Grycuk, R., Scherer, M., Voloshynovskiy, S.: Local keypoint-based image detector with object detection. In: Rutkowski, L., Korytkowski, M., Scherer, R., Tadeusiewicz, R., Zadeh, L.A., Zurada, J.M. (eds.) ICAISC 2017. LNCS (LNAI), vol. 10245, pp. 507–517. Springer, Cham (2017). https://doi.org/10.1007/978-3-319-59063-9_45

18. Grycuk, R., Scherer, R., Gabryel, M.: New image descriptor from edge detector and blob extractor. J. Appl. Math. Comput. Mech. **14**(4), 31–39 (2015)
19. Huang, J., Kumar, S., Mitra, M., Zhu, W.J., Zabih, R.: Image indexing using color correlograms. In: Proceedings of 1997 IEEE Computer Society Conference on Computer Vision and Pattern Recognition, pp. 762–768, June 1997
20. Iakovidou, C., Bampis, L., Chatzichristofis, S.A., Boutalis, Y.S., Amanatiadis, A.: Color and edge directivity descriptor on GPGPU. In: 2015 23rd Euromicro International Conference on Parallel, Distributed and Network-Based Processing (PDP), pp. 301–308. IEEE (2015)
21. Karczmarek, P., Kiersztyn, A., Pedrycz, W., Dolecki, M.: An application of chain code-based local descriptor and its extension to face recognition. Pattern Recogn. **65**, 26–34 (2017)
22. Kumar, P.P., Aparna, D.K., Rao, K.V.: Compact descriptors for accurate image indexing and retrieval: FCTH and CEDD. Int. J. Eng. Res. Technol. (IJERT) **1** (2012). ISSN 2278–0181
23. Lavoué, G.: Combination of bag-of-words descriptors for robust partial shape retrieval. Vis. Comput. **28**(9), 931–942 (2012)
24. Liu, Y., Zhang, D., Lu, G., Ma, W.Y.: A survey of content-based image retrieval with high-level semantics. Pattern Recogn. **40**(1), 262–282 (2007)
25. Lowe, D.G.: Distinctive image features from scale-invariant keypoints. Int. J. Comput. Vis. **60**(2), 91–110 (2004)
26. Meskaldji, K., Boucherkha, S., Chikhi, S.: Color quantization and its impact on color histogram based image retrieval accuracy. In: First International Conference on Networked Digital Technologies, NDT 2009, pp. 515–517, July 2009
27. Riid, A., Preden, J.S.: Design of fuzzy rule-based classifiers through granulation and consolidation. J. Artif. Intell. Soft Comput. Res. **7**(2), 137–147 (2017)
28. Sadiqbatcha, S., Jafarzadeh, S., Ampatzidis, Y.: Particle swarm optimization for solving a class of type-1 and type-2 fuzzy nonlinear equations. J. Artif. Intell. Soft Comput. Res. **8**(2), 103–110 (2018)
29. Śmietański, J., Tadeusiewicz, R., Łuczyńska, E.: Texture analysis in perfusion images of prostate cancer-a case study. Int. J. Appl. Math. Comput. Sci. **20**(1), 149–156 (2010)
30. Valle, E., Cord, M.: Advanced techniques in CBIR: local descriptors, visual dictionaries and bags of features. In: 2009 Tutorials of the XXII Brazilian Symposium on Computer Graphics and Image Processing (SIBGRAPI TUTORIALS), pp. 72–78. IEEE (2009)
31. Veltkamp, R.C., Tanase, M.: Content-based image retrieval systems: a survey, pp. 1–62. Utrecht University, Department of Computing Science (2002)
32. Wang, J.Z., Boujemaa, N., Del Bimbo, A., Geman, D., Hauptmann, A.G., Tesić, J.: Diversity in multimedia information retrieval research. In: Proceedings of the 8th ACM International Workshop on Multimedia Information Retrieval, pp. 5–12. ACM (2006)

Symmetry of Hue Distribution
in the Images

Piotr Milczarski[(✉)] [ID]

Faculty of Physics and Applied Informatics, University of Lodz,
Pomorska street 149/153, 90-236 Lodz, Poland
piotr.milczarski@uni.lodz.pl
http://www.wfis.uni.lodz.pl

Abstract. In the paper, a new symmetry measure is proposed to evaluate the symmetry/asymmetry of the hue distribution within the segmented part of the image. A new symmetry/asymmetry area measure (*ASM*) as well as their parts: the asymmetry measures of: the shape distribution (*ASMShape*), hue distribution (*ASMHue*) and structures distribution (*ASMStuct*) are proposed and discussed. In the paper, a dermatological asymmetry measure in shape (*DASMShape*) and hue (*DASMHue*) are presented and discussed thoroughly as well as their *ASMShape* and *ASMHue* applications. The hue distribution of the symmetry/asymmetry of the segmented skin lesion is discussed. One of the DASMHue measures is thoroughly presented. The results of the DASMHue algorithm based on the threshold binary masks using PH2 dataset shows stronger overestimating results but the total ratio 95.8% of correctly and overestimated cases is better than the ratio which takes into account only shape alone.

Keywords: Asymmetry area measure of the hue distribution
(*ASMHue*) · Dermatological symmetry and asymmetry of skin lesion
Dermatological asymmetry measure of hue distribution
Pattern symmetry assessment · Texture symmetry

1 Introduction

Symmetry and asymmetry of hue or color distribution in a given object can be one of the properties allowing to distinguish between the objects.

Object symmetry is discussed in many papers (see Table 1). It may be defined as a mathematical, physics, medical or even interdisciplinary term. [14] It can be even defined as an abstract term. In image processing we can define 2D or 3D symmetry of the image objects e.g. as an axial, rotational or reflectional one, or as a whole symmetry consisting of different types. The symmetry in 2D can refer to the whole image or a part/parts of the segmented image. In the paper, the symmetry/asymmetry is defined for 2D images.

In the paper, general symmetry/asymmetry measure of the object shape *ASMShape* and the symmetry of the hue/color distribution, *ASMHue*, are proposed and discussed.

© Springer International Publishing AG, part of Springer Nature 2018
L. Rutkowski et al. (Eds.): ICAISC 2018, LNAI 10842, pp. 48–61, 2018.
https://doi.org/10.1007/978-3-319-91262-2_5

The paper is organized as follows. In Sect. 2 object symmetry and its different definition and approaches are presented widely. A new symmetry/asymmetry area measure (ASM) as well as their parts: the asymmetry measures of: the shape distribution ($ASMShape$), hue distribution ($ASMHue$) and structures distribution ($ASMStuct$) are proposed and discussed in Sect. 3. A dermatological asymmetry measure in shape ($DASMShape$) and hue ($DASMHue$) are also presented. Two different applications of the measures are shown in Sect. 4. The results and conclusions of asymmetry in hue distribution presented are shown in Sect. 5.

2 Object Symmetry

The symmetry axial transform (SAT) can be considered one of the first approaches for the detection of symmetry, but the first works of the symmetry as an computer science instance appeared with the rise of the computers epoch in 40's and 50's of 20 century. In Table 1 the short summary of the objects symmetry approaches is given. The examples of the methods of symmetry in the images can base on textures rotation invariance [24, 39].

Table 1. The examples of symmetry definitions and approaches

Authors	Approach/methods	Description/idea/limitations
Blum and Nagel [4]	The symmetry axial transform (SAT); weighted symmetric axis, center of maximal circles for inclusion	Shape description using weighted symmetric axis features. The symmetry axial transform (SAT). SAT has an ability to retrieve only maximal axes of symmetry
Bay et al. [2]	Speeded up robust features (SURF)	SURF is a patented local feature detector and descriptor, applied in object recognition, image registration, classification or 3D reconstruction. SURF was conceptually based on scale-invariant feature transform (SIFT) descriptor
Brady and Asada [5]	Smoothed Local Symmetry (SLS)	Retrieving of a global symmetry (if it exists) from the local curvature of contours, through the locus of mid-edge-point pairs
Cross and Hancock [9]	Curl of the vector potential	A vector potential is constructed from the gradient field extracted from filtered images. Edge and symmetry lines are extracted through a topographical analysis of the vector field
Di Gesù and Valenti [12]	Symmetry operators, discrete symmetry transform (DST)	Symmetries are based on the evaluation of the axial moment around its center of gravity. Gray levels are considered the point masses. The descriptor has been applied at a local level to define DST. Object symmetries are studied with axial moments of regions previously selected

(*continued*)

Table 1. (*continued*)

Authors	Approach/methods	Description/idea/limitations
Lowe [20, 21]	Scale-invariant feature transform, SIFT	The scale-invariant feature transform (SIFT) is an algorithm in computer vision to detect and describe local features in images
Manmatha and Sawhney [22]	Reflectional symmetry	A "measure" of symmetry and an axis orientation are provided at each point. It is computed in convolving with the first and second derivative of Gaussians
Marola [23]	Gray level information, symmetry descriptor	The symmetry descriptor of a given object is based on a cross correlation operator evaluated on the gray levels
Menzies et al. [26]	The symmetry for ABCD rule than Stolz[]	Axial symmetry of pigmentation refers to pattern symmetry around any axis through the center of the lesion. This does not require the lesion to have symmetry of shape. Different definition of symmetry for ABCD rule than in Stolz [38]
Shen et al. [33], Bigun et al. [3]	Fourier or Gabor transforms	Complex moments - Fourier or Gabor transforms of the images
Shen et al. [34]	Symmetric and asymmetric energy	A measure is built of two terms: symmetric and asymmetric energy. Minimizing the asymmetric term of the energy over an image
Sirakov et al. [35]	Active contour (AC) evolution	The automatic extraction of skin lesion's boundary to measure symmetry applying minimal boundary box
Soyer et al. [36, 37]	Dermoscopy; symmetry/asymmetry of the lesion	The axial symmetry of pigmentation refers to pattern symmetry around any axis through the center of the lesion
Shen et al. [33]	Three-point checklist of dermoscopy	Axial symmetry of pigmentation refers to shape, hue/color and structure distributed on symmetry around any axis through the center of the lesion or two perpendicular axes
Stolz et al. [38]	ABCD rule. Symmetry takes into account the contour, colors, and structures within the lesion	The lesion is bisected by two lines that are placed 90° to each other. The first line attempts to bisect the lesion at the division of most symmetry and the other one is placed 90° to it
Zabrodsky et al. [42]	The measure considered is the point to point L2-distance from the pattern to its nearest symmetric version	Based on selection of equidistant points along contours or equiangular edge points around the patterns centre-of-mass (centroid). From these n points the nearest C_n-*symmetric* pattern is built in rotating the average of the counter rotation version of the points by $2i\pi/n$, $[0 \leq i \leq n-1]$
Zavidovique and Di Gesù [43]	A measure of symmetry based on "symmetry kernel" (SK)	Given any symmetry transform S, SK of a pattern P is the maximal included symmetric sub-set of P for all directions and shifts. The associated symmetry measure is a modified difference between the respective surfaces of a pattern and its kernel

As it can be seen from Table 1 there are several different approaches to symmetry of the objects. They are also define differently in different scientific fields e.g. General approaches to dermatological image processing are presented in [10,11,13,15–19,30–32,39–41].

3   A New Symmetry/Asymmetry Area Measure (*ASM*)

3.1   Asymmetry Measure – Definition List

In the paper, apart from the definitions and abbreviations defined above the following definitions and abbreviations are defined:

- *mNA* - maximum number of symmetry axes for the given problem. The value of *mNA* will depend on the method accuracy, e.g. if rotational symmetry is checked every 10 then *mNA* should be regarded as 16.
- AS – Asymmetry, it is asymmetry of the shape, hue and structure distribution. The values of the asymmetry AS are discreet: 0 for fully symmetric shapes; 1 for symmetric ones in one axis or 2 for asymmetric ones.
- ASM – Asymmetry Measure. It depends on ASMShape, ASMHue and ASMStruct.
- GSSPT – a geometrical shape symmetry precision threshold.
- VoSS – a vector of shape symmetry. Its coefficients are equal to the number of symmetry axes for the given thresholds.
- GCL – a geometric center of the lesion.
- LSA – list of symmetry axes.
- ST – shape thresholds: ST = {*lst, ust*}, where

$$ST(ASMShape(W)) = \begin{matrix} 0 \Leftrightarrow ASMShape(W) < lst \\ 1 \Leftrightarrow lst \leq ASMShape(W) < ust \\ 2 \Leftrightarrow ASMShape(W) > ust \end{matrix} \qquad (1)$$

The values of *lst* and *ust* will depend on the ASMShape function type and will be derived after optimization of results.

3.2   Asymmetry Measure of Shape, ASMShape – Method Description

There are discussed two types of functions, Eqs. 3–4, with different vectors of shape symmetry (VoSS) **W**:

$$W = [n(t_1), n(t_2), \ldots, n(t_k)], \qquad (2)$$

where $k \geq 2$.

The first, exponential type is defined as:

$$ASMShape(W) = mNA * \exp(-f(W)), \qquad (3)$$

where $f: R^k \ R^+ \bigcup \{0\}$.

The second type, a rational one is defined as:

$$ASMShape(W) = \frac{mNA}{f(W)} \tag{4}$$

In the experimental research several versions of function $f(W)$ in (2) and (3) with different coefficients and for a different subset of GSSPT thresholds were tested. The smaller the number of thresholds values the faster deriving of VoSS vector W defined as in (1). We have achieved the best results for the following subset of threshold values: $\{0.8, 0.81, \ldots, 0.97\}$. i.e. some of these thresholds were used for finding building symmetry, e.g. for dermoscopic images it is $\{0.9, 0.93, 0.94, 0.95, 0.97\}$.

In the exponential case of the DASMShape, the inner function $f(W)$ was proposed as follows:

$$f(W) = \sum_{i=1}^{5} a_i n_i^2 \tag{5}$$

where values $n(t)$ defined for (4) are $n_1 = n(t_1) = n(0.9)$, $\ldots$, $n_5 = n(t_5) = n(0.97)$.

In the research, several vectors of coefficients $\mathbf{a}$ have been tested:

$$\mathbf{a} = [a_1, \ldots, a_5] \tag{6}$$

Apart from estimating shape symmetry/asymmetry of the lesion there are also derived the list of the symmetry axes and the geometric center of the lesion. That list of axes is a starting point for estimating asymmetry measure of hue ASMHue.

3.3   Asymmetry Measure of Hue, ASMSHue

The method of deriving and estimating the value of new asymmetry measure of hue can be described as follows:

1. Use the algorithm of the asymmetry measure of shape ASMShape described in the Subsect. 3.2 to derive the list of the symmetry axes, LSA and the geometric center of the lesion, GCL. If the list is empty (there are no symmetry axes regardless the similarity threshold values) then start form the horizontal line crossing the GCL.
2. Using the image and the binary mask of the lesion extract the lesion as an image (ELI).
3. Derive the histograms for ELI image in gray scale, red, green blue channels. Find local minima of the histograms.
4. Estimate a set of at least 2 vector thresholds $hTLow$ and $hTHigh$ (it will depend on the histograms; it shouldn't be more than 4 of them).

$$\begin{aligned} histThresh = \{hTLow, hTHigh\} = \\ f(hist(Grey), hist(Red), hist(Green), hist(Blue)) \end{aligned} \tag{7}$$

$$\overrightarrow{\text{histTresh}} = \{\overrightarrow{\text{hTLow}}, \overrightarrow{\text{hTHigh}}\} = \\ \text{f(hist(Grey), hist(Red), hist(Green), hist(Blue))} \tag{8}$$

where

$$\overrightarrow{hTLow} = [hLGrey, hLR, hLG, hLB]$$
$$\overrightarrow{hTHigh} = [hHGrey, hHR, hHG, hHB] \tag{9}$$

$$0 < hL < hH < 255, \tag{10}$$

for each color Gray, R, G, B separately. While deriving the lower (hL) and upper (hH) threshold from the local minima they should separate at least 15% of the ELI color pixels. If there is only one single narrow peak in a given histogram then choose lower and upper limits so they omit around 1–2% from each side of the peak.

5. Derive binary masks of the ELI with the defined in the step 4 thresholds. In the result, 3 binary images of the lesion are obtained for the given color scale image (at the beginning for the gray scale image):

$$BIM_1 = BIM(pix < hTLow)$$
$$BIM_2 = BIM(hTLow \leq pix < hTHigh) \tag{11}$$
$$BIM_3 = BIM(pix \geq hTHigh)$$

After the step 5 the result is a vector

$$\mathbf{w} = [w_1, w_2, w_3], \tag{12}$$

where each w_i is defined as equal to the ratio of the extracted lesion pixels from BIM_i to the total number of pixels in the lesion, where $i = 1, 2, 3$,

6. In the first step, use the procedure of ASHShape to derive symmetry axes LSA(BIM_i), for binary images BIM_1, BIM_2 and BIM_3 for Gray scale images and the corresponding geometric centers GCL(BIM_i), where $i = 1, 2, 3$;

7. Compare the axes LSA and LSA(BIM_i) as well as geometric centers GCL and GCL(BIM_i), where $i = 1, 2, 3$.

(a) If the axes

   i. differ less than $10°$–$20°$ from each other;
   ii. cross the other binary images centers GCL(BIM_i),
   they can be estimated as the same axis.

(b) If there are 3 axes and they differ by $60°$ we have 3 separate symmetry axes in general. But in some cases e.g. in dermatology, the number of dermatological symmetry axes equals 1. Hence, for $mNA = 2$, DASMHue $= 1$.

After that procedure, build a symmetry number vector

$$\mathbf{T} = f(LSA(BIM_i), GCL(BIM_i)) = [t_1, t_2, t_3], \tag{13}$$

where t_i might be equal the number of symmetry axes: 0,1,..., mNA. The conditions for the values t_i are defined below.

8. The coefficient t_i is incremented by 1 under conditions that
   (a) geometrical centers GCL(BIM_i) are bound in a circle with a radius less than 10% of the radius of the lesion, and
   (b) there exists 2 vertical symmetry axes in all LSA(BIM_i) that differs less than 15° from each other for the given threshold.

9. The coefficient t_i equals 1, if it is not equal 2 or more, and
   (a) geometrical centers GCL(BIM_i) are **not** bound in a circle with a radius less than 10% of the radius of the lesion,
   (b) the tangent of the lines defined by each two geometrical points GCL(BIM_i) gives the angle that they do not differ less than 15 from each other,
   (c) there exists 1 symmetry axis in all LSA(BIM_i) that is parallel to the line going through GCL(BIM_i) that they differ less than 15 from each other for the given threshold.

10. If any of the conditions in step 8 and 9 is not fulfilled then t_i equals 0.

11. After the above procedure for Gray scale image we might obtain:
    (a) the ratio vector **w** defined in the step 5;
    (b) the subset of LSA(BIM_i) that consists of 2,1 or none of the symmetry axes;
    (c) the vector $\mathbf{T} = [\mathbf{t}_1, \mathbf{t}_2, \mathbf{t}_3]$

12. The final value of AMHue can be defined as

$$ASMHue = mNA - w * T \tag{14}$$

13. If the $ASMHue$ is not equal or less than 0 (i.e. it is fully symmetric) or close to it, then calculate $ASMHueR, ASMHueG$ and $ASMHueB$ in the same way as for the Gray scale image repeating steps 5–12. As a result we obtain a vector **ASMHueRGB** defined as

$$ASMHueRGB = [ASMHueR, ASMHueG, ASMHueB] \tag{15}$$

The final value of hue asymmetry measure is a set of **ASMHue** for the Gray scale image and the vector **ASMHueRGB** for the given R, G and B scale images.

4   Examples of the Method Application

The described above method is working for the images as a whole or segmented parts of the image. Below, there are presented the examples of the method for 167 dermatological skin lesions, segmented from the dataset PH2 [25] and a block of flats.

4.1 Dermatological Asymmetry Measure, DASM and Dermatological Asymmetry Measure of Hue Distribution, DASMHue

In the papers [28,29], Dermatological Asymmetry Measure (DASM) have been introduced and asymmetry of shape was discussed thoroughly. There have been also introduced separate measures of shape (DASMShape), hue/color distribution (DASMHue) and structure distribution (DASMStruct) with their continuous values from the subset $<0,2>$, i.e. $mNA = 2$. For each symmetry value 0 is given for symmetric in two perpendicular axes; value 1 for symmetric in one axis and 2 for completely asymmetric ones [1,6–8]. In this case ASM measure is named dermatological asymmetry measure DASM, so the parts of ASM.

The values in the middle column (DAS(PH2)) shows the values for dermatological asymmetry as obtained by PH2 dataset experts. The first three vectors of coefficients **a** are chosen for the exponential type of DASMShape function in (3) and thresholds equal {0.9, 0.93, 0.94, 0.95, 0.97}: $\mathbf{a}_x = [0.01, 0.025, 1/15, 1/3, 1.0]$, $\mathbf{a}_y = [0.01, 0.02, 0.04, 0.2, 2.0]$, $\mathbf{a}_z = [0.004, 0.01, 1/6, 1/3, 0.5]$. The last two vectors $\mathbf{a}_k = [0.004, 0.01, 0.17, 0.33, 0.5]$ and $\mathbf{a}_m = [0.1, 0.2, 0.3, 0.5, 0.9]$ are chosen for the rational type of DASMShape function (Table 2).

The dermatological asymmetry measure of hue/color distribution for the chosen 5 lesions from PH2 there is presented in Table 3.

Table 2. The examples of **W** vector for images from PH2 dataset with DASMShape values

Image ID from PH2	VoSS vector W coefficient values for n(t), where t					DAS PH2	DASMShape values for f(W) and coefficients a and n(t)				
	0.9	0.93	0.94	0.95	0.97		a_x	a_y	a_z	a_k	a_m
IMD003	10	4	3	0	0	0	0.27067	0.37275	0.25491	1.43885	0.54054
IMD035	1	0	0	0	0	2	1.98010	1.98010	1.99202	1.99800	1.81818
IMD002	8	3	3	1	0	1	0.33115	0.50316	0.22623	1.13122	0.52632
IMD075	1	1	1	0	0	2	1.80666	1.86479	1.66943	1.78412	1.25000
IMD155	2	2	2	2	1	0	0.12914	0.09192	0.15523	0.53447	0.39216
IMD339	12	12	7	6	0	2	0.00000	0.00000	0.00000	0.12231	0.08097
IMD211	2	1	1	1	0	1	1.25627	1.48164	1.18193	1.31406	0.90909
IMD405	2	1	1	1	0	2	1.25627	1.48164	1.18193	1.31406	0.90909
IMD406	13	6	5	2	0	2	0.00747	0.02969	0.00290	0.61862	0.28571

We have achieved the best results for DASMHue and DASMHueRGB using the following subset of threshold values for the exponential type of DASMShape function in (3) and thresholds equal {0.9, 0.93, 0.94, 0.95, 0.97}.

The discussion about the segmentation of the skin lesions and symmetry of the hue and structures distribution can be also found in [15–19, 27].

Table 3. Asymmetry values for the chosen images from PH2 dataset

Image ID from PH2	DAS	DASM Shape	Color scale	hL	hH	Ratio vector w = [w₁, w₂, w₃]	Tangent GCL (BIMᵢ) (if applicable)	Subset LSA (BIMi) (degrees)	Symmetry vector T = [t₁,t₂,t₃]	DASM Hue	Hue symmetry axes (°)	Shape symmetry axes (°)
IMD002			Gray	85	120	[0.447, 0.369, 0.185]		-	[0, 0, 0]	2	-	36, 98, 160
	1	0.22623	R	132	180	[0.421, 0.384, 0.195]		{2}	[1, 0, 0]	1.579	2	
			G	81	107	[0.525, 0.293, 0.202]			[0, 0, 0]	2		
			B	24	80	[0.015, 0.805, 0.180]		{29, 39, 92, 153}	[0, 1, 0]	1.195	39, 92, 153	
IMD010			Gray	70	135	[0.001, 0.998, 0.001]		{4, 94, 27, 117, 16, 106}	[0, 2, 0]	0.002	Symmetric	Completely symmetric
	0	0.00000	R			[0.015, 0.977, 0.015]		{4, 94, 27, 117, 16, 106}	[0, 2, 0]	0.03	Symmetric	
			G			[0.013, 0.977, 0.017]		{4, 94, 27, 117, 16, 106}	[0, 2, 0]	0.03	Symmetric	
			B			[0.015, 0.977, 0.015]		{4, 94, 27, 117, 16, 106}	[0, 2, 0]	0.03	Symmetric	
IMD168			Gray	48	91	[0.469, 0.342, 0.189]		{16, 178}	[0, 0, 1]	1.811	-	67, 160
	2	1.23757	R	67	100	[0.482, 0.238, 0.28]		-	[0, 0, 0]	2	-	
			G	67	100	[0.536, 0.244, 0.22]		-	[0, 0, 0]	2	-	
			B	42	92	[0.552, 0.351, 0.097]			[0, 0, 0]	2	-	
IMD211			Gray	48	95	[0.01, 0.71, 0.28]	Tg = 0.25, 165	{8, 55, 78, 169, 176}	[0, 1, 1]	1.01	169	78, 171
	1	1.48164	R	55	135	[0.01, 0.752, 0.238]	Tg = 0.21, 167	{4, 58, 155, 165, 172}	[0, 1, 1]	1.01	165	
			G	39	91	[0.01, 0.754, 0.236]	Tg = 0.236, 167	{5, 42, 78, 159, 172}	[0, 1, 1]	1.01	172	
			B	34	85	[0.01, 0.788, 0.202]	-	{4, 77, 161, 174}	[0, 1, 0]	1.212	174	
IMD406			Gray	46	113	[0.01, 0.492, 0.498]		-	[0, 0, 0]	2	-	{40, 127}
	2	0.02969	R	92	145	[0.147, 0.327, 0.516]		-	[0, 0, 0]	2	-	
			G	42	105	[0.001, 0.491, 0.508]		-	[0, 0, 0]	2	-	
			B	31	85	[0.001, 0.522, 0.477]		-	[0, 0, 0]	2	-	

4.2   Image Object Symmetries

In the following case an image of block of flats was used. Figure 1, and one of the binary masks with thresholds $(0, 120)$ for the selected segment is shown at Fig. 2. The image thresholds used for finding building symmetry are $\{0.8, 0.82, 0.84, 0.9, 0.95\}$ and the vector VoSS is $W = [2, 1, 1, 0, 0]$; the ratio division $w = [0.22, 0.52, 0.24]$ for $hTLow = 120$ and $hTHigh = 230$.

Fig. 1. Original picture of the block of flats taken from Internet (Color figure online)

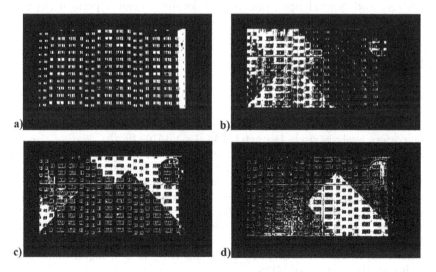

Fig. 2. Binary image for the image at Fig. 1. (a) Gray scale with thresholds $(0, 120)$; (b) gray scale with $(189, 255)$; (c) red channel with $(114, 170)$; (d) blue channel with $(150, 172)$.

One can assume nMA = 4 (square shape), but the shape symmetry ASMShape goes to 1. In color segmentation the images show symmetry equal 1 for the color distribution only using the (0, $hTLow$) thresholds in horizontal axis but only for the resemblance threshold 0.8. That can be seen at Fig. 2b–d. The image was taken on purpose because of the way it is painted. It results in all ASMHue symmetry coefficients with values close to 0.

5   Results and Conclusions

In the paper, the analysis of the object symmetry was conducted and a new general asymmetry measure ASM of the objects was defined and tested. After general definition of ASM and their parts shape, hue and structures, some practical application were shown. One of them is dermatological asymmetry measure DASM.

In the research, the images the classification results for the 167 cases from PH2 dataset out of 200. The difference of 33 of the excluded cases results from the fact that images contain only part of the lesion. That is why in that cases asymmetry of their shape or hue distribution are impossible to derive automatically. Using DASMHue 139 out of 167 cases were correctly classified using thresholds {0.791, 1.367} applying quantified values {0, 1, 2} to the DASMHue value. Additionally 7 cases were underestimated, in comparison to 21 overestimated ones. The difference is a result of the definition of the asymmetry (shape, hue, structures) in three-point checklist that is used in PH2 dataset. By using only DASMShape the accuracy of 70% was achieved.

The method defined in the paper will be used in the tool built for supporting the doctors and general practitioners non-dermatological experts using three-point checklist of dermoscopy.

References

1. Argenziano, G., Fabbrocini, G., Carli, P., De Giorgi, V., Sammarco, E., Delfino, M.: Epiluminescence microscopy for the diagnosis of doubtful melanocytic skin lesions. Comparison of the ABCD rule of dermatoscopy and a new 7-point checklist based on pattern analysis. Arch. Dermatol. **134**, 1563–1570 (1998)
2. Bay, H., Ess, A., Tuytelaars, T., Van Gool, L.: SURF: speeded up robust features. Comput. Vis. Image Underst. (CVIU) **110**(3), 346–359 (2008)
3. Bigun, J., DuBuf, J.M.H.: N-folded symmetries by complex moments in Gabor space and their application to unsupervized texture segmentation. IEEE Pattern Anal. Mach. Intell. **16**(1), 80–87 (1994)
4. Blum, H., Nagel, R.N.: Shape description using weighted symmetric axis features. Pattern Recogn. **10**, 167–180 (1978)
5. Brady, M., Asada, H.: Smoothed local symmetries and their implementation. Int. J. Robot. Res. **3**(3), 36–61 (1984)
6. Cardili, R.N., Roselino, A.M.: Elementary lesions in dermatological semiology: literature review. Anais brasileiros de dermatologia **91**(5), 629–633 (2016)

7. Chummun, S., McLean, N.R.: The management of malignant skin cancers. Surgery **29**(10), 529–533 (2011)
8. Celebi, M.E., Wen, Q., Iyatomi, H., Shimizu, K., Zhou, H., Schaefer, G.: A state-of-the-art survey on lesion border detection in dermoscopy images. In: Celebi, M.E., Mendonca, T., Marques, J.S. (eds.) Dermoscopy Image Analysis, pp. 97–129. CRC Press, Boca Raton (2015)
9. Cross, A.D.J., Hancock, E.R.: Scale space vector fields for symmetry detection. Image Vis. Comput. **17**(5–6), 337–345 (1999)
10. Deserno, T.M.: Biomedical Image Processing. Springer, Heidelberg (2011). https://doi.org/10.1007/978-3-642-15816-2
11. Esteva, A., Kuprel, B., Novoa, R.A., Ko, J., Swetter, S.M., Blau, H.M., Thrun, S.: Dermatologist-level classification of skin cancer with deep neural networks. Nature **542**(7639), 115–118 (2017)
12. Di Gesù, V., Valenti, C.: Symmetry operators in computer vision. Vistas Astronom. **40**(4), 461–468 (1996)
13. Henning, J.S., et al.: The CASH (color, architecture, symmetry, and homogeneity) algorithm for dermoscopy. J. Am. Acad. Dermatol. **56**(1), 45–52 (2007)
14. Jain, A.K.: Fundamentals of Digital Image Processing. Prentice Hall of India, New Delhi (2002)
15. Jaworek-Korjakowska, J., Kłeczek, P., Tadeusiewicz, R.: Detection and classification of pigment network in dermoscopic color images as one of the 7-point checklist criteria. In: Augustyniak, P., Maniewski, R., Tadeusiewicz, R. (eds.) PCBBE 2017. AISC, vol. 647, pp. 174–181. Springer, Cham (2018). https://doi.org/10.1007/978-3-319-66905-2_15
16. Jaworek-Korjakowska, J., Kłeczek, P., Grzegorzek, M., Shirahama, K.: Automatic detection of blue-whitish veil as the primary dermoscopic feature. In: Rutkowski, L., Korytkowski, M., Scherer, R., Tadeusiewicz, R., Zadeh, L.A., Zurada, J.M. (eds.) ICAISC 2017. LNCS (LNAI), vol. 10245, pp. 649–657. Springer, Cham (2017). https://doi.org/10.1007/978-3-319-59063-9_58
17. Jaworek-Korjakowska, J., Tadeusiewicz, R.: Assessment of asymmetry in dermoscopic colour images of pigmented skin lesions. In: Proceedings of IASTED International Conference on Biomedical Engineering, BioMed 2013, pp. 368–375 (2013)
18. Jaworek-Korjakowska, J., Tadeusiewicz, R.: Determination of border irregularity in dermoscopic color images of pigmented skin lesions. In: 36th Annual International Conference of the IEEE Engineering in Medicine and Biology Society, EMBC 2014, pp. 6459–6462 (2014)
19. Jaworek-Korjakowska, J., Tadeusiewicz, R.: Assessment of dots and globules in dermoscopic color images as one of the 7-point check list criteria. In: Proceedings of IEEE International Conference on Image Processing, ICIP 2013, pp. 1456–1460 (2013)
20. Lowe, D.G.: Object recognition from local scale-invariant features. In: Proceedings of International Conference on Computer Vision, vol. 2, pp. 1150–1157 (1999)
21. Lowe, D.G.: Distinctive image features from scale-invariant keypoints. Int. J. Comput. Vis. **60**(2), 91–110 (2004)
22. Manmatha, R., Sawhney, H.: Finding symmetry in intensity images. Technical report (1997)
23. Marola, G.: On the detection of the axes of symmetry of symmetric and almost symmetric planar images. IEEE Trans. Pattern Anal. Mach. Intell. **11**, 104–108 (1989)
24. Mehta, R., Egiazarian, K.O.: Rotation invariant texture description using symmetric dense microblock difference. IEEE Sig. Process. Lett. **23**(6), 833–837 (2016)

25. Mendoncca, T., Ferreira, P.M., Marques, J.S., Marcal, A.R.S., Rozeira, J.: PH2 - a dermoscopic image database for research and benchmarking. In: 35th Annual International Conference of the IEEE Engineering in Medicine and Biology Society (EMBC), Osaka, pp. 5437–5440 (2013)
26. Menzies, S.W., Crotty, K.A., Ingvar, C., McCarthy, W.H.: An Atlas of Surface Microscopy of Pigmented Skin Lesions: Dermoscopy, 2nd edn. McGrawHill, Roseville (2003)
27. Milczarski, P.: Skin lesion symmetry of hue distribution. In: Proceedings of IEEE 9th International Conference on Intelligent Data Acquisition and Advanced Computing Systems: Technology and Applications, IDAACS 2017, pp. 1006–1013 (2017)
28. Milczarski, P., Stawska, Z., Maślanka, P.: Skin lesions dermatological shape asymmetry measures. In: Proceedings of IEEE 9th International Conference on Intelligent Data Acquisition and Advanced Computing Systems: Technology and Applications, IDAACS 2017, pp. 1056–1062 (2017)
29. Milczarski, P., Stawska, Z., Was, L., Wiak, S., Kot, M.: New dermatological asymmetry measure of skin lesions. Int. J. Neural Netw. Adv. Appl. **4**, 32–38 (2017). (Prague)
30. Pathan, S., et al.: Biomed. Sig. Process. Control **39**, 237–262 (2018). Elsevier
31. Rosendahl, C., Cameron, A., McColl, I., Wilkinson, D.: Dermatoscopy in routine practice "Chaos and Clues". Aust. Fam. Phys. **41**(7), 482487 (2012)
32. Schmid, P.: Segmentation of digitized dermatoscopic images by two-dimensional color clustering. IEEE Trans. Med. Imaging **18**(2), 164–171 (1999)
33. Shen, D., Ip, H., Cheung, K.T., Teoh, E.K.: Symmetry detection by generalized complex moments: a close-form solution. IEEE Pattern Anal. Mach. Intell. **21**(5), 466–476 (1999)
34. Shen, D., Ip, H., Teoh, E.K.: An energy of assymmetry for accurate detection of global reflexion axes. Image Vis. Comput. **19**, 283–297 (2001)
35. Sirakov, N.M., Mete, M., Chakrader, N.S.: Automatic boundary detection and symmetry calculation in dermoscopy images of skin lesions. In: 18th IEEE International Conference on Image Processing, Brussels, pp. 1605–1608 (2011)
36. Soyer, H.P., Argenziano, G., Hofmann-Wellenhof, R., Zalaudek, I.: Dermoscopy: The Essentials, 2nd edn. Saunders Ltd., Philadelphia (2011)
37. Soyer, H.P., Argenziano, G., Zalaudek, I., Corona, R., Sera, F., Talamini, R., et al.: Three-point checklist of dermoscopy. A new screening method for early detection of melanoma. Dermatology **208**(1), 27–31 (2004)
38. Stolz, W., Riemann, A., Cognetta, A.B., Pillet, L., Abmayr, W., Hölzel, D., et al.: ABCD rule of dermatoscopy: a new practical method for early recognition of malignant melanoma. Eur J. Dermatol. **4**, 521–527 (1994)
39. Was, L., Milczarski, P., Stawska, Z., Wyczechowski, M., Kot, M., Wiak, S., Wozniacka, A., Pietrzak, L.: Analysis of dermatoses using segmentation and color hue in reference to skin lesions. In: Rutkowski, L., Korytkowski, M., Scherer, R., Tadeusiewicz, R., Zadeh, L.A., Zurada, J.M. (eds.) ICAISC 2017. LNCS (LNAI), vol. 10245, pp. 677–689. Springer, Cham (2017). https://doi.org/10.1007/978-3-319-59063-9_61
40. Wighton, P., Lee, T.K., Lui, H., McLean, D.I., Atkins, M.S.: Generalizing common tasks in automated skin lesion diagnosis. IEEE Trans. Inf Technol. Biomed. **15**, 622–629 (2011)

41. Xie, F., Bovik, A.C.: Automatic segmentation of dermoscopy images using self-generating neural networks seeded by genetic algorithm. Pattern Recognit. **46**, 1012–1019 (2013)
42. Zabrodsky, H., Peleg, S., Avnir, D.: Symmetry as a continuous feature. IEEE Pattern Anal. Mach. Intell. **17**(12), 1154–1166 (1995)
43. Zavidovique, B., Di Gesù, V.: The S-kernel: ameasure of symmetry of objects. Pattern Recogn. **40**, 839–852 (2007)

Image Completion with Smooth Nonnegative Matrix Factorization

Tomasz Sadowski[✉] and Rafał Zdunek

Faculty of Electronics, Wroclaw University of Science and Technology,
Wybrzeze Wyspianskiego 27, 50-370 Wroclaw, Poland
{tomasz.sadowski,rafal.zdunek}@pwr.edu.pl

Abstract. Nonnegative matrix factorization is an unsupervised learning method for part-based feature extraction and dimensionality reduction of nonnegative data with a variety of models, algorithms, structures, and applications. Smooth nonnegative matrix factorization assumes the estimated latent factors are locally smooth, and the smoothness is enforced by the underlying model or the algorithm. In this study, we extended one of the algorithms for this kind of factorization to an image completion problem. It is the B-splines ADMM-NMF (Nonnegative Matrix Factorization with Alternating Direction Method of Multipliers) that enforces smooth feature vectors by assuming they are represented by a linear combination of smooth basis functions, i.e. B-splines. The numerical experiments performed on several incomplete images show that the proposed method outperforms the other algorithms in terms of the quality of recovered images.

1 Introduction

Nonnegative Matrix Factorization (NMF) [1,2] is a commonly-used method for feature extraction and dimensionality reduction of nonnegative data with many applications in machine learning and signal processing. The examples include: spectral unmixing problems [3–5], textual document analysis [6,7], image classification [8], etc. A certain class of input data demonstrates intrinsic smoothness, e.g. spatial smoothness in hyperspectral imaging, and hence, the features extracted from such data may also reflect smoothness properties. Hence, there is a need to impose the smoothness onto the estimated latent factors for the selected applications.

The smoothness in NMF can be enforced in many ways – typically, by adding the smoothness penalty term to an objective function [9–11]. The smoothness term may result from the Gaussian priors or the Markov Random Field (MRF) models [10].

Another possibility is to enforce smoothness in the feature vectors by a linear combination of some smooth piecewise or unimodal nonnegative basis functions, e.g. Gaussian Radial Basic Functions (GRBF) [12] or B-splines [13]. The latter are more flexible in choosing their parameters (order, knots, etc.), and they

© Springer International Publishing AG, part of Springer Nature 2018
L. Rutkowski et al. (Eds.): ICAISC 2018, LNAI 10842, pp. 62–72, 2018.
https://doi.org/10.1007/978-3-319-91262-2_6

work very efficiently for spectra as well as image modeling. In the paper [14], the smooth NMF model with the B-splines was further extended by using a different computational strategy for updating the factors in the analyzed model. Instead of the multiplicative updates, Alternating Direction Method of Multipliers (ADMM) [15] was used, which considerably improves the convergence properties. ADMM, known also as the Douglas-Rachford splitting, become recently very popular in machine learning and image processing [16,17], including NMF problems [14,18–20], matrix completion [21,22], and low-rank matrix decomposition [23].

In this study, we extend the ADMM-NMF algorithm [14] to a hybrid version that combines it with the concept of the image completion that was proposed in [24]. This approach assumes that an incomplete image is iteratively approximated by a lower-rank approximation, where the known entries are fixed in each iteration. The lower-rank approximation was obtained with the fast Hierarchical Alternating Least-Squares (HALS) algorithm [1]. This idea was also explored in [25] but mostly in the context of a matrix decomposition. In this study, we assume the similar approach to an image completion problem but instead of the HALS-based updates, the current approach is motivated by the model and the algorithm from the paper [14]. It means that we assume that the feature vectors in a lower-rank approximation of a completed image should be locally smooth. The numerical experiments demonstrate that this approach is very efficient for solving an image completion problem with a highly incomplete input image. Furthermore, the proposed algorithm combined with the right image segmentation method is suitable for reducing snowflakes in images with snow occlusion.

The remainder of this paper is organized as follows. The model of a matrix completion problem, the smooth NMF model with B-splines, and the proposed method for solving an image completion problem are presented in Sect. 2. The numerical experiments performed for various image completion problems are presented in Sect. 3. The last section contains the summary and conclusions.

2   Image Completion with Smooth NMF

The aim of NMF is to find such lower-rank nonnegative matrices $\boldsymbol{A} = [a_{ij}] \in \mathbb{R}_+^{I \times J}$ and $\boldsymbol{X} = [x_{jt}] \in \mathbb{R}_+^{J \times T}$ that $\boldsymbol{Y} = [y_{it}] \cong \boldsymbol{AX} \in \mathbb{R}_+^{I \times T}$, given the data matrix $\boldsymbol{Y}$, the lower rank J, and possibly some prior knowledge on the matrices $\boldsymbol{A}$ or $\boldsymbol{X}$. The set of nonnegative real numbers is denoted by $\mathbb{R}_+$. When NMF is applied for model dimensionality reduction, we usually assume: $J \ll \frac{IT}{I+T}$ or at least: $J \leq \min\{I, T\}$.

The columns of $\boldsymbol{A}$ represents the features, and the columns of $\boldsymbol{X}$ contains coefficients of a conic combination of the features. In smooth NMF, each column of $\boldsymbol{A}$ represents a locally smooth profile, which can be obtained in various ways.

2.1   Matrix Completion Problem

Let $\boldsymbol{M} = [m_{it}] \in \mathbb{R}_+^{I \times T}$ be an original incomplete matrix, and Ω be the set of indices of known entries in $\boldsymbol{M}$. The aim of the matrix completion problem [26] is to find the minimum-rank matrix $Y = [y_{it}] \in \mathbb{R}_+^{I \times T}$ that has the same entries as the matrix M in the items indicated by the set Ω. Such a problem can be formulated as follows:

$$\min_{Y} \text{rank}(Y), \quad \text{s.t.} \quad y_{it} = m_{it}, \forall (i, t) \in \Omega. \tag{1}$$

The above approach has been widely used for various applications, especially for image completion [24,27,28] (including fingerprints restoration [29]) as well as recommendation systems. It can be applied for recovering missing pixels in gray scale (2D matrix) or color (3D matrix or tensor). It can be also used for estimating users' preference in an user-item matrix, especially for large-scale problems, e.g. for the MovieLens dataset (22,884,377 ratings and 586,994 tag applications for 33,670 movies evaluated by 247,753 users) [25].

2.2 Smooth NMF

In this study, we focus on the smooth NMF model that assumes that each feature vector a_j from A is a linear combination of locally smooth nonnegative basis vectors $\{s_n^{(r)}\}$. Thus:

$$\forall j : a_j = \sum_{n=1}^{N} w_{nj} s_n^{(r)}, \tag{2}$$

where $\{w_{nj}\}$ are coefficients of a linear combinations of the basis vectors. In [13,14], the basis functions are expressed by r-th order B-splines, i.e. each $s_n^{(r)} = \left[S_n^{(r)}(\xi)\right] \in \mathbb{R}_+^J$ is determined uniformly in the interval $[\xi_{min} \leq \xi < \xi_{max}]$. The points $\{\xi_1, \ldots, \xi_N\}$ are known as knots. The B-splines can be obtained by the "Cox-DeBoor" recursive formula. For our applications, we set $r = 4$. Taking into account (2), the model for smooth NMF takes the form:

$$Y = SWX, \tag{3}$$

where $S = [s_1^{(r)}, \ldots, s_N^{(r)}] \in \mathbb{R}_+^{I \times N}$, $W = [w_{nj}] \in \mathbb{R}^{N \times J}$ and $X = [x_{jt}] \in \mathbb{R}_+^{J \times T}$.

Let $A = SW \in \mathbb{R}_+^{I \times J}$. To estimate W in (3), the following minimization problem is formulated:

$$\min_{W, A} \Psi(W) + \Phi(A), \quad \text{s.t.} \quad SW = A, \tag{4}$$

where $\Psi(W) = \frac{1}{2}\|Y - SWX\|_F^2$ is a closed convex function w.r.t. W, and $\Phi(A) = \sum_{i,j} \mathcal{I}_\Gamma(a_{ij})$ is a non-differentiable convex function, where

$$\mathcal{I}_\Gamma(a_{ij}) = \begin{cases} 0 & \text{if } a_{ij} \in \Gamma, \\ \infty & \text{else} \end{cases} \tag{5}$$

is an indicator function for the set $\Gamma = \{\xi : \xi \geq 0\}$.

To estimate the factors W and A in (4), we used the ADMM from [14]. The factor X in (3) is updated with the fast HALS algorithm [1].

2.3   Image Completion with Smooth NMF

The image completion can be expressed by the problem (1), which belongs to a class of NP-hard problems. Following [1, 24, 25], the matrix Y can be approximated by the product of lower-rank nonnegative factors A and X obtained by solving the constrained NMF problem:

$$\min_{A,X} ||P_\Omega(AX) - P_\Omega(M)||_F, \quad \text{s.t. } A \in \mathbb{R}_+^{I \times J} \text{ and } X \in \mathbb{R}_+^{J \times T}. \qquad (6)$$

where $P_\Omega(.)$ stands for the projection onto Ω, and $J \ll \min\{I, T\}$. Thus $Y \approx P_\Omega(AX) \in \mathbb{R}_+^{I \times T}$. Considering the above approach, the recursive rule for updating Y can be expressed by Algorithm 1, where the function $[A, X] =$ ADMM-NMF(Y, J) is executed by the ADMM-NMF with the B-splines. Note that

---

Algorithm 1. B-Splines-based Algorithm for Image Completion (BSA-IC)

Input : $M \in \mathbb{R}^{I \times T}$ – incomplete matrix, Ω – set of indices of known entries,
δ – threshold for stagnation, J – rank of factorization
Output: $Y \in \mathbb{R}_+^{I \times T}$ – completed matrix

1 Initialize: $Z = 0 \quad \in \mathbb{R}^{I \times T}$, $\epsilon_1 > \delta$, $\epsilon_2 = 0$;
2 **while** $|\epsilon_1 - \epsilon_2| > \delta$ **do**

/* Update for Y */

3 $\quad$ $\tilde{Y} = Z + Y$, $\epsilon_1 = ||Z||_F^2$;

4 $\quad$ $Y = [y_{ij}]$, where $y_{ij} = \begin{cases} m_{ij} & \text{if } (i,j) \in \Omega \\ \tilde{y}_{ij} & \text{otherwise} \end{cases}$

5 $\quad$ $[A, X] = $ ADMM-NMF(Y, J);   // Smooth NMF

6 $\quad$ $\tilde{Y} = AX$;

7 $\quad$ $\tilde{Z} = Y - \tilde{Y}$;

8 $\quad$ $z_{ij} = \begin{cases} \tilde{z}_{ij} & \text{if } (i,j) \in \Omega \\ 0 & \text{otherwise} \end{cases}$;

9 $\quad$ $Z = [z_{ij}] \in \mathbb{R}_+^{I \times T}$;

10 $\quad$ $\epsilon_2 = ||Z||_F^2$;

---

the ADMM-NMF aims to provide locally smooth features, hence the matrix Y should demonstrate locally vertical smoothness. To enforce local smoothness in both orthogonal directions, the step 5 of Algorithm 1 should be executed alternatingly for Y and Y^T. When applied to Y^T, then $[X^T, A^T] =$ ADMM-NMF(Y^T, J).

The **while** loop in Algorithm 1 should be repeated until the residual error determined by the objective function in (6) drops below a given threshold δ. The computational complexity of one iterative step of Algorithm 1 amounts to $O(IJT)$, provided that the computational complexity for the ADMM-NMF can be also upper-bounded by $O(IJT)$.

3   Experiments

The proposed algorithm is evaluated for various image completion problems using the natural images that are illustrated in Fig. 1.

Fig. 1. Original images: (left) boat; (middle) mountain; (right) snowfall night-view city

The experiments were performed using the PLGRID[1] queues on the distributed cluster server in Wroclaw Center for Networking and Supercomputing (WCSS)[2] using the nodes with 8 cores (ncpus) and 8GB RAM (mem), and also on the workstation with the following parameters: Windows 10, CPU Intel i7-4790K 4.00 GHz, 8 GB RAM, using Matlab 2016a and its parallelization pool.

In the first experiment, the tests are performed on 2 images: boat (512×512 pixels, gray scale) and mountain (384×254 pixels, color). The incomplete images are obtained by removing from the original ones: (a) randomly selected 50%, 70% and 90% of pixels, (b) single lines of pixels forming a regular grid of 10 pixels wide. In this experiment, the following algorithms are compared: the HALS for image completion [1,25], SVT [27], LMaFit [28], and BSA-IC (Algorithm 1).

Due to the non-convexity property of NMF algorithms, each algorithm is re-run 100 times for various random initializations. The recovered images are validated quantitatively using the Signal-to-Interference Ratio (SIR) measure [1]. The boxplots of SIR values are demonstrated in Figs. 2 and 3, together with the selected completed images. The averaged runtime [in seconds] of each algorithm is listed in Table 1.

In the second experiment, we test the usefulness of the proposed algorithm for tackling the rain or snow removal problem [30,31]. The aim is to remove the snowflakes from the image snowfall night-view city that is shown in Fig. 1. To use image completion algorithms for this problem, the snowflakes need to be first marked by the set Ω, and we achieved this task with the image segmentation method based on the Markov Random Field (MRF) [32]. For this test, we selected 2 best algorithms for image completion: the HALS and BSA-IC. The results are depicted in Fig. 4. The SIR-values and the runtime are shown in Table 2.

---

[1] http://www.plgrid.pl/en.

[2] https://www.wcss.pl/en/.

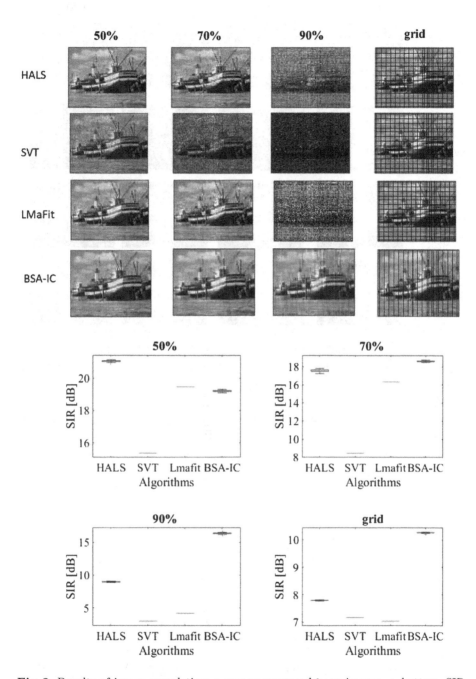

Fig. 2. Results of image completion: • *upper:* recovered `boat` images, • *bottom:* SIR statistics for the `boat` image.

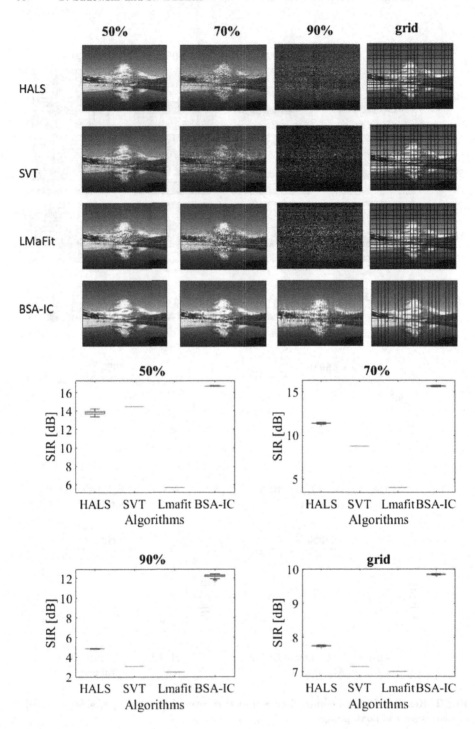

Fig. 3. Results of image completion: • *upper:* recovered mountain images, • *bottom:* SIR statistics for the mountain image.

Table 1. Averaged runtime [sec.] and the standard deviation obtained with the tested algorithms for various image completion problems.

	HALS	SVT	LMaFit	BSA-IC
Boat				
50%	422.6 ± 36.15	145.11 ± 7.7	21.39 ± 3.04	592.27 ± 372.0
70%	431.68 ± 22.03	84.73 ± 7.28	14.97 ± 1.62	944.18 ± 364.53
90%	442.02 ± 32.42	62.79 ± 9.44	12.68 ± 1.99	2843.95 ± 568.47
Grid%	246.87 ± 83.36	418.76 ± 30.91	30.1 ± 4.93	527.55 ± 281.57
Mountain				
50%	378.58 ± 22.24	199.45 ± 8.49	13.6 ± 1.78	408.64 ± 95.18
70%	295.62 ± 11.12	147.64 ± 8.28	11.58 ± 1.3	753.22 ± 138.84
90%	118.75 ± 3.75	82.06 ± 3.75	8.99 ± 0.6	1728.07 ± 246.29
Grid	97.58 ± 14.85	191.48 ± 7.99	17.98 ± 2.23	252.26 ± 66.81

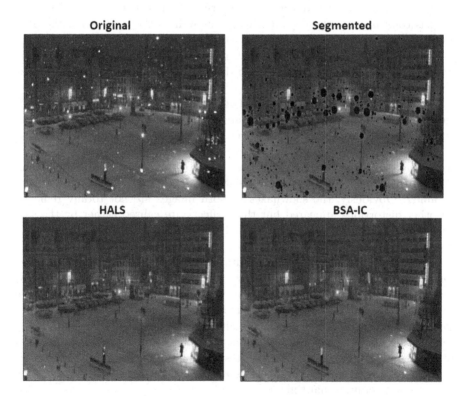

Fig. 4. Snow removal results: • *upper left:* input incomplete image, • *upper right:* segmented image with marked snowflakes, • *bottom left:* HALS-based estimate, • *bottom right:* BSA-IC-based estimate

Table 2. SIR-values [dB] and runtime [sec.] obtained with the HALS and BSA-IC for the snow removing problem.

	HALS	BSA-IC
SIR [dB]	15.67 ± 0.07	16.40 ± 0.08
time [s]	494.57 ± 34.40	885.99 ± 242.52

4   Conclusions

In this study, we demonstrated experimentally that the smooth NMF model with B-splines works more efficiently for the selected image completion problems than the other tested algorithms, including the HALS-NMF model. The proposed algorithm is particularly efficient for highly incomplete images. Figures 2 and 3 show that the images recovered by the BSA-IC have the highest quality. Only the boat image, which reconstructed from 50% pixels by the HALS, seems to have better quality than with the BSA-IC. In all other cases, the BSA-IC significantly (considering the standard deviation) outperforms the other tested algorithms, which is confirmed by the depicted images as well as the boxplots of the SIR-values. Even the images reconstructed from only 10% of available pixels have quite good quality (especially the boat image). The BSA-IC is also the only method that partially removes the grid distortion, leading to the highest SIR-values.

The results listed in Table 1 show that the BSA-IC is slower than the other algorithms, especially when the images are highly incomplete. Further works are necessary to optimize the Matlab code of the proposed algorithm.

The BSA-IC also works very well for the snow removal problem. Figure 4 shows that the image recovered with the BSA-IC has better quality than obtained with the HALS. This statement is also confirmed by the SIR values in Table 2. Obviously, the snowflakes in both images are not totally removed and other distortions appeared but these effects are caused rather by the segmentation method. The analyzed original image is very challenging because it presents the night-view city and the snowflakes are poorly visible and strongly colored by the lantern light. The segmentation of such an image is very difficult. Moreover, the lantern light points are very bright, and hence they are indexed by the segmentation methods similarly as the snowflakes. As a result, the image completion algorithms remove these areas from the analyzed image. Further research is also needed to choose more suitable image segmentation method.

Summing up, the proposed method is suitable for solving image completion problems, and it outperforms the other tested methods with respect to the quality of the recovered images. Further researches are needed to optimize it with respect to a computational time.

Acknowledgment. This work was partially supported by the grant 2015/17/B/ST6/01865 funded by National Science Center (NCN) in Poland. Calculations were performed at the Wroclaw Center for Networking and Supercomputing under grant no. 127.

References

1. Cichocki, A., Zdunek, R., Phan, A.H., Amari, S.I.: Nonnegative Matrix and Tensor Factorizations: Applications to Exploratory Multi-way Data Analysis and Blind Source Separation. Wiley, Hoboken (2009)
2. Lee, D.D., Seung, H.S.: Learning the parts of objects by non-negative matrix factorization. Nature **401**, 788–791 (1999)
3. Zhang, H., He, W., Zhang, L., Shen, H., Yuan, Q.: Hyperspectral image restoration using low-rank matrix recovery. IEEE Trans. Geosci. Remote Sens. **52**(8), 4729–4743 (2014)
4. Miao, L., Qi, H.: Endmember extraction from highly mixed data using minimum volume constrained nonnegative matrix factorization. IEEE Trans. Geosci. Remote Sens. **45**(3), 765–777 (2007)
5. Jia, S., Qian, Y.: Constrained nonnegative matrix factorization for hyperspectral unmixing. IEEE Trans. Geosci. Remote Sens. **47**(1), 161–173 (2009)
6. Xu, W., Liu, X., Gong, Y.: Document clustering based on non-negative matrix factorization. In: SIGIR 2003: Proceedings of 26th Annual International ACM SIGIR Conference on Research and Development in Information Retrieval, pp. 267–273. ACM Press, New York (2003)
7. Pauca, V.P., Shahnaz, F., Berry, M.W., Plemmons, R.J.: Text mining using nonnegative matrix factorizations. In: Proceedings of SIAM Interernational Conferene on Data Mining, Orlando, FL, pp. 452–456 (2004)
8. Kersting, K., Wahabzada, M., Thurau, C., Bauckhage, C.: Hierarchical convex NMF for clustering massive data. In: Proceedings of 2nd Asian Conference on Machine Learning, ACML, Tokyo, Japan, pp. 253–268 (2010)
9. Virtanen, T.: Monaural sound source separation by nonnegative matrix factorization with temporal continuity and sparseness criteria. IEEE Trans. Audio Speech Lang. Process. **15**(3), 1066–1074 (2007)
10. Zdunek, R., Cichocki, A.: Blind image separation using nonnegative matrix factorization with Gibbs smoothing. In: Ishikawa, M., Doya, K., Miyamoto, H., Yamakawa, T. (eds.) ICONIP 2007. LNCS, vol. 4985, pp. 519–528. Springer, Heidelberg (2008). https://doi.org/10.1007/978-3-540-69162-4_54
11. Févotte, C., Bertin, N., Durrieu, J.L.: Nonnegative matrix factorization with the Itakura-Saito divergence: with application to music analysis. Neural Comput. **21**(3), 793–830 (2009)
12. Zdunek, R.: Approximation of feature vectors in nonnegative matrix factorization with Gaussian radial basis functions. In: Huang, T., Zeng, Z., Li, C., Leung, C.S. (eds.) ICONIP 2012. LNCS, vol. 7663, pp. 616–623. Springer, Heidelberg (2012). https://doi.org/10.1007/978-3-642-34475-6_74
13. Zdunek, R., Cichocki, A., Yokota, T.: B-spline smoothing of feature vectors in nonnegative matrix factorization. In: Rutkowski, L., Korytkowski, M., Scherer, R., Tadeusiewicz, R., Zadeh, L.A., Zurada, J.M. (eds.) ICAISC 2014. LNCS (LNAI), vol. 8468, pp. 72–81. Springer, Cham (2014). https://doi.org/10.1007/978-3-319-07176-3_7
14. Zdunek, R.: Alternating direction method for approximating smooth feature vectors in nonnegative matrix factorization. In: IEEE International Workshop on Machine Learning for Signal Processing, MLSP 2014, Reims, France, 21–24 September 2014, pp. 1–6 (2014)
15. Boyd, S., Parikh, N., Chu, E., Peleato, B., Eckstein, J.: Distributed optimization and statistical learning via the alternating direction method of multipliers. In: Foundations and Trends in Machine Learning. vol. 3, pp. 1–122. NOW Publishers (2011)

16. Goldstein, T., O'Donoghue, B., Setzer, S., Baraniuk, R.G.: Fast alternating direction optimization methods. SIAM J. Imaging Sci. **7**, 1588–1623 (2014)
17. Hajinezhad, D., Chang, T., Wang, X., Shi, Q., Hong, M.: Nonnegative matrix factorization using ADMM: algorithm and convergence analysis. In: ICASSP, pp. 4742–4746. IEEE (2016)
18. Sun, D.L., Fevotte, C.: Alternating direction method of multipliers for non-negative matrix factorization with the beta-divergence. In: Proceedings of IEEE International Conference on Acoustics, Speech and Signal Processing (ICASSP), Florence, Italy (2014)
19. Esser, E., Möller, M., Osher, S., Sapiro, G., Xin, J.: A convex model for nonnegative matrix factorization and dimensionality reduction on physical space. IEEE Trans. Image Process. **21**(7), 3239–3252 (2012)
20. Xu, Y.: Alternating proximal gradient method for nonnegative matrix factorization. CoRR abs/1112.5407 (2011)
21. Xu, Y., Yin, W., Wen, Z., Zhang, Y.: An alternating direction algorithm for matrix completion with nonnegative factors. Front. Math. China **7**(2), 365–384 (2012)
22. Sun, D.L., Mazumder, R.: Non-negative matrix completion for bandwidth extension: a convex optimization approach. In: Proceedings of IEEE International Workshop on Machine Learning for Signal Processing (MLSP 2013), Southampton, UK (2013)
23. Yuan, X., Yang, J.F.: Sparse and low rank matrix decomposition via alternating direction method. Pac. J. Optim. **9**(1), 167–180 (2013)
24. Yokota, T., Zhao, Q., Cichocki, A.: Smooth PARAFAC decomposition for tensor completion. IEEE Trans. Sig. Process. **64**(20), 5423–5436 (2016)
25. Sadowski, T., Zdunek, R.: Modified HALS algorithm for image completion and recommendation system. In: Świątek, J., Borzemski, L., Wilimowska, Z. (eds.) ISAT 2017. AISC, vol. 656, pp. 17–27. Springer, Cham (2018). https://doi.org/10.1007/978-3-319-67229-8_2
26. Guo, X., Ma, Y.: Generalized tensor total variation minimization for visual data recovery. In: Proceedigngs of IEEE Conference on Computer Vision and Pattern Recognition (CVPR) (2015)
27. Cai, J.F., Candès, E.J., Shen, Z.: A singular value thresholding algorithm for matrix completion. SIAM J. Optim. **20**(4), 1956–1982 (2010)
28. Wen, Z., Yin, W., Zhang, Y.: Solving a low-rank factorization model for matrix completion by a nonlinear successive over-relaxation algorithm. Math. Program. Comput. **4**(4), 333–361 (2012)
29. Chugh, T., Cao, K., Zhou, J., Tabassi, E., Jain, A.K.: Latent fingerprint value prediction: crowd-based learning. IEEE Trans. Inf. Forensics Secur. **13**(1), 20–34 (2018)
30. Pei, S.C., Tsai, Y.T., Lee, C.Y.: Removing rain and snow in a single image using saturation and visibility features. In: IEEE International Conference on Multimedia and Expo Workshops (ICMEW), pp. 1–6 (2014)
31. Barnum, P., Kanade, T., Narasimhan, S.G.: Spatio-temporal frequency analysis for removing rain and snow from videos. In: Workshop on Photometric Analysis For Computer Vision (PACV), in Conjunction with ICCV, Pittsburgh, PA (2007)
32. Demirkaya, O., Asyali, M., Sahoo, P.: Image Processing with MATLAB: Applications in Medicine and Biology, 2nd edn. Taylor and Francis, Oxford (2015)

A Fuzzy SOM for Understanding Incomplete 3D Faces

Janusz T. Starczewski[1,2(✉)], Katarzyna Nieszporek[1], Michał Wróbel[1], and Konrad Grzanek[3,4]

[1] Institute of Computational Intelligence, Częstochowa University of Technology, Częstochowa, Poland
`janusz.starczewski@iisi.pcz.pl`
[2] Radom Academy of Economics, Radom, Poland
[3] Information Technology Institute, University of Social Sciences, Lódz, Poland
[4] Clark University, Worcester, MA 01610, USA

Abstract. This paper presents a new recognition method for three-dimensional geometry of the human face. The method measures biometric distances between features in 3D. It relies on the common self-organizing map method with fixed topological distances. It is robust to missing parts of the face due to the introduction of an original fuzzy certainty mask.

Keywords: Biometric · 3D face · Self-organizing map
Fuzzy certainty map

1 Introduction

Currently, many studies are devoted to biometric face identification and verification. The number of approaches to face recognition has been raised significantly for the last decade [2,5,7,10]. Numerous methods have the potential to be adapted to face recognition [1,3,8,13].

In this paper we present our subsequent method addressed to 3D Face Recognition (3DFR). It has been demonstrated that 3D face recognition methods gain higher accuracy with respect to 2D methods due to its ability to measure precisely the face geometry as intervals between the characteristic points (features) on the face (see e.g. [5]). Moreover, 3DFR is resistant to variable lighting conditions, make-up, glasses, changes in beards and facial hair or finally, different face orientation. An interesting side effect is a possibility to read and interpret facial expressions and intentions or truthfulness of analyzed people.

Our contribution to this field is a novel 3DFR method based on self-organized maps (continuation of our previous research [12]) and fuzzy measures, which is robust to missing parts of the face. We propose a method that marks and labels characteristic points of faces. Since the marked features are identified and labeled, we refer to as the understanding of faces rather than feature extraction. Understood features are i.a. the corners of eyes, lips, eyebrow. A complete set of considered features is presented in Fig. 1a.

© Springer International Publishing AG, part of Springer Nature 2018
L. Rutkowski et al. (Eds.): ICAISC 2018, LNAI 10842, pp. 73–80, 2018.
https://doi.org/10.1007/978-3-319-91262-2_7

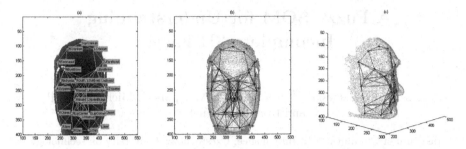

Fig. 1. Face feature understanding: (a) initial SOM with labeled features, (b) and (c) expansion of the SOM on a 3D face to be understood (actual features indicated by plus signs)

2 A SOM Approach to 3DFR

As a direct consequence of our previous research [12], we have applied the SOM with labeled nodes to understand face characteristic features. The topology of the proposed SOM has been fitted to the features on the face. The method has two stages:

- generation of an initial SOM,
- application of the SOM to new faces without modification of neighborhood coefficients.

I stage

The initial SOM has been expanded on a generic 3D face model with prior identification of characteristic points by cluster analysis according to the following steps:

1. We have made use of the Sobel gradient detector to find more significant changes in the Zdimension of the face coordinates according to X- and Y-axes separately.
2. Both gradients have been used to calculate a magnitude of the resultant vector, hence, we have unified both positive and negative gradients along the two dimensions. In detail,

$$I = \sqrt{\left(conv\left(\mathbf{h}, \mathbf{I}\right)^2 + conv\left(\mathbf{h}', \mathbf{I}\right)\right)^2} \tag{1}$$

where the power has been calculated element-wise and the Sobel mask for the convolutions has been chosen as

$$\mathbf{h} = \begin{bmatrix} 1 & 0 & -1 \\ 2 & 0 & -2 \\ 1 & 0 & -1 \end{bmatrix} \tag{2}$$

3. In order to make SOMs be sensitive to the magnitude of gradients, we have been choosing the training points randomly with the probability proportional to the gradient magnitudes.
4. Such points have been clustered by the standard Fuzzy C-Means algorithm with a number of clusters validated by apparent their utility as characteristic points. During multiple runs of the algorithm, we have decided to limit the number of centers to 27 characteristic points. The averaged labeled points are illustrated in the Fig. 2.
5. The real distances between the 3D characteristic points on the generic model surface have been used as lateral distances. For simplicity of calculations, we have omitted the distances of 3rd and the higher level of the neighborhood. Consequently, the Gaussian neighborhood grades could be stored in an array for calculations in the next stage. The resultant SOM has been set as the initial map for further identification of 3D face characteristic points.

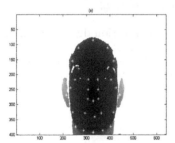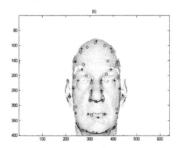

Fig. 2. Interpretable characteristic points (indicated by stars) obtained by FCM clustering: (a) shaded representation of the 3D generic model, (b) reference to single-run FCM clusters (indicated by circles) on the face model after the Sobel transformation

II stage

In the second stage, we have trained quasi-standard SOM on 3D faces with the difference that the neighborhood coefficients had been fixed in the previous stage. We have made use of an exponentially decreasing learning factor.

3   A New Fuzzy Certainty Mask for SOM

A problem of dead neurons is commonly known in SOMs. Namely, the node which is closer to the winner neuron learns more efficiently than neurons which are farther according to the vanishing neighbor function centered at the winner neuron. Such units that have not been the winner at least once in the whole presentation of the dataset are not able to organize themselves properly. They are so-called dead neurons. Nevertheless, we use this drawback as a profit to the uncertain treatment of recognition.

In cases of partially lacking faces (e.g. with lost fragments during the acquisition process), the standard SOM moves all its nodes to the existing fragment of the face. Our motivation is to "kill" such neurons that are not affected directly by the face model and their only possibility to learn results from the neighborhood of other winner neurons. We check such uncommitted neurons within a single run of the SOM algorithm nod modifying locations of neurons, hence called a zero-run. Nodes that never wined in the zero-run have to be blocked from any modification and treated consequently as dead. To this purpose, we assign to such nodes a zero membership grade in an introduced fuzzy certainty mask. Actually, at this moment our certainty mask is hard instead of fuzzy and we are in need of fuzzification of the border between dead neurons and active neurons in order that active neurons close to the borderline are not moved to far toward the center of the face. Ergo we can make use of the percentages of winning given by counter $winner selected$, i.e.,

$$certainty_n = \left(\frac{winner selected_n}{\max \left(winner selected_n \right)} \right)^m \tag{3}$$

for each active n-th neuron, where m is the concentration degree empirically set as close to 8.

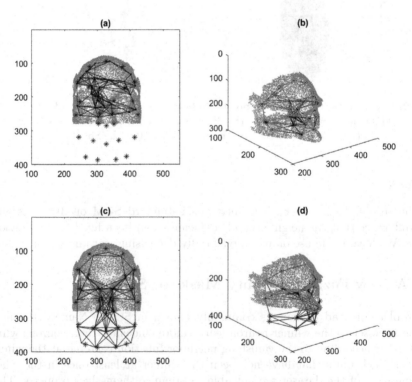

Fig. 3. Face understanding with asymmetrically missing part: (a, b) SOM, (c, d) SOM with the fuzzy certainty mask applied; desired and labeled features indicated by asterisks

Consequently, we use a modified learning formula for SOMs:

$$\Delta\mathbf{w}_n = \eta\left(t\right) h_{i(\mathbf{x}),n} certainty_n \left(\mathbf{x} - \mathbf{w}_n\right) \tag{4}$$

where

- η is a time-decaying learning rate,
- $h_{i(\mathbf{x}),n}$ is the neighbor function for the topological neighborhood of the winner node $i\left(\mathbf{x}\right)$,
- $certainty_n$ is a fuzzy grade of the certainty mask,
- $\mathbf{x}$ is a 3D vector for the face.

The behavior of our method is demonstrated (and compared with the ordinary SOM) for two cases of incomplete 3D faces in Figs. 3 and 4. We may notice that the standard map tries to extend entirely in the available surface of the face, while the map using fuzzy certainties leave the dead neurons in the starting (neutral) position.

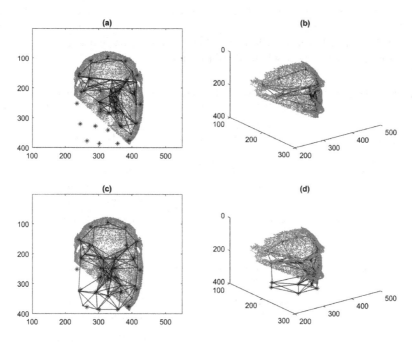

Fig. 4. Face understanding with symmetrically missing part: (a, b) SOM, (c, d) SOM with the fuzzy certainty mask applied; desired and labeled features indicated by asterisks

4   Experimental Results

The comparative study was carried out on a set of biometric three-dimensional images NDOff-2007 [4]. We selected 5 first faces from the whole collection of 6940 3D images gathered for 387 human faces (for a single person, there are several variants of face orientation; however we did not take advantage of multiple variants). We removed parts of the body not belonging to faces in preprocessing. Initial nodes for the SOM algorithm were translated according to the mean and scaled according to the variance of the cloud of facial points.

We performed 10 iterations of the SOM algorithm for each case. The learning factor decayed from 0.3 to 0.1 with the time constant equal to 3. The results for three cases of face completeness are presented in Table 1.

In all cases, the use of fuzzy certainty map resulted with usually multiple Root Mean Square Error (RMSE) reduction. Only in cases of complete face presentation, the profit coming from the certainty map was around 50% counted for all nodes including the dead ones to 65% considering certainty degrees of only active nodes. In the case of missing bottom parts of faces, we observed 4 times lower RMSE while using the certainty map calculated for all nodes (3.4 times lower for only active nodes). In the case of missing triangular parts of faces, we observed 3.2 times lower RMSE in the fuzzy approach when calculated for all nodes (and 2.6 times lower when calculated for only active nodes).

Table 1. RMSE for SOM and fuzzy-SOM algorithms in case of complete and incomplete faces

Face available	Face no.	SOM	FuzzyCertaintySOM	SOM	FuzzyCertaintySOM
		Including dead neurons		For only active neurons	
Complete	1	10.5893	7.8782	6.2506	4.6312
	2	15.3386	10.1428	5.0752	2.305
	3	13.9167	8.3723	4.978	3.4467
	4	23.6253	14.4362	10.8655	4.4711
	5	9.7819	8.4643	4.635	4.4462
	Average	12.2086	8.2156	5.3007	3.2167
Missing bottom	1	50.4051	6.4774	15.6969	2.4066
	2	17.5421	10.0572	6.4704	3.1328
	3	38.952	9.0042	5.013	3.2977
	4	39.4753	14.5856	15.594	4.0839
	5	42.1151	7.3578	7.7177	1.9245
	Average	31.4149	7.9137	8.4153	2.4743
Missing triangular	1	41.2353	9.735	13.8217	3.7895
	2	33.5679	10.4203	7.0264	2.6572
	3	25.4928	8.5424	3.8218	2.9305
	4	42.2614	15.8118	17.5471	4.6325
	5	26.1938	8.3608	3.2972	3.8349
	Average	28.1252	8.8117	7.5857	2.9741

5   Conclusion

We have demonstrated that SOMs together with fuzzy certainty masks are robust to incomplete face patterns. Due to the faithful assignment of characteristic points to labeled marks, the method is specially dedicated for automatic understanding of 3D human faces. It has been observed that variability in poses adversely affects the construction of maps. We suppose that the use of surface normals or the Laplacian in the face analysis will reduce this drawback. The model should be robust to each kind of uncertainties, i.e. that localized in the envelope face features. We are positive to solve such problems by introducing higher order uncertainty in the SOM model and processing it with the aid of fuzzy logic and the rough set theory [6,9,11].

References

1. Beg, I., Rashid, T.: Modelling uncertainties in multi-criteria decision making using distance measure and topsis for hesitant fuzzy sets. J. Artif. Intell. Soft Comput. Res. **7**(2), 103–109 (2017)
2. Bronstein, A.M., Bronstein, M.M., Kimmel, R.: Three-dimensional face recognition. Int. J. Comput. Vis. **64**(1), 5–30 (2005)
3. Chang, O., Constante, P., Gordon, A., Singaña, M.: A novel deep neural network that uses space-time features for tracking and recognizing a moving object. Artif. Intell. Soft Comput. **7**, 125–136 (2017). (LNCS, Springer)
4. Faltemier, T., Bowyer, K., Flynn, P.: Rotated profile signatures for robust 3D feature detection. In: 8th IEEE International Conference on Automatic Face Gesture Recognition, FG 2008, pp. 1–7, September 2008
5. Gupta, S., Markey, M.K., Bovik, A.C.: Anthropometric 3D face recognition. Int. J. Comput. Vis. **90**(3), 331–349 (2010)
6. Nowicki, R.: On combining neuro-fuzzy architectures with the rough set theory to solve classification problems with incomplete data. IEEE Trans. Knowl. Data Eng. **20**, 1239–1253 (2008)
7. Okuwobi, I.P., Chen, Q., Niu, S., Bekalo, L.: Three-dimensional (3D) facial recognition and prediction. SIViP **10**(6), 1151–1158 (2016)
8. Prasad, M., Liu, Y.T., Li, D.L., Lin, C.T., Shah, R.R., Kaiwartya, O.P.: A new mechanism for data visualization with TSK-type preprocessed collaborative fuzzy rule based system. J. Artif. Intell. Soft Comput. Res. **7**(1), 33–46 (2017)
9. Rivero, C.R., Pucheta, J., Laboret, S., Sauchelli, V., Patiño, D.: Energy associated tuning method for short-term series forecasting by complete and incomplete datasets. J. Artif. Intell. Soft Comput. Res. **7**(1), 5–16 (2017)
10. Spreeuwers, L.: Breaking the 99% barrier: optimisation of 3D face recognition. IET Biometr. **4**(3), 169–177 (2015)
11. Starczewski, J.T.: Advanced Concepts in Fuzzy Logic and Systems with Membership Uncertainty. Studies in Fuzziness and Soft Computing, vol. 284. Springer, Heidelberg (2013). https://doi.org/10.1007/978-3-642-29520-1

12. Starczewski, J.T., Pabiasz, S., Vladymyrska, N., Marvuglia, A., Napoli, C., Woźniak, M.: Self organizing maps for 3D face understanding. In: Rutkowski, L., Korytkowski, M., Scherer, R., Tadeusiewicz, R., Zadeh, L.A., Zurada, J.M. (eds.) ICAISC 2016. LNCS (LNAI), vol. 9693, pp. 210–217. Springer, Cham (2016). https://doi.org/10.1007/978-3-319-39384-1_19
13. Villmann, T., Bohnsack, A., Kaden, M.: Can learning vector quantization be an alternative to SVM and deep learning? - recent trends and advanced variants of learning vector quantization for classification learning. J. Artif. Intell. Soft Comput. Res. **7**(1), 65–81 (2017)

Feature Selection for 'Orange Skin' Type Surface Defect in Furniture Elements

Bartosz Świderski[1](✉)(iD), Michał Kruk[1](iD), Grzegorz Wieczorek[1](iD), Jarosław Kurek[1](iD), Katarzyna Śmietańska[2](iD), Leszek J. Chmielewski[1](iD), Jarosław Górski[2](iD), and Arkadiusz Orłowski[1](✉)(iD)

[1] Faculty of Applied Informatics and Mathematics – WZIM, Warsaw University of Life Sciences – SGGW, ul. Nowoursynowska 159, 02-776 Warsaw, Poland
{bartosz_swiderski,arkadiusz_orlowski}@sggw.pl
[2] Faculty of Wood Technology – WTD, Warsaw University of Life Sciences – SGGW, ul. Nowoursynowska 159, 02-776 Warsaw, Poland
http://www.wzim.sggw.pl, http://www.wtd.sggw.pl

Abstract. The surfaces of furniture elements having the *orange skin* surface defect were investigated in the context of selecting optimum features for surface classification. Features selected from a set of 50 features were considered. Seven feature selection methods were used. The results of these selections were aggregated and found consistently positive for some of the features. Among them were primarily the features based on local adaptive thresholding and on Hilbert curves used to evaluate the image brightness variability. These types of features should be investigated further in order to find the features with more significance in the problem of surface quality inspection. The groups of features which appeared least profitable in the analysis were the two features based on percolation, and the one based on Otsu global thresholding.

Keywords: Feature selection · Surface defect · Orange skin
Detection · Furniture · Feature selection · Brightness variability

1 Introduction

Visual inspection seems to remain the main method of assessing quality in the furniture industry. Despite that the machine vision methods have definitely attained their maturity and became the part of routine industrial processes, in the furniture industry this is still not the case.

To our best knowledge, the literature status for orange skin or orange peel did not change since our previous review [1]. In brief, we have stated that what concerns furniture elements quality inspection, the image processing methods are very rarely if not at all mentioned in the literature. The existing references mention the defects of furniture only in the context of more general domains [2], or the raw material quality is of main concern [3]. In fact, in very few papers the *orange skin* called also *orange peel* as a surface defect is considered directly.

© Springer International Publishing AG, part of Springer Nature 2018
L. Rutkowski et al. (Eds.): ICAISC 2018, LNAI 10842, pp. 81–91, 2018.
https://doi.org/10.1007/978-3-319-91262-2_8

Konieczny and Meyer [4] generate the images of *orange peel* artificially and study the visibility of this defect for humans in various conditions. Armesto et al. [5] describe a system of moving lighting and cameras. It is designed to improve the results of quality inspection of painted surfaces in the car industry. The target of this system is to enable the *defect augmentation phenomena* as the authors call the processes the light causes on surface features of various kinds. Among the broad class of surface defects the *orange peel* is present. The system makes it possible to use local adaptive thresholding as the only detection method.

Besides the papers, there are numerous patents in which methods are shown to avoid or remove orange skin in the painting process; let us name just one by Allard et al. [6] as an example. In none of these patents the image analysis methods are recalled.

On the contrary to the furniture industry, in the timber industry the image-based analysis of structural and anatomical defects is a well established technology with broad literature (cf. the reviews [7,8]).

It seems reasonable that in the preliminary stage in which the surface inspection in furniture industry is now, one of the main questions is the issue of finding proper image features which could capture the visual phenomena that make the surface look *good* or *bad*, in the context of *orange skin*. It is possible to take generic features like local spectral features, wavelet features or others, like for example the Histogram of Oriented Gradients, which we did in one of our previous papers [1]. What is more interesting, however, is to learn what features of the surface are important in the problem of our concern, and the real challenge would be to discover the meaning of these features in the problem.

In this paper we shall consider a set of features already found effective in a series of various demanding applications: classification of dermoscopic images of melanoma [9], regions in mammographic images [10,11] and images of microorganisms in soil [10]. Preliminary tests with feature selection made on this set have been previously made [12], and one of the methods of finding some order in the set which could make it easier to perform feature selection in a bottom-up manner was used. The features were ordered according to their Fisher measure of information content. They were added sequentially and the process was stopped when the classification accuracy attained its maximum. It must be admitted that drawing any conclusion from the result of this single experiment would be premature.

We shall use a set of seven feature selection methods chosen from those described by Pohjalainen et al. [13]. We shall check which features were selected the most frequently and finally we shall try to look at those features more carefully to discover how their design made them useful in the application of our interest. This will go far beyond the preliminary experiments presented previously [14] where we found that *orange skin* can be detected with simple differentiation and thresholding of the image intensity function.

The remaining part of this paper is organized as follows. In the next section the surface defect considered will be recalled, and the way the defect is seen in the images and the method with which the objects for classification are formed will be presented. In Sect. 3 the features and the classifier will be described. Section 4 will be devoted to the problem of feature selection. The results from seven feature selection methods will be presented in a combined way. These results will be briefly discussed in Sect. 5. Conclusions and outlook for further work will close the paper.

2   Classified Objects

2.1   The Defect: *Orange Skin*

The defect called *orange skin* or otherwise *orange peel* can appear on lacquered surfaces. In furniture elements it is one of the reasons for reduced esthetic quality of the product. It can emerge as small hollows, that is, an uneven structure of the hardened surface. The depth of hollows is smaller than the thickness of one lacquer layer so its order of magnitude is tenths parts of a millimeter. Numerous reasons can cause the defect to appear: insufficient quantity or bad quality of dilutent, large difference of temperatures between the lacquer and the surface, improper pressure or distance of spraying, excessive air circulation during spraying or drying, and insufficient air humidity. The structure of wood is hidden under the lacquer in the analysis.

Because the reason for classification of the surface as good or bad is of esthetic nature and because not only the presence of hollows but also their relations are important, it is not possible to indicate a well-defined *defect* on the surface, like it would be for example in the case of a crack or scratch. The parts of the surface which *look good* gradually pass to those which *look bad*. The *good* surface is not free from texture and has some deviations from planarity. Example images will be shown in the next section.

2.2   Images and Objects

In the present paper we have used the same set of images we analyzed in a previous paper [12]. The images were taken with the Nikon D750 24 Mpix camera equipped with the Nikon lens F/2.8, 105 mm. The distance from the focal plane to the object surface was 1 m and the optical axis of the camera was normal to the surface. The lighting was provided by a flash light with a typically small light emitting surface, located at 80 cm from the object, with the axis of the light beam deflected by 70° from the normal to the surface. In this way, the light came from a direction close to parallel to the surface, to emphasize the surface profile. The camera was fixed on a tripod and it was fired remotely to avoid vibration.

The objects were painted with white lacquer in a typical technological process. The surfaces imaged belonged to several different objects. The surfaces were classified by the furniture quality expert into three classes: *very good*, *good* and *bad* in the terms of the orange skin defect, before the experiment. The photographs were made of a part of the object which was not farther than 30 cm from the center of the image. An image of a part of an elongated objects was taken once at a time, then the object was moved an a next image was taken, to include all of the surface of the objects in the experiment. The images were made in color mode and stored losslessly.

From these images, small non-overlapping images were cut, each of them of size 300 × 300 pixels. There were 900 such images total. Each of these images was treated as a separate object and was classified independently of the other images.

From these objects, the training and testing sets were chosen for cross-validation. Ten cross-validation rounds were planned. In each round, there were 90 images in the testing set, selected randomly from the set of images, with equal numbers of images belonging to each class. The remaining 810 images formed the training set in the given round. The numbers of images belonging to the classes resulted from the classification made by the human expert and were close to equal, but not precisely equal.

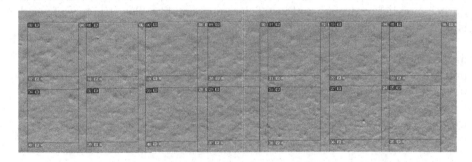

Fig. 1. Example of images of the surface of furniture elements. Small images of size 300 × 300 like those outlined with blue lines and marked with small dark blue icons were cut for the training and testing processes. (Color figure online)

The way the small images were cut can be seen in Fig. 1. The examples of images belonging to the three classes selected in the experiment are shown in Fig. 2. A very good surface has a fine and even texture. A bad surface has an uneven and strong texture. A good surface is everything in the middle. Note that a good surface can differ from a very good one by that its texture is less even, although it can be weaker, like that in the images in the lowest row of Fig. 2.

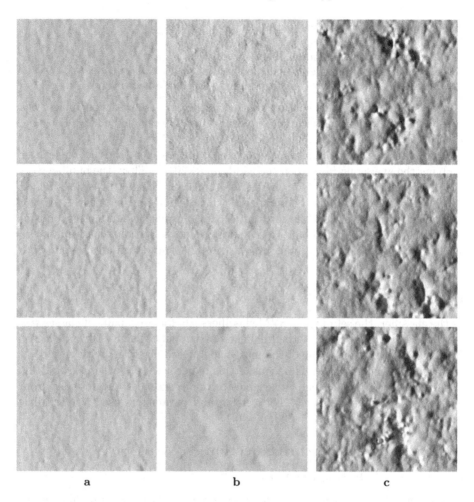

a   b   c

Fig. 2. Examples of images of the surfaces belonging to three classes: (a) very good, (b) good and (c) bad.

3   Features and Classification

3.1   Features

All the features were generated from the luminance component Y of the YIQ color model, $Y \in \{0, 1, \ldots, 255\}$.

The features for each small image were formed with the methods listed below. There were 50 features in total. The ranges of indexes of features are given in boldface:

01–01: number of black fields after thresholding with Otsu method – 1 feature;
02–08: Kolmogorow-Smirnow features [15] – 7 features;

09–14: maximum subregions features [9] – 6 features;
15–16: features based on the percolation [9] – 2;
17–32: features based on the Hilbert curve [10, 11] – 16;
33–41: features from single-valued thresholding [12] (explained below) – 9;
42–50: features from adaptive thresholding [12] (explained below) – 9.

The single-valued thresholding was performed as follows. The image was thresholded, in sequence, with thresholds: $i/10 \times 255, i = 1, 2, \ldots, 9$. The nine features are the numbers of black regions after each thresholding.

The adaptive thresholding was performed as follows. Let A be the image after applying the averaging filter with the window 20×20 pix. Then, the number I_2 is calculated as $I_2 = A - Y - C$, where C is a constant. The result is thresholded at 0.1×255. The feature is the number of white regions in the image I_2, giving 9 features. It remains to set the constant C. Setting the constant corresponds in fact to modifying the threshold. To scan a range of thresholds, nine values were taken, $C = i, i = 1, 2, \ldots, 9$, giving nine features. A white region corresponds to a dark blob in the image.

3.2 Classifier

As the classifier, the set of three SVM classifiers [16] for three pairs of classes was used. The classifiers voted for the final class assignment. The SVM was selected for this study due to it is one of the most widely used classifiers with very good performance in many cases. The focus of this paper is set on features, so at this stage we reduced the number of variables in the experiment and resigned from considering multiple classifiers. The version and parameters used were: radial-basis function kernel, cost $c = 300$, $\sigma = 0.1$.

4 Training and Feature Selection

4.1 Feature Selection Methods

To find the globally optimal set of features, an exhaustive search within the set of available features should be performed, which is totally impractical (in the set of 50 features there are $2^{50} - 1 > 10^{16}$ different nonempty subsets). Therefore, we used seven feature selection methods chosen from those previously described [13]. The software for these methods is publicly available [17]. The methods are:

1. method based on Chi square test (Chi2) [18],
2. Correlation-based feature selection (CFS) [19],
3. Fast Correlation-based Filter (FCBF) [20, 21],
4. method based on Fisher score (FS) [22],
5. method based on Information Gain (IG) [23],
6. Sparse Multinomial Logistic Regression via Bayesian L1 Regularization (SMLR) [24],
7. Kruskal-Wallis variance test [25].

With each method, the features were selected on the basis of just one of the ten training sets. Then, these features were used in the cross-validation process to assess the accuracy of classification.

4.2   Accuracy of Classification

Accuracies of classification attained with the methods are shown in Fig. 3. They are sorted according to the average accuracy attained. It happened that the standard deviation of errors increased nearly monotonically together with the decrease of accuracy. Consequently, the method Chi square appeared the best both in relation to its high accuracy as well as low accuracy deviation.

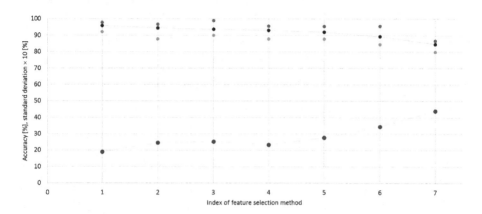

Fig. 3. Accuracy of classification for feature selection methods attained in the cross-validation process, sorted according to descending average accuracy. Indexes of feature selection methods comply with those in Table 1. Upper graphs: average accuracy (black ■ dots with line); maximum (blue ■) and minimum (red ■) errors attained in the cross-validation series; lower graph: standard deviation (green ■ dots with line) of these results, 10× enlarged for better visibility. Lines connecting points related to methods have no meaning except for indicating the correspondence between these points. (Color figure online)

The values of average, maximum and minimum accuracies obtained in the ten cross-validation rounds are shown in Table 1. The best average accuracy slightly exceeds 95% and the minimum accuracy, which is the pesimistic estimate of actual accuracy, is slightly larger than 92%. This indicates that indeed some more work should be done to look for better features.

4.3   Feature Selection Results

In Table 1, alongside with the accuracies, the five best features are shown for those methods in which the single features are assigned a measure of quality in a natural way[1]. More insight in the results of feature selection can be gained by examining the histograms in Fig. 4 which show the number of times a given feature was selected in the seven feature selection methods used. It can be seen

---

[1] This does not concern CFS, where features are not sequenced; in this method, the following features were selected: $\{2, 3, 13, 14, 23, 24, 28, 34, 39, 40, 41, 43, 45, 47\}$.

Table 1. Results of feature selection and accuracy measures attained. Rows sorted according to descending average accuracy (in **bold font**).

#	Method	Acronym	Accuracy [%]				No. fea.	5 best features (if applicable)				
			Avg	Min	Max	Sdv						
1	Chi Square	Chi2	**95.88**	92.22	97.78	1.90	41	29	28	43	18	41
2	Information gain	IG	**94.43**	87.78	96.67	2.46	26	43	29	28	41	18
3	Fisher score	FS	**93.65**	90.00	98.89	2.53	26	29	24	18	20	13
4	Correlation-based feature selection	CFS	**93.09**	87.78	95.56	2.33	14					
5	Kruskal-Wallis variance test	KWVT	**91.98**	87.78	95.51	2.76	17	43	41	39	6	5
6	Sparse multinomial logistic regression	SMLR	**89.31**	84.44	95.56	3.45	10	43	13	10	40	38
7	Fast correlation-based filter	FCBF	**84.75**	80.00	86.67	4.40	5	43	28	23	14	36

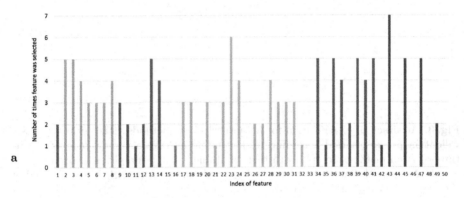

a

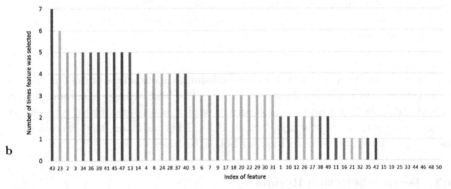

b

Fig. 4. Cumulated results of feature selection. (a) Features sorted according to the groups of features; (b) features sorted according to the number of times of being selected. Groups of features marked with colors (cf. enumeration in Sect. 3.1): **01–01:** ■ dark green, Otsu, 1 feature; **02–08:** ■ green, Kolmogorow-Smirnow, 7 features; **09–14:** ■ red, maximum subregions, 6; **15–16:** ■ blue, percolation, 2; **17–32:** ■ yellow, Hilbert, 16; **33–41:** ■ violet, single-valued thresholding, 9; **42–50:** ■ dark blue, adaptive thresholding, 9 features. (Color figure online)

that some features were selected in more feature selection algorithm, and some in less of them. Some features were not selected at all.

5 Discussion

The graphics in Fig. 4 indicate that the feature 43 was always chosen (7 times) and the next most frequently chosen feature was 23. Other features belonging to nearly all groups of features were chosen 5 times. On the opposite end, features like 15, 19, 25, etc. were not chosen at all. Both the most frequently chosen features and the least frequently chosen ones belong to various groups. No general conclusion concerning the groups of features can be drawn.

The most frequently chosen feature with the index 43 is the second feature from the set of adaptive thresholding-based features. This indicates that adaptive thresholding is a viable method in looking for good features in our task. Other features from this group were chosen as well, but the thresholds used in them differed by more than one (except index 1 and 2). This means that serial thresholding with different threshold has its merit.

The second best feature has the index 23 and it is the sevenths feature based on Hilbert curves. The features from this group are using the following scheme. Two strings of pixels located on subsequent fragments of the Hilbert curve are compared. In the comparison, the Kolmogorow-Smirnow statistics S_{KS} is used together with its minimum significance level p. The larger S_{KS} and p, the more the pixels belonging to the two strings are different. The strings are moved along the curve. In each location a new pair (S_{KS}, p) is found. Several measures of the amount of variation in the image can be built for these pairs. The sevenths Hilbert feature is $\mathrm{std}(p)/\mathrm{mean}(p)$. The fact that this feature was selected so frequently seems to indicate that the statistical measures of brightness variability in which the mapping between the image surface and some function with which this surface can be mapped into a 1-dimensional curve perform well in texture analysis. This can be interpreted as an encouragement to investigate more the features based on such a concept.

The group of features which should probably not be investigated more are the two features based on percolation, and the one based on Otsu global thresholding.

The results were obtained with a single classifier.

6 Summary and Prospects

Features selected from a set of 50 features were used to classify the surfaces affected by the *orange skin* surface defect. The features were selected with seven methods. The results of these selections were consistently positive for some of the features. Among them were primarily the features based on local adaptive thresholding and on Hilbert curves used to evaluate the image brightness variability. These types of features should be investigated further in order to find the features with more significance in the problem of surface quality inspection.

It is planned to extend the experiments with the *orange skin* defect in furniture by studying more images taken in slightly varying conditions to further test the stability of the obtained results. The extension to more than one classifier is also considered.

References

1. Chmielewski, L.J., Orłowski, A., Wieczorek, G., Śmietańska, K., Górski, J.: Testing the limits of detection of the orange 'skin' defect in furniture elements with the HOG features. In: Nguyen, N.T., Tojo, S., Nguyen, L.M., Trawiński, B. (eds.) ACIIDS 2017. LNCS (LNAI), vol. 10192, pp. 276–286. Springer, Cham (2017). https://doi.org/10.1007/978-3-319-54430-4_27
2. Karras, D.A.: Improved defect detection using support vector machines and wavelet feature extraction based on vector quantization and SVD techniques. In: Proceedings of International Joint Conference on Neural Networks, vol. 3, pp. 2322–2327, July 2003. https://doi.org/10.1109/IJCNN.2003.1223774
3. Musat, E.C., Salca, E.A., Dinulica, F., et al.: Evaluation of color variability of oak veneers for sorting. BioResources **11**(1), 573–584 (2016). https://doi.org/10.15376/biores.11.1.573-584
4. Konieczny, J., Meyer, G.: Computer rendering and visual detection of orange peel. J. Coat. Technol. Res. **9**(3), 297–307 (2012). https://doi.org/10.1007/s11998-011-9378-2
5. Armesto, L., Tornero, J., Herraez, A., Asensio, J.: Inspection system based on artificial vision for paint defects detection on cars bodies. In: 2011 IEEE International Conference on Robotics and Automation, pp. 1–4, May 2011. https://doi.org/10.1109/ICRA.2011.5980570
6. Allard, M., Jaecques, C., Kauffer, I.: Coating material which can be thermally cured and hardened by actinic radiation and use thereof. US Patent 6,949,591, 27 September 2005
7. Bucur, V.: Techniques for high resolution imaging of wood structure: a review. Meas. Sci. Technol. **14**(12), R91 (2003). https://doi.org/10.1088/0957-0233/14/12/R01
8. Longuetaud, F., Mothe, F., Kerautret, B., et al.: Automatic knot detection and measurements from X-ray CT images of wood: a review and validation of an improved algorithm on softwood samples. Comput. Electron. Agric. **85**, 77–89 (2012). https://doi.org/10.1016/j.compag.2012.03.013
9. Kruk, M., Świderski, B., Osowski, S., Kurek, J., et al.: Melanoma recognition using extended set of descriptors and classifiers. EURASIP J. Image Video Process. **2015**(1) (2015). https://doi.org/10.1186/s13640-015-0099-9
10. Kurek, J., Świderski, B., Dhahbi, S., Kruk, M., et al.: Chaos theory-based quantification of ROIs for mammogram classification. In: Tavares, J.M.R.S., Natal, J.R.M. (eds.) Computational Vision and Medical Image Processing V. Proceedings of 5th ECCOMAS Thematic Conference on Computational Vision and Medical Image Processing VipIMAGE 2015, pp. 187–191. CRC Press, Tenerife, 19–21 October 2015. https://doi.org/10.1201/b19241-32
11. Świderski, B., Osowski, S., Kurek, J., Kruk, M., et al.: Novel methods of image description and ensemble of classifiers in application to mammogram analysis. Expert Syst. Appl. **81**, 67–78 (2017). https://doi.org/10.1016/j.eswa.2017.03.031

12. Kruk, M., Świderski, B., Śmietańska, K., Kurek, J., Chmielewski, L.J., Górski, J., Orłowski, A.: Detection of 'orange skin' type surface defects in furniture elements with the use of textural features. In: Saeed, K., Homenda, W., Chaki, R. (eds.) CISIM 2017. LNCS, vol. 10244, pp. 402–411. Springer, Cham (2017). https://doi.org/10.1007/978-3-319-59105-6_34

13. Pohjalainen, J., Räsänen, O., Kadioglu, S.: Feature selection methods and their combinations in high-dimensional classification of speaker likability, intelligibility and personality traits. Comput. Speech Lang. **29**(1), 145–171 (2015). https://doi.org/10.1016/j.csl.2013.11.004

14. Chmielewski, L.J., Orłowski, A., Śmietańska, K., Górski, J., Krajewski, K., Janowicz, M., Wilkowski, J., Kietlińska, K.: Detection of surface defects of type 'orange skin' in furniture elements with conventional image processing methods. In: Huang, F., Sugimoto, A. (eds.) PSIVT 2015. LNCS, vol. 9555, pp. 26–37. Springer, Cham (2016). https://doi.org/10.1007/978-3-319-30285-0_3

15. Świderski, B., Osowski, S., Kruk, M., Kurek, J.: Texture characterization based on the Kolmogorov-Smirnov distance. Expert Syst. Appl. **42**(1), 503–509 (2015). https://doi.org/10.1016/j.eswa.2014.08.021

16. Cortes, C., Vapnik, V.: Support-vector networks. Mach. Learn. **20**(3), 273–297 (1995). https://doi.org/10.1007/BF00994018

17. Pohjalainen, J.: Feature selection code (2015). http://users.spa.aalto.fi/jpohjala/featureselection/. Accessed 25 Apr 2017

18. Liu, H., Setiono, R.: Chi2: feature selection and discretization of numeric attributes. In: Vassilopoulos, J. (ed.) Proceedings of 7th IEEE International Conference on Tools with Artificial Intelligence, pp. 388–391. IEEE Computer Society, Herndon, 5–8 November 1995. https://doi.org/10.1109/TAI.1995.479783

19. Hall, M.A., Smith, L.A.: Feature selection for machine learning: comparing a correlation-based filter approach to the wrapper. In: Proceedings of 12th International Florida AI Research Society Conference FLAIRS 1999, AAAI, 1–5 May 1999

20. Liu, H., Yu, L.: Feature selection for high-dimensional data: a fast correlation-based filter solution. In: Proceedings of 20th International Conference on Machine Leaning ICML2003, pp. 856–863. ICM, Washington, D.C. (2003)

21. Liu, H., Hussain, F., Tan, C.L., Dash, M.: Discretization: an enabling technique. Data Min. Knowl. Disc. **6**(4), 393–423 (2002). https://doi.org/10.1023/A:1016304305535

22. Duda, R., Hart, P., Stork, D.: Pattern Classification, 2nd edn. Wiley, New York (2001)

23. Cover, T.M., Thomas, J.A.: Elements of Information Theory. Wiley, New York (1991)

24. Cawley, G.C., Talbot, N.L.C.: Gene selection in cancer classification using sparse logistic regression with Bayesian regularization. Bioinformatics **22**(19), 2348–2355 (2006). https://doi.org/10.1093/bioinformatics/btl386

25. Wei, L.J.: Asymptotic conservativeness and efficiency of Kruskal-Wallis test for K dependent samples. J. Am. Stat. Assoc. **76**(376), 1006–1009 (1981). https://doi.org/10.1080/01621459.1981.10477756

Image Retrieval by Use of Linguistic Description in Databases

Krzysztof Wiaderek[1(✉)], Danuta Rutkowska[1,2],
and Elisabeth Rakus-Andersson[3]

[1] Institute of Computer and Information Sciences,
Czestochowa University of Technology, 42-201 Czestochowa, Poland
{krzysztof.wiaderek,danuta.rutkowska}@icis.pcz.pl
[2] Information Technology Institute, University of Social Sciences,
90-113 Lodz, Poland
[3] Department of Mathematics and Natural Sciences,
Blekinge Institute of Technology, 37179 Karlskrona, Sweden
elisabeth.andersson@bth.se

Abstract. In this paper, a new method of image retrieval is proposed. This concerns retrieving color digital images from a database that contains a specific linguistic description considered within the theory of fuzzy granulation and computing with words. The linguistic description is generated by use of the CIE chromaticity color model. The image retrieval is performed in different way depending on users' knowledge about the color image. Specific database queries can be formulated for the image retrieval.

Keywords: Image retrieval · Image recognition
Information granulation · Linguistic description
Fuzzy sets · Computing with words · Image databases
CIE chromaticity color model · Knowledge-based system

1 Introduction

There are many publications concerning image retrieval that is a significant research area since collections of color digital images have been rapidly increasing; see e.g. a survey on image retrieval methods [8]. However, our approach differs from those presented in the literature. The main issue is the goal of image recognition and retrieval. Our aim is not to precisely recognize an object or a scene in an image but only a color that can be described within the framework of fuzzy set theory [21]. This may concern the color as well as other attributes such as amount of the color in an image (color participation or size of the fuzzy color cluster) and optionally its fuzzy location and shape. The problem formulated in this way allows to quickly retrieve an image (or images) corresponding to a fuzzy description that a user introduces into an image retrieval system. An intelligent pattern recognition system that generates linguistic description of color digital images is proposed and developed in authors' previous papers [15–20].

© Springer International Publishing AG, part of Springer Nature 2018
L. Rutkowski et al. (Eds.): ICAISC 2018, LNAI 10842, pp. 92–103, 2018.
https://doi.org/10.1007/978-3-319-91262-2_9

In this article, we use the method of generating the linguistic description of images in order to create specific databases that allow to quickly retrieve images responding to fuzzy queries. In addition, further image analysis can be performed by an intelligent knowledge-based system. As a result, such a system may be able to realize fuzzy inference in the direction to image understanding.

2   Color Model for Image Processing

In our approach, we employ the CIE chromaticity color model [6] in the image processing in order to produce the linguistic description (see [14–20]). Figure 1 presents the CIE chromaticity diagram (triangle) where the color areas, considered as fuzzy sets, are depicted and denoted by numbers 1, 2, ..., 23, associated with the following colors (hues): white, yellowish green, yellow green, greenish yellow, yellow, yellowish orange, orange, orange pink, reddish orange, red, purplish red, pink, purplish pink, red purple, reddish purple, purple, bluish purple, purplish blue, blue, greenish blue, bluegreen, bluish green, green. The CIE chromaticity diagram shows the range of perceivable hues for the normal human eye.

It is worth emphasizing that chromaticity is an objective specification of the quality of a color regardless of its luminance. This means that the CIE diagram removes all intensity information, and uses its two dimensions to describe hue and saturation. The CIE color model is a color space that separates the three dimensions of color into one luminance dimension and a pair of chromaticity dimension. For simplicity, in our considerations we ignore the luminance but it can be taken into account in further, more detailed research.

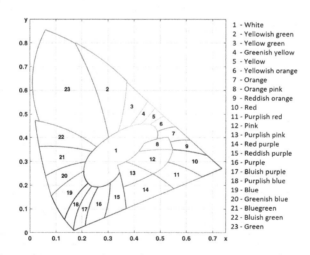

Fig. 1. The CIE chromaticity diagram (Color figure online)

The main advantage of using the CIE color model is the fuzzy granulation of the color space, so we can employ the granular recognition system introduced in

[17] and developed in [18]. The CIE color model is suitable from artificial intelligence point of view because the intelligent recognition system should imitate the way of human perception of colors.

3   Linguistic Description of Color Images

The color granules presented in Fig. 1 are viewed as fuzzy sets with membership functions defined in [14]. Of course, according to [21] different shapes of membership functions can be employed (see e.g. [11–13]). The granular recognition system that produces the linguistic description of input images is a rule-based system (knowledge-based system) with inference using fuzzy logic [22], like e.g. [2,7,10]. In our approach, fuzzy granulation [23] concerns the color granules as well as location granules within an image. With regard to the shape attribute, we also consider rough granulation based on rough sets [9]; see our previous papers, e.g. [17].

In [19,20], the process of producing the linguistic descriptions of color digital images based on the fuzzy color granules, determined in the CIE color model, is explained. Figures 2 and 3 present results of classification pixels of two images into fuzzy color granules of the CIE diagram.

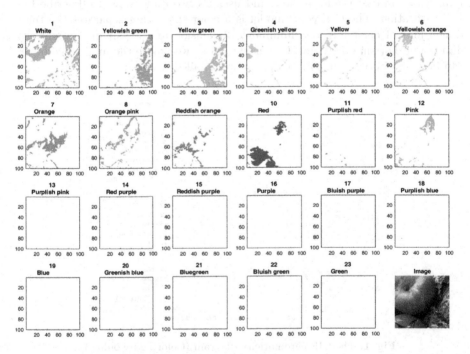

Fig. 2. Color granules in input image 1 (Color figure online)

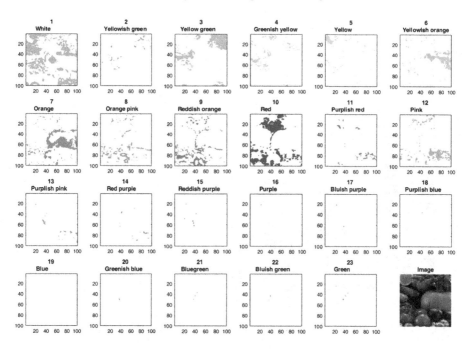

Fig. 3. Color granules in input image 2 (Color figure online)

Then, histograms portrayed in Figs. 4 and 5, respectively, illustrate participation rates of particular colors (fuzzy color granules) in both images. It should be emphasized that values of the participation rates are viewed as fuzzy numbers, and measured by use of the fuzzy unit P (fuzzy set with the membership equal 1 only for p); see e.g. [13]. Thus, value p is the kernel of the fuzzy number P that denotes the participation rate of every color granule, assuming that each of them participates in the image with the same rate. This fuzzy number is applied as the unit of participation of particular colors in an image.

Figure 6 presents trapezoidal membership functions of fuzzy sets VS, S, M, B, VB denoting *Very Small, Small, Medium, Big, Very Big*, respectively, as linguistic values of color participation (p_rate) in an input image. It is obvious that the p_rate axis corresponds to the vertical axes in the histograms (Figs. 4 and 5); the unit value p is employed in all the axes.

In the next section, the database table (Table 1) that shows the participation of colors in different images has been produced by use of the fuzzy unit P. In this table, the linguistic values depend on the fuzzy numbers expressed by means of the unit P. Fuzzy numbers are described in [5], and applied in many problems (see e.g. [3]). Values indicated by the histograms and expressed by the unit P are described by linguistic labels according to the membership functions of fuzzy sets presented in Fig. 6.

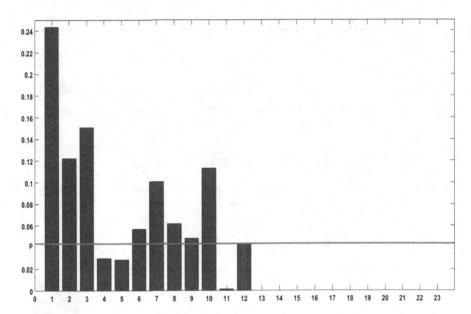

Fig. 4. Histogram of color participation in input image 1 (Color figure online)

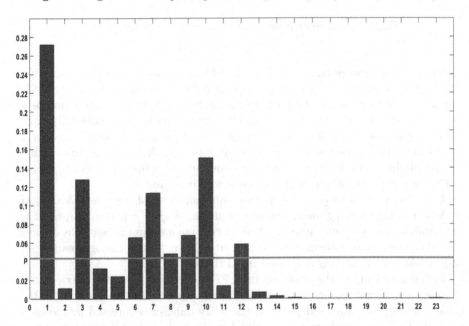

Fig. 5. Histogram of color participation in input image 2 (Color figure online)

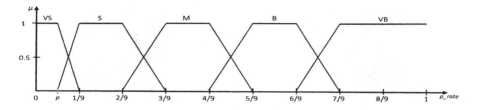

Fig. 6. Fuzzy sets of color participation rate

4   Databases for the Linguistic Description of Images

Table 1 is a database table that includes data concerning participation of particular colors, C_1, C_2, ..., C_{23}, in images from an image collection. Values of the data corresponds to the percentage of pixels belonging to the color granules.

More detailed data, concerning the color participation in particular locations of the images can be included in a hierarchical database and also in the form of a multidimensional cube. This means that Table 1, in addition to the values that denote the participation rates of colors in the images, may also contain two-dimensional tables of the color participations in parts of the images. This refers to the macropixels, introduced and employed in [15–20]. The fuzzy macropixels indicate locations within an image LU, ..., RD (see Fig. 7). The macropixels can be of different size, as Fig. 8 illustrates. Semantic meaning of the location names is explained later in this section, when referred to Fig. 10.

Table 1. Database table: participation of color

File of image	Participation of C_1	Participation of C_2	...	Participation of C_{23}
Image 1	0.24	0.12	...	0.00
...	...	...	...	...
Image 2	0.27	0.01	...	0.00
...	...	...	...	...

Such a multidimensional model of the data table can be considered as an OLAP cube (see e.g. [4]); OLAP stands for OnLine Analytical Processing. As a matter of fact, in our case, this multidimensional cube is viewed as a fuzzy data model (see e.g. [1]).

Figure 9 illustrates how to create a three-dimensional cube that represents an image. The cube is composed of every matrix $M_{C\_1}, M_{C\_2}, ..., M_{C\_23}$, of membership values of particular color granules from the CIE diagram (Fig. 1). Based on the matrix cube the visualizations shown in Fig. 2 have been generated and also put in form of the corresponding cube as we see in this figure.

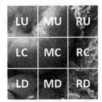

Fig. 7. Locations in image 1 (Color figure online)

Fig. 8. Locations in image 2 (Color figure online)

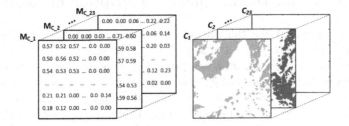

Fig. 9. Multidimensional cubes of color granules in image 1 (Color figure online)

It should be emphasized that OLAP cubes are used in data warehouses for analytical processing of the data. OLAP cubes consist of facts, also called measures, categorized by dimensions (in general, it can be more than three dimensions). An OLAP cube provides a convenient way of collecting measures of the same dimensionality. The useful feature of an OLAP cube is that the data can be contained in an aggregated form. Special operations allow to slice, dice, drill down/up, roll-up, and rotate an cube in order to navigate, select, and view particular subsets of the data. In this way we can easily analyze specific parts od the data. Especially, the drill down and up enable to navigate among levels of data ranging from the most summarized (up) to the most detailed (down). A slice is a subset of a multidimensional cube corresponding to single value for one or more numbers of the dimensions.

In case of our example, we can slice a particular matrix M_{C_i}, for i = 1, 2, ..., 23, from the cube presented in Fig. 9. In addition, we can drill down (or up) to analyze specific regions of an image (location indicated by macropixels of different size); see Fig. 10. It should be emphasized that in our case, the dimensions: color granules and two-dimensional space of an image (composed of pixels) have values considered as fuzzy sets, i.e. fuzzy color granules and fuzzy macropixels (see [15–20]).

Figure 10 presents two fuzzy histograms that portray participation of pixels of the same color granule (C_2 – yellowish green) in locations determined by macropixels of different size (big and small). This concerns image 1; see Figs. 2, 7, and 9. Let us notice that color C_2 is mostly visible at the right side of the picture (Righ Upper - RU, Right Central - RC, and Right Down - RD macropixels), and

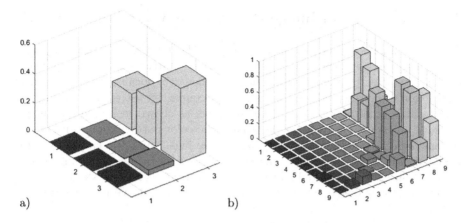

a)   b)

Fig. 10. Color C_2 in image 1, in locations of macropixels: (a) big size (b) small size (Color figure online)

also (but much less) in the Left Down - LD and Middle Down - MD macropixels. This corresponds to Fig. 7 where color Yellowish green does not exist in locations LU - Left Upper, MU - Middle Upper, LC - Left Central, and MC - Middle Central.

The three-dimensional cube considered so far represents a single image described by data concerning participation of particular colors in the image and smaller regions (location determined by fuzzy macropixels). This data model can be viewed as a part of an OLAP cube that contains data concerning a collection of images. Figure 11 illustrates the multi-dimensional cube composed of the data of the form depicted in Fig. 9, for many digital color pictures. By use of the OLAP operations, it is easy to analyze an image base with regard to colors and locations in a set of the pictures. An example of image retrieval based on the data that can be aggregated in such a cube is considered in Sect. 5.

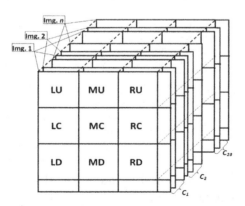

Fig. 11. Multidimensional data model of a collection of images

5   An Example of Image Retrieval by Use of the Database

Figure 12 portrays several images from the base of color digital pictures employed to illustrate our approach to image retrieval. As a matter of fact, only data concerning the color participation in images without the localization as shown in Figs. 7 and 8 is presented in this section. Of course, by use of the multidimensional data model of the form of OLAP cube, as shown in Fig. 11, the image retrieval procedure can be extended to analyze color participation in particular regions indicated by the fuzzy macropixels (see Figs. 7 and 8). Data in the cube depicted in Fig. 11 are viewed as hierarchical granulated cubes, enabling to navigate within more aggregated and more detailed levels. This refers to the deeper granulation of an image area by smaller macropixels as shown in Fig. 8 and also in Fig. 10b).

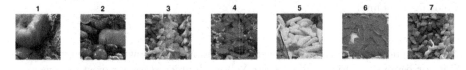

Fig. 12. Part of the image base used in the example of image retrieval (Color figure online)

Table 2 contains linguistic values obtained according to the membership functions depicted in Fig. 6 that describe color participation in the images included in this base. Two first rows of this table contain the linguistic values describing participation of particular colors in images 1 and 2. Of course, the table includes linguistic values for every image from the collection, much more than only seven presented.

Table 2. Database table with linguistic values of color participation in images

Im.	C_1	C_2	C_3	C_4	C_5	C_6	C_7	C_8	C_9	C_{10}	C_{11}	C_{12}	C_{13}	C_{14}	C_{15}	C_{16}	C_{17}	C_{18}	C_{19}	C_{20}	C_{21}	C_{22}	C_{23}
1	S	S	S	VS	VS	VS	S	VS	VS	S	VS	VS	VS	VS	VS	VS	VS	VS	VS	VS	VS	VS	VS
2	S	VS	S	VS	VS	VS	S	VS	VS	S	VS	VS	VS	VS	VS	VS	VS	VS	VS	VS	VS	VS	VS
3	M	VS	S	VS	VS	VS	VS	VS	VS	VS	VS	VS	VS	VS	VS	VS	VS	VS	VS	VS	VS	VS	VS
4	VS	M	S	VS	VS	VS	VS	VS	VS	S	VS	VS	VS	VS	VS	VS	VS	VS	VS	VS	VS	VS	VS
5	S	VS	VS	B	VS	VS	VS	VS	VS	VS	VS	VS	VS	VS	VS	VS	VS	VS	VS	VS	VS	VS	VS
6	S	VS	VS	VS	VS	VS	VS	S	VS	S	VS	B	VS	VS	VS	VS	VS	VS	VS	VS	VS	VS	VS
7	M	VS	S	VS	VS	VS	VS	VS	VS	VS	VS	S	VS	VS	VS	VS	VS	VS	VS	VS	VS	VS	VS
...	...	...	...	...	...	...	...	...	...	...	...	...	...	...	...	...	...	...	...	...	...	...	...

The image retrieval is performed by use of the data in Table 2 and fuzzy IF-THEN rules of the following form, e.g.:

$$\textbf{IF } c_1 \text{ is S } \textbf{AND } c_2 \text{ is VS } \textbf{AND } c_3 \text{ is S } \textbf{AND} \dots \textbf{ THEN } \text{Im. 2} \qquad (1)$$

$$\textbf{IF } c_4 \text{ is B } \textbf{THEN } \text{Im.5} \tag{2}$$

$$\textbf{IF } c_{12} \text{ is B } \textbf{THEN } \text{Im.6} \tag{3}$$

where c_1, c_2,..., c_{12} are linguistic variables corresponding to fuzzy color granules C_1, C_2,..., C_{12}, respectively.

An inference process that employs fuzzy logic and the fuzzy IF-THEN rules produces outputs (image or images) matching an user's query. Since descriptions of images are included in database tables the SELECT instruction used in the SQL language can be employed, e.g. SELECT FROM Table 1 WHERE Color is greenish yellow AND participation is Big. It is worth emphasizing that in our approach the fuzzy queries are employed (see e.g. [24]). In our example, an answer to this query is image 5 as the output.

6   Conclusions and Final Remarks

The image retrieval approach presented in this paper is very useful when a problem is formulated as follows: Find a picture (or pictures), from an image collection, including color described with regard to names of the color granules (see Figs. 1, 2 and 3) located in regions (indicated by names as in Figs. 7, 8, and 11) of size defined by fuzzy linguistic values (as shown in Fig. 6). There are situations requiring quick retrieval of pictures including an object that can be recognized by its color, size, and (optionally) location, approximately defined. For example – a wanted person who escapes with a yellow bag.

When the data describing images by use of the linguistic values are contained in the form of a multidimensional cube (OLAP), as illustrated in Fig. 11, we can analyze the color pictures in the direction of image understanding. It is worth noticing that deeper granulation of an image area (as shown in Fig. 8) allows to inference concerning shapes of objects by means of macropixels of various size.

References

1. Alain, K.M., Nathanael, K.M., Rostin, M.M.: Integrating fuzzy concepts to design a fuzzy data warehouse. Int. J. Comput. **27**(1), 112–132 (2017)
2. Almohammadi, K., Hagras, H., Alghazzawi, D., Aldabbagh, G.: A survey of artificial intelligence techniques employed for adaptive educational systems within e-learning platforms. J. Artif. Intell. Soft Comput. Res. **7**(1), 47–64 (2017)
3. Beg, I., Rashid, T.: Modelling uncertainties in multi-criteria decision making using distance measure and TOPSIS for hesitant fuzzy sets. J. Artif. Intell. Soft Comput. Res. **7**(2), 103–109 (2017)
4. Biere, M.: Business Intelligence for the Enterprise. Prentice Hall, Upper Saddle River (2003)
5. Dubois, D., Prade, H.: Fuzzy Sets and Systems: Theory and Applications. Academic Press, New York (1980)
6. Fortner, B., Meyer, T.E.: Number by Color. A Guide to Using Color to Undersdand Technical Data. Springer, Heidelberg (1997). https://doi.org/10.1007/978-1-4612-1892-0

7. Liu, H., Gegov, A., Cocea, M.: Rule based networks: an efficient and interpretable representation of computational models. J. Artif. Intell. Soft Comput. Res. **7**(2), 111–1239 (2017)
8. Marshall, A.M., Gunasekaran, S.: Image retrieval - a review. Int. J. Eng. Res. Technol. **3**(5), 1128–1131 (2014)
9. Pawlak, Z: Granularity of knowledge, indiscernibility and rough sets. In: IEEE International Conference Fuzzy Systems Proceedings. IEEE World Congress on Computational Intelligence, vol. 1, pp. 106–110 (1998)
10. Prasad, M., Liu, Y.-T., Lin, C.-T., Shah, R.R., Kaiwartya, O.P.: A new mechanism for data visualization with TSK-type preprocessed collaborative fuzzy rule based system. J. Artif. Intell. Soft Comput. Res. **7**(1), 33–46 (2017)
11. Rakus-Andersson, E.: Fuzzy and Rough Techniques in Medical Diagnosis and Medication. Springer, Heidelberg (2007). https://doi.org/10.1007/978-3-540-49708-0
12. Riid, A., Preden, J.-S.: Design of fuzzy rule-based classifiers through granulation and consolidation. J. Artif. Intell. Soft Comput. Res. **7**(2), 137–147 (2017)
13. Rutkowska, D.: Neuro-Fuzzy Architectures and Hybrid Learning. Springer, Heidelberg (2002). https://doi.org/10.1007/978-3-7908-1802-4
14. Wiaderek, K.: Fuzzy sets in colour image processing based on the CIE chromaticity triangle. In: Rutkowska, D., Cader, A., Przybyszewski, K. (eds.) Selected Topics in Computer Science Applications. Academic Publishing House EXIT, Warsaw, Poland, pp. 3–26 (2011)
15. Wiaderek, K., Rutkowska, D.: Fuzzy granulation approach to color digital picture recognition. In: Rutkowski, L., Korytkowski, M., Scherer, R., Tadeusiewicz, R., Zadeh, L.A., Zurada, J.M. (eds.) ICAISC 2013. LNCS (LNAI), vol. 7894, pp. 412–425. Springer, Heidelberg (2013). https://doi.org/10.1007/978-3-642-38658-9_37
16. Wiaderek, K., Rutkowska, D., Rakus-Andersson, E.: Color digital picture recognition based on fuzzy granulation approach. In: Rutkowski, L., Korytkowski, M., Scherer, R., Tadeusiewicz, R., Zadeh, L.A., Zurada, J.M. (eds.) ICAISC 2014. LNCS (LNAI), vol. 8467, pp. 319–332. Springer, Cham (2014). https://doi.org/10. 1007/978-3-319-07173-2_28
17. Wiaderek, K., Rutkowska, D., Rakus-Andersson, E.: Information granules in application to image recognition. In: Rutkowski, L., Korytkowski, M., Scherer, R., Tadeusiewicz, R., Zadeh, L.A., Zurada, J.M. (eds.) ICAISC 2015. LNCS (LNAI), vol. 9119, pp. 649–659. Springer, Cham (2015). https://doi.org/10.1007/978-3-319-19324-3_58
18. Wiaderek, K., Rutkowska, D., Rakus-Andersson, E.: New algorithms for a granular image recognition system. In: Rutkowski, L., Korytkowski, M., Scherer, R., Tadeusiewicz, R., Zadeh, L.A., Zurada, J.M. (eds.) ICAISC 2016. LNCS (LNAI), vol. 9693, pp. 755–766. Springer, Cham (2016). https://doi.org/10.1007/978-3-319-39384-1_67
19. Wiaderek, K., Rutkowska, D., Rakus-Andersson, E.: Linguistic description of color images generated by a granular recognition system. In: Rutkowski, L., Korytkowski, M., Scherer, R., Tadeusiewicz, R., Zadeh, L.A., Zurada, J.M. (eds.) ICAISC 2017. LNCS (LNAI), vol. 10245, pp. 603–615. Springer, Cham (2017). https://doi.org/ 10.1007/978-3-319-59063-9_54
20. Wiaderek, K., Rutkowska, D.: Linguistic description of images based on fuzzy histograms. In: Choraś, M., Choraś, R. (eds.) IP&C 2017. AISC, vol. 681, pp. 27–34. Springer, Cham (2017). https://doi.org/10.1007/978-3-319-68720-9_4
21. Zadeh, L.A.: Fuzzy sets. Inf. Control **8**, 338–353 (1965)
22. Zadeh, L.A.: Fuzzy logic = computing with words. IEEE Trans. Fuzzy Syst. **4**, 103–111 (1996)

23. Zadeh, L.A.: Toward a theory of fuzzy information granulation and its centrality in human reasoning and fuzzy logic. Fuzzy Sets Syst. **90**, 111–127 (1997)
24. Zadrozny, S., De Tre, G., De Caluve, R., Kacprzyk, J.: An overview of fuzzy approaches to flexible database querying. In: Galindo, J. (ed.) Handbook of Research on Fuzzy Information Processing in Databases, vol. I, pp. 34–54. Information Science Reference (2008)

Bioinformatics, Biometrics and Medical Applications

On the Use of Principal Component Analysis and Particle Swarm Optimization in Protein Tertiary Structure Prediction

Óscar Álvarez[1], Juan Luis Fernández-Martínez[1],
Celia Fernández-Brillet[1], Ana Cernea[1], Zulima Fernández-Muñiz[1],
and Andrzej Kloczkowski[2,3(✉)]

[1] Department of Mathematics, University of Oviedo,
C. Calvo Sotelo S/N, 33007 Oviedo, Spain
oscalmac@gmail.com, celia.fernandez.98@gmail.com,
{jlfm,cerneadoina,zulima}@uniovi.es
[2] Battelle Center for Mathematical Medicine, Nationwide Children's Hospital,
Columbus, OH, USA
Andrzej.Kloczkowski@nationwidechildrens.org
[3] Department of Pediatrics, The Ohio State University, Columbus, OH, USA

Abstract. We discuss applicability of Principal Component Analysis and Particle Swarm Optimization in protein tertiary structure prediction. The proposed algorithm is based on establishing a low-dimensional space where the sampling (and optimization) is carried out via Particle Swarm Optimizer (PSO). The reduced space is found via Principal Component Analysis (PCA) performed for a set of previously found low-energy protein models. A high frequency term is added into this expansion by projecting the best decoy into the PCA basis set and calculating the residual model. Our results show that PSO improves the energy of the best decoy used in the PCA considering an adequate number of PCA terms.

Keywords: Principal component analysis · Particle swarm optimization
Tertiary protein structure · Conformational sampling
Protein structure refinement

1 Introduction

The problem of protein tertiary structure prediction consists of determining the unique three-dimensional conformation of protein (corresponding to the lowest energy) from its amino acid sequence. Currently, this problem represents one of the biggest challenges for biomedicine and biotechnology since it is of utter relevance in areas such as drug design or design and synthesis of new enzymes with desired properties that have not yet been appeared naturally by evolution, that fold to a desired target protein structure [1, 2].

Despite the constantly growing number of protein structures deposited in the Protein Data Bank (PDB), there is a rapidly increasing gap between the number of protein sequences obtained from large-scale genome and transcriptome sequencing and

© Springer International Publishing AG, part of Springer Nature 2018
L. Rutkowski et al. (Eds.): ICAISC 2018, LNAI 10842, pp. 107–116, 2018.
https://doi.org/10.1007/978-3-319-91262-2_10

the number of PDB structures. Currently PDB contains over 130,000 macromolecular structures, while the UniProt Knowledge base contains around 50 million sequences (after recent redundancy reduction). Thus, less than 1% of protein sequences have the native structures in the PDB database. Therefore, accurate computational methods for protein tertiary structure prediction, which are much cheaper and faster than experimental techniques, are needed [1, 2].

The main methodologies to generate protein tertiary structure models are divided into two categories: template-based and template-free modeling. Template-based homology modeling allows building a model of the target protein based on a template structure of a homologue (protein with known structure and high (at least 30%) sequence identity to the target protein), by simulating the process of evolution - i.e. introducing amino acid substitutions as well as insertions and deletions, while maintaining the same fold.

Template-free methods predict the protein tertiary structure from physical principles based on optimizing the energy function that describes the interaction between the protein residues to find the global minimum without using any template information. Some well-known programs in the literature use template-free modeling [2–4] mainly when no structural homologs exist in the PDB. Template-based modeling methods use the known structures (as templates) of the proteins that are analogous to the target protein to construct structural models [5].

Regardless the method utilized, the tertiary structure protein prediction is hampered by the curse of the dimensionality, since these prediction methods are unable to explore the whole conformational space. The curse of dimensionality describes how the ratio of the volume of the hyper sphere enclosed by the unit hypercube becomes irrelevant for higher dimensionality (more than 10 dimensions). Therefore, there is a need to simplify the protein tertiary structure prediction problem by using model reduction techniques to alleviate its ill-posed character [1].

Protein refinement methods are a good alternative to approximate the native structure of a protein using template-based approximate models. Some of these methods use molecular dynamics, coarse-grained models and also spectral decomposition. In our earlier work, [6] we applied Elastic Network Models to protein structure refinement. This mathematical model provides a reliable representation of the fluctuational dynamics of proteins and explains various conformational changes in protein structures. In this article, we use the tertiary structure information provided by other decoys to reduce the dimensionality of the protein tertiary structure prediction problem. We were able to accomplish this task by constraining the sampling within the subspace spanned by the largest principal components of a series of templates. These low-energy protein models (or templates) are previously found using different optimization techniques, or performing local optimization and using different initial and reference models, via template-free methods. In the present study we used as templates, models submitted by the different prediction groups during the CASP experiment.

This methodology allows the sampling of the lowest-energy models in a low dimensional space close to the native conformation. Due to the fact that, the native structure is unknown for most cases, the refined protein structure requires its uncertainty assessment in order to gain a deeper understanding of the protein and its alternate states [7]. The number of PCA terms (PCAs) used to construct the reduced search space for energy optimization and sampling affects the refined structure. Therefore, in this paper we try to understand the effect of PCA dimensionality in the protein tertiary structure prediction problem.

The main conclusions are that the dimensionality reduction alleviates the ill-posed character of this high-dimensional optimization problem, as well as the possibility to increase the uncertainty of the predicted backbone structure. Therefore, a tradeoff is required since, determining the minimum number of PCA terms is a crucial step for achieving a successful refinement.

2 Computational Methods

2.1 Protein Energy Function Landscape

In the tertiary structure protein prediction problem the model parameters are the proteins coordinates determined by n_a atoms, $\mathbf{m} = (m_1, m_2, \ldots, m_n) \in \mathbf{M} \subset \mathbf{R}^n$, with $n = 3n_a$, being $\mathbf{M}$ the set of admissible protein models elaborated taking into account their biological consistency. The tertiary structure of a given protein is defined by knowing the free-energy function, $E(\mathbf{m}) : \mathbf{R}^n \to \mathbf{R}$ and finding the modal that minimizes that free energy function, $\mathbf{m}_p = \min_{\mathbf{m} \in \mathbf{M}} E(\mathbf{m})$ [8].

The main issue with this problem is its high dimensionality. That implies that the optimization algorithm utilized in this problem needs to tackle the high dimension of the model space consisting of thousands of atoms and also the landscape of the energy function.

Also, assuming that $\mathbf{m}_p$ is the global optimum for the energy function satisfying the condition, $\nabla E(\mathbf{m}_p) = \mathbf{0}$, there exist a set of models $M_{tol} = \{\mathbf{m} : E(\mathbf{m}) \le E_{tol}\}$ whose energy is lower than a given energy cut-off E_{tol}. These models, in the neighborhood of $\mathbf{m}_p$, belongs to the linear hyperquadric [9]:

$$\frac{1}{2}(\mathbf{m} - \mathbf{m}_p)^T HE(\mathbf{m}_p)(\mathbf{m} - \mathbf{m}_p) \le E_{tol} - E(\mathbf{m}_p) \tag{1}$$

where $HE(\mathbf{m}_p)$ is the Hessian matrix calculated in $\mathbf{m}_p$. Nevertheless, the linear hyper quadric only describes locally in the neighborhood of $\mathbf{m}_p$ the global complexity of the energy landscape with one or more flat curvilinear elongated valleys with almost null gradients where the local optimization methods might get trapped.

2.2 Protein Model Reduction via Principal Component Analysis

Principal component analysis is mathematical model reduction technique that transforms a set of correlated variables into a smaller number of uncorrelated ones known as principal components. The resulting transformation has the advantage of being smaller and being more computationally advantageous while maintaining as much as possible the previous variability. This procedure has been applied in several fields but, in protein tertiary structure, it was carried out a preliminary application utilizing the three largest PCs while optimizing via the Powel method [3]. However, in this paper, we perform stochastic sampling in higher dimensions using a member of the family of Particle Swarm Optimizers (RR-PSO) [10, 11]. We study the protein structure prediction and how the number of PCA terms affects the final protein structure obtained via RR-PSO. This PCA is of great relevance in protein structure prediction as it aids us sampling the parameters when a correlation among exists, it also avoids the issue of a high dimensional problem and alleviates the ill-posed character of the tertiary structure optimization problem as the solutions are found in a smaller dimensional space: finding

$$\mathbf{a}_k \in \mathbb{R}^d : E(\hat{\mathbf{m}}_k) = E(\boldsymbol{\mu} + \mathbf{V}_d\,\mathbf{a}_k) \leq E_{tol}, \tag{2}$$

where $\boldsymbol{\mu}, \mathbf{V}_d$ are provided by the model reduction technique that it is used.

The PCA dimensionality reduction is carried out as follows [11]:

An ensemble of l decoys $\mathbf{m}_i \in R^n$ is selected and arranged column wise into a matrix: $\mathbf{X} = (\mathbf{m}_1, \mathbf{m}_2, \ldots, \mathbf{m}_l) \in M(n, l)$. The problem consists of finding a set of protein patterns $\mathbf{V}_d = (\mathbf{v}_1, \mathbf{v}_2, \ldots, \mathbf{v}_d)$ that provides an accurate low dimensional representation of the original set with $d << l$. This carried out by diagonalization the matrix X as follows:

$$C_{prior} = (\mathbf{X} - \boldsymbol{\mu})(\mathbf{X} - \boldsymbol{\mu})^T \in M(n, n), \tag{3}$$

where $\boldsymbol{\mu}$ is either the experimental mean of the decoys, the median, or any other decoy around we desire to perform the search as a backbone structure.

Matrix C_{prior} has a maximum rank of $l - 1$, therefore, as a maximum $l - 1$ eigenvectors of C_{prior} are require to expand the whole prior variability. Thus, it is easier to diagonalize $C_{prior}^T \in M(l, l)$ and to obtain the $l - 1$ first eigenvectors of C_{prior} as follows:

$$\begin{aligned}
&\mathbf{X} - \boldsymbol{\mu} = V\Sigma U^T, \\
&C_{prior}^T = U\Sigma\Sigma^T U^T \Rightarrow B = V\Sigma = (\mathbf{X} - \boldsymbol{\mu})U, \\
&\mathbf{v}_k = \frac{B(:, k)}{\|B(:, k)\|_2}, k = 1, \ldots, l - 1.
\end{aligned} \tag{4}$$

The centered character of the experimental covariance C_{prior} is crucial to maintain consistency with the centroid model $\boldsymbol{\mu}$.

Ranking the eigenvalues of C_{prior}^T in decreasing order allows us to select a certain number of PCA terms ($d << l - 1 << n$) to match most of the variability in the model ensemble. Additionally, a high frequency term is included within the PCA in order to consider the model with the lowest energy, and projecting it into the PCA basis as follows:

$$\mathbf{v}_{d+1} = \mathbf{m}_{BEST} - \boldsymbol{\mu} + \sum_{i=1}^{d} a_i\mathbf{v}_i. \tag{5}$$

Consequently, any protein model in the reduced base is represented as a unique linear combination of the eigen-modes:

$$\hat{\mathbf{m}}_k = \boldsymbol{\mu} + \sum_{i=1}^{d+1} a_i \mathbf{v}_i = \boldsymbol{\mu} + \mathbf{V}\mathbf{a}_k. \tag{6}$$

The projection of any decoy $\hat{\mathbf{m}}_k$ is very fast, since matrix $\mathbf{V}$ is orthogonal:

$$\mathbf{a}_k = \mathbf{V}^T(\hat{\mathbf{m}}_k - \boldsymbol{\mu}). \tag{7}$$

This technique allows global optimization methods to perform efficiently the required sampling in the reduced search space. The PCA procedure helps alleviating the ill-posed character of any highly dimensional problem and we look to study how the number of PCA terms affects the final predicted configuration.

2.3   The Particle Swarm Optimizer

For each backbone conformation, we have performed the optimization via Particle Swarm Optimization (PSO). This methodology is a stochastic and evolutionary optimization technique, which is inspired in individual's social behavior (particles) [12–14]. The sampling problem consists of finding an appropriate sample of protein models $\hat{\mathbf{m}}_k = \boldsymbol{\mu} + \mathbf{V} \cdot \mathbf{a}_k$, such as $E(\hat{\mathbf{m}}_k) \leq E_{tol}$. Although the search is carried out in the reduced search space (PCA), the sampled proteins must be reconstructed in the original atom space in order to correctly evaluate their energy. The PSO algorithm is as follows:

We define a prismatic space of admissible protein models, $\mathbf{M}$:

$$l_j \leq a_{ji} \leq u_j, \quad 1 \leq j \leq n, \quad 1 \leq i \leq n_{size}, \tag{8}$$

where l_j, u_j, are the lower and upper limits for the j-th coordinate for each model. Each plausible model is a particle that is represented by a vector whose length is the number of PCA terms. Each model has its own position in the search space. The perturbations we produced in the PCA search space required in order to carry out he sampling and explore the solutions are represented by the particle velocities. In our case, the search space is designed by projecting back all the decoys to the reduced PCA space and finding the lower and upper limits that expand the variability in each PCA coordinate.

At each iterations, the algorithm updates the positions, $\mathbf{a}_i(k)$, and the velocities, $\mathbf{v}_i(k)$ of each particle swarm. The velocity of each particle, i, at each iteration, k, is a function of three major components:

The inertia term, a real constant, w that modifies the velocities.

The social term, the difference between the global best position found thus far in the entire swarm, $\mathbf{g}(k)$ and the particle's current position, $\mathbf{a}_i(k)$.

The cognitive term, the difference between the particle's best position found $\mathbf{l}_i(k)$ and the particle's current position, $\mathbf{a}_i(k)$.

Thus, the algorithm is written as follows: [14]

$$\mathbf{v}_i(k+1) = \omega\mathbf{v}_i(k) + \phi_1(\mathbf{g}(k) - \mathbf{a}_i(k)) + \phi_2(\mathbf{l}_i^k - \mathbf{a}_i(k))$$
$$\mathbf{a}_i(k+1) = \mathbf{a}_i(k) + \mathbf{v}_i(k+1), \tag{9}$$
$$\phi_1 = r_1 a_g, \quad \phi_2 = r_2 a_l, \quad r_1, r_2 \in U(0,1), \quad \omega, a_g, a_l \in \mathbf{R}.$$

r_1, r_2 are vectors of random numbers uniformly distributed in (0,1) to weight the global and local acceleration constants, a_g, a_l. $\overline{\phi} = \frac{a_g + a_l}{2}$ is the total mean acceleration, crucial in determining the algorithm's stability and convergence [12].

Protein structure calculations are performed via the Bioshell computational package [15–17]. Additionally, Bioshell was considered and essential tool in our research as it was used to carry out tertiary structure in the different PCA basis dimensions, that is, it enabled us to eliminate the distortion of bond angles and lengths accompanying the displacement of protein coordinates when we sample moving along the PCA terms. Furthermore, Bioshell package help us maintaining the structure unchanged and, ultimately, obtaining a backbone structure closer to the determined structures via experiments. Finally, Bioshell also evaluates at each time step each protein conformation, calculating its residues and performing energy minimization to evaluate the energy conformation.

3   Results

In this section we look to study how different PCA dimension affect the prediction capabilities of the PSO algorithm when applied to different predictions found on the CASP database. We consider protein predictions whose native structures are known in order to assess how our prediction differs from the native structure. As it has been explained previously in the methodology section we utilize different decoys from proteins found in the CASP experiment, we randomly selected the protein T0545 to show the energy values of 185 different decoys and plotted in Fig. 1A. If we select every single decoys that is in the 30th energy percentile, that is, those with an energy less than this −300, we are capable of constructing a reliable PCA base (see Fig. 1B). In this sense, it is possible to describe the vast majority of the backbone conformational variation, a fact that has also been reported by Baker et al. (3) However, we were able to further tune the methodology in order to account for the highest energy details by adding an additional term, known as the high frequency term. This study suggests that we can efficiently sample and optimize a great number of conformational variations in tertiary protein structures by selecting the few first decoys.

The search space utilized is based on the PCA expansion. It is observed that, regardless the PCA coordinates we consider, the width of the first PCA coordinate interval is bigger and, afterwards, it starts getting narrower as the PCA index increases. Additionally, we consider another PCA with eleven terms plus the High Frequency term, in this case, a higher variability within the protein decoys is considered.

Once the PCAs are determined, we perform the PSO search and optimization by adopting a swarm of 40 particles and 100 iterations. To carry out the PSO sampling and optimization, we used the RR-PSO family member while its exploration capabilities were monitored in order to ensure that a good exploration of the PCA search space is

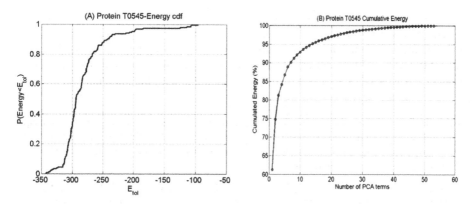

Fig. 1. Energy values of 185 different decoys for protein T0545 (A), is used to construct a reliable PCA base (B)

performed. The monitoring is, then, carried out by measuring the median distance for each particles and the center of gravity and, normalizing it with respect to the first iteration, considered to be a 100%. When the median dispersion falls below 3%, we can assume that the swarm has collapsed towards the global best, and we can either stop sampling or increase the exploration utilizing steps much greater than 1. When the collapse happens, all the particles of the same iteration will be considered as a unique particle in the posterior sampling.

As shown in Table 1, the predictions utilizing three PCA terms are not of good quality with the majority of the predictions with energies far from the native structures. On the other hand, those predictions carried out with a higher dimensionality yield to lower energies. This point is due to the fact that the explorative character of the PSO is strongly correlated with the number of dimensions utilized in constructing the search space. That is, the more dimensions we use, the better the exploration of the protein structure conformational variations and, as a consequence, the final energy predicted.

Table 1. Summary of the computational experiments performed in this paper, via Principal Component Analysis and Particle Swarm Optimization. Energy of the best decoy used in the PCA and lower energy found after PSO optimization. Bold faces indicate the cases where the energy after optimization improved.

Protein CASP9 code	Native structure	Best decoy	3 PCA terms	5 PCA terms	7 PCA terms	9 PCA terms	11 PCA terms
T0545	−348.8	−342.1	−256.8	−299.0	**−343.5**	**−344.6**	**−345.5**
T0557	278.9	−273.7	**−275.3**	**−275.2**	**−275.4**	**−277.2**	**−277.6**
T0555	−389.4	−370.6	23.67	18.68	**−370.9**	**−370.9**	**−371.3**
T0561	−483.6	−448.6	13.28	−400.8	−447.7	**−449.4**	**−450.2**
T0580	−258.3	**−253.8**	−196.4	−250.8	−249.7	−249.5	−250.8
T0635	−466.5	−462.8	−43.7	−324.1	−361.7	**−463.1**	**−463.6**
T0637	−384.5	−372.0	−46.7	−103.7	−369.2	−371.4	**−372.4**
T0639	−380.6	−343.6	−102.3	−335.5	**−345.4**	**−345.7**	**−345.4**
T0643	−234.3	**−209.4**	−138.9	−209.2	**−209.5**	**−210.0**	**−210.0**

The point remarked by the energy predictions is further confirmed when the Root Mean Squared (RMS) distance is scrutinized in Table 2. Predictions obtained with a PCA with low dimensions are structurally far from the native structures as shown by the RMS, whose values are extremely high. However, when we increase the dimensionality, it is possible to obtain better RMS closer to the native structure.

Table 2. Summary of the computational experiments performed in this paper, via Principal Component Analysis and Particle Swarm Optimization. RMSD of the best decoy used in the PCA and lower energy found after PSO optimization. Bold faces indicate the cases where the RMSD after optimization improved.

Protein CASP9 code	Best decoy	3 PCA terms	5 PCA terms	7 PCA terms	9 PCA terms	11 PCA terms
T0545	1.942	9.231	**1.931**	**1.923**	**1.919**	**1.889**
T0555	8.566	14.411	8.568	**8.566**	**8.522**	**8.516**
T0557	1.617	1.696	**1.606**	**1.596**	**1.024**	**0.780**
T0561	5.898	14.156	5.941	5.899	**5.895**	**5.892**
T0580	**1.284**	1.716	1.331	1.303	1.304	1.291
T0635	2.450	12.520	9.238	6.388	**2.225**	**2.222**
T0637	4.961	12.610	7.468	4.966	**4.964**	**4.286**
T0639	7.944	13.390	10.310	8.967	**6.068**	**4.693**
T0643	3.882	20.670	19.800	**3.728**	**3.432**	**2.915**

It can be observed, when three PCA terms are considered, the structure is not well defined compared to the native structure, on the other hand, considering 11 PCA terms, the structure is better defined and closer to the native structure, as expected based on the previous analysis of the RMS and the energy function optimization results.

We computed the median coordinates of the sampled protein decoys that fullfil that the energy is below -200 for each PCA search space case. For each case, we presented the protein as a matrix with rows containing the coordinates x, y and z and the columns containing the atoms in the protein. This way of representing the protein helps us visualizing better the uncertainty behind the coordinates. We observed, that larger variations ins the coordinates occurs in the protein borders. Additionally, as the number of PCA terms decreases, the variations are observed to be smaller, a possible confirmation that, as the terms gets reduced, the ill-conditioned character of the tertiary protein structure prediction problem is reduced. On the other hand, the more PCA terms, the more ill-conditioned the optimization problem is as it is considering more information. As it can be observed, there is a trade-off between the ill-conditioned character and the prediction capability of the model. This is due to the fact that, as we reduce the PCA Search Space, some crucial information required to get a good prediction is lost in the model reduction procedure when accounting for fewer structural variations.

4 Conclusions

In this study, we present an study of the Principal Component Analysis dimensionality and how this can affect the energy prediction and tertiary structure of proteins from the CASP9 competition. The algorithm utilized successfully establishes a low dimensional space in order to apply the energy optimization procedure via a member of the family of Particle Swarm Optimizers. This model reduction has been performed in order to obtain four different search spaces (3, 5, 7, 9 and 11 dimensions plus a high frequency term) to perform the energy optimization later on. The optimizer was capable o modelling the protein sequence and sample the selected decoys projected over the four different PCA Search spaces. Different energy optimum values were obtained depending on the dimensions of the PCA Search Space. It was concluded that as the number of PCA terms increases, it is possible to obtain a better refinement of both the protein energy and the backbone structure of the native protein and its alternative states. As the number of PCA increases, a greater level of information of the decoys utilized to construct the PCA is included and, a lower energy and uncertainty is obtained in the predictions.

Finally, this paper serves to explain how the model reduction technique serves to alleviate the ill-posed character of this high-dimensional optimization problem and how to choose an appropriate.

Acknowledgements. A. K. acknowledges financial support from NSF grant DBI 1661391 and from The Research Institute at Nationwide Children's Hospital.

References

1. Zhang, Y.: Progress and challenges in protein structure prediction. Curr. Opin. Struct. Biol. **18**, 342–348 (2008)
2. Bonneau, R., Strauss, C.E., Rohl, C.A., Chivian, D., Bradley, P., Malmstrom, L., Robertson, T., Baker, D.: De novo prediction of three-dimensional structures for major protein families. J. Mol. Biol. **322**, 65–78 (2002)
3. Bradley, P., Chivian, D., Meiler, J., Misura, K., Rohl, C., Schief, W.W.W., Schueler-Furman, O., Murphy, P., Schonbrun, J., Rosetta predictions in: CASP5: successes, failures, and prospects for complete automation. Proteins **53**, 457–468 (2003)
4. Chivian, D., Kim, D.E., Malmstrom, L., Bradley, P., Robertson, T., Murphy, P., Strauss, C. E., Bonneau, R., Rohl, C.A., Baker, D.: Automated prediction of CASP-5 structures using the Robetta server. Proteins **53**, 524–533 (2003)
5. Sen, T.Z., Feng, Y., Garcia, J.V., Kloczkowski, A., Jernigan, R.L.: The extent of cooperativity of protein motions observed with elastic network models is similar for atomic and coarser-grained models. J. Chem. Theory Comput. **2**, 696–704 (2006)
6. Gniewek, P., Kolinski, A., Jernigan, R.L., Kloczkowski, A.: Elastic network normal modes provide a basis for protein structure refinement. J. Chem. Phys. **136**, 195101 (2012)
7. Fernández-Martínez, J.L.: Model reduction and uncertainty analysis in inverse problems. Lead. Edge **34**, 1006–1016 (2015)
8. Price, S.L.: From crystal structure prediction to polymorph prediction: interpreting the crystal energy landscape. Phys. Chem. Chem. Phys. **10**, 1996–2009 (2008)

9. Fernández-Martínez, J.L., et al.: On the topography of the cost functional in linear and nonlinear inverse problems. Geophysics **77**, W1–W15 (2012)

10. Fernández-Martínez, J.L., García-Gonzale, E.: Stochastic stability analysis of the linear continuous and discrete PSO models. Trans. Evol. Comp. **15**, 405–423 (2011)

11. Fernández-Martínez, J.L., García-Gonzalo, E.: Stochastic stability and numerical analysis of two novel algorithms of the PSO family: PP-PSO and RR-PSO. Int. J. Artif. Intell. Tools **21**, 1240011 (2012)

12. Jolliffe, I.T.: Principal Component Analysis. Springer, Heidelberg (2002). https://doi.org/10.1007/b98835

13. Kennedy, J., Eberhart, R.: A new optimizers using particle swarm theory. In: Proceedings of Sixth International Symposium Micromachine Human Science, vol. 1, pp. 39–46 (1995)

14. Fernández-Martínez, J.L., García-Gonzalo, E.: The generalized PSO a new door to PSO evolution. J. Artif. Evol. Appl. **2008**, 861275 (2008)

15. Fernández-Martínez, J.L., García-Gonzalo, E.: The PSO family: deduction, stochastic analysis and comparison. Swarm Intell **3**, 245–273 (2009)

16. Gront, D., Kolinski, A.: BioShell – A package of tools for structural biology prediction. Bioinformatics **22**, 621–622 (2006)

17. Gront, D., Kolinski, A.: Utility library for structural bioinformatics. Bioinformatics **24**, 584–585 (2008)

18. Gniewek, P., Kolinski, A., Jernigan, R.L., Kloczkowski, A.: BioShell - threading: a versatile monte carlo package for protein threading. BMC Bioinform. **22**, Article no. 22 (2014)

19. Aramini, J.M., et al.: Solution NMR structure of a putative Uracil DNA glycosylase from Methanosarcina acetivorans. Northeast Structural Genomics Consortium Target MvR76 (2010)

20. Ramelot, T.A., et al.: Solution NMR structure of the PBS linker Polypeptide domain (fragment 254-400) of Phycobilisome linker protein ApcE from Synechocystis sp. PCC 6803. Northeast Structural Genomics Consortium Target SgR209C

21. Eletsky, A., et al.: Solution NMR structure of the N-terminal domain of putative ATP-dependent DNA Helicase RecG-related Protein from Nitrosomonas europaea. Northeast Structural Genomics Consortium Target NeR70A (2010)

22. Heidebrecht, T., et al.: The structural basis for recognition of J-base containing DNA by a Novel DNA-binding domain in JBP1. Northeast Structural Genomics Consortium and others (2010)

23. Cuff, M.E., et al.: The lactose-specific IIB component domain structure of the phosphoenolpyruvate: carbohydrate phosphotransferase system (PTS) from Streptococcus pneumoniae. Midwest Center for Structural Genomics Target TIGR4 (2010)

24. Ramagopal, U.A. et al.: Structure of putative HAD superfamily (subfamily III A) hydrolase from Legionella pneumophila. 3N1U, New York Structural Genomics Research Center Target (2010)

25. Oke, M., et al.: Crystal structure of the hypothetical protein PA0856 from Pseudomonas Aeruginosa. Joint Center for Structural Genomics NP_249547.1 (2010)

26. Zhang, R., et al.: The crystal structure of functionally unknown protein from Neisseria Meningitidis MC58. Midwest Center for Structural Genomics Target 3NYM (2008)

27. Forouhar, F., et al.: Crystal structure of the N-terminal domain of DNA-binding protein SATB1 from Homo Sapiens. Northeast Structural Genomics Consortium Target HR4435B (2010)

The Shape Language Application to Evaluation of the Vertebra Syndesmophytes Development Progress

Marzena Bielecka[1(✉)], Rafał Obuchowicz[2], and Mariusz Korkosz[3]

[1] Chair of Geoinformatics and Applied Computer Science, Faculty of Geology,
Geophysics and Environmental Protection,
AGH University of Science and Technology,
Mickiewicza 30, 30-059 Cracow, Poland
bielecka@agh.edu.pl
[2] Department of Radiology, Jagiellonian University Medical College, Cracow, Poland
r.obuchowicz@gmail.com
[3] Division of Rheumatology, Departement of Internal Medicine and Gerontology,
Jagiellonian University Hospital, Śniadeckich 10, 31-531 Cracow, Poland
mariuszk@mp.pl

Abstract. In this paper, a measure for assessment the progress of pathological changes in spine bones is introduced. The definition of the measure is based on a syntactic description of geometric features of the bone contours. The proposed approach is applied for analysis of vertebra syndesmophytes in X-ray images of the spine. It turns out that the proposed measure assesses the progress of the disease effectively. The results obtained by the algorithm based on the introduced measure are consistent with the assessment done by an expert.

Keywords: Vertebrae radiographs · Shape language
Geometric features · Syntactic description

1 Introduction

X-ray images play a crucial role in the diagnosis of many inflammatory diseases including the ones of the spine. The early diagnosis and, as a consequence, good chances of effective therapy are crucial for spine diseases the more so because they usually limit the patient mobility and functionality significantly. Because X-ray imaging is cheap and commonly performed, there is a great demand for tools of automatic analysis of such images. Therefore, the studies concerning the topic are conducted intensively [2–5,7,11,14,17,20,24,26,27]. Bone contours analysis, based on the contour geometrical description by using syntactic methods, is one

This paper was supported by the AGH - University of Science and Technology, Faculty of Geology, Geophysics and Environmental Protection as a part of the statutory project.

© Springer International Publishing AG, part of Springer Nature 2018
L. Rutkowski et al. (Eds.): ICAISC 2018, LNAI 10842, pp. 117–126, 2018.
https://doi.org/10.1007/978-3-319-91262-2_11

of the applied tools [6,7]. It was used to analysis of various bones - the palm and the spine can be put as examples.

So far the studies were conducted only in the context of detecting pathological changes in bones [6,7]. In this paper, the method of assessment of the progress of the disease is proposed. Such tool would be useful, among others, for making an appraisal whether the applied therapy is effective. In such case, a sequence of X-ray images of the same patient would be analysed. The proposed method is based on a geometric description of the contour based on the syntactic approach and is a continuation of the previous studies conducted by the authors [6,7]. The paper is organized in the following way. In the next section, the clinical backgrounds are discussed. Then, in Sect. 3 theoretical basis is described. The measure that allows the physicians to assess the degree of the disease advance is introduced in Sect. 4. Results are put forward in the same section.

2   Clinical Background

Inflammatory spondyloarthropaties present clinically with joints swelling but also back pain related to spine involvement [15]. Inflammatory disease of the spine can be diagnosed by the different imaging modalities with most important role of radiography (CR) and magnetic resonance (MR) [1]. As MR has been proposed as a gold standard for the assessment of the inflammation of the medulla, present in the vertebral bodies of the spine, the shape of the vertebra can be effectively assessed with the use the CR lateral and AP views [16]. Crosstalk between immune system cells present in the medulla in the bone namely osteoblasts and osteoclasts which control bone tissue turnover was proven on molecular cellular and anatomical levels [19]. It is well proven that the morphological effect of immunological activation of the bone is a buildup of the syndesmophytes which form at the vertebral corners. Their shape and size reflect the activity of the disease, and they are known as an important predictive factor used for the assessment of the disease change and its dynamics. Careful monitoring of the shape of the corner of vertebral bodies has important clinical value [8,13,22]. This involves observing two features the cranio-caudal and antero-posterior dimensions of the osteophytes - see Figs. 1 and 2.

One of the important limitation of the CR is restricted accessibility of the thoracic spine (important site of osteophyte formation) where a shade of the vertebra limits thoracic spine assessment [10]. The diversity of radiological interpretation of the subtle changes in the shape of the osteophytes is another factor that limits reliable interpretation of the progression of the disease. The detection of the cranio-caudal and antero-posterior dimensions of the osteophytes is ambiguous because of small dimensions of the specimen, different quality of the radiographic pictures and lastly various habits of medical imaging professionals [25]. Careful analysis of the horizontal and vertical alignment of the bone protrusion is crucial for its assessment and classification. Reliable diagnosis of those discrete changes is crucial for the classification of the response to the treatment [9,10,12,21,23]. Therefore a semi-automatic or an automatic system addressed

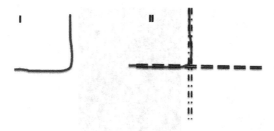

Fig. 1. I - The fragment of the anterior lower outline of the healthy vertebral. II - The spacing between two vertical lines is the antero-posterior dimension AP, the spacing between two horizontal lines represents the height of the change, this is the cranio-caudal dimension CC.

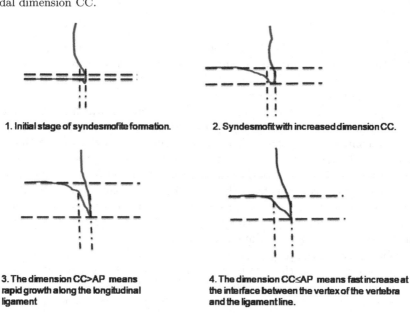

Fig. 2. The stages of syndesmofite formation.

to shape and size recognition of the syndesmophytes is highly expected by the medical community - both by diagnostic imaging specialists and rheumatologists.

3   The Generalized Shape Language

The contours of vertebrae are the objects of our interest. They have been received by using the Statistical Dominance Algorithm (SDA), which was developed for preprocessing of X-ray images with various dynamics and noise levels [18]. Next, the received contours were described by primitives that contain information

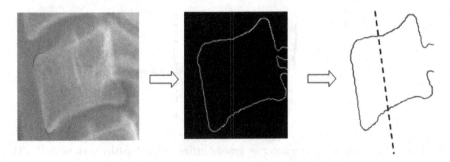

Fig. 3. The example of a vertebrae radiograph and the output image of SDA (R = 25, t = 100).

related to their shapes. Since in a given contour we focus on finding syndesmo-phytes, the area of our interest is limited to the places where they can occur. A received X-ray image of a vertebra, its contour and the contour section that is being examined are shown in Fig. 3.

The proposed primitives are fragments of a contour which have the same values of four characteristics $[c_t, c_c, c_x, c_y]$, which are calculated at each point (x, y) of a contour. The tangent line c_t, the contour convexity c_c and the signs c_x and c_y of the increment of the x-values and y-values are the characteristics. It should be mentioned that although theoretically, there is an infinite number of the points in the contour, from the computational point of view, the contour consists of the finite number of pixels. All the characteristics are calculated numerically. Primitives are marked by $p_{ij}, i, j \in \{1, 2, 3, 4\}$, where index i denotes geometrical features of the primitives, i.e. whether the fragment is a straight line, concave or convex as well as whether it is increasing or decreasing. The index j corresponds to the number of a quadrant of the Cartesian plane [7]. Each of the primitives p_{ij} is an equivalence class. It means that all fragments of a contour that can be described by p_{ij} belongs either to one of the semi-axes of the coordinate system or to one quadrant of the Cartesian plane. Next, the received string of primitives is converted into a sequence of sinquads. The *l-sinquad* is a string of subsequent primitives which belong to the same quadrant of the Cartesian plane [7]. A contour of a healthy vertebra, divided into sinquads by using the introduced primitives, is presented in Fig. 4. The points marked in red are transitions between sinquads. They provide information about basic properties of the analyzed shape. The string of sinquads, in turn, creates so-called biquads [7]. For the fragment *AB* of the contour shown in Fig. 4 the biquad has the form: 34.41. If the syndesmophyte lesions occur, they appear in the transition points between *3-sinquad* and *4-sinquad* and between *4-sinquad* and *1-sinquad* [8].

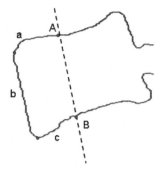

Fig. 4. The fragment AB of a contour of a healthy vertebra with marked sinquads. The label a describes *3-sinquad*, b it is *4-sinquad* and c it is *1-sinquad*.

4   Construction of Fuzzy Measure for the Syndesmophytes

To asses the progress of a syndesmophyte for a given patient, a measure is proposed. Two features are important in the process of assessing the rate of the disease progression [8,22]. One of them is the angle between $4 - sinquad$ and $1 - sinquad$ - see Fig. 5, while the other is related to the cranio-cadual size CC. Formally, this feature is the relative distance d between the points that belong to $1 - sinquad$ - see Fig. 6.

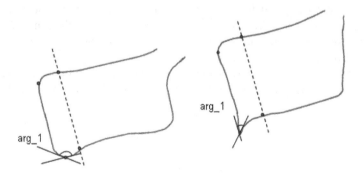

Fig. 5. The angle between $4 - sinquad$ and $1 - sinquad$. On the left there is a contour of a healthy vertebra with marked arg_1, on the right there is a contour with a syndesmophyte with marked arg_1.

The first feature, arg_1, takes small value if the syndesmophyte occurs. Generally, the larger syndesmophyte, the smaller value of arg_1. The second feature, arg_2, takes large value for a major syndesmophyte. The first one indicates that the syndesmophyte exists. There exists a limit, however, beyond which the arg_1 stays constant. Therefore the second feature arg_2 is needed. This feature, in turn, is not sufficient to measure the growth of syndesmophyte in the early stage of

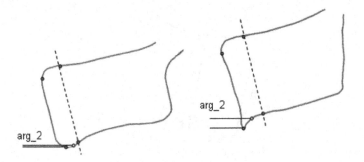

Fig. 6. The feature related with the cranio-cadual size CC. On the left there is a contour of a healthy vertebra with marked arg_2, on the right there is a contour with a syndesmophyte with marked arg_2.

disease. Thus, only a combination of these two features gives a reliable measure. Both of them can be easily determined on the basis of the introduced shape language [7]. Application of this language allows us to receive a description of a given contour of vertebra by primitives that, in turn, create sinquads. For a given vertebra a string of sinquads determines unequivocally the place where a syndesmophyte can occur. Thus, the arg_1 is the angle between $4 - sinquad$ and $1 - sinquad$ and it can be computed on the basis of the component c_t of the last primitive which belongs to $4 - sinquad$ and the first primitive which belongs to $1 - sinquad$. The arg_2 is computed on the basis of $1 - sinquad$. The first point of $1 - sinquad$ and the point for which the component c_t starts to be equal to the typical values for a healthy vertebra are determined. The difference d between values of y of these two points is the basis to define the arg_2. The value d is normalized by n which, for a given patient, is a maximal space between vertebrae. Both features are treated as fuzzy ones and their combination creates the argument t of the sigmoidal function μ, which defines the proposed measure:

$$\mu(t) = \frac{1}{1 + e^{-\beta t}},$$

where

$$t = \frac{180 - arg_1}{180} * arg_2$$

and

$$arg_2 = \frac{d}{n}.$$

For a healthy vertebra

$$100° \leq arg_1 \leq 180°$$

and arg_2 is small. This means that the argument t takes values around 0, which results in small value of function μ. The bigger syndesmophyte, the bigger value of t and, as the result, the bigger value of μ. In this paper, n takes the value equal to 50 which is the maximum distance between vertebrae observed in the analyzed sample. The sample contained 166 examples of vertebrae, 33 of them were diagnosed as affected by syndesmophytes.

In Table 1 the values of μ for seven chosen vertebrae with syndesmophytes for which a radiologist established the progress of the disease are presented. The contours of the chosen vertebrae are shown in Fig. 7.

Table 1. Values of function $\mu(t)$ with parameter $\beta = 5$ for some cases of vertebrae with syndesmophytes

No.	arg_1	arg_2	t	$\mu(t)$	Progress of the disease
1	50,4	0,16	0,113	0,115	2
2	56,52	0,7	0,48	0,416	4
3	90,08	0,42	0,21	0,24	2
4	75,63	0,46	0,267	0,291	2
5	74,28	0,4	0,235	0,264	2
6	65,12	0,5	0,319	0,331	3
7	56,19	0,46	0,316	0,329	3

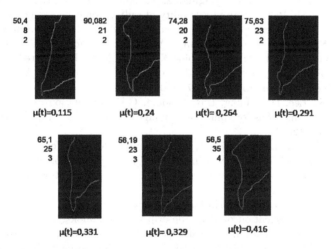

Fig. 7. Examples of vertebrae with syndesmophytes. Next to each contour there are values of arg_1, d and the progress of the disease. Under each contour, the values of function μ are shown.

5   Concluding Remarks

In this paper the measure for assessment the progress of the pathological changes in spine bones is introduced. Such assessment is crucial for verification whether the applied therapy is effective. Since a lot of X-ray images is taken every day, there is a great demand for the tools of automatic analysis of such images. The introduced measure is based on two geometric features of the bone contour. It turns out that they are sufficient to differentiate three stages of the development of syndesmophytes that were specified by an expert - the radiologist. It should be stressed that the results are preliminary. Although the authors had the access to the base of X-ray images at the Collegium Medicum of the Jagiellonian University, it was difficult to find sufficiently many images with clear examples for the third and the fourth stage of the syndesmophyte development. Values of the proposed measure were calculated for all 166 examples. It turned out that the measure correctly separated the contours of healthy vertebrae from the contours of the vertebrae affected by syndesmophytes.

It should be mentioned that the creation of a good base of X-ray images for this studies is planned. To sum up, the proposed measure turned to be an effective tool for the assessment of the syndesmophytes development. The value of the measure increases if the changes are more advanced.

References

1. Aydin, S.Z., Kasapoglu Gunal, E., Kurum, E., Akar, S., Mungan, H.E., Alibaz-Oner, F., Lambert, R.G., Atagunduz, P., Marzo Ortega, H., McGonagle, D., Maksymowych, W.P.: Limited reliability of radiographic assessment of spinal progression in ankylosing spondylitis. Rheumatology **56**, 2162–2169 (2017)
2. Antani, S., Long, L.R., Thoma, G.R.: A biomedical information system for combined content-based retrieval of spine X-ray images and associated text information. In: Proceedings of the 3rd Indian Conference on Computer Vision, Graphics and Image Processing, pp. 242–247 (2002)
3. Antani, S., Lee, D.J., Long, L.R., Thoma, G.R.: Evaluation of shape similarity measurement methods for spine X-ray images. J. Vis. Commun. Image Represent. **15**, 285–302 (2004)
4. Benerjee, S., Bhunia, S., Schaefer, G.: Osteophyte detection for hand osteoarthritis identification in X-ray images using CNNs. In: Proceedings of the Annual International Conference of the IEEE Engineering in Medicine and Biology Society, EMBS pp. 6196–6199 (2011)
5. Bielecka, M., Bielecki, A., Korkosz, M., Skomorowski, M., Wojciechowski, W., Zieliński, B.: Application of shape description methodology to hand radiographs interpretation. In: Bolc, L., Tadeusiewicz, R., Chmielewski, L.J., Wojciechowski, K. (eds.) ICCVG 2010. LNCS, vol. 6374, pp. 11–18. Springer, Heidelberg (2010). https://doi.org/10.1007/978-3-642-15910-7_2
6. Bielecka, M., Piórkowski, A.: Optimization of numerical calculations of geometric features of a curve describing preprocessed X-Ray images of bones as a starting point for syntactic analysis of finger bone contours. In: Chmielewski, L.J., Datta, A., Kozera, R., Wojciechowski, K. (eds.) ICCVG 2016. LNCS, vol. 9972, pp. 365–376. Springer, Cham (2016). https://doi.org/10.1007/978-3-319-46418-3_32

7. Bielecka, M., Korkosz, M.: Generalized shape language application to detection of a specific type of bone erosion in X-ray images. In: Rutkowski, L., Korytkowski, M., Scherer, R., Tadeusiewicz, R., Zadeh, L.A., Zurada, J.M. (eds.) ICAISC 2016. LNCS (LNAI), vol. 9692, pp. 531–540. Springer, Cham (2016). https://doi.org/10. 1007/978-3-319-39378-0_45

8. Baraliakos, X., Listing, J., Rudwaleit, M., Haibel, H., Brandt, J., Sieper, J., et al.: Progression of radiographic damage in patients with ankylosing spondylitis: defining the central role of syndesmophytes. Ann. Rheum. Dis. **66**, 910–915 (2007)

9. Creemers, M., Franssen, M.J., van't Hof, M.A., Gribnau, F.W., van de Putte, L.B., van Riel, P.L.: Assessment of outcome in ankylosing spondylitis: an extended radiographic scoring system. Ann. Rheum. Dis. **64**, 127–129 (2005)

10. El Maghraoui, A., Bensabbah, R., Bahiri, R., Bezza, A., Guedira, N., Hajjaj-Hassouni, N.: Cervical spine involvement in ankylosing spondylitis. Clin. Rheumatol. **22**, 94–98 (2003)

11. Howe, B., Gururajan, A., Sari-Sarraf, A., Long, L.R.: Hierarchical segmentation of cervical and lumbar vertebrae using a customized generalized Hough transform and extensions to active appearance models. In: Proceedings of the 6th IEEE Southwest Symposium on Image Analysis and Interpretation, pp. 182–186 (2004)

12. Heuft-Dorenbosch, L., Landewe, R., Weijers, R., Wanders, A., Houben, H., van der Linden, S., et al.: Combining information obtained from magnetic resonance imaging and conventional radiographs to detect sacroiliitis in patients with recent onset inflammatory back pain. Ann. Rheum. Dis. **65**, 804–808 (2006)

13. Lee, H.S., Kim, T.H., Yun, H.R., Park, Y.W., Jung, S.S., Bae, S.C., et al.: Radiologic changes of cervical spine in ankylosing spondylitis. Clin. Rheumatol. **20**, 262–266 (2001)

14. Long, L.R., Thoma, G.R.: Use of shape models to search digitized spine X-rays. In: Proceedings of the 13th IEEE Symposium on Computer-Based Medical Systems, pp. 255–260 (2000)

15. Mandl, P., Navarro-Compn, V., Terslev, L., Aegerter, P., et al.: EULAR recommendations for the use of imaging in the diagnosis and management of spondyloarthritis in clinical practice. Ann. Rheum. Dis. **74**, 1327–1339 (2015)

16. Maas, F., Spoorenberg, A., Brouwer, E., van der Veer, E., Bootsma, H., Bos, R., Wink, F.R., Arends, S.: Radiographic damage and progression of the cervical spine in ankylosing spondylitis patients treated with TNF-a inhibitors: facet joints vs. vertebral bodies. Semin. Arthritis. Rheum. **46**(5), 562–568 (2017)

17. Nurzynska, K., Piórkowski, A., Bielecka, M., Obuchowicz, R., Taton, G., Sulicka, J., Korkosz, M.: Automatical syndesmophyte contour extraction from lateral C spine radiographs. In: Augustyniak, P., Maniewski, R., Tadeusiewicz, R. (eds.) PCBBE 2017. AISC, vol. 647, pp. 164–173. Springer, Cham (2018). https://doi. org/10.1007/978-3-319-66905-2_14

18. Piórkowski, A.: A statistical dominance algorithm for edge detection and segmentation of medical images. In: Piętka, E., Badura, P., Kawa, J., Wieclawek, W. (eds.) Information Technologies in Medicine. AISC, vol. 471, pp. 3–14. Springer, Cham (2016). https://doi.org/10.1007/978-3-319-39796-2_1

19. Geusens, P., Lems, W.F.: Osteoimmunology and osteoporosis. Arthritis Res. Ther. **13**, 242 (2011)

20. Ogiela, M.R., Tadeusiewicz, R., Ogiela, L.: Image languages in intelligent radiological palm diagnostics. Pattern Recogn. **39**, 2157–2165 (2006)

21. Stolwijk, C., van Tubergen, A., Castillo-Ortiz, J.D., Boonen, A.: Prevalence of extra-articular manifestations in patients with ankylosing spondylitis: a systematic review and meta-analysis. Ann. Rheum. Dis. **74**, 65–73 (2015)

22. Spoorenberg, A., de Vlam, K., van der Linden, S., Dougados, M., Mielants, H., van de Tempel, H., et al.: Radiological scoring methods in ankylosing spondylitis. Reliability and change over 1 and 2 years. J. Rheumatol. **31**, 125–132 (2004)
23. Tan, S., Wang, R., Ward, M.M.: Syndesmophyte growth in ankylosing spondylitis. Curr. Opin. Rheumatol. **27**, 326–332 (2015)
24. Tezmol, A., Sari-Sarraf, H., Mitra, S., Long, R., Gururajan, A.: Customized Hough transform for robust segmentation of cervical vertebrae from X-ray images. In: Proceedings of the 5th IEEE Southwest Symposium on the Image Analysis and Interpretation, pp. 224–228 (2002)
25. Wanders, A.J., Landewe, R.B., Spoorenberg, A., Dougados, M., van der Linden, S., Mielants, H., et al.: What is the most appropriate radiologic scoring method for ankylosing spondylitis? A comparison of the available methods based on the outcome measures in rheumatology clinical trials filter. Arthritis Rheum. **50**, 2622–2632 (2004)
26. Xu, X., Lee, D.J., Antani, S., Long, L.R.: A spine X-ray image retrieval system using partial shape matching. IEEE Trans. Inf. Technol. Biomed. **12**, 100–108 (2008)
27. Zamora, G., Sari-Sarraf, H., Long, R.: Hierarchical segmentation of vertebrae from X-ray images. In: Proceedings of SPIE, Medical Imaging 2003: Image Processing, vol. 5032, p. 631 (2003)

Analytical Realization of the EM Algorithm for Emission Positron Tomography

Robert Cierniak[1(✉)], Piotr Dobosz[1], Piotr Pluta[1], and Zbigniew Filutowicz[2,3]

[1] Institute of Computational Intelligence, Czestochowa University of Technology,
Armii Krajowej 36, 42-200 Czestochowa, Poland
robert.cierniak@iisi.pcz.pl
[2] Information Technology Institute, University of Social Science,
90-113, Lodz, Poland
[3] Clark University, Worcester, MA 01610, USA

Abstract. The presented paper describes an analytical iterative approach to reconstruction problem for positron emission tomography (PET). The reconstruction problem is formulated taking into consideration the statistical properties of signals obtained by PET scanner and the analytical methodology of image processing. Computer simulations have been performed which prove that the reconstruction algorithm described here, does indeed significantly outperform conventional analytical methods on the quality of the images obtained.

Keywords: Image reconstruction from projections
Positron emission tomography
Statistical iterative reconstruction algorithm

1 Introduction

Medical imaging is one of the most useful diagnostic tools available to medicine. The algorithm presented here relates to one of the most popular imaging techniques belonging to the emission tomography category: positron emission tomography (PET). This medical imaging technique allows us to look inside a person and obtain images that illustrate various biological processes and functions. In this technique, a patient is initially injected with a radiotracer, which contains biochemical molecules. These molecules are tagged with a positron emitting radioisotope and can participate in physiological processes in the body. After the decay of these radioisotope molecules, positrons are emitted from the various tissues of the body which have absorbed the molecules. As a consequence of the annihilation of the positrons, pairs of gamma photons are produced and are

The original version of this chapter was revised: The name of the 4th author was corrected. The correction to this chapter is available at https://doi.org/10.1007/978-3-319-91262-2_71

released in opposite directions. In PET scanners, these pairs of photons are registered by detectors and counted. A pair of detectors detecting a pair of gamma photons at the same time constitutes a line of response (LOR). A count of photons registered on a certain LOR will be called a projection. The goal of the PET is to reconstruct the distribution of the radiotracer in the tissues of the investigated cross-sections of the body based on a set of projections from various LORs obtained by the PET scanner. The problem formulated in this way is called an image reconstruction from projections problem and is solved using various reconstruction methods. Because of the relatively small number of annihilations observed in a single LOR, the statistical nature of the measurements performed has a strong influence and must be taken into account. Recently, some new concepts regarding reconstruction algorithms have been applied to emission tomography techniques, with statistical approaches to image reconstruction being particularly preferred (see e.g. [1,2]). The standard reconstruction method used in PET is the maximum likelihood - expectation maximization (ML-EM) algorithm, as described for example in [3,4]. In this algorithm an iterative procedure is used in the reconstruction process, as follows:

$$f_l^{t+1} = f_l^t \frac{1}{\sum_k a_{kl}} \sum_k a_{kl} \frac{\lambda_k}{\sum_{\bar{l}} a_{k\bar{l}} f_{\bar{l}}^t} \tag{1}$$

where: f_l is an estimate of the image representing the distribution of the radiotracer in the body; $l = 1, \ldots, L$ is an index of pixels; t is an iteration index; λ_k is the number of annihilation events detected along the k-th LOR; a_{kl} is an element of the system matrix.

The image processing methodology used in this algorithm is consistent with the algebraic image reconstruction scheme, where the reconstructed image is conceptually divided into homogeneous blocks representing pixels. In this algebraic conception, the elements of the system matrix a_{kl} are determined for every pixel l separately, for every annihilation event λ_k detected along the k-th LOR. Unfortunately, algebraic reconstruction problems are formulated using matrices with very large dimensionality. Algebraic reconstruction algorithms are thus much more complex than analytical methods.

In this paper, a new statistical approach to the image reconstruction problem is proposed, which is consistent with the analytical methodology of image processing during the reconstruction process. The problem can be defined as an approximate discrete 2D reconstruction problem (see e.g. [5]). It takes into consideration a form of the interpolation function used in back-projection operations. The preliminary conception of this kind of image reconstruction from projections strategy for transmission tomography, i.e. x-ray computed tomography (CT), is represented in the literature only in the original works published by the authors of this paper, for parallel scanner geometry (see e.g. [6]), for fan-beam geometry (see e.g. [5]) and for spiral cone-beam tomography (see e.g. [7]). Thanks to the analytical origins of the reconstruction method proposed in the above papers, most of the above-mentioned difficulties connected with using algebraic methodology can be avoided. Although the proposed reconstruction method has to establish certain coefficients, these can be pre-calculated and, because of the small memory requirements, can be stored in memory. Generally, in algebraic methods, the coefficients a_{kl} are calculated dynamically during the reconstruction process, because of the huge dimensionality of the matrix containing these elements of the system. The analytical reconstruction problem is formulated as a shift-invariant system, which allows the application of an FFT algorithm during the most demanding calculations, and in consequence, significantly accelerates the image reconstruction process. In emission tomography, e.g. the PET, the measurements obtained from the scanner are subject directly to statistics consistent with the Poisson distribution. This means that the preferred approaches for this imaging technique (using the ML method) are based on the Kullback-Leibler divergence, and the EM algorithm associated with it. This conception is strictly concerned with the analytical reconstruction approach previously devised for the transmission CT technique, and the adoption of this solution to the emission PET imaging technique, using an EM algorithm with an analytical scheme of image processing.

2   Analytical Statistical Iterative Reconstruction Algorithm for PET Technique

The analytical approximate reconstruction problem was originally formulated for a parallel CT scanner [5,6,8–12]. However, the concept can also form the starting point for the design of a reconstruction algorithm for PET technique. The general scheme of the reconstruction procedure we propose is depicted in Fig. 1.

Firstly, the direct measurements using a PET scanner are obtained, and then statistically processed. In this way, the input signals for the reconstruction procedure are obtained, denoted below as $\lambda(s, \alpha)$, where s and α are parameters of a given LOR in the rotated coordinate system $x - y$, as is depicted in Fig. 2.

Having all the values $\lambda(s, \alpha)$, the reconstruction algorithm can be started, as specified in the described below steps, for the reconstruction of the cross-section with its center located at a fixed position regarding the axis z which is perpendicular to the pane $x - y$.

Before the main reconstruction procedure is started, the $h_{\Delta i, \Delta j}$ coefficients matrix is established. All of the calculations in this step of the reconstruction procedure can be pre-calculated, i.e. they can be carried out before the scanner performs any measurements. We make the simplification that the coefficients are the same for all pixels of the reconstructed image, and they can be calculated numerically, as follows:

$$h_{\Delta i, \Delta j} = \Delta_\alpha \sum_{\psi=0}^{\Psi-1} int\left(\Delta i \cos \psi \Delta_\alpha + \Delta j \sin \psi \Delta_\alpha\right), \tag{2}$$

where: Δi (Δj) is the difference between the index of pixels in the x-direction (y- direction); Δ_{xy} is the distance between the pixels in the reconstructed image;

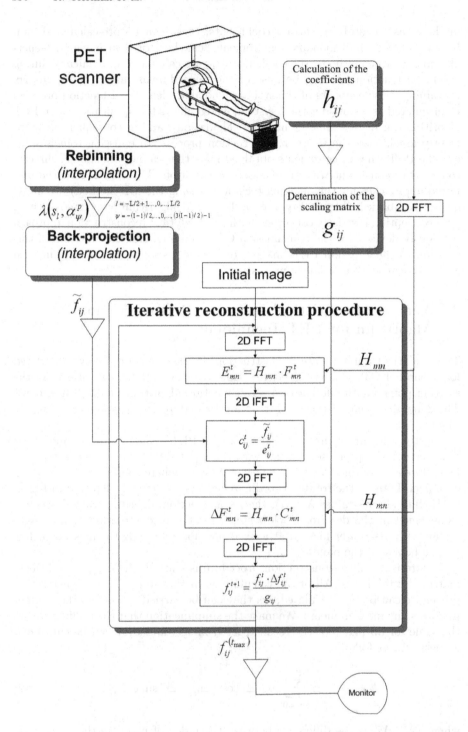

Fig. 1. An image reconstruction algorithm for PET technique

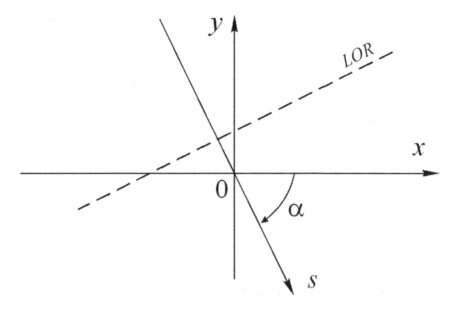

Fig. 2. The parameters of a line of response related to the reconstruction plane

Δ_α is the raster of angles of rotation; $\theta = 0, 1, \ldots, \frac{2\pi}{\Delta_\alpha} - 1$; Int is an interpolation function.

Then, the matrix of the coefficients $h_{\Delta i, \Delta j}$ is transformed into the frequency domain using a 2D FFT transform. The output of this step is a matrix of the coefficients H_{kl} with dimensions $2I \times 2I$ (if the reconstructed image has dimensions $I \times I$).

Also pre-calculated can be a scaling matrix g_{ij} which is determined based on the matrix of the coefficients $h_{\Delta i, \Delta j}$. If the reconstructed image has dimensions $I \times I$ then this operation is performed according to the relation:

$$g_{ij} = \sum_{\Delta i=-i+1}^{I-i} \sum_{\Delta j=-j+1}^{I-j} h_{\Delta i, \Delta j}. \tag{3}$$

2.1   Rebinning Operation

The reconstruction procedure presented below relates to the so-called rebinning methodology, where the image is reconstructed from a set of virtual parallel projections and the calculation is based on real measurements. In this rebinning operation, we will, first of all, consider the parallel-beam raster determined by the pair (s_l, α_ψ), where: $s_l = (l - 0.5)\,\Delta_{xy}$; $l = -L/2, \ldots, L/2$ is the sample index of the detectors in a hypothetical parallel-beam system; L is an even number of virtual detectors, and $\alpha_\psi = \psi \Delta_\alpha$; $\psi = 0, 1, \ldots, \Psi - 1$ is the index of the individual projections in the parallel-beam system; Ψ is the maximum number of projections; Δ_α is the angular distance between projections. In order to convert the real measurement values to the parallel system we interpolate parallel projection values from the immediate neighborhood of the determined pair (s_l, α_ψ), based on a group of four projection values: $\lambda\left(s_l^\uparrow, \alpha_\psi^\uparrow\right)$, $\lambda\left(s_l^\uparrow, \alpha_\psi^\downarrow\right)$, $\lambda\left(s_l^\downarrow, \alpha_\psi^\uparrow\right)$, $\lambda\left(s_l^\downarrow, \alpha_\psi^\downarrow\right)$, where $s_l^\downarrow$ is the next value below s_l, $s_l^\uparrow$ is the next value above s_l, $\alpha_\psi^\downarrow$ is the next value below α_ψ, $\alpha_\psi^\uparrow$ is the next value above α_ψ. We can use bilinear interpolation, for instance, to estimate the projection value of the hypothetical ray, according to the following relation:

$$\dot\lambda\left(s_l, \alpha_\psi\right) = \frac{\alpha_\psi - \alpha_\psi^\downarrow}{\alpha_\psi^\uparrow - \alpha_\psi^\downarrow}\left(\frac{s_l - s_l^\downarrow}{s_l^\uparrow - s_l^\downarrow}\lambda\left(s_l^\uparrow, \alpha_\psi^\uparrow\right) + \right. \tag{4}$$

$$\left.\frac{s_l^\uparrow - s_l}{s_l^\uparrow - s_l^\downarrow}\lambda\left(s_l^\downarrow, \alpha_\psi^\uparrow\right)\right) + \frac{\alpha_\psi^\uparrow - \alpha_\psi}{\alpha_\psi^\uparrow - \alpha_\psi^\downarrow}\left(\frac{s_l - s_l^\downarrow}{s_l^\uparrow - s_l^\downarrow}\lambda\left(s_l^\uparrow, \alpha_\psi^\downarrow\right) + \frac{s_l - s_l^\uparrow}{s_l^\uparrow - s_l^\downarrow}\lambda\left(s_l^\downarrow, \alpha_\psi^\downarrow\right)\right),$$

where $\dot\lambda\left(s_l, \alpha_\psi\right)$ is the interpolated value of the hypothetical parallel system.

2.2   Back-Projection Operation

The next part of the reconstruction algorithm begins by performing the back-projection operation. This operation is described by the following relation:

$$\tilde f_{ij} = \Delta_\alpha \sum_\psi \bar\lambda\left(s_{ij}, \alpha_\psi\right), \tag{5}$$

where: $\tilde f_{ij}$ is the image of the cross-section obtained after the back-projection operation at position z_p, for voxels described by coordinates (i, j, z_p); $i = 1, 2, \ldots, I$; $j = 1, 2, \ldots, I$, and the measurements $\bar\lambda\left(s_{ij}, \alpha_\psi\right)$ are interpolated for all pixels (i, j) in the reconstructed image at every angle α_ψ using the following interpolation formula:

$$\bar\lambda\left(s_{ij}, \alpha_\psi\right) = \frac{s_{ij} - s^\downarrow}{\Delta_{xy}}\dot\lambda\left(s^\uparrow, \alpha_\psi\right) + \frac{s^\uparrow - s_{ij}}{\Delta_{xy}}\dot\lambda\left(s^\downarrow, \alpha_\psi\right), \tag{6}$$

where $s^\downarrow$ is the next value below s_{ij}, $s^\uparrow$ is the next value above s_{ij}, and

$$s_{ij} = \left(i - \frac{I}{2}\right)\Delta_{xy}\cos\psi\Delta_\alpha + \left(j - \frac{I}{2}\right)\Delta_{xy}\sin\psi\Delta_\alpha. \tag{7}$$

2.3   Iterative Reconstruction Procedure

Before the iterative reconstruction procedure is started the initial image have to be determined. It can be any image f_{ij}^0 but in order to accelerate the reconstruction process, it is determined using a standard reconstruction method based on the same set of the measurements $\lambda\left(s_l, \alpha_\psi\right)$, for instance, the well-known FBP method.

The proposed in this paper reconstruction method used in PET is the maximum likelihood - expectation maximization (ML-EM) algorithm, and the used there image processing methodology with the analytical scheme is consistent. In this algorithm an iterative procedure is used in the reconstruction process, as follows:

$$f_{ij}^{t+1} = f_{ij}^{t} \frac{1}{g_{ij}} \sum_{i} \sum_{j} \frac{\tilde{f}_{ij}}{\sum_{\bar{i}} \sum_{\bar{i}} f_{i,j}^{t} h_{\Delta i, \Delta j}} h_{\Delta i, \Delta j} \tag{8}$$

where: f_{ij}^{t} is an estimate of the image representing the distribution of the radio-tracer in the body; $i = 1, 2, \ldots, I$ and $j = 1, 2, \ldots, I$ are indexes of pixels; t is an iteration index; $h_{\Delta i, \Delta j}$ are elements of the coefficients matrix determined according to the relation (2).

The EM formula from (8) is formulated using a shift-invariant system, which allows the application of an FFT algorithm during the most demanding calculations (as is shown in Fig. 1), and in consequence, significantly accelerates the image reconstruction process.

3   Experimental Results

In our experiments, we have adapted the well-known Shepp-Logan mathematical phantom of the head. During the simulations, for parallel projections, we fixed L = 512 virtual measurement points (detectors) on the screen. The number of parallel views was chosen as $\Psi = 1610$ per half-rotation and the size of the processed image was fixed at $I \times I = 512 \times 512$ pixels.

After making these assumptions, it is then possible to conduct the virtual measurements and complete all the required parallel projections which relate to the LORs. Then, through suitable rebinning operations, the back-projection operation can be carried out to obtain an image $\tilde{f}_{ij}$ which can be used as a referential image for the reconstruction procedure to be realized iteratively. There is depicted the reconstructed image after 30000 iterations in the Table 1C.

Coefficients $h_{\Delta i, \Delta j}$ necessary for the optimization procedure can be pre-calculated before the reconstruction process is started, and in our experiments, these coefficients were fixed for the subsequent processing. The image obtained after the back-projection operation was then subjected to a process of reconstruction, whose procedure is described by relations (8), wherein convolution operations were performed in the frequency domain. For comparison, a view of the reconstructed image using a traditional FBP algorithm is also presented (see Table 1B).

Table 1. Views of the images (window centre $C = 1.05 \cdot 10^{-3}$, window width $W = 0.1 \cdot 10^{-3}$): original image (A); reconstructed image using the standard FBP method with Shepp-Logan kernel (B); reconstructed image using the method described in this paper after 30000 iterations (C).

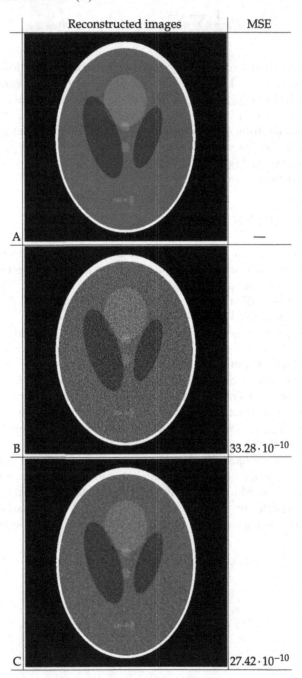

Reconstructed images	MSE
A	—
B	$33.28 \cdot 10^{-10}$
C	$27.42 \cdot 10^{-10}$

4 Conclusion

In this paper, it has been proven that this statistical approach, which was originally formulated for CT scanner with parallel beam geometry, can be adapted for PET technique. We have presented a fully feasible statistical reconstruction ML-EM algorithm. Simulations have been performed, which prove that our reconstruction method is very fast (thanks to the use of FFT algorithms) and gives satisfactory results with suppressed noise, even without the introduction of any additional regularization term. The computational complexity for 2D reconstruction geometries (e.g. parallel rays), is proportional to $I^2 \times \Psi \times L$ for each iteration of the algebraic reconstruction procedure described by the relation (1), but our original analytical approach only needs approximately $8I^2 \log_2 (2I)$ operations. However, soft computing techniques can find their application in reconstruction techniques, such as described e.g. in [13–25].

Acknowledgments. This work was partly supported by The National Centre for Research and Development in Poland (Research Project POIR.01.01.01-00-0463/17).

References

1. Sauer, K., Bouman, C.: A local update strategy for iterative reconstruction from projections. IEEE Trans. Signal Process. **41**(3), 534–548 (1993)
2. Fessler, J.A.: Penalized weighted least-squares image reconstruction for positron emission tomography. IEEE Trans. Med. Imaging **13**(2), 290–300 (1994)
3. Shepp, L.A., Vardi, Y.: Maximum likelihood reconstruction for emission tomography. IEEE Trans. Med. Imaging **MI–1**(2), 113–122 (1982)
4. Green, P.J.: Bayesian reconstructions from emission tomography data using a modified EM algorithm. IEEE Trans. Med. Imaging **9**(1), 84–93 (1990)
5. Cierniak, R.: New neural network algorithm for image reconstruction from fan-beam projections. Neurocomputing **72**, 3238–3244 (2009)
6. Cierniak, R.: A new approach to tomographic image reconstruction using a Hopfield-type neural network. Int. J. Artif. Intell. Med. **43**(2), 113–125 (2008)
7. Cierniak, R.: A three-dimentional neural network based approach to the image reconstruction from projections problem. In: Rutkowski, L., Scherer, R., Tadeusiewicz, R., Zadeh, L.A., Zurada, J.M. (eds.) ICAISC 2010. LNCS (LNAI), vol. 6113, pp. 505–514. Springer, Heidelberg (2010). https://doi.org/10.1007/978-3-642-13208-7_63
8. Cierniak, R.: A novel approach to image reconstruction from discrete projections using Hopfield-type neural network. In: Rutkowski, L., Tadeusiewicz, R., Zadeh, L.A., Żurada, J.M. (eds.) ICAISC 2006. LNCS (LNAI), vol. 4029, pp. 890–898. Springer, Heidelberg (2006). https://doi.org/10.1007/11785231_93
9. Cierniak, R.: A new approach to image reconstruction from projections problem using a recurrent neural network. Int. J. Appl. Math. Comput. Sci. **183**(2), 147–157 (2008)
10. Cierniak, R.: A novel approach to image reconstruction problem from fan-beam projections using recurrent neural network. In: Rutkowski, L., Tadeusiewicz, R., Zadeh, L.A., Zurada, J.M. (eds.) ICAISC 2008. LNCS (LNAI), vol. 5097, pp. 752–761. Springer, Heidelberg (2008). https://doi.org/10.1007/978-3-540-69731-2_72

11. Cierniak, R.: Neural network algorithm for image reconstruction using the grid-friendly projections. Australas. Phys. Eng. Sci. Med. **34**, 375–389 (2011)
12. Cierniak, R.: An analytical iterative statistical algorithm for image reconstruction from projections. Appl. Math. Comput. Sci. **24**(1), 7–17 (2014)
13. Chu, J.L., Krzyźak, A.: The recognition of partially occluded objects with support vector machines, convolutional neural networks and deep belief networks. J. Artif. Intell. Soft Comput. Res. **4**(1), 5–19 (2014)
14. Bas, E.: The training of multiplicative neuron model artificial neural networks with differential evolution algorithm for forecasting. J. Artif. Intell. Soft Comput. Res. **6**(1), 5–11 (2016)
15. Chen, M., Ludwig, S.A.: Particle swarm optimization based fuzzy clustering approach to identify optimal number of clusters. J. Artif. Intell. Soft Comput. Res. **4**(1), 43–56 (2014)
16. Aghdam, M.H., Heidari, S.: Feature selection using particle swarm optimization in text categorization. J. Artif. Intell. Soft Comput. Res. **5**(4), 231–238 (2015)
17. El-Samak, A.F., Ashour, W.: Optimization of traveling salesman problem using affinity propagation clustering and genetic algorithm. J. Artif. Intell. Soft Comput. Res. **5**(4), 239–245 (2015)
18. Leon, M., Xiong, N.: Adapting differential evolution algorithms for continuous optimization via greedy adjustment of control parameters. J. Artif. Intell. Soft Comput. Res. **6**(2), 103–118 (2016)
19. Miyajima, H., Shigei, N., Miyajima, H.: Performance comparison of hybrid electromagnetism-like mechanism algorithms with descent method. J. Artif. Intell. Soft Comput. Res. **5**(4), 271–282 (2015)
20. Rutkowska, A.: Influence of membership function's shape on portfolio optimization results. J. Artif. Intell. Soft Comput. Res. **6**(1), 45–54 (2016)
21. Bologna, G., Hayashi, Y.: Characterization of symbolic rules embedded in deep DIMLP networks: a challenge to transparency of deep learning. J. Artif. Intell. Soft Comput. Res. **7**(4), 265–286 (2017)
22. Notomista, G., Botsch, M.: A machine learning approach for the segmentation of driving maneuvers and its application in autonomous parking. J. Artif. Intell. Soft Comput. Res. **7**(4), 243–255 (2017)
23. Rotar, C., Iantovics, L.B.: Directed evolution - a new metaheuristc for optimization. J. Artif. Intell. Soft Comput. Res. **7**(3), 183–200 (2017)
24. Chang, O., Constante, P., Gordon, A., Singana, M.: A novel deep neural network that uses space-time features for tracking and recognizing a moving object. J. Artif. Intell. Soft Comput. Res. **7**(2), 125–136 (2017)
25. Liu, H., Gegov, A., Cocea, M.: Rule based networks: an efficient and interpretable representation of computational models. J. Artif. Intell. Soft Comput. Res. **7**(2), 111–123 (2017)

An Application of Graphic Tools and Analytic Hierarchy Process to the Description of Biometric Features

Paweł Karczmarek[1(✉)], Adam Kiersztyn[1], and Witold Pedrycz[2,3,4]

[1] Institute of Mathematics and Computer Science,
The John Paul II Catholic University of Lublin,
ul. Konstantynów 1H, 20-708 Lublin, Poland
{pawelk,adam.kiersztyn}@kul.pl
[2] Department of Electrical and Computer Engineering, University of Alberta,
Edmonton, AB T6R 2V4, Canada
wpedrycz@ualberta.ca
[3] Department of Electrical and Computer Engineering,
Faculty of Engineering, King Abdulaziz University, Jeddah 21589, Saudi Arabia
[4] Systems Research Institute, Polish Academy of Sciences, Warsaw, Poland

Abstract. AHP is a well-known method supporting decision-making based on a pairwise comparison process. Previous results of our research show that this tool can be effectively used to describe biometric features, in particular facial parts. In this paper, we present an original and innovative development of this approach augmented by a graphical interface that allows the user to get rid of restrictions in the form of certain numerical (linguistic) values, which were adapted beforehand, answering questions about comparisons of individual features. The presented results of experiments show the efficiency and ease of use of AHP based on a graphical interface in a context of description of biometric features. An application a proper non-linear transformation which parameters can be found on a basis of Particle Swarm Optimization can significantly improve the consistency of expert's evaluation.

Keywords: Analytic Hierarchy Process (AHP)
Decision-making theory · Particle Swarm Optimization
Facial features · Biometric description

1 Introduction

A problem of describing biometric features has been one of the most widely examined and thoroughly discussed topics in the literature of biometrics, criminology, and forensic science. In [2,3] the initial studies of the human being features including the facial ones in the context of forensic science and crime description were considered. Currently, the main approaches to describing features of the human face are sketch-based methods and memory portrait (e.g., Evofits and

© Springer International Publishing AG, part of Springer Nature 2018
L. Rutkowski et al. (Eds.): ICAISC 2018, LNAI 10842, pp. 137–147, 2018.
https://doi.org/10.1007/978-3-319-91262-2_13

IdentiKit [5,23]), methods based on expert knowledge (description provided by a qualified expert) with the use of linguistic descriptors [4,11,12,17,18], fuzzy logic and linguistic modeling [6,26–28], sketching with words [29], a describable attributes and similes obtained using MTurk [19,20], Granular Computing [21], and others. A survey can be found in [13]. Moreover, an extensive research has been conducted to investigate the way people describe other people on the ground of psychological experiments [24,34]. An important issue that still remains incompletely solved is the accuracy of the description of the face (and possibly other traits) by the expert and the behavior of the expert (his/her consistency of judgements), but also the proportion and the interdependence of different parts of the face, which could be particularly confusing in the context of recognition of people. In particular, unqualified witnesses describing the face may be susceptible to making a mistake in drawing up the description. For example, eyes that can be objectively staged, that they are large, embedded in a wide face can give the impression of being small. Similarly, comparing different faces with each other may give rise to similar difficulties. One of the reasons may be, for instance, an incorrectly scaled photo. Despite the decisive development of automated methods such as deep machine learning [35], sparse representation [36], etc., the final decision about the classification of a person must be made by the system operator. One of the forms of operator support can be an adequate method based on a graphical interface. The well-known tool in the decision making theory, AHP (Analytic Hierarchy Process) [31,33], becomes essential here.

Our ultimate objective of this study is to design and use a graphical interface to the AHP method in order to provide a convenient, effective description of individual parts of the face, both by a qualified expert, as well as by an unqualified witness involved in the description process. Traditionally, AHP is based on answering questions about comparisons realized between pairs of different attributes of a given feature in order to select or create a ranking of these attributes. The expert answers the questions, whose answers are usually expressed in a scale 1–9 or 1–7. In our proposal, this numeric scale is replaced by a simple graphical tool such as a slider. The user can easily adjust the position of the slider in accordance with his/her preferences. The numeric values of the slider (its position) are not presented to the user in order to make the decision based only on his/her own preferences and not feel limited by the values imposed in advance, which is often discouraging for participants of various surveys. The slider is obviously designed so that one cannot see the values used in the scale and the values are transformed to the traditional reciprocal matrix used in AHP. Then, in order to maintain a proper level of consistency in the users assessments, the matrix is optimized in order to minimize the inconsistency index using a well-known Particle Swarm Optimization method [16]. It is one of the possible methods of decreasing the level of inconsistency in an expert's choices. In detail, this process is carried out by finding the appropriate parameters of a piecewise linear function, which transforms the values of individual cells of the reciprocal matrix and gives the possibility to reduce the inconsistency index of this matrix. A general procedure of graphical approach to the APH was thoroughly described in the previous study [14]. Here, we discuss an application of the slider-based approach to the problem of facial description by the experts. It is worth to note that the witnesses of a crime or sometimes even specialists in the biometrics have difficulties in describing the facial features. Therefore, we present a possible application of AHP which can let the experts depart from typical explanations of linguistic or numerical values.

In this paper, we present preliminary promising results of experimental studies with the participation of an expert illustrating how the effectiveness and rationality of expert judgments can be improved using our approach. These results are obtained using the PUT Face Database [15] and related to selected part of the face such as the eyebrows. The originality of our approach is due to the fact that the graphic approach to Analytic Hierarchy Process, albeit intuitively appealing, has not been studied in the literature. Moreover, the results of experiments with apparently measurable facial features show potential applicability of the method in the novel applications such as description of features.

The paper is structured as follows. In Sect. 2 we show the main properties underlying the use of AHP. Our proposal is discussed in depth in Sect. 3. The results of experiments are shown in Sect. 4 while Sect. 5 is devoted to the conclusions and future works.

2   Analytic Hierarchy Process

Let us briefly recall the most important aspects of the Analytic Hierarchy Process [31,33]. It is a method of pairwise comparisons proceeded by one or more experts in order to obtain the importance, ranking, or priorities of the compared features. Usually, in order to obtain comparable answers, the following scale is used: equal importance (1), weak importance (2), moderate importance (3), moderate plus (4), essential (strong) importance (5), strong plus (6), demonstrated (very strong) importance (7), very, very strong (8), extreme importance (9). The scale is often modified, namely the range is reduced, sometimes real values, interval values, or fuzzy values [22], etc. are considered. The answers are recorded in the so-called reciprocal matrix A. This matrix is built as follows: The a_{ij} elements has the property $a_{ij} = 1/a_{ji}$, $i, j = 1, \ldots, n$, while $a_{ii} = 1$. In the theory as well as in practice, the value of the coefficient $\nu = (\lambda_{max} - n) / (n - 1)$ has the most significant meaning since it allows to assess the consistency of an expert ratings. Here $\lambda_{max} \geq n$ is the maximal eigenvalue of the reciprocal matrix A, and the value of $\mu = \nu/r$, where $r = 0, 0, 0.52, 0.89, 1.11, 1.25, 1.35, 1.40, 1.45, 1.49$ for $n = 1, , \ldots, 10$ compared elements, respectively. The values of r were obtained in the series of experiments as the mean consistency indices of 500 random reciprocal matrices [32]. The coefficients ν, μ, and r are called inconsistency index, consistency ratio, and random inconsistency index, respectively. It should be noted that in case of $n > 10$, the way of obtaining the r-values were discussed, for instance, in [1,30]. The final results of pairwise comparisons are expressed as the values of the elements of the eigenvector associated with the λ_{max} parameter.

3   Our Proposal

People often encounter difficulties with adapting to a predetermined scale, forms of answering questions, especially when the problem is not trivial. To such a category of problems, one can certainly include the issue of the description of the face and its parts and their relevance in the identification or classification process. Therefore, as an alternative, we propose an approach with a slider (a tool present in most graphical programming environments) whose values on a numerical scale are hidden from the user in such a way that he/she does not feel too large "jumps", and the result of his/her response is linearly transformed into the classical AHP scale (1/9, 9). An example of a transformation of this form may have the following form:

$$p(x) = \begin{cases} \frac{8}{9v}x + 1, & x \in [-v, 0), \\ \frac{8}{v}x + 1, & x \in [0, v]. \end{cases} \tag{1}$$

where $-v$ is the smallest and v is the largest position of the slider.

However, such a transformation is insufficient to ensure adequate consistency, coherence, reliability and comparability of expert assessments with assessments of other experts. Therefore, it is necessary to introduce a non-linear transformation, the task of which will be to map values from the AHP scale to new one being more in line with the real preferences of the expert. One of such proposals may be the piecewise linear function, whose formula can be explicitly given as follows:

$$f(x) = \begin{cases} \frac{\left(\frac{1}{b_{p-1}}-\frac{1}{9}\right)\left(x-\frac{1}{9}\right)}{\frac{1}{a_{p-1}}-\frac{1}{9}} + \frac{1}{9}, & x \in \left[\frac{1}{9}, \frac{1}{a_{p-1}}\right), \\ \cdots \\ \frac{\left(\frac{1}{b_{i-1}}-\frac{1}{b_i}\right)\left(x-\frac{1}{a_i}\right)}{\frac{1}{a_{i-1}}-\frac{1}{a_i}} + \frac{1}{b_i}, & x \in \left[\frac{1}{a_i}, \frac{1}{a_{i-1}}\right), \\ \cdots \frac{\left(1-\frac{1}{b_2}\right)\left(x-\frac{1}{a_2}\right)}{1-\frac{1}{a_2}} + 1, & x \in \left[\frac{1}{a_2}, 1\right), \\ \frac{(b_2-1)(x-1)}{a_2-1} + 1, & x \in [1, a_2), \\ \cdots \\ \frac{(b_i-b_{i-1})(x-a_{i-1})}{a_i-a_{i-1}} + b_{i-1}, & x \in [a_{i-1}, a_i), \\ \cdots \\ \frac{(9-b_{p-1})(x-a_{p-1})}{9-a_{p-1}} + b_{p-1}, & x \in [a_{p-1}, 9]. \end{cases} \tag{2}$$

The value of a coefficient p is the p-th broken cutoff point starting from the coordinate $(1, 1)$. The coefficients $a_2, a_3, \ldots, a_{p-1}$ are cutoff points lying on the x-axis, while the coefficients $b_2, b_3, \ldots, b_{p-1}$ are its points lying on the y-axis. Note that $a_1 = b_1 = 1$ and $a_p = b_p = 9$. These coefficients can be found on the basis of the minimization of the sum of inconsistency indices related to each reciprocal matrix. In particular, one reciprocal matrix can occur when only one expert considers only one facial feature. In this work, we use the well-known PSO (Particle Swarm Optimization) [10,16] method, which is a sociologically inspired optimization algorithm.

In the process of PSO, particles forming a certain swarm are randomly initialized. It is done by setting their initial positions and velocities. Next, when the successive generations of the algorithm are generated, the values of positions and velocities are obtained with the use of the following formulae:

$$\overline{\mathbf{v}}_i = \overline{\mathbf{v}}_i + 2\mathbf{r}_1 \otimes (\mathbf{p}_i - \mathbf{y}_i) + 2\mathbf{r}_2 \otimes (\mathbf{p}_g - \mathbf{y}_i), \tag{3}$$

$$\mathbf{y}_i = \mathbf{y}_i + \overline{\mathbf{v}}_i. \tag{4}$$

Here, $\mathbf{y}_i$ stands for a particle (corresponding to an i-th vector of cutoff points $a_2 < a_3 < \ldots < a_{p-1}, b_1 < b_2 < \ldots < b_{p-1}$) which are the parameters of the function f. $\overline{\mathbf{v}}_i$ stands for its velocities vector, its personal vest value is stored in the variable $\mathbf{p}_i$, while global best value is $\mathbf{p}_g$. A symbol $\otimes$ is used to denote an operation of element-wise vector multiplication, while $\mathbf{r}_1$ and $\mathbf{r}_2$ are randomly chosen values belonging to $[0, 1]$.

4   Experimental Studies

In the series of experiments, we involved an experienced expert in the field of face recognition to evaluate a subset of 20 faces from the PUT Face Database [15]. We have chosen this set of photos since the face images are of high resolution and selected facial features, such as eye positions, are determined manually with high precision. The example images are presented in Fig. 1. Therefore, it is possible to check the dependency between the experts assessments concerning the faces and the real values corresponding to measureable features according to their lengths. An example of applying the slider instead of constructing a AHP matrix with manually inserted numeric entries is shown in Fig. 2.

Fig. 1. PUT example faces

Let us consider the eyebrow width. We asked the expert to evaluate 20 first persons from the dataset PUT. We have chosen their pictures with central position of a head with at least slight variations. In the optimization process (repeated 10 times to get reliable averaged results) we have obtained the 10-segment piecewise linear function f as presented in Fig. 3. Figure 4 presents the maximal eigenvectors coefficient values of the AHP reciprocal matrices before

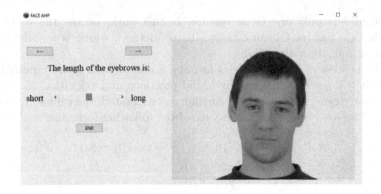

Fig. 2. The application form. The image comes from the PUT dataset [15]

running the optimization process while at Fig. 5 depicted are the results obtained after the optimization, namely after the transformation of the reciprocal matrix values through the non-linear function f. Figure 5 shows a slight improvement of the results which can be particularly seen when observing the trend lines corresponding to the linguistic variables *short, average,* and *long*. It is worth noting that the average values of the inconsistency index before and after the optimization were 0.072 and 0.007, respectively. Figure 6 depicts similar results in relation to the facial width feature. The function f for this case is presented in Fig. 7, while the inconsistency index decreased from 0.048 to 0.007. These results show the efficiency of the method and its applicability in the context of biometric features description.

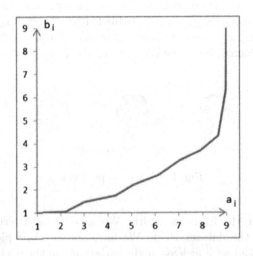

Fig. 3. Transformation function f (eyebrows width)

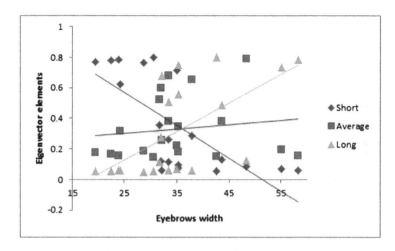

Fig. 4. The set of AHP results including trend lines (before the optimization process)

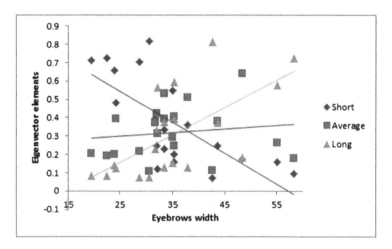

Fig. 5. The AHP results after PSO-based optimization with respect to the sum of maximal eigenvalues

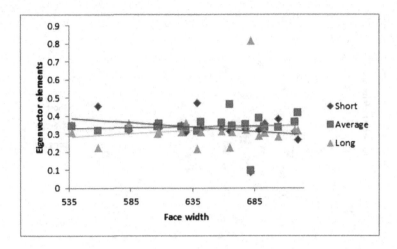

Fig. 6. Results related to the face width feature (after the optimization procedure)

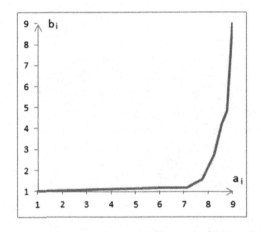

Fig. 7. Transformation function f (face width)

5   Conclusions and Future Work

In the study, we have proposed a novel and intuitive approach to the problem of
facial features description which is based on an application of graphical interface
to the well-known decision-making method, namely Analytic Hierarchy Process.
A presented series of experiments confirm the efficiency of the method when
applied to the description of facial feature such as eyebrow width. Among all
of the future work directions the most interesting seems to be a combination of
fuzzy AHP and graphical programming tools. Furthermore, an interesting aspect
of the future investigation may be a comparison of various object detection
algorithms [7–9, 25] and experts assessments.

Acknowledgements. The authors are supported by National Science Centre, Poland (grant no. 2014/13/D/ST6/03244). Support from the Canada Research Chair (CRC) program and Natural Sciences and Engineering Research Council is gratefully acknowledged (W. Pedrycz).

References

1. Alonso, J.A., Lamata, M.T.: Consistency in the analytic hierarchy process: a new approach. Int. J. Uncertain. Fuzz. **14**, 445–459 (2006)
2. Bertillon, A.: La photographie judiciaire: avec un appendice sur la classification et l'identification anthropométriques. Gauthier-Villars, Paris (1890)
3. Bertillon, A.: Identification anthropométrique: instructions signaltiques. Imprimerie administrative, Melun (1983)
4. Dolecki, M., Karczmarek, P., Kiersztyn, A., Pedrycz, W.: Face recognition by humans performed on basis of linguistic descriptors and neural networks. In: Proceedings of the 2016 International Joint Conference on Neural Networks (IJCNN 2016), pp. 5135–5140 (2016)
5. Frowd, C.D., Hancock, P.J.B., Carson, D.: EvoFIT: a holistic, evolutionary facial imaging technique for creating composites. ACM Trans. Appl. Percept. **1**, 19–39 (2004)
6. Fukushima, S., Ralescu, A.L.: Improved retrieval in a fuzzy database from adjusted user input. J. Intell. Inf. Syst. **5**, 249–274 (1995)
7. Grycuk, R., Gabryel, M., Korytkowski, M., Scherer, R.: Content-based image indexing by data clustering and inverse document frequency. In: Kozielski, S., Mrozek, D., Kasprowski, P., Małysiak-Mrozek, B., Kostrzewa, D. (eds.) BDAS 2014. CCIS, vol. 424, pp. 374–383. Springer, Cham (2014). https://doi.org/10.1007/978-3-319-06932-6_36
8. Grycuk, R., Gabryel, M., Korytkowski, M., Scherer, R., Voloshynovskiy, S.: From single image to list of objects based on edge and blob detection. In: Rutkowski, L., Korytkowski, M., Scherer, R., Tadeusiewicz, R., Zadeh, L.A., Zurada, J.M. (eds.) ICAISC 2014. LNCS (LNAI), vol. 8468, pp. 605–615. Springer, Cham (2014). https://doi.org/10.1007/978-3-319-07176-3_53
9. Grycuk, R., Gabryel, M., Nowicki, R., Scherer, R.: Content-based image retrieval optimization by differential evolution. In: 2016 IEEE Congress on Evolutionary Computation (CEC), pp. 86–93 (2016)
10. Kacprzyk, J., Pedrycz, W.: Springer Handbook of Computational Intelligence. Springer, Heidelberg (2015). https://doi.org/10.1007/978-3-662-43505-2
11. Karczmarek, P., Kiersztyn, A., Pedrycz, W., Dolecki, M.: Linguistic descriptors and analytic hierarchy process in face recognition realized by humans. In: Rutkowski, L., Korytkowski, M., Scherer, R., Tadeusiewicz, R., Zadeh, L.A., Zurada, J.M. (eds.) ICAISC 2016. LNCS (LNAI), vol. 9692, pp. 584–596. Springer, Cham (2016). https://doi.org/10.1007/978-3-319-39378-0_50
12. Karczmarek, P., Kiersztyn, A., Pedrycz, W., Rutka, P.: A study in facial features saliency in face recognition: an analytic hierarchy process approach. Soft. Comput. **21**, 7503–7517 (2017)
13. Karczmarek, P., Kiersztyn, A., Rutka, P., Pedrycz, W.: Linguistic descriptors in face recognition: a literature survey and the perspectives of future development. In: SPA 2015 Signal Processing, Algorithms, Architectures, Arrangements, and Applications, Conference Proceedings, pp. 98–103 (2015)

14. Karczmarek, P., Pedrycz, W., Kiersztyn, A.: Graphic interface to analytic hierarchy process and its optimization. IEEE Trans. Fuzzy Syst. (submitted)

15. Kasiński, A., Florek, A., Schmidt, A.: The PUT face database. Image Process. Commun. **13**, 59–64 (2008)

16. Kennedy, J.F., Eberhart, R.C., Shi, Y.: Swarm Intelligence. Academic Press, San Diego (2001)

17. Kiersztyn, A., Karczmarek, P., Dolecki, M., Pedrycz, W.: Linguistic descriptors and fuzzy sets in face recognition realized by humans. In: Proceedings of the 2016 IEEE International Conference on Fuzzy Systems (FUZZ-IEEE 2016), pp. 1120–1126 (2016)

18. Kiersztyn, A., Karczmarek, P., Rutka, P., Pedrycz, W.: Quantitative methods for linguistic descriptors in face recognition. In: Zapała, A. (ed.) Recent Developments in Mathematics and Informatics, Contemporary Mathematics and Computer Science, vol. 1, pp. 123–138. The John Paul II Catholic University of Lublin Press, Lublin (2016)

19. Kumar, N., Berg, A.C., Belhumeur, P.N., Nayar, S.K.: Attribute and simile classifiers for face verification. In: Proceedings of IEEE 12th International Conference on Computer Vision, pp. 365–372 (2009)

20. Kumar, N., Berg, A.C., Belhumeur, P.N., Nayar, S.K.: Describable visual attributes for face verification and image search. IEEE Trans. Pattern Anal. Mach. Intell. **33**, 1962–1977 (2011)

21. Kurach, D., Rutkowska, D., Rakus-Andersson, E.: Face classification based on linguistic description of facial features. In: Rutkowski, L., Korytkowski, M., Scherer, R., Tadeusiewicz, R., Zadeh, L.A., Zurada, J.M. (eds.) ICAISC 2014. LNCS (LNAI), vol. 8468, pp. 155–166. Springer, Cham (2014). https://doi.org/10.1007/978-3-319-07176-3_14

22. van Laarhoven, P.J.M., Pedrycz, W.: A fuzzy extension of Saaty's priority theory. Fuzzy Sets Syst. **11**, 199–227 (1983)

23. Laughery, K.R., Fowler, R.H.: Sketch artists and Identi-kit, procedure for recalling faces. J. Appl. Psychol. **65**, 307–316 (1980)

24. Matthews, M.L.: Discrimination of Identikit constructions of faces: evidence for a dual processing strategy. Percept. Psychophys. **23**, 153–161 (1978)

25. Moreira, J.L., Braun A., Musse, S.R.: Eyes and eyebrows detection for performance driven animation. In: 2010 23rd SIBGRAPI Conference on Graphics, Patterns and Images, pp. 17–24 (2010)

26. Nakayama, M., Miyajima, K., Iwamoto, H., Norita, T.: Interactive human face retrieval system based on linguistic expression. In: Proceedings of 2nd International Conference on Fuzzy Logic and Neural Networks, IIZUKA 1992, vol. 2, pp. 683–686 (1992)

27. Nakayama, M., Norita, T., Ralescu, A.: A fuzzy logic based qualitative modeling of image data. In: Proceedings of IPMU 1992, pp. 615–618 (1992)

28. Norita, T.: Fuzzy theory in an image understanding retrieval system. In: Ralescu, A.L. (ed.) Applied Research in Fuzzy Technology. International Series in Intelligent Technologies, vol. 1, pp. 215–251. Springer Science+Business Media, New York (1994). https://doi.org/10.1007/978-1-4615-2770-1_6

29. Rahman, A., Sufyan Beg, M.M.: Face sketch recognition using sketching with words. Int. J. Mach. Learn. Cyber. **6**, 597–605 (2015)

30. Saaty, T.L.: Fundamentals of Decision Making and Priority Theory with the Analytic Hierarchy Process. Analytic Hierarchy Process Series, vol. 6. RWS Publications, Pittsburgh (2000)

31. Saaty, T.L.: The Analytic Hierarchy Process. McGraw-Hill, New York (1980)
32. Saaty, T.L., Mariano, R.S.: Rationing energy to industries: priorities and input-output dependence. Energy Syst. Policy **3**, 85–111 (1979)
33. Saaty, T.L., Vargas, L.G.: Models, Methods, Concepts & Applications of the Analytic Hierarchy Process. Springer, New York (2012). https://doi.org/10.1007/978-1-4614-3597-6
34. Sadr, J., Jarudi, I., Sinha, P.: The role of eyebrows in face recognition. Perception **32**, 285–293 (2003)
35. Sun, Y., Wang, X., Tang, X.: Deep learning face representation from predicting 10,000 classes. In: 2014 IEEE CVPR, pp. 1891–1898 (2014)
36. Wright, J., Yang, A.Y., Ganesh, A., Sastry, S.S., Ma, Y.: Robust face recognition via sparse representation. IEEE Trans. Pattern Anal. Mach. Intell. **31**, 210–227 (2009)

On Some Aspects of an Aggregation Mechanism in Face Recognition Problems

Paweł Karczmarek[1]([✉]), Adam Kiersztyn[1], and Witold Pedrycz[2,3,4]

[1] Institute of Mathematics and Computer Science, The John Paul II Catholic University of Lublin, ul. Konstantynów 1H, 20-708 Lublin, Poland
{pawelk,adam.kiersztyn}@kul.pl
[2] Department of Electrical and Computer Engineering, University of Alberta, Edmonton, AB T6R 2V4, Canada
wpedrycz@ualberta.ca
[3] Department of Electrical and Computer Engineering, Faculty of Engineering, King Abdulaziz University, Jeddah 21589, Saudi Arabia
[4] Systems Research Institute, Polish Academy of Sciences, Warsaw, Poland

Abstract. In the paper, we investigate the problem of an aggregation of classifiers based on numerical and linguistic values of facial features. In the literature, there are many reports of the studies discussing the aggregation or information fusion, however in the situation when the specific classification methods utilize numeric, not linguistic values. Here, we examine the well-known methods (Eigenfaces, Fisherfaces, LBP, MB-LBP, CCBLD) supported by the linguistic values of the measurable facial segments. The detailed results of experiments on the MUCT and PUT facial databases show which of the common aggregation functions and methods have a significant potential to improve the classification process.

Keywords: Classifiers aggregation · Clustering · FCM
Face recognition · Eigenfaces · Fisherfaces · Local descriptors

1 Introduction

Face recognition has been one of the most challenging problems in computer vision and machine learning community for over 20 years. It is caused by various and commonly encountered applications such as surveillance systems, border control, passport verification, drivers licenses verification, and many others. Most of the biometric systems produce results of acceptable accuracy. However, at least from the theoretical point of view, some aspects of the facial recognition problems have not been fully developed as of now. One of them is an aggregation of classifiers based on various sources of information about a subject to be classified or verified. In particular, there is one important question to be addressed. Can the merging of the results of classifiers based on classic face recognition algorithms and linguistic (or, directly, geometric) measures be made efficient? There are a few important results in the area of aggregation techniques

© Springer International Publishing AG, part of Springer Nature 2018
L. Rutkowski et al. (Eds.): ICAISC 2018, LNAI 10842, pp. 148–156, 2018.
https://doi.org/10.1007/978-3-319-91262-2_14

concerning face recognition or, most broadly, biometric recognition systems. For instance, in [6] scoring and template matching strategy for four facial regions were successively applied. In [44] the eigenfaces algorithm [46] for various facial regions was applied and the results were aggregated. In [15] RBF neural networks and majority rule were used. A method of combination of classifiers based on different image transformations was discussed in [17]. In [35] a weighted sum rule and similarity matrices were proposed. T-norms regarded as an aggregation mechanism were comprehensively described in [16]. Utility functions viewed as aggregation operators were discussed in [9]. In [2] a three-valued logic and fuzzy set-based decision making mechanism to aggregate the classifiers were proposed. In the context of deep learning the fusion of colors was discussed in [33]. Finally, there were many studies exploring the applications of fuzzy measure sought as a way of classifier aggregation, see [19–21,26,31,32,36–38,40]. A class of methods, which are closely related to aggregation of classification based on facial features, is the expert-oriented, and more precisely, linguistic descriptors-based approach. This class of methods is derived from the observation that computers can recognize faces in a manner similar to people who are extremely effective in recognizing others and that they can work on the basis of similar premises, i.e. a linguistic description of the face and its parts, and more generally, related with linguistic modeling and Granular Computing [43]. Examples of works in this field are, among others [10,22,23,28–30,42,45]. Finally, a comprehensive discussion of various aggregation techniques and their applications can be found, among others, in [4,7,11,18]. The main objective of this paper is to explore the fundamental relationships between the classic face recognition algorithms such as Principal Component Analysis [46], Linear Discriminant Analysis [3], and newer ones such as Local Binary Patterns [1], Multi-scale Block LBP [8,34], and Chain Code-based Local Descriptor, CCBLD [24,25] and the pixel lengths of facial features. Since the clustering method [5] leads to linguistic descriptions of facial features and their membership grades to the linguistic descriptors such as *short*, *quite short, average, quite long, long* these two approaches, namely geometric and linguistic, can be seen here as identical. Our aim is to investigate how the above-mentioned algorithms work supported by the information such as the lengths of features and which of commonly used aggregation functions (such as minimum, maximum, etc.) are the best proposals for these combinations of methods. The paper is organized as follows. Section 2 briefly covers the aggregation technique discussed in this study. In Sect. 3, we present experimental results. The last section is devoted to conclusions and future directions.

2 Aggregation and Feature Extraction

The model we consider here can be outlined as follows. Consider two classifiers. The first is based on the nearest neighbour algorithm applied to the distances between the vector representing images after the well-known transformations such as PCA, LDA, LBP, MB-LBP, and CCBLD. The second one is based on the following scheme. If considered are n facial features (expressed in terms of their pixel lengths) then we apply the Fuzzy C-Means algorithm to each of them to find the clusters representing the linguistic terms *short, quite short, average, quite long,* and *long.* In this manner, for every image I in the dataset, we form the vector $V_I = (v_{1,1}, v_{1,2}, v_{1,3}, v_{1,4}, v_{1,5}, \ldots, v_{n,1}, v_{n,2}, v_{n,3}, v_{n,4}, v_{n,5})$ representing the face as a set of $5n$ membership grades. For such vectors, we apply the nearest neighbour classifier. Since there are only two classifiers, i.e., the first which is based on a numerical method and the second one, which is based on FCM-generated features, the set of possible aggregation functions is not large. Those most intuitive and easy to apply are the following: minimum, maximum, average, geometric mean, harmonic mean, median, and voting. Of course, the aggregation operator is applied to the normalized values of two distances, namely the one coming from the numerical method (e.g., PCA) and the second being the distance between the vectors of memberships.

3   Experimental Studies

In this section, we present the results of a series of experiments with two datasets for which a very detailed coordinates of chosen facial features were provided by the creators to fully address the precision task regarding to particular lengths of features. The datasets being used in the study are PUT Face Database [27] and The MUCT Face Database [39], respectively.

3.1   The MUCT Face Database

The MUCT database consists of 3,755 faces with 76 manual landmarks. We selected 15 photos for each of 199 individuals. Only those peoples who had 15 images were selected. Based on 76 available landmarks we are able to identify 14 facial features. For further research, the top 10 were selected, which are also available in the second of the analysed bases: length of mouth; width of the upper lip; width of the lower lip; eyebrow length; width of eyebrows; width at the bottom of the nose; the distance between the pupils; length of eye socket; width of eye socket; width of face. An example of a picture from the MUCT database with the landmarks is shown in Fig. 1. All the files were preprocessed (cropped and converted to the grayscale). During the analysis, the data set was randomly divided into a learning set containing two images of each person in the first series of experiments and five images per person in the second series, respectively, and a testing set containing remaining images. Based on data coming from the learning set, for each of the discussed features, five clusters were defined. In the next step, for each feature and each of the available photos, the degree of membership to each of the clusters was determined. The description of each image was obtained by using the real numbers vector $v = [v_1, v2, \ldots, v_5]$. In the series of experiments we have tested the accuracy of individual classifiers as PCA (with a Canberra distance between the vectors after the transformation), LDA (with a cosine distance), local binary pattern (LBP with a partition of the image onto 7×7 subregions), multi-scale block LBP (MBLBP with a partition

Fig. 1. MUCT landmarks

of the image onto 7×7 subregions and the blocks of pixels of a size 3×3), and chain code-based local descriptor (CCBLD with a partition of the image onto 3×3 subregions and the blocks of pixels of a size 3×3). As the aggregation function, we have chosen a few most popular and intuitive two-argument operators, namely minimum, maximum, average, geometric and harmonic means, median, and voting. Here, the result of a method based on FCM clustering is very poor (21.61%) and, therefore, it is difficult to obtain satisfactory classification level with using of this method. However, particularly in case of local descriptors LBP, MBLBP, and CCBLD the maximum function helps improving the final result of classification after an aggregation. A relatively good choice was the use of the average, median, or harmonic mean, see Table 1 for details.

Table 1. Results obtained for the MUCT. The consecutive columns are the average recognition rate of the method (PCA, LDA, etc.,), vector comparison-based recognition and the combination of two classifiers with different aggregation functions. The results where the improvement took place are written bold

	Method	Vector	Min.	Max.	Average	Geometric mean	Harmonic mean	Median	Voting
PCA	74.12	21.61	24.92	73.72	63.12	55.03	46.28	63.12	48.89
LDA	90.80	21.61	**90.8**	31.71	80.3	90.5	**91.36**	80.3	57.54
LBP	53.17	21.61	21.61	**57.84**	49.55	38.79	31.36	49.55	39.25
MBLBP	63.57	21.61	21.61	**65.33**	57.74	46.48	35.93	57.74	43.92
CCBLD	52.31	21.61	26.78	**53.72**	**53.67**	48.89	44.27	**53.67**	38.74

3.2   PUT Face Database

The PUT database contains 2,200 photos: 22 photos for each of 100 people. For further analysis, 11 photos for each person were selected. For each photo there is a very large collection of key points and contours of key facial features. Sample photo with selected key points presents Fig. 2. Based on the available information was selected a set of ten most important facial features: length of mouth; width of the upper lip; width of the lower lip; eyebrow length; width of eyebrows; width at the bottom of the nose; the distance between the pupils; length of eye socket; width of eye socket; width of face. As in the case of MUCT, the available set of photos was randomly divided into two parts: The learning set containing five images of each person and the testing set containing the remaining images. Next, centres of clusters and degrees of membership for each image and for each feature were determined. Here, we have conducted similar series of experiments as in the previous case. In case of LBP and MBLBP the images were divided onto 3×3 subregions. Moreover, the pixel block sizes in the case of MBLBP were 3×3. The results are significantly better in relation to both accuracy (96.17% by vectors obtained in FCM method) and, consequently, in relation to potential improvement of the method by an application of aggregation operators, see Table 2. The PCA and FCM-based vectors classifiers can be aggregated efficiently by almost all the aggregators excluding voting. Similarly, the LDA accuracy can be improved by vectors with help of minimum, average, geometric and harmonic means, and median. All the means and the median are efficient for the LBP and MBLBP cases. From the other hand, CCBLD could be improved by maximum function.

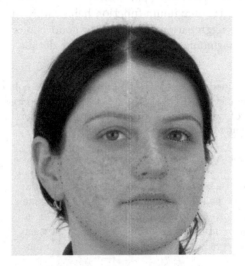

Fig. 2. Sample photo from PUT database with selected key points

Table 2. Results obtained for the PUT database

	Method	Vector	Min.	Max.	Average	Geometric mean	Harmonic mean	Median	Voting
PCA	96	96.17	**96.67**	98	**99.67**	**99.33**	**99**	**99.67**	95.67
LDA	99.83	96.17	**99.83**	98.67	100	100	**99.83**	100	97.67
LBP	99.17	96.17	99.00	97.83	100	100	100	100	97.67
MBLBP	100	96.17	99.83	98.5	100	100	100	100	97.83
CCBLD	100	96.17	97.67	**100**	99.83	99.83	99.5	99.83	97.83

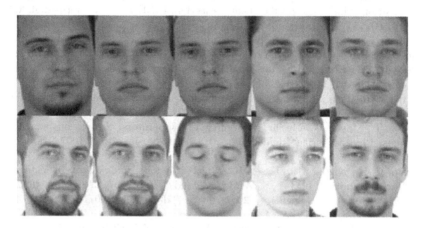

Fig. 3. Example faces not classified by LBP (first row) and distance measures-based method (second row)

4   Conclusions and Future Studies

In this study, we have thoroughly studied the problem of aggregation of classifiers based on the well-known numerical algorithms and the values of facial features coming in the form of membership grades to the linguistic values typically used by humans describing a face. A series of experiments have shown the potential applicability of the method, specifically when the dataset of considered images is relatively small and there is a possibility to determine the precise lengths of particular facial features. We have found the best two-argument aggregation functions suitable for the aggregation process. Future studies may focus on a comprehensive examination of other classification methods and the usage of larger datasets to facilitate the fully automation of the method and an improvement by an application of algorithms of content-based image retrieval, e.g., [12–14, 41].

Acknowledgements. The authors are supported by National Science Centre, Poland (grant no. 2014/13/D/ST6/03244). Support from the Canada Research Chair (CRC) program and Natural Sciences and Engineering Research Council is gratefully acknowledged (W. Pedrycz).

References

1. Ahonen, T., Hadid, A., Pietikäinen, M.: Face recognition with local binary patterns. In: Pajdla, T., Matas, J. (eds.) ECCV 2004. LNCS, vol. 3021, pp. 469–481. Springer, Heidelberg (2004). https://doi.org/10.1007/978-3-540-24670-1_36
2. Al-Hmouz, R., Pedrycz, W., Daqrouq, K., Morfeq, A.: Development of multimodal biometric systems with three-way and fuzzy set-based decision mechanisms. Int. J. Fuzzy Syst. **20**, 128–140 (2018). https://doi.org/10.1007/s40815-017-0299-9
3. Belhumeur, P.N., Hespanha, J.P., Kriegman, D.J.: Eigenfaces vs. Fisherfaces: recognition using class specific linear projection. IEEE Trans. Pattern Anal. Mach. Intell. **19**, 711–720 (1997)
4. Beliakov, G., Pradera, A., Calvo, T.: Aggregation Functions: A Guide for Practitioners. Springer, Heidelberg (2007). https://doi.org/10.1007/978-3-540-73721-6
5. Bezdek, J.C., Ehrlich, R., Full, W.: FCM: the fuzzy c-means clustering algorithm. Comput. Geosci. **10**, 191–203 (1984)
6. Brunelli, R., Poggio, T.: Face recognition: features versus templates. IEEE Trans. Pattern Anal. Mach. Intell. **15**, 1042–1052 (1993)
7. Calvo, T., Mayor, G., Mesiar, R.: Aggregation Operators. New Trends and Applications. Physica-Verlag, Heidelberg (2014). https://doi.org/10.1007/978-3-7908-1787-4
8. Chan, C.-H., Kittler, J., Messer, K.: Multi-scale local binary pattern histograms for face recognition. In: Lee, S.-W., Li, S.Z. (eds.) ICB 2007. LNCS, vol. 4642, pp. 809–818. Springer, Heidelberg (2007). https://doi.org/10.1007/978-3-540-74549-5_85
9. Dolecki, M., Karczmarek, P., Kiersztyn, A., Pedrycz, W.: Utility functions as aggregation functions in face recognition. In: 2016 IEEE Symposium Series on Computational Intelligence (SSCI), Athens, pp. 1–6 (2016)
10. Fukushima, S., Ralescu, A.L.: Improved retrieval in a fuzzy database from adjusted user input. J. Intell. Inf. Syst. **5**, 249–274 (1995)
11. Grabisch, M., Marichal, J.-L., Mesiar, R., Pap, E.: Aggregation Functions. Cambridge University Press, Cambridge (2009)
12. Grycuk, R., Gabryel, M., Korytkowski, M., Scherer, R.: Content-based image indexing by data clustering and inverse document frequency. In: Kozielski, S., Mrozek, D., Kasprowski, P., Małysiak-Mrozek, B., Kostrzewa, D. (eds.) BDAS 2014. CCIS, vol. 424, pp. 374–383. Springer, Cham (2014). https://doi.org/10.1007/978-3-319-06932-6_36
13. Grycuk, R., Gabryel, M., Korytkowski, M., Scherer, R., Voloshynovskiy, S.: From single image to list of objects based on edge and blob detection. In: Rutkowski, L., Korytkowski, M., Scherer, R., Tadeusiewicz, R., Zadeh, L.A., Zurada, J.M. (eds.) ICAISC 2014. LNCS (LNAI), vol. 8468, pp. 605–615. Springer, Cham (2014). https://doi.org/10.1007/978-3-319-07176-3_53
14. Grycuk, R., Gabryel, M., Nowicki, R., Scherer, R.: Content-based image retrieval optimization by differential evolution. In: 2016 IEEE Congress on Evolutionary Computation (CEC), pp. 86–93 (2016)
15. Haddadnia, J., Ahmadi, M.: N-feature neural network human face recognition. Image Vis. Comput. **22**, 1071–1082 (2004)
16. Hu, X., Pedrycz, W., Wang, X.: Comparative analysis of logic operators: a perspective of statistical testing and granular computing. Int. J. Approx. Reason. **66**, 73–90 (2015)

17. Jarillo, G., Pedrycz, W., Reformat, M.: Aggregation of classifiers based on image transformations in biometric face recognition. Mach. Vis. Appl. **19**, 125–140 (2008)
18. Kacprzyk, J., Pedrycz, W.: Springer Handbook of Computational Intelligence. Springer, Heidelberg (2015). https://doi.org/10.1007/978-3-662-43505-2
19. Karczmarek, P., Kiersztyn, A., Pedrycz, W.: An evaluation of fuzzy measure for face recognition. In: Rutkowski, L., Korytkowski, M., Scherer, R., Tadeusiewicz, R., Zadeh, L.A., Zurada, J.M. (eds.) ICAISC 2017. LNCS (LNAI), vol. 10245, pp. 668–676. Springer, Cham (2017). https://doi.org/10.1007/978-3-319-59063-9_60
20. Karczmarek, P., Kiersztyn, A., Pedrycz, W.: On developing Sugeno fuzzy measure densities in problems of face recognition. Int. J. Mach. Intell. Sens. Signal Process. **2**, 80–96 (2017)
21. Karczmarek, P., Kiersztyn, A., Pedrycz, W.: Generalized Choquet integral for face recognition. Int. J. Fuzzy Syst. **20**, 1047–1055 (2018). https://doi.org/10.1007/s40815-017-0355-5
22. Karczmarek, P., Kiersztyn, A., Pedrycz, W., Rutka, P.: A study in facial features saliency in face recognition: an analytic hierarchy process approach. Soft. Comput. **21**, 7503–7517 (2017)
23. Karczmarek, P., Kiersztyn, A., Rutka, P., Pedrycz, W.: Linguistic descriptors in face recognition: a literature survey and the perspectives of future development. In: SPA 2015 Signal Processing, Algorithms, Architectures, Arrangements, and Applications, Conference Proceedings, pp. 98–103 (2015)
24. Karczmarek, P., Pedrycz, W., Kiersztyn, A., Dolecki, M.: An application of chain code-based local descriptor and its extension to face recognition. Pattern Recognit. **65**, 26–34 (2017)
25. Karczmarek, P., Kiersztyn, A., Pedrycz, W., Rutka, P.: Chain code-based local descriptor for face recognition. In: Burduk, R., Jackowski, K., Kurzyński, M., Woźniak, M., Żołnierek, A. (eds.) Proceedings of the 9th International Conference on Computer Recognition Systems CORES 2015. AISC, vol. 403, pp. 307–316. Springer, Cham (2016). https://doi.org/10.1007/978-3-319-26227-7_29
26. Karczmarek, P., Pedrycz, W., Reformat, M., Akhoundi, E.: A study in facial regions saliency: a fuzzy measure approach. Soft. Comput. **18**, 379–391 (2014)
27. Kasiński, A., Florek, A., Schmidt, A.: The PUT face database. Image Process. Commun. **13**, 59–64 (2008)
28. Kumar, N., Berg, A.C., Belhumeur P.N., Nayar, S.K.: Attribute and simile classifiers for face verification. In: Proceedings of IEEE 12th International Conference on Computer Vision, pp. 365–372 (2009)
29. Kumar, N., Berg, A.C., Belhumeur, P.N., Nayar, S.K.: Describable visual attributes for face verification and image search. IEEE Trans. Pattern Anal. Mach. Intell. **33**, 1962–1977 (2011)
30. Kurach, D., Rutkowska, D., Rakus-Andersson, E.: Face classification based on linguistic description of facial features. In: Rutkowski, L., Korytkowski, M., Scherer, R., Tadeusiewicz, R., Zadeh, L.A., Zurada, J.M. (eds.) ICAISC 2014. LNCS (LNAI), vol. 8468, pp. 155–166. Springer, Cham (2014). https://doi.org/10.1007/978-3-319-07176-3_14
31. Kwak, K.-C., Pedrycz, W.: Face recognition using fuzzy integral and wavelet decomposition method. IEEE Trans. Syst. Man Cybern. B Cybern. **34**, 1666–1675 (2004)
32. Kwak, K.-C., Pedrycz, W.: Face recognition: a study in information fusion using fuzzy integral. Pattern Recognit. Lett. **26**, 719–733 (2005)

33. Li, L., Alrjebi, M., Liu, W.: Face recognition against pose variations using multi-resolution multiple colour fusion. Int. J. Mach. Intell. Sens. Signal Process. **1**, 304–320 (2016)
34. Liao, S., Zhu, X., Lei, Z., Zhang, L., Li, S.Z.: Learning multi-scale block local binary patterns for face recognition. In: Lee, S.-W., Li, S.Z. (eds.) ICB 2007. LNCS, vol. 4642, pp. 828–837. Springer, Heidelberg (2007). https://doi.org/10.1007/978-3-540-74549-5_87
35. Liu, Z., Liu, C.: Fusion of color, local spatial and global frequency information for face recognition. Pattern Recognit. **43**, 2882–2890 (2010)
36. Martínez, G.E., Melin, P., Mendoza, O.D., Castillo, O.: Face recognition with cho-quet integral in modular neural networks. In: Castillo, O., Melin, P., Pedrycz, W., Kacprzyk, J. (eds.) Recent Advances on Hybrid Approaches for Designing Intelligent Systems. SCI, vol. 547, pp. 437–449. Springer, Cham (2014). https://doi.org/10.1007/978-3-319-05170-3_30
37. Martínez, G.E., Melin, P., Mendoza, O.D., Castillo, O.: Face recognition with a Sobel edge detector and the Choquet integral as integration method in a modular neural networks. In: Melin, P., et al. (eds.) Design of Intelligent Systems Based on Fuzzy Logic, Neural Networks and Nature-Inspired Optimization, pp. 59–70. Springer, Part I (2015)
38. Melin, P., Felix, C., Castillo, O.: Face recognition using modular neural networks and the fuzzy Sugeno integral for response integration. Int. J. Intell. Syst. **20**, 275–291 (2005)
39. Milborrow, S., Morkel, J., Nicolls, F.: The MUCT landmarked face database. In: Pattern Recognition Association of South Africa (2010)
40. Mirhosseini, A.R., Yan, H., Lam, K.-M., Pham, T.: Human face image recognition: an evidence aggregation approach. Comput. Vis. Image Underst. **71**, 213–230 (1998)
41. Moreira, J.L., Braun, A., Musse, S.R.: Eyes and eyebrows detection for performance driven animation. In: 2010 23rd SIBGRAPI Conference on Graphics, Patterns and Images, pp. 17–24 (2010)
42. Nakayama, M., Miyajima, K., Iwamoto, H., Norita, T.: Interactive human face retrieval system based on linguistic expression. In: Proceedings of 2nd International Conference on Fuzzy Logic and Neural Networks, IIZUKA 1992, vol. 2, pp. 683–686 (1992)
43. Pedrycz, W.: Granular Computing: Analysis and Design of Intelligent Systems. CRC Press, Boca Raton (2013)
44. Pentland, A., Moghaddam, B., Starner, T.: View-based and modular eigenspaces for face recognition. In: Proceedings of 1994 IEEE Computer Society Conference on Computer Vision and Pattern Recognition, CVPR 1994, pp. 84–91 (1994)
45. Rahman, A., Sufyan Beg, M.M.: Face sketch recognition using sketching with words. Int. J. Mach. Learn. Cyber. **6**, 597–605 (2015)
46. Turk, M., Pentland, A.: Eigenfaces for recognition. J. Cogn. Neurosci. **3**, 71–86 (1991)

Nuclei Detection in Cytological Images Using Convolutional Neural Network and Ellipse Fitting Algorithm

Marek Kowal$^{(\boxtimes)}$, Michał Żejmo, and Józef Korbicz

Institute of Control and Computation Engineering, University of Zielona Góra,
Zielona Góra, Poland
M.Kowal@issi.uz.zgora.pl

Abstract. Morphometric analysis of nuclei play an essential role in cytological diagnostics. Cytological samples contain hundreds or thousands of nuclei that need to be examined for cancer. The process is tedious and time-consuming but can be automated. Unfortunately, segmentation of cytological samples is very challenging due to the complexity of cellular structures. To deal with this problem, we are proposing an approach, which combines convolutional neural network and ellipse fitting algorithm to segment nuclei in cytological images of breast cancer. Images are preprocessed by the colour deconvolution procedure to extract hematoxylin-stained objects (nuclei). Next, convolutional neural network is performing semantic segmentation of preprocessed image to extract nuclei silhouettes. To find the exact location of nuclei and to separate touching and overlapping nuclei, we approximate them using ellipses of various sizes and orientations. They are fitted using the Bayesian object recognition approach. The accuracy of the proposed approach is evaluated with the help of reference nuclei segmented manually. Tests carried out on breast cancer images have shown that the proposed method can accurately segment elliptic-shaped objects.

Keywords: Deep learning · Convolutional neural network
Ellipse fitting · Bayesian object recognition · Nuclei detection
Breast cancer

1 Introduction

Recently, cancer diagnostics is based heavily on the results of cytological and histological examinations. Biological material necessary for cytological examination is taken from affected tissue using needle biopsy. Next, cellular material is fixed and stained. Finally, slide glass with cells is examined by a pathologist. He evaluates morphometric parameters of nuclei or cells to diagnose cancer cases. Therefore, nuclei segmentation is critical to performance of Computer-Aided Cytology (CAC). The most common approaches of nuclei segmentation are based on the image thresholding, watershed transform, region growing, level

© Springer International Publishing AG, part of Springer Nature 2018
L. Rutkowski et al. (Eds.): ICAISC 2018, LNAI 10842, pp. 157–167, 2018.
https://doi.org/10.1007/978-3-319-91262-2_15

sets, graph cuts, morphological mathematics and deep learning [1–6]. However, the problem of nuclei segmentation is challenging and remains open.

The aim of our study is to develop and test the new method of nuclei segmentation based on convolutional neural network (CNN) and ellipse fitting. CNN seems to be a promising technique for semantic segmentation of cytological images. The well-known property of CNN is the ability to learn features invariant to object scaling, rotation, and shifting. We expect that it will be able to extract nuclei even thought staining of nuclei is strongly heterogeneous. Difficulties may arise due to overlapping or touching nuclei which may be hard to separate and count. Of course, CNN may be trained to separate overlapping objects, but this requires a large number of training samples representing areas of the image on which objects overlap [7,8]. Thus, additional effort must be taken to generate ground truth images. Unfortunately, the natural imbalance in the frequency of the overlapping nuclei cases makes it difficult to build larger datasets [9]. Trained CNN will be able to detect overlapping regions but to extract consistent objects from pixelwise segmentation post-processing is necessary.

To overcome the problem, we propose to use Bayesian object recognition framework to fit a set of ellipses to the nuclei mask generated by the CNN. This method belongs to the branch of stochastic geometry, which deals with the analysis of random spatial patterns [10]. The idea of our approach is to construct the marked point process which with high probability generates ellipses consistent with the image and with a priori constraints. The problem of finding most likely configuration of ellipses can be formulated as maximum a posterior estimation issue and solved by using the steepest ascent optimization algorithm.

In order to verify the accuracy of the proposed method, it was used to segment nuclei in breast cancer cytological images. The obtained results were compared with the manual segmentation and with the segmentation results produced by intensity thresholding combined with ellipse fitting.

The remainder of this paper is organized as follows. In Sect. 2, procedures used to image preprocessing are presented. Sections 3 and 4 present methods used to detect nuclei. Results of experiments are presented in Sect. 5. Concluding remarks are given in Sect. 6.

2   Image Preprocessing

Cytological images of breast cancer were collected by pathologists from University Hospital in Zielona Góra, Poland. Cellular material was acquired from affected breast tissue using fine needle biopsy under the control of an ultrasonograph. The set contains 25 benign and 25 malignant cases. Next, the material was fixed with fixative spray and dyed with hematoxylin (blue color) and eosin (red color). Cytological preparations were then digitized into virtual slides using the Olympus VS120 Virtual Microscopy System. For experimental studies, we used only small, selected fragments of these slides of the size 500 × 500 pixels. The cytological sample usually consists of cell nuclei, cytoplasm, and red blood cells. Hematoxylin is mainly absorbed by nuclei. Cytoplasm and red blood cells absorb eosin. As the result, nuclei have blue color and cytoplasm and red blood cells are red. We need for further processing only nuclei structures. Unfortunately, nuclei can deposit eosin to some extent and cytoplasm can deposit hematoxylin to some extent. Moreover, absorption spectra of hematoxylin and eosin are overlapping in RGB space, thus we cannot easily separate them. To quantify the contribution of hematoxylin and eosin, we can use color deconvolution method [11]. Three separate intensity images are created as a result, the first represents the hematoxylin density, second eosin density, and third residuals. For further processing, we are using the image of hematoxylin density.

3   Semantic Segmentation

Images obtained as a result of deconvolution are subjected to semantic segmentation to determine the location of nuclei. Semantic segmentation is realized using CNN classifier. It is used to predict classes for all pixels in the image. Each pixel can belong to one of four pixel categories: nuclei, cytoplasm, nuclei edge and background. The result of classification is a semantic map of the size of processed image. It stores class labels for all pixels. In fact, the output of CNN is the probability distribution over these four classes. Therefore each pixel is always labeled by the class which gained the highest probability. CNN is classifying the pixel based on the patch centered in that pixel. For each pixel a single patch must be prepared as an input load for CNN. Therefore, to segment, the whole image, such classification procedure must be repeated for each pixel in the processed image.

In this work we used patches of the size 43×43 pixels. Thus, all training and testing images must be cut into bunch of patches before they are processed by the CNN classifier. Sizes and distributions of training and testing subsets are presented in Table 1.

Table 1. Collections of patches

	Training	Testing
Number of images	20	20
Total patches	18,315,912	18,961,569
- nuclei border	1,739,200	1,968,232
- nuclei	5,105,059	5,788,305
- cytoplasm	5,588,398	5,770,039
- background	5,883,255	5,434,993

We can observe that the class of patches describing nuclei borders is underrepresented. To tackle this problem, classes were artificially balanced by augmenting and preprocessing some patches. To generate artificial patches each original nuclei border patch was subjected to randomized scaling, rotation or flipping.

Typically, CNN is constructed of several convolutional layers interleaved by pooling layers [12]. On the top of the network, there is at least one fully connected layer. Convolutional layer is composed of a set of learnable filters. Each filter extracts different features from the input patch. During the learning procedure, parameters of filters (weights) are tuned to minimize error (loss function). Pooling layer is used to progressively reduce the spatial size of the input patch and to extract invariant features. Spatial reduction in the size of the patch is usually realized by max-pooling operation applied to the window of predefined size. The fully-connected layer is connected to flattened activations of the last convolutional-pooling ensemble. The task of fully connected layer is to capture the complex relationships between high-level features and output labels. The size of the output layer is equal to the number of classes.

The structure of our CNN is presented in Table 2. The network is composed of four convolutional layers, which are separated by two max-pooling layers. There is one fully-connected layer (512 neurons) and output layer (4 neurons) at the top of the network. All convolutional layers are followed by rectified linear units (ReLU) to deal with gradient vanishing problem. Training was conducted using stochastic gradient descent, mini-batch was set to 256 and training process took 20 epochs. To prevent over-fitting, we applied dropout technique. This allowed us to achieve the 91.33% classification accuracy for the test patches.

The semantic mask generated by the CNN classifier is turned into nuclei mask. In nuclei mask, pixels belonging to nuclei are labelled by 1, while others by 0. In Fig. 1 we can compare the nuclei mask generated by CNN with the mask generated by Otsu thresholding. We can visually asses that CNN is much more accurate. However, further processing is necessary to locate all nuclei, determine their shape and separate touching and overlapping nuclei. This task is realized by ellipse fitting applied to nuclei mask.

Table 2. The architecture of used CNN. The first row indicates the type of layer (C - convolutional layer, MP - max-pooling layer, FC - fully-connected layer). The second row contains information about kernel sizes, stride and padding, respectively.

C32 →	C32 →	MP →	C64 →	C64 →	MP →	FC512 →	FC4
3x3x1x1	3x3x1x1	2x2x2x0	3x3x1x1	3x3x2x0	2x2x2x0	—	—

4   Ellipse Fitting

In the second step of the proposed method, we are trying to find objects in the nuclei mask which resemble the nuclei. This is done by fitting ellipses to nuclei mask using Bayesian inference. The method is looking for the configuration of ellipses that will cover as many as possible nuclei pixels in binary nuclei mask. It is a loose criterion because a lot of different ellipse configurations with similar silhouettes exist. To select the realistic configuration from candidate configurations, we must define a priori model to penalize configurations consisting of overwhelming numbers of overlapping objects. Of course, overlapping of ellipses can not be completely banned because we need them to model scenarios with overlapping nuclei.

Therefore, we merely assume that configurations with many overlapping ellipses are less likely than configurations with fewer overlapping ellipses.

4.1 Bayesian Object Recognition

The task at hand is to reconstruct unknown configuration of ellipses $\tilde{\mathbf{x}}$ based on the nuclei mask $\mathbf{y}$ generated by color deconvolution and CNN classifier. The binary image of nuclei mask is defined on finite pixel lattice S and $y_t \in \{0,1\}$ denotes the pixel value at position t. Configuration of ellipses is described by $\mathbf{x} = \{x_1, x_2, \ldots, x_n\}$, where $x_i \in U$ is a single ellipse. Atlas of ellipses is defined by specifying ranges for their parameters: major axis length $r_M \in [15, \ldots, 45]$, the ratio of minor axis length to major axis length $r_R \in [0.5, 0.65, 0.8, 1]$ and rotation angle $o \in [0°, 30°, 60°, 90°, 120°, 150°]$.

Unknown configuration $\tilde{\mathbf{x}}$ cannot be determined with certainty but it is described probabilistically by the conditional probability mass function $p(\mathbf{x}|\mathbf{y})$:

$$p(\mathbf{x}|\mathbf{y}) \propto f(\mathbf{y}|\mathbf{x})p(\mathbf{x}), \tag{1}$$

where likelihood $f(\mathbf{y}|\mathbf{x})$ evaluates the fitting of ellipse configuration $\mathbf{x}$ to nuclei mask $\mathbf{y}$ and a priori term $p(\mathbf{x})$ restrain the number of overlapping ellipses. Given nuclei mask $\mathbf{y}$, we must find most likely configuration $\hat{\mathbf{x}}$:

$$\hat{\mathbf{x}} = \arg \max_x f(\mathbf{y}|\mathbf{x})p(\mathbf{x}). \tag{2}$$

Assuming that variables representing nuclei mask y_t are conditionally independent given configuration $\mathbf{x}$, we can present likelihood function $f(\mathbf{y}|\mathbf{x})$ in the following form:

$$f(\mathbf{y}|\mathbf{x}) = \prod_{t \in S(\mathbf{x})} b(y_t; p_N) \prod_{t \in S \setminus S(\mathbf{x})} b(y_t; p_B), \tag{3}$$

where $S(\mathbf{x}) \subseteq S$ is a silhouette of configuration $\mathbf{x}$:

$$S(\mathbf{x}) = \bigcup_{i=1}^{n} S(x_i), \tag{4}$$

where $S(x_i)$ is the part of the silhouette mask $\mathbf{s}$ occupied by the disk x_i. Probability mass functions of Bernoulli distributions $b(y_t; p_N)$ and $b(y_t; p_B)$ are used to evaluate the likelihood of pixels on nuclei mask $\mathbf{y}$ within nuclei region $S(x)$ and background region $S \setminus S(x)$ respectively:

$$b(y_t; p_N) = \begin{cases} 1 - p_N & \text{if } y_t = 0 \\ p_N & \text{if } y_t = 1, \end{cases} \tag{5}$$

$$b(y_t; p_B) = \begin{cases} 1 - p_B & \text{if } y_t = 1 \\ p_B & \text{if } y_t = 0, \end{cases} \tag{6}$$

where p_N is the probability that the pixel belong to nuclei will be classified by the CNN as a nuclei pixel and p_B is the probability that the pixel representing background will be classified by the CNN as a background pixel. The values of these probabilities are estimated using manually segmented images.

The goal of the liklihood term $f(\mathbf{y}|\mathbf{x})$ is to reward configurations $\mathbf{x}$ which fit the nuclei mask $\mathbf{y}$ best. However, some of them can be completely unrealistic due to an excessive number of overlapping disks [13]. To prevent this, we used pairwise interaction model proposed by Strauss:

$$p(\mathbf{x}) = \alpha \beta^{n(\mathbf{x})} \gamma^{u(\mathbf{x})}, \tag{7}$$

where $u(x)$ is the number of pairwise overlaps in configuration $\mathbf{x}$ [13,14] and γ is the parameter controlling interactions between disks [13,14].

4.2 Optimization

Our aim is to find the configuration of ellipses $\hat{\mathbf{x}}$ which maximizes $p(\mathbf{x}|\mathbf{y})$ given nuclei mask $\mathbf{y}$. Therefore the problem reduces to optimization task. To solve this problem we have used well-known steepest ascent procedure. The algorithm starts with the provisional configuration $\hat{\mathbf{x}}_0$ and then begins updating it by slight changes to the current configuration like adding the new ellipse or deleting/shifting existed one. So, the procedure generates collection of candidate configurations x_k by some basic changes to the current configuration $\hat{\mathbf{x}}_k$ and scans them in some predefined order. If the tested candidate configuration $\mathbf{x}_c$ is better than current one $\hat{\mathbf{x}}_k$ then algorithm accepts it as the new current configuration $\hat{\mathbf{x}}_{k+1} = \mathbf{x}_c$. Such deterministic approach ensures that probability of configuration never decreases at any stage so the procedure convergence is guaranteed. However, we have to take into account the fact that algorithm can stuck in the local maxima. The crucial step of the algorithm is to determine if the tested configuration $\mathbf{x}_c$ is better than the current one $\hat{\mathbf{x}}_k$. In the proposed approach, log posterior likelihood ratio was used for this purpose:

$$w(\mathbf{x}_c) = \ln \left(\frac{f(\mathbf{y}|\mathbf{x}_c) p(\mathbf{x}_c)}{f(\mathbf{y}|\hat{\mathbf{x}}_k) p(\hat{\mathbf{x}}_k)} \right)$$

$$= \sum_{t \in S_N} z(y_t) + \ln\left(\gamma\right)\left(u(\mathbf{x}_c) - u(\hat{\mathbf{x}}_k)\right)$$

$$+ \ln\left(\beta\right)\left(n(\mathbf{x}_c) - n(\hat{\mathbf{x}}_k)\right), \tag{8}$$

where $z(y_t) = \ln\left(b(y_t; p_N)\right) - \ln\left(b(y_t; p_B)\right)$, $S_N = \left(S(\mathbf{x}_c) \cup S(\hat{\mathbf{x}}_k)\right) \setminus \left(S(\mathbf{x}_c) \cap S(\hat{\mathbf{x}}_k)\right)$, $\ln(\gamma) = -200$ and $\ln(\beta) = -50$. Although we do not know a posteriori probability for either $\mathbf{x}_c$ or $\hat{\mathbf{x}}_k$, we are able to evaluate whether the tested configuration is better than the current configuration based on the $w(\mathbf{x}_c)$ ratio.

5 Experimental Results

In order to verify the effectiveness of the proposed approach, it was applied to detect nuclei in 50 test images of the size 500×500 pixels (see Fig. 1).

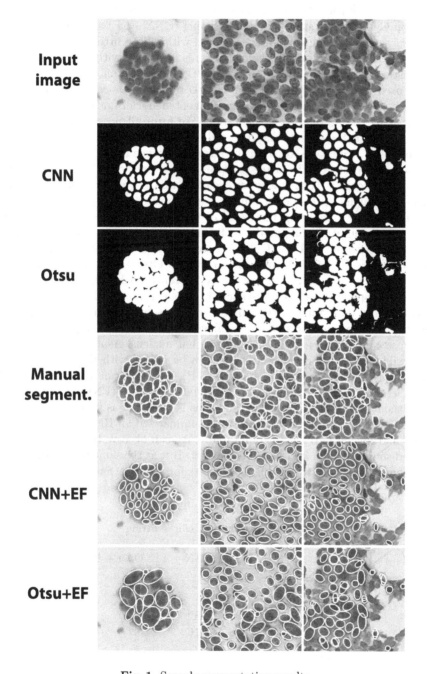

Fig. 1. Sample segmentation results

The set contains 25 benign and 25 malignant cases. Test images were manually segmented to get the reference database of nuclei. The accuracy of the approach based on CNN classification and ellipse fitting (CNN + EF) was compared with the accuracy of Otsu thresholding combined with ellipse fitting (Otsu + EF).

The accuracy of automatic segmentation is measured with the help of Hausdorff distance (HD) and Jaccard distance (JD). These distance metrics are commonly used to measure the similarity of 2D objects. For each test image, we are given a list of masks of manually segmented nuclei and masks of ellipses generated by the automatic segmentation. We form all possible pairs between manually segmented objects and detected ellipses. Then, we compute distances for these pairs, which are stored in the form of distance matrices. For each manually segmented object, the nearest ellipse is determined, and the corresponding distance is recorded. That is why for each test image we get as many distances as manually segmented objects. Finally, we can evaluate the accuracy of nuclei segmentation by computing the mean distance and distance standard deviation. These values were computed for 1405 manually segmented nuclei. Obtained results are presented in Table 3. We can see that CNN + EF segmentation generates more precise approximation of nuclei than Otsu + EF segmentation.

Moreover, we can calculate how many reference nuclei were properly identified with respect to a chosen distance threshold. Distance threshold is chosen arbitrarily and determines if the ellipse can represent given nucleus. If the distance is lower than the threshold then ellipse is matching the nucleus. Otherwise, the ellipse cannot be used to approximate that nucleus. As a result 3 scenarios are possible: manually segmented nuclei can be paired with the nearest ellipse and such case is classified as true positive (TP), no ellipse can be found to match the nuclei and such case is classified to be false negative (FN), and ellipse can stay without corresponding nucleus and thus is classified as false positive (FP). The results of segmentation accuracy were summarized for Hausdorff distance in Table 4 and for Jaccard distance in Table 5. We can observe that both methods achieved similar results for the FP coefficient. But at the same time it should be noted that CNN+EF achieved higher TP rate for all cases. Sample segmentation results are presented in Fig. 1.

Table 3. Accuracy of ellipse fitting

		CNN + EF	Otsu + EF
Hausdorff distance	mean	**14**	21
	std.	**16**	23
Jaccard distance	mean	**0.45**	0.48
	std.	**0.16**	0.23

Table 4. Accuracy of segmentation (Hausdorff distance)

Segmentation method	Accuracy measures	T_{HD}		
		25	**50**	**75**
CNN + EF	TP	**1246**	**1327**	**1332**
(num. nuclei)	FP	**359**	278	273
Otsu + EF	TP	981	1131	1133
(num. nuclei)	FP	387	**237**	**235**
Manual segmentation	num. nuclei	1405		

Table 5. Accuracy of segmentation (Jaccard distance)

Segmentation method	Accuracy measures	T_{JD}		
		0.25	**0.5**	**0.75**
CNN + EF	TP	**262**	**995**	**1294**
(num. nuclei)	FP	**1143**	610	311
Otsu + EF	TP	195	813	1092
(num. nuclei)	FP	1173	**555**	**276**
Manual segmentation	num. nuclei	1405		

6   Conclusions

The content of cytological images is highly complex and their analysis is difficult in an automated way. Generally, such methods of segmentation as intensity thresholding, edge detection, watershed transform and active contours are not able to extract nuclei with satisfactory accuracy. This paper presents an alternative way of nuclei segmentation which is based on CNN and Bayesian ellipse recognition. Method is able to locate nuclei and roughly approximate their shape. Presented results are promising because the method is able to detect nuclei even from a clumped cytological material with many overlapping nuclei.

It was observed that CNN performed very well with the semantic segmentation of cytological images. The obtained nuclei masks were very accurate and did not contain too many overlapping objects. In the second stage algorithm locates nuclei and determines their shape by approximating them by ellipses. Fitting algorithm prefers to add large ellipses over small ones due to the method of ellipse evaluation. For this reason, the algorithm is sensitive to the upper limit of the size of ellipses which has to be chosen with great care. It was observed that steepest ascent optimization usually converges very fast to the local maxima. At the beginning, the algorithm mostly is adding ellipses to the configuration, then switch to shifting existed ellipses, very rarely is deleting the ellipse from the configuration.

Presented framework is not limited to model nuclei by ellipses. In practice, we can compose an atlas of prospective objects of any shape without increasing the computational complexity.

Acknowledgement. The research was supported by National Science Centre, Poland (2015/17/B/ST7/03704).

References

1. Spanhol, F.A., Oliveira, S.L.E., Petitjean, C., Heutte, L.: Breast cancer histopathological image classification using convolutional neural networks. In: Proceedings of International Conference on Neural Networks (IJCNN 2016), Vancouver, Canada (2016)
2. Bembenik, R., Jóźwicki, W., Protaziuk, G.: Methods for mining co-location patterns with extended spatial objects. Int. J. Appl. Math. Comp. Sci. **27**(4), 681–695 (2017)
3. Qi, J.: Dense nuclei segmentation based on graph cut and convexity–concavity analysis. J. Microsc. **253**(1), 42–53 (2014)
4. Kłeczek, P., Dyduch, G., Jaworek-Korjakowska, J., Tadeusiewicz, R.: Automated epidermis segmentation in histopathological images of human skin stained with hematoxylin and eosin. In: Proceedings of SPIE Medical Imaging, vol. 10140, pp. 10140–10140–19 (2017)
5. Nurzynska, K., Mikhalkin, A., Piórkowski, A.: CAS: cell annotation software - research on neuronal tissue has never been so transparent. Neuroinformatics **15**, 365–382 (2017)
6. Kowal, M., Filipczuk, P.: Nuclei segmentation for computer-aided diagnosis of breast cancer. Int. J. Appl. Math. Comp. Sci. **24**(1), 19–31 (2014)
7. Chu, J.L., Krzyżak, A.: The recognition of partially occluded objects with support vector machines, convolutional neural networks and deep belief networks. J. Artif. Intell. Soft Comput. Res. **4**(1), 5–19 (2014)
8. Surya, S., Babu, R.V.: TraCount: a deep convolutional neural network for highly overlapping vehicle counting. In: Proceedings of 10th Indian Conference on Computer Vision, Graphics and Image Processing, ICVGIP 2016, pp. 46:1–46:6, New York, NY, USA (2016)
9. Hu, R.L., Karnowski, J., Fadely, R., Pommier, J.P.: Image segmentation to distinguish between overlapping human chromosomes. In: Proceedings of 31st Conference on Neural Information Processing Systems (NIPS 2017), Long Beach, CA, USA (2017)
10. Descombes, X.: Multiple objects detection in biological images using a marked point process framework. Methods **115**(Supplement C), 2–8 (2017). Image Processing for Biologists
11. Ruifrok, A.C., Johnston, D.A.: Quantification of histochemical staining by color deconvolution. Anal. Quant. Cytol. Histol. **23**(4), 291–299 (2001)
12. LeCun, Y., Huang, F.J., Bottou, L.: Learning methods for generic object recognition with invariance to pose and lighting. In: Proceedings of IEEE Conference on Computer Vision and Pattern Recognition, CVPR 2004, vol. 2, pp. II-97. IEEE (2004)

13. van Lieshout, M.N.M.: Markov point processes and their applications in high-level imaging. Bull. Int. Stat. Inst. **56**, 559–576 (1995)
14. Kowal, M., Korbicz, J.: Marked point process for nuclei detection in breast cancer microscopic images. In: Augustyniak, P., Maniewski, R., Tadeusiewicz, R. (eds.) PCBBE 2017. AISC, vol. 647, pp. 230–241. Springer, Cham (2018). https://doi. org/10.1007/978-3-319-66905-2_20

Towards the Development of Sensor Platform for Processing Physiological Data from Wearable Sensors

Krzysztof Kutt, Wojciech Binek, Piotr Misiak, Grzegorz J. Nalepa, and Szymon Bobek[✉]

AGH University of Science and Technology,
al. Mickiewicza 30, 30-059 Krakow, Poland
{kkutt,wojciech.binek,pmisiak,gjn,sbobek}@agh.edu.pl

Abstract. The paper outlines a mobile sensor platform aimed at processing physiological data from wearable sensors. We discuss the requirements related to the use of low-cost portable devices in this scenario. Experimental analysis of four such devices, namely Microsoft Band 2, Empatica E4, eHealth Sensor Platform and BITalino (r)evolution is provided. Critical comparison of quality of HR and GSR signals leads to the conclusion that future works should focus on the BITalino, possibly combined with the MS Band 2 in some cases. This work is a foundation for possible applications in affective computing and telemedicine.

1 Introduction

Recent rapid development of wearable devices equipped with physiological sensors is an opportunity for new systems in the areas of telemedicine, quantified self, and affective computing (AfC). Our recent works [14] focus on the last of these areas. There are two main aspects of AfC [18]. The first one is related to the detection of emotional responses of humans. The second one is related to the simulation of emotional responses in artificial systems. We are interested in the first aspect, which recently can largely benefit from wearable devices.

In AfC appropriate identification of the affective condition of a person requires a certain model of emotions. There are multiple models considered in psychology, philosophy and cognitive science. William James was the precursor of the appraisal theory which is among most popular in the community of computational emotional modeling [2,11]. One of the most popular appraisal theories is OCC (Ortony et al.) [17] which categorizes emotion on basis of appraisal of pleasure/displeasure (valence) and intensity (arousal). Research indicates that they can be measured by the use of Autonomic Nervous System (ANS) activity, including the use of Heart Rate (HR) and Skin Conductance/Galvanic Skin Response (GSR) signals (for meta-analysis see [4]).

The ANS measures can be accurately acquired in laboratory experiments. However, a practical challenge is the quality of measurement provided by field

© Springer International Publishing AG, part of Springer Nature 2018
L. Rutkowski et al. (Eds.): ICAISC 2018, LNAI 10842, pp. 168–178, 2018.
https://doi.org/10.1007/978-3-319-91262-2_16

devices, such as wearables, e.g. wristbands. These devices are low-cost and accessible thus creating opportunity for real life applications. On the other hand they use lower quality sensors, often applied in a non-optimal way. Recently there has been growing interest in assessing the quality of such devices, e.g. [5].

The original contribution of our work presented in this paper is the critical evaluation of several wearable devices delivering physiological data monitoring. We aim at comparing these devices considering future applications in AfC and in telemedicine. As such, we focus on the continous monitoring of the HR and GSR signals. The rest of the paper is organized as follows. We begin with the overall design of the sensor platform in Sect. 2, then moving to the discussion of selected devices in Sect. 3. Based on this, we present the measurement procedures in Sect. 4, along with the detailed signal processing in Sect. 5. We then move to the evaluation of results in Sect. 6 and conclusions in the final section.

2   Outline of a Mobile Sensor Platform

Our proposal of the mobile platform aimed at Affective Computing applications supports the affective data flow through several interconnected modules:

1. The person is experiencing emotion which is connected with the reactions of person's ANS.
2. Mobile sensor monitors these signals (e.g. HR and GSR).
3. Data is transmitted to the processing device (e.g. smartphone) via bluetooth or other interface.
4. The processing device reads the data using API provided by the mobile sensor distributor.
5. Statistical or machine learning model is used to transform sensor data into emotion values (e.g. in Valence x Arousal dimensions or nominal values defining the names of emotions).
6. All data, i.e. gathered from sensors and outputs from model, may be saved in CSV files, broadcasted to other applications or combined with other data streams (e.g. GPS signal and network connection usage) to provide more reliable contextual information[1].
7. Finally, the data may be used by number of applications, including: affect identification, context processing and adaptation, and health monitoring.

Our work presented in this paper is focused on steps 2–4, i.e. on gathering sensory data from user. With this in mind, one can specify requirements for mobile sensors to be fit in the presented platform:

1. The ANS measurement should be *accurate*. As we aim at low-cost devices available for almost everyone, it cannot be very precise. There is only a need for differentation of various valence and arousal levels and their changes, what makes the devices usable from the AfC applications point of view.

---

[1] With the use of e.g. AWARE framework, see http://www.awareframework.com/.

2. Collecting affective information should be done on a continuous basis, which is related to: (a) platform *mobility*, as it will assist user everywhere, (b) reliable sensor *contact*, as it will assist user during various activites, (c) being *comfortable* for user, as it should not distract her in regular life, (d) sufficient *battery* capacity that lasts at least one working day without recharging.
3. Data should be processed live to provide an affective feedback loop, so there is a need for: (a) *connection* with mobile device, e.g. through Bluetooth, (b) *raw signal* access, as each filtering done by the sensor or API results in data loss, what makes further processing difficult, (c) clearly defined *unit of measurement*, to provide a possibility of comparison with other devices and to allow the use of general model that gathers signals in some specified units, (d) *open API* that will allow the access to current sensor readings.

Low-cost mobile sensors that we consider are described in the next section.

3   Overview of Selected Devices

Wristbands. Empatica E4 [6] is a research- and clinical-oriented sensory wristband based on the technologies previously developed in the Affective Computing division of MIT Media Lab. The band has a photoplethysmography sensor for blood volume pulse measurements, as well as galvanic skin response sensor, infrared thermopile, 3-axis accelerometer and event mark button.

Microsoft Band 2 is a health and fitness tracking-oriented wristband. Equipped with optical heart rate and galvanic skin response sensors, as well as skin temperature, ambient light and UV sensors, 3-axis accelerometer, GPS and barometer.

Signals from both Empatica E4 and MS Band 2 were obtained and recorded via a custom dedicated application for Android devices created by authors [10].

BITalino. The BITalino (r)evolution kit[2] is a complete platform designed to deal with the body signals. It is ready to use out-of-the-box and allows the user to acquire biometric data using included software, which also enables the full control of the device. The device sends raw signals produced by the analog-to-digital converter. These can be converted to the correct physical units using the right transfer function for the sensor that produced the data. Communication is available via Bluetooth or any UART-compatible device (e.g. ZigBee, WiFly or FTDI). There is also set of APIs for many platforms including Arduino, Android, Python, Java, iOS and many more, which let the user create custom software.

e-Health. The e-Health Sensor Platform V2.0[3] is an open-source platform that allows users to develop biometric and medical applications where body monitoring is used. It is possible to perform real time monitoring or to get sensitive

---

[2] For details see http://bitalino.com/.

[3] For details see https://www.cooking-hacks.com/documentation/tutorials/ehealth-biometric-sensor-platform-arduino-raspberry-pi-medical.

data, which will be subsequently analysed for medical diagnosis. Our e-Health kit consists of the following components useful for AfC experiments:

- e-Health PCB, which can be connected to Arduino or Raspberry Pi; all sensors are connected to this PCB,
- heart rate/blood saturation sensor – easy to use on-finger sensor,
- electrocardiogram (ECG) – monitors the electrical and muscular functions of the heart. It consists of 3 electrodes and 3 leads (positive, negative, neutral),
- body temperature sensor,
- galvanic skin response (GSR/EDA) – the sensor consists of two metallic electrodes and is a type of ohmmeter with human body being a resistor,
- electromyograph (EMG) – measures the electrical activity of skeletal muscles by detecting the electrical potential generated by muscle cells when these cells are activated. It consists of 3 electrodes (MID, END and GND).

The e-Health library methods are needed to get measured data. In most cases the value is passed using Arduino analog input pins connected to the responding output pins of the e-Health PCB. These methods read the voltage and use it to calculate the value returned by the method. Serial communication is the best way to transfer data, but wireless channels like WiFi, Bluetooth are also available.

4   Evaluation and Measurement Procedures

To achieve the goal of devices comparison with consideration of future AfC applications, experimental procedure consisting of 3 parts was designed. It was aimed at data acquisition in various settings similar to target Affective Computing setting, including various affective states that cover wide range of valence and arousal values. Affect-related physiological signals, Heart Rate (HR) and Galvanic Skin Response (GSR), were collected using four low-cost wearable devices described in Sect. 3 and Polar H6 strap as a reference. It is a professional fitness device used for HR tracking.

The first part was designed using the PsychoPy 2^4 environment, a standard software framework in Python to support psychological experiments. Subjects were asked to watch affective pictures. Each of them was presented for 3 s, then it disappears and subject had 5 s for valence evaluation on 7-levels scale [1, 7]. The set of 60 pictures was grouped into training session (6 images) and three experimental sessions (each of them with 18 images). To provide valid emotional descriptions of images, in terms of valence and arousal scores, a subset of Nencki Affective Picture System was used [12].

In the second part, the "London Bridge" platform game was run. The task was to collect points as you go through the 2D world. The gameplay incorporates current score and remaining time indicators, and random events of current score reduction or remaining time shortening. Full game design is discussed in [13].

Finally, the accuracy of acquired data during the physical activity was examined. After finishing the game, participants were asked to do 20 squats and rest for a minute. This should induce significant changes in both HR and GSR.

---

[4] See: http://psychopy.org.

5 Analysis Workflow

Performed experiment was focused on two basic parameters: HR and GSR. To evaluate HR signals we used Polar H6 chest strap as a reference. For GSR no reference data was available therefore different processing strategy was applied.

HR Processing. MS Band 2, eHealth and reference Polar H6 provide direct HR measurements. Polar H6 and MS Band 2 record HR with 1 Hz sampling rate, eHealth uses 32 Hz sampling rate. Because upsampling would not introduce any new information, we decided to downsample all signals to 1 Hz. E4 and BITalino do not provide direct HR information. The devices record blood volume pressure (BVP), and electrocardiography (ECG) signals that we used to calculate HR.

Proposed algorithm for HR detection is based on primary tone extraction methods. It combines autocorrelation and frequency domain analysis. Input signal is filtered using FIR bandpass filter preserving frequency range from 1.1 to 5 Hz. Filtered data is windowed using 8 s triangular, asymmetric window with 0.875 overlap producing an output of one sample per second. For each window position we calculate an autocorrelation and power spectrum. The first HR candidate is calculated from the delay between first and second maximum of autocorrelation result. Furthermore three candidates are calculated from position of first peak in power spectrum and distances between first, second and third peak. Both operations give total of 4 HR candidates. If previous HR sample is available, candidates that differ more than a defined threshold (7 BPM) are discarded. The remaining values are averaged to obtain HR sample. If all candidates are rejected, extrapolated values from last 3 s and 10 s are added to candidate set. The most extreme values (without comparison to previous sample) are discarded and the remaining set is averaged to produce HR. The final result is filtered using lowpass IIR filter. In order to compare HR signals from different devices we calculated the root mean squared error (RMSE) between the signals and a reference as well as a Pearson correlation coefficient between data gathered from all devices for one participant.

GSR Processing. GSR contains two components, the skin conductance level (SCL) (tonic level) and skin conductance response (SCR) known as phasic response. SCL is slowly changing signal dependent on individual factors such as hydration level or skin dryness. It is considered not to be informative on its own [9]. SCR is a set of alterations on top of SCL usually correlated with emotional responses to presented stimuli [7,15]. Galvanic skin response can be described in terms of skin resistance or conductance and can be expressed using different units. The first analysis step was to convert all the data to conductance in μS. The optimal GSR sampling rate is not clearly defined. Schmidt [19] recommends 15–20 Hz sampling rate and low pass filtering to 5 Hz, Ohme [16] proposes downsampling to 32 Hz without specifying the initial sampling rate, Nourbakhsh in his research used sampling rate of 10 Hz [15]. Sampling rates of MS Band 2 (5 Hz), Empatica E4 (4 Hz) and BITalino (1000 Hz) are fixed by the manufacturers so it was impossible adjust the values. E-Health acquired data using 32 samples per second. Finally, for data comparison we resampled all signals to 10 Hz.

To extract SCR, we estimated the tonic level and subtracted it from original signal. The SCL was obtained by extracting and interpolating local skin conductance minima. Finally, the SCL was smoothed using IIR lowpass filter [7]. To compare devices we calculated the Pearson correlation coefficient between recorded signals. Because each measurement was taken in the same time, on the same person the results should be highly correlated. This however does not provide an answer to the question which signal is better.

In order to assess which device produces better results we used following procedure. The SCR signal was deconvolved with single phasic response model [1, 3] defined by equation $b[t] = e^{\frac{-t}{\tau_0}} - e^{\frac{-t}{\tau_1}}$. According to [1] recommended values for τ_0 and τ_1 are 2 and 0.75 respectively. The resulting signal is called driver function. It is smoothed using lowpass IIR filter. If SCR signal contains fragments similar to model response, corresponding high amplitude positive peaks appear in smoothed driver function, therefore if SCR signal contains data similar to model response, the driver function should be mostly positive. To examine it we calculate ratio of energy of positive values to whole signal (PTR):

$$PTR = 10 \log_{10} \left(\frac{RMS(max(0, DF))}{RMS(DF)} \right), \tag{1}$$

where DF is a driver function.

The SCR signal is reconstructed by convolving smoothed driver with model response. For noisy SCR signal smoothing the driver should alter the reconstructed signal significantly. To examine it we calculate signal-to-noise ratio:

$$SNR = 10 \log_{10} \left(\frac{SCR_r}{SCR_o - SCR_r} \right), \tag{2}$$

where SCR_r is the reconstructed signal and SCR_o is the original SCR data.

6   Results of Experiments

The experiment was conducted on 7 participants. We selected the fragments were data from all the devices was available, obtaining more than 2 hours of overlapping signals. Initial evaluation revealed that data gathered during a physical activity vary significantly more from the reference than the data from the initial part of the experiment. As our main interest is the quality of signals gathered while the participant is stationary, we decided to analyse both parts separately. The presented results refer to stationary part unless stated otherwise.

Correlation of HR signals is presented in Table 1. Each cell contains maximum, average and minimum value from the whole dataset. Average correlation between measured data and reference (Polar H6) is the best for MS Band 2. High spread for BITalino and Empatica E4 indicate that they are capable of producing better results, but they are very sensitive to device placement and measurement conditions. Moreover, the correlation factor for BITalino was decreasing with each experiment. Most likely it is caused by reusing the ECG electrodes, which should be replaced more frequently. The differences in correlation are mirrored in RMSE parameter (Table 2, lower values are better). Additionally, we observed that MS Band 2 and eHealth record HR with clearly visible quantization (Fig. 1).

Table 1. Heart rate signals correlation ($_{max}$/average/$_{min}$)

	BITalino	eHealth	Empatica E4	MS Band 2	Polar H6
BITalino	$_{1.00}/1.00/_{1.00}$	$_{0.76}/0.13/_{-0.34}$	$_{0.92}/0.48/_{-0.17}$	$_{0.88}/0.17/_{-0.35}$	$_{0.97}/0.18/_{-0.40}$
eHealth	$_{0.76}/0.13/_{-0.34}$	$_{1.00}/1.00/_{1.00}$	$_{0.78}/0.28/_{-0.21}$	$_{0.68}/0.42/_{0.19}$	$_{0.84}/0.55/_{0.14}$
Empatica E4	$_{0.92}/0.48/_{-0.17}$	$_{0.78}/0.28/_{-0.21}$	$_{1.00}/1.00/_{1.00}$	$_{0.86}/0.32/_{-0.25}$	$_{0.93}/0.41/_{-0.26}$
MS Band 2	$_{0.88}/0.17/_{-0.35}$	$_{0.68}/0.42/_{0.19}$	$_{0.86}/0.32/_{-0.25}$	$_{1.00}/1.00/_{1.00}$	$_{0.88}/0.66/_{0.37}$
Polar H6	$_{0.97}/0.18/_{-0.40}$	$_{0.84}/0.55/_{0.14}$	$_{0.93}/0.41/_{-0.26}$	$_{0.88}/0.66/_{0.37}$	$_{1.00}/1.00/_{1.00}$

Table 2. RMSE [dB] for heart rate signals ($_{max}$/average/$_{min}$)

	BITalino	eHealth	Empatica E4	MS Band 2
RMSE [dB]	$_{27.37}/13.67/_{1.70}$	$_{8.53}/4.82/_{2.81}$	$_{18.11}/6.97/_{1.53}$	$_{7.81}/4.06/_{2.54}$

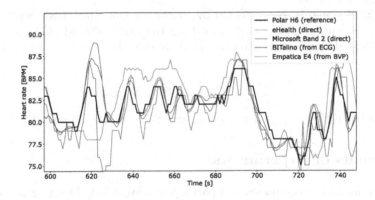

Fig. 1. Fragment of HR signals gathered using different devices

While working on the HR extraction algorithm we noticed that using shorter analysis window and skipping final smoothing reveals HR variability (HRV). This is a great advantage of BITalino and Empatica E4 over the remaining devices. According to [8], during inhale the HR is growing and while exhale it is decreasing. This information may be used to extract breathing frequency and respiratory sinus arrhythmia (RSA). In future experiments both parameters may be used to examine the emotional state of the subjects.

GSR signals correlation indicate high similarity of BITalino and eHealth (Table 3). The devices provide data with distinguishable phasic response and similar peak locations (Fig. 2). Signals from eHealth contain more noise than from BITalino, but they can be easily filtered due to high frequency and low amplitude of distortions. The amplitude of skin conductance response from Empatica E4 is lower than from eHealth and BITalino, but individual peaks are recognizable. Signals from Empatica have highest SNR and PTR (similarity to theoretical response) levels (Table 4). Unfortunately not all the peaks observed in results from BITalino and eHealth are present what is reflected in lower correlation factor. It may be caused by different sensor location (Empatica measures conductance GSR on wrist, eHealth and BITalino on fingers). GSR from MS Band 2 is uncorrelated with other devices and we were unable to extract any useful data. No phasic response can be observed therefore there is no point in analysing the remaining parameters.

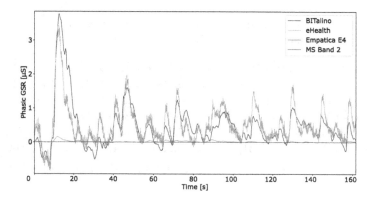

Fig. 2. Comparison of GSR response signals gathered using different devices

Table 3. Skin conductance response correlation ($_{max}$/average/$_{min}$)

	BITalino	eHealth	Empatica E4	MS Band 2
BITalino	$_{1.00}$/1.00/$_{1.00}$	$_{0.84}$/0.70/$_{0.41}$	$_{0.55}$/0.30/$_{0.05}$	$_{0.19}$/0.06/$_{-0.03}$
eHealth	$_{0.84}$/0.70/$_{0.41}$	$_{1.00}$/1.00/$_{1.00}$	$_{0.64}$/0.32/$_{0.02}$	$_{0.06}$/0.04/$_{0.02}$
Empatica E4	$_{0.55}$/0.30/$_{0.05}$	$_{0.64}$/0.32/$_{0.02}$	$_{1.00}$/1.00/$_{1.00}$	$_{0.09}$/0.01/$_{-0.09}$
MS Band 2	$_{0.19}$/0.06/$_{-0.03}$	$_{0.06}$/0.04/$_{0.02}$	$_{0.09}$/0.01/$_{-0.09}$	$_{1.00}$/1.00/$_{1.00}$

Table 4. Signal parameters for extracted skin conductance response ($_{max}$/average/$_{min}$)

	BITalino	eHealth	Empatica E4	MS Band 2
PTR [dB]	$_{-0.19}$/-0.76/$_{-2.80}$	$_{-0.42}$/-0.73/$_{-1.63}$	$_{-0.14}$/-0.55/$_{-1.68}$	$_{0.00}$/-0.63/$_{-1.58}$
SNR [dB]	$_{23.13}$/18.07/$_{13.32}$	$_{16.51}$/12.59/$_{4.84}$	$_{26.80}$/19.99/$_{3.52}$	$_{20.48}$/9.23/$_{0.22}$

The physical activity test was taken only by 4 out of 7 experiment participants. The results are compared to the stationary part of the experiment. Obtained parameters are expressed as a difference between the active and stationary part. In some cases the recorded HR was correlated with subject movement instead of the real HR. For BITalino and Emapatica, where HR was calculated by our own script it can probably be fixed by combining the ECG/BVP data with signals from acceleration sensor, but it was not tested yet. This problem affects results from MS Band 2 the most and eHealth the least (Table 5). Other issue occurs in GSR measurements, where during movement the contact between body and sensor was not constant what leads to sudden conductance changes. Both problems result in lower signal correlation for GSR and HR (Tables 5 and 6). The only case were the correlation has significantly grown is the GSR signal from MS Band 2 and Empatica E4, however it should be noted that the change was from correlation factor $-0,02$ to $0,20$, therefore the data is still weekly correlated.

Table 5. Difference between HR signals correlation during the activity test and the first part of the experiment (max/average/min)

	BITalino	eHealth	Empatica E4	MS Band 2	Polar H6
BI[a]	0.00/0.00/0.00	-0.17/-0.14/-0.21	-0.58/-0.23/0.14	-0.48/-0.21/-0.03	-0.31/0.02/0.41
eH	-0.17/-0.14/-0.21	0.00/0.00/0.00	-0.38/-0.33/0.21	0.13/-0.14/-0.23	0.00/-0.31/-0.35
E4	-0.58/-0.23/0.14	-0.38/-0.33/0.21	0.00/0.00/0.00	-0.11/-0.17/0.21	0.03/-0.22/0.14
B2	-0.48/-0.21/-0.03	0.13/-0.14/-0.23	-0.11/-0.17/0.21	0.00/0.00/0.00	-0.41/-0.54/-0.60
H6	-0.31/0.02/0.41	0.00/-0.31/-0.35	0.03/-0.22/0.14	-0.41/-0.54/-0.60	0.00/0.00/0.00

[a]BI – BITalino, eH – eHealth, E4 – Empatica E4, B2 – MS Band 2, H6 – Polar H6

Table 6. Difference between skin conductance response correlation during the activity test and the first part of the experiment (max/average/min)

	BITalino	eHealth	Empatica E4	MS Band 2
BITalino	0.00/0.00/0.00	-0.05/-0.20/-0.32	0.13/-0.02/0.08	0.13/0.00/-0.14
eHealth	-0.05/-0.20/-0.32	0.00/0.00/0.00	-0.03/-0.02/0.16	0.15/0.00/-0.13
Empatica E4	0.13/-0.02/0.08	-0.03/-0.02/0.16	0.00/0.00/0.00	0.34/0.22/0.15
MS Band 2	0.13/0.00/-0.14	0.15/0.00/-0.13	0.34/0.22/0.15	0.00/0.00/0.00

Our critical analysis demonstrates, that Bitalino remains the most prospective platform for both HR and GSR measurements, especially that the technical support for the e-Health is being phased out. As for secondary HR readings MS Band can be used. While the Bitalino kit does not have the wristband form, it can be turned into a wearable using the Bitalino Freestyle kit and 3D-printed boxes. It is worth emphasizing, that the devices we selected in this paper offer real-time HR and GSR monitoring, as opposed to the vast majority of fitness trackers, e.g. from Fitbit, that offer only highly filtered and averaged data.

7   Conclusions and Future Work

This paper discusses the practical aspects of the construction of a measurement framework for affective computing and telemedicine based on low-cost, portable devices. We provide analysis of results of experiments aimed at critical comparison of quality of HR and GSR signals in selected devices.

In our future works on the framework we will consider focusing on the Bitalino, possibly combined with the MS Band 2 in some cases. Using the more reliable affective data acquired from these devices we will work on developing effective classification methods for the emotional condition of the user. These methods will be implemented on mobile devices, such as smartphones.

References

1. Alexander, D.M., Trengove, C., Johnston, P., Cooper, T., August, J.P., Gordon, E.: Separating individual skin conductance responses in a short interstimulus-interval paradigm. J. Neurosci. Methods **146**(1), 116–123 (2005)
2. Arnold, M.B.: Emotion and Personality. Columbia University Press, New York (1960)
3. Benedek, M., Kaernbach, C.: Decomposition of skin conductance data by means of nonnegative deconvolution. Psychophysiology **47**(4), 647–658 (2010)
4. Cacioppo, J.T., Berntson, G.G., Larsen, J.T., Poehlmann, K.M., Ito, T.A.: The psychophysiology of emotion. In: Handbook of Emotions, pp. 173–191. Guildford Press, New York (2000)
5. Düking, P., Hotho, A., Holmberg, H.C., Fuss, F.K., Sperlich, B.: Comparison of non-invasive individual monitoring of the training and health of athletes with commercially available wearable technologies. Front. Physiol. **7**, 71 (2016)
6. Garbarino, M., Lai, M., Bender, D., Picard, R., Tognetti, S.: Empatica E3 - a wearable wireless multi-sensor device for real-time computerized biofeedback and data acquisition. In: 2014 EAI 4th International Conference on Wireless Mobile Communication and Healthcare (Mobihealth), pp. 39–42 (2014)
7. Grundlehner, B., Brown, L., Penders, J., Gyselinckx, B.: The design and analysis of a real-time, continuous arousal monitor. In: Proceedings of the Sixth International Workshop on Wearable and Implantable Body Sensor Networks, pp. 156–161, June 2009
8. Hirsch, J.A., Bishop, B.: Respiratory sinus arrhythmia in humans: how breathing pattern modulates heart rate. Am. J. Physiol. **241**(4), H620–H629 (1981). https://pdfs.semanticscholar.org/48fa/f00ce055ae1bfc5535dc446037b1d9aacf89.pdf, http://www.ncbi.nlm.nih.gov/pubmed/7315987
9. IMotions Biometric Research Platform: GSR Pocket Guide. IMotions Biometric Research Platform (2016)
10. Kutt, K., Nalepa, G.J., Giżycka, B., Jemioło, P., Adamczyk, M.: Bandreader - a mobile application for data acquisition from wearable devices in affective computing experiments. In: ICAISC 2018 (2018, submitted)
11. Lazarus, R.S.: Psychological Stress and the Coping Process. McGraw-Hill, New York (1966)
12. Marchewka, A., Żurawski, Ł., Jednoróg, K., Grabowska, A.: The nencki affective picture system (NAPS): introduction to a novel, standardized, wide-range, high-quality, realistic picture database. Behav. Res. Methods **46**(2), 596–610 (2014)

13. Nalepa, G.J., Gizycka, B., Kutt, K., Argasinski, J.K.: Affective design patterns in computer games. Scrollrunner case study. In: Communication Papers of the 2017 Federated Conference on Computer Science and Information Systems, FedCSIS 2017, pp. 345–352 (2017). https://doi.org/10.15439/2017F192
14. Nalepa, G.J., Kutt, K., Bobek, S., Lepicki, M.Z.: AfCAI systems: affective computing with context awareness for ambient intelligence. Research proposal. In: Ezquerro, M.T.H., Nalepa, G.J., Mendez, J.T.P. (eds.) Proceedings of the Workshop on Affective Computing and Context Awareness in Ambient Intelligence (AfCAI 2016). CEUR Workshop Proceedings, vol. 1794 (2016). http://ceur-ws.org/xxx-1794/
15. Nourbakhsh, N., Wang, Y., Chen, F., Calvo, R.A.: Using galvanic skin response for cognitive load measurement in arithmetic and reading tasks. In: Proceedings of the 24th Conference on Australian Computer-Human Interaction, OzCHI 2012, pp. 420–423 (2012)
16. Ohme, R., Reykowska, D., Wiener, D., Choromanska, A.: Analysis of neurophysiological reactions to advertising stimuli by means of EEG and galvanic skin response measures. J. Neurosci. Psychol. Econ. **2**(1), 21–31 (2009)
17. Orthony, A., Clore, G., Collins, A.: The Cognitive Structure of Emotions. Cambridge University Press, Cambridge (1988)
18. Picard, R.W.: Affective Computing. MIT Press, Cambridge (1997)
19. Schmidt, S., Walach, H.: Electrodermal activity (EDA) - state-of-the-art measurement and techniques for parapsychological purposes. J. Parapsychol. **64**, 139–163 (2000)

Severity of Cellulite Classification Based on Tissue Thermal Imagining

Jacek Mazurkiewicz[1]([✉]), Joanna Bauer[2], Michal Mosion[3],
Agnieszka Migasiewicz[4], and Halina Podbielska[2]

[1] Department of Computer Engineering, Faculty of Electronics,
Wroclaw University of Science and Technology, ul. Wybrzeze Wyspianskiego 27,
50-370 Wroclaw, Poland
Jacek.Mazurkiewicz@pwr.edu.pl
[2] Department of Biomedical Engineering, Faculty of Fundamental Problems
of Technology, Wroclaw University of Science and Technology,
ul. Wybrzeze Wyspianskiego 27, 50-370 Wroclaw, Poland
{Joanna.Bauer,Halina.Podbielska}@pwr.edu.pl
[3] Comarch S.A., ul. Dlugosza 2-6, 51-162 Wroclaw, Poland
mosion93@gmail.com
[4] Department of Cosmetology, Faculty of Physiotherapy,
Wroclaw University School of Physical Education,
al. Ignacego Jana Paderewskiego 35, 51-612 Wroclaw, Poland
Agnieszka.Migasiewicz@awf.wroc.pl

Abstract. In this article we present a novel approach to cellulite classification that can be personlised based on non-contact thermal imaging using IR thermography. By analysing the superficial temperature distribution of the body it is possible to diagnose the stages of cellulite development. The study investigates thermal images of posterior of thighs of female volunteers and identifies cellulite areas in an automatic way using image processing. The Growing Bubble Algorithm has been used for thermal picture conversion into valid input vector for a neural network based classifier scheme. Using machine learning process of training the input database was prepared as the stage of cellulite classifier according to the state of the art Nürnberger-Müller diagnosis scheme. Our work demonstrates that it is possible to diagnose the cellulite with over 70% accuracy using a cost-effective, simple and unsophisticated classifier which operates on low-definition pictures. In essence, our work shows that IR-thermography, when coupled with computer aided image analysis and processing, can be a very convenient and effective tool to enable personalized diagnosis and preventive medicine to improve the quality of life of women suffering from cellulite problems.

Keywords: Thermal imaging · Infrared thermography · Cellulite
MLP · Image processing

© Springer International Publishing AG, part of Springer Nature 2018
L. Rutkowski et al. (Eds.): ICAISC 2018, LNAI 10842, pp. 179–190, 2018.
https://doi.org/10.1007/978-3-319-91262-2_17

1   Introduction

Cellulite, also known as edematous fibrosclerotic panniculopathy or gynoid lipodystrophy [1,9], is a disorder of subcutaneous layer, which belongs to a wide group of skin dysfunctions. In many cases, it is only a cosmetic defect, however, it is often mistaken as a cellulitis, which is an inflammation of connective tissues. Cellulite affects about 85% of women in the age over 20 [3,8,10]. Women are more likely to develop cellulite than men, and, it is usually diagnosed in more advanced stage. It is caused mainly by the differences between females and males in the anatomical structure of their respective skins e.g. thickness, divergent cross-linking of the connective tissue of the dermis and different fat contents. Cellulite appears as unnatural folds of fat fabric resulting from hormonal changes, especially an estrogen level decrease, which may lead to changes in blood circulation and reduction of collagen production. Blood and lymphatic vessels, located in the middle layer of the skin structure, can have their structure changed by an adipocytes, leading to an insufficiency in the microcirculation system. For this reason, some cells can have improper level of nutrition that can result in metabolic disorders (fat cellulite) or swellings due to water accumulation (water cellulite) [6]. As a result of the above changes, orange-colored swelling and imbalances (known as orange peel) can appear on the surface of the skin. Origins of the disorders of microcirculation and skin irregularities discussed above are complex. One of the theories holds responsible adipocyte, where visible lumps on the surface are the results of degeneration of cells fibrosis under the skin. This can occur as a result of excess calcium and magnesium in the blood thereby causing an increased osmotic pressure in blood vessels. The transportation of oxygen to all cells becomes difficult which limits cellular respiration and nutrition. Toxic metabolic wastes produced are stored in the body. The accumulation of these wastes can cause damages such as swelling and microcracking of cells, thereby increasing the hydrostatic pressure in the remaining vessels and intercellular fluid. It comes to the distortion of the skin structures, where the rigid collagen fibres begin to squeeze. This action has negative effects, because it reduces the flexibility of the tissues [6]. Another theory of the fibrosclerotic panniculopathy holds responsible the impairment of the hormonal economy in the body. This mainly concerns the excess of estrogens, which are essential of regulation of fertility in both sexes. In the female body, the concentration of estrogen is higher, especially during the adolescence, pregnancy, hormone therapy or even due to contraception [8]. These hormones are responsible for the expansion of the blood and lymph vessels, what cause the bulges and increasing permeability of toxins through the capsules to the cells. The individual sensitivity of hormonal receptors occurring inside adipose cells is also an important factor. Activation of them under the binding of endogenous estrogens determines the gene expression in these cells leading to overactivity of the adipogenesis [8]. Fibroblast activity is regulated by sympathetic fibers, which are responsible for the innervation of the body integument. Fibroblasts produce collagen and fibers of the intracellular substance as well as determine the rigidity of the skin membrane. The nervous system controls the metabolism of adipocytes and the intensity of the microcirculation [8]. The theory linking an unbalanced fat and carbohydrates containing diet with lipodystrophy is, widely known in the public. Excessive consumption of these unbalanced foods promotes lipogenesis and dysregulation of the pancreatic hormone management, which is responsible for regulation of blood sugar level and pressure in blood vessels. Cellulites (mainly water cellulites) can appear due to the excess of salt, which stops water in the body. Cellulites may also have a genetic background that has not been fully investigated yet. So far, based on the available literature, it has been determined that chromosome 17, which codes angiotensin converting enzyme (ACE) and transcriptional factors (HIF1A), can be responsible. ACE is responsible for blood pressure regulation and conversion of angiotensin that leads to the contraction of capillaries. HIF1A is associated with cellular hypoxia. The distribution of both genes differs significantly in women with cellulite and in healthy women [4]. Cellulite examination can be done in many different ways. In the case of manual tests (e.g. unaided visible inspection and visible inspection with a measuring tool) the diagnoses are not comprehensive and could be incorrect because of measurement errors. Here we propose an original approach of cellulite examination using a combination of infrared thermographic imaging and neural network classifier scheme that may potentially revolutionize the personalized diagnoses of cellulite at its different stages of development. The advantage of the our proposed methodology of diagnoses is that it is contactless, fast, suitable for practical usage and thus may lead to the personalized lipodystrophy diagnosis and monitoring of the therapy progress. In this article, after an introduction to scope of the study in Sect. 1, we presents a short overview of state of the art methods of cellulite diagnosis (Sect. 2). Section 3 of the article then describes the details of our proposed approach of combining IR thermography and neural network classifier in cellulite diagnosis. We then report the input data from clinical investigations and the discuss new methods in the personalized diagnoses of cellulites at different stages of development (Sect. 4).

2   Cellulite Examination

The basic grouping of paniculopathy symptoms consists of distinguishing the subjective symptoms (felt by the patient herself), and the objective symptoms (observed by the specialist). Subjective symptoms are most often associated with the feeling of heaviness, ramps and muscle tensions in the limbs. In advanced stages, it can even come to tingling and pain. Objective symptoms include excessive skin pigmentation, discoloration and stretch marks. Symptoms depend on the form and stage of the disease. For women who do not practice sport or lead a sedentary lifestyle or for women who have lost weight within a short period, the characteristic type of cellulite is soft cellulite. For this kind of cellulite, there is a specific "mattress" effect, where changes of skins surface (hollows and protuberances) are very well visible during movement. Beads and grains are easily feel under the fingers while testing. However, women, who take care of their figure, can have hard cellulite, which becomes apparent when the skin is grasped. The symptoms of hard cellulite are also stretch marks. The form of water cellulite is accompanied with an edema and large surfaces of folds whereas in the case of lipid cellulite surface irregularities called the "orange peel" can appear. A hybrid form of above types of paniculopathy is also possible. That is why the recognition of different classes and stages of cellulite without any specialised equipment is very uncertain and should be carried out by a specialist. Any assessment of the cellulite stage is a complicated process as the pathology (hard, soft, lipid, water, mixed) should be determined first. Each of these pathologies, at the end, can differ in the results of the measurement. Initially, the smooth skin, due to the influence of dystrophy, can become distorted, which is visible only when it is examined carefully. Further development of cellulite causes skin folds with uneven hollows. After some time, this effect becomes visible in the standing position and the beads are noticeable under examining fingers [11]. Advanced stages of pathological changes in the tissues can be easily observed. There are also palpation methods, which uses some predefined scales available in the literature. Primary signs of cellulite are recognized only with the methods based on measurements of characteristic parameters of pathology. As a result, the diagnosis of cellulite's type and severity becomes more reliable in these measurements, which are independent of factors such as age, type or thickness of the skin. One of the most popular and often used scale to detect the severity of the cellulite is Nürnberger-Müller's scale [11]. Examinations using palpation method along with a visible observation assign the observations into one of the following four groups:

1. No dimpling or apparent visible alterations to the skin surface upon standing or lying down or upon pinching the skin.
2. No dimpling or apparent visible alterations to the skin surface upon standing or lying down. Dimpling appears with the pin h test or muscular contraction.
3. Dimpling appears spontaneously when standing but not when lying down. The orange peel appearance of the skin is evident to the naked eye, without need for manipulation.
4. Dimpling is spontaneously present when both standing and lying down, evident to the naked eye without need for manipulation, orange peel skin surface appearance with raised areas and nodules.

3   Proposed Solution

Conventional thermographic methods use special liquid crystal film, which, in contact with the body, changes colour in response to the skin temperature. The film is applied to places either where cellulite is present or places where there are visible changes in blood vessels. The colour scales applied in the investigation is similar to a rainbow - the warmest areas are marked with white and yellow, while the coolest colours are either blue or purple. The temperature range applied on the film is from 28.5 °C to 31 °C. The type of pathological changes and severity is determined by the contrast between the adjacent colours, which develops as a result of the differences in blood microcirculation in the targeted area [7]:

1. Grade 0 applies to people, who exhibits no visible pathology - image presented on fillm contains small, unevenly distributed spots.
2. Grade 1 occurs in the cases of enlarged pathologies of the focal point, which are surrounded by significantly cooler tissue as a result of ischemia.
3. Grade 2 represents images with speckles that resemble to a leopard skin pattern. This effect is caused by alternating beads and nodules.
4. Grade 3 is found when there are large, black holes showing in hemispherical areas. A black colour represents that the temperature was below the measurement scale.

In contrast, we take a non-contact thermal imaging approach as an alternative method for cellulite stage classification. Our approach allows a remote assessment of the surface temperature distribution of the examined body similar to what has been described in references [2,12]. Dermatological effects such as cellulite manifest as superficial temperature changes, which can be conveniently detected, quantified and machine-analyzed using IR thermography for automatic decision making. The proposed approach is presented in Fig. 1.

Fig. 1. Flow chart of the proposed approach

3.1   Input Images

The group of female volunteers, aged 19–22 with different stages of cellulite has been diagnosed a'priori by a licensed cosmetologist using the Nürnberger-Müller scale [11]. In order to maintain the constant ambient conditions, measurements were made in one room at a specific time of the day. The air temperature was also fixed between 22–24 °C and the humidity set to 35–40%. Volunteers were asked to expose their thighs. Then, they stayed in the standing position for 20 min to adopt to the conditions of the experiment. The body temperature we found was stable, because the volunteers were not involved in any physical activity. Thermal images of the backside (posterior) of the thigh of each volunteer were recorded using a thermographic camera FLIR T335 with a IR spectral range of 7.5–13 μm, a temperature sensitivity of 50 mK at 30 °C. The images were taken from a fixed distance of 1.2 m from the volunteer. Figure 2 shows typical thermal images taken on volunteers.

3.2   Image Processing

Due to the quality differences among all the input images, each of the image was processed in the following steps to obtain the clearest data: Gaussian blur

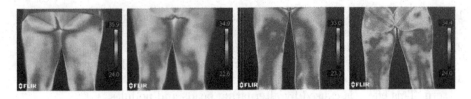

Fig. 2. Typical thermal images of thighs of volunteers

filter - if necessary, sharpness enhancement and - finally - colour balance [5]. The Gaussian blur (also known as Gaussian smoothing) is a filter that blurs an image using a Gaussian function using one component: radius, which determines the size of the area to be taken into a count for blurring action. Such blurring is used to reduce image noise and details. As the result, the image is smoother and blurred. The 'unsharp' mask is a filter to achieve the same contrast for the lower frequency of image. This is a high-pass filter that detects edges between the bright and the dark fields. Detected edges are then sharpened by the earlier created mask. The filter has two components. Firstly, the sigma component determines the size of the area that is to be taken into account during Gaussian blurring. It means that finer details are sharpened if a smaller radius is used. Secondly, the weight component determines the minimum changes in brightness that must occur for the filter to be applied. If the smaller value is fixed, the lower values are strengthened. The colour balance algorithm is described as a sequence of the following steps: create a 256-element array (colours are in range 0–255), get each pixel value and according to the value, increment appropriate position in the array to create an array with a histogram of the image; histogram array is the source of most frequent values of colour, threshold each pixel in a colour layer (red, green and blue) (Fig. 3). In the final step, useless colours are eliminated. For example magenta colour is converted into white, while blue and cyan colours are converted into black (Fig. 4) [13].

3.3 Classifier

The cellulite stage classification is based on the total size of the white, or, in general, the lightest fields visible in the thermal pictures after the end of the processing. If the size of the region is larger, the cellulite problem is more serious. The lightest regions in the pictures are usually separable and irregular. This is the reason why we propose to use the Growing Bubble Algorithm to estimate the total size of them [9]. The following observation can be converted to a solution. The proportion of darker and lighter areas analysis and the frequency of appearance can potentially create the valid input vector for the classifier. Additionally, the place and the rotation of patterns visible in the pictures are ignored.

The idea here is to fill the lighter and darker areas, to count them and to measure its sizes. We have to limit an expansion of the filling areas and this is exactly what bubbles in the Growing Bubble Algorithm do. The areas are filled by the bubbles instead of using a simple colour scale. The bubbles cannot grow

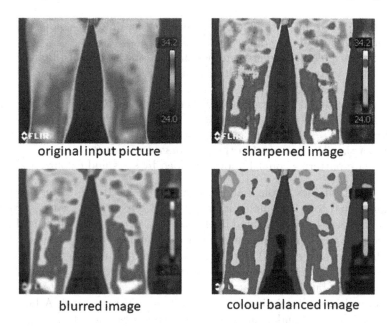

Fig. 3. Input picture preprocessing (Color figure online)

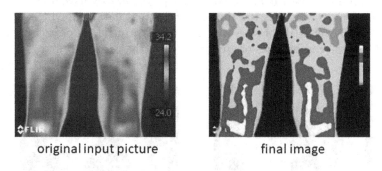

Fig. 4. Input picture and final image (Color figure online)

bigger than the colour borders and consequently are able to preserve its sizes. The bubbles size for darker and brighter area analyses can be a useful tool in quantitatively pointing to the stage of a given cellulite (Fig. 5). This is achieved in the following steps, for example:

1. Set R - radius of bubble to 4.
2. Iterate through all of the pixels.
3. If actual pixel is white then take lighter colour, If it is black then take the darker colour, otherwise go to Point 2.
4. Set error variable to zero.
5. In place pointed by the actual pixel draw the circle of radius R pixels with a chosen colour.

6. For each pixel out of the picture or overlapping different colour increment error variable and save average error place.
7. If error variable not larger then increment R and go back to Point 4.
8. If error is to big then move circle in an opposite side to the average error place and draw again.
9. If error variable is less than before then increment R and go to Point 4.
10. If error variable is bigger than draw the previous circle and fill it with chosen colour.
11. If it was not the last pixel in the picture go to Point 2.
12. Calculate pattern - count bubbles grouped by sizes and colours.

The algorithm goes through all pixels in the image. If the lighter pixels area of a radius equal to 4 is available it draws an IR bubble there. Next, it tries to enlarge and move the circle around the lighter area to possibly best fit in the solid colour space. It stops enlarging when the circle is going to overlay the darker zone or the already drawn circle. The process is repeated until all lighter regions are touched. Meanwhile, in an analogical way, the darker areas are processed where the second set of bubbles is created (Fig. 5). In fact, the process of enlargement is not stopped immediately if the circle overlaps the single pixel. A little bit of overlapping is allowed and thanks to that the circles can fit better ragged edges. As the consequence, the circles are filled with colour. Later on, when new circle is looking for a place, it calculates how many pixels already have overlaps. When all bubbles are drawn, it is the time for the actual data retrieving. All circles are grouped by intervals of sizes and colours. The members of each group are counted and the cardinalities of groups becomes the classifier input.

 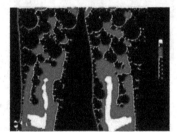

final image   Growing Bubble Algorithm

Fig. 5. Input image at the end of Growing Bubble Algorithm

The classifier was created in the neural network as a three-layer Multilayer Perceptron: 4–16-input layer neurons depending on the input vector size, 8 neurons in hidden layer, 4 neurons in output layer to point the one of 4 possible stages of the cellulite [9]. Hidden layer neurons were tested for the following activation functions: Gate, Linear, LSTM, Sigmoid, Softmax, and Tanh. The neural network was trained using backpropagation algorithm by 25000 epochs, The learning error was finally less than 1%.

4   Experimental Data and Results

Input database consists of a total of 140 thermal images of 320 × 240 pixels: 20 images without cellulite, 67 images with 1st level degree of cellulite, 43 images with 2nd level degree of cellulite, 10 images with 3rd level degree of cellulite. The database was divided into two separated groups: 90% of images from each category were used for learning procedure, 10% - for testing. The images were taken with auto assigned temperature values where these values had been assigned by the thermal camera. It can be compared to auto white balance in an ordinary camera. Such solution is correct in general usage but is insufficient for severity of cellulite classification. For this reason we decided to normalize the temperature scale using FLIR Tool v2.1 application software provided by the manufacturer of the thermal camera, FLIR. This tool allows the user to open an image that has been taken by the thermal camera and adjust the temperature range and colour pallets ex post. With the help of this tool, the temperature of each of the images taken in the current investigation was set to an appropriate test temperature range as it was otherwise difficult to fix the minimum and maximum value of temperature to border the scale of normalisation. Finally, seven temperature ranges were created based on the minimum and maximum temperature recorded by the camera and by the average temperature stored during the experiment:

1. auto assigned - values left 'unhanged',
2. 28.5–31.0 °C - temperature range based on liquid crystal thermographic film method of classification,
3. 24.6–36.2 °C - from max temperature recorded as lower limit to max temperature recorded as upper limit,
4. 20.6–30.8 °C - from min temperature recorded as lower limit to min temperature recorded as upper limit,
5. 24.6–30.8 °C - from max temperature recorded as lower limit to reduced max temperature recorded as upper limit,
6. 22.6–33.5 °C - from average of max and min temperature recorded as lower limit, to average of max and min temperature recorded as upper limit,
7. 24.1–34.1 °C - from average of all images temperature recorded as lower limit, to average of all images temperature recorded as upper limit.

The effect of temperature-ranges creation is similar to the manual white-balance tuning used in conventional photography. The severity of cellulite classification tests were carried out for the each temperature-range separately. This way we tried to avoid the negative influence of camera feature for the final results. We also checked the best activation function for hidden neurons. Table 1 shows the correct stage of cellulite classification obtained in this way. We can notice the Sigmoid function came out to be the best activation function for the hidden layer neurons. The results presented in this article are preliminary but very promising especially given the fact that the machine-learning algorithm was built on assumptions that are simple and not very sophisticated so that the classifier scheme could operate on low-definition pictures. Our proposed approach is thus highly advantageous for building commercial systems that could be affordable and used not only in medical outpatient clinics but also in many beautician practices and be operated by relatively less-skilled operators. It is easy to find that the correct temperature scale tuning for the thermal pictures that would have an impact on the final classification. We have obtained the best results for the severity of cellulite classification using the temperature range based on minimum and maximum temperatures recorded by thermal camera during the experiment. This finding has a very significant practical relevance as it signifies that there is no need to make any sophisticated preprocessing of the thermal images for the purpose of the severity classifications. The same quality of the classification was obtained for the temperature range fixed by the average values of the temperature. Here the image preprocessing is a little more complex, but still is not so much time-consuming. It is however evident that auto assigned values of the temperature are relatively less useful for obtaining a proper classification basis. It means that we have to care about the temperature range tuning to create the valid input data from the input pictures. The Growing Bubbles Algorithm appeared to be a robust method to exchange the picture into valuable input vector for the classifier. It is easy for implementation and can be tuned, if necessary, by the radius of the bubbles if we have different shapes of input pictures. These different shapes can be the result of distance from the patient to the camera or different types of the thermal cameras.

Table 1. Stage of cellulite classification results

| Activation function | Gate | Linear | LSTM | Sigmoid | Softmax | Tanh |
Temperature range	[%]	[%]	[%]	[%]	[%]	[%]
AUTO	43	50	50	46	50	50
28.5–31.0 °C	50	36	50	67	29	36
24.6–36.2 °C	43	50	50	74	57	50
20.6–30.8 °C	50	36	36	67	36	43
24.6–30.8 °C	57	50	50	74	50	43
22.6–33.5 °C	50	50	29	74	50	43
24.1–34.1 °C	36	36	50	67	50	50

For the cellulite stage classification problem the typical confusion matrix has no real sense. This is the reason why during the second part of tests we try to find how far is the pointed cellulite stage from the standard Nürnberger - Müller's scale output if the classification result is incorrect. This way we estimate the kind of "functional error" measured as "the number of grades difference". We check how significant is wrong classification. The same temperatures ranges are preserved and we use the sigmoid activation function - because this configuration provides the best results of classification. We have only two possibilities: "one grade difference" and "two grades difference". It is easy to notice (Table 2) that over 81% of wrong classifier answers point the next stage instead of the correct one and used temperature range has not significant influence for the result. The highest output we find for the same temperature ranges which guarantee the best correct answers. It means if the classification result is wrong it can be still useful in practical point of view. One grade of difference is acceptable for cellulite medical practices.

Table 2. Wrong classification - functional error - how far is the result from the standard Nürnberger - Müller's scale output

Functional error	One grade difference	Two grades difference
Temperature range	[%]	[%]
AUTO	84.2	15.8
28.5–31.0 °C	84.6	15.4
24.6–36.2 °C	90.0	10.0
20.6–30.8 °C	92.3	7.7
24.6–30.8 °C	81.8	18.2
22.6–33.5 °C	84.4	15.6
24.1–34.1 °C	84.6	15.4

While the quality of the recognition presented here is reasonable, it is still lower than what expected. This could be due to the way of preliminary data classification is conducted based on the Nürnberger-Müller scale, which is manual and thus very subjective. The borders among the cellulite stages are sometimes very subtle and quite difficult to perform any contradistinction. This could cause some mistakes in the stage of learning database creation. To avoid this problem in the future the preliminary examination should use more reliable diagnostic tools such as standard or high frequency ultrasonography. An enlargement of the learning database for training procedure is also necessary to improve the accuracy of recognition. Another issue for consideration is setting the temperature ranges for the pictures normalization. From our study, it appears to have a significant impact on the final results.

5   Conclusions

We have successfully demonstrated an original approach of cellulite examination that takes benefit of neural network based classifier scheme applied to infrared thermographic imaging. The approach presented here is fast, non-contact and, when further developed, can be used as an alternative method for many classical examinations related to cellulite. We showed the feasibility of IR thermography as a potential tool for personalized diagnosis of cellulite. Presented classifier operates on low-definition pictures and can provide recognition accuracy higher than manual methods which are commonly used nowadays. Preliminary data obtained from such classifiers applied to the cellulite stage classification provided an accuracy over 70% which is highly promising for the implementation of our approach of cellulite severity classification. Proposed methodology may lead to the development of affordable and effective commercially available system which will support aesthetic or cosmetic specialists in every day practice, particularly in prevention of cellulite, as well as objective and personalised therapy.

References

1. Avram, M.M.: Cellulite: a review of its physiology and treatment. J. Cosmet. Laset. Ther. **6**(4), 181–185 (2004)
2. Bauer, J., Deren, E.: Standardization of infrared thermal imaging in medicine and physiotherapy. Acta. Bio. Opt. Inform. Med. **20**(1), 11–20 (2014)
3. Cellulite Statistics (2006). http://www.worldvillage.com/cellulite-statistics/. Accessed 3 July 2017
4. Emanuele, E.: A multilocus candidate approach identifies ACE and HIF1A as susceptibility genes for cellulite. J. Eur. Acad. Dermatol. Venereol. **24**(8), 930–935 (2010)
5. Faundez-Zanuy, M., Mekyska, J., Espinosa-Duro, V.: On the focusing of thermal images. Pattern Recogn. Lett. **32**(11), 1548–1557 (2011)
6. Galazka, M., Galeba, A., Nurein, H.: Cellulite as a medical and aesthetic problem - etiopathogenesis, symptoms, diagnosis and treatment. Hygeia Public Health **49**(3), 425–430 (2014)
7. Goldman, M.P., Hexsel, D.: Cellulite Pathophysiology and Treatment. CRC Press, Boca Raton (2010)
8. Janda, K., Tomikowska, A.: Cellulite - causes, prevention. Treatment. Ann. Acad. Med. Stettin. **60**(1), 29–38 (2014)
9. Jankowski, M., Mazurkiewicz, J.: Road surface recognition system based on its picture. In: Rutkowski, L., Korytkowski, M., Scherer, R., Tadeusiewicz, R., Zadeh, L.A., Zurada, J.M. (eds.) ICAISC 2013. LNCS (LNAI), vol. 7894, pp. 548–558. Springer, Heidelberg (2013). https://doi.org/10.1007/978-3-642-38658-9_50
10. Junqueira, J.P., Alfonso, M., de Mello Tucunduva, T.C., Bussamara Pinheiro, M.V., Bagatin, E.: Cellulite: a review. Surg. Cosmet. Dermatol. **2**(3), 214–219 (2010)
11. Nürnberger, F., Müller, G.: So-called cellulite - an invented disease. J. Dermatol. Surg. Oncol. **4**(3), 221–229 (1978)
12. Ring, F.: The historical development of thermometry and thermal imaging in medicine. J. Med. Eng. Technol. **30**(4), 192–198 (2006)
13. Toet, A.: Natural colour mapping for multiband nightvision imagery. Inf. Fusion **4**(3), 155–166 (2003)

Features Selection for the Most Accurate SVM Gender Classifier Based on Geometrical Features

Piotr Milczarski[1]([✉]) [iD], Zofia Stawska[1] [iD], and Shane Dowdall[2]

[1] Faculty of Physics and Applied Informatics, University of Lodz,
Pomorska Street 149/153, 90-236 Lodz, Poland
{piotr.milczarski,zofia.stawska}@uni.lodz.pl
[2] Department of Visual and Human Centred Computing,
Dundalk Institute of Technology, Dundalk, Country Louth, Ireland
shane.dowdall@dkit.ie
http://www.wfis.uni.lodz.pl

Abstract. In the paper, we have focused on the problem of choosing the best set of features in the task of gender classification/recognition. Choosing a minimum set of features, that can give satisfactory results is also important in the case where only a part of the face is visible. The minimum set of features can simplify the classification process to make it useful for mobile applications. Many authors have used SVM in facial classification and recognition problems, but there are not many works using facial geometry features in the classification neither in SVM. Almost all works are based on the appearance-based methods. In the paper, we show that the classifier constructed on the base of only two or three geometric facial features can give satisfactory (though not always optimal) results with accuracy 82% and positive predictive value 87%, also in incomplete facial images. We show that Matlab and Mathematica can produce very different SVMs given the same data.

Keywords: Geometric facial features · Biometrics
Gender classification · Support Vector Machine

1 Introduction

Recognition of human biometric features is a very current issue. It is increasingly used, e.g. in various types of identification systems. One of the human basic features is gender. Although the first gender recognition algorithms were developed many years ago, they are still being researched because their results are not always satisfactory.

One of the objectives of our research is to classify faces when only a part of the face is visible. We search for points of the face that are the best for gender classification. We show the conditions for facial features to achieve higher accuracy in case of whole face and partial face visibility. In the papers [4,11,18,

© Springer International Publishing AG, part of Springer Nature 2018
L. Rutkowski et al. (Eds.): ICAISC 2018, LNAI 10842, pp. 191–206, 2018.
https://doi.org/10.1007/978-3-319-91262-2_18

19,22,26,37] authors showed the results of gender classification using only a part of the face. The authors used lower part of the face [18], top half of the face [4], veiled faces [19], periocular region [11,26] or they taking into account multiple facial parts such as lip, eyes, jaw, etc. [22].

Gender can be recognized using many different human biometric features such as silhouette, gait, voice, etc. However, the most-used feature is human face [21, 29]. We can distinguish two basic approaches for the gender recognition problem [7,19]. The first one takes into account the full facial image (set of pixels). Then, after pre-processing, that image is a training set for the classifier (appearance-based methods). In the feature-based methods, the set of face characteristic points is a training set.

Appearance based methods are based on the values of image pixels that were previously transformed on the local or global level, e.g. at the local level, the image can be divided into lower windows or specific face regions such as mouth, nose or eyes. This approach preserves natural geometric relationships which can be used as naïve features. This solution does not require any image characteristics to be detected before the learning process starts, but its disadvantage is a relatively large set of features.

Feature-based methods require finding the facial characteristic points such as nose, mouth, eyes, ears or hair, called fiducial points [27,28]. The geometric relations between these points (fiducial distances) are used as a feature vector in the classification process. The importance of these distances in the gender discrimination/classification task are confirmed by the psychophysical studies [23,27]. This approach requires pre-processing of the image to determine the characteristic points, but in return the classifier is based on a small set of features. An approach based on characteristic points is rarely used. It is probably due to the need to take an additional step, which is to extract relevant points from the image. Nonetheless, its application can give very good results [27].

In our research, we decided to use geometric face features to limit computational complexity.

Many various classification methods can be used in a gender recognition task. The most popular classification methods include neural networks [13,17], radial basis function networks (RBF) [1], Gabor wavelets [36], Adaboost [5,32,35], Support Vector Machines (SVM) [2,8,10] or Bayesian classifiers [16,33]. All of them give comparable results (see Table 1) [19]. Comparison of different gender classification methods (Table 1) leads to the conclusion that the differences between them are minimal. Authors, regardless of the classifier used, report results at the level of 90%. Taking this into account, for our research we chose one of the most frequently used classification method - SVM.

Gender classification using geometrical features can give similar accuracy about 90% [16] to 94% [9].

To analyze a classifier's accuracy, we need a set of face examples. We can prepare database ourselves or use an existing one. Most of researchers exploit the generally available FERET database. Makinen shows that the best results have been obtained by authors using the same database to train and to test the

Table 1. Comparison of various classification methods [19]

Author	Classifier	Training data	Test data	Result [%]
Baluja [5]	Adaboost	FERET	Cross validation	94.3
Fok [17]	Neural network	FERET	Cross validation	97.2
Demirkus [14]	Bayesian	FERET	Video seqs.	90
Wang [35]	Adaboost	Mix (FERET, CAS-EAL, Yale)	Cross validation	~97
Alexandre [2]	SVM-linear	FERET	FERET	99.07
Buchala [8]	SVM-RBF	Mix (FERET, AR, BioID)	Cross validation	92.25

classifier [19]. High classification rate can be connected with similarity of the training and testing facial photos parameters.

There are several publicly available database that have been used for experiments. The most popular is the FERET database [31]. Publicly available datasets examples are AR, BioID, CAS-PEAL-R1, MORPH-2, LFW.

In our research we decided to use a part of AR face database [25] containing frontal facial images without expressions and a part of face dataset prepared by Angélica Dass in humanæ project [20].

The AR face database was prepared by Aleix Martinez and Robert Benavente in the Computer Vision Center (CVC) at the U.A.B. It contains over 4,000 color images of 126 people's faces (70 men and 56 women). Images show frontal view faces with different illumination conditions and occlusions (sun glasses and scarfs). The pictures were taken at the CVC under strictly controlled conditions, they are of 768×576 pixel resolution and of 24 bits of depth.

Humanæ project face dataset contains over 1500 color photos of different faces (men, women and children). There are only frontal view faces, prepared in the same conditions, with resolution 756×756 pixels.

The paper is organized as follows. In Sect. 2 we discuss the various strategies of SVM classifier construction. In Sect. 3 a description of facial geometrical features is presented. Section 4 describes the methodology and results of the research. A deeper analysis of the obtained results can be found in Sect. 5 as well as the paper conclusions.

2   Support Vector Machines

In this section we provide some preliminary information on Support Vector Machines [34]: how to create them, measure their accuracies as well as comparing between SVM accuracies.

2.1   Kernels

A Support Vector Machine is a type of classifier that takes input vectors and maps them non-linearly to a higher dimensional feature space. A linear decision surface is then constructed in the feature space thus allowing for the input vectors to be classified into one of two classes [6, 12]. The accuracy of an SVM is usually highly dependent on the choice of the Kernel and its parameters. Common choices for Kernels include linear, polynomial, radial basis and sigmoid functions.

2.2   Sample Data and Features

It is a goal of this paper to determine a minimal set of features to be extracted from an image of a human face so that gender can then be classified. Section 3 below gives a full description of the 9 features that were chosen for our work. 120 images were selected: 60 using female subjects and 60 using male subjects.

2.3   Cross-validation

There are several methods that can be employed to predict the accuracy of a Support Vector Machine when it is applied to an independent data set. One method is to simply split the data into training data and test data. However, this method does not, in general, give reliable results when dealing with a small set of sample data. Hence, two standard approaches were used: Leave-One-Out and k-fold cross-validation. When using k-fold cross-validation it is best to employ stratified sampling i.e. to ensure an equal number of male and female examples are in each test set.

2.4   MATLAB, Mathematica and Preliminary Settings

It was decided to create the SVMs using two different programs: MATLAB and Mathematica. This allowed for the comparison of different, kernels, optimization techniques, parameters and cross-validation methods. Preliminary experiments took place on a subset of the final sample data that consisted of 110 elements and using Radial Basis function (RBF), linear, polynomial and sigmoid kernel functions. Based on these preliminary experiments it was decided to use the Radial Basis function as the Kernel as these gave SVMs with the highest average accuracy when tested using Leave-One-Out and k-fold cross-validation. Furthermore, Bayesian Optimization is regarded as more accurate, hence we tested that in our research. In general, accuracy tested using Leave-One-Out cross-validation was higher than when 12-fold cross-validation was used. As a result, it was decided that k-fold cross-validation is accurate enough to allow the best feature set to be determined.

2.5   Comparison Tests

After approximating the accuracy rates of classifiers it is desired to determine which are the best feature sets to use. Given the same accuracy rate, an SVM that used a smaller number of features is preferable. Hence there is a need for a statistical test to determine if indeed one SVM can be assumed to have a different accuracy level to another. Several such tests exist including: tests for the difference of proportions, paired t-tests, paired differences t-test based of 10-fold cross-validation, McNemar's Test, and 5×2cv t-test. Dieterich [15] determined that some of these tests have unacceptable Type I errors and concluded the 5×2cv t-test would be the most appropriate choice. However, in a later paper Alpaydın [3] proposed a variant of this test, known as the 5×2cv F-test which gives an even lower Type I error.

To apply the 5×2cv F-test we first choose two feature sets, A & B, that we wish to compare. Next, we perform 5 iterations of 2-fold cross-validation. In each iteration, the data is randomly split into two equal sized sets: S_1 & S_2. Two SVMs are then created for feature set A and B respectively, both created using training set S_1 and the test set S_2. The error rates of these SVMs are denoted $p_A^{(1)}$ and $p_B^{(1)}$. Then two more SVMs and error rates are created where S_2 is used as the training set and S_1 as the test set. These are denoted $p_A^{(2)}$ & $p_B^{(2)}$ respectively. The difference in corresponding error rates is recorded: $p^{(1)} = p_A^{(1)} - p_B^{(1)}$ and $p^{(2)} = p_A^{(2)} - p_B^{(2)}$ and a variance, s^2, is calculated using

$$s^2 = (p^{(1)} - p^*)^2 + (p^{(2)} - p^*)^2, \tag{1}$$

where $p^* = (p^{(1)} + p^{(2)})/2$. We repeat this process 5 times giving $p_i^{(1)}$, $p_i^{(2)}$ & s_i^2 where i denotes the iteration. Finally, we produce a statistic f using the following formula

$$f = \frac{\sum_{i=1}^{5} \sum_{j=1}^{2} \left(p_i^{(j)}\right)^2}{2 \sum_{i=1}^{5} s_i^2}. \tag{2}$$

Note f is approximately F distributed with 10 and 5 degrees of freedom. The 5×2cv t-test is calculated in a similar fashion where the t statistic is calculated using the following formula

$$t = \frac{p_1^{(1)}}{\sqrt{\sum_{i=1}^{5} s_i^2/5}} \tag{3}$$

and the t follows a t distribution with 5 degrees of freedom.

3   Facial Geometrical Features - A Facial Model Using Muld

In the training process, we used a database of images originating from two different available face databases. The AR database, which we initially used contains a small number of faces. It has been extended by a number of cases from a different dataset - Humanæ project. As Makinen pointed out in [24], training the classifier on photos from only one database, made in the same, controlled conditions, adjusts the classifier to a certain type of picture. As a result, it can transpose into a very good classification results testing classifier with a part of training database and at the same time significantly worse in the case of testing with a set of photos from another source, e.g. from the Internet. It seems that constructing training datasets using more diverse photos gives better results.

In our research we took into account 11 facial characteristic points (Fig. 1): **RO** - right eye outer corner; **RI** – right eye inner corner; **LI** – left eye inner corner, **LO** – left eye outer corner; **RS** and **LS** – right and left extreme face point at eyes level; **MF** – forehead point – in the direction of facial vertical axis defined as in [21] or in [28]; **M** – nose bottom point/philtrum point; **MM** – mouth central point; **MC** – chin point; **Oec**, the anthropological facial point, that has coordinates derived as an arithmetical mean value of the points **RI** and **LI** [28].

Points were marked manually on each image. These features were described in [21,28] and are only a part of facial geometric features described in [14]. The coordinates are bounded with the anthropological facial **Oec** point. The points and distance values are recalculated in Muld [28] unit equal to the diameter of the eye. The diameter of the eye does not change in a person older than 4–5 years [30] and it measures in reality 1 Muld $= 10 \pm 0.5$ [mm].

The chosen points allow us to define 9 distances which are used as the features in the classification process. The name and ordinal number are used interchangeably. The names of the distances are identical with the name of the point not to complicate the issue, and they are:

1. MM – distance between anthropological point and mouth center.
2. MC – distance between anthropological point and chin point.
3. MC-MM – chin/jaw height.
4. MC-M – distance between nose-end point and chin point.
5. RSLS – face width at eye level.
6. ROLO – distance between outer eye corners.
7. MF-MC – face height.
8. M – distance between anthropological point and nose bottom point/philtrum point.
9. MF - distance between anthropological point and forehead point.

All the facial characteristic points were taken manually, in the same conditions, using the same feature point definitions. The accuracy of the measurements is ± 1px. It results in accuracy $\leq 5\%$ taking into account the images resolution and the eyes' sizes/ diameters.

The above 9 features have been chosen taking into account the average value and variance for males and females are distinguished. The second reason is that the chosen set of features have some anthropological invariance i.e. the outer eyes corners cannot be expanded, you can move the jaw but we took closed-mouth faces only.

In our experiments, we test classification efficiency using subsets of the set of features described above. We look for a minimal set of features that give the best classification results. We also want to check which of partial view areas A1, A2, A3 or A4 from Fig. 1 gives comparable accuracy with the full view area. The classification efficiency is the ratio of correctly classified test examples to the

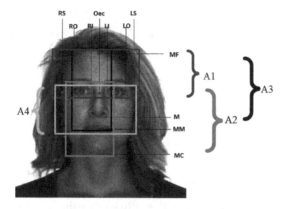

Fig. 1. Face characteristic points [23,28] (image from AR database).

total number of test examples. We train and test the classifier on the different subsets. We use cross validation as a method of result testing because the facial set consists only of 120 images.

4   Results

At the beginning we have conducted several calculations to choose a proper kernel function, as we mentioned in 2.3. We tested our dataset using SVM with: RBF, linear, polynomial and sigmoid kernel functions. There were small differences between the results, but the RBF-kernel gave always the best results approx. at least 2% better than other kernels.

4.1   Method Description

The data set used consists of 120 elements:

- 60 females – 49 from AR database and 11 from Humanæ project dataset;
- 60 males – 43 from AR database and 17 from Humanæ project dataset;

We build classifiers on j out of 9 features, where $1 \leq j \leq 9$ and systematically tried every combination of j features (the feature sets).

We also use either Leave-One-Out cross-validation or k-fold cross-validation with $k = 12$. The following describes the k-fold cross-validation method used:

1. Take 5 female and 5 male cases from the entire data set and use these as the test set.
2. Use the remaining 110 cases (55 females and 55 males) as a training set.
3. A SVM classifier is then trained using the training set with the particular j features chosen and its Classification Rate, CR, is measured using the following:
   CR = (number of correctly classified cases in the test set)/10.
   In MATLAB one of the SVMs is trained using Bayes optimization.

4. Steps 1, 2 and 3 are then repeated 12 times, each time with different elements in the test set. As a result, each element of the data-set is used in exactly one test-set.
5. The overall accuracy for a feature set is taken as the average of the 12 classification rates.

Leave-One-Out cross-validation is done in a similar way except the test set consists of just one element. Hence 120 SVMs are trained with 119 elements in the training set and the remaining 1 in the test set.

4.2   Results of Classification

In our experiments, we test classification efficiency using the subsets taken from the set of features described above. Table 2 displays the best and worst accuracies obtained by SVMs for different sized feature sets. Using MATLAB the highest accuracy found was 81.67% and positive predictive value 89%. using the three features: (1, 4, 7) and positive predictive value 89%. The highest accuracies got poorer as more features were added.

Table 2. The best and worst sets of features (MATLAB results using 12-fold cross-validation with Bayes optimization).

No. feat.	The best set	Acc. (%)	The worst set	Acc. (%)
1	(2)	75.0 A	(9)	54.2
2	(1, 2), (1, 3), (1, 9), (2, 8)	75.8 AR * A * A	(6, 9)	57.5
3	(1, 4, 7)	81.7 R*	(3, 5, 9), (6, 8, 9)	58.3
4	(1, 3, 4, 9)	80.8 A	(5, 6, 8, 9)	62.5
5	(1, 3, 4, 6, 9)	80.8 A	(3, 4, 5, 6, 9)	64.2
6	(1, 3, 4, 5, 7, 9)	80.0 A	(1, 3, 5, 6, 7, 8)	69.2
7	(1, 3, 4, 5, 7, 8, 9)	78.3 A	(1, 2, 4, 5, 6, 7, 8) (2, 3, 4, 5, 6, 8, 9) (2, 4, 5, 6, 7, 8, 9)	71.7
8	(1, 3, 4, 5, 6, 7, 8, 9) (1, 2, 3, 4, 5, 7, 8, 9)	77.5 AA	(2, 3, 4, 5, 6, 7, 8, 9) (1, 2, 3, 4, 5, 6, 8, 9)	72.5
9	(1, 2, 3, 4, 5, 6, 7, 8, 9)	77.5 A	(1, 2, 3, 4, 5, 6, 7, 8, 9)	77.5

Table 3 shows the results generated in Mathematica using 12-fold cross-validation. A SVMs with four features, (1, 2, 5, 7), produced the highest accuracy of 78.8%. Again, accuracies got poorer as more features were added. It is noteworthy that the accuracies produced by Mathematica were generally poorer that those produced by MATLAB. This is possibly due to the different optimization strategies used, because of the Bayes optimization used in the later tool.

Table 3. The best and worst sets of features (Mathematica results using 12-fold cross-validation).

No. feat.	The best set	Acc. (%)	The worst set	Acc. (%)
1	(2), (1)	72.1A, 71.8A	(1)	48.1
2	(2, 9), (2, 7), (1, 9)	72.4A, 72.2A, 72.0A*	(6, 9)	58.3
3	(1, 2, 9), (1, 4, 7), (1, 2, 7)	77.0A, 76.5R*, 76.2A	(6, 8, 9)	58.7
4	(1, 2, 5, 7), (1, 2, 5, 9), (1, 5, 7, 9)	78.8A, 78.5A, 77.9A	(3, 6, 8, 9)	64.1
5	(1, 2, 5, 7, 9), (1, 2, 3, 5, 9)	78.2A, 77.2A	(3, 5, 6, 7, 8)	66.3
6	(1, 2, 3, 5, 7, 9), (1, 2, 4, 5, 8, 9), (1, 2, 3, 6, 7, 9)	77.4A, 76.5A, 76.5A	(2, 3, 4, 5, 6, 8)	70.4
7	(1, 2, 3, 4, 5, 7, 9), (1, 2, 3, 4, 5, 8, 9)	77.2A, 76.2A	(2, 4, 5, 6, 7, 8, 9)	70.7
8	(1, 2, 3, 4, 5, 7, 8, 9)	76.0A	(1, 2, 3, 4, 5, 6, 7, 8)	73.1
9	(1, 2, 3, 4, 5, 6, 7, 8, 9)	73.5A	(1, 2, 3, 4, 5, 6, 7, 8, 9)	73.5

Table 3 shows the results generated in Mathematica using Leave-One-Out cross-validation. An SVM using four features, (1, 5, 7, 9) gave the highest accuracy of 79.2%. We note that this feature set measured an accuracy of 77.9% when using 12-fold cross-validation (Table 3).

It is noted that the accuracies found using the Leave-One-Out method are generally higher that those measured using k-fold. These discrepancies may be due to the fact that SVMs trained using Leave-One-Out are trained with more training data.

The main objective of our research was to find which features are most useful in the classification process. As we can see in Tables 2, 3 and 4 we obtained various best set of features using different training methods. However, some features are repeated in many sets. Statistically, in the best sets, the features that occur most often are 1, 3 and 9 which are measured in the vertical direction of facial axis.

We also presented the sets of features giving the worst results to show that in this case feature 6 is most common. It suggests that some features, which seems to be important, are not useful for the gender recognition, because these features might be the same for both genders.

From the Tables 3, 4, 5 and 6 it can be seen that high accuracy features that are common to all methods: (2), (1), (1, 9), (1, 2, 7). Apart from these - high accuracy features that are common to LOO CV methods (1, 2, 9), (1, 4, 7, 9); high accuracy features that are common to KFold CV methods (2, 7), (1, 4, 7), (1, 2, 9). The deeper analysis of all results shows rather significant differences in the results (up to 6–7%).

Tables 7, 8 and 9 breakdown the accuracies of the SVMs by gender. Often two feature sets are displayed; the first is the feature set that produced the highest accuracy for male inputs and the second for female inputs. If only one is displayed, then this feature set produced the highest accuracy for both genders.

Table 4. The best and worst sets of features (Mathematica results using Leave-One-Out cross-validation).

No. feat.	The best set	Acc. (%)	The worst set	Acc. (%)
1	(2), (1)	72.9A, 72.1A	(9)	37.1
2	(2, 6), (2, 3), (7, 9)	72.9A, 72.5A, 72.1A	(6, 9)	52.5
3	(1, 5, 9), (1, 2, 7), (1, 2, 9)	76.2A, 75.8A, 75.4A	(6, 8, 9)	54.2
4	(1, 5, 7, 9), (1, 3, 5, 9), (3, 4, 7, 9), (1, 4, 8, 9), (1, 4, 7, 9)	79.2A, 77.5A, 77.5A, 77.1R, 77.1A	(3, 6, 8, 9)	62.9
5	(1, 2, 5, 7, 9), (1, 3, 4, 6, 9), (3, 4, 5, 7, 9)	79.2A, 77.5A, 77.5R*	(3, 5, 6, 7, 8)	64.2
6	(1, 3, 5, 7, 8, 9), (1, 3, 4, 5, 7, 9)	78.3A, 78.3A	(1, 2, 5, 6, 7, 9)	69.6
7	(1, 2, 3, 4, 5, 7, 9)	77.9A	(2, 3, 4, 5, 7, 8, 9)	70.4
8	(1, 3, 4, 5, 6, 7, 8, 9)	76.2A	(1, 2, 3, 4, 6, 7, 8, 9)	71.7
9	(1, 2, 3, 4, 5, 6, 7, 8, 9)	70.8A	(1, 2, 3, 4, 5, 6, 7, 8, 9)	70.8

Table 5. The best and worst sets of features (MATLAB results using 12-fold cross-validation only).

No. feat.	The best set	Acc. (%)	The worst set	Acc. (%)
1	(2), (4), (1)	68.3, 65.0, 64.1	(9)	40
2	**(1, 9)** (1, 2) (2, 7) (1, 4) (3, 7)	72.5, 71.7, 70.8, 70.0, 70.0	(7, 8) (8, 9)	51.7
3	(1, 4, 9), **(1, 2, 7)**, (1, 2, 9), (1, 4, 7)	80.0, 78.3, 77.5, 77.5	(3, 4, 9) (3, 4, 6) (6, 8, 9) (3, 6, 8)	54.2, 55, 55.8
4	(1, 2, 4, 9) (1, 4, 7, 9) (1, 2, 8, 9)	80.0, 79.2, 78.3	(3, 4, 6, 7)	58.3
5	(1, 4, 7, 8, 9) (1, 2, 4, 7, 9) (1, 2, 4, 8, 9) (1, 3, 4, 8, 9)	80.0, 79.2, 78.3, 78.3	(1, 2, 4, 5, 6)	60.8
6	(1, 2, 3, 4, 8, 9) (1, 2, 4, 7, 8, 9) (1, 3, 4, 7, 8, 9)	80.8, 79.2, 79.2	(3, 4, 5, 6, 7, 8) (1, 2, 3, 5, 6, 7) (1, 2, 4, 5, 6, 8)	65.0, 65.8, 65.8
7	(1, 2, 3, 4, 7, 8, 9)	80.8 A	(1, 2, 3, 4, 5, 6, 8) (1, 2, 3, 4, 5, 6, 7)	65.0, 66.7
8	(1, 2, 3, 4, 5, 7, 8, 9)	74.2 AA	(1, 3, 4, 5, 6, 7, 8, 9) (1, 2, 3, 4, 5, 6, 8, 9)	69.2
9	(1, 2, 3, 4, 5, 6, 7, 8, 9)	81.7	(1, 2, 3, 4, 5, 6, 7, 8, 9)	81.7

Table 6. The best and worst sets of features (MATLAB results using Leave-One-Out cross-validation only).

No. feat.	The best set	Acc. (%)	The worst set	Acc. (%)
1	(3), (2), (1)	65.0, 64.1, 64.1	(9)	40.8
2	(1, 2) (1, 4) (1, 9) (1, 3)	73.3, 72.5, 72.5, 71.7	(7, 8)	49.1
3	(1, 4, 9) (1, 2, 9) (1, 4, 7) (1, 2, 7) (2, 3, 7)	81.7, 79.2, 79.2, 78.3, 78.3	(3, 4, 9)	53.3
4	(1, 4, 7, 9) (1, 2, 4, 7) (1, 2, 4, 9) (1, 2, 3, 9) (1, 2, 7, 9) (1, 3, 4, 9) (1, 3, 8, 9) (2, 3, 4, 7)	80.8, 80.0, 80.0, 79.2, 79.2, 79.2, 79.2, 79.2	(3, 4, 6, 7)	56.7
5	(1, 2, 4, 7, 9) (1, 2, 3, 4, 9) (1, 3, 4, 7, 9) (1, 4, 7, 8, 9)	80.8, 80.0, 80.0, 80.0	(3, 4, 5, 6, 7)	60.0
6	(1, 2, 4, 7, 8, 9) (1, 2, 3, 7, 8, 9) (1, 2, 3, 4, 7, 9) (1, 2, 3, 4, 8, 9) (1, 3, 4, 7, 8, 9)	81.7, 80.8, 80.0, 80.0, 80.0	(1, 2, 3, 4, 5, 6)	62.5
7	(1, 2, 3, 4, 7, 8, 9)	80.8	(1, 2, 3, 4, 5, 6, 8) (1, 2, 3, 4, 5, 6, 7)	65.0, 66.7
8	(1, 2, 3, 4, 6, 7, 8, 9) (2, 3, 4, 5, 6, 7, 8, 9)	75.8	(1, 2, 3, 4, 5, 6, 7, 8)	70.8
9	(1, 2, 3, 4, 5, 6, 7, 8, 9)	84.2	(1, 2, 3, 4, 5, 6, 7, 8, 9)	84.2

Table 7. The best results for male and female (MATLAB results using 12-fold cross-validation with Bayes optimization).

No. feat.	Set of features	Acc. male (%)	Acc. fem. (%)	Set of features	Acc. fem. (%)	Acc. male (%)
1	(2)	83.3	66.7	(8)	76.7	53.3
2	(1, 2), (1, 3), (2, 8) (2, 9)	85.0	66.765.0	(1, 9)	75.0	76.7
3	(1, 2, 4) (1, 4, 7)	88.3	61.775.0	(1, 4, 7)	75.0	88.3
4	(1, 2, 5, 7)	91.7	68.3	(1, 4, 7, 9)	78.3	80.0
5	(1, 5, 7, 8, 9) (2, 3, 4, 5, 9) (2, 3, 4, 6, 9)	90.0	70.070.0 65.0	(1, 3, 4, 6, 9)	78.3	83.3
6	(1, 2, 3, 6, 8, 9)	91.7	66.7	(1, 3, 4, 5, 7, 9)	76.7	83.3
7	(1, 2, 3, 4, 5, 7, 9) (1, 2, 4, 6, 7, 8, 9)	88.3	61.765.0	(1, 2, 3, 4, 6, 8, 9) (1, 2, 3, 4, 7, 8, 9)	75.0	80.080.0
8	(1, 3, 4, 5, 6, 7, 8, 9)	86.7	68.3	(1, 2, 3, 4, 5, 6, 7, 9)	70.0	81.7
9	(1, 2, 3, 4, 5, 6, 7, 8, 9)	88.3	66.7	(1, 2, 3, 4, 5, 6, 7, 8, 9)	88.3	66.7

Table 8. The best results for male and female (Mathematica results using Leave-One-Out cross-validation).

No. feat.	Set of features	Acc. male (%)	Acc. fem. (%)	Set of features	Acc. fem. (%)	Acc. male (%)
1	(2), (1)	81.7, 72.5	64.2, 71.7	(9)	34.2	40
2	(2, 6), (5, 8)	80.0, 63.3	65.8, 70.0	(6, 9), (6, 8)	53.3, 56.7	51.7, 50.0
3	(1, 5, 9), (1, 2, 7)	81.7, 78.3	70.8, 73.3	(6, 8, 9)	54.2	54.2
4	(1, 5, 7, 9), (1, 3, 7, 9)	85.8, 80.0	72.5, 73.3	(5, 6, 8, 9), (3, 6, 8, 9)	60.8, 65.0	65.8, 60.8
5	(1, 2, 5, 7, 9) (3, 4, 5, 7, 9)	85.8, 80.8	72.5, 74.2	(1, 5, 6, 7, 8), (3, 5, 6, 7, 8)	65.8, 68.3	65.0, 60.0
6	(1, 3, 5, 7, 8, 9), (1, 3, 4, 5, 7, 9)	84.2, 83.3	72.5, 73.3	(2, 3, 5, 6, 7, 9), (1, 2, 5, 6, 7, 9)	73.3, 74.2	67.5, 65.0
7	(1, 2, 3, 4, 5, 7, 9)	85.0	70.8	(2, 3, 4, 5, 7, 8, 9) (1, 2, 3, 4, 5, 6, 8)	73.3, 76.7	67.5, 65.0
8	(1, 2, 3, 4, 5, 6, 8, 9), (1, 3, 4, 5, 6, 7, 8, 9)	82.5, 80.0	66.7, 72.5	(1, 2, 3, 4, 6, 7, 8, 9), (1, 2, 3, 4, 5, 6, 8, 9)	75.0, 82.5	68.3, 66.7
9	(1, 2, 3, 4, 5, 6, 7, 8, 9)	75.0	66.7	(1, 2, 3, 4, 5, 6, 7, 8, 9)	75.0	66.7

Table 9. The best sets of features for male and female (Mathematica results using 12-fold cross-validation).

No. feat.	Set of features	Acc. male (%)	Acc. fem. (%)	Set of features	Acc. fem. (%)	Acc. male (%)
1	(2), (1)	79.4, 72.4	64.9, 71.2	(9)	47.8	48.3
2	(1, 4), (1, 9)	79.0, 74.1	64.7, 69.9	(3, 8), (6, 8)	59.0, 63.1	59.4, 56.0
3	(2, 5, 7), (3, 7, 9)	82.9, 75.8	67.6, 73.8	(6, 7, 8), (6, 8, 9)	59.6, 61.7	62.6, 55.8
4	(1, 2, 5, 7), (1, 5, 7, 9)	85.4, 80.9	72.2, 74.9	(3, 6, 7, 8), (3, 5, 6, 9)	63.8, 66.9	66.3, 62.4
5	(1, 2, 5, 7, 9)	84.0,	72.4	(1, 5, 6, 7, 8), (3, 5, 6, 7, 8)	67.8, 70.5	68.2, 62.1
6	(1, 2, 4, 6, 8, 9), (1, 2, 3, 5, 7, 9)	82.9, 82.3	69.2, 72.6	(2, 5, 6, 7, 8, 9), (2, 3, 4, 5, 6, 8)	74.7, 75.3	67.8, 65.5
7	(1, 2, 3, 4, 5, 7, 9)	81.4	73.1	(2, 4, 5, 6, 7, 8, 9), (1, 2, 3, 4, 5, 6, 8)	75.6, 76.5	65.8, 65.6
8	(1, 2, 3, 4, 6, 7, 8, 9), (1, 2, 3, 4, 5, 7, 8, 9)	81.3, 80.6	69.7, 71.4	(1, 2, 3, 5, 6, 7, 8, 9), (1, 2, 3, 4, 5, 6, 8, 9)	77.2, 79.7	70.1, 67.9
9	(1, 2, 3, 4, 5, 6, 7, 8, 9)	77.7	69.2	(1, 2, 3, 4, 5, 6, 7, 8, 9)	77.7	69.2

We first note that nearly all SVMs were better at classifying males. If one focuses on the SVMs that produce the highest accuracy for females, then we see that the accuracy for males with these SVMs is still usually higher but the gap between them is lower.

5 Discussion of the Results and Conclusions

It is noted that the same feature sets produce different accuracies depending on whether MATLAB or Mathematica was used and also whether Leave-One-Out or k-fold cross-validation were used. This suggests that there is high variance in the accuracy rates and it would be better to have more data to work with.

In general, MATLAB found more accurate SVMs than Mathematica (when given the same dataset) which suggests that a researcher should ensure they try many different tools when finding SVMs on a dataset. As a consequence, the features found, that give high accuracy SVMs, differed depending on whether MATLAB or Mathematica was used.

It was decided to test whether the discrepancy in accuracy levels found could be put down to chance. As a result, the accuracy levels found using all feature sets were compared to the results found using the feature set (1, 5, 7, 9). These comparisons were made using the 5×2cv F-test and the 5×2cv t-test. The results for the F-test are also displayed in Tables 2, 3 and 4 as 'A' or 'R' beside the accuracy rates. An 'A' implies that one should accept the Null Hypothesis at the 5% Level and an 'R' implies you should reject it. The Null Hypothesis states that the error rate is the same for the two SVMs. The results imply that you should accept that the accuracy levels of most of the top performing SVMs should be considered the same. If the t-test gave the opposite result to the F-test then a '*' was placed beside the letter. In most cases the two tests give the same result. These tests imply that there is no significant difference in the accuracy levels produced by most feature sets.

As we can see in Tables 7, 8 and 9, the classification accuracy for male and female is not symmetrical. We obtained better results for the male in all cases. However, SVMs that produce the highest accuracy for females have an even higher accuracy for males, but lower gap between the accuracy rates. This has a bearing on how one should choose an overall best SVM as there appears to be a trade-off between choosing an SVM that has the best chance of classifying males, females or both. So, when it comes to choosing the best feature set, it may be better to choose an SVM that has the highest accuracy in classifying females – this will depend on the relative sizes of the genders in the underlying population.

In our research we noticed, that there are cases poorly classified always or almost always, independently of the SVM construction method. This applies to a greater extent to female than male. It leads to the conclusion that some females have many male traits, and vice versa a number of men exhibit a lot of feminine features. We obtained 3 females and 1 male cases having 0% classification rate. There are also several cases where classification rate is under 5%. It seems that such examples should be added to the training set to facilitate the subsequent classification of such "difficult" faces.

In the paper, we show that the classifier constructed on the base of only two or three geometric facial features can give satisfactory (though not always optimal) results with accuracy 81.7% and positive predictive value 86.5%. We show that classification using a part of the face is also possible. In our methodology, the iris of one eye must be visible in order to correctly scale the remaining values. However, the rest of the face does not have to be visible at the picture entirety. For example, facial width does not have a significant impact on the classification accuracy. Good results were given by sets of features using only the lower or upper part of the face (e.g. (1,9) partial area A3). This means that with a certain area of the face covering the eye, we can find a set of features that despite the incompleteness of the data will be able to classify the photo with accuracy 75.8% (lower by 5.9% than for the full facial view), and positive predictive value 76.3%.

References

1. Abdi, H., Valentin, D., Edelman, B., O'Toole, A.J.: More about the difference between men and women: evidence from linear neural network and principal component approach. Neural Comput. **7**(6), 1160–1164 (1995)
2. Alexandre, L.A.: Gender recognition: a multiscale decision fusion approach. Pattern Recogn. Lett. **31**(11), 1422–1427 (2010)
3. Alpaydin, E.: Combined 5 × 2cv F test for comparing supervised classification learning algorithms. Neural Comput. **11**(8), 1885–1892 (1999)
4. Andreu, Y., Mollineda, R.A., Garcia-Sevilla, P.: Pattern Recognition and Image Analysis. LNCS, vol. 5524. Springer, Heidelberg (2009). https://doi.org/10.1007/978-3-642-02172-5
5. Baluja, S., Rowley, H.A.: Boosting sex identification performance. Int. J. Comput. Vis. **71**(1), 111–119 (2007)
6. Boser, B.E., Guyon, I.M., Vapnik, V.N.: A training algorithm for optimal margin classifiers. In: Proceedings of 5th Annual Workshop on Computational Learning Theory COLT-1992, p. 144 (1992)
7. Brunelli, R., Poggio, T.: Face recognition: features versus templates. IEEE Trans. Pattern Anal. Mach. Intell. **15**(10), 1042–1052 (1993)
8. Buchala, S., Loomes, M.J., Davey, N., Frank, R.J.: The role of global and feature based information in gender classification of faces: a comparison of human performance and computational models. Int. J. Neural Syst. **15**, 121–128 (2005)
9. Burton, A.M., Bruce, V., Dench, N.: What's the difference between men and women? Evidence from facial measurements. Perception **22**, 153–176 (1993)
10. Castrillon, M., Deniz, O., Hernandez, D., Dominguez, A.: Identity and gender recognition using the encara real-time face detector. In: Conferencia de la Asociacin Espaola para la Inteligencia Artificial, vol. 3 (2003)
11. Castrillon-Santana, M., Lorenzo-Navarro, J., Ramon-Balmaseda, E.: On using periocular biometric for gender classification in the wild. Pattern Recogn. Lett. **82**, 181–9 (2016)
12. Cortes, C., Vapnik, V.: Support-vector network. Mach. Learn. **20**(3), 273–297 (1995)
13. Cottrell, G.W., Metcalfe, J.: EMPATH: face, emotion, and gender recognition using holons. In: Lippmann, R., Moody, J.E., Touretzky, D.S. (eds.) Proceedings of Advances in Neural Information Processing Systems (NIPS), vol. 3, pp. 564–571. Morgan Kaufmann (1990)
14. Demirkus, M., Toews, M., Clark, J.J., Arbel, T.: Gender classification from unconstrained video sequences. In: 2010 IEEE Computer Society Conference on Computer Vision and Pattern Recognition Workshops (CVPRW), pp. 55–62 (2010)
15. Dietterich, T.G.: Approximate statistical tests for comparing supervised classification learning algorithms. Neural Comput. **10**, 1895–1923 (1998)
16. Fellous, J.M.: Gender discrimination and prediction on the basis of facial metric information. Vis. Res. **37**(14), 1961–1973 (1997)
17. Fok, T.H.C., Bouzerdoum, A.: A gender recognition system using shunting inhibitory convolutional neural networks. In: 2006 IEEE International Joint Conference on Neural Network Proceedings, pp. 5336–5341 (2006)

18. Hasnat, A., Haider, S., Bhattacharjee, D., Nasipuri, M.: A proposed system for gender classification using lower part of face image. In: Proceedings of International Conference on Information Processing, pp. 581–585 (2015)
19. Hassanat, A.B., Prasath, V.B.S., Al-Mahadeen, B.M., Alhasanat, S.M.M.: Classification and gender recognition from veiled-faces. Int. J. Biometr. 9(4), 347–364 (2017)
20. Humanæ Project. http://humanae.tumblr.com. Accessed 15 Nov 2017
21. Jain, A., Huang, J., Fang, S.: Gender identification using frontal facial images. In: IEEE International Conference on Multimedia and Expo, ICME 2005, p. 4 (2005)
22. Kawano, T., Kato, K., Yamamoto, K.: An analysis of the gender and age differentiation using facial parts. In: IEEE International Conference on Systems Man and Cybernetics, vol. 4, pp. 3432–3436, 10–12 October 2005
23. Kompanets, L., Milczarski, P., Kurach, D.: Creation of the fuzzy three-level adapting brainthinker. In: 6th International Conference on Human System Interaction (HSI), pp. 459–465 (2013). https://doi.org/10.1109/HSI.2013.6577865
24. Mäkinen, E., Raisamo, R.: An experimental comparison of gender classification methods. Pattern Recogn. Lett. **29**, 1544–56 (2008)
25. Martinez, A.M., Benavente, R.: The AR face database. CVC Technical report #24 (1998)
26. Merkow, J., Jou, B., Savvides, M.: An exploration of gender identification using only the periocular region. In: Proceedings of 4th IEEE International Conference on Biometrics Theory Applications and Systems BTAS, pp. 1–5 (2010)
27. Milczarski, P.: A new method for face identification and determining facial asymmetry. In: Katarzyniak, R., et al. (eds.) Semantic Methods for Knowledge Management and Communication. SCI, vol. 381, pp. 329–340. Springer, Heidelberg (2011). https://doi.org/10.1007/978-3-642-23418-7_29
28. Milczarski, P., Kompanets, L., Kurach, D.: An approach to brain thinker type recognition based on facial asymmetry. In: Rutkowski, L., Scherer, R., Tadeusiewicz, R., Zadeh, L.A., Zurada, J.M. (eds.) ICAISC 2010. LNCS (LNAI), vol. 6113, pp. 643–650. Springer, Heidelberg (2010). https://doi.org/10.1007/978-3-642-13208-7_80
29. Moghaddam, B., Yang, M.H.: Learning gender with support faces. IEEE Trans. Pattern Anal. Mach. Intell. **24**(5), 707–711 (2002)
30. Muldashev, E.R.: Whom Did We Descend From? OLMA Press, Moscow (2002). (in Russian)
31. Phillips, P.J., Moon, H., Rizvi, S.A., Rauss, P.J.: The FERET evaluation methodology for face-recognition algorithms. IEEE Trans. Pattern Anal. Mach. Intell. **22**(10), 1090–1104 (2000)
32. Shakhnarovich, G., Viola, P.A., Moghaddam, B.: A unified learning framework for real time face detection and classification. In: Proceedings of International Conference on Automatic Face and Gesture Recognition (FGR 2002), pp. 14–21. IEEE (2002)
33. Sun, Z., Bebis, G., Yuan, X., Louis, S.J.: Genetic feature subset selection for gender classification: a comparison study. In: Proceedings of IEEE Workshop on Applications of Computer Vision (WACV 2002), pp. 165–170 (2002)
34. Vapnik, V.N., Kotz, S.: Estimation of Dependences Based on Empirical Data. Springer, New York (2006). https://doi.org/10.1007/0-387-34239-7
35. Wang, J.G., Li, J., Lee, C.Y., Yau, W.Y.: Dense SIFT and Gabor descriptors-based face representation with applications to gender recognition. In: 11th International Conference on Control Automation Robotics & Vision (ICARCV), no. December, pp. 1860–1864 (2010)

36. Wiskott, L., Fellous, J.-M., Krüger, N., von der Malsburg, C.: Face recognition by elastic bunch graph matching. In: Sommer, G., Daniilidis, K., Pauli, J. (eds.) CAIP 1997. LNCS, vol. 1296, pp. 456–463. Springer, Heidelberg (1997). https://doi.org/10.1007/3-540-63460-6_150

37. Yamaguchi, M., Hirukawa, T., Kanazawa, S.: Judgment of gender through facial parts. Perception **42**, 1253–1265 (2013)

Parallel Cache Efficient Algorithm and Implementation of Needleman-Wunsch Global Sequence Alignment

Marek Pałkowski[✉], Krzysztof Siedlecki, and Włodzimierz Bielecki

Faculty of Computer Science and Information Systems,
West Pomeranian University of Technology in Szczecin,
Zolnierska 49, 71210 Szczecin, Poland
{mpalkowski,ksiedlecki,wbielecki}@wi.zut.edu.pl
http://www.wi.zut.edu.pl

Abstract. An approach allowing us to improve the locality of a parallel Needleman-Wunsch (NW) global sequence alignment algorithm is proposed. The original NW algorithm works with an arbitrary gap penalty function and examines all possible gap lengths. To compute the score of an element of an NW array, cells gap symbols are looked back over entire row and column as well as one adjacent cell. We modified the NW algorithm so to read cells only with the row-major order by means of forming a copy of the transposed scoring array. The loop skewing technique is used to generate parallel code. A formal parallel NW algorithm is presented. Experimental results demonstrate super-linear speed-up factor of the accelerated code due to considerable increasing code locality on a studied modern multi-core platform.

Keywords: Needleman-Wunsch algorithm
Global sequence alignment · Cache efficiency · Loop skewing
Bioinformatics

1 Introduction

Sequence alignment is a fundamental and well studied problem in bioinformatics. The score of an alignment is determined using a matching (or scoring) matrix that assigns a score to each pair of characters from the alphabet in use as well as a gap penalty model that determines the penalty associated with a gap sequence. The first dynamic programming algorithm for global sequence alignment was introduced in 1970 by Needleman and Wunsch [7]. The original algorithm works with an arbitrary gap penalty function $\gamma(n)$ and has much better sensitivity and specificity, but requires cubic computation time. Hence, a sequential version of the algorithm is impractical for long queries and/or database sequences.

Optimization and parallelization of computational biology dynamic programming algorithms and applications is still a challenging task for developers and

© Springer International Publishing AG, part of Springer Nature 2018
L. Rutkowski et al. (Eds.): ICAISC 2018, LNAI 10842, pp. 207–216, 2018.
https://doi.org/10.1007/978-3-319-91262-2_19

researchers. Increase in computing power depends not only on the number of cores available but also on organization of cores and cache in multi-processors.

In this paper, we propose optimization and evaluation of a parallel version of the Needleman-Wunsch algorithm (NW) that is cache efficient and scalable. The presented approach introduces a copy of a transposed scoring matrix to replace column-major order array access by only row-major one. Next, so modified algorithm is parallelized by means of loop skewing. A formal parallel algorithm increasing code locality is presented.

Experimental results demonstrate that increasing cache efficiency significantly reduces computation time in comparison to that of the corresponding serial code. We compare also the performance of parallel versions of the NW algorithm based on loop skewing with that of a program without the proposed locality improvement. Code speed-up is measured for a multi-core platform Intel Xeon 2699 v.3.

The rest of the paper is organized as follows. Section 2 introduces global sequence alignment realized with the NW algorithm. Section 3 presents techniques exposing parallelism and increasing locality in this algorithm. Implementation details and memory models are discussed. Section 4 presents results of experiments, which demonstrate that proposed modifications make the algorithm dramatically faster than the original one for modern processors due to significant reducing cache misses. This section discusses the following factors: time of code execution, code speed-up, and code scalability. Related work is discussed in Sect. 5. Section 6 concludes this paper considering future work.

2   The Needleman-Wunsch Algorithm Description

The Needleman-Wunsch algorithm is typically used to perform global alignment of protein or nucleotide sequences. The algorithm was developed by Needleman and Wunsch and published in 1970 [7]. It was one of the first applications of dynamic programming to determine similarity between sequences. Dynamic programming is a method for solving complex problems by breaking them down into simpler sub-problems. The NW algorithm is still widely used for optimal global alignment, particularly when the quality of the global alignment is fundamental.

Algorithm 1 formalizes the NW algorithm which aligns two sequences a and b of length M and N, respectively. A scoring matrix F is first initialized: (i) cell $(0,0)$ is zeroed and (ii) the first row and first column are subject to gap penalty.

The next step is carrying out a recursion filling matrix F (see lines 9 to 14). A score is calculated as the best possible score (i.e. highest) from existing scores to the left, top or top-left (diagonal). The scoring system $s(a_i, b_j)$ is a similarity score of elements a_i, b_j that constitute the two sequences. It describes scores when two letters are the same, differential, or one letter aligns to a gap in the other string. Scores can be negative. A penalty of a gap, γ_k, has length k and is a component of scores from upper and side cells.

The final step of the NW algorithm is traced back to generate the best global alignment. The step starts with the cell at the lower right of the matrix and

---

Algorithm 1. The original Needleman-Wunsch algorithm with arbitrary gap penalties

---

Input: Two sequences a and b of length M and N, respectively; scoring matrix $\sigma(a,b)$; gap penalty function $\gamma(n)$.
Output: Dynamic programming matrix F.
1: **Initialization:**
2:    $F(0,0) = 0$.
3:    **for** i=1 **to** M **do**
4:        $F(i,0) = \gamma(i)$.
5:    **end for**
6:    **for** i=1 **to** N **do**
7:        $F(0,i) = \gamma(i)$.
8:    **end for**
9: **Recursion:**
10:    **for** i=1 **to** M **do**
11:        **for** j=1 **to** N **do**

12:            $F(i,j) = \max \begin{cases} F(i-1,j-1) + \sigma(a_i, b_j), \\ \max_{1 \leq k < i} (F(i-k,j) - \gamma(k)), \\ \max_{1 \leq k < j} (F(i,j-k) - \gamma(k)). \end{cases}$

13:        **end for**
14:    **end for**

---

continues to the top left cell. Because the optimal path tracing is of linear-time complexity, we do not consider any optimization of the corresponding code in this paper.

Algorithm 1 presents the recursion for arbitrary gap penalties, called also general gap penalty functions which are useful when sequence alignment is penalised too much with the linear gap penalty function [11]. Hence, the gap weight function can be anything (this is the most general case) and allows us to incorporate more complicated and physically realistic user-defined functions into the algorithm in a facile manner. However, the computations require cubic time and makes the original NW algorithm not useful in practice. This motivates us to develop a way of an efficient acceleration of the general NW recursion algorithm.

3   Methods

To carry out a (data) cache miss analysis, we focus on read and write misses of array F and ignore misses due to the reads of sequences. Computing each cell of array F goes through iterations within column-major and row-major to look back previous gaps (line 12 of Algorithm 1). Furthermore, adjacent cell $F(i-1, j-1)$ is required. Figure 1 presents those cells in green used for computing yellow cell (i,j).

To improve code locality, we form another array (see Fig. 1a) in a such a manner that the column-order major (red rectangle) is changed to the row-order

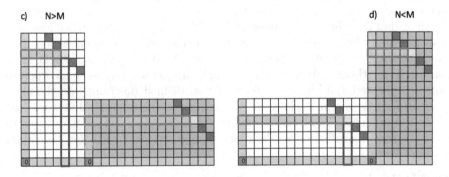

Fig. 1. Parallel filling the NW array with the row-major order using a copy of the transposed array and loop skewing. (Color figure online)

major in the copy of an original array F, i.e., the transposition of the array is carried out. A value of cell (i,j) must be also put into the cell in the transposed array with coordinate $(M + 1 + j, i)$. Hence, we can read side cells for next ones (Fig. 1b).

Algorithm 2. Cache efficient recursion of the NW algorithm

9: **Recursion:**
10: **for** i=1 **to** M **do**
11:     **for** j=1 **to** N **do**

12:     $$F(i,j) = \max \begin{cases} F(i-1, j-1) + \sigma(a_i, b_j), \\ \max_{1 \le k < i}(F(M+j, i-k) - \gamma(k)), \\ \max_{1 \le k < j}(F(i, j-k) - \gamma(k)), \end{cases}$$

13:         $F(M+j, i) = F(i,j)$.
14:     **end for**
15: **end for**

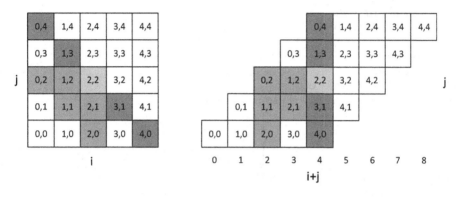

Fig. 2. Loop skewing of the NW filling

The transposed array is flipped left-right and up-down. The first cell $(M + 1, 0)$ is zeroed with $(0,0)$. The column $(M+1, i)$ is assigned as row $(i, 0)$ with $\gamma(i)$ for $1 \le i \le M$ whereas the row $(M+1+i, 0)$ is assigned as column $(0, i)$ for $1 \le i \le N$.

Algorithm 2 shows a modified recursion of the NW algorithm providing cache improvements.

The total memory cost is defined with the $2 * ((N+1) \times (M+1))$ array size. Figures 1(c) and (d) present array F when $N > M$ and $M > N$, respectively. The shape of the NW array is formed with the two rectangular arrays. However, it can be allocated dynamically. It is worth noting that a dynamic memory implementation with pointers is easier when $N \ge M$ (see Fig. 1(c)). In such a case, the first column includes pointers indicating rows with $N+M$ and M sizes. The algorithm allows for $N \ge M$ swapping sequences if it is required.

Next, we apply the well-known affine transformation, loop skewing [12], to parallelize the algorithm. We can "skew" the NW loop-nest so that the inner-loop iterates over the statement instances along the array diagonal (see Fig. 2). So, blue statement instances get executed in one iteration of the outer loop and the outer loop proceeds at the 45° angle direction. Now the inner loop can be parallelized. One way to think about loop-skewing is that we align the iteration

---

Algorithm 3. The cache efficient Needleman-Wunsch algorithm with loop skewing applied

---

Input: Two sequences a and b of length M and N, respectively; gap penalty function $\gamma(n)$.

Output: Dynamic programming matrix F.

1: **Initialization:**
2:    $F(0,0) = 0$.
3:    $F(M+1,0) = 0$.
4:    **for** i=1 **to** M **do**
5:       $F(i,0) = \gamma(i)$.
6:       $F(M+1, i) = \gamma(i)$.
7:    **end for**
8:    **for** i=1 **to** N **do**
9:       $F(0,i) = \gamma(i)$.
10:      $F(M+1+i,0) = \gamma(i)$.
11:   **end for**
12: **Recursion:**
13:    **for** i=1 **to** $N+M-1$ **do**
14:       **parfor** j=max(0, i-M) **to** min(N-1, i-1) **do**
15:
$$F(i-j,j+1) = \max \begin{cases} F(i-j,j) + \sigma(a_{i-j}, b_{j+1}), \\ \max_{1 \le k < i-j} (F(M+j+1, i-j-k) - \gamma(k)), \\ \max_{1 \le k < j+1} (F(i-j, j+1-k) - \gamma(k)). \end{cases}$$
16:          $F(M+j+1, i-j) = F(i-j, j+1)$.
17:       **end parfor**
18:    **end for**

---

space coordinates and re-label the statement instances in the new coordinates as follows

$$\begin{bmatrix} 1 & 1 \\ 0 & 1 \end{bmatrix} \begin{bmatrix} i \\ j \end{bmatrix} = \begin{bmatrix} i+j \\ j \end{bmatrix}.$$

It is worth noting that the iterator of the outer loop is formed as $i + j$ and it enumerates time partitions $1, 2, \ldots, (N + M - 1)$, while the inner loop scans the iterations belonging to the $i + j$ diagonal. Based on this observation, we can modify the NW recursion being careful about the loop-bounds since the iteration space of a new loop nest is no longer "rectangular". Summing up, we may conclude that loop skewing serializes the outermost loop, which satisfies all dependences available in the original loop nest, leaving the inner loop free to be parallelized.

Applying loop skewing is demonstrated in Fig. 1: the independent cells placed in the two blue diagonals, they read cells and write values safely (see Fig. 1(b)) within the original and transposed arrays F for the both conditions $N > M$ and $M < N$ (Fig. 1(c) and (d), respectively).

Algorithm 3 presents a formalized NW procedure with using loop skewing and locality improvement. In the loop nest, which implements the recursion (Lines

12 to 18), the outer loop is serial, while the inner one is parallel. It is worth observing that the statements shown in Lines 15 and 16 are the same as those in the code presented in Algorithm 2. In the target loop nest, the loop bounds are modified and variables (i, j) are replaced with expressions $(i - j, j + 1)$ according to the loop skewing technique.

4 Related Work

To accelerate sequence alignment, fast sequence-alignment heuristics have been developed, for example BLAST [2] and FASTA [10]. However, the speed-up that is obtained with such methods is often provided at the cost of the algorithm sensitivity. Another direction of research, also aimed at speeding sequence alignment, has been the development of parallel and cache efficient algorithms.

A Parallel programming framework for multi-core DNA sequence alignment was discussed in paper [1]. The presented methodology is based on a slicing procedure of both the query and database sequence. Zao and Sahni [14] consider best alignment using a single array $F[0:n]$. They introduce score, swap and strip algorithms to reduce cache misses and computing complexity. However, arbitrary gap penalties are not analysed in those papers and studied alignments are limited to a linear model.

Well-known cache optimization techniques, like loop permutation or vectorization [8], cannot be applied to the NW loop nest. Polyadic cells reading and non-uniform dependences patterns limit many classical transformations. Loop tiling [13], implemented in modern compilers [3,4,9], is able to increase cache efficiency for dynamic programming kernels. However, tile size searching (TSS) is required to expose the best available cache efficiency.

The Nussinov RNA folding algorithm is another dynamic programming kernel of bioinformatics with a similar dependence pattern. Chang et al. [5] accelerated that algorithm by using the lower triangle of the recurrence array to store the transpose of computed values in the upper triangle. Li et al. [6] improved cache locality for the Chang recurrence by locating values in the same row instead of the same column. Reading values in a row is more cache efficient than reading values in a column. In this paper, we accelerated the NW algorithm in the similar manner. However, the proposed NW algorithm uses all cells in the NW array (not only upper triangle as the Nussinov RNA folding) thanks to the use a copy of the transposed NW array.

5 Experiments

This section presents results of an experimental study aimed at estimation of speed-up of codes implementing the NW algorithm with and without locality improvement. To carry out experiments, we have used a machine with two processors Intel Xeon E5-2699 v.3 (2.3 GHz base and 3.6 GHz turbo, 18 cores/36 threads, 45 MB Cache) and 128 GB RAM. All codes were compiled with the -O3 flag of optimization and auto-vectorization by means of the Intel C++ Compiler (*icc* 17.0.1). Times are measured in seconds. We consider two randomly generated sequences of the same length, $N = M$.

First, we carried out evaluations using from one to 64 threads for sequences of length 3000 (see Fig. 3(a)). We compared execution time of parallel original and optimized codes implementing the NW algorithm. Parallelization was carried out by means of loop skewing. Obtained results show that optimized code is cache efficient and scalable, it overcomes the performance of the original parallel code. Optimized code demonstrates super-linear speed-up even for one thread. Figure 3(b) presents speed-up as the ratio of the execution time of the original serial code to that of parallel codes with and without locality improvement (blue and orange curves, respectively).

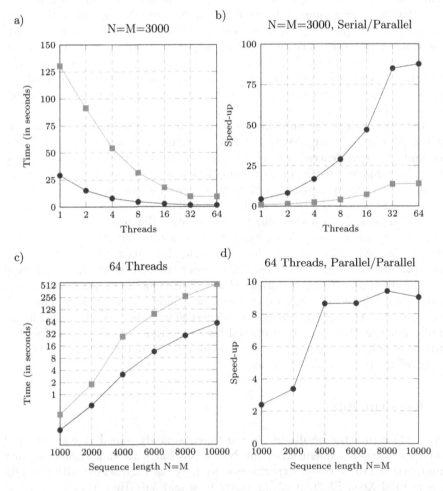

Fig. 3. Times and speed-ups of the accelerated NW algorithm.

Experiments were carried out also for six different lengths (from 1000 to 10000) using 64 threads (see Fig. 3(c)[1]). We can observe a dramatic code acceleration, particularly for longer sequences. Figure 3(d) depicts speed-up as the ratio of the original parallel NW code execution time to that of the cache efficient parallel code.

Summing up, we conclude that the presented approach successfully reduces time of execution and cache-misses for the NW recursion. The presented experiments show that computation complexity can be significantly reduced for the NW algorithm with arbitrary gap penalties and its practical usage should be reconsidered for modern multi-core architectures.

6   Conclusion

In this paper, we presented a cache efficient NW algorithm. Loop skewing was used to parallelize that algorithm. To compute a score of the NW array, cells gap symbols are looked back only within the row major order. For generated optimized code, we demonstrated that additional memory cost is acceptable in practice and super-linear code speed-up is obtained. We carried out evaluations for 64 threads on modern processors Intel Xeon 2699 v.3.

The contribution of this paper over the closely related work [6,15] is that the suggested algorithm is parallel while the related ones are serial. This allows us to increase algorithm and its implementation performance.

In future, we are going to develop a source-to-source compiler framework allowing for automatic improving similar codes according to the presented way. The following tasks should be resolved: (i) row and column major order detection, (ii) transformation legality checkness, (iii) procedure formalization, (iv) valid code generation. We strongly believe that such a solution would be useful to automatically optimize other computationally intensive kernels in bioinformatics. We are going also to study applying the presented approach to multiple sequence alignment solutions.

Acknowledgments. Thanks to the Miclab Team (miclab.pl) from the Technical University of Czestochowa (Poland) that provided access to high performance multi-core machines for the experimental study presented in this paper.

References

1. de Almeida, T.J.B.M., Roma, N.F.V.: A parallel programming framework for multi-core DNA sequence alignment. In: International Conference on Complex, Intelligent and Software Intensive Systems, pp. 907–912, February 2010
2. Altschul, S.F., Gish, W., Miller, W., Myers, E., Lipman, D.: A basic local alignment search tool. J. Mol. Biol. **215**, 403–410 (1990)
3. Bielecki, W., Palkowski, M.: Perfectly nested loop tiling transformations based on the transitive closure of the program dependence graph. Soft Comput. Comput. Inf. Sci. **342**, 309–320 (2015)

---

[1] To depict two time lines with large disparities, we used a logarithmic time axis.

4. Bondhugula, U., Hartono, A., Ramanujam, J., Sadayappan, P.: A practical automatic polyhedral parallelizer and locality optimizer. SIGPLAN Not. **43**(6), 101–113 (2008)
5. Chang, D.J., Kimmer, C., Ouyang, M.: Accelerating the Nussinov RNA folding algorithm with CUDA/GPU. In: 10th IEEE International Symposium on Signal Processing and Information Technology, pp. 120–125, December 2010
6. Li, J., Ranka, S., Sahni, S.: Multicore and GPU algorithms for Nussinov RNA folding. BMC Bioinform. **15**(8), S1 (2014). https://doi.org/10.1186/1471-2105-15-S8-S1
7. Needleman, S.B., Wunsch, C.D.: A general method applicable to the search for similarities in the amino acid sequence of two proteins. J. Mol. Biol. **48**(3), 443–453 (1970)
8. Padua, D.A. (ed.): Encyclopedia of Parallel Computing. Springer, New York (2011)
9. Palkowski, M., Klimek, T., Bielecki, W.: TRACO: an automatic loop nest parallelizer for numerical applications. In: Federated Conference on Computer Science and Information Systems, FedCSIS 2015, Lódz, Poland, 13–16 September 2015, pp. 681–686 (2015). https://doi.org/10.15439/2015F34
10. Pearson, W.R., Lipman, D.J.: Rapid and sensitive protein simlarity searches. Science **227**, 1435–1441 (1985)
11. Smith, T., Waterman, M.: Identification of common molecular subsequences. J. Mol. Biol. **147**(1), 195–197 (1981)
12. Wolfe, M.: Loops skewing: the wavefront method revisited. Int. J. Parallel Prog. **15**(4), 279–293 (1986)
13. Xue, J.: Loop Tiling for Parallelism. The Springer International Series in Engineering and Computer Science, vol. 575. Springer, New York (2000). https://doi.org/10.1007/978-1-4615-4337-4. https://books.google.pl/books?id=DPJNwR2SBF0C
14. Zhao, C., Sahni, S.: Cache and energy efficient alignment of very long sequences. In: IEEE 5th International Conference on Computational Advances in Bio and Medical Sciences (ICCABS), pp. 1–6 (2015)
15. Zhao, C., Sahni, S.: Cache and energy efficient algorithms for Nussinov's RNA folding. BMC Bioinform. **18**(15), 518 (2017)

Using Fuzzy Numbers for Modeling Series of Medical Measurements in a Diagnosis Support Based on the Dempster-Shafer Theory

Sebastian Porebski$^{(\boxtimes)}$ ⑩ and Ewa Straszecka ⑩

Faculty of Automatic Control, Electronics and Computer Science,
Institute of Electronics, Silesian University of Technology, Gliwice, Poland
{sebastian.porebski,ewa.straszecka}@polsl.pl

Abstract. This work concern attempts to model imprecise symptoms in the medical diagnosis support tools. Patient's self-check is very important, particularly in chronic diseases. In hypertension or diabetes patients record measurements. Still, these measurements are made in different circumstances, thus they are imprecise. A physician takes into account rather a trend in a series of measurements to diagnose a patient. Till now, knowledge engineers' approach is different since they often use a single value as input information of a decision support system. In this work, a series of measurements is modeled as a fuzzy number. The main purpose of the presented approach is to check whether it is possible to replace a single measurement with a series of measurements in the diagnosis support system and to examine the impact of this change on the diagnosis process. Preliminary results show that use of the fuzzy number in determining the diagnosis may increase its certainty and can be profitable when used in real medical problems.

Keywords: Medical diagnosis support · Series of measurements
Imprecise information · Dempster-Shafer theory · Fuzzy numbers

1 Introduction

One of the crucial challenges in medical diagnosis support systems is an information imperfection [3]. Significant part of the imperfection is associated with symptom imprecision. We deal with imprecise symptoms by carrying out measurements that are always performed with finite precision but also when the patient states his condition with imprecise terms, e.g. *"frequent fainting"* or *"severe dizziness"*. There is no doubt that modeling of such statements implies using the theory of fuzzy sets [8]. Moreover, diagnostic tests or measurements are imprecise also when performed by a patient himself. We can observe this phenomenon, especially when repeating blood pressure or glucose level measurements at short time intervals. It would be desirable to adjust the diagnosis support tools to deal with series of measurements using an appropriate model.

© Springer International Publishing AG, part of Springer Nature 2018
L. Rutkowski et al. (Eds.): ICAISC 2018, LNAI 10842, pp. 217–228, 2018.
https://doi.org/10.1007/978-3-319-91262-2_20

The Dempster-Shafer theory (DST) [12] makes it possible to use uncertain knowledge and to support a decision with the belief measure [17]. It allows capturing the uncertainty of knowledge by modeling a focal element, which is a premise of a diagnostic rule. The focal element can be described by a fuzzy membership function. Hence, the fuzzy focal element represents imprecision of a medical symptom [13]. The mechanism of diagnosis support using the DST is related to a calculation of the belief measure about a given diagnosis based on the available knowledge [13]. More precisely, the belief measure equals the sum of the basic probability of every focal element which is a characteristic symptom for the diagnosis and which is confirmed for the diagnosed patient. This approach can be successfully used in the medical diagnosis support [10]. There are many approaches to joining the Dempster-Shafer and the fuzzy sets theories (e.g. [1,2,4,14,15]). The method that is used in the present problem is particularly medical diagnosis-oriented. In this problem, symptoms may be imprecise i.e. not exactly matching that formulated in medical guidelines. They can also be fraught with measurement errors. On the other hand, diagnostic rules are evaluated by the basic probability assignment as an uncertain measure. This evaluation allows for tuning significance of rules to population data.

Some previous studies present a calculation of the belief measure when both an input (observation) and a focal element (knowledge) are modeled with fuzzy sets [5,6,9,16]. Yet, they do not concern measurements. In this work, simulation results will be presented, which prove that processing series of diagnostic measurements as a piece of fuzzy information allows strengthening the diagnosis confidence. It is a good reason for modeling diagnostic measurements as imprecise information using fuzzy numbers [7].

This work is organized in the following way. The first part of the work presents our approach to modeling a series of measurements by a fuzzy number in detail. Then, synthetic patients data set is used to define diagnostic rules that are fuzzy focal elements related to one symptom. The use of one symptom allows tracking of calculations. After that, three ways of fuzzy set inclusion calculation will be presented since they are necessary for use when symptoms are now fuzzy numbers. Simulation results allow comparing the belief measure when it is calculated for a single patient data and for a modeled series of measurements. Moreover, the belief measure is also used to determine the diagnosis based on a pair of measurements (systolic and diastolic pressure), which directly relates to real-life problems, where the diagnosis is based on set of symptoms.

2 Materials and Methods

2.1 Data Notation

Let us denote by $D^{(l)}$ the label of the l-th diagnosis, $l = 1, \cdots, C$. In the simplest case, the diagnoses are: "the patient needs the medicine" (or a dose increase) and "the patient does not need any medicament" (or the dose is accurate). If X is the medical data set containing information about N patients it can be defined in the following way:

$$X = \begin{bmatrix} x_1 \ x_2 \ \cdots \ x_i \ \cdots \ x_N \\ d_1 \ d_2 \ \cdots \ d_i \ \cdots \ d_N \end{bmatrix}, \tag{1}$$

where x_i is the measurement (or a diagnostic test result) of the i-th patient and d_i is the diagnosis index assigned to this patient. For C different diagnoses in the considered medical problem, $d_i \in \{1, \cdots, C\}$. We can denote the set of measurements concerning the l-th diagnosis ($l = 1, \cdots, C$) in the following way:

$$X^{(l)} = \{x_i \in X \,|\, d_i = l\}. \tag{2}$$

Let us suppose that we dispose K repeated measurements (or diagnostic test results) of x_i and denote them as the vector:

$$\hat{x}_i = [x_{i1}, x_{i2}, \cdots, x_{ik}, \cdots, x_{iK}], \tag{3}$$

where x_{ik} is the k-th evaluation of the x_i, $k = 1, \cdots, K$.

2.2 Modeling of Imprecise Information

A fuzzy number $\tilde{x}_i$ which represents $\hat{x}_i$ can be defined by the following triangular membership function:

$$\tilde{\mu}_i(x) = \begin{cases} 0, & \text{if } x \le a, \\ \frac{x-a}{b-a}, & \text{if } a < x \le b, \\ \frac{c-x}{c-b}, & \text{if } b < x \le c, \\ 0, & \text{if } x > c, \end{cases} \tag{4}$$

where a, b, c are parameters and x is the measurement domain. We can use the mean and the standard deviation of $\hat{x}_i$ (denoted as $\bar{x}_i$ and σ_i, respectively) to obtain a, b, c:

$$a = \bar{x}_i - \beta \cdot \sigma_i, \quad b = \bar{x}_i, \quad c = \bar{x}_i + \beta \cdot \sigma_i \tag{5}$$

where β is a coefficient that let us to change the support of $\tilde{\mu}_i(x)$. An exemplar fuzzy number created according to the $\hat{x}_i$ (3) is presented together with the histogram of its data in Fig. 1.

The $\tilde{\mu}_i(x)$ membership function models a series of measurements made by a patient. In a similar way a membership function that model knowledge may be designed using training data [13]. For instance, for the blood pressure three fuzzy sets "low", "normal" and "high" can be determined, with membership functions $\mu_j^{(l)}(x)$, presented in Fig. 2. In this approach one fuzzy membership function is defined for each diagnosis. Most often, two competing diagnoses are considered, but in this work, three diagnoses are taken into account (hence $l = 1, 2, 3$) which also confirms the real-life diagnosis [11].

These sets concern diagnoses e.g.: "patient needs treatment against hypotension" (or "a dose of a medicine is too high"), "patient does not need treatment" (or "a dose of a medicine is accurate") and "patient needs treatment against

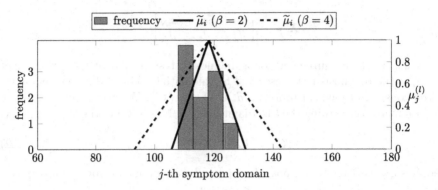

Fig. 1. A histogram of 10 measurements sample and fuzzy numbers $\widetilde{\mu}_i$ that model the sample (two different β parameters)

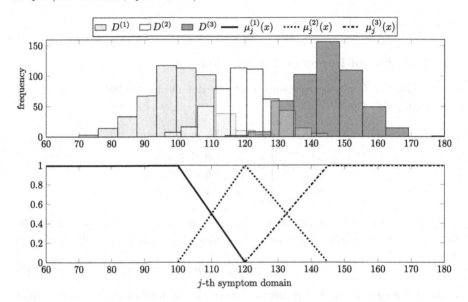

Fig. 2. Histogram of generated data related to three diagnoses $D^{(l)}$, $l = 1, \cdots, 3$ and fuzzy membership functions $\mu_j^{(l)}$ for fuzzy focal elements created according to the generated data

hypertension" (or "a dose of a medicine is too low"). The $\mu_j^{(l)}(x)$, $l = 1, 2, 3$ membership functions can be also designed by an expert [13], unlike the $\widetilde{\mu}_i(x)$, which always results from measurements. Three membership functions $\mu_j^{(l)}$, $l = 1 \cdots, 3$ are ordered to the mean values of each $\boldsymbol{X}^{(l)}$. Let us suppose that mean values of $\boldsymbol{X}^{(l)}$ ($\overline{x}^{(l)}$) are ordered in such a way that $\overline{x}^{(1)} < \overline{x}^{(2)} < \overline{x}^{(3)}$. Hence the $\mu_j^{(l)}(x)$ can be defined in the following way:

$$\mu_j^{(1)}(x) = \begin{cases} 1, & \text{if} \quad x \le \overline{x}^{(1)}, \\ \frac{\overline{x}^{(2)} - x}{\overline{x}^{(2)} - \overline{x}^{(1)}}, & \text{if} \quad \overline{x}^{(1)} < x \le \overline{x}^{(2)}, \\ 0, & \text{otherwise}, \end{cases} \tag{6}$$

$$\mu_j^{(2)}(x) = \begin{cases} \frac{x - \overline{x}^{(1)}}{\overline{x}^{(2)} - \overline{x}^{(1)}}, & \text{if} \quad \overline{x}^{(1)} < x \le \overline{x}^{(2)}, \\ \frac{\overline{x}^{(3)} - x}{\overline{x}^{(3)} - \overline{x}^{(2)}}, & \text{if} \quad \overline{x}^{(2)} < x \le \overline{x}^{(3)}, \\ 0, & \text{otherwise}, \end{cases} \tag{7}$$

$$\mu_j^{(3)}(x) = \begin{cases} \frac{x - \overline{x}^{(3)}}{\overline{x}^{(3)} - \overline{x}^{(2)}}, & \text{if} \quad \overline{x}^{(2)} < x \le \overline{x}^{(3)}, \\ 1, & \text{if} \quad x > \overline{x}^{(3)}, \\ 0, & \text{otherwise}. \end{cases} \tag{8}$$

The main difference between the membership functions for a fuzzy number $\tilde{\mu}_i(x)$ (4) and for focal elements $\mu_j^{(l)}(x)$ (6–8) is that the focal elements can be defined by functions with infinite support (for extreme symptom values – $\mu_j^{(1)}(x)$ and $\mu_j^{(3)}(x)$ in Fig. 2), therefore they require a separate definition.

2.3 Diagnosis Support Based on the Belief Measure

In DST [12], the knowledge consists of focal elements. The set of focal elements can be denoted as $S^{(l)}$. Each fuzzy focal element $s_j^{(l)} \in S^{(l)}$ is a logical statement regarding the symptom characteristic for the l-th diagnosis. A single focal element, referring to one symptom and the l-th diagnosis, can be for instance: $s_j^{(l)} \equiv$ "systolic pressure is high", with the $\mu_j^{(3)}(x)$ membership function.

The $s_j^{(l)}$ is assigned with the basic probability value $m_j^{(l)}$ that expresses uncertainty that $s_j^{(l)}$ supports the l-th diagnosis. The basic probability assignment for the whole $S^{(l)}$ must fulfill the following constraints [12]:

$$m(f) = 0 \qquad \sum_{s_j^{(l)} \in S^{(l)}} m_j^{(l)} = 1, \tag{9}$$

The first condition of (9) requires that the symptom f which is not related to the diagnosis obtain zero probability value. The second condition ensures that the sum of the basic probability of all focal elements related to one diagnosis is equal to one.

The belief measure of the l-th diagnosis, according to the input information and knowledge of fuzzy focal elements, is a sum of the following form [9]:

$$Bel^{(l)}(s^{(*)}) = \sum_{s_j^{(l)} \in S^{(l)}} I(s^{(*)} \subset s_j^{(l)}) \cdot m_j^{(l)}, \tag{10}$$

where $s^{(*)}$ can be a single value (a singleton corresponding a measurement result) or a fuzzy set (corresponding a series of measurements) in the general case. Intuitively, the belief measure sums up the probability value of all confirmed symptoms (focal elements) that are considered for l-th diagnosis. In the first case, the $s^{(*)}$ is modeled by the singleton $\mu_s(x; x_i) = 1$, while in the second by $\tilde{\mu}_i(x)$ (4). The both cases, generally, can be denoted as $\mu^{(*)}(x)$.

The $I(s^{(*)} \subset s_j^{(l)})$ is the inclusion measure. If DST is generalized for fuzzy focal elements $s_j^{(l)}$, different inclusion measures can be proposed:

$$I_I(s^{(*)} \subset s_j^{(l)}) = \frac{\min\limits_{x \in supp(s^{(*)})} [1, 1 + (\mu_j^{(l)}(x) - \mu^{(*)}(x))]}{\max\limits_{x \in supp(s^{(*)})} \mu^{(*)}(x)}, \tag{11}$$

$$I_Y(s^{(*)} \subset s_j^{(l)}) = \min\limits_{x \in supp(s^{(*)})} [\max(\overline{\mu}^{(*)}(x), \mu_j^{(l)}(x))], \tag{12}$$

$$I_O(s^{(*)} \subset s_j^{(l)}) = \frac{\sum\limits_{x \in supp(s^{(*)})} \min[\mu_j^{(l)}(x), \mu^{(*)}(x)]}{\sum\limits_{x \in supp(s^{(*)})} \mu^{(*)}(x)}, \tag{13}$$

where $supp(s^{(*)})$ is the set of x values for which $\mu^{(*)}(x) > 0$. The inclusions I_I, I_Y and Y_O are proposed in [6], [16] and [9], respectively.

Three inclusion measures process fuzzy input $s^{(*)}$ and the fuzzy focal element $s_j^{(l)}$ in different ways. They influence Bel values, so their performance should be tested. When $s^{(*)}$ is a single value, the singleton $\mu_s(x; x_i) = 1$ is used in (10) in the place of $\mu^{(*)}(x)$. It implies the same result for each proposed inclusion measure (11)–(13). This can be proved in the following way. Let us denote the single measurement as x_i. Then: $\mu^{(*)}(x_i) = 1, \overline{\mu}^{(*)}(x_i) = 0$, and $0 \leq \mu_j^{(l)}(x_i) \leq 1$. In this way:

$$I_I(x_i \subset s_j^{(l)}) = \frac{\min[1, 1 + (\mu_j^{(l)}(x_i) - \mu^{(*)}(x_i))]}{\mu^{(*)}(x_i)} \tag{14}$$

$$= \min[1, \mu_j^{(l)}(x_i)] = \mu_j^{(l)}(x_i),$$

$$I_Y(x_i \subset s_j^{(l)}) = \max(\overline{\mu}^{(*)}(x), \mu_j^{(l)}(x_i)) = \max(0, \mu_j^{(l)}(x_i)) = \mu_j^{(l)}(x_i), \tag{15}$$

$$I_O(x_i \subset s_j^{(l)}) = \frac{\min[\mu_j^{(l)}(x_i), \mu^{(*)}(x_i)]}{\mu^{(*)}(x_i)} = \min[\mu_j^{(l)}(x_i), 1] = \mu_j^{(l)}(x_i). \tag{16}$$

According to (14)–(16), we can see that:

$$Bel^{(l)}(x_i = \{x_i\}) = \sum\limits_{s_j^{(l)} \in S^{(l)}} \mu_j^{(l)}(x_i) \cdot m_j^{(l)}. \tag{17}$$

In the first step of the study, we do not discuss the influence of the basic probability assignment (9) on the diagnosis. Therefore, we will further consider the trivial case of only one symptom affecting the Bel value, which implies $m_j = 1$, $j = 1, 2, 3$. Next, an experiment for assumed values of the probability is performed. A wider discussion of the basic probability can be found in other authors' works, e.g. [10, 13].

3   Experiments

In this point, the goal is to check how different $Bel^{(l)}(s^{(*)})$ measure values can be obtained when $s^{(*)}$ is a single measurement value x_i (1) and when it is the vector of measurements $\hat{x}_i$ modeled by (4). These input data values will be changed to observe $Bel^{(l)}(s^{(*)})$ values for the whole symptom domain and each of the proposed inclusion measures (11)–(13). Since belief measure calculated for single input information x_i is the same (17), it is not illustrated separately (their shapes are the same as the $\mu_j^{(l)}$ presented in Fig. 2).

3.1   Simulation Procedure

Simulations are performed in the following steps:

1. Three data sets $\boldsymbol{X}^{(l)}$ are generated as 500 realizations of random variable of normal distribution for each diagnosis ($l = 1, \cdots, 3$). Probability density function parameters (mean and standard deviation) are set to generate non-trivial data (they cannot be linearly separated). Hence, mean values are 100, 120 and 145 for $l = 1, \cdots, 3$. Standard deviation value is equal to three for each probability density function.
2. According to (6–8) three membership functions $\mu_j^{(l)}(x)$ are defined. They are used to represent $s_j^{(l)} \in S^{(l)}$. Number of considered symptoms is one in the study, hence $j = 1$ and there is one focal element in each $S^{(l)}$ and $m_j^{(l)} = 1$, $l = 1, \cdots, 3$.
3. A single measurement is modeled as a singleton. Its position is changed in the whole symptom domain to obtain Bel value for the whole symptom domain.
4. Series of measurements $\hat{x}_i$ (3) is the set of 10 randomly chosen measurement values of one diagnosis and modeled by the fuzzy number $\tilde{x}^{(*)}$ using the membership function (4). The fuzzy number position is also changed in the whole range of the symptom domain.
5. Kołmogorow-Smirnov test (K-S) is performed to find out matches between measurements $\hat{x}_i$ and distributions modeling blood pressures for three diagnoses. This is done to obtain the intervals, for which $\hat{x}_i$ can be considered as coming from these distributions. In the intervals, Bel values for different inclusions are compared.

4   Results

Results are illustrated by figures. Thick lines (solid, dotted and dotted-dashed) in Figs. 3, 4 and 5 are belief measure values calculated for the fuzzy number $\tilde{x}_i$ defined by $\tilde{\mu}_i(x)$ for $\beta = 2$ (see (4)–(5)). This fuzzy number is well-fitted to the sample data (see Fig. 1). On the other hand belief measure values are also calculated when series of measurements (3) is modeled by not very accurate fuzzy number (Fig. 1 for $\beta = 4$). When results of the measurement series are less consistent then the fuzzy number has a very wide support, in worst cases even covering supports of both diagnoses membership functions. In the latter case, both diagnoses will be almost equally supported and the final diagnosis cannot be elaborated. If measurements are not so inconsistent, the fuzzy number can be modeled by the (Fig. 1 for $\beta = 2$) membership function. Belief measure values are presented by thin lines (solid, dotted and dotted-dashed) in Figs. 3, 4 and 5. The fuzzy numbers are moved along the whole range of the symptom value. In this way values of belief measures for different matching cases of data and knowledge are obtained. These values are represented by thin and thick lines equivalent to the input membership functions. Grey areas in Figs. 3, 4 and 5 indicate the intervals for which $\hat{x}_i$ pass the K-S test and come from normal distribution as the generated data of three diagnoses (Fig. 2).

Figures 3, 4 and 5 allow us comparing effects of different inclusion measures (11)–(13) on the $Bel^{(l)}(s^{(*)})$ calculation. Since we use only one focal element in each diagnosis, each belief measure value is equal to the appropriate inclusion measure value (11)–(13). Hence, they can be directly compared. When the fuzzy number $\widetilde{x}_i$ is well-fitted to the sample data (Fig. 1) for Ishizuka's measure (11), changes of belief measures are the same as the shape of the appropriate fuzzy membership function $\mu_j^{(l)}(x)$. The inclusion measure proposed by Yager (12) does not seem to be very useful since $Bel^{(l)}(s^{(*)})$ is never equal to one. The reason is the property of the Yager's inclusion: it can be equal to one only when fuzzy number $\widetilde{x}_i$ is entirely included in the focal element defined by $\mu_j^{(l)}(x) = 1$ for $x = supp(\widetilde{x}_i)$. Ogawa's measure (13) obtains the greatest belief values for intervals where samples are considered as consistent with normally distributed data (when they fit K-S test).

Belief calculation is also compared when fuzzy number rather poorly models the series of measurements (Fig. 1, for $\beta = 4$). All thin lines in (Figs. 3, 4 and 5) indicate that the fuzzy number $\widetilde{x}_i$, when it is not matched properly to $\hat{x}_i$, it results in the belief measure value reduction. Moreover, significant shift of the highest $Bel^{(l)}(s^{(*)})$ value is observed for Ishizuka's measure (Fig. 3). Nonetheless, Ogawa's measure (Fig. 5) still provides the highest $Bel^{(l)}$ values for the intervals indicated by the K-S test.

The last experiment is performed for a complex problem. Let us suppose that we dispose of a blood pressure result which is related to the systolic and diastolic pressure value. Assume that we obtain diastolic blood pressure values in the same way as systolic blood pressure values explained in the point Sect. 3.1. This time 500 values are generated as normally distributed data for three diagnoses when mean values are equal 55, 70 and 95 for $l = 1, 2, 3$ and systolic blood pressure data are not changed. In this way, we can create the fuzzy focal element set $S^{(l)}$ that includes not only two single focal elements $(s_1^{(l)}, s_2^{(l)})$ related individually to the systolic and diastolic blood pressure, but also the complex focal element $(s_3^{(l)})$ that is related to a combination of these symptoms, e.g. "systolic pressure is low and diastolic pressure is low". Basic probability value is assigned to each focal element $j = 1, 2, 3$ for each diagnosis $l = 1, 2, 3$. We can assign following values of the basic probability $m_j^{(l)}$ (9):

$$m_1^{(1)} = 0.53,\ m_2^{(1)} = 0.33,\ m_3^{(1)} = 0.14,$$
$$m_1^{(2)} = 0.34,\ m_2^{(2)} = 0.37,\ m_3^{(2)} = 0.29, \tag{18}$$
$$m_1^{(3)} = 0.31,\ m_2^{(3)} = 0.55,\ m_3^{(3)} = 0.24.$$

In Figs. 3, 4 and 5 different belief shapes are presented. When we consider two symptom domains, belief measure shape should be illustrated in three dimensions. In this paper we only present the difference between diagnosis decision performed when $Bel^{(l)}$ is calculated for two symptoms represented by two single values and two fuzzy numbers. These are compared in the Fig. 6. Diagnoses for the input given as a single value and as a fuzzy set are similar. Thus, we cannot spoil the diagnosis by using series of data as the input. Still, if measurements are imprecise, the fuzzy input, which rather resembles a median than a mean, can be of value. We can also observe a softer change among diagnoses.

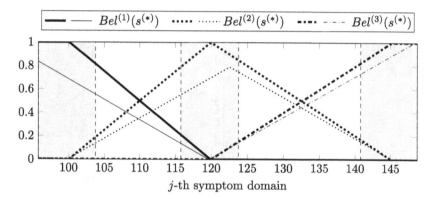

Fig. 3. Belief measure values calculated for three diagnoses when input data is modeled by two different fuzzy numbers (Ishizuka's inclusion measure). See description in text.

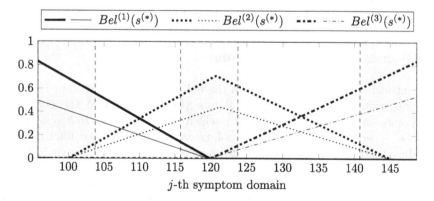

Fig. 4. Belief measure values calculated for three diagnoses when input data is modeled by two different fuzzy numbers (Yager's inclusion measure). See description in text.

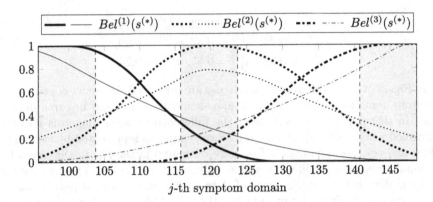

Fig. 5. Belief measure values calculated for three diagnoses when input data is modeled by two different fuzzy numbers (Ogawa's inclusion measure). See description in text.

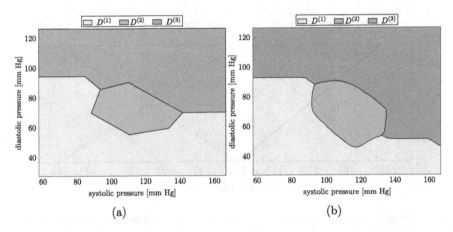

Fig. 6. Diagnosis decision based on belief calculation when blood pressure measurements are: (a) single values, (b) fuzzy numbers (Ogawa's inclusion measure)

5   Discussion and Conclusions

Different inclusion measures influence the belief measure calculation. However, when observing the belief measure shapes (Figs. 3, 4 and 5) the characteristic points can be noticed (these for which $Bel^{(1)}(s^{(*)}) = Bel^{(2)}(s^{(*)})$ and $Bel^{(2)}(s^{(*)}) = Bel^{(3)}(s^{(*)})$). These points are common, hence final decision according to the maximal belief measure value is the same for each inclusion measure.

Nonetheless, Ogawa's measure (13) seems to be the most suitable to handle a fuzzy input information since it does not reduce the belief measure value when the fuzzy number accurately matches knowledge. It is clear that when the fuzzy input and fuzzy focal element match to each other, we should obtain rather high confidence. Only Ishizuka's measure (11) results in the maximal belief value, still it occurs for a narrow symptom value interval and this result is not different from using the single input value. Three inclusion measures influence the calculation cost. Ogawa's inclusion calculation time is ca. 60–50% faster than Ishizuka's but it is ca. 20% longer than Yager's.

Simulation results let us judge that modeling the series of measurements with the fuzzy number is profitable if an imprecise symptom is obtained as a result of repeated diagnostic procedures. This approach is ready to use in the real medical data problems. Data of real measurements will be soon gathered. However, the simulations are based on our experience in biomedical electronics and partly collected data. A suitable representation of a symptom by a fuzzy number must be carefully performed. The membership function ill-fitting the series of measurements spoils the results.

When more symptoms influence diagnosis, i.e. $j > 1$, then the Bel value is the sum of their $m_j^{(l)}$ weighted by the inclusion of inputs in knowledge ($\mu_j^{(l)}$ in $s_j^{(l)}$). This may improve the diagnosis which is based on more extended information than just single measurement or average. Simultaneously, properties of inclusion measures have even a stronger impact on the final conclusion than membership functions. We hope that observations will help to choose the right method of modeling the diagnosis.

Acknowledgements. This research is financed from the statutory funds (BKM-510/Rau-3/2017 & BK-232/Rau-3/2017) of the Institute of Electronics of the Silesian University of Technology, Gliwice, Poland.

References

1. Casanovas, M., Merigo, J.M.: Fuzzy aggregation operators in decision making with Dempster-Shafer belief structure. Expert Syst. Appl. **39**(8), 7138–7149 (2012)
2. Chai, K.C., Tay, K.M., Lim, C.P.: A new method to rank fuzzy numbers using Dempster-Shafer theory with fuzzy targets. Inf. Sci. **346**, 302–317 (2016)
3. Esfandiari, N., Babavalian, M.R., Moghadam, A.-M.E., Tabar, V.K.: Knowledge discovery in medicine: current issue and future trend. Expert Syst. Appl. **41**(9), 4434–4463 (2014)
4. Ghasemini, J., Ghaderi, R., Mollaei, M.R.K., Hojjatoleslami, S.A.: A novel fuzzy Dempster-Shafer inference system for brain MRI segmentation. Inf. Sci. **223**, 205–220 (2013)
5. Hwang, C.M.: Belief and plausibility functions on intuitionistic fuzzy sets. Int. J. Intell. Syst. **31**(6), 556–568 (2016)
6. Ishizuka, M.: Inference procedures under uncertainty for the problem-reduction method. Inf. Sci. **28**(3), 179–206 (1982)
7. Jiang, W., Yang, W., Luo, Y., Qin, X.Y.: Determining basic probabilisty assignment based on the improved similarity measures of generalized fuzzy numbers. Int. J. Comput. Commun. Control **10**(3), 333–347 (2015)
8. Liao, H., Xu, Z., Zeng, X.-J., Merigo, J.M.: Qualitative decision making with correlation coefficients of hesitant fuzzy linguistic term sets. Knowl. Based Syst. **76**, 127–138 (2015)
9. Ogawa, H., Fu, K.S., Yao, J.T.P.: An inexact inference for damage assessment of existing structures. Int. J. Man-Mach. Stud. **22**(3), 295–306 (1985)

10. Porebski, S., Straszecka, E.: Extracting easily interpreted diagnostic rules. Inf. Sci. **426**, 19–37 (2018)
11. Porwik, P., Orczyk, T., Lewandowski, M., Cholewa, M.: Feature projection k-NN classifier model for imbalanced and incomplete medical data. Biocybern. Biomed. Eng. **36**(4), 644–656 (2016)
12. Shafer, G.: A Mathematical Theory of Evidence. Princeton University Press, New Jersey (1976)
13. Straszecka, E.: Combining knowledge from different sources. Expert Syst. **27**(1), 40–52 (2010)
14. Tang, H.: A novel fuzzy soft set approach in decision making based on grey relational analysis and Dempster-Shafer theory of evidence. Appl. Soft Comput. **31**, 317–325 (2015)
15. Wang, J., Hu, Y., Xiao, F., Deng, X., Deng, Y.: A novel method to use fuzzy soft sets in decision making based on ambiguity measure and Dempster-Shafer theory of evidence: an application in medical diagnosis. Artif. Intell. Med. **69**, 1–11 (2016)
16. Yager, R.R.: Generalized probabilities of fuzzy events from fuzzy belief structures. Inf. Sci. **28**(192), 45–62 (1982)
17. Yager, R.R.: On the fusion of imprecise uncertainty measures using belief structures. Inf. Sci. **181**(15), 3199–3209 (2011)

Averaged Hidden Markov Models in Kinect-Based Rehabilitation System

Aleksandra Postawka$^{(\boxtimes)}$ and Przemysław Śliwiński

Faculty of Electronics, Wroclaw University of Science and Technology,
Wrocław, Poland
{aleksandra.postawka,przemyslaw.sliwinski}@pwr.edu.pl

Abstract. In this paper the Averaged Hidden Markov Models (AHMMs) are examined for the upper limb rehabilitation purposes. For the data acquisition the Microsoft Kinect 2.0 sensor is used. The system is intended for low-functioning autistic children whose rehabilitation is often based on sequences of images presenting the subsequent gestures. The number of such training sets is limited and the preparation of a new one is not available for everyone, whereas each child requires the individual therapy. The advantage of the presented system is that new activities models could be easily added.

The conducted experiments provide satisfactory results, especially in the case of single hand rehabilitation and both hands rehabilitation based on asymmetric gestures.

Keywords: Autistic children · Rehabilitation
Hidden Markov Models · Averaged Hidden Markov Models
Microsoft Kinect 2.0 · Depth sensor

1 Introduction

The necessity of rehabilitation is usually considered to be useful only after accidents or injuries. However, it is also needed in some cognitive impairments, such as autism. The rehabilitation of low-functioning autistic children is often based on sequences of images presenting the subsequent gestures to be performed. The number of such rehabilitation sets is limited and the preparation of a new one is not available for everyone. It is a well known fact that every autistic person is different and needs the individual approach [1–3]. Hence, the individuals with autism may need significantly different rehabilitation exercises.

In the literature there are numerous examples of rehabilitation systems. The one of mostly considered issues is a post-stroke rehabilitation as that strokes are the often a cause of motor deficits. Regenbrecht *et al.* proposed the system ART for the treatment of upper limb dysfunctions [4]. The system uses the augmented reality and computer games based approach in order to increase patients motivation. The post-stroke rehabilitation system has been also developed by Kuttuva *et al.* [5]. In this tool called Rutgers Arm the virtual reality (VR) technology

© Springer International Publishing AG, part of Springer Nature 2018
L. Rutkowski et al. (Eds.): ICAISC 2018, LNAI 10842, pp. 229–239, 2018.
https://doi.org/10.1007/978-3-319-91262-2_21

is used. The another example is a game-based system for upper limbs rehabilitation presented by Pastor *et al.* [6]. For post-stroke patients rehabilitation the wearable wireless sensors have been used as well [7].

Along with the development of the depth sensors technology the Microsoft Kinect became an increasingly used device in rehabilitation. Clark *et al.* have confirmed Kinect to be successfully used for assessing the postural control [8]. Scherer *et al.* developed the Kinect-based system for injured athletes [9]. Another example for using Kinect in rehabilitation is a system Kinere, which has been applied for patients suffering from (1) severe cerebral palsy and (2) acquired muscle atrophy [10]. Kusaka et al. used Kinect to develop a rehabilitation system for patients with hemiplegia [11]. The Kinect-based rehabilitation system for home usage was designed by Su *et al.* [12]. Kinect was also used in the system intended for patients with body scheme dysfunctions and left-right confusion [13].

Despite the great number of research dealing with rehabilitation, we believe that there is still a lack of the system which would make it possible to easily add a new rehabilitation task. Such a feature would be very useful in the very individualized in their nature therapies for autistic persons.

The other motivation for this research is that Hidden Markov Models (HMMs) have been never used, to the best of authors knowledge, in the task of rehabilitation, in spite of the fact that the left-to-right HMM [14] models, which preserve the information about the observation symbols order, seem to be a valuable tool for the purpose of motion tracking and evaluation. The Averaged HMMs [15], composed of multiple left-to-right HMMs, combine the features of all component models, thus the most commonly occurring features could be retrieved from such a final model. Moreover, the observation symbol distribution in states is a property that describes movements' noise and uncertainty in a very natural way.

- First of all the symbol distribution in AHMM contains only motions that really occurred in the therapists movements.
- Secondly, the multitude of learning sequences ensures that a wide variety of proper movements is included.

In this paper the usefulness of Averaged Hidden Markov Models (AHMMs) in the rehabilitation exercises have been examined. The 13 activity AHMM models have been used for the upper limbs rehabilitation. For the data acquisition the Microsoft Kinect 2.0 depth sensor is used.

The paper is organized as follows. The notation is introduced in Sect. 2. The Sect. 3 contains the description of methods used in the research. The application and usage examples have been presented in Sect. 4. The Sect. 5 contains the overall conclusions and plans for the future.

2 Notation

In the paper the following notation is used for HMMs:

$\lambda = \{A, B, \pi\}$ - the complete parameter set for HMM,
N - the number of states,
M - the number of observation symbols,
T - the length of observation sequence,
$A = \{a_{ij} : i, j \in \{1, \ldots, N\}\}$ - the state transition matrix,
$B = \{b_{ij} : i \in \{1, \ldots, N\}, j \in \{1, \ldots, M\}\}$ - the probability distribution matrix for observed symbols,
$\pi = \{\pi_i : i \in \{1, \ldots, N\}\}$ - the initial state distribution vector,
$O = O_1, O_2, \ldots, O_T$ - the observation sequence,
$O_{j:k}$ - the part of observation sequence including symbols from j-th to k-th, inclusive.

For AHMMs the following notation is used:

D - the number of component models,
$x^{(d)} : d \in \{1, \ldots, D\}$ - the value x from d-th component model.

3 Methods

The rehabilitation module is based on Averaged Hidden Markov Models (AHMMs) [15], briefly described below. In order to decide whether the new motion is correct or not, the rehabilitation models extend the methods for action recognition in HMM. These algorithms are listed in Sect. 3.2. The other methods used for rehabilitation are described in Sects. 3.3 and 3.4.

3.1 Averaged Hidden Markov Models

Each activity model is created from multiple learning sequences. In the later stages of AHMM generation each single learning sequence becomes the base for a one component HMM model. Thus the number D of learning sequences is equal to the number of component models.

One of the learning sequences is chosen as a *pattern sequence* and defines all models structure, i.a. the number of states. The pattern sequence is used for the *base model* definition. The base model is a left-to-right HMM with states matched to subsequent observation symbols, thus in this stage the model could be reduced to the simple Markov chain.

Each of the rest of component models is computed based on the base model and the corresponding learning sequence. Such a *child model* has a structure *similar* to the base model:

- the child model is also a left-to-right HMM model,
- the base and child models have the same number of states,
- the same observation symbols in base and child models occur in the same states, taking under consideration the symbol order.

The detailed algorithm for base and child models parameters computation is described in the previous work [15]. At this stage there is D similar left-to-right HMMs.

Finally, all the component models are simply averaged using the Eqs. (1), (2) and (3). In consequence, we obtain one resultant Averaged HMM model which generates each of its learning sequences $O^{(i)}$ with the probability $P(O^{(i)}|\lambda) > 0$.

$$\bar{a}_{ij} = \sum_{d=1}^{D} \frac{1}{D} \cdot a_{ij}^{(d)} \tag{1}$$

$$\bar{b}_{ij} = \sum_{d=1}^{D} \frac{1}{D} \cdot b_{ij}^{(d)} \tag{2}$$

$$\bar{\pi}_{i} = \sum_{d=1}^{D} \frac{1}{D} \cdot \pi_{i}^{(d)} \tag{3}$$

3.2   Action Recognition

The task of rehabilitation could be considered as the subproblem of the real-time recognition problem. In both cases the information whether the activity is completed or not is crucial. The posterior probability $P(O|\lambda)$ is not a sufficient indicator, while each beginning part of the recognized sequence is also recognized, i.e.

$$\forall_{1 \leq t < T} P(O_{1:T}|\lambda) > 0 \implies P(O_{1:t}|\lambda) > 0.$$

Therefore the method developed for real-time recognition [16] has been used. The N_R value (the real number of states, which is different for each of the component models) has been added to the model during the learning phase as the id of the last state that is always accessed while recognizing any of learning sequences. It means that the sequence ending in the state with lower id than N_R is not complete. The last state is estimated by the Viterbi algorithm [14] based on the observation symbol sequence.

Because of short activities, which consist of less than 4 symbol changes and where noise might change the recognized class (activity id), the additional condition has been added. The last symbol O_T in the considered sequence has to be probable to occur in the last state in the model, i.e. $b_{N_O T} > 0$.

The complete sequence recognized by model λ fulfills all the three conditions for this model.

3.3   Rehabilitation

The idea of rehabilitation using HMMs is based on displaying the most probable symbol in the next most probable state based on the actual state. The actual state is estimated using the Viterbi algorithm. In general the rehabilitation problem could be stated as a set of following equations. The first state is chosen as the most probable one due to the initial state distribution using the Eq. (4). Each next state is chosen based on the probability transition matrix and the actual state (Eq. (5)). While having the most probable state, the next symbol to be displayed is calculated by Eq. (6).

$$\text{Find state } i \in \{1, \dots, N\} : \forall_{j \in \{1,\dots,N\}} \pi_i \geq \pi_j \tag{4}$$

$$\text{For the state } j \text{ find state } i \in \{1, \dots, N\} : i \neq j \wedge \forall_{k \in \{1,\dots,N\}} a_{ji} \geq a_{jk} \tag{5}$$

$$\text{For the state } i \text{ find symbol } k \in \{1, \dots, M\} : \forall_{l \in \{1,\dots,M\}} b_{ik} \geq b_{il} \tag{6}$$

Because of the fact that in evaluation problem [14] the posterior probability tends to zero exponentially along with the increase of the number of observation symbols, the logarithmized value is calculated instead. However, during the long rehabilitation process (e.g. the motions are very slow) the range of double is also exceeded and $log(P(O|\lambda)) = -\infty$. Therefore, in the system only the symbol changes are registered. Sometimes the list of symbols to be displayed calculated by Eqs. (4–6) may consist of series of the same symbol - the most probable symbol in the next most probable state may be the same symbol as the most probable symbol in the previously most probable state. While we register only the symbol changes, such situation detection had to be introduced. In order not to reduce the intelligibility of the algorithm this case has been marked in red in the Fig. 1. The variable *skippedSameSymbol* is set to the number of next equal symbols. If this variable is greater than zero then exceptionally the adjacent equal symbols (in a number equal to this variable) are added to the symbol history. In further discussion this case will be omitted.

The diagram showing the complete rehabilitation algorithm is presented in Fig. 1. First of all the HMM symbol is calculated depending on the chosen motion (right–, left– or both–handed model) according to the algorithm described in [17]. The information about the previous symbol (performed motion) and previously displayed symbol is saved from the previous iteration. Also the zero-one information whether the motion was correct or not (due to the value $log(P(O|\lambda))$) is remembered.

Secondly, if the actual symbol is the same as in the previous iteration, then the old values of (1) the next symbol to be displayed and (2) correctness of motion are returned. Otherwise, the symbol is added to the history and remembered as the previous symbol for the next iteration. The model is evaluated using the forward algorithm [14] based on the actual observation symbol history.

Next, if $log(P(O|\lambda)) \neq -\infty$ then the actual motion is marked as correct and the state sequence is estimated using the Viterbi algorithm. The next state is estimated based on the last decoded state using the Eq. (5). Next symbol to be displayed is calculated based on the estimated future state using the Eq. (6). Otherwise, if $log(P(O|\lambda)) = -\infty$ then the actual motion is marked as incorrect, the symbol is rolled back from the history and the algorithm returns the same values as in the previous iteration.

3.4 Hand Coordinates Estimation

The only information about the next movement obtained from the HMM is the observation symbol, i.e. the natural number in the range of $\{1; M\}$. For the rehabilitation module the visualization is the one of the most important features, thus the hands coordinates have to be calculated. In order to estimate these

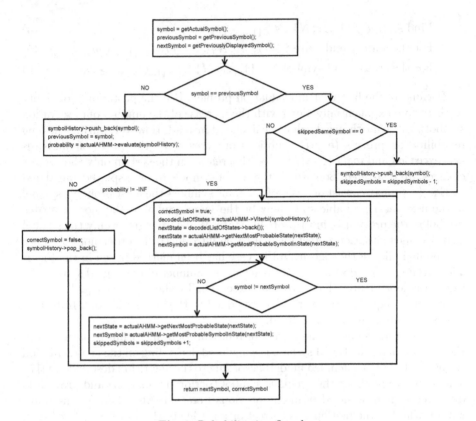

Fig. 1. Rehabilitation flowchart

parameters the inverse operations than in the hand position classifier described in [17] (Eqs. (1) and (2)) are needed.

Firstly, for each feature h_i and the HMM hand symbol s_N the id of the including interval $z_i \in \{0; m_i - 1\}$ (m_i is the number of intervals that the range of values for h_i had been divided into) is calculated using the Eq. (7). The r_i is an auxiliary variable.

$$\begin{cases} r_0 = s_N \\ z_i = r_{i-1} div \prod_{k=i+1}^{N} m_k \\ r_i = r_{i-1} mod \prod_{k=i+1}^{N} m_k \end{cases} \tag{7}$$

Secondly, based on the minimum $h_{i_{MIN}}$ and maximum $h_{i_{MAX}}$ value in the range, the value of h_i is estimated (Eq. 8). In order to minimize the error for the singular feature value estimation the middle of the range is chosen.

$$h_i = h_{i_{MIN}} + (z_i + 0,5) \cdot \frac{h_{i_{MAX}} - h_{i_{MIN}}}{m_i} \tag{8}$$

Finally, based on the estimated feature values and the length of the arm, the hand coordinates are calculated. The features used in the classification were as follows:

- h_1 - an angle between projection of the vector v to the OXZ plane and the x axis,
- h_2 - an angle between the vector v and the y axis,
- h_3 - a relative length of the vector v (the quotient of $|v|$ and the length of the whole arm),

where v is the radius vector connecting the shoulder and the hand.
Therefore, in order to calculate the hand's coordinates, the point $(|v|, 0, 0)$ is rotated by the calculated angles in the OXY and OXZ planes.

4   Application

The rehabilitation module is a part of the more complex system designed for children with autism [17]. The application is designed to track one- or both-hands movements. The list of AHMM activities models chosen for rehabilitation is presented in Table 1. The list has been divided into left-, right- and both-handed activities. The system is fully scalable, while the files with models could be easily added or deleted from the models directory.

Table 1. The list of modeled activities used for rehabilitation

Left-handed activities	Right-handed activities	Both-handed activities
Left arm twisting forward	Right arm twisting forward	Both hands twisting forward
Left arm twisting backward	Right arm twisting backward	Both hands twisting backward
Raising and lowering left hand	Raising and lowering right hand	Raising and lowering both hands
		Clapping hands
		Clapping hands over the head
		Crawl forward
		Crawl backward

In the rehabilitation mode the two skeletons are displayed - the pattern to follow and the actual motion. The skeletons coordinates are normalized as in [17], so that the *Spine Shoulder* joint overlaps in both cases. The complete body coordinates set is retained in the file, but the hand coordinates are calculated based on the next symbol to be displayed (algorithm described in Sect. 3.4). The elbow coordinates for the symbols are tabulated. The actual hand position is additionally surrounded by a circle which changes the color depending on whether the motion is correct (green) or not (red) according to the model. Since in the 2D picture it is difficult to guess the limb distance from the camera, the displayed color changes. If the joint is further from the camera than the *Spine Shoulder* joint (greater z value) then the joint connection is painted in blue.

After choosing the rehabilitation mode and the activity AHMM model, the first motion is displayed. Depending on the patient's movements the next symbols (hands positions) are estimated and displayed. An example for the rehabilitation is presented in Fig. 2. In the picture the most important fragments for the recording of twisting left arm forward have been included. An example of the incorrect motion is presented in Fig. 3(a) - the hand is surrounded with the red circle.

Fig. 2. Rehabilitation with the exercise: twisting left arm forward (Color figure online)

In the case of single hand rehabilitation, the future movements estimated by the AHMM model coincide with expectances. The speed of motions does not affect the result of action recognition, however it has the influence on the motion evaluation $P(O|\lambda)$.

The issue of both hands rehabilitation is much more complicated. The data taken from the real world abound in noises, for example it is nearly impossible to perform the strictly symmetric motion by each of the hands. Therefore, in such a symmetric movements like twisting arms forward, the most probable future symbol chosen by the algorithm often corresponds to asymmetric hands position. In such a case even if in the real motion (learning sequences) the hands positions did not differ much, the algorithm (Sect. 3.4) estimates the coordinates strongly asymmetric. The example of asymetric hands position prediction is presented in Fig. 3(b), while the Fig. 3(c) presents the next stage of the same activity which is symmetric. The problem is not visible if there is no symmetry in both hands movement. The problems with visualization however do not affect the motion assessment.

Summarizing, the applied algorithms meet the requirements, especially when it comes to the one–handed motions. The bottleneck of this rehabilitation system is the lack of legible visualization, as that the position of the 3D point presented on the 2D screen is ambiguous.

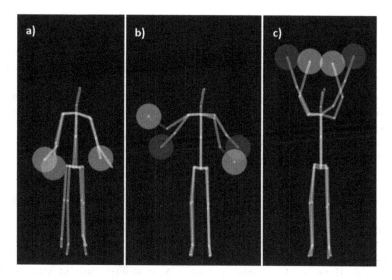

Fig. 3. (a) The one–hand incorrect motion (red circle) (b) The both–hand asymmetric movement (c) The both–hand symmetric movement (Color figure online)

5   Conclusions and Future Work

In this paper the Averaged Hidden Markov Models were examined for the rehabilitation purposes. The one– and two–handed activity models were taken under the consideration.

The conducted experiments indicate that AHMMs provide the satisfactory results for the rehabilitation purpose as the motion is tracked and assessed properly. In the case of single hands rehabilitation and both hands rehabilitation based on motions without symmetry, the next movement prediction coincides with the expectations.

The visualization of the future movement sometimes could be unintuitive, as that the predicted next hands positions for the symmetric both hands motion could have no symmetry. Also the position of the 3D point presented on the 2D screen is ambiguous. The symbolic representation (skeleton) also seems to be too abstract for children with autism. For the practical usage of the examined algorithms the visualization methods need to be improved. On the other hand, such a symbolic representation respects the privacy as that only the skeleton joints are taken under consideration.

The advantage of the system is that the new activities could be easily added. This function could be especially important in the case of autistic children rehabilitation, as that each child with autism needs individual therapy.

Acknowledgment. This work was supported by the statutory funds of the Faculty of Electronics 0401/0159/17, Wroclaw University of Science and Technology, Wroclaw, Poland.

References

1. Seach, D., Lloyd, M., Preston, M.: Supporting Children with Autism in Mainstreem Schools. The Questions Publishing Company Ltd., Birmingham (2003). ISBN 83-60215-17-0
2. Barry, A.: Some people think that every person with autism is like Rain Man, or a wizard at maths. Thejournal (2017). http://www.thejournal.ie/autism-aspergers-ireland-3297234-Mar2017/
3. Autism Awareness - Frequently Asked Questions About Autism. Staffordshire Adults Autistic Society. http://www.saas.uk.com/p/autism-awareness-questions. php
4. Regenbrecht, H., Hoermann, S., McGregor, G., Dixon, B., Franz, E., Ott, C., Hale, L., Schubert, T., Hoermann, J.: Visual manipulations for motor rehabilitation. Comput. Graph. (Pergamon) 36(7), 819–834 (2012)
5. Kuttuva, M., Boian, R., Merians, A., Burdea, G., Bouzit, M., Lewis, J., Fensterheim, D.: The rutgers arm, a rehabilitation system in virtual reality: a pilot study. CyberPsychol. Behav. 9(2), 148–152 (2006)
6. Pastor, I., Hayes, H.A., Bamberg, S.J.M.: A feasibility study of an upper limb rehabilitation system using Kinect and computer games. In: 2012 Annual International Conference of the IEEE Engineering in Medicine and Biology Society, pp. 1286–1289 (2012)
7. Chee, K.L., Chen, I.M., Zhiqiang, L., Yeo, S.H.: A low cost wearable wireless sensing system for upper limb home rehabilitation. In: 2010 IEEE Conference on Robotics, Automation and Mechatronics, pp. 1–8 (2010)
8. Clark, R.A., Pua, Y.H., Fortin, K., Ritchie, C., Webster, K.E., Denehy, L., Bryant, A.L.: Validity of the Microsoft Kinect for assessment of postural control. Gait Posture 36(3), 372–377 (2012)
9. Scherer, M., Unterbrunner, A., Riess, B., Kafka, P.: Development of a system for supervised training at home with Kinect V2. Procedia Eng. 147, 466–471 (2016)
10. Chang, Y.J., Chen, S.F., Huang, J.D.: A Kinect-based system for physical rehabilitation: a pilot study for young adults with motor disabilities. Res. Dev. Disabil. 32(6), 2566–2570 (2011)
11. Kusaka, J., Obo, T., Botzheim, J., Kubota, N.: Joint angle estimation system for rehabilitation evaluation support. In: 2014 IEEE International Conference on Fuzzy Systems (FUZZ-IEEE), pp. 1456–1462 (2014)
12. Su, Ch.J., Chiang, Ch.Y., Huang, J.Y.: Kinect-enabled home-based rehabilitation system using dynamic time warping and fuzzy logic. Appl. Soft Comput. 22, 652–666 (2014). Elsevier B.V
13. González-Ortega, D., Díaz-Pernas, F.J., Martínez-Zarzuela, M., Antón-Rodríguez, M.: A Kinect-based system for cognitive rehabilitation exercises monitoring. Comput. Methods Programs Biomed. 113, 620–631 (2014)
14. Rabiner, L., Juang, B.: An introduction to hidden Markov models. IEEE ASSP Mag. 3, 4–16 (1986)
15. Postawka, A.: Exercise recognition using averaged hidden Markov models. In: Rutkowski, L., Korytkowski, M., Scherer, R., Tadeusiewicz, R., Zadeh, L.A., Zurada, J.M. (eds.) ICAISC 2017 Part II. LNCS (LNAI), vol. 10246, pp. 137–147. Springer, Cham (2017). https://doi.org/10.1007/978-3-319-59060-8_14

16. Postawka, A.: Real-time monitoring system for potentially dangerous activities detection. In: Proceedings of the 22nd International Conference on Methods and Models in Automation and Robotics (MMAR), pp. 1005–1008. IEEE Xplore Digital Library (2017)
17. Postawka, A., Śliwiński, P.: A Kinect-based support system for children with autism spectrum disorder. In: Rutkowski, L., Korytkowski, M., Scherer, R., Tadeusiewicz, R., Zadeh, L.A., Zurada, J.M. (eds.) ICAISC 2016 Part II. LNCS (LNAI), vol. 9693, pp. 189–199. Springer, Cham (2016). https://doi.org/10.1007/978-3-319-39384-1_17

Genome Compression: An Image-Based Approach

Kelvin Vieira Kredens[1]([✉]), Juliano Vieira Martins[1], Osmar Betazzi Dordal[1], Edson Emilio Scalabrin[1], Roberto Hiroshi Herai[2], and Bráulio Coelho Ávila[1]

[1] Graduate Program in Computer Science – PPGIa,
Pontifical Catholic University of Paraná – PUCPR, Curitiba, Brazil
kredens@pucpr.edu.br,
{julianovmartins,osmarbd,scalabrin,avila}@ppgia.pucpr.br
[2] Graduate Program in Health Sciences – PPGCS,
Pontifical Catholic University of Paraná – PUCPR, Curitiba, Brazil
roberto.herai@pucpr.br

Abstract. With the advent of Next Generation Sequencing Technologies, it has been possible to reduce the cost and time of genome sequencing. Thus, there was a significant increase in demand for genomes that were assembled daily. This demand requires more efficient techniques for storing and transmitting genomic data. In this research, we discussed the horizontal compression of lossless genomic sequences, using two image formats, WEBP, and FLIF. For this, the genomic sequence is transformed into a matrix of colored pixels, where an RGB color is assigned to each symbol of the A, T, C, G alphabet at a position x-y. The WEBP format showed the best data-rate saving (76.15%, SD = 0.84) when compared to FLIF. In addition, we compared the data-rate savings of two specialized DELIMINATE and MPCompress genomic data compression tools with WEBP. The results obtained show that the WEBP is close to DELIMINATE (76.03%, SD = 2.54%) and MFCompress (76.97%). SD = 1.36%). Finally, we suggest using WEBP for genomic data compression.

Keywords: Data compression · Genome compression
Assembled genomic sequence · Lossless compression · Image file format

1 Introduction

The development of the next-generation sequencing technologies [1,2] has reduced the cost and sequencing time of complete genomes. This fact led to an expressive demand for the use of the many sequenced genomes. According to [3] in two decades, the number of sequenced individuals reach 1 billion people. The amount of sequenced genomes has challenged the development of more efficient ways to process large amounts of genetic data [4,5]. Data compression is one of the alternatives to reduce the amount of genomic data to store and transmit.

© Springer International Publishing AG, part of Springer Nature 2018
L. Rutkowski et al. (Eds.): ICAISC 2018, LNAI 10842, pp. 240–249, 2018.
https://doi.org/10.1007/978-3-319-91262-2_22

Assembled genomic sequences are usually stored in FASTA format files. This type of file uses ASCII characters to represent the genetic bases defined as A (Adenine), T (Thymine), C (Cytosine) and G (Guanine). Thus, the format is not optimized for saving the data space. This explains why several specialized tools for genomic data compression have been proposed. According to [6], the compression process can take a Vertical or Horizontal approach. Both, during compression, use one or more genomic sequences as a source of information, whose the purpose is to identify repeated segments or genomic grammar rules [7].

The vertical compression approach is advantageous when the input is a collection of genomic sequences, mainly from organisms of the same species. The advantage is in the strong similarity between the genomes of these organisms, which for some species can reach 99.5% [8]. Thus, in theory, it is necessary to store only 0.5% of the differences [9]. Similarly, in this approach, all sequences used as sources of information for compression must be accessible to the decompression stage. In the horizontal compression approach, each genomic sequence is compressed using the sequence itself as an information source [6]. In this research, we are only exploring the horizontal approach.

The compression of genetic data was examined in [10–17]. In [18], a set of genomic sequences has been proposed to evaluate the performance of the compression tools that follow the horizontal approach, where specialized tools were tested: DELIMINATE, MFCOMPRESS, and COMRAD. In this paper, the COMRAD tool was not taken into account because it does not work with the alphabet A, T, C, G, N.

Here, our research efforts are to show the use of image format as a viable method for lossless compression of genomic sequences. In this sense, the image formats that have been evaluated – in decreasing order of space saving – are *WEBP*, and *FLIF*. These image formats were compared to the values of the specialized genomic compression tools DELIMINATE and MFCOMPRESS. The last two tools used less than 1% of the storage space compared to the *WEBP* and *FLIF* image formats. Thus, in what follows this article, we will examine only *WEBP* and *FLIF* image formats. Other image formats have had less expressive results.

The following sections describe our proposal and how we evaluate it.

2   Materials and Methods

A compression of genomic sequences using image formats is viable and to make it useful one does not have much effort to do. In this sense, we propose to show that with little effort the image formats *WEBP* and *FLIF* can produce good results compared to other methods specialized. The idea is simple. A square matrix Q of pixels is created. And each pixel of Q will encode a nucleotide. The Algorithm 1 will be explained in Sect. 2.1.

2.1 Proposal

In this section, we will describe the compression process, summarized by Algorithm 1. The compression process begins with the line-by-line reading of the input FASTA file [19]. Two parts compose this text-based file format: sequence header and biological sequence (DNA, RNA or Protein), having the nucleotides letter A, T, C, and G to represent a series of biological sequence. The non-ATCG symbols and the *FASTA* header are stored into a separated temporary file, called codebook. Due to the current sequencing technology and genomic data assembly software, a *FASTA* file could present long scratches of n's corresponding to inaccessible genomic locations or highly redundant repetitive regions. For both cases, it is used a sequence simplification strategy to reduce the required space for data representation. The non-ATCG symbols are stored along with their positions in the original sequence. Therefore, repeated strings of symbols are represented as a single entry in the codebook. Each non-ATCG symbol or each string of repetitive non-ATCG symbols is represented in the codebook as a triad, one on each line. This triad is composed by the values (p, s, l), where p is the position in which the symbol or chain occurs in the original sequence, s is the symbol and l is the chain length. At the end of processing, the codebook is stored in a plain text-based file to be compressed.

Algorithm 1. cpAg

1: $maxSize \leftarrow ceiling\langle\sqrt{sizeOf(inputFile)}\rangle$
2: $Q \leftarrow newPixel[maxSize][maxSize]$
3: $codebook \leftarrow newList\langle triad\rangle$ ▷ p=position, s=symbol/chain, l=lenght
4: $header \leftarrow getHeader(inputFile)$
5: $codebook.add(header)$
6: $sequence \leftarrow getSequence(inputFile)$
7: **for** $p \leftarrow sizeOf(sequence)$ **do** $p < sizeOf(sequence)$
8:     $c \leftarrow sequence[p]$
9:     **if** $c \in alphabet$ **then**
10:         $Q.add(f(c))$ ▷ f(c) assigns an RGB color to each nucleotide
11:     **else**
12:         $l \leftarrow 0$
13:         **while** $sequence[p] \notin alphabet \land sequence[p] \equiv sequence[p-1]$ **do**
14:             $l \leftarrow l + 1$
15:             $p \leftarrow p + 1$
16:         $codebook.add(p, c, l)$
17: $write(Q, inOut)$
18: $Text\ outTxt \leftarrow newText()$
19: $write(codebook, outTxt)$
20: **return** $outTxt$

The next step is to create a square matrix of pixels. It was created square because some image formats have limitations in row/column length such as

WEBP (discussed later), that could limit the sequence length to be used. The matrix order was defined from the length of the target genomic sequence to be converted, and calculated according to $MatrixSize = \lceil\sqrt{SequenceSize}\rceil$. Once the matrix is created, the next step is to code each A, T, C, G symbol into a pixel (see Algorithm 1). The basic idea is to convert each symbol into a pixel with a different grayscale in RGB encoding. And, if the sequence size does not allow an exact square root – empty pixels are added to ensure a complete fill of Q.

The final step of sequence compression generates two files. The first one stores the matrix of pixels of the selected image format. The second stores the codebook which contains the following information: (a) the header from the FASTA file; (b) the length of the genomic sequence; (c) the number of columns used to represent the sequence in the FASTA file; and (d) the list of symbols or non-ATCG chains of symbols removed from the genomic sequence. The decompression process is simple. First, it is necessary to read the image file and the codebook. Using a specific codec for the selected image format, the image file is converted to a matrix of pixels. Next, the matrix is traversed, converting each pixel into a nucleotide symbol, where C (Cytosine), T (Thymine), A (Adenine) and G (Guanine) are respectively associated with the gray variations in RGB: (0, 0, 0), (1, 0, 0), (2, 0, 0), and (3, 0, 0). For the reconstruction of the FASTA file we use the information stored previously in the codebook, FASTA header, column length and non-ATCG symbols if any. In this way, we process the sequence by inserting the line breaks and the non-ATCG symbols into their respective positions.

2.2   Performance Evaluation

The evaluation procedure is based on genomic sequences written in *FASTA* files, which are compressed and encoded into files whose formats are images. The image formats used were: *WEBP*[1] and *FLIF*[2]. Here, it should be noted that when the codebook is larger than 200 bytes, it is compressed using 7zip with the algorithm *PPMD*[3]. Thus, the number of bytes of each compressed *FASTA* file is given by the sum of the number of bytes of the resulting image file plus the number of bytes of the codebook.

The data set used for performance evaluation was the same as defined by Biji et al. [18], composed of the genomic sequences of 1,547 organisms. However, we do not use any virus genomes or multi-FASTA files. We do not use virus sequences because they are small in size. Already, the multi-FASTA files have not been considered as a way to simplify the experimentation procedures. We also do not use genomic sequences larger than 268,435,456 nucleotides. This limitation, although circumvented by dividing the sequence into more than one file, occurs because of the *WEBP* image format. In the version used, WEB format does

---

[1]   https://developers.google.com/speed/webp/.

[2]   http://flif.info/.

[3]   https://www.dotnetperls.com/ppmd - parameter -m0=PPMd".

Table 1. Distribution of genomic sequences by kingdoms. In some situations, the amount of sequences is greater than the amount of organism since the genome of such organisms are divided into distinct FASTA files.

Kingdoms	Number of organisms	Number of sequences (*)
Animalia	3	14
Archaea	24	24
Bacteria	1,101	1,101
Fungi	27	275
Plant	3	23
Protist	5	110
Total	**1,163**	**1,547**

(*) Each sequence represents a distinct FASTA file

not allow images with the matrix of pixels greater than 16,384 x 16,384 pixels. Thus, the final set of data used in the evaluation consists of the genome of 1,163 organisms, ranging from the following biological kingdoms: *Animalia, Archaea, Bacteria, Fungi, Plant* and *Protist* (Table 1). As in some cases, the genome of an organism is divided into more than one file; in the end, the *FASTA* file set consists of 1,547 distinct files.

$$DR = 1 - \left(\frac{1}{CR}\right), \text{ such that } CR = \frac{UFS}{CFS} \tag{1}$$

where, UFS is the Uncompressed File Size and CFS is the Compressed File Size.

The obtained results using the image formats were compared with the results of the following specialized genomic data compression tools: DELIMINATE [20] and MFCompress [21]. To analyze the results we used the Data Rate savings DR, given by Eq. 1, where CR is defined as the Compressed Ratio.

The following set of features formed the computational configuration used in the tests: INTEL Xeon 24 Cores 2.40 GHz (64 bits) and 182 GB of RAM; and operating environment Linux CENTOS 6.

3   Results

Next, we present the performance evaluation results of the proposed method. For standardization purposes, the execution of tested methods, as well as the collection of performance metrics, were automated within a software framework. It was implemented using the Python language, in which the methods and the sequences are initially configured for automatic execution. Overall, each of the 1,547 *FASTA* files was individually compressed nine times, generating a total of 13,923 file compressions.

After the tests, for the WEBP and FLIF formats, it was observed that the use of some different RGB colors, that is closer to black color, impacted on the

final size. That is because the WEBP and FLIF formats implement optimization methods for space savings this way, we tested two configurations. In the
first, the symbols A, T, C, and G were converted to RGB using blue, green,
red and white. In the second, gray variations were used, with only the red value
from RGB varying between 0 to 3. In this configuration, the value of green and
blue was kept equal to 0. We tested the two configurations and found that the
gray variation configuration obtained slightly better data rate savings, on average, 1.03% for WEBP and 0.02% for FLIF when compared to the color version.
Thus, for the ratings, we focused on the configuration that worked with gray
variation. Figure 1B shows the average data rate savings achieved by the specialized genomic data compression tools, DELIMINATE and MFCompress, and two
image formats that generate the best data rate savings, *WEBP(g)* and *FLIF(g)*.
Table 2 shows the result of all evaluated specialized genomic data compression
tools and image formats.

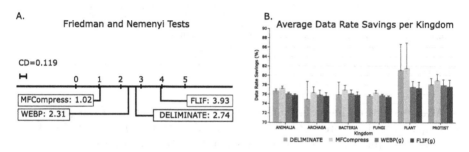

Fig. 1. (A) Friedman and Nemenyi Tests considering the average data rate savings
obtained by the methods in the individual compression of all 1,547 FASTA files.
(B) Average data rate saving by grouping the genomic sequences by the Kingdom of
each Organism. In this graph, we are presenting the result of the two best-specialized
genomic data compression tools and the two best image formats.

As can be seen, Table 2 shows the average data rate saving achieved by each
tool. In this table, we are presenting the values of all the specialized genomic compression tools Deliminate and MFCompress and for the image formats WEB and
FLIF. The tool MFCompress presented the best data rate savings with 76.98%,
followed by the WEBP(g) image format (gray variation) of 76.15%. To validate
the results, two non-parametric tests, Friedman and Nemenyi, were applied.
The Friedman test [22,23] is a nonparametric test equivalent to ANOVA [24] for
repeated measures. This test sorts the algorithms for each data set separately.
With the obtained result from the Friedman test, it was possible to reject the
null hypothesis since the p-value was below 0.01 ($p < 0.01$) for 1,547 observations and four different methods (see Fig. 1B). Thus, it was necessary to apply
a posthoc test to identify which methods generated differentiated results. The
Nemenyi test [25] is used when all algorithms are compared on a peer-by-peer
basis. The performance of two algorithms is significantly different if the corresponding mean compressions differ by at least the critical distance. The set of

tests showed that the critical distance is 0.119 Thus, no tool was statistically similar concerning the average compression ratio. (see Fig. 1A).

Concerning data-rate saving versus kingdoms, on Table 2 shows the Protist kingdom with the best average compression results (average data rate savings of 77.37%) with the lower standard deviation (SD = 1.87). While for the Plant kingdom, despite the high data rate savings (78.70%), the variation was higher (SD = 4.26).

Table 2. The values average of data-rate saving of all evaluated tools specialized in genomic compression and image formats. The column Average per Kingdom presents the average data-rate saving achieved by all the methods for a given kingdom and the column Average per method displays the average of each tool/image format for all kingdoms.

Methods	Animalia (%)	Archaea (%)	Bacteria (%)	Fungi (%)	Plant (%)	Protist (%)	Average per Kingdom (%)
DELIMINATE*	76.75 ±0.44	74.90 ±3.82	75.94 ±2.63	75.64 ±0.51	80.89 ±4.92	77.17 ±2.48	76.04 ±2,55
MFCompress*	77.43 ±0.54	76.35 ±1.20	76.92 ±0.98	76.33 ±0.80	81.69 ±4.51	78.26 ±1.68	76.98 ±1,36
WEBP(g)**	76.09 ±0.31	75.86 ±0.98	76.14 ±0.76	75.78 ±0.50	76.23 ±0.71	77.20 ±1.36	76.15 ±0,84
FLIF(g)**	75.76 ±0.31	75.58 ±0.71	75.81 ±0.70	75.42 ±0.48	75.99 ±0.64	76.85 ±1.43	75.81 ±0,81
Average per Method	76.51 ±0.76	75.67 ±2.16	76.20 ±1.56	75.79 ±0.67	78.70 ±4.26	77.37 ±1.87	--

*Tools and **Image Format

After evaluating the performance of the methods, an analysis was performed to compare the data rate savings of the methods with the characteristics of the subjected genomic sequences for compression. Based on the results (Table 2), we calculated the Pearson correlation. This investigation had the aim to evaluate in which situations each tool or image format stands out, and it was based in the following features: size (in bytes) of the *FASTA* file before compression, entropy and repetitiveness index. Table 3 shows the results of this evaluation.

To evaluate the positive or negative impact on saving space, we evaluate the data rate saving versus the size of the sequences. After discarding the

Table 3. Pearson correlation coefficient between the features: Size, Information Entropy and Repetitiveness Index. Such features were extracted from each genomic sequence and used to calculate the correlation coefficient concerning the compression ratio.

Kingdom	Sequence Size (%)	Information Entropy (%)	Repetitiveness Index (%)
Animalia	−0.5131	−0.3560	−0.0677
Archaea	0.5757	−0.4270	0.1859
Bacteria	0.1433	−0.3549	0.0673
Fungi	−0.0164	−0.0038	0.1823
Plant	0.9617	0.0712	−0.1411
Protist	0.3936	−0.5475	0.1172
All	0.3287	−0.2971	0.1093

FASTA sequence header, the size of each sequence was calculated by counting the number of symbols, including the non-ATCG. The data in Table 3 shows that there is a strong correlation between the Plants kingdom. However, this correlation is positive for the two specialized tools for genomic compression (0.97 and 0.94), respectively for DELIMINATE and MFCompress, and negative for image formats (−0.67 and −0.63), respectively for *WEBP* and *FLIF*. These results (detailed in Table 4) are because the image formats were not useful in compressing the large genomes of the Plant kingdom. In general, except for the Plant kingdom, the sequence size does not present a substantial correlation for the image formats.

Another evaluated characteristic was the Repetitiveness Index (RI). As described by the authors in [26], RI is a measure that attempts to gauge the amount of intra-repeatability present in a DNA sequence. An RI is expected to be zero in random DNA sequence of any G/C content. The higher value is greater than zero for sequences that contain expressive repeats. As can be seen in Table 3, this characteristic does not present strong correlation with any of the evaluated kingdoms.

The last evaluated characteristic was the Information Entropy, calculated as defined by Shanon [27]. Thereunto, the present information entropy in each genomic sequence was calculated. In this process, we used as input the genomic sequence, discarding the non-ATCG symbols. A random sequence restricted to the alphabet of ATCG symbols has the entropy value equal to 2. That is, 2 bits are required to represent each symbol of the DNA sequence. In the case of a genomic DNA sequence restricted to the ATCG symbols, the smaller the entropy, the more unbalanced it's the distribution of these symbols. Using the data rate savings and the entropy of each genomic sequence were used to calculate the Pearson correlation. In this way, we verified that there is a strong correlation between the Plant kingdom with values 0.72 and 0.76, respectively for *WEBP(g)* and *FLIF(g)*, whereas there is no correlation for specialized genomic compression tools (detailed in Table 4). These results can be explained by the fact that the specialized tools apply different transformations on the data before the compression. Thus, the coding step is not based solely on the original distribution of the symbols. Since the image formats are based on symbol distribution, there is a strong correlation between data rate savings and entropy.

Table 4. Pearson correlation coefficients between size and entropy information of genetic sequences with the tools and image formats evaluated – only to the kingdom Plant.

Kingdom	*Deliminate*	*MFCompress*	*WEBP(g)*	*FLIF(g)*
Plant - size	0.9726	0.9470	−0.6737	−0.6314
Plant - entropy	−0.0070	−0.0815	0.7282	0.7627

4   Conclusion

This research aimed to demonstrate the viability of storing assembled genomic sequences, usually represented as *FASTA* files, as image files under different formats. We also showed that resulting image files applied over genomic sequences presented similar data rate savings when compared to specialized tools for genomic data compression. Also, the results showed that entropy is correlated with gains in compression rates when we evaluate specialized genomic compression tools because they apply techniques of transforming the data before encoding, especially for the kingdom whose genomic sequences are larger in size and repeatability. Furthermore, precedents are opened for the use of techniques of manipulation of images for the search for patterns and similarities that can be analysed directly in the image and without there being decompression for the original *FASTA* file. In this way, the genomic analysis programs will work directly with the data in the image format of these experiments. Based on these results, our next step is focused on applying transformation techniques to reduce the entropy of information from genomic sequences and then use image-based compression.

Acknowledgments. We thank Biji Christopher Leela for her help, sharing with us the sequences that compound the dataset she created.

Funding. This work was partially supported by CAPES-Brazilian Federal Agency for Support and Evaluation of Graduate Education-scholarship. That provided Master Fellowship to JVM. Ph.D. Fellowship to KVK. and Postdoctoral Fellowship to OBD. The computational infrastructure for data analysis of this manuscript was supported by Fundação Araucária (grant #CP09/2016) and Graduate Program in Computer Science (PPGIa) from PUCPR.

References

1. Schuster, S.C.: Next-generation sequencing transforms today's biology. Nat. Methods **5**, 16–18 (2008)
2. Reuter, J.A., Spacek, D.V., Snyder, M.P.: High-throughput sequencing technologies. Mol. Cell **58**(4), 586–597 (2015)
3. Stephens, Z.D., Lee, S.Y., Faghri, F., Campbell, R.H., Zhai, C., Efron, M.J., Iyer, R., Schatz, M.C., Sinha, S., Robinson, G.E.: Big data: astronomical or genomical? PLoS Biol. **13**, e1002195 (2015)
4. Hsi-Yang Fritz, M., Leinonen, R., Cochrane, G., Birney, E.: Efficient storage of high throughput DNA sequencing data using reference-based compression. Genome Res. **21**, 734–740 (2011)
5. Hayden, E.C.: Genome researchers raise alarm over big data. Nature (2015)
6. Grumbach, S., Tahi, F.: Compression of DNA sequences. In: Data Compression Conference DCC 1993, pp. 340–350 (1993)
7. Yamagishi, M.E.B., Herai, R.H.: Chargaff's "Grammar of Biology": New Fractal-Like Rules. Quantitative Biology, Arxiv preprint arXiv, p. 17 (2011)

8. Levy, S., Sutton, G., Ng, P.C., Feuk, L., Halpern, A.L., Walenz, B.P., Axelrod, N., Huang, J., Kirkness, E.F., Denisov, G., Lin, Y., MacDonald, J.R., Pang, A.W.C., Shago, M., Stockwell, T.B., Tsiamouri, A., Bafna, V., Bansal, V., Kravitz, S.A., Busam, D.A., Beeson, K.Y., McIntosh, T.C., Remington, K.A., Abril, J.F., Gill, J., Borman, J., Rogers, Y.-H., Frazier, M.E., Scherer, S.W., Strausberg, R.L., Venter, J.C.: The diploid genome sequence of an individual human. PLoS Biol. **5**, e254 (2007)

9. Giancarlo, R., Rombo, S.E., Utro, F.: Compressive biological sequence analysis and archival in the era of high-throughput sequencing technologies. Brief. Bioinform. **15**, 390–406 (2013)

10. Giancarlo, R., Scaturro, D., Utro, F.: Textual data compression in computational biology: algorithmic techniques. Comput. Sci. Rev. **6**(1), 1–25 (2012)

11. Nalbantoglu, Ö.U., Russell, D.J., Sayood, K.: Data compression concepts and algorithms and their applications to bioinformatics. Entropy **12**, 34–52 (2009)

12. Bhattacharyya, M., Bhattacharyya, M., Bandyopadhyay, S.: Recent directions in compressing next generation sequencing data. CBIO **7**, 2–6 (2012)

13. Deorowicz, S., Grabowski, S.: Data compression for sequencing data. Algorithms Mol. Biol. **8**, 25 (2013)

14. Giancarlo, R., Rombo, S.E., Utro, F.: Compressive biological sequence analysis and archival in the era of high-throughput sequencing technologies. Brief. Bioinform. **15**, 390–406 (2014)

15. Bakr, N.S., Sharawi, A.A.: DNA lossless compression algorithms: review. Am. J. Bioinf. Res. **3**(3), 72–81 (2013)

16. Wandelt, S., Bux, M., Leser, U.: Trends in genome compression. Curr. Bioinform. **9**, 315–326 (2014)

17. Hosseini, M., Pratas, D., Pinho, A.J.: A survey on data compression methods for biological sequences. Information **7**, 56 (2016)

18. Biji, C.L., Nair, A.S.: Benchmark dataset for whole genome sequence compression. IEEE/ACM Trans. Comput. Biol. Bioinform. **14**, 1228–1236 (2017)

19. Nomenclature for incompletely specified bases in nucleic acid sequences. Recommendations 1984. Nomenclature committee of the international union of biochemistry (NC-IUB). Proc. Natl. Acad. Sci. U.S.A. **83**, 4–8 (1986)

20. Mohammed, M.H., Dutta, A., Bose, T., Chadaram, S., Mande, S.S.: DELIMINATE-a fast and efficient method for loss-less compression of genomic sequences: sequence analysis. Bioinformatics **28**, 2527–2529 (2012)

21. Pinho, A.J., Pratas, D.: MFCompress: a compression tool for FASTA and multi-FASTA data. Bioinformatics **30**, 117–118 (2014)

22. Mann, H.B., Whitney, D.R.: Institute of mathematical statistics is collaborating with JSTOR to digitize, preserve, and extend access to the annals of mathematical statistics. Ann. Stat. 50–60. ® https://www.jstor.org/

23. Friedman, M.: The use of ranks to avoid the assumption of normality implicit in the analysis of variance. J. Am. Stat. Assoc. **32**(200), 675–701 (1937)

24. Fisher, R.: Statistical methods and scientific induction (1955)

25. Nemenyi, P.: Distribution-Free Multiple Comparisons (1963)

26. Haubold, B., Wiehe, T.: How repetitive are genomes? BMC Bioinf. **7**(1), 541 (2006)

27. Shannon, C.E.: A mathematical theory of communication. Bell Syst. Tech. J. **27**, 379–423 (1948)

Stability of Features Describing the Dynamic Signature Biometric Attribute

Marcin Zalasiński[1]($\boxtimes$), Krzysztof Cpałka[1], and Konrad Grzanek[2,3]

[1] Institute of Computational Intelligence, Częstochowa University of Technology, Częstochowa, Poland
{marcin.zalasinski,krzysztof.cpalka}@iisi.pcz.pl
[2] Information Technology Institute, University of Social Sciences, 90-113 Lodź, Poland
kgrzanek@spoleczna.pl
[3] Clark University, Worcester, MA 01610, USA

Abstract. Behavioral biometric attributes tend to change over time. Due to this, analysis of their changes is an important issue in the context of identity verification. In this paper, we present an evaluation of stability of features describing the dynamic signature biometric attribute. The dynamic signature is represented by nonlinear waveforms describing dynamics of the signing process. Our analysis takes into account a set of features extracted using a partitioning of the signature in comparison to so-called global features of the signature. It shows which features change more and how it is associated with identification efficiency. Our simulations were performed using ATVS-SLT DB dynamic signature database.

Keywords: Biometrics · Dynamic signature
Evaluation of signature stability

1 Introduction

Biometrics is a science of recognizing the identity of a person based on some kind of unique personal attributes [6,9]. A signature is one of these attributes. In the literature we can find two types of the signature biometric attribute-dynamic (on-line, [18,29,43]) and static (off-line, [5,21,41]).

Methods used for identity verification on the basis of the dynamic signature can be divided into few groups. One of them uses e.g. features describing whole signature (so-called global features, [14]). Others compare directly waveforms describing signatures [28]. The third group of methods used for the signature verification uses characteristic features of the signature extracted from its regions [17]. In this paper, we focus on two kinds of features: global features and features extracted from the signature regions.

Behavioral biometric attributes tend to change over time, so analysis of the features' changes is an important issue in the context of identity verification.

© Springer International Publishing AG, part of Springer Nature 2018
L. Rutkowski et al. (Eds.): ICAISC 2018, LNAI 10842, pp. 250–261, 2018.
https://doi.org/10.1007/978-3-319-91262-2_23

The purpose of this paper is an evaluation of stability of features describing the dynamic signature biometric attribute. It is realized using properly defined measures of change. Moreover, we also show how changes of features are associated with the identification efficiency.

Structure of the paper is as follows: Sect. 2 contains a description of the signature features considered in the paper, Sect. 3 presents a description of adopted criteria for evaluating the dynamic signature variability over time, Sect. 4 shows simulation results, conclusions are drawn in Sect. 5.

2   Introduction to the Dynamic Signature Verification Methods and Features

In this paper, we focus on two types of methods used for identity verification based on a signature and two types of features describing the signature.

First of them is a method based on the dynamic signature partitioning which uses templates created in the selected partitions of the signature. It consists of the following steps: **1. Creation of the signature partitions.** The partitions are created individually for each user on the basis of his/her reference signatures. In this process values of pen velocity and pressure signals and values of time moments of the signing process are used. **2. Determination of templates in the partitions.** Each partition contains signals of trajectories x and y, for which templates $\mathbf{tc}_{i,p,r}^{\{s,a\}}$ are created, where i is the user index, $\{p,r\}$ are indices indicating the partition (p is the vertical section index, r is the horizontal section index), s is the type of signal used to create partition (velocity v or pressure z) and a is the type of trajectory used to create the template (x or y). The templates are average values of trajectory signals a of the reference signatures of the user i in the partition denoted by indices $\{p,r\}$. **3. Creation of the classifier.** For each considered user a flexible neuro-fuzzy one-class classifier is created. Neuro-fuzzy systems [11–13,35] combine the natural language description of fuzzy systems [1–4,8,16,31,33,36–38] and the learning properties of neural networks [7,19,20,22,27,30,32,34,39]. Parameters of the classifier are determined on the basis of the values of reference signatures' signals in partitions. **4. Identity verification.** This process is performed by the classifier using distance values $dp_{i,j,p,r}^{\{s,a\}}$ between trajectory signals of the signature j and the template, determined for each partition.

The second method is based on so-called global features, which contain information about signature characteristic, e.g. signature length, number of pen-ups, etc. It consists of the following steps: **1. Extraction of global features values.** Values of all global features $g_{i,n,j}$, where n is the number of the feature considered in the method have to be determined. **2. Determination of global features templates.** The templates $\overline{g}_{i,n}$ are average values of global features extracted from all reference signatures. **3. Creation of the classifier.** For each considered user a flexible neuro-fuzzy one-class classifier is created. Parameters of the classifier are determined on the basis of the values of reference signatures' global features. **4. Identity verification.** This process is performed by the classifier using distance values $dg_{i,n,jr}$ between global features of the signature j and the template, determined for each global feature. In this paper, we consider a subset of the least variable global features defined in [15]. However, the subset of features can be selected in a different way, e.g. using population-based algorithms [10, 23, 25, 26].

A more detailed description of the methods mentioned above can be found in our previous works [12, 42, 44]. The remainder of this article presents a way of variability analysis of the features describing the dynamic signature (Sect. 3).

3   Description of the Adopted Criteria for Evaluation of the Dynamic Signature Features' Stability

Analysis of the stability of the dynamic signature features is based on the defined criteria. Values of the criteria are determined for each acquisition session nS. The criteria can be described as follows:

Average Distance Between Feature and Its Reference Value. It is used to determine variability level of the dynamic signature features in subsequent acquisition sessions in relation to the reference feature. For features used by method based on partitioning this coefficient is defined as follows [45]:

$$\overline{dp}_{i,p,r,nS}^{\{s,a\}} = \frac{1}{J} \sum_{j=1}^{J} dp_{i,j,p,r,nS}^{\{s,a\}}. \tag{1}$$

This coefficient can be defined analogously for global features. It has the following form:

$$\overline{dg}_{i,n,nS} = \frac{1}{J} \sum_{j=1}^{J} dg_{i,n,j,nS}. \tag{2}$$

The Standard Deviation of Distances Between Feature and Its Reference Value. It is used to determine dispersion level of the dynamic signature features in subsequent acquisition sessions. For features used by method based on partitioning this coefficient is defined as follows [45]:

$$\sigma p_{i,p,r,nS}^{\{s,a\}} = \sqrt{\frac{1}{J} \sum_{j=1}^{J} \left(\overline{dp}_{i,p,r,nS}^{\{s,a\}} - dp_{i,j,q,nS}^{\{s,a\}} \right)^2}. \tag{3}$$

This coefficient can be defined analogously for global features. It has the following form:

$$\sigma g_{i,n,j,nS} = \sqrt{\frac{1}{J} \sum_{j=1}^{J} \left(\overline{dg}_{i,n,j,nS} - dg_{i,n,j,nS} \right)^2}. \tag{4}$$

The Product of the Average and Variance Relative Variation of the Mentioned Distances Between Two Acquisition Sessions. It has been proposed in the paper [15]. It is used to determine the most stable features if the signature. For features used by method based on partitioning this coefficient is defined as follows [45]:

$$VCp_{i,p,r,nS}^{\{s,a\}} = \left| \overline{dp}_{i,p,r,nS}^{\{s,a\}} - \overline{dp}_{i,p,r,nS=1}^{\{s,a\}} \right| \cdot \left| \frac{\sigma p_{i,p,r,nS}^{\{s,a\}}}{\overline{dp}_{i,p,r,nS}^{\{s,a\}}} - \frac{\sigma p_{i,p,r,nS=1}^{\{s,a\}}}{\overline{dp}_{i,p,r,nS=1}^{\{s,a\}}} \right|. \tag{5}$$

This coefficient can be defined analogously for global features. It has the following form:

$$VCg_{i,n,nS} = \left| \overline{dg}_{i,n,nS} - \overline{dg}_{i,n,nS=1} \right| \cdot \left| \frac{\sigma g_{i,n,nS}}{\overline{dg}_{i,n,nS}} - \frac{\sigma g_{i,n,nS=1}}{\overline{dg}_{i,n,nS=1}} \right|. \tag{6}$$

Next sections of the paper present simulations scenario, simulation results, and conclusions.

4   Simulation Results

Simulations were performed using authorial test environment written in C# for ATVS-SLT database [15] which contains signatures of 27 users, created in 6 sessions. First 4 sessions contain 4 signatures of each user and 2 last sessions contain 15 signatures of each user.

During simulations, we used two methods described in Sect. 2-one based on the templates extracted from partitions and the second based on global features. For each method training phase was performed individually for each user, taking into account 4 signatures from the first session ($nS = 1$). In the method using partitioning, we assumed that each signature was partitioned into 2 vertical sections and 2 horizontal sections ($p = 2$, $r = 2$). In the method using global features we use 10 the least variable features described in [15]. All features values used in both methods were normalized to the range $[0,1]$. In the test phase, we used 5 remaining sessions to evaluate the efficiency of the signature verification, which is expressed using coefficients FAR, FRR, and ERR commonly used in biometrics [40]. These results are shown in Tables 1 and 2.

Moreover, we also determined values of stability criteria presented in Sect. 3. They were determined for $nS = [2, 3, \ldots, 6]$ and they are presented in Tables 3, 4, 5, 6, 7 and 8. It should be noted that the simulations were carried out five times and the results were averaged.

Conclusions from the simulations can be summarized as follows:

- Values of all defined criteria for evaluation of stability of the dynamic signature features tend to increase over time.

Table 1. Identity verification errors for method using templates created in the partitions averaged in the context of all users from the database ATVS-SLT DB.

nS	FAR	FRR	EER
2	3.70%	2.78%	3.24%
3	4.63%	4.63%	4.63%
4	4.63%	4.63%	4.63%
5	4.20%	12.84%	8.52%
6	6.17%	11.36%	8.77%

Table 2. Identity verification errors for method using global features averaged in the context of all users from the database ATVS-SLT DB.

nS	FAR	FRR	EER
2	0.10%	0.14%	0.24%
3	0.33%	0.36%	0.69%
4	0.56%	0.76%	1.32%
5	0.74%	1.12%	1.86%
6	0.76%	1.16%	1.92%

- The lowest variability level associated with the value of coefficient $\overline{dp}_{i,p,r,nS}^{\{s,a\}}$ is related to the trajectory x from partition denoted by indices $\{p = 0, r = 0\}$, created on the basis of the signal v (see Table 3). The lowest variability level associated with the value of coefficient $\overline{dg}_{i,n,nS}$ is related to the feature number 97 (see Table 4).
- The lowest dispersion level associated with the value of coefficient $\sigma p_{i,p,r,nS}^{\{s,a\}}$ is related to the trajectory x from partition denoted by indices $\{p = 0, r = 0\}$, created on the basis of the signal v (see Table 5). The lowest dispersion level associated with the value of coefficient $\sigma g_{i,n,nS}$) is related to the feature number 97 (see Table 6).
- The most stable feature determined on the basis of the value of coefficient $VCp_{i,p,r,nS}^{\{s,a\}}$ is the trajectory y from partition denoted by indices $\{p = 1, r = 0\}$, created on the basis of the signal v (see Table 7). The most stable feature determined on the basis of the value of coefficient $VCg_{i,n,nS}$ is the feature number 93 (see Table 8).
- The system used for identity verification tends to decrease the verification accuracy over time (see Tables 1 and 2). This situation can be observed in case of both types of features. The trend of increasing verification error over time is consistent with the trend of increasing values of the coefficients used for evaluation of the dynamic signature stability.
- Taking into account values of parameters used for evaluation of the dynamic signature stability we can see that changes in global features values are higher than changes of templates created in the partitions. However, it doesn't affect verification efficiency. It seems that this is not a key parameter in the evaluation of the feature usefulness in the verification process.

Table 3. Values of the coefficient $\overline{dp}_{i,p,r,nS}^{\{s,a\}}$ averaged in the context of all users from the database ATVS-SLT DB.

Template (s, a, p, r)	$nS = 2$	$nS = 3$	$nS = 4$	$nS = 5$	$nS = 6$	Average
$v, x, 0, 0$	0.2648	0.2518	0.2494	0.2921	0.2861	0.2688
$v, x, 0, 1$	0.2896	0.2715	0.2512	0.3366	0.3120	0.2922
$v, x, 1, 0$	0.5346	0.5079	0.5535	0.4897	0.5333	0.5238
$v, x, 1, 1$	0.4115	0.3739	0.4212	0.3955	0.4406	0.4085
$v, y, 0, 0$	0.2823	0.2732	0.2720	0.3460	0.3289	0.3005
$v, y, 0, 1$	0.2981	0.2682	0.2807	0.3675	0.3713	0.3171
$v, y, 1, 0$	0.6070	0.5700	0.5680	0.5545	0.5483	0.5696
$v, y, 1, 1$	0.4356	0.4457	0.4413	0.4729	0.4950	0.4581
$z, x, 0, 0$	0.3094	0.2920	0.3164	0.3387	0.3434	0.3200
$z, x, 0, 1$	0.3585	0.3423	0.3390	0.4035	0.4140	0.3715
$z, x, 1, 0$	0.6346	0.5861	0.5909	0.5725	0.5895	0.5947
$z, x, 1, 1$	0.5118	0.4839	0.4699	0.4670	0.5002	0.4866
$z, y, 0, 0$	0.3286	0.3023	0.2991	0.3620	0.3514	0.3287
$z, y, 0, 1$	0.4041	0.3795	0.3685	0.4509	0.4586	0.4123
$z, y, 1, 0$	0.6125	0.5878	0.5687	0.5580	0.5871	0.5828
$z, y, 1, 1$	0.4964	0.4529	0.4778	0.5042	0.5222	0.4907
Average	0.4237	0.3993	0.4042	0.4320	0.4426	-

Table 4. Values of the coefficient $\overline{dg}_{i,n,nS}$ averaged in the context of all users from the database ATVS-SLT DB.

Feature number (n)	$nS = 2$	$nS = 3$	$nS = 4$	$nS = 5$	$nS = 6$	Average
3	1.6620	1.7361	2.4676	2.7043	2.7698	2.2680
7	0.0073	0.0175	0.0190	0.0211	0.0193	0.0169
17	0.0383	0.0431	0.0389	0.0490	0.0460	0.0430
38	0.0319	0.0347	0.0357	0.0579	0.0567	0.0434
45	0.0179	0.0182	0.0202	0.0223	0.0242	0.0205
58	0.0445	0.0760	0.0524	0.0811	0.0779	0.0664
59	0.0538	0.0653	0.0465	0.0751	0.0710	0.0623
72	0.2482	0.3917	0.4058	0.4524	0.4504	0.3897
93	0.0044	0.0047	0.0042	0.0049	0.0047	0.0046
97	0.0033	0.0037	0.0040	0.0043	0.0045	0.0040
Average	0.2112	0.2391	0.3094	0.3472	0.3525	-

Table 5. Values of the coefficient $\sigma p_{i,p,r,nS}^{\{s,a\}}$ averaged in the context of all users from the database ATVS-SLT DB.

Template (s, a, p, r)	$nS = 2$	$nS = 3$	$nS = 4$	$nS = 5$	$nS = 6$	Average
$v, x, 0, 0$	0.0763	0.0533	0.0659	0.0993	0.0633	0.0716
$v, x, 0, 1$	0.1044	0.0904	0.0844	0.1456	0.0811	0.1012
$v, x, 1, 0$	0.1422	0.1304	0.1563	0.1707	0.1930	0.1585
$v, x, 1, 1$	0.1252	0.0922	0.1170	0.1500	0.1819	0.1333
$v, y, 0, 0$	0.0711	0.0581	0.0830	0.1019	0.1000	0.0828
$v, y, 0, 1$	0.0896	0.0681	0.0941	0.1393	0.1107	0.1004
$v, y, 1, 0$	0.1567	0.1711	0.1507	0.1904	0.2000	0.1738
$v, y, 1, 1$	0.1393	0.1159	0.1422	0.1556	0.1822	0.1470
$z, x, 0, 0$	0.0900	0.0570	0.0700	0.0993	0.0967	0.0826
$z, x, 0, 1$	0.1104	0.0893	0.0970	0.1267	0.1111	0.1069
$z, x, 1, 0$	0.1300	0.1174	0.1393	0.1678	0.1600	0.1429
$z, x, 1, 1$	0.1159	0.1144	0.1174	0.1219	0.1378	0.1215
$z, y, 0, 0$	0.0807	0.0800	0.0919	0.1222	0.0907	0.0931
$z, y, 0, 1$	0.0993	0.1037	0.1267	0.1422	0.1263	0.1196
$z, y, 1, 0$	0.1274	0.1500	0.1393	0.1630	0.1807	0.1521
$z, y, 1, 1$	0.1356	0.1111	0.1307	0.1433	0.1600	0.1361
Average	0.1121	0.1002	0.1129	0.1399	0.1360	-

Table 6. Values of the coefficient $\sigma g_{i,n,nS}$ averaged in the context of all users from the database ATVS-SLT DB.

Feature number (n)	$nS = 2$	$nS = 3$	$nS = 4$	$nS = 5$	$nS = 6$	Average
3	1.1689	0.9483	1.3955	1.5958	1.5144	1.3246
7	0.0044	0.0089	0.0078	0.0083	0.0078	0.0074
17	0.0223	0.0184	0.0248	0.0347	0.0333	0.0267
38	0.0202	0.0197	0.0194	0.0311	0.0289	0.0239
45	0.0100	0.0114	0.0132	0.0146	0.0162	0.0131
58	0.0234	0.0612	0.0320	0.0440	0.0375	0.0396
59	0.0505	0.0535	0.0253	0.0484	0.0357	0.0427
72	0.1581	0.2038	0.2007	0.2057	0.1875	0.1912
93	0.0027	0.0029	0.0026	0.0035	0.0033	0.0030
97	0.0020	0.0022	0.0024	0.0028	0.0028	0.0024
Average	0.1462	0.1330	0.1724	0.1989	0.1867	-

Table 7. Values of the coefficient $VCp_{i,p,r,nS}^{\{s,a\}}$ averaged in the context of all users from the database ATVS-SLT DB.

Template (s,a,p,r)	$nS = 2$	$nS = 3$	$nS = 4$	$nS = 5$	$nS = 6$	Average
$v, x, 0, 0$	0.0003	0.0010	0.0002	0.0035	0.0027	0.0015
$v, x, 0, 1$	0.0012	0.0001	0.0002	0.0085	0.0030	0.0026
$v, x, 1, 0$	0.0002	0.0005	0.0001	0.0023	0.0006	0.0007
$v, x, 1, 1$	0.0009	0.0001	0.0000	0.0018	0.0085	0.0023
$v, y, 0, 0$	0.0003	0.0005	0.0010	0.0047	0.0045	0.0022
$v, y, 0, 1$	0.0024	0.0001	0.0021	0.0152	0.0067	0.0053
$v, y, 1, 0$	0.0003	0.0012	0.0003	0.0009	0.0005	0.0006
$v, y, 1, 1$	0.0002	0.0007	0.0003	0.0014	0.0047	0.0015
$z, x, 0, 0$	0.0006	0.0008	0.0000	0.0014	0.0014	0.0008
$z, x, 0, 1$	0.0011	0.0009	0.0019	0.0026	0.0016	0.0016
$z, x, 1, 0$	0.0001	0.0006	0.0014	0.0061	0.0035	0.0023
$z, x, 1, 1$	0.0000	0.0004	0.0012	0.0018	0.0009	0.0009
$z, y, 0, 0$	0.0007	0.0002	0.0000	0.0073	0.0019	0.0020
$z, y, 0, 1$	0.0003	0.0009	0.0027	0.0072	0.0029	0.0028
$z, y, 1, 0$	0.0003	0.0001	0.0005	0.0023	0.0002	0.0007
$z, y, 1, 1$	0.0017	0.0008	0.0004	0.0026	0.0052	0.0021
Average	0.0007	0.0006	0.0008	0.0044	0.0030	-

Table 8. Values of the coefficient $VCg_{i,n,nS}$ averaged in the context of all users from the database ATVS-SLT DB.

Feature number (n)	$nS = 2$	$nS = 3$	$nS = 4$	$nS = 5$	$nS = 6$	Average
3	1.2640	1.6422	2.6229	2.8082	3.0417	2.2758
7	0.0021	0.0054	0.0046	0.0047	0.0040	0.0042
17	0.0103	0.0103	0.0115	0.0222	0.0157	0.0140
38	0.0101	0.0089	0.0113	0.0185	0.0157	0.0129
45	0.0036	0.0042	0.0058	0.0048	0.0066	0.0050
58	0.0110	0.0435	0.0148	0.0240	0.0191	0.0225
59	0.0358	0.0391	0.0129	0.0266	0.0175	0.0264
72	0.0306	0.0447	0.0379	0.0702	0.0673	0.0501
93	0.0010	0.0012	0.0011	0.0016	0.0013	0.0012
97	0.0013	0.0013	0.0013	0.0015	0.0016	0.0014
Average	0.1370	0.1801	0.2724	0.2982	0.3191	-

5   Conclusions

In this paper, we analyzed the stability of features describing the dynamic signature biometric attribute. The analysis was performed taking into account signatures acquired in different sessions. Between them, there was a minimum two-month interval. It was assumed that the basic shape of the signature of each user is not radically changing but it only evolves. Our approach has only an informative meaning and can be used to select the most stable features describing the dynamic signature or other biometric feature.

References

1. Bartczuk, Ł., Dziwiński, P., Starczewski, J.T.: New method for generation type-2 fuzzy partition for FDT. In: Rutkowski, L., Scherer, R., Tadeusiewicz, R., Zadeh, L.A., Zurada, J.M. (eds.) ICAISC 2010. LNCS (LNAI), vol. 6113, pp. 275–280. Springer, Heidelberg (2010). https://doi.org/10.1007/978-3-642-13208-7_35
2. Bartczuk, Ł., Łapa, K., Koprinkova-Hristova, P.: A new method for generating of fuzzy rules for the nonlinear modelling based on semantic genetic programming. In: Rutkowski, L., Korytkowski, M., Scherer, R., Tadeusiewicz, R., Zadeh, L.A., Zurada, J.M. (eds.) ICAISC 2016. LNCS (LNAI), vol. 9693, pp. 262–278. Springer, Cham (2016). https://doi.org/10.1007/978-3-319-39384-1_23
3. Bartczuk, Ł., Przybył, A., Cpałka, K.: A new approach to nonlinear modelling of dynamic systems based on fuzzy rules. Int. J. Appl. Math. Comput. Sci. **26**(3), 603–621 (2016)
4. Beg, I., Rashid, T.: Modelling uncertainties in multi-criteria decision making using distance measure and TOPSIS for hesitant fuzzy sets. J. Artif. Intell. Soft Comput. Res. **7**(2), 103–109 (2017)
5. Batista, L., Granger, E., Sabourin, R.: Dynamic selection of generative discriminative ensembles for off-line signature verification. Pattern Recogn. **45**, 1326–1340 (2012)
6. Bobulski, J.: 2DHMM-based face recognition method. In: Choraś, R.S. (ed.) Image Processing and Communications Challenges 7. AISC, vol. 389, pp. 11–18. Springer, Cham (2016). https://doi.org/10.1007/978-3-319-23814-2_2
7. Bologna, G., Hayashi, Y.: Characterization of symbolic rules embedded in deep DIMLP networks: a challenge to transparency of deep learning. J. Artif. Intell. Soft Comput. Res. **7**(4), 265–286 (2017)
8. Chang, O., Constante, P., Gordon, A., Singana, M.: A novel deep neural network that uses space-time features for tracking and recognizing a moving object. J. Artif. Intell. Soft Comput. Res. **7**(2), 125–136 (2017)
9. Connor, P., Ross, A.: Biometric recognition by gait: a survey of modalities and features. Comput. Vis. Image Underst. **167**, 1–27 (2018)
10. Cpałka, K., Łapa, K., Przybył, A.: A new approach to design of control systems using genetic programming. Inf. Technol. Control **44**(4), 433–442 (2015)
11. Cpałka, K., Rebrova, O., Nowicki, R., Rutkowski, L.: On design of flexible neuro-fuzzy systems for nonlinear modelling. Int. J. Gen. Syst. **42**(6), 706–720 (2013)
12. Cpałka, K., Zalasiński, M., Rutkowski, L.: A new algorithm for identity verification based on the analysis of a handwritten dynamic signature. Appl. Soft Comput. **43**, 47–56 (2016)

13. Dziwiński, P., Avedyan, E.D.: A new method of the intelligent modeling of the non-linear dynamic objects with fuzzy detection of the operating points. In: Rutkowski, L., Korytkowski, M., Scherer, R., Tadeusiewicz, R., Zadeh, L.A., Zurada, J.M. (eds.) ICAISC 2016. LNCS (LNAI), vol. 9693, pp. 293–305. Springer, Cham (2016). https://doi.org/10.1007/978-3-319-39384-1_25

14. Faundez-Zanuy, M.: On-line signature recognition based on VQ-DTW. Pattern Recogn. **40**, 981–992 (2007)

15. Galbally, J., Martinez-Diaz, M., Fierez, J.: Aging in biometrics: an experimental analysis on on-line signature. PLoS One **8**(7), e69897 (2013)

16. Grycuk, R., Gabryel, M., Scherer, R., Voloshynovskiy, S.: Multi-layer architecture for storing visual data based on WCF and microsoft SQL server database. In: Rutkowski, L., Korytkowski, M., Scherer, R., Tadeusiewicz, R., Zadeh, L.A., Zurada, J.M. (eds.) ICAISC 2015. LNCS (LNAI), vol. 9119, pp. 715–726. Springer, Cham (2015). https://doi.org/10.1007/978-3-319-19324-3_64

17. Ibrahim, M.T., Khan, M.A., Alimgeer, K.S., Khan, M.K., Taj, I.A., Guan, L.: Velocity and pressure-based partitions of horizontal and vertical trajectories for on-line signature verification. Pattern Recogn. **43**, 2817–2832 (2010)

18. Jeong, Y.S., Jeong, M.K., Omitaomu, O.A.: Weighted dynamic time warping for time series classification. Pattern Recogn. **44**, 2231–2240 (2011)

19. Ke, Y., Hagiwara, M.: An English neural network that learns texts, finds hidden knowledge, and answers questions. J. Artif. Intell. Soft Comput. Res. **7**(4), 229–242 (2017)

20. Khan, N.A., Shaikh, A.: A smart amalgamation of spectral neural algorithm for nonlinear Lane-Emden equations with simulated annealing. J. Artif. Intell. Soft Comput. Res. **7**(3), 215–224 (2017)

21. Kumar, R., Sharma, J.D., Chanda, B.: Writer-independent off-line signature verification using surroundedness feature. Pattern Recogn. Lett. **33**, 301–308 (2012)

22. Laskowski, Ł., Laskowska, M., Jelonkiewicz, J., Boullanger, A.: Spin-glass implementation of a hopfield neural structure. In: Rutkowski, L., Korytkowski, M., Scherer, R., Tadeusiewicz, R., Zadeh, L.A., Zurada, J.M. (eds.) ICAISC 2014. LNCS (LNAI), vol. 8467, pp. 89–96. Springer, Cham (2014). https://doi.org/10.1007/978-3-319-07173-2_9

23. Łapa, K., Cpałka, K.: On the application of a hybrid genetic-firework algorithm for controllers structure and parameters selection. In: Borzemski, L., Grzech, A., Świątek, J., Wilimowska, Z. (eds.) ISAT 2015. AISC, vol. 429, pp. 111–123. Springer, Cham (2016). https://doi.org/10.1007/978-3-319-28555-9_10

24. Łapa, K., Cpałka, K., Galushkin, A.I.: A new interpretability criteria for neuro-fuzzy systems for nonlinear classification. In: Rutkowski, L., Korytkowski, M., Scherer, R., Tadeusiewicz, R., Zadeh, L.A., Zurada, J.M. (eds.) ICAISC 2015. LNCS (LNAI), vol. 9119, pp. 448–468. Springer, Cham (2015). https://doi.org/10.1007/978-3-319-19324-3_41

25. Łapa, K., Szczypta, J., Saito, T.: Aspects of evolutionary construction of new flexible PID-fuzzy controller. In: Rutkowski, L., Korytkowski, M., Scherer, R., Tadeusiewicz, R., Zadeh, L.A., Zurada, J.M. (eds.) ICAISC 2016. LNCS (LNAI), vol. 9692, pp. 450–464. Springer, Cham (2016). https://doi.org/10.1007/978-3-319-39378-0_39

26. Szczypta, J., Łapa, K., Shao, Z.: Aspects of the selection of the structure and parameters of controllers using selected population based algorithms. In: Rutkowski, L., Korytkowski, M., Scherer, R., Tadeusiewicz, R., Zadeh, L.A., Zurada, J.M. (eds.) ICAISC 2014. LNCS (LNAI), vol. 8467, pp. 440–454. Springer, Cham (2014). https://doi.org/10.1007/978-3-319-07173-2_38

27. Liu, H., Gegov, A., Cocea, M.: Rule based networks: an efficient and interpretable representation of computational models. J. Artif. Intell. Soft Comput. Res. **7**(2), 111–123 (2017)
28. Maiorana, E.: Biometric cryptosystem using function based on-line signature recognition. Expert Syst. Appl. **37**, 3454–3461 (2010)
29. Manjunatha, K.S., Manjunath, S., Guru, D.S., Somashekara, M.T.: Online signature verification based on writer dependent features and classifiers. Pattern Recogn. Lett. **80**, 129–136 (2016)
30. Minemoto, T., Isokawa, T., Nishimura, H., Matsui, N.: Pseudo-orthogonalization of memory patterns for complex-valued and quaternionic associative memories. J. Artif. Intell. Soft Comput. Res. **7**(4), 257–264 (2017)
31. Nowicki, R., Scherer, R., Rutkowski, L.: A method for learning of hierarchical fuzzy systems. In: Intelligent Technologies-Theory and Applications, pp. 124–129 (2002)
32. Prasad, M., Liu, Y.-T., Li, D.-L., Lin, C.-T., Shah, R.R., Kaiwartya, O.P.: A new mechanism for data visualization with TSK-type preprocessed collaborative fuzzy rule based system. J. Artif. Intell. Soft Comput. Res. **7**(1), 33–46 (2017)
33. Riid, A., Preden, J.-S.: Design of fuzzy rule-based classifiers through granulation and consolidation. J. Artif. Intell. Soft Comput. Res. **7**(2), 137–147 (2017)
34. Rutkowski, L.: Adaptive probabilistic neural networks for pattern classification in time-varying environment. IEEE Trans. Neural Netw. **15**(4), 811–827 (2004)
35. Rutkowski, L., Cpałka, K.: Compromise approach to neuro-fuzzy systems. In: Proceedings of the 2nd Euro-International Symposium on Computation Intelligence. Frontiers in Artificial Intelligence and Applications, vol. 76, pp. 85–90 (2002)
36. Scherer, R.: Multiple Fuzzy Classification Systems. Springer, Heidelberg (2012). https://doi.org/10.1007/978-3-642-30604-4
37. Scherer, R., Rutkowski, L.: A fuzzy relational system with linguistic antecedent certainty factor. In: Rutkowski, L., Kacprzyk, J. (eds.) Neural Networks and Soft Computing. AINSC, vol. 19, pp. 563–569. Physica, Heidelberg (2003). https://doi.org/10.1007/978-3-7908-1902-1_86
38. Scherer, R., Rutkowski, L.: Connectionist fuzzy relational systems. In: Halgamuge, S.K., Wang, L. (eds.) Computational Intelligence for Modelling and Prediction. SCI, vol. 2, pp. 35–47. Springer, Heidelberg (2005). https://doi.org/10.1007/10966518_3
39. Tezuka, T., Claramunt, C.: Kernel analysis for estimating the connectivity of a network with event sequences. J. Artif. Intell. Soft Comput. Res. **7**(1), 17–31 (2017)
40. Yeung, D.-Y., Chang, H., Xiong, Y., George, S., Kashi, R., Matsumoto, T., Rigoll, G.: SVC2004: first international signature verification competition. In: Zhang, D., Jain, A.K. (eds.) ICBA 2004. LNCS, vol. 3072, pp. 16–22. Springer, Heidelberg (2004). https://doi.org/10.1007/978-3-540-25948-0_3
41. Yilmaz, M.B., Yanikoglu, B.: Score level fusion of classifiers in off-line signature verification. Inf. Fusion **32**(Part B), 109–119 (2016)
42. Zalasiński, M., Cpałka, K., Hayashi, Y.: New fast algorithm for the dynamic signature verification using global features values. In: Rutkowski, L., Korytkowski, M., Scherer, R., Tadeusiewicz, R., Zadeh, L.A., Zurada, J.M. (eds.) ICAISC 2015. LNCS (LNAI), vol. 9120, pp. 175–188. Springer, Cham (2015). https://doi.org/10.1007/978-3-319-19369-4_17
43. Zalasiński, M., Cpałka, K.: A new method for signature verification based on selection of the most important partitions of the dynamic signature. Neurocomputing **289**, 13–22 (2018)

44. Zalasiński, M., Cpałka, K., Rakus-Andersson, E.: An idea of the dynamic signature verification based on a hybrid approach. In: Rutkowski, L., Korytkowski, M., Scherer, R., Tadeusiewicz, R., Zadeh, L.A., Zurada, J.M. (eds.) ICAISC 2016. LNCS (LNAI), vol. 9693, pp. 232–246. Springer, Cham (2016). https://doi.org/10. 1007/978-3-319-39384-1_21

45. Zalasiński, M., Cpałka, K., Er, M.J.: Stability evaluation of the dynamic signature partitions over time. In: Rutkowski, L., Korytkowski, M., Scherer, R., Tadeusiewicz, R., Zadeh, L.A., Zurada, J.M. (eds.) ICAISC 2017. LNCS (LNAI), vol. 10245, pp. 733–746. Springer, Cham (2017). https://doi.org/10.1007/978-3-319-59063-9_66

Data Mining

Text Categorization Improvement
via User Interaction

Jakub Atroszko[1], Julian Szymański[1(✉)], David Gil[2], and Higinio Mora[2]

[1] Department of Computer Systems Architecture,
Gdansk University of Technology, Gdańsk, Poland
`jakub.atroszko@gmail.com, julian.szymanski@eti.pg.gda.pl`
[2] Department of Computer Science Technology and Computation,
University of Alicante, Alicante, Spain
`{david.gil,mora}@ua.es`

Abstract. In this paper, we propose an approach to improvement of text categorization using interaction with the user. The quality of categorization has been defined in terms of a distribution of objects related to the classes and projected on the self-organizing maps. For the experiments, we use the articles and categories from the subset of Simple Wikipedia. We test three different approaches for text representation. As a baseline we use Bag-of-Words with weighting based on Term Frequency-Inverse Document Frequency that has been used for evaluation of neural representations of words and documents: Word2Vec and Paragraph Vector. In the representation, we identify subsets of features that are the most useful for differentiating classes. They have been presented to the user, and his or her selection allow increase the coherence of the articles that belong to the same category and thus are close on the SOM.

Keywords: Text representation · Document categorization
Wikipedia · Word2Vec · Paragraph vector · Self-organizing maps

1 Introduction

Continuous growth of textual data creates demands on effective processing methods that enable key information retrieval. [1] The fundamental concept of reflecting relations in data in terms of the Vector Space Model, especially grouping and dispersing characteristic points that represent objects in order to present existing patterns provide benefits explained by authors in A Vector Space Model for Automatic Indexing. There are known models that provide meaningful representations such as Bag-of-Words with weighting based on Term Frequency-Inverse Document Frequency [2], distributed representation of words and documents also known as Word2Vec [3] and Paragraph Vector [4]. They were used with success to solve numerous problems using machine learning algorithms [5–7] but it is a human sole responsibility to evaluate the quality of the obtained results. One of

© Springer International Publishing AG, part of Springer Nature 2018
L. Rutkowski et al. (Eds.): ICAISC 2018, LNAI 10842, pp. 265–275, 2018.
https://doi.org/10.1007/978-3-319-91262-2_24

the possibilities to achieve this is to compare them to the results obtained by a human in the same task [8,9].

Text categorization is an example of such task and it is defined in terms of assignment of predefined categories to text documents [10]. Human expertise in it is usually limited to defining correct labels for whole documents [11].

Another great example of such task is clustering. One of the most popular data mining algorithms that can be used in categorization and visualization. The clustering is the task of finding groups of similar documents in a collection of documents [12]. It might be improved by eg.: by adding information from external lexical resources [13] or including user participation in the selection of the features that are used to distinguish between documents [14]. Such feature feedback can find its application in tasks related to filtering, personalization, recommendation [15].

Visual data mining applied to natural language domain can make finding unknown patterns much easier [16]. There are known methods such as Multidimensional scaling [17] which allow analyzing data in terms of distances among points that represents data in a geometric space. Visualizing data using t-SNE [18] is another one and it is based on the same fundamental concept that multidimensional data can be presented in a meaningful way in a two or three-dimensional map. Among different methods, there are self-organizing maps [19] which are similar to those discovered in the brains. They have the ability to clustering and highlighting abstract relationships [20].

Work presented in Clustering Wikipedia Search Results [21] and Self Organizing Maps for Visualization of Categories [22] already have shown the purposefulness of using the self-organizing maps [19] in the context of semantic text processing.

In this paper, we additionally test three different data representation models including neural embeddings as an input to the self-organizing maps and feature selection method. We define text categorization quality, based on the self-organizing maps, which is measurable but also intuitive for human via clustering and visualization. We present text categorization quality assessment methodology and show its usability with the experimental results obtained by the user.

2   Feature Selection

The selection of a smaller number of the representative features in some cases allows sufficient representation of multidimensional data. Feature selection methods can be categorized in different ways. Methods can be generally assigned to one of three models: filter, wrapper, and embedded [23] or distinguished as feature extraction and feature selection. Features can be extracted for example as in methods: Principle Component Analysis (PCA), Linear Discriminant Analysis (LDA) and Canonical Correlation Analysis (CCA) or selected as in Information Gain, Relief, Fisher Score or Lasso [24].

There are known several methods to evaluate single features or feature subsets such as inconsistency rate, inference correlation, classification error, fractal dimension, distance measure, etc. [25].

3 Our Approach

In this paper, we present an interactive approach to text categorization that can be defined as combinations of three components: **representation, evaluation, optimization** [26].

3.1 Representation

We took keywords from the Simple Wikipedia articles and let us denote a set of them as a dictionary D. Originally the words in the articles are ordered, therefore, let us call article's keywords as a vector of words or article vector a. Article vectors with the same category labels assigned by a human were concatenated by us into a new vector which we call category vector c. Order of the article vectors in the category vectors was chosen arbitrarily. The category label of category vector was set to be the same as the concatenated article vectors. Because both types of vectors are described by the same dictionary D, we use one common term for them - document vectors d. Weights of words or in the other words coordinates of such document vectors can be learned with a chosen data representation model.

As a baseline we used Bag-of-Words with weighting based on Term Frequency - Inverse Document Frequency (**TFIDF**) [2] that has been used for evaluation of a neural representations of words and documents: Word2Vec [3] (**W2V**) and Paragraph Vector [4] (**D2V**). Let us denote document vector d with weights learned using $R \in \{\mathbf{TFIDF}, \mathbf{W2V}, \mathbf{D2V}\}$ as $R[d]$ then:

$TFIDF[d]$ - vector representation of document d based on tf-idf weighting,
$D2V[d]$ - distributed representation of document d with Paragraph Vector,
$W2V[d]$ - vector representation of document d based on distributed representation of words, created with formula 1.

$$W2V[d] = [W2V[0][0], W2V[1][0], ..., W2V[i][0]..., W2V[|D| - 1][0]] \quad (1)$$

where:

$|D|$ - dictionary size,
$W2V[i][0]$ - the only weight (index 0) taken from 1-dimensional i-th word embedding learned with Word2Vec model. If i-th word was not present in document d then we used $W2V[i][0] = 0$.

Paragraph vector dimensionality was set to be equal $|D|$ consequently each representation had the same size. Vectors represented in the model R can be treated as rows of the matrix $\mathbf{R}$ where $R[d][i]$ is i-th weight in vector representation of document d.

3.2 Evaluation

Vector representations as defined in the previous Sect. 3.1 were used as an input to the self-organizing maps [19]. Hence each document d had its position on the resulting map associated with Cartesian coordinates x and y. Formula 2 presents this idea.

$$SOM(R[d]) = [x_d, y_d] \tag{2}$$

where:

$SOM(R[d])$ - document's d position on the self organizing map,
x_d - the x-coordinate of document d,
y_d - the y-coordinate of document d,

The coordinates were used to carry out categorization based on the process of matching points that represent the articles to the nearest points that represent the categories – similarly to the method known as k-Nearest Neighbour (kNN) [27]. The Euclidean distance metric 3 [28] presented below was used to carry out categorization.

$$d(SOM(R[a])), SOM(R[c]))) = \sqrt{|x_a - x_c|^2 + |y_a - y_c|^2} \tag{3}$$

where:

d - Euclidean distance between article a and category c on the SOM,
x_a - the x-coordinate of article a,
y_a - the y-coordinate of article a,
x_c - the x-coordinate of category c,
y_c - the y-coordinate of category c.

Cosine metric usually used in the natural language domain was also tested but the results for the Euclidean metric were better and allowed us to use formula 3 with success in our application.

Classification rate obtained with the above method due to the usage of self-organizing maps can be visually perceived by a human, therefore, it can be called text categorization quality. The formula 4 allows measuring the percentage result of achieving the expected grouping of objects near their classes and dispersing classes between each other.

$$Q = \frac{s}{a} \times 100 \tag{4}$$

where:

Q - categorization quality,
s - number of successfully categorized articles. Obtained by comparing the results of carried out classification to categories assigned by human.
a - total number of articles.

The higher is Q the more articles are grouped in relation to categories and categories are dispersed between each other.

3.3   Optimization

In order to see if we are able to increase text categorization quality (4) we decided to properly modify weights of selected features in vector representations (Sect. 3.1). Below we are defining used in interactive experiments a feature selection method Sect. 3.3.

1. Select the feature subset size $k \in N$ and $k \in <1, \frac{c}{t}>$. Where c is the number of columns in the matrix $\mathbf{R}$ (Sect. 3.1) and t is the number of top subsets of features in the ranking. If the remainder of the division is different than 0 then add columns at the end of matrix $\mathbf{R}$ filled with zeros to align. If $k \neq 1$ sort the columns of matrix $\mathbf{R}$ relative to the recommendations obtained by this algorithm with $k = 1$ then continue.
2. By transposing and collapsing every next k columns of sorted matrix $\mathbf{R}$ create matrix $\mathbf{F}$ where row $n \in <0, \frac{c}{k})$ is a vector representation of subset of features $[n * k, n * k + 1, ..., n * k + k - 1]$ and is in the form of $F[n] = [R[0][n * k], R[0][n * k + 1], ..., R[0][n * k + k - 1],$
$R[1][n * k], R[1][n * k + 1], ..., R[1][n * k + k - 1],$
...
$R[r - 1][n * k], R[r - 1][n * k + 1], ..., R[r - 1][n * k + k - 1]].$
where r is the number of rows in the matrix $\mathbf{R}$
If $k = 1$ then $\mathbf{F} = \mathbf{R}^T$.
3. Compute correlation matrix $\mathbf{C}$ using matrix $\mathbf{F}$.
4. For each row in $\mathbf{C}$ compute the average.
5. Sort the rows of the matrix $\mathbf{F}$ relative to the increasing average correlation value calculated in the previous step.
6. Recommended features are those related to the first t vectors from the sorted matrix $\mathbf{F}$.

4   Experiments and Results

4.1   Parameters Tuning

Parameter t of the feature selection method Sect. 3.3 was arbitrary set to be equal 10 due to the size of the graphical user interface. Different values of k were tested and we found out that each model $R \in \{\mathbf{TFIDF}, \mathbf{W2V}, \mathbf{D2V}\}$ used in the method behaves similarly after reaching some thresholding value of k when further increase does not change features already selected at first position. All chosen models had that point at $k = \frac{|D|}{10}$ for the input data presented below and taken from Simple Wikipedia.

- Domesticated animals
  - Pets
    * Cats
      · Cat breeds
    * Dogs
      · Dog breeds
    * Hamsters

4.2   Evaluation of Optimization Method

We decided to combine the evaluation method (Sect. 3.2) with the optimization method (Sect. 3.3) into an experimental methodology Sect. 4.2 presented below.

1. Input matrix **R** (Sect. 3.1) to the self-organizing map and to the feature selection algorithm (Sect. 3.3) with parameters $t = 10$ and $k = \frac{|D|}{10}$.
2. Select features recommended at first position and multiply their initial weight in vector representations by the power with base 10 and the exponent 256.
3. Input modified vectors to the self-organizing map.
4. Compare text categorization quality Q (4) for both maps. If $Q_1 - Q_0 < 0$ return to the second step and use the exponent 128.

Presented above scheme (Sect. 4.2) was applied to the input data T presented below which we divided into the subsets.

- Carnivores
  - Felines
    * Cats
      · Cat breeds
  - Canids
    * Dogs
      · Dog breeds

The first subset of the above collection used in the experiments contained articles related to two categories Cat breeds and Dog breeds alternatively formulated $T_1 = Catbreeds \cup Dogbreeds$. Then it was extended with parent categories Cats and Dogs so $T_2 = T_1 \cup Cats \cup Dogs$. Similarly, the third set included the parent categories Felines and Canids, $T_3 = T_2 \cup Felines \cup Canids$. The last collection extended with root category Carnivores combined all categories into one hierarchical structure - $T_4 = T_3 \cup Carnivores$. Due to computational cost, the collections were reduced to only three articles from each category.

The results of all conducted experiments based on the presented approach (Sect. 4.2) in this paper can be reviewed in the Table 1. At the intersection of the row and the column there is the difference between Q_1 and Q_0 let us denote it as **Improved** if $Q_1 - Q_0 > 0$ and **No change** in case of $Q_1 - Q_0 = 0$.

Table 1. The results of the experiments

T	BOW	W2V	D2V
1	Improved	Improved	No change
2	Improved	Improved	Improved
3	Improved	Improved	No change
4	No change	Improved	Improved

Making use of the fact that each result has its visual interpretation we can compare the resulting maps and see what has changed with optimization method. Figures 1 and 2 below present visualizations of self-organizing maps obtained in the experiments conducted on the fourth collection T_4 represented with Word2Vec model. They were created with The Databionic ESOM Tools [29]. The Fig. 1 presents the visualization of the self-organizing map on which the $Q_0 = 25.00\%$ was obtained. Similarly, the Fig. 2 presents the visualization of the self-organizing map after optimization process. In this case $Q_1 = 43.75\%$.

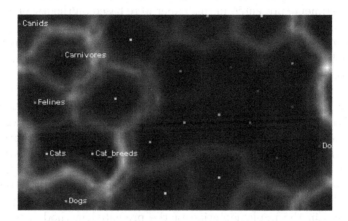

Fig. 1. Visualization of self-organizing map with text categorization quality $Q_0 = 25.00\%$ for T_4 and Word2Vec

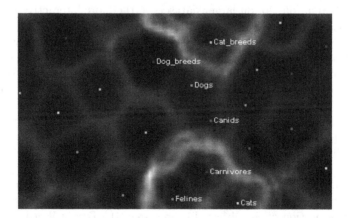

Fig. 2. Visualization of self-organizing map with text categorization quality $Q_1 = 43.75\%$ after the optimization process Sect. 3.3 for T_4 and Word2Vec

The first thing to notice is categories placement on the maps which have changed. In the Fig. 2 they are placed in the center of the map. The second thing is the highlighted distinction between articles. This can be noticed on the basis of the disappearance of the characteristic lowland from the first map on the second map.

5   Discussion and Conclusion

Two main perspectives can be distinguished. The first concerning the user and the second related to the potential applications of the presented methodology.

The experiments presented in this paper were based on the use of a graphical user interface in the program, which allowed to improve the quality of the text categorization. The user could choose selected by the recommendation algorithm features and adjust their weights as he wishes. The result of weighting could be viewed with two visualizations. The first one with unmodified vectors and the second one with modified weights of the selected words by the user. What is more, both visualizations had corresponding text categorization quality computed and displayed to the user. So he could judge if chosen weighting direction brings him to his goal. Therefore participation of human in the process of the text categorization allowed to improve the performance of machine learning algorithms. The weighting scheme applied in the experiments was established as a part of those experiments. Clearly and precisely defined goal in both qualitative and quantitative terms led to the promising results. Despite the small size of the datasets used in the implementation of the presented methodology numerous potential uses of the described method for big data exists. The task of designing the optimal feature weighing scheme was so simplified that the user was able to conduct all needed experiments in a straightforward way to establish the method of the quality improvement. We also plan to extend the experiments and perform their evaluation [30] on larger datasets in the more interactive way similar to the approach presented in [31]. To do this first we need to construct scalable hardware architecture that can be designed using our modeling software [32].

An interactive approach to the problem allows drawing conclusions about the properties of the individual data representation models. BOW was the simplest model to implement. D2V was the most difficult one but at the same time if the size of the vector was relatively small compared to other representations it allowed to accelerate the achievement of results with this representation used as the input to the self-organizing maps. W2V had the biggest vectors thus the time needed for self-organization was longest but the obtained results were the best.

If we consider the process of the implementation itself, it is worth noting the problem of mapping names understood by people into individual features in the vector representation of the object. In the case of the Bag-of-Words model, this is trivial. For embedded representations, it becomes more complicated because of their dispersed nature. The W2V model, due to the fine granularity could be described in terms of words understood by people. However, D2V operates on concepts of entire documents, therefore, the task of matching the words building the document with the features representing the object is significantly impeded.

In addition, weighing features for this model did not always affect the results. Lack of names understandable to people, as well as not always visible impact of weighing on the maps allow thinking that it is a representation that requires much more to do in order to adapt it to the interactive mode.

Comparing the result in the visual form gives us intuition about the interpretation of the corresponding classification rate in the form of the single number. Sometimes the visualizations of the resulting maps were incomprehensible and the number was the only way to know what exactly has happened. It can, therefore, be seen as a precise numerical measure with a visual interpretation.

The methodology presented in this paper is a tool for research in the area of representation learning, feature selection and quality assessment. The next steps that should be taken in the future may concern the use of much larger datasets. Implementing solutions based on effective big data processing such as parallel and distributed processing. Trying different visualization models and distance metrics. Testing precisely selected weights and different feature selection methods. Redefining method E to incorporate multi-label categorization in order to obtain the hierarchical categories structure quality assessment.

References

1. Tayal, S., Goel, S.K., Sharma, K.: A comparative study of various text mining techniques. In: 2015 2nd International Conference on Computing for Sustainable Global Development (INDIACom), pp. 1637–1642 (2015)
2. Schütze, H., Manning, C.D., Raghavan, P.: Introduction to Information Retrieval, pp. 117–119. Cambridge University Press, New York (2008)
3. Mikolov, T., Chen, K., Corrado, G., Dean, J.: Efficient estimation of word representations in vector space. CoRR abs/1301.3781 (2013)
4. Le, Q.V., Mikolov, T.: Distributed representations of sentences and documents. CoRR abs/1405.4053 (2014)
5. Mujtaba, G., Shuib, L., Raj, R.G., Rajandram, R., Shaikh, K.: Automatic text classification of ICD-10 related CoD from complex and free text forensic autopsy reports. In: 2016 15th IEEE International Conference on Machine Learning and Applications (ICMLA), pp. 1055–1058 (2016)
6. Mikolov, T., Sutskever, I., Chen, K., Corrado, G.S., Dean, J.: Distributed representations of words and phrases and their compositionality. In: Advances in Neural Information Processing Systems, pp. 3111–3119 (2013)
7. Bengio, Y., Courville, A., Vincent, P.: Representation learning: a review and new perspectives. IEEE Trans. Pattern Anal. Mach. Intell. **35**, 1798–1799 (2013)
8. Resnik, P.: Semantic similarity in a taxonomy: an information-based measure and its application to problems of ambiguity in natural language. CoRR abs/1105.5444, p. 95 (2011)
9. Finkelstein, L., Gabrilovich, E., Matias, Y., Rivlin, E., Solan, Z., Wolfman, G., Ruppin, E.: Placing search in context: the concept revisited. In: Proceedings of the 10th International Conference on World Wide Web, pp. 406–414. ACM (2001)
10. Yang, Y., Pedersen, J.O.: A comparative study on feature selection in text categorization. In: ICML, vol. 97, pp. 412–420 (1997)

11. Godbole, S., Harpale, A., Sarawagi, S., Chakrabarti, S.: Document classification through interactive supervision of document and term labels. In: Boulicaut, J.-F., Esposito, F., Giannotti, F., Pedreschi, D. (eds.) PKDD 2004. LNCS (LNAI), vol. 3202, pp. 185–196. Springer, Heidelberg (2004). https://doi.org/10.1007/978-3-540-30116-5_19

12. Allahyari, M., Pouriyeh, S., Assefi, M., Safaei, S., Trippe, E.D., Gutierrez, J.B., Kochut, K.: A brief survey of text mining: classification, clustering and extraction techniques. arXiv preprint arXiv:1707.02919 (2017)

13. Stanković, R., Krstev, C., Obradović, I., Kitanović, O.: Improving document retrieval in large domain specific textual databases using lexical resources. In: Nguyen, N.T., Kowalczyk, R., Pinto, A.M., Cardoso, J. (eds.) TCCI XXVI. LNCS, vol. 10190, pp. 162–185. Springer, Cham (2017). https://doi.org/10.1007/978-3-319-59268-8_8

14. Hu, Y., Milios, E.E., Blustein, J.: Interactive feature selection for document clustering. In: Proceedings of the 2011 ACM Symposium on Applied Computing, SAC 2011, pp. 1143–1150. ACM, New York (2011)

15. Raghavan, H., Madani, O., Jones, R.: Interactive feature selection. In: IJCAI, vol. 5, pp. 841–846 (2005)

16. Dzemyda, G., Kurasova, O., Žilinskas, J.: Multidimensional Data Visualization. SOIA, vol. 75. Springer, New York (2012). https://doi.org/10.1007/978-1-4419-0236-8

17. Borg, I., Groenen, P.J.F.: Modern Multidimensional Scaling: Theory and Applications. SSS. Springer, New York (2005). https://doi.org/10.1007/0-387-28981-X

18. van der Maaten, L., Hinton, G.: Visualizing data using t-SNE. J. Mach. Learn. Res. 9, 2579–2605 (2008)

19. Kohonen, T.: The self-organizing map. Proc. IEEE 78, 1464–1465, 1474 (1990)

20. Ultsch, A.: Emergence in self-organizing feature maps. University Library of Bielefeld (2007)

21. Szymański, J.: Self-organizing map representation for clustering Wikipedia search results. In: Nguyen, N.T., Kim, C.-G., Janiak, A. (eds.) ACIIDS 2011. LNCS (LNAI), vol. 6592, pp. 140–149. Springer, Heidelberg (2011). https://doi.org/10.1007/978-3-642-20042-7_15

22. Szymański, J., Duch, W.: Self organizing maps for visualization of categories. In: Huang, T., Zeng, Z., Li, C., Leung, C.S. (eds.) ICONIP 2012. LNCS, vol. 7663, pp. 160–167. Springer, Heidelberg (2012). https://doi.org/10.1007/978-3-642-34475-6_20

23. Zhao, Z., Morstatter, F., Sharma, S., Alelyani, S., Anand, A., Liu, H.: Advancing feature selection research. ASU feature selection repository, pp. 1–28 (2010)

24. Tang, J., Alelyani, S., Liu, H.: Feature selection for classification: a review. In: Data Classification: Algorithms and Applications, p. 37 (2014)

25. Vergara, J.R., Estévez, P.A.: A review of feature selection methods based on mutual information. Neural Comput. Appl. 24, 175–186 (2014)

26. Domingos, P.: A few useful things to know about machine learning. Commun. ACM 55, 78–87 (2012)

27. Kotsiantis, S.B., Zaharakis, I.D., Pintelas, P.E.: Machine learning: a review of classification and combining techniques. Artif. Intell. Rev. 26, 159–190 (2006)

28. Cha, S.H.: Comprehensive survey on distance/similarity measures between probability density functions. Int. J. Math. Models Methods Appl. Sci. 1, 300–302, 306 (2007)

29. Ultsch, A., Mörchen, F.: ESOM-maps: tools for clustering, visualization, and classification with emergent SOM. Technical report, Department of Mathematics and Computer Science, University of Marburg, Germany (2005)
30. Draszawka, K., Szymański, J.: External validation measures for nested clustering of text documents. In: Ryżko, D., Rybiński, H., Gawrysiak, P., Kryszkiewicz, M. (eds.) Emerging Intelligent Technologies in Industry. SCI, vol. 369, pp. 207–225. Springer, Heidelberg (2011). https://doi.org/10.1007/978-3-642-22732-5_18
31. Szymański, J., Duch, W.: Semantic memory knowledge acquisition through active dialogues. In: 2007 International Joint Conference on Neural Networks, IJCNN 2007, pp. 536–541. IEEE (2007)
32. Czarnul, P., Rościszewski, P., Matuszek, M., Szymański, J.: Simulation of parallel similarity measure computations for large data sets. In: 2015 IEEE 2nd International Conference on Cybernetics (CYBCONF), pp. 472–477. IEEE (2015)

Uncertain Decision Tree Classifier for Mobile Context-Aware Computing

Szymon Bobek$^{(\boxtimes)}$ (ID) and Piotr Misiak

AGH University of Science and Technology,
al. Mickiewicza 30, 30-059 Krakow, Poland
szymon.bobek@agh.edu.pl

Abstract. Knowledge discovery from uncertain data is one of the major challenges in building modern artificial intelligence applications. One of the greatest achievements in this area was made with a usage of machine learning algorithms and probabilistic models. However, most of these methods do not work well in systems which require intelligibility, efficiency and which operate on data are not only uncertain but also infinite. This is the most common case in mobile contex-aware computing. In such systems data are delivered in streaming manner, requiring from the learning algorithms to adapt their models iteratively to changing environment. Furthermore, models should be understandable for the user allowing their instant reconfiguration. We argue that all of these requirements can be met with a usage of incremental decision tree learning algorithm with modified split criterion. Therefore, we present a simple and efficient method for building decision trees from infinite training sets with uncertain instances and class labels.

Keywords: Decision trees · Uncertainty · Machine learning

1 Introduction

Introducing uncertainty into the learning process is an important research topic in the field of knowledge discovery across different areas of application. This is due to rapid development of technology such as mobile and wearable devices and cognitive services that altogether deliver huge amount of data to be processed. Such data is not only characterized by high volume, but also by high heterogeneity, as it comes from many different sources called context providers [2]. These sources, may vary with respect to the quality and certainty of delivered information requiring form mechanisms that processes it to handle both of these issues. Furthermore, the underlying model that these data is sampled from may change over time, as the user preferences and habits evolves and the environment conditions change.

In this paper we focus on context-aware [19] systems which operation strictly depends on large (possibly infinite) volumes of uncertain and heterogeneous data.

© Springer International Publishing AG, part of Springer Nature 2018
L. Rutkowski et al. (Eds.): ICAISC 2018, LNAI 10842, pp. 276–287, 2018.
https://doi.org/10.1007/978-3-319-91262-2_25

This data can be directly obtained from the mobile device sensors (so called low-level context), or that can be inferred based on this data (high-level context). In both cases the mechanism for modeling and processing such uncertain data is required.

Context-aware systems (CAS) are class of artificial intelligence hybrid solutions, that base their reasoning on large volumes of heterogeneous data. Mobile context-aware systems form a subclass of CAS that has been extensively developed over the last decade, along with rapid development of mobile and wearable devices [15]. The distinctive features about mobile CAS is that they operate in highly dynamic environments which may change rapidly or gradually, evolving over time [6,7]. Such feature, makes usage of static models impractical as they have to follow the changes that are present in the environment or in user needs, emotional state, preferences and goals.

In our previous works we identified requirements that mobile CAS system have to fulfill to assure its high quality [3,5]. These requirements are:

1. Intelligibility – mobile context-aware system should allow the user to understand and modify its performance.
2. Robustness – mobile context-aware system should provide adaptability with respect to the changing user habits or environment conditions, and should be able to model and process uncertain and incomplete data.
3. Privacy – sensitive data should be secured and not accessible by the third party.
4. Efficiency – mobile context-aware system should be efficient both in terms of resource efficiency and responsiveness.

In this paper we focus on two of the above aspects, namely: robustness (understood both as adaptability and uncertainty handling) and intelligibility in terms of providing a solution that will allow for semi-automatic knowledge discovery from data streams with uncertain or missing class labels that will be understandable for the end user. We used decision trees generation algorithm based on modified information gain split criterion, which takes into account uncertainty of data. Such an approach allows fro application of our algorithm to a wide range of methods for building decision trees, that are based on entropy measures (e.g. CVFDT [12], VFDT [9], etc.).

The rest of the paper is organised as follows. In Sect. 2 methods for handling uncertainty in training datasets was presented and motivation was stated. Description of the algorithm for generating uncertain decision trees was given in Sect. 3. Evaluation of the algorithm and comparison with selected state-of-the-art classifiers was presented in Sect. 4. In Sect. 5, we summarized the work and presented future work plans.

2 Approaches for Handling Uncertain Training Data

There are several approaches that allows to handle uncertain training data:

1. (DU) Discarding information about uncertainty and taking all training instances as they are.
2. (DI) Discarding instances, or attributes values that fall below the certainty threshold.
3. (PB) Using probability theory or statistics to handle missing or incomplete training instances.

While the first approach may appear very naive, this is actually what most of the machine learning algorithms do. The implicit assumption about the training data is that the uncertainty (if even exists) oscillates around the mean of the normal distribution. This allows to treat most of the data as certain, while only a fraction as incorrect. Due to the generalisation nature of machine learning algorithms, this incorrect data will be excluded from the model as a minority. However, in mobile context-aware systems a lot of data is of very low quality. Therefore, leaving the data as it is, may end up in models that have very low accuracy.

The second approach for handling uncertain training data assumes that there is a constant ϵ below which the instance (or only a value of an attribute which delivered uncertain information) is discarded from the training set. However, choosing the value of the ϵ has to be done empirically, and depends on the quality of sensors that the mobile device is equipped with. This makes the task non trivial, as there is a high variety of different devices and different hardware. Choosing wrong ϵ value may end up in a lot of data missing, or a lot of mistaken data in training sets. The example of the algorithm that is able to handle missing values of attributes is C4.5, which simply do not take into account instances with missing attributes while calculating split measures [18]. Although C4.5 propagates instances with missing labels down to the tree child nodes multiplying them with an appropriate weighting factor, the information about the uncertainty associated with each value in the instance is lost.

The third group form solutions based on probability theory and statistics. This includes UK-means [8] algorithm, a modification of k-means handles uncertain objects whose locations are represented by probability density functions. Hoverer it efficiency is limited, as it computes expected distances between objects and cluster representatives using costly numerical integrations. Uncertain Decision Trees (UDT) [21] rather than abstracting uncertain data by statistical derivatives (such as mean and median), the complete information of a data the probability density function is utilized. This results in better performance, but suffers from resource inefficiency and limited intelligibility, as the UDT tend to be complex structures. In [13] authors propose a UCVFDT algorithm that has the ability to handle examples with uncertain attributes with nominal values. In their work, uncertainty is represented by a probability degree on the set of possible values considered in the classification problem. For handling ambiguous data both in the feature set and in class labels, a fuzzy decision trees were developed [22]. A comprehensive survey of uncertain data algorithms and applications is provided in [1,11].

Despite the variety of existing solutions, there are still lack of methods that generates intelligible models, which allow for handling uncertain knowledge, but also can be verified and modified directly by the non-expert end user. The following section provides more detail on the motivation for our work.

2.1 Motivation

Among the solution that learn models from uncertain data described in previous paragraphs, we mainly focused on the third group, which are based on probabilistic models. Only these methods transfer the knowledge about the uncertainty of training data into the model. Such knowledge can later be utilized by the user to improve his or her understanding of the model and thus to improve the overall system intelligibility.

However, none of the aforementioned solutions tackle this issue directly. Thus, the models generated with such algorithms as [13,21] tend to be overcomplicated and hardly understandable by non technical users. On the other hand, methods such as [22] are difficult and inefficient to implement as an on-line algorithms, which is important in case of mobile systems, which operate in evolving environments.

Therefore, the primary motivation for our work was threefold:

1. Provisioning a method for transferring uncertainty from data to model in a compact, easily maintainable form
2. Development of an algorithm that can be instantly applied to existing learning methods, including on line learning solutions, such as VFDT, or CVFDT
3. Provisioning of the uncertain model representation in a way understandable and modifiable by the end user.

As a result of our work, the uncertain decision tree generation algorithm was developed and a translator that converts these threes into human readable, visual representation. The original contribution includes development of a method for building statistics of the certainty of training instances and using it as a heuristic in building a decision tree model. It adapts the classical information gain split criterion to take into consideration uncertain data while selecting splitting attribute, which makes the method fast and simple to implement. Additionally, by using simple statistics rather than complex probability estimation methods, it can be efficiently used in on line learning algorithms, such as VFDT or CVFDT. It does not discard any data, but takes the most probable measure for calculation and includes the certainty information in this calculation. Additionally, such an approach allows for handling uncertain class labels, being at the same time fast and easy to implement. Finally, it solves the *cold start* problem [20], for which most of the probabilistic approaches suffer. This problem appears, where there is not enough data to build the model, yet the system needs the model to work. Because the user has immediate insight into the model, at every stage of learning, the model can be instantly modified by ones, allowing the system to make use of it, even on very early stage.

3   Uncertain Decision Trees

The main goal of the work presented in this paper is focused on finding the heuristic that will be based on the information gain measure, but at the same time will include uncertainty of training data into the calculation. This allows to apply it to a variety of algorithms that are based on it, such as classic ID3 algorithm, or more complex, incremental versions such as VFDT or CVFDT.

The classic information gain formula for the attribute A and a training set S is defined as follows:

$$Gain(A) = H(S) - \sum_{v \in Domain(A)} \frac{|S_v|}{|S|} H(S_v) \tag{1}$$

Where S_v is a subset of S, such that for every $s \in S$ value of $A = v$. The entropy for the training set S is defined as follows:

$$H(S) = - \sum_{v \in Domain(C)} p(v) \log_2 p(v) \tag{2}$$

Where $Domain(C)$ is a set of all classes in S and $p(v)$ is a ratio of the number of elements of class v to all the elements in S.

In case of the uncertain data, the $p(v)$ from the Eq. (2) has to be defined as a probability of observing element of class v in the dataset S. This probability will be denoted further as capital $P_{total}(C = v)$. Similarly, a fraction $\frac{|S_v|}{|S|}$ from Eq. (1) has to be redefined as a probability of observing value v of attribute A in the dataset S. This probability will be referred later as $P_{total}(A = v)$ and is defined as probability of observing a value v_j^i of an attribute A_i in the set S_t that contains k training instances. This can be defined as follows:

$$P_{total}(A_i = v_j^i) = \frac{1}{k} \sum_{S_t \ni P_{j=1 \ldots n}} P_j(A_i = v_j^i) \tag{3}$$

Similarly $P_{total}(C = v_j)$ can be defined, which represents probability of observing class label in a set. Having that, the uncertain information gain measure can be defined as shown in the Eq. (4).

$$Gain^U(A) = H^U(S) - \sum_{v \in Domain(A)} P_{total}(A = v) H^U(S_v) \tag{4}$$

Where the H^U is the uncertain entropy measure defined as:

$$H^U(S) = - \sum_{v \in Domain(C)} P_{total}(C = v) \log_2 P_{total}(C = v) \tag{5}$$

Measures represented by the Eqs. (4) and (5) can be used to build decision tree, using any algorithm that is based on the information gain heuristic. Such a tree contains information about the classification accuracy in its leaves. However, this accuracy does not take into account the uncertainty associated with it. It only denotes how many instances from the training set are covered with particular leaf. Therefore, although the uncertainty information was taken into consideration while building a tree, it is no longer used during the classification.

The complete procedure of generating the uncertain tree was presented in Algorithm below.

Algorithm uID3(S, A) – grow a decision tree from uncertain data.

Input   : data S; set of attributes A.
Output: uncertain decision tree uT.

1 **if** Homogeneous(S) **then return** MajorityClass(S);
2 $R \leftarrow$ Best split using $Gain^U(S)$ split S into subsets S_i according to Domain(R);
3 **for** each i **do**
4 | **if** $S_i \neq \emptyset$ **then** $uT_i \leftarrow$ uID3(S_i, A) ;
5 | **else** uT_i is a leaf labeled with MajorityClass(S);
6 **return** a root R of the decision tree

This loss of information was solved by including statistics about the sensor accuracy for particular branches, as depicted in Fig. 1. Specifically, the $P_{total}(A = v)$ was included for each branch. Such information can be useful while translating decision tree into rule-based knowledge representation. In such a translation uncertain branches can be verified by the user or skipped, keeping the size of the knowledge small.

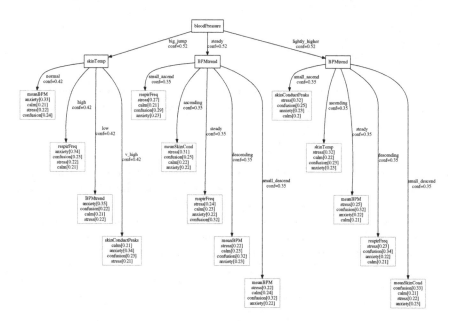

Fig. 1. Decision tree generated with uncertain data.

4   Evaluation

The evaluation of the uID3 algorithm described in this paper was performed using artificially created dataset related to the area of affective computing [16,17]. The created dataset describes dependency between certain physiological features and emotional states associated with these features. The following features are taken into account: mean heart rate, average heart rate trend during some period of time (e.g. during an experiment), skin temperature on some chosen body part, skin conduction peaks (such peaks may indicate sudden scare), mean skin conductance during some period of time, respiration depth (whether someone is breathing deeply or not), respiration frequency, blood pressure, **class attribute:** emotion.

4.1   Prepared Input Data and uARFF Format

The custom uARFF format used in the uID3 algorithm is the extension of traditional ARFF format and allows to keep information about certainty level of the class assigned to given case.

The process of preparing data for the experiments consisted of the following steps:

1. name of the dataset, set of feature attributes and domains of the feature attributes and the class attribute were established (as in the previous section). These assumptions were encoded in the following way:

```
1   @relation affective_state
2   @attribute meanBPM {low, normal, high}
3   @attribute BPMtrend {descending, small_descend, steady, small_ascend, ascending}
4   @attribute skinTemp {low, normal, high, v_high}
5   @attribute skinConductPeaks {0,1,2, more}
6   @attribute meanSkinCond {below_normal, normal, above_normal}
7   @attribute respirDepth {shallow, deep}
8   @attribute respirFreq {slow, typical, fast}
9   @attribute bloodPressure {steady, lightly_higher, big_jump}
10  @attribute emotion {anxiety, calm, confusion, stress}
```

2. set of data cases were created (let us name it *original* dataset). About 12,000 cases were generated. The following listing presents a few first data cases:

```
1   @data
2   low, descending, low, 0, below_normal, shallow, slow, lightly_higher, confusion
3   low, descending, low, 0, below_normal, shallow, slow, big_jump, anxiety
4   low, descending, low, 0, below_normal, shallow, typical, lightly_higher, confusion
```

3. the new dataset with certainty factors was generated from the existing dataset (let us name it *noisy* dataset). In this case the class attribute is replaced by the list of all classes from the domain with certainty factors assigned to them. These factors sum up to 1 for each case. It is important that this dataset contains some noise, i.e. in some cases the class with the highest probability is not the proper class assigned in the previous dataset:

```
1 || @data
2 || low , descending , low ,0 , below_normal , shallow , slow , lightly_higher , confusion [0.49];
        calm [0.3]; anxiety [0.17]; stress [0.04]
3 || low , descending , low ,0 , below_normal , shallow , slow , big_jump , anxiety [0.40]; calm [0.38];
        stress [0.12]; confusion [0.09]
4 || low , descending , low ,0 , below_normal , shallow , typical , lightly_higher , confusion [0.37];
        stress [0.36]; calm [0.20]; anxiety [0.07]
```

4. the dataset without certainty factor is created the way that for each case the class with the highest probability is chosen (let us name it *simple* dataset):

```
1 || low , descending , low ,0 , below_normal , shallow , slow , lightly_higher , confusion
2 || low , descending , low ,0 , below_normal , shallow , slow , big_jump , anxiety
3 || low , descending , low ,0 , below_normal , shallow , typical , lightly_higher , confusion
```

Finally, the generated datasets were divided into train and test set applying the common 66%/33% ratio. In total, 12958 cases were created, of which 8455 were assigned to the train set and 4503 were assigned to the test set. The classes distribution in the original dataset is as follows: 4320 have the anxiety class, 328 cases have the calm class, 4266 cases have the confusion class and 4044 cases have the stress class. All three datasets were saved to the separate files, which were loaded as training sets and test sets for different classifiers, respectively.

4.2 Tests

The uID3 algorithm is the only one that takes data uncertainty into account, that is why it is not possible to compare the accuracy and efficiency of all algorithms using prepared uncertain data. In order to overcome this obstacle we proposed the following evaluation process:

1. the *noisy* dataset will be put as the input data to the uID3 classifier in order to demonstrate performance of this algorithm on uncertain data
2. the *simple* dataset will be used in standard classifiers to test their performance on our data

In order to describe and compare the accuracy of tested algorithms, the following evaluation metrics were used [10]:

- accuracy – number of correctly classified instances to all instances
- true positive rate/recall – number of true positives (i.e. cases which were labeled as positive and were classified as positive) to the number of all positives (i.e. cases which were labeled as positive)
- false positive rate – number of false positives (i.e. negative cases classified as positive) to the number of all negatives (i.e. cases labeled as negative)
- area under the ROC curve – area under the curve splitting correctly and incorrectly classified instances.

Concluding, the following algorithm were tested: uID3, J48, Hoeffding Tree, Random Forest, Naive Bayes and ZeroR as baseline to compare all the algorithms.

4.3 Evaluation Results

The following tables present evaluation results. Table 1 shows results, where only class label was uncertain. It can be noticed that the uID3 performed no worse than the majority of the classifiers.

Table 1. Evaluation results of the tested algorithms on the *noisy* (uID3) and *simple* (rest) datasets

uID3, accuracy: 59.73%				ZeroR, accuracy: 21.0%		
TP rate	FP rate	ROC area	**Class**	TP rate	FP rate	ROC area
0.855	0.096	0.877	**Anxiety**	0.0	0.0	0.304
0	0	0.455	**Calm**	0.0	0.0	0.085
0.482	0.125	0.733	**Stress**	0.0	0.0	0.400
0.683	0.330	0.712	**Confusion**	1.0	1.0	0.211
J48, accuracy: 60.42%				HoeffdingTree, accuracy: 59.73%		
TP rate	FP rate	ROC area	**Class**	TP rate	FP rate	ROC area
0.689	0.170	0.573	**Anxiety**	0.855	0.069	0.742
0.006	0.007	0.096	**Calm**	0.0	0.0	0.120
0.336	0.135	0.554	**Stress**	0.483	0.587	0.676
0.668	0.387	0.274	**Confusion**	0.684	0.330	0.352
Naive Bayes, accuracy: 57.52%				Random Forest, accuracy: 58.92%		
TP rate	FP rate	ROC area	**Class**	TP rate	FP rate	ROC area
0.855	0.096	0.752	**Anxiety**	0.845	0.102	0.764
0.073	0.074	0.083	**Calm**	0.076	0.022	0.145
0.500	0.099	0.729	**Stress**	0.429	0.075	0.698
0.516	0.293	0.303	**Confusion**	0.732	0.348	0.380

Table 2 shows results for dataset, where all values were noisy, including features. It can be noticed, that uID3 has the best performance among all tested algorithms along with Naive Bayes classifier. This is because in uID3, the best candidate for split is chosen based on most certain attributes that at the same time reduces the entropy of dataset. This results in decrease importance of attributes that exhibit high variance in probability distribution of their values.

Another important observation is that the decrease in accuracy along with decrease of certainty of data is relatively small for uID3 (about 10%) comparing to other algorithms (12% for J48, 14% for HoeffdingTree, 16% for Random Forest). Only Naive Bayes had lower accuracy drop (8%).

Table 2. Evaluation results of the tested algorithms on the *noisy* (uID3) and *simple* (rest) datasets

uID3, accuracy: 49.71%				ZEROR, accuracy: 21.0%		
TP rate	FP rate	ROC area	Class	TP rate	FP rate	ROC area
0.689	0.162	0.763	**Anxiety**	0.0	0.0	0.312
0	0	0.552	**Calm**	0.0	0.0	0.080
0.360	0.146	0.645	**Stress**	0.0	0.0	0.398
0.662	0.384	0.629	**confusion**	1.0	1.0	0.5
J48, accuracy: 48.96%				HOEFFDINGTREE, accuracy: 46.42%		
TP rate	FP rate	ROC area	Class	TP rate	FP rate	ROC area
0.689	0.170	0.573	**Anxiety**	0.689	0.162	0.576
0.006	0.007	0.096	**Calm**	0.0	0.0	0.094
0.336	0.135	0.554	**Stress**	0.225	0.074	0.547
0.668	0.387	0.274	**Confusion**	0.760	0.481	0.279
NAIVE BAYES, accuracy: 49.55%				RANDOM FOREST, accuracy: 42.91%		
TP rate	FP rate	ROC area	Class	TP rate	FP rate	ROC area
0.689	0.162	0.591	**Anxiety**	0.594	0.204	0.536
0.0	0.0	0.101	**Calm**	0.069	0.046	0.087
0.358	0.148	0.581	**Stress**	0.308	0.162	0.516
0.658	0.385	0.310	**Confusion**	0.551	0.367	0.265

5   Summary and Future Works

In this paper we presented an algorithm for generating decision trees from uncertain datasets. Proposed method uses probability distribution of features values to select optimal split attribute, that not only reduces entropy of dataset, but also reduces the uncertainty of decision process. We provided an evaluation of our method with comparison to selected classifiers from Weka[1].

Although uID3 shown to be not worse than other methods, exposing lowest drop in accuracy on highly uncertain data, our main goal was to provide method that will generate knowledge easily understandable and editable for the user. The algorithm provides a fair trade-off between complex probabilistic approaches and simple state-of-the-art classifiers. It allows to achieve reasonable accuracy on uncertain data, providing simple and human-readable format.

Our future work will focus on implementing our method in on-line learning algorithms such as CVFDT to improve their interpretability and intelligibility. Furthermore, we will incorporate this work in our decision support system based on rules [14] with uncertainty [4].

---

[1] Weka is a collection of machine learning algorithms for data mining tasks. See: https://www.cs.waikato.ac.nz/ml/weka.

References

1. Aggarwal, C.C., Yu, P.S.: A survey of uncertain data algorithms and applications. IEEE Trans. Knowl. Data Eng. **21**(5), 609–623 (2009)
2. Bobek, S.: Methods for modeling self-adaptive mobile context-aware sytems. Ph.D. thesis, AGH University of Science and Technology, April 2016. Supervisor: G.J. Nalepa
3. Bobek, S., Nalepa, G.J.: Uncertain context data management in dynamic mobile environments. Future Gener. Comput. Syst. **66**(Jan), 110–124 (2017). https://doi. org/10.1016/j.future.2016.06.007
4. Bobek, S., Nalepa, G.J.: Uncertainty handling in rule-based mobile context-aware systems. Pervasive Mob. Comput. **39**(Aug), 159–179 (2017). https://doi.org/10. 1016/j.pmcj.2016.09.004
5. Bobek, S., Nalepa, G.J., Ślażyński, M.: Challenges for migration of rule-based reasoning engine to a mobile platform. In: Dziech, A., Czyżewski, A. (eds.) MCSS 2014. CCIS, vol. 429, pp. 43–57. Springer, Cham (2014). https://doi.org/10.1007/ 978-3-319-07569-3_4
6. Bobek, S., Porzycki, K., Nalepa, G.J.: Learning sensors usage patterns in mobile context-aware systems. In: Ganzha, M., Maciaszek, L.A., Paprzycki, M. (eds.) Proceedings of the Federated Conference on Computer Science and Information Systems - FedCSIS 2013, Krakow, Poland, 8–11 September 2013, pp. 993–998. IEEE, September 2013
7. Bobek, S., Ślażyński, M., Nalepa, G.J.: Capturing dynamics of mobile context-aware systems with rules and statistical analysis of historical data. In: Rutkowski, L., Korytkowski, M., Scherer, R., Tadeusiewicz, R., Zadeh, L.A., Zurada, J.M. (eds.) ICAISC 2015. LNCS (LNAI), vol. 9120, pp. 578–590. Springer, Cham (2015). https://doi.org/10.1007/978-3-319-19369-4_51
8. Chau, M., Cheng, R., Kao, B., Ng, J.: Uncertain data mining: an example in clustering location data. In: Ng, W.-K., Kitsuregawa, M., Li, J., Chang, K. (eds.) PAKDD 2006. LNCS (LNAI), vol. 3918, pp. 199–204. Springer, Heidelberg (2006). https://doi.org/10.1007/11731139_24
9. Domingos, P., Hulten, G.: Mining high-speed data streams. In: Proceedings of the Sixth ACM SIGKDD International Conference on Knowledge Discovery and Data Mining, KDD 2000, pp. 71–80. ACM, New York (2000). https://doi.org/10.1145/ 347090.347107
10. Flach, P.: Machine Learning: The Art and Science of Algorithms that Make Sense of Data. Cambridge University Press, New York (2012)
11. Goyal, N., Jain, S.K.: A comparative study of different frequent pattern mining algorithm for uncertain data: a survey. In: 2016 International Conference on Computing, Communication and Automation (ICCCA), pp. 183–187, April 2016
12. Hulten, G., Spencer, L., Domingos, P.: Mining time-changing data streams. In: Proceedings of the Seventh ACM SIGKDD International Conference on Knowledge Discovery and Data Mining, KDD 2001, pp. 97–106. ACM, New York (2001). https://doi.org/10.1145/502512.502529
13. Liang, C., Zhang, Y., Song, Q.: Decision tree for dynamic and uncertain data streams. In: Sugiyama, M., Yang, Q. (eds.) Proceedings of 2nd Asian Conference on Machine Learning. Proceedings of Machine Learning Research, 08–10 November 2010, vol. 13, pp. 209–224. PMLR, Tokyo (2010). http://proceedings.mlr.press/ v13/liang10a.html

14. Nalepa, G.J.: Architecture of the HeaRT hybrid rule engine. In: Rutkowski, L., Scherer, R., Tadeusiewicz, R., Zadeh, L.A., Zurada, J.M. (eds.) ICAISC 2010. LNCS (LNAI), vol. 6114, pp. 598–605. Springer, Heidelberg (2010). https://doi.org/10.1007/978-3-642-13232-2_73
15. Nalepa, G.J., Bobek, S.: Rule-based solution for context-aware reasoning on mobile devices. Comput. Sci. Inf. Syst. **11**(1), 171–193 (2014)
16. Nalepa, G.J., Kutt, K., Bobek, S.: Mobile platform for affective context-aware systems. Future Gener. Comput. Syst. (2018). https://doi.org/10.1016/j.future.2018.02.033
17. Picard, R.W.: Affective Computing. MIT Press, Cambridge (1997)
18. Quinlan, J.R.: C4.5: Programs for Machine Learning. Morgan Kaufmann Publishers Inc., San Francisco (1993)
19. Salber, D., Dey, A.K., Abowd, G.D.: The context toolkit: aiding the development of context-enabled applications. In: Proceedings of the SIGCHI Conference on Human Factors in Computing Systems, CHI 1999, pp. 434–441. ACM, New York (1999)
20. Schein, A.I., Popescul, A., Ungar, L.H., Pennock, D.M.: Methods and metrics for cold-start recommendations. In: Proceedings of the 25th Annual International ACM SIGIR Conference on Research and Development in Information Retrieval, SIGIR 2002, pp. 253–260. ACM, New York (2002)
21. Tsang, S., Kao, B., Yip, K.Y., Ho, W.S., Lee, S.D.: Decision trees for uncertain data. IEEE Trans. Knowl. Data Eng. **23**(1), 64–78 (2011)
22. Yuan, Y., Shaw, M.J.: Induction of fuzzy decision trees. Fuzzy Sets Syst. **69**(2), 125–139 (1995). https://doi.org/10.1016/0165-0114(94)00229-Z

An Efficient Prototype Selection Algorithm Based on Dense Spatial Partitions

Joel Luís Carbonera[1](✉) and Mara Abel[2]

[1] IBM Research, Rio de Janeiro, Brazil
joelc@br.ibm.com
[2] UFRGS, Porto Alegre, Brazil
marabel@inf.ufrgs.br

Abstract. In order to deal with big data, techniques for prototype selection have been applied for reducing the computational resources that are necessary to apply data mining approaches. However, most of the proposed approaches for prototype selection have a high time complexity and, due to this, they cannot be applied for dealing with big data. In this paper, we propose an efficient approach for prototype selection. It adopts the notion of spatial partition for efficiently dividing the dataset in sets of similar instances. In a second step, the algorithm extracts a prototype of each of the densest spatial partitions that were previously identified. The approach was evaluated on 15 well-known datasets used in a classification task, and its performance was compared to those of 6 state-of-the-art algorithms, considering two measures: accuracy and reduction. All the obtained results show that, in general, the proposed approach provides a good trade-off between accuracy and reduction, with a significantly lower running time, when compared with other approaches.

Keywords: Prototype selection · Data reduction · Data mining
Machine learning · Big data

1 Introduction

Prototype selection is a data-mining (or machine learning) pre-processing task that consists in producing a smaller representative set of instances from the total available data, which can support a data mining task *with no performance loss* (or, at least, a reduced performance loss) [8]. Thus, every prototype selection strategy faces a *trade-off* between the *reduction rate* of the dataset and the resulting *classification quality* [7].

Most of the proposed algorithms for prototype selection, such as [2,12–16] have a *high time complexity*, which is an undesirable property for algorithms that should deal with big volumes of data. In this paper, we propose an algorithm for prototype selection, called PSDSP (*P*rototype *S*election based on *D*ense *S*patial

© Springer International Publishing AG, part of Springer Nature 2018
L. Rutkowski et al. (Eds.): ICAISC 2018, LNAI 10842, pp. 288–300, 2018.
https://doi.org/10.1007/978-3-319-91262-2_26

*P*artitions)[1]. The algorithm has two main steps: (I) it uses the notion of spatial partition for efficiently dividing the dataset in sets of instances that are similar to each other, and (II) it extracts prototypes from the densest sets of instances identified in the first step. This simple strategy is remarkably efficient, and results in a linear time complexity on the number of instances. Also, it allows the user to define the desired number of prototypes that will be extracted from the dataset. This level of control is not common in most of the approaches for prototype selection.

Our approach was evaluated on 15 well-known datasets and its performance was compared with the performance of 6 important algorithms provided by the literature, according to 2 different performance measures: *accuracy* and *reduction*. The accuracy was evaluated considering two classifiers: SVM and KNN. The results show that, when compared to the other algorithms, PSDSP provides a good trade-off between accuracy and reduction, while presents a significantly low time complexity. The PSDSP algorithm has *linear* time complexity on the number of objects, while most of the prototype selection algorithms have a quadratic time complexity. Due to this, the proposed algorithm is very fast. These results suggest that PSDSP is a promising algorithm for scenarios in which the running time is important.

Section 2 presents some related works. Section 3 presents the notation that will be used throughout the paper. Section 4 presents our approach. Section 5 discusses our experimental evaluation. Finally, Sect. 6 presents our main conclusions and final remarks.

2   Related Works

The *Condensed Nearest Neighbor* (CNN) algorithm [11] and *Reduced Nearest Neighbor* algorithm (RNN) [9] are some of the earliest proposals for instance selection. Both can assign noisy instances to the final resulting set, are dependent on the order of the instances and have a high time complexity. The *Edited Nearest Neighbor* (ENN) algorithm [16] removes every instance that does not agree with the label of the majority of its k nearest neighbors. This strategy is effective for removing noisy instances, but it does not reduce the dataset as much as other algorithms. In [15], the authors present 5 approaches, named the *Decremental Reduction Optimization Procedure* (DROP). These algorithms assume that those instances that have x as one of their k nearest neighbors are called the *associates* of x. Among the proposed algorithms, DROP3 has the best trade-off between the reduction of the dataset and the classification accuracy. It applies a noise-filter algorithm such as ENN. Then, it removes an instance x if its associates in the original training set can be correctly classified without x. The main drawback of DROP3 is its high time complexity. The *Iterative Case Filtering algorithm* (ICF) [2] is based on the notions of *Coverage set* and *Reachable set*. The coverage set of an instance x is the set of instances in T whose distance from x is less than

---

[1] The source code of the algorithm is available in https://www.researchgate.net/publication/323701200_PSDSP_algorithm.

the distance between x and its nearest enemy (instance with a different class). The Reachable set of an instance x, on the other hand, is the set of instances in T that have x in their respective coverage sets. In this method, a given instance x is removed from S if $|Reachable(x)| > |Coverage(x)|$. This algorithm also has a high running time. In [12], the authors adopted the notion of *local sets* for designing complementary methods for instance selection. In this context, the local set of a given instance x is the set of instances contained in the largest hypersphere centered on x such that it does not contain instances from any other class. The first algorithm, called *Local Set-based Smoother* (LSSm) uses two notions for guiding the process: *usefulness* and *harmfulness*. The usefulness $u(x)$ of a given instance x is the number of instances having x among the members of their local sets, and the harmfulness $h(x)$ is the number of instances having x as the nearest enemy. For each instance x in T, the algorithm includes x in S if $u(x) \geq h(x)$. Since the goal of LSSm is to remove harmful instances, its reduction rate is lower than most of the instance selection algorithms. The author also proposed the *Local Set Border selector* (LSBo). Firstly, it uses LSSm to remove noise, and then, it computes the local set of every instance $\in T$. Then, the instances in T are sorted in the ascending order of the cardinality of their local sets. In the last step, LSBo verifies, for each instance $x \in T$ if any member of its local set is contained in S, thus ensuring the proper classification of x. If that is not the case, x is included in S to ensure its correct classification. The time complexity of the two approaches is $O(|T|^2)$. In [4], the authors proposed the *Local Density-based Instance Selection* (LDIS) algorithm. This algorithm selects the instances with the highest density in their neighborhoods. It provides a good balance between accuracy and reduction and is faster than the other algorithms discussed here. The literature provide some extensions to the basic LDIS algorithm, such as [3,5]. Other approaches can be found in surveys such as [8,10].

3   Notations

In this section, we introduce a notation adapted from [4] that will be used throughout the paper.

- $T = \{o_1, o_2, ..., o_n\}$ is the non-empty set of n instances (or data objects), representing the original dataset to be reduced in the prototype selection process.
- $D = \{d_1, d_2, ..., d_m\}$ is a set of m dimensions (that represent features, or attributes), where each $d_i \subseteq \mathbb{R}$.
- Each $o_i \in T$ is an $m - tuple$, such that $o_i = (o_{i1}, o_{i2}, ..., o_{im})$, where o_{ij} represents the value of the j-th feature (or dimension) of the instance o_i, for $1 \leq j \leq m$.
- $val: T \times D \rightarrow \mathbb{R}$ is a function that maps a data object $o_i \in T$ and a dimension $d_j \in D$ to the value o_{ij}, which represents the value in the dimension d_j for the object o_i.

- $L = \{l_1, l_2, ..., l_p\}$ is the set of p class labels that are used for classifying the instances in T, where each $l_i \in L$ represents a given class label.
- $l: T \rightarrow L$ is a function that maps a given instance $x_i \in T$ to its corresponding class label $l_j \in L$.
- $c: L \rightarrow 2^T$ is a function that maps a given class label $l_j \in L$ to a given set C, such that $C \subseteq T$, which represents the set of instances in T whose class is l_j. Notice that $T = \bigcup_{l \in L} c(l)$. In this notation, 2^T represents the *powerset* of T, that is, the set of all subsets of T, including the empty set and T itself.

4   The PSDSP Algorithm

In this paper, we propose the PSDSP (*P*rototype *S*election based on *D*ense *S*patial *P*artitions) algorithm, which can be viewed as a specific variation of the general schema represented in [6].

Before discussing the PSDSP algorithm, it is important to consider the notion of *spatial partition*.

Definition 1. *A spatial partition of a spatial region that contains a given set of objects $H \subseteq T$ is a set $\mathcal{SP}_H = \{s_1, s_2, ..., s_m\}$, where:*

$$\forall s_i \in \mathcal{SP}_H \rightarrow \exists d_i \in D \wedge s_i \subseteq d_i \tag{1}$$

$$\forall d_i \in D \rightarrow \exists s_i \in \mathcal{SP}_H \tag{2}$$

$$\forall s_i \in \mathcal{SP}_H \rightarrow s_i = [x, y] \wedge$$
$$x \geq min(d_i, H) \wedge \tag{3}$$
$$y \leq max(d_i, H)$$

, considering $min(d_i, H)$ as the lowest value of the dimension d_i within the set H, and $max(d_i, H)$ as the greatest value of the dimension d_i within the set H.

Thus, a *spatial partition* of the spatial region that contains a given set of objects H is intuitively a set of intervals, one for each dimension $d_i \in D$, defining a specific multidimensional region (a hyperrectangle) in the spatial region containing the objects in H. Figure 1 presents an example of a dataset with 100 data objects in a 2D space with 12 spatial partitions.

Considering the notion of *spatial partition*, the PSDSP algorithm (formalized in Algorithm 1) firstly identifies a set of spatial partitions for each class of objects of the dataset and, in a second step, it selects the prototypes from the *densest* spatial partitions previously identified. Thus, it can be viewed as a combination of different aspects of the algorithms proposed in [4,6].

The PSDSP algorithm takes as input a set of data objects T, a value[2] $n \in \mathbb{N}^*$, which determines the number of intervals in which each dimension of the dataset will be divided; and a value $p \in [0, 1]$, which determines the expected number

---

[2] Notice that we are assuming in this paper that the set of *natural numbers* ($\mathbb{N}$) is the set of non-negative integers and, due to this, it includes zero. When we are referring to the set of natural numbers excluding zero, we use $\mathbb{N}^*$.

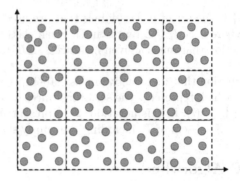

Fig. 1. Representation of a 2D space, with 12 *spatial partitions*.

of prototypes that should be selected, as a percentage of the total number of instances ($|T|$) in the dataset. After, the algorithm initializes P as an empty set and, for each $l \in L$, it:

1. Determines the number k of objects in $c(l)$ that should be included in P.
2. Determines the set R of sets of objects within $c(l)$, such that each set $r_i \in R$ is a set of objects contained within a specific *spatial partition* of the spatial region that contains the objects in $c(l)$. The set R is produced by the function *partitioning*, represented in Algorithm 2, which takes as input the set $c(l)$ of instances classified by l and the number n of intervals in which each dimension will be divided.
3. Sorts the set R in descending order, according to the number of instances included by each set $r_i \in R$. That is, after this step, the first set in R would represent the set of instances included by the *densest spatial partition* of $c(l)$. The density of a given spatial partition is the number of instances that it includes.
4. Extracts k prototypes, from the first k sets in R, and includes them in the resulting set Ps. When $|R| < k$, only $|R|$ prototypes are included.

The function *partitioning*, on the other hand, is formalized by the Algorithm 2. This algorithm takes as input a set of instances H and a number $n \in \mathbb{N}^*$ of intervals in which each dimension will be divided. It results in a set R of sets of instances, where each set $r_i \in R$ is the set of instances contained in some *spatial partition* of the spatial region that contains H. Initially, the algorithm defines R as an empty set. After, for each dimension $d_i \in D$, the algorithm defines $DRange$, which represents the range of the dimension d_i, as the absolute difference between the highest and the lowest value of d_i in H; and $range_i$, which represents the range of an interval of d_i, as $\frac{DRange}{n}$. Notice that the intervals of each dimension are homogeneous. After, the algorithm considers $region$ as a hash table whose keys are $|D|$-tuples in the form of $(x_1, x_2, ..., x_j)$, where each $x_i \leq n$ identify one of the intervals of the $i - th$ dimension in D. Also, for each key x, $region$ stores a set objects, such that $region[x] \subseteq H$. Notice that a key

---

Algorithm 1. PSDSP algorithm

Input: A set instances T, a number $n \in \mathbb{N}^*$ of intervals, and a value $p \in [0, 1]$, which is the number of prototypes, as a percentage of $|T|$.
Output: A set P of prototypes.
begin

   $P \leftarrow \emptyset$;
   foreach $l \in L$ **do**
      $k \leftarrow p \cdot |c(l)|$;
      $R \leftarrow partitioning(c(l), n)$;
      Sorting R in descending order, according to the number of instances included by each set $r_i \in R$;
      $i \leftarrow 0$;
      while $i \leq |R|$ and $i \leq k$ **do**
         $prot \leftarrow extractsPrototype(r_i)$;
         $P \leftarrow P \cup \{prot\}$;
         $i \leftarrow i + 1$;
   return P;

---

of this hash table can be viewed as the identification of a given *spatial partition* and, therefore, $region[x]$ represents the set of objects located within the spatial partition identified by x. After, for each object $o \in H$, the algorithm:

1. Considers x as an empty $|D|$-tuple.
2. Defines, for each dimension $d_i \in D$, the value of x_i, as the identification of the interval that contains the value $val(o, d_i)$, such that $x_i \leftarrow \lfloor \frac{val(o,d_i) - min(d_i, H)}{range_i} \rfloor$. In this way, x determines the identification of the spatial partition that contains o.
3. Includes o as an element of $region[x]$.

After, for each key x of $region$, the algorithm includes the set $region[x]$ in R as an element. Thus, each element of R is a set of objects located in some spatial partition defined by the algorithm. Finally, the algorithm returns R.

Algorithm 2. partitioning

Input: A set instances H and a number $n \in \mathbb{N}^*$ of intervals.
Output: A set R of sets of instances.
begin

   $R \leftarrow \emptyset$;
   Let $range_i$ be the range of an interval of the dimension $d_i \in D$.;
   foreach $d_i \in D$ **do**
      $DRange \leftarrow abs(max(d_i, H) - min(d_i, H))$;
      $range_i \leftarrow \frac{DRange}{n}$;
   Let $region$ be a hash table whose keys are $|D|$-tuples in the form of $(x_1, x_2, ..., x_j)$, where each x_i identifies one of the intervals of the $i - th$ dimension within D. Also, for each key x, $region$ stores a set of objects, such that $region[x] \subseteq H$.;
   foreach $o \in H$ **do**
      Let x be an empty $|D|$-tuple.;
      foreach $d_i \in D$ **do**
         $x_i \leftarrow \lfloor \frac{val(o,d_i) - min(d_i, H)}{range_i} \rfloor$;
      $region[x] \leftarrow region[x] \cup \{o\}$;
   foreach key x of $region$ **do**
      R includes $region[x]$ as its element;
   return R;

Finally, the function *extractsPrototype*, adopted by the Algorithm 1, takes as input a set of instances $H \subseteq T$ and produces a $|D|$-tuple that represents the centroid of the instances in H. This is the same strategy used by [6] for extracting prototypes.

It is important to notice that the Algorithm 2 uses the notion of spatial partition in an *implicit* way for identifying the set of instances that are included in each spatial partition. Also, it assumes that the spatial region that included all the elements in H is divided in a set of non-overlapping spatial partitions, which covers all the set H.

In addition, the algorithm extracts prototypes of the k *densest* spatial partitions because it assumes that the density of a spatial partition indicates the amount of information that it represents, and that the resulting set of prototypes should include the prototypes that abstract a richer amount of information.

Moreover, since the notion of *spatial partition* defined in this work can be applied only to datasets with quantitative (numerical) dimensions, the PSDSP can be applied only to datasets with this kind of dimensions.

Each of the steps of the algorithm has, at most, a time complexity that is *linear* on the number of instances. This is a remarkable feature of the PSDSP algorithm.

5   Experiments

For evaluating our approach, we compared the PSDSP algorithm in a classification task, with 6 important prototype selection algorithms[3] provided by the literature: DROP3, ENN, ICF, LSBo, LSSm and LDIS. We considered 15 well-known datasets with numerical dimensions: cardiotocography, diabetes, E. Coli, glass, heart-statlog, ionosphere, iris, landsat, letter, optdigits, page-blocks, parkinson, segment, spambase and wine. All datasets were obtained from the UCI Machine Learning Repository[4]. Table 1 presents the details of the datasets that were used.

We use two standard measures to evaluate the performance of the algorithms: *accuracy* and *reduction*. Following [4,12], we assume: *accuracy* $=$ $|Success(Test)|/|Test|$ and *reduction* $= (|T| - |S|)/|T|$, where $Test$ is a given set of instances that are selected for being tested in a classification task, and $|Success(Test)|$ is the number of instances in $Test$ correctly classified in the classification task.

For evaluating the classification *accuracy* of new instances in each respective dataset, we adopted a SVM and a KNN classifier. For the KNN classifier, we considered $k = 3$, as assumed in [4,12]. For the SVM, following [1], we adopted the implementation provided by Weka 3.8, with the standard parametrization ($c = 1.0$, *toleranceParameter* $= 0.001$, *epsilon* $= 1.0E - 12$, using a polynomial kernel and a multinomial logistic regression model with a ridge estimator as calibrator).

---

[3] All algorithms were implemented by the authors.

[4] http://archive.ics.uci.edu/ml/.

Table 1. Details of the datasets used in the evaluation process.

Dataset	Instances	Attributes	Classes
Cardiotocography	2126	21	10
Diabetes	768	9	2
E. Coli	336	8	8
Glass	214	10	7
Heart-statlog	270	14	2
Ionosphere	351	35	2
Iris	150	5	3
Landsat	4435	37	6
Letter	20000	17	26
Optdigits	11240	65	10
Page-blocks	5473	11	5
Parkinson	195	23	2
Spambase	9544	58	2
Segment	2310	20	7
Wine	178	14	3

Besides that, following [4], the accuracy and reduction were evaluated in an *n-fold cross-validation* scheme, where $n = 10$. Thus, firstly a dataset is randomly partitioned in 10 equally sized subsamples. From these subsamples, a single subsample is selected as validation data (*Test*), and the union of the remaining 9 subsamples is considered the *initial training set* (*ITS*). Next, a prototype selection algorithm is applied for reducing the *ITS*, producing the *reduced training set* (*RTS*). At this point, we can measure the *reduction* of the dataset. Finally, the *RTS* is used as the training set for the classifier, which is used for classifying the instances in *Test*. At this point, we can measure the accuracy achieved by the classifier, using *RTS* as the training set. This process is repeated 10 times, with each subsample used once as *Test*. The 10 values of accuracy and reduction are averaged to produce, respectively, the *average accuracy* (*AA*) and *average reduction* (*AR*). Tables 2, 3 and 4 report, respectively, for each combination of dataset and prototype selection algorithm: the resulting *AA* achieved by the SVM classifier, *AA* achieved by the KNN classifier, and the *AR*. The best results for each dataset is marked in bold typeface.

In all experiments, following [4], we adopted $k = 3$ for DROP3, ENN, ICF, and LDIS. For the PSDSP algorithm, we adopted $n = 5$ and $p = 0.1$, since this parametrization provides a good balance between accuracy and reduction. Besides that, for the algorithms that use distance (dissimilarity) function, we adopted the standard *Euclidean* distance.

Tables 2 and 3 show that LSSm achieves the highest *accuracy* in most of the datasets, for both classifiers. This is expected, since that LSSm was designed

Table 2. Comparison of the *accuracy* achieved by the training set produced by each algorithm, for each dataset, adopting a SVM classifier.

Algorithm	DROP3	ENN	ICF	LSBo	LSSm	LDIS	PSDSP	Average
Cardiotocography	0.64	**0.67**	0.64	0.62	**0.67**	0.62	0.59	0.64
Diabetes	0.75	**0.77**	0.76	0.75	**0.77**	0.75	0.71	0.75
E. Coli	0.81	0.82	0.78	0.74	**0.83**	0.77	0.81	0.79
Glass	0.47	0.49	0.49	0.42	**0.55**	0.50	0.48	0.49
Heart-statlog	0.81	0.83	0.79	0.81	**0.84**	0.81	0.82	0.82
Ionosphere	0.81	0.87	0.58	0.45	**0.88**	0.84	0.86	0.76
Iris	0.94	**0.96**	0.73	0.47	**0.96**	0.81	0.84	0.82
Landsat	0.86	**0.87**	0.85	0.85	**0.87**	0.84	0.84	0.85
Letter	0.80	**0.84**	0.75	0.73	**0.84**	0.75	0.74	0.78
Optdigits	0.98	0.98	0.97	0.98	**0.99**	0.96	0.97	0.98
Page-blocks	0.93	**0.94**	0.93	0.92	**0.94**	0.94	0.91	0.93
Parkinsons	0.85	**0.87**	0.85	0.82	**0.87**	0.82	0.85	0.85
Segment	0.91	**0.92**	0.91	0.80	0.91	0.89	0.87	0.89
Spambase	**0.90**	**0.90**	**0.90**	**0.90**	**0.90**	0.89	0.87	0.89
Wine	0.93	0.95	0.94	0.96	**0.97**	0.94	0.95	0.95
Average	0.83	0.84	0.79	0.75	**0.85**	0.81	0.81	0.81

Table 3. Comparison of the *accuracy* achieved by the training set produced by each algorithm, for each dataset, adopting a KNN classifier.

Algorithm	DROP3	ENN	ICF	LSBo	LSSm	LDIS	PSDSP	Average
Cardiotocography	0.63	0.64	0.57	0.55	**0.67**	0.54	0.50	0.59
Diabetes	0.72	0.72	0.72	**0.73**	0.72	0.68	0.69	0.71
E. Coli	0.84	0.84	0.79	0.79	**0.86**	0.82	0.79	0.82
Glass	0.63	0.63	0.64	0.54	**0.71**	0.62	0.51	0.61
Heart-statlog	**0.67**	0.64	0.63	0.66	0.66	**0.67**	0.65	0.66
Ionosphere	0.82	0.83	0.82	**0.88**	0.86	0.85	0.86	0.85
Iris	**0.97**	**0.97**	0.95	0.95	0.96	0.95	0.94	0.96
Landsat	0.88	**0.90**	0.83	0.86	**0.90**	0.87	0.85	0.87
Letter	0.88	0.92	0.80	0.73	**0.93**	0.79	0.73	0.83
Optdigits	0.97	**0.98**	0.91	0.91	**0.98**	0.95	0.94	0.95
Page-blocks	0.95	**0.96**	0.93	0.94	**0.96**	0.94	0.72	0.92
Parkinsons	0.86	**0.88**	0.83	0.85	0.85	0.74	0.76	0.82
Segment	0.92	**0.94**	0.87	0.83	**0.94**	0.88	0.86	0.89
Spambase	0.79	0.81	0.79	0.81	**0.82**	0.75	0.74	0.79
Wine	0.69	0.66	0.66	**0.74**	0.71	0.69	0.67	0.69
Average	0.82	0.82	0.78	0.79	**0.83**	0.78	0.75	0.80

Table 4. Comparison of the *reduction* achieved by each algorithm, for each dataset.

Algorithm	DROP3	ENN	ICF	LSBo	LSSM	LDIS	PSDSP	Average
Cardiotocography	0.70	0.32	0.71	0.69	0.14	0.86	**0.90**	0.62
Diabetes	0.77	0.31	0.85	0.76	0.13	**0.90**	**0.90**	0.66
E. Coli	0.72	0.17	0.87	0.83	0.09	**0.92**	0.90	0.64
Glass	0.75	0.35	0.69	0.70	0.13	**0.90**	**0.90**	0.63
Heart-statlog	0.74	0.35	0.78	0.67	0.15	**0.93**	0.90	0.65
Ionosphere	0.86	0.15	0.96	0.81	0.04	**0.91**	0.90	0.66
Iris	0.70	0.04	0.61	0.92	0.05	0.87	**0.90**	0.58
Landsat	0.72	0.10	0.91	0.88	0.05	**0.92**	0.90	0.64
Letter	0.68	0.05	0.80	0.84	0.04	0.82	**0.90**	0.59
Optdigits	0.72	0.01	**0.93**	0.92	0.02	0.92	0.90	0.63
Page-blocks	0.71	0.04	**0.95**	0.96	0.03	0.87	0.90	0.64
Parkinsons	0.72	0.15	0.80	0.87	0.11	0.83	**0.90**	0.63
Segment	0.68	0.05	0.79	**0.90**	0.05	0.83	**0.90**	0.60
Spambase	0.74	0.19	0.79	0.82	0.10	0.82	**0.90**	0.62
Wine	0.80	0.30	0.82	0.75	0.11	0.88	**0.90**	0.65
Average	0.73	0.17	0.82	0.82	0.08	0.88	**0.90**	0.58

for removing noisy instances and does not provide high reduction rates. Besides that, for most of the datasets, the difference between the accuracy of PSDSP and the accuracy achieved by the other algorithms is not big. The average accuracy achieved by PSDSP is equivalent to the average accuracy of LDIS, and similar to the average accuracy of DROP3. In cases where the achieved accuracy is lower than the accuracy provided by other algorithms, this can be compensated by a higher reduction produced by PSDSP and by a much lower running time. Table 4 shows that PSDSP achieves the highest *reduction* in most of the datasets, and achieves also the highest average reduction rate. This table also shows that, in some datasets (such as parkinson and segment), the PSDSP algorithm achieved a reduction rate that is significantly higher than the reduction achieved by other algorithms, with a similar accuracy.

We also carried out experiments for evaluating the impact of the parameters n and p in the performance of PSDSP. The Table 5 represents the accuracy achieved by an SVM classifier (with the standard parametrization of Weka 3.8), as a function of the parameters n and p, with n assuming the values 2, 5, 10 and 20; and p assuming the values 0.05, 0.1 and 0.2. In this experiment, we also considered the 10-fold cross validation schema. Notice that the tables present the measures grouped primarily by the parameter n, and within each value of n, they present the results for each value of p.

The experiment show that with $n = 2$, the algorithm achieves the poorer performance. On the other hand, with $n = 5$, $n = 10$ and $n = 20$, there is no significant differences in the accuracy achieved by PSDSP. The small differences that we can identify in the results achieved with different values of n cannot be explained solely in terms of the change of n. This suggest that this parameter interacts with the structure of the dataset in a complex way. Further investigations should identify the properties and constraints regarding this interaction.

On the other hand, as the value of p increases, considering a fixed value of n, the accuracy increases. This is expected, since as p increases, the total number of prototypes selected by the algorithm also increases. Since each prototype abstracts the local information of its spatial partition, in most of the cases, increasing the value of p allows the resulting set of prototypes to capture more local information of the dataset. These additional prototypes allows the classifier to use the additional information to make more fine-grained distinctions in the classification process.

Table 5. Comparing the accuracy achieved by an SVM classifier trained with prototype selected by PSDSP with different values of n and p.

Algorithm	$n=2$			$n=5$			$n=10$			$n=20$			Average
	$p=0.05$	$p=0.1$	$p=0.2$	$p=0.05$	$p=0.1$	$p=0.2$	$p=0.05$	$p=0.1$	$p=0.2$	$p=0.05$	$p=0.1$	$p=0.2$	
Cardiotocography	0.58	**0.60**	**0.60**	0.57	0.59	**0.64**	0.57	0.61	**0.62**	0.58	0.59	**0.63**	0.60
Diabetes	0.69	**0.71**	**0.71**	0.71	0.71	**0.73**	0.70	0.74	**0.75**	0.72	0.71	**0.74**	0.72
E. Coli	**0.73**	0.71	0.71	0.76	**0.81**	**0.81**	0.74	0.75	**0.80**	0.64	0.75	**0.81**	0.75
Glass	0.52	**0.54**	0.51	0.48	0.48	**0.50**	0.49	0.46	**0.49**	**0.52**	0.47	0.50	0.50
Heart-statlog	0.77	**0.81**	0.80	0.72	0.82	**0.83**	0.76	**0.81**	0.80	0.77	0.78	**0.81**	0.79
Ionosphere	0.84	0.85	**0.87**	0.84	**0.86**	0.85	0.74	0.78	**0.84**	0.81	0.82	**0.86**	0.83
Iris	**0.87**	0.82	0.85	0.85	0.84	**0.92**	0.83	0.85	**0.90**	0.81	0.80	**0.93**	0.86
Landsat	0.83	0.83	**0.84**	0.76	0.84	**0.86**	0.83	**0.85**	**0.85**	0.84	0.85	**0.86**	0.84
Letter	0.67	**0.70**	0.67	0.68	0.74	**0.79**	0.64	0.72	**0.78**	0.65	0.71	**0.77**	0.71
Optdigits	0.95	**0.97**	**0.97**	0.95	0.97	**0.98**	0.96	**0.97**	**0.97**	0.95	0.97	**0.98**	0.97
Page-blocks	**0.92**	**0.92**	0.91	0.91	0.91	**0.92**	**0.92**	0.91	0.91	**0.93**	**0.93**	0.92	0.92
Parkinsons	0.78	0.83	**0.84**	0.75	**0.85**	**0.85**	0.81	0.82	**0.85**	0.78	**0.83**	**0.83**	0.82
Segment	0.82	**0.83**	0.81	0.78	0.87	**0.90**	0.84	0.88	**0.90**	0.84	0.87	**0.90**	0.85
Spambase	**0.71**	0.65	0.65	**0.88**	0.87	**0.88**	0.88	**0.89**	**0.89**	0.87	0.88	**0.89**	0.83
Wine	0.88	0.94	**0.98**	0.92	0.95	**0.96**	0.93	0.95	0.95	0.89	0.94	**0.96**	0.94
Average	0.77	**0.78**	**0.78**	0.77	0.81	**0.83**	0.78	0.80	**0.82**	0.78	0.79	**0.83**	0.80

We also carried out a comparison of the running times of the prototype selection algorithms considered in our experiments. In this comparison, we applied the 7 prototype selection algorithms to reduce the 3 biggest datasets considered in our tests: *letter*, *optdigits* and *spambase*. We adopted the same parametrizations that were adopted in the first experiment. We performed the experiments in an Intel® Core™ i5-5200U laptop with a 2.2 GHz CPU and 8 GB of RAM. The Fig. 2 shows that, considering these datasets, the PSDSP algorithm achieves the lowest running time, in comparison to the other algorithms. This result is a consequence of the *linear* time complexity of PSDSP. Notice that the Fig. 2

uses a *logarithmic scale* in the time axis, since there are big differences in the running time of the algorithms.

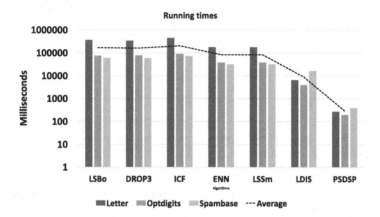

Fig. 2. Comparison of the running times of 7 prototype selection algorithms, considering the three biggest datasets. Notice that the time axis uses a logarithmic scale.

In summary, the experiments show that the PSDSP algorithm has the lowest running time, in comparison with other state-of-the-art algorithms. Besides that, PSDSP also presents the highest reduction rates, while preserves a good accuracy, which is similar to the accuracy achieved by other algorithms, such as DROP3 and LDIS. These features suggest that PSDSP is a promising algorithm for prototype selection. It is indicated in scenarios that a lower running time is critical.

6   Conclusion

In this paper, we proposed an efficient algorithm for prototype selection, called PSDSP (Prototype Selection based on Dense Spatial Partitions). It adopts the notion of *spatial partition*, which, in an overview, is a multidimensional region in the dataset. The PSDSP algorithm defines a set of spatial partitions for each class of objects in the dataset. After, it identifies the sets of instances that are contained by the k densest spatial partitions, and extracts prototypes from these sets. The algorithm takes as input the value n, which represents the number of intervals in which the dimensions of the dataset will be divided; and the value p, which determines the desired number of prototypes, as a percentage of the original number of instances in the dataset. Thus, the algorithm provides control to the user regarding the size of the reduced dataset. This is not a common feature in the prototype selection algorithms provided in the literature.

Our experiments show that PSDSP provides a good balance between accuracy and reduction, with the lowest time complexity, when compared with other algorithms available in the literature. The empirical evaluation of running times showed that PSDSP algorithm has a running time that is significantly lower than the running times of other stated-of-the-art algorithms. These features make PSDSP a promising algorithm for dealing with big volumes of data, in scenarios that the running time is critical.

In future works, we plan to investigate how to allow the algorithm to automatically identify the best way of dividing the dataset in a set of spatial partitions, without user intervention.

References

1. Anwar, I.M., Salama, K.M., Abdelbar, A.M.: Instance selection with ant colony optimization. Procedia Comput. Sci. **53**, 248–256 (2015)
2. Brighton, H., Mellish, C.: Advances in instance selection for instance-based learning algorithms. Data Min. Knowl. Disc. **6**(2), 153–172 (2002)
3. Carbonera, J.L.: An efficient approach for instance selection. In: Bellatreche, L., Chakravarthy, S. (eds.) DaWaK 2017. LNCS, vol. 10440, pp. 228–243. Springer, Cham (2017). https://doi.org/10.1007/978-3-319-64283-3_17
4. Carbonera, J.L., Abel, M.: A density-based approach for instance selection. In: 2015 IEEE 27th International Conference on Tools with Artificial Intelligence (ICTAI), pp. 768–774. IEEE (2015)
5. Carbonera, J.L., Abel, M.: A novel density-based approach for instance selection. In: 2016 IEEE 28th International Conference on Tools with Artificial Intelligence (ICTAI), pp. 549–556. IEEE (2016)
6. Carbonera, J.L., Abel, M.: Efficient prototype selection supported by subspace partitions. In: 2017 IEEE 29th International Conference on Tools with Artificial Intelligence (ICTAI), pp. 921–928. IEEE (2017)
7. Chou, C.H., Kuo, B.H., Chang, F.: The generalized condensed nearest neighbor rule as a data reduction method. In: 2006 18th International Conference on Pattern Recognition, ICPR 2006, vol. 2, pp. 556–559. IEEE (2006)
8. García, S., Luengo, J., Herrera, F.: Data preprocessing in data mining. Springer, Switzerland (2015). https://doi.org/10.1007/978-3-319-10247-4
9. Gates, G.W.: Reduced nearest neighbor rule. IEEE Trans. Inf. Theory **18**(3), 431–433 (1972)
10. Hamidzadeh, J., Monsefi, R., Yazdi, H.S.: IRAHC: instance reduction algorithm using hyperrectangle clustering. Pattern Recogn. **48**(5), 1878–1889 (2015)
11. Hart, P.E.: The condensed nearest neighbor rule. IEEE Trans. Inf. Theory **14**, 515–516 (1968)
12. Leyva, E., González, A., Pérez, R.: Three new instance selection methods based on local sets: a comparative study with several approaches from a bi-objective perspective. Pattern Recogn. **48**(4), 1523–1537 (2015)
13. Lin, W.C., Tsai, C.F., Ke, S.W., Hung, C.W., Eberle, W.: Learning to detect representative data for large scale instance selection. J. Syst. Softw. **106**, 1–8 (2015)
14. Nikolaidis, K., Goulermas, J.Y., Wu, Q.: A class boundary preserving algorithm for data condensation. Pattern Recogn. **44**(3), 704–715 (2011)
15. Wilson, D.R., Martinez, T.R.: Reduction techniques for instance-based learning algorithms. Mach. Learn. **38**(3), 257–286 (2000)
16. Wilson, D.L.: Asymptotic properties of nearest neighbor rules using edited data. IEEE Trans. Syst. Man Cybern. SMC **2**(3), 408–421 (1972)

Complexity of Rule Sets Induced by Characteristic Sets and Generalized Maximal Consistent Blocks

Patrick G. Clark[1], Cheng Gao[1], Jerzy W. Grzymala-Busse[1,2(✉)], Teresa Mroczek[2], and Rafal Niemiec[2]

[1] Department of Electrical Engineering and Computer Science, University of Kansas, Lawrence, KS 66045, USA
patrick.g.clark@gmail.com, {cheng.gao,jerzy}@ku.edu
[2] Department of Expert Systems and Artificial Intelligence, University of Information Technology and Management, 35-225 Rzeszow, Poland
{tmroczek,rniemiec}@wsiz.rzeszow.pl

Abstract. We study mining incomplete data sets with two interpretations of missing attribute values, lost values and "do not care" conditions. For data mining we use characteristic sets and generalized maximal consistent blocks. Additionally, we use three types of probabilistic approximations, lower, middle and upper, so altogether we apply six approaches to data mining. Since it was shown that an error rate, associated with such data mining is not universally smaller for any approach, we decided to compare complexity of induced rule sets. Therefore, our objective is to compare six approaches to mining incomplete data sets in terms of complexity of induced rule sets. We conclude that there are statistically significant differences between these approaches.

Keywords: Incomplete data · Lost values · "do not care" conditions
Characteristic sets · Maximal consistent blocks
MLEM2 rule induction algorithm · Probabilistic approximations

1 Introduction

We study mining incomplete data sets with two interpretations of missing attribute values, lost values and "do not care" conditions. A missing attribute value is interpreted as lost if the original value existed but currently is unavailable, for example it is forgot or erased. A "do not care" condition means that the missing attribute value may be replaced by any value from the attribute domain. A "do not care" condition may occur as a result of a refusal to answer a question during the interview.

For data mining we use probabilistic approximations, a generalization of the lower and upper approximations, well known in rough set theory. A probabilistic approximation is associated with a parameter α, interpreted as a probability. When $\alpha = 1$, a probabilistic approximation becomes the lower approximation; if

© Springer International Publishing AG, part of Springer Nature 2018
L. Rutkowski et al. (Eds.): ICAISC 2018, LNAI 10842, pp. 301–310, 2018.
https://doi.org/10.1007/978-3-319-91262-2_27

α is small positive number, e.g., 0.001, a probabilistic approximation is the upper approximation. Initially, probabilistic approximations were applied to completely specified data sets [9,12–19]. Probabilistic approximations were generalized to incomplete data sets in [8].

Characteristic sets, for incomplete data sets with any interpretation of missing attribute values, were introduced in [7]. Maximal consistent blocks, restricted only to data sets with "do not care" conditions, were introduced in [11]. Additionally, in [11] maximal consistent blocks were used as granules to define only ordinary lower and upper approximations. A definition of the maximal consistent block was generalized to cover lost values and probabilistic approximations in [1]. The applicability of characteristic sets and maximal consistent blocks for mining incomplete data, from the view point of an error rate, was studied in [1]. As it happened, there is a small difference in quality of rule sets induced either way. Thus, we decided to compare characteristic sets with generalized maximal consistent blocks in terms of complexity of induced rule sets. In our experiments, the Modified Learning from Examples Module, version 2 (MLEM2) was used for rule induction [6].

2   Incomplete Data

In this paper, the input data sets are presented in the form of a decision table. An example of a decision table is shown in Table 1. Rows of the decision table represent cases, while columns are labeled by variables. The set of all cases will be denoted by U. In Table 1, $U = \{1, 2, 3, 4, 5, 6, 7, 8\}$. Independent variables are called *attributes* and a dependent variable is called a *decision* and is denoted by d. The set of all attributes is denoted by A. In Table 1, $A = \{$ *Temperature*, *Headache*, *Cough*$\}$. The value for a case x and an attribute a is denoted by $a(x)$.

We distinguish between two interpretations of missing attribute values: lost values, denoted by "?" and "do not care" conditions, denoted by "*". Table 1 presents an incomplete data set with both lost values and "do not care" conditions.

The set X of all cases defined by the same value of the decision d is called a *concept*. For example, a concept associated with the value *yes* of the decision *Flu* is the set $\{1, 2, 3, 4\}$.

For a completely specified data set, let a be an attribute and let v be a value of a. A *block* of (a, v), denoted by $[(a, v)]$, is the set $\{x \in U \mid a(x) = v\}$ [4].

For incomplete decision tables the definition of a block of an attribute-value pair (a, v) is modified in the following way.

- If for an attribute a and a case x we have $a(x) = ?$, the case x should not be included in any blocks $[(a, v)]$ for all values v of attribute a,
- If for an attribute a and a case x we have $a(x) = *$, the case x should be included in blocks $[(a, v)]$ for all specified values v of attribute a.

For the data set from Table 1 the blocks of attribute-value pairs are:
[(Temperature, normal)] = $\{3, 6, 8\}$,

Table 1. A decision Table

	Attributes			Decision
Case	Temperature	Headache	Cough	Flu
1	high	yes	?	yes
2	high	no	*	yes
3	*	?	yes	yes
4	high	no	?	yes
5	?	no	*	no
6	normal	*	no	no
7	high	no	yes	no
8	*	no	?	no

$[(\text{Temperature, high})] = \{1, 2, 3, 4, 7, 8\}$,
$[(\text{Headache, no})] = \{2, 4, 5, 6, 7, 8\}$,
$[(\text{Headache, yes})] = \{1, 6\}$,
$[(\text{Cough, no})] = \{2, 5, 6\}$,
$[(\text{Cough, yes})] = \{2, 3, 5, 7\}$.

For a case $x \in U$ and $B \subseteq A$, the *characteristic set* $K_B(x)$ is defined as the intersection of the sets $K(x, a)$, for all $a \in B$, where the set $K(x, a)$ is defined in the following way:

- If $a(x)$ is specified, then $K(x, a)$ is the block $[(a, a(x))]$ of attribute a and its value $a(x)$,
- If $a(x) = ?$ or $a(x) = *$, then $K(x, a) = U$.

For Table 1 and $B = A$,
$K_A(1) = \{1\}$,
$K_A(2) = \{2, 4, 7, 8\}$,
$K_A(3) = \{2, 3, 5, 7\}$,
$K_A(4) = \{2, 4, 7, 8\}$,
$K_A(5) = \{2, 4, 5, 6, 7, 8\}$,
$K_A(6) = \{6\}$,
$K_A(7) = \{2, 7\}$,
$K_A(8) = \{2, 4, 5, 6, 7, 8\}$.

A binary relation $R(B)$ on U, defined for $x, y \in U$ in the following way

$$(x, y) \in R(B) \ if \ and \ only \ if \ y \in K_B(x)$$

will be called the *characteristic relation*. In our example $R(A) = \{(1, 1), (2, 2), (2, 4), (2, 7), (2, 8), (3, 2), (3, 3), (3, 5), (3, 7), (4, 2), (4, 4), (4, 7), (4, 8), (5, 2), (5, 4), (5, 5), (5, 6), (5, 7), (5, 8), (6, 6), (7, 2), (7, 7), (8, 2), (8, 4), (8, 5), (8, 6), (8, 7), (8, 8)\}$.

We quote some definitions from [1]. Let X be a subset of U. The set X is B-*consistent* if $(x, y) \in R(B)$ for any $x, y \in X$. If there does not exist a B-consistent subset Y of U such that X is a proper subset of Y, the set X is called a *generalized maximal B-consistent block*. The set of all generalized maximal B-consistent blocks will be denoted by $\mathscr{C}(B)$. In our example, $\mathscr{C}(A) = \{\{1\}, \{2, 4, 8\}, \{2, 7\}, \{3\} \{5, 8\}, \{6\}\}$.

Let $B \subseteq A$ and $Y \in \mathscr{C}(B)$. The set of all generalized maximal B-consistent blocks which include an element x of the set U, i.e. the set

$$\{Y | Y \in \mathscr{C}(B), x \in Y\}$$

will be denoted by $\mathscr{C}_B(x)$.

For data sets in which all missing attribute values are "do not care" conditions, an idea of a maximal consistent block of B was defined in [10]. Note that in our definition, the generalized maximal consistent blocks of B are defined for arbitrary interpretations of missing attribute values. For Table 1, the generalized maximal A-consistent blocks $\mathscr{C}_A(x)$ are

$\mathscr{C}_A(1) = \{\{1\}\}$,
$\mathscr{C}_A(2) = \{\{2, 4, 8\}, \{2, 7\}\}$,
$\mathscr{C}_A(3) = \{\{3\}\}$,
$\mathscr{C}_A(4) = \{\{2, 4, 8\}\}$,
$\mathscr{C}_A(5) = \{\{5, 8\}\}$,
$\mathscr{C}_A(6) = \{\{6\}\}$,
$\mathscr{C}_A(7) = \{\{2, 7\}\}$,
$\mathscr{C}_A(8) = \{\{2, 4, 8\}, \{5, 8\}\}$.

3   Probabilistic Approximations

In this section, we will discuss two types of probabilistic approximations: based on characteristic sets and on generalized maximal consistent blocks.

3.1   Probabilistic Approximations Based on Characteristic Sets

In general, probabilistic approximations based on characteristic sets may be categorized as singleton, subset and concept [3,7]. In this paper we restrict our attention only to concept probabilistic approximations, for simplicity calling them probabilistic approximations based on characteristic sets.

A *probabilistic approximation based on characteristic sets* of the set X with the threshold α, $0 < \alpha \leq 1$, denoted by $appr_\alpha^{CS}(X)$, is defined as follows

$$\cup \{K_A(x) \mid x \in X, \ Pr(X | K_A(x)) \geq \alpha\}.$$

For Table 1 and both concepts $\{1, 2, 3, 4\}$ and $\{5, 6, 7, 8\}$, all distinct probabilistic approximations, based on characteristic sets, are

$appr_{0.5}^{CS}(\{1,2,3,4\}) = \{1,2,3,4,5,7,8\},$

$appr_1^{CS}(\{1,2,3,4\}) = \{1\},$

$appr_{0.667}^{CS}(\{5,6,7,8\}) = \{2,4,5,6,7,8\},$

$appr_1^{CS}(\{5,6,7,8\}) = \{6\}.$

If for some β, $0 < \beta \leq 1$, a probabilistic approximation $appr_\beta^{CS}(X)$ is not listed above, it is equal to the probabilistic approximation $appr_\alpha^{CS}(X)$ with the closest α to β, $\alpha \geq \beta$. For example, $appr_{0.2}^{CS}(\{1,2,3,4\}) = appr_{0.5}^{CS}(\{1,2,3,4\})$.

3.2 Probabilistic Approximations Based on Generalized Maximal Consistent Blocks

By analogy with the definition of a probabilistic approximation based on characteristic sets, we may define a probabilistic approximation based on generalized maximal consistent blocks as follows:

A *probabilistic approximation* based on generalized maximal consistent blocks of the set X with the threshold α, $0 < \alpha \leq 1$, and denoted by $appr_\alpha^{MCB}(X)$, is defined as follows

$$\cup\{Y \mid Y \in \mathscr{C}_x(A),\ x \in X,\ Pr(X|Y) \geq \alpha\}.$$

All distinct probabilistic approximations based on generalized maximal consistent blocks are

$appr_{0.5}^{MCB}(\{1,2,3,4\}) = \{1,2,3,4,7,8\},$

$appr_{0.667}^{MCB}(\{1,2,3,4\}) = \{1,2,3,4,8\},$

$appr_1^{MCB}(\{1,2,3,4\}) = \{1,3\},$

$appr_{0.333}^{MCB}(\{5,6,7,8\}) = \{2,4,5,6,7,8\},$

$appr_{0.5}^{MCB}(\{5,6,7,8\}) = \{2,5,6,7,8\},$

$appr_1^{MCB}(\{5,6,7,8\}) = \{5,6,8\}.$

4 Experiments

Our experiments were conducted on eight data sets that are available in the University of California at Irvine *Machine Learning Repository*. For any such data set a template was created by replacing (randomly) 5% of existing specified

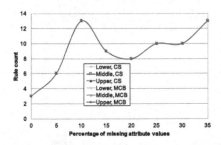

Fig. 1. Error rate for the *bankruptcy* data set with lost values

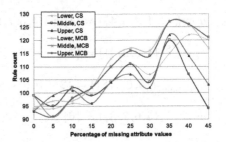

Fig. 2. Error rate for the *breast cancer* data set with lost values

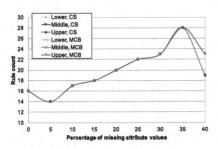

Fig. 3. Error rate for the *echocardiogram* data set with lost values

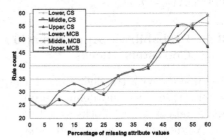

Fig. 4. Error rate for the *hepatitis* data set with lost values

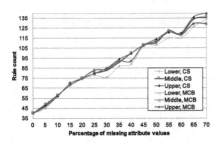

Fig. 5. Error rate for the *image segmentation* data set with lost values

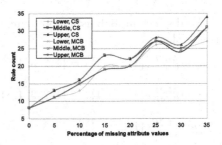

Fig. 6. Error rate for the *iris* data set with lost values

attribute values by *lost values*, then adding another 5% of lost values, and so on, until an entire row was full of lost values. The same templates were used for constructing data sets with "do not care" conditions, by replacing "?"s with "*"s, so we created 16 families of incomplete data sets.

In our experiments we used the MLEM2 rule induction algorithm of the LERS (Learning from Examples using Rough Sets) data mining system [2, 5, 6]. We used characteristic sets and generalized maximal consistent blocks for mining incomplete datasets. Additionally, we used three different probabilistic

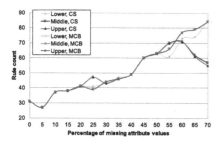

Fig. 7. Error rate for the *lymphography* data set with lost values

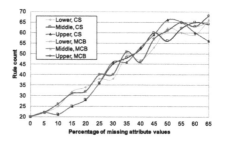

Fig. 8. Error rate for the *wine recognition* data set with lost values

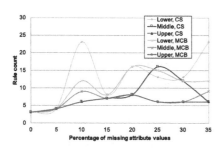

Fig. 9. Number of rules for the *bankruptcy* data set with "do not care" conditions

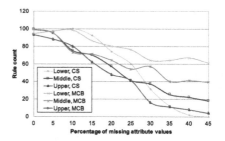

Fig. 10. Error rate for the *breast cancer* data set with "do not care" conditions

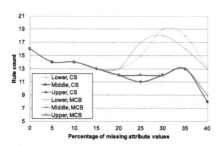

Fig. 11. Error rate for the *echocardiogram* data set with "do not care" conditions

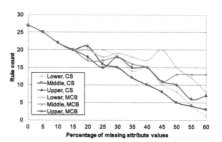

Fig. 12. Error rate for the *hepatitis* data set with "do not care" conditions

approximations, lower ($\alpha = 1$), middle ($\alpha = 0.5$) and upper ($\alpha = 0.001$). Thus our experiments were conducted on six different approaches to mining incomplete data sets. These six approaches were compared by applying the Friedman rank sum test combined with multiple comparisons, with a 5% level of significance. We applied this test to all 16 families of data sets, eight with lost values and eight with "do not care" conditions.

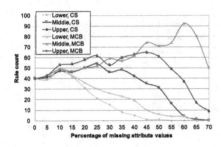

Fig. 13. Error rate for the *image segmentation* data set with "do not care" conditions

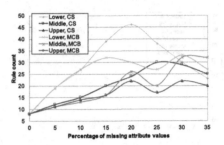

Fig. 14. Error rate for the *iris* data set with "do not care" conditions

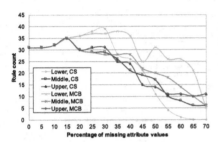

Fig. 15. Error rate for the *lymphography* data set with "do not care" conditions

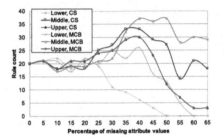

Fig. 16. Error rate for the *wine recognition* data set with "do not care" conditions

Results of our experiments are presented in Figs. 1, 2, 3, 4, 5, 6, 7, 8, 9, 10, 11, 12, 13, 14, 15 and 16, where "CS" denotes a characteristic set and "MCB" denotes a maximal consistent block. For eight data sets with lost values, the null hypothesis H_0 of the Friedman test saying that differences between these approaches are insignificant was rejected for four families of data sets (*breast cancer*, *hepatitis*, *image recognition* and *iris*). However, the post-hoc test (distribution-free multiple comparisons based on the Friedman rank sums) indicated that the differences between all six approaches were statistically insignificant for *breast cancer* and *hepatitis*. Results for *image recognition* and *iris* are listed in Table 2.

For eight data sets with "do not care" conditions, the null hypothesis H_0 of the Friedman test was rejected for all eight families of data sets. Additionally, for three families of data sets (*bankruptcy*, *echocardiogram* and *hepatitis* families of data sets the post-hoc test shown that the differences between all six approaches were insignificant. Results for the remaining five data sets are presented in Table 2. Obviously, for data sets with "do not care" conditions, *concept* upper approximations based on characteristic sets are identical with upper approximations based on maximal consistent blocks [11].

Table 2. Results of statistical analysis

Data set	Friedman test results (5% significance level)
Image recognition, ?	Lower, CS is better than Middle, CS and Upper, CS
	Lower, CS is better than all three approaches with MCB
Iris, ?	Lower, CS is better than Upper, CS
Breast cancer, *	Upper, CS is better than Lower, MCB
	Upper, MCB is better than Lower, MCB
Image recognition, *	Lower, CS is better than Upper, CS; Middle, MCB and Upper MCB
	Lower, MCB is better than Upper, CS; Middle, MCB and Upper MCB
Iris, *	Upper, CS is better than Lower, CS and Lower, MCB
	Upper, MCB is better than Lower, CS and Lower, MCB
Lymphography, *	Middle, CS is better than Lower, MCB
Wine recognition, *	Lower, CS is better than Middle, CS and Middle, MCB

5   Conclusions

Our objective was to compare six approaches to mining incomplete data sets (combining characteristic sets and generalized maximal consistent blocks with three types of probabilistic approximations). Our conclusion is that the choice between characteristic sets and generalized maximal consistent blocks and between types of probabilistic approximation is important, since there are statistically significant differences in complexity of induced rule sets. However, for every data set all six approaches should be tested and the best one should be selected. There is no universally best approach.

References

1. Clark, P.G., Gao, C., Grzymala-Busse, J.W., Mroczek, T.: Characteristic sets and generalized maximal consistent blocks in mining incomplete data. In: Polkowski, L., Yao, Y., Artiemjew, P., Ciucci, D., Liu, D., Ślęzak, D., Zielosko, B. (eds.) IJCRS 2017. LNCS (LNAI), vol. 10313, pp. 477–486. Springer, Cham (2017). https://doi.org/10.1007/978-3-319-60837-2_39
2. Clark, P.G., Grzymala-Busse, J.W.: Experiments on probabilistic approximations. In: Proceedings of the 2011 IEEE International Conference on Granular Computing, pp. 144–149 (2011)
3. Clark, P.G., Grzymala-Busse, J.W.: Experiments using three probabilistic approximations for rule induction from incomplete data sets. In: Proceeedings of the MCCSIS 2012, IADIS European Conference on Data Mining ECDM 2012, pp. 72–78 (2012)

4. Grzymala-Busse, J.W.: LERS–a system for learning from examples based on rough sets. In: Slowinski, R. (ed.) Intelligent Decision Support. Handbook of Applications and Advances of the Rough Set Theory, pp. 3–18. Kluwer Academic Publishers, Dordrecht (1992)

5. Grzymala-Busse, J.W.: A new version of the rule induction system LERS. Fundam. Inform. **31**, 27–39 (1997)

6. Grzymala-Busse, J.W.: MLEM2: a new algorithm for rule induction from imperfect data. In: Proceedings of the 9th International Conference on Information Processing and Management of Uncertainty in Knowledge-Based Systems, pp. 243–250 (2002)

7. Grzymala-Busse, J.W.: Rough set strategies to data with missing attribute values. In: Notes of the Workshop on Foundations and New Directions of Data Mining, in Conjunction with the Third International Conference on Data Mining, pp. 56–63 (2003)

8. Grzymała-Busse, J.W.: Generalized parameterized approximations. In: Yao, J.T., Ramanna, S., Wang, G., Suraj, Z. (eds.) RSKT 2011. LNCS (LNAI), vol. 6954, pp. 136–145. Springer, Heidelberg (2011). https://doi.org/10.1007/978-3-642-24425-4_20

9. Grzymala-Busse, J.W., Ziarko, W.: Data mining based on rough sets. In: Wang, J. (ed.) Data Mining: Opportunities and Challenges, pp. 142–173. Idea Group Publishing, Hershey (2003)

10. Leung, Y., Li, D.: Maximal consistent block technique for rule acquisition in incomplete information systems. Inf. Sci. **153**, 85–106 (2003)

11. Leung, Y., Wu, W., Zhang, W.: Knowledge acquisition in incomplete information systems: a rough set approach. Eur. J. Oper. Res. **168**, 164–180 (2006)

12. Pawlak, Z., Skowron, A.: Rough sets: some extensions. Inf. Sci. **177**, 28–40 (2007)

13. Pawlak, Z., Wong, S.K.M., Ziarko, W.: Rough sets: probabilistic versus deterministic approach. Int. J. Man Mach. Stud. **29**, 81–95 (1988)

14. Ślęzak, D., Ziarko, W.: The investigation of the bayesian rough set model. Int. J. Approx. Reason. **40**, 81–91 (2005)

15. Wong, S.K.M., Ziarko, W.: INFER–an adaptive decision support system based on the probabilistic approximate classification. In: Proceedings of the 6-th International Workshop on Expert Systems and their Applications, pp. 713–726 (1986)

16. Yao, Y.Y.: Probabilistic rough set approximations. Int. J. Approx. Reason. **49**, 255–271 (2008)

17. Yao, Y.Y., Wong, S.K.M.: A decision theoretic framework for approximate concepts. Int. J. Man Mach. Stud. **37**, 793–809 (1992)

18. Ziarko, W.: Variable precision rough set model. J. Comput. Syst. Sci. **46**(1), 39–59 (1993)

19. Ziarko, W.: Probabilistic approach to rough sets. Int. J. Approx. Reason. **49**, 272–284 (2008)

On Ensemble Components Selection in Data Streams Scenario with Gradual Concept-Drift

Piotr Duda[✉]

Institute of Computational Intelligence, Czestochowa University of Technology,
Al. Armii Krajowej 36, 42-200 Czestochowa, Poland
piotr.duda@iisi.pcz.pl

Abstract. In the paper we study the issue of components selection of an ensemble for data stream classification. Decision about adding or removing single component has significant meaning not only for an accuracy in the current instant, but can be also significant for the further stream processing. The algorithm proposed in this paper is an enhanced version of the ASE (Automatically Sized Ensemble) algorithm which guarantees that a new component will be added to the ensemble only if it increases the accuracy not only for the current data chunk but also for the whole data stream. The algorithm is designed to improve data stream processing in the case when one concept is gradually replaced by the other. The Hellinger distance is applied to allow adding a new component, if its predictions differ significantly from the rest of the ensemble, even though that component does not increase accuracy of the whole ensemble.

Keywords: Ensemble methods · Data streams · Gradual concept drift

1 Introduction

Currently, machine learning methods find more and more interesting applications, see e.g. [3,6,8,9,21,30,32]. One of the most recent and difficult problems in machine learning is the analysis of a huge amount of data which comes to the system continuously during a learning process. The information incorporated into every data element has to be included in the model as fast as it is possible. This limitation is caused by the huge volume of data, which cannot be stored in a system at a one time. Moreover, the fact that data come to the system in a continuous manner entails that the model must be able to respond at any time. In consequence, the processing time of every instance needs to be minimized. An additional challenge related to the assumed data characteristics is the possibility of changing the data distribution. This feature is called a concept drift. In consequence, the method has to be able to adjust to the non-stationary environment. The algorithms that meet all of these assumptions are called the Data Stream Analysis (DSA) methods.

© Springer International Publishing AG, part of Springer Nature 2018
L. Rutkowski et al. (Eds.): ICAISC 2018, LNAI 10842, pp. 311–320, 2018.
https://doi.org/10.1007/978-3-319-91262-2_28

In the literature, there are plenty methods which try to solve such data stream analysis tasks as classification [11,26,27], regression [12,13,16] density estimation [7], clustering [1], etc. DSA is currently a popular field of study in view of its many practical application, e.g. in sensors data analysis, in financial data analysis, in environment monitoring, and in network security. These methods use different approaches to deal with streaming data. The most popular solutions are incremental learning, sliding windows, and ensemble methods. In the case of incremental algorithms, each data element is processed as soon as it comes to the system and it is forgotten as soon as it is processed. The other approach is presented by the algorithms using sliding windows. In this case, the model stores some number of the most recent data elements. When a new data element enters the system, the oldest data is deleted and the new is stored in a window. The traditional ensemble methods work on data chunks. In this case, the system waits until a certain number of data will be gathered. Then based on the data, stored in the chunk, it builds the so-called weak learner. The more detailed review of the classification ensemble methods is presented in Sect. 3.

The choice of the proper strategy to deal with data stream has a crucial impact on the performance of DSA algorithms especially taking into consideration a concept-drift. The changes in a data distribution can occur at any time and in any manner. Intuitively the types of the non-stationary can be seen as one of the following cases or as their combinations: the abrupt changes, the incremental changes, the gradual changes, and the reoccurring changes. In consequence, there are not the one, best approach. Every type of changes should be analyzed separately. Among aforementioned types, the gradual changes deserve a special attention, see e.g. [2,15,31].

Based on the above motivation we decided to propose a novel method to deal with gradual changes in data streams. It is an updated version of the SEM algorithm developed in [23]. In our approach, we will use the Hellinger distance to make the SEM algorithm more sensitive to the gradual changes in data streams.

The rest of the paper is organized as follows. The basic definition and notations are introduced in Sect. 2. The related works are summarized in Sect. 3. The main contribution is presented in Sect. 4. Experimental evaluation of the proposed method is described in Sect. 5. Finally, Sect. 6 presents conclusions and suggestions for future works.

2 Preliminaries

In a classification task, we want to find the best approximation of the function $\psi : \mathbb{X} \to \mathbb{Y}$, where $\mathbb{X}$ is a d dimensional space of attributes, and $\mathbb{Y}$ is a finite set of the labels. Every single element of the space $\mathbb{X}$ takes continuous or discrete values. An approximation $\hat{\psi}$ of function ψ is built based on a training set.

In a data stream scenario, the training set takes the following form

$$S = \{s_1, s_2, \dots\} \tag{1}$$

where subsequent s_i, $i = 1, 2, \dots$ are generated continuously, $s_i = (X_i, Y_i)$, $X_i \in \mathbb{X}$ and $Y_i \in \mathbb{Y}$. If the stream is stationary, all the data are generated from the same probability distribution, however, in a general case, such assumption cannot be made. The concept drift phenomenon is commonly described on the background of the Bayesian Decision Theory. In this case, we assign the attribute vector X to the class that maximizes the conditional probability

$$P(Y|X) = \frac{P(Y)P(X|Y)}{P(X)}. \tag{2}$$

In a non-stationary environment all the probabilities on the right-hand side of (2) can change over time and in consequence, the joint probability distribution of X and Y can differ in time, i.e.

$$P_i(X,Y) \neq P_{i+1}(X,Y). \tag{3}$$

In the case of gradual concept-drift, we assume that the data coming from the stream are generated from two different probability distributions, given in time i by the density functions $f_i(x,y)$ and $g_i(x,y)$, and the contribution of each distribution is changing.

$$X_i \sim \begin{cases} f_i(x,y) & \text{with probability } w_i \\ g_i(x,y) & \text{with probability } 1 - w_i \end{cases} \tag{4}$$

where $0 \leq w_i \leq 1$. The method dealing with gradual changes has to be able to keep information about both distributions and allow to update the model during processing the data stream. The natural solution seems to be an ensemble approach. In data stream scenario, the ensemble $\varGamma$ is a set of static classifiers $hi(\cdot)$ (weak learners) that are learned based on the subsequent chunks of data. It is a big challenge to propose a method to add and remove the components from the ensemble. A lot of work has been done to solve this problem, however, most of the algorithm are destined to work in a general case and an application of ensemble methods, especially to gradual changes, has not been investigated and justified sufficiently.

3   Related Works

The ensemble methods use the effective pre-processing method (chunk based) to adjust classifier to changes in an environment. One of the first works in this field was the Streaming Ensemble Algorithm (SEA) [28]. The authors proposed to create a new classifier based on every chunk of data. The components are stored in a memory until their number does not exceed some assumed limit. After that, a newly created component (also called a weak learner or a base classifier) can replace one of the current components only if its accuracy is higher than the weakest component of the ensemble. Otherwise, the newly created component is discarded. The label for a new instance is established on a base of majority voting. The Accuracy Weighted Ensemble (AWE) algorithm was proposed in [29]. The authors proposed to improve the SEA algorithm by weighting the power of a vote of each component according to its accuracy. Additionally, the authors proved that the decision made by the ensemble will be always at least as good as made by a single classifier. The Learn++ algorithm was proposed in [25]. In fact, the authors proposed a procedure to construct an ensemble in a case of the non-stationary environment, however, they did not use the term 'stream data'. The weights for the weak classifiers were established in a new way and additionally the resampling method, inspired by AdaBoost, was introduced. This idea was adapted to the data stream scenario in [14], and further was extended to the imbalanced data, in [10]. The online version of Bagging and Boosting algorithms was proposed in [22] and this approach was extended in [4]. The Diversity for Dealing with Drifts (DDD) algorithm [20] merged the method of ensemble construction with a drift detector. In [23] the authors proposed a procedure for determining the proper ensemble size automatically, based on an appropriate statistical test. Next, its special version, dedicated to the decision trees, was proposed in [24]. Instead of assigning a weight to the whole tree, the authors proposed to determine weights on the level of leaves. The ensemble methods dedicated to gradual concept drift can be found in [17–19].

4   The ASE-GD Algorithm

In this section, the proposed procedure is presented in details. The described algorithm is an improvement of the method introduced in our previous paper [23] called the Automatically Sized Ensemble (ASE) algorithm.

Let the set $S^t = (s_{t_1}, \ldots, s_{t_n})$ be a sequence of data elements coming from the stream in the t-th data chunk. A new weak learner $h_t(\cdot) : \mathbb{X} \to \mathbb{Y}$ is created, based on the dataset S^t. At the beginning, the first learned classifier determines the whole ensemble. For a subsequent data chunk t, a temporal ensemble $\Gamma^+ = \Gamma \cup \{h_t\}$ is created. Then the decision about adding a newly created weak learner is determined based on a statistical test. This test ensures that the additional component significantly improves the accuracy of the whole ensemble. To fulfill this task, after gathering the new chunk of data, the accuracies of both ensembles Γ and Γ^+ are computed (denoted by $Acc(\Gamma, S^{t+1})$ and $Acc(\Gamma^+, S^{t+1})$, respectively). Next, the following condition is checked:

$$Acc(\Gamma^+, S^{t+1}) - Acc(\Gamma, S^{t+1}) > z_{1-\alpha} \frac{1}{\sqrt{n}}, \tag{5}$$

where $z_{1-\alpha}$ is the $(1 - \alpha)$ quantile of the standard normal distribution $\mathbf{N}(0, 1)$ and α is a significance level of the test (fixed by the user).

The more complicated issue is to decide when an existing component should be removed from the ensemble. In [23] we proposed to remove component if the following inequality is true

$$Acc(\Gamma, S^{t+1}) - Acc(\Gamma^-, S^{t+1}) < z_{1-\alpha} \frac{1}{\sqrt{n}}, \tag{6}$$

where Γ^- includes every component of Γ without a currently investigated one.

4.1   Proposed Method

In the case of a gradual concept drift, every chunk of data contains some number of elements generated from the first distribution and some number generated from the other one. In such a case, we may want to force an algorithm to store of an 'unimportant' (in that moment) component to better adjust the whole ensemble in the future. For this purpose we propose to apply the Hellinger distance:

$$H^2(P, Q) = 1 - \sum_{i=1}^{k} \sqrt{p_i q_i} \tag{7}$$

where $P = (p_1, \ldots, p_k)$ and $Q = (q_1, \ldots, q_k)$ are the discrete probability distributions. If the considered distributions are similar, the Hellinger distance will be close to zero. The value close to 1 indicates that the distributions differ significantly.

To decide what should be done with the considered component h_t, we will always compare the outputs of the ensemble Γ with the outputs of the component h_t. If we check that a weak learner should be incorporated into the ensemble then h_t is a newly created component. To check which component can be removed from the ensemble, every single component is considered separately. This decision depends on the inequalities (5) or (6) and on the distribution of the outputs. Particularly, in the case of the adding a new component, we have

$$p_1 = P(\Gamma^+(X) = 1), \ p_2 = P(\Gamma^+(X) = 0) \tag{8}$$
$$q_1 = P(h_t(X) = 1), \ q_2 = P(h_t(X) = 0) \tag{9}$$

The obtained value of the Hellinger distance (7) is compared with the previously fixed threshold $\Theta > 0$. The pseudo-code of the proposed procedure, called ASE-GD, is presented in Fig. 1.

5   Experimental Results

In this section, the performance of the ASE-GD algorithm is investigated. The experiments are conducted to demonstrate the influence of the parameters of the ASE-GD algorithm on its performances. The two distributions (RT1 and RT2) were generated by the Random Tree Generator [11]. Then stream data were generated from these distribution. The ith element of the stream is coming from the RT2 distribution with probability

$$P(X_i \sim RT2) = \left(\tanh\left(\frac{i - 50000}{2}\right) + 1\right) \cdot 0.5, \tag{10}$$

and from the RT1 distribution with probability $1 - P(X_i \sim RT2)$.

To conduct the experiments we generated $50'000$ data elements. To generate the random trees we applied the Massive Online Analysis (MOA) framework [5].

```
Inputs: α > 0, Θ > 0 ,
01. t:=1
02. Gather S^t
03. Train h_t
04. Γ = {h_1}
05. t++
06. Until data are coming from the stream
07.     Gather S^t
08.        Γ^+ := Γ ∪ h_t
09.        Compute Acc(Γ, S^{t+1}) and Acc(Γ^-, S^{t+1})
10.        If inequality (5) is satisfied then
11.           Γ := Γ^+
12.        else
13.           Compute statistics (8) and (9)
14.           If H^2(P,Q) > Θ
15.              Γ := Γ^+
16.        For each h in Γ
17.           Γ^- := Γ\h
18.           Compute statistics (8) and (9)
19.           If Inequality (6) satisfied then and H^2(P,Q) < Θ
20.              Γ := Γ^-
21.        t++
22.        Go back to the line 06.
```

Fig. 1. The ASE-GD algorithm

The generated data have 15 binary attributes. Every instance belongs to one of two classes. The first leaves were allowed to appear beginning from the 3rd level of the tree and the maximum depth of the tree was set to 10.

The presented results were obtained using the prequential strategy. The performance of the ASE-GD algorithm is compared with the ASE algorithm [23]. The weak learners were established in the form of the ID3 decision tree.

In the first experiment, the dependences between the data chunk size and the accuracies are presented. Figure 2 presents prequential accuracies obtained for size of the chunk equal to $200, 300, \ldots, 2500$. The experiments were conducted with maximal depth of the tree fixed to 15 and parameter $\Theta = 0.1$. The accuracies of the ASE-GD algorithm are marked as a purple line and of the ASE algorithm as a green line.

One can see that a proper choice of the data chunk size has a crucial meaning. The chunk has to be big enough to allow a weak learner to properly develop. For the small size of the chunk (200–400), both algorithms present similar results. For bigger values, from 500 to 1200, the ASE-GD algorithm is significantly better. For the data chunk bigger than 1200 the improvement is negligible.

Next, the significance of the parameter Θ is investigated. The chunk size was fixed to 1000 data elements. The results obtained for $\Theta = 0.01, 0.02, \ldots, 0.2$ are presented in Fig. 3. The algorithm ASE takes the constant value because it does not depend on this parameter. If the value of Θ is set to zero, both algorithms provide the same results. The best result was achieved for $\Theta = 0.1$, and the higher its values reduced the improvement. That indicates that a proper determination of Θ is an important issue and a non-trivial task.

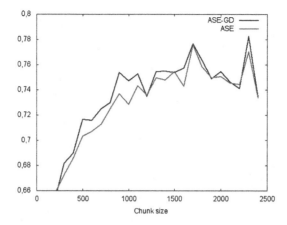

Fig. 2. Influence of the chunk size (Color figure online)

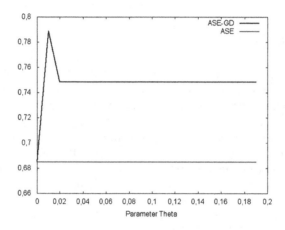

Fig. 3. Influence of the parameter Θ

The last experiment investigates the influence of the maximal depth of the trees, which varies from 3 to 15. The obtained accuracies are presented in Fig. 4.

The results of this experiment are consistent with our predictions. Increasing value of the examined parameter allowed for getting better accuracy. When the maximum depth of the tree reaches the maximum depth of random trees RT1 and RT2, it stops to affect accuracy. The ASE-GD algorithm presents better results during the whole experiment.

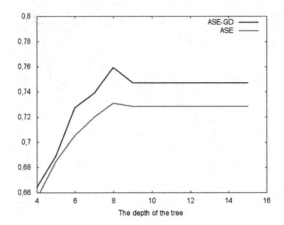

Fig. 4. The influence of the maximal depth of the trees on the performance of the ensemble

6   Conclusions

The selection of the ensemble components based only on the accuracy of the classification is not the optimal solution in the face of the occurrence of the concept drift. Incorporation of an additional measure, ensuring greater diversification of the components, has a positive effect on the performance of ensemble algorithms in a non-stationary environment. In this paper, we examined the utility of the Hellinger distance to the gradual changes. The presented experimental results confirm its usefulness. In the future work, we plan to propose a method for adjusting parameter Θ to the changes in a stream.

Acknowledgments. This work was supported by the Polish National Science Centre under Grant No. 2014/15/B/ST7/05264.

References

1. Amini, A., Wah, T.Y., Saboohi, H.: On density-based data streams clustering algorithms: a survey. J. Comput. Sci. Technol. **29**(1), 116–141 (2014)
2. Andressian, V., Parent, E., Claude, M.: A distributions free test to detect gradual changes in watershed behavior. Water Resour. Res. 39(9) (2003). https://doi.org/10.1029/2003WR002081
3. Ayadi, N., Derbel, N., Morette, N., Novales, C., Poisson, G.: Simulation and experimental evaluation of the ekf simultaneous localization and mapping algorithm on the wifibot mobile robot. J. Artif. Intell. Soft Comput. Res. **8**(2), 91–101 (2018). https://doi.org/10.1515/jaiscr-2018-0006
4. Beygelzimer, A., Kale, S., Luo, H.: Optimal and adaptive algorithms for online boosting. In: Proceedings of the 32nd International Conference on Machine Learning (ICML-15), pp. 2323–2331 (2015)
5. Bifet, A., Holmes, G., Kirkby, R., Pfahringer, B.: MOA: massive online analysis. J. Mach. Learn. Res. **11**(May), 1601–1604 (2010)

6. Bustamam, A., Sarwinda, D., Ardenaswari, G.: Texture and gene expression analysis of the MRI brain in detection of Alzheimers disease. J. Artif. Intell. Soft Comput. Res. **8**(2), 111–120 (2018). https://doi.org/10.1515/jaiscr-2018-0008
7. Cao, Y., He, H., Man, H.: SOMKE: Kernel density estimation over data streams by sequences of self-organizing maps. IEEE Trans. Neural Netw. Learn. Syst. **23**(8), 1254–1268 (2012)
8. Davis, J.J.J., Lin, C.T., Gillett, G., Kozma, R.: An integrative approach to analyze EEG signals and human brain dynamics in different cognitive states. J. Artif. Intell. Soft Comput. Res. **7**(4), 287–299 (2017)
9. Devi, V.S., Meena, L.: Parallel MCNN (PMCNN) with application to prototype selection on large and streaming data. J. Artif. Intell. Soft Comput. Res. **7**(3), 155–169 (2017)
10. Ditzler, G., Polikar, R.: Incremental learning of concept drift from streaming imbalanced data. IEEE Trans. Knowl. Data Eng. **25**(10), 2283–2301 (2013)
11. Domingos, P., Hulten, G.: Mining high-speed data streams. In: Proceedings of 6th ACM SIGKDD International Conference on Knowledge Discovery and Data Mining, pp. 71–80 (2000)
12. Duda, P., Jaworski, M., Rutkowski, L.: Knowledge discovery in data streams with the orthogonal series-based generalized regression neural networks. Inf. Sci. (2017). https://doi.org/10.1016/j.ins.2017.07.013
13. Duda, P., Jaworski, M., Rutkowski, L.: Convergent time-varying regression models for data streams: tracking concept drift by the recursive parzen-based generalized regression neural networks. Int. J. Neural Syst. **28**(02), 1750048 (2018)
14. Elwell, R., Polikar, R.: Incremental learning of concept drift in nonstationary environments. IEEE Trans. Neural Netw. **22**(10), 1517–1531 (2011)
15. Hoffmann, M., Vetter, M., Dette, H.: Nonparametric inference of gradual changes in the jump behaviour of time-continuous processes. Stoch. Process. Appl. (2018). https://doi.org/10.1016/j.spa.2017.12.005
16. Ikonomovska, E., Gama, J., Džeroski, S.: Online tree-based ensembles and option trees for regression on evolving data streams. Neurocomputing **150**, 458–470 (2015)
17. Jaworski, M., Duda, P., Rutkowski, L.: On applying the restricted Boltzmann machine to active concept drift detection. In: 2017 IEEE Symposium Series on Computational Intelligence (SSCI), pp. 1–8. IEEE (2017)
18. Liu, A., Zhang, G., Lu, J.: Fuzzy time windowing for gradual concept drift adaptation. In: 2017 IEEE International Conference on Fuzzy Systems (FUZZ-IEEE), pp. 1–6. IEEE (2017)
19. Mahdi, O.A., Pardede, E., Cao, J.: Combination of information entropy and ensemble classification for detecting concept drift in data stream. In: Proceedings of the Australasian Computer Science Week Multiconference, p. 13. ACM (2018)
20. Minku, L., Yao, X.: DDD: a new ensemble approach for dealing with concept drift. IEEE Trans. Knowl. Data Eng. **24**(4), 619–633 (2012)
21. Notomista, G., Botsch, M.: A machine learning approach for the segmentation of driving maneuvers and its application in autonomous parking. J. Artif. Intell. Soft Comput. Res. **7**(4), 243–255 (2017)
22. Oza, N.C.: Online bagging and boosting. In: 2005 IEEE International Conference on Systems, Man and Cybernetics, vol. 3, pp. 2340–2345. IEEE (2005)
23. Pietruczuk, L., Rutkowski, L., Jaworski, M., Duda, P.: A method for automatic adjustment of ensemble size in stream data mining. In: 2016 International Joint Conference on Neural Networks (IJCNN), pp. 9–15. IEEE (2016)
24. Pietruczuk, L., Rutkowski, L., Jaworski, M., Duda, P.: How to adjust an ensemble size in stream data mining? Inf. Sci. **381**, 46–54 (2017)

25. Polikar, R., Upda, L., Upda, S.S., Honavar, V.: Learn++: an incremental learning algorithm for supervised neural networks. IEEE Trans. Syst. Man Cybern. Part C (Appl. Rev.) **31**(4), 497–508 (2001)
26. Rutkowski, L., Jaworski, M., Pietruczuk, L., Duda, P.: Decision trees for mining data streams based on the Gaussian approximation. IEEE Trans. Knowl. Data Eng. **26**(1), 108–119 (2014)
27. Rutkowski, L., Pietruczuk, L., Duda, P., Jaworski, M.: Decision trees for mining data streams based on the McDiarmid's bound. IEEE Trans. Knowl. Data Eng. **25**(6), 1272–1279 (2013)
28. Street, W.N., Kim, Y.: A streaming ensemble algorithm (sea) for large-scale classification. In: Proceedings of the Seventh ACM SIGKDD International Conference on Knowledge Discovery and Data Mining, pp. 377–382. ACM (2001)
29. Wang, H., Fan, W., Yu, P.S., Han, J.: Mining concept-drifting data streams using ensemble classifiers. In: Proceedings of the Ninth ACM SIGKDD International Conference on Knowledge Discovery and Data Mining, pp. 226–235. ACM (2003)
30. Woźniak, M., Połap, D., Napoli, C., Tramontana, E.: Graphic object feature extraction system based on cuckoo search algorithm. Expert Syst. Appl. **66**, 20–31 (2016)
31. Zalasiński, M., Cpałka, K., Er, M.J.: Stability evaluation of the dynamic signature partitions over time. In: Rutkowski, L., Korytkowski, M., Scherer, R., Tadeusiewicz, R., Zadeh, L.A., Zurada, J.M. (eds.) ICAISC 2017. LNCS (LNAI), vol. 10245, pp. 733–746. Springer, Cham (2017). https://doi.org/10.1007/978-3-319-59063-9_66
32. Zalasiński, M., Cpałka, K., Rakus-Andersson, E.: An idea of the dynamic signature verification based on a hybrid approach. In: Rutkowski, L., Korytkowski, M., Scherer, R., Tadeusiewicz, R., Zadeh, L.A., Zurada, J.M. (eds.) ICAISC 2016. LNCS (LNAI), vol. 9693, pp. 232–246. Springer, Cham (2016). https://doi.org/10.1007/978-3-319-39384-1_21

An Empirical Study of Strategies Boosts Performance of Mutual Information Similarity

Ole Kristian Ekseth[(✉)] and Svein-Olav Hvasshovd

Department of Computer Science (IDI), NTNU, Trondheim, Norway
oekseth@gmail.com

Abstract. In the recent years, the application of *mutual information* based measures has received broad popularity. The *mutual information* MINE measure is asserted to be the best strategy for identification of relationships in challenging data sets. A major weakness of the MINE similarity metric concerns its high execution time. To address the performance issue numerous approaches are suggested both with respect to improvement of software implementations and with respect to the application of simplified heuristics. However, none of the approaches manage to address the high execution-time of MINE computation.

In this work, we address the latter issue. This paper presents a novel MINE implementation which manages a 530x+ performance increase when compared to established approaches. The novel high-performance approach is the result of a structural evaluation of 30+ different MINE software implementations, implementations which do not make use of simplified heuristics. Hence, the proposed strategy for computation of MINE *mutual information* is both accurate and fast. The novel *mutual information* MINE software is available at https://bitbucket. org/oekseth/mine-data-analysis/downloads/. To broaden the applicability the high-performance MINE metric is integrated into the *hpLysis machine learning library* (https://bitbucket.org/oekseth/hplysis-cluster-analysis-software).

1 Introduction

The advent and application of analysis software which are both generic and accurate are believed to significantly improve the utilization of research efforts, *e.g.*, with respect to knowledge discovery [1–3]. An established approach to identification of complex relationships is the application of *mutual information*. In life science the use of *mutual information* [4] is seen as an accurate strategy to address the latter issue, *e.g.*, with respect to the analysis of datasets with different topological traits.

A recent contribution is the *MINE* similarity metric [5]. The *MINE* method describes a dynamic programming approach to the recursive *Mutual Information* metric specified in [4]. However, software which supports the computation of *MINE* is constrained by high execution time, as exemplified in Fig. 1.

© Springer International Publishing AG, part of Springer Nature 2018
L. Rutkowski et al. (Eds.): ICAISC 2018, LNAI 10842, pp. 321–332, 2018.
https://doi.org/10.1007/978-3-319-91262-2_29

When evaluating existing *MINE* implementation strategies, such as with respect to [5–15], we observe how existing software are constrained by low utilization of computer hardware. While fast execution and accurate predictions are important in data mining [16], Fig. 1 demonstrates how established approaches result in high-performance penalties.

An important motivation in research is to gain insight into complex relationships, *e.g.*, in the context of epigenetics [17,18] and inference of transcriptional networks [19–21]. A use case is to realize why Rheumatoid arthritis (RA) synovitis [are] causing pain, swelling and loss of function [22]. Ramifications of addressing performance issues in data-mining software are therefore expected to be a significant boost to low-cost high-inventive approaches for knowledge discovery [23].

A major challenge in the application of established similarity metrics concerns how to correctly choose similarity metrics which captures both application centered use cases and data topology [5, 24–28]. The latter challenge motivates the use of generalized similarity metrics. The *MINE* similarity metric [5] is demonstrated to accurately capture similarities in challenging feature data vectors. In the recent years, the *MINE* similarity metric has seen numerous permutations, *e.g.*, with respect to the optimization efforts of [8–10]. The established optimization strategies for *MINE* computation suffers from high execution time performance. To optimize the *MINE* computation the work of [7] unsuccessfully tries to improve implementation efficiency. In contrast [8,10] present a modified version of the *MINE* metric. The work of [7] asserts that their heuristic *MINE* permutation results in a lower degree of prediction accuracy.

The major performance bottleneck in the *MINE* metric, and *mutual information* strategies in general concerns the computation of logarithms. Therefore, optimization of *MINE* is strongly overlapping with optimization of entropy metrics. While there are numerous proposed strategies for optimization of logarithm optimization they all suffer from simplified heuristics known to reduce prediction accuracy.

1.1   Our Proposal

In this paper, we describe a methodology to address performance issues in the computation of *"mutual information"*. Importantly, while we from Fig. 2 observe how our approach enables a 500x+ reduction in execution time, our approach produces by definition exactly the same prediction results as non-optimized approaches. We have constructed a new software for computation of *mutual information* based measures, both as stand-alone tools, and integrated into the hpLysis machine learning software [29]. Through an optimized implementation of the *MINE* similarity metric [4,5], we achieve a 530x+ performance increase. Therefore, our approach enables a significant performance boost to established software approaches such the MIDER software [7].

This article addresses the issues of:

1. *MINE* optimization: Fig. 1 discuss the time-cost benefit of *MINE* based implementations: the results demonstrates how our novel entropy strategy manages a 533x+ reduction in execution time;

2. entropy computations: the quantification of entropy overhead Fig. 1 captures how a naive approach for logarithm computation results in a 50x+ overall performance lag;
3. computer hardware: Fig. 2 depicts the time benefit of the different logarithm optimization strategies: the figure identify how the proposed optimization ustrategy both provides the fastest results and enable accurate predictions;

The remainder of this paper is organized as follows. Section 2 evaluate approaches for optimization of *MINE* and logarithm computations. Section 3.1 list a subset of the implementation strategies we evaluate, *i.e.*, for computation of *mutual information* based metrics. Section 3 describes how an evaluation of 30+ implementation strategies the significant reduction in execution time without reducing the correctness of the *MINE* algorithm (Fig. 1). A subset of the evaluated implementation strategies are listed in Subsect. 3.1, From the empirical evaluation of implementation strategies we identify approaches which are both accurate and fast. An example concerns the approach to pre-compute logarithms. Figure 1 demonstrates how the latter results in a 50x+ performance increase. Section 4 summarizes the observations derived from an extensive empirical evaluation, before presenting conclusions and future work in Sect. 5.

2   Related Work

In this section, we evaluate the established approaches for improving the performance of "*Mutual Information (MI)*" based approaches. To reduce the time cost of the of *MINE* algorithm it is necessary to understand the use cases and applications of *MI*. This section evaluates:

1. *MINE* application: why improvements to *MI*-based software significantly improve quality of research;
2. *MINE* optimization: current *MI* software approaches and why these have high execution time cost;
3. generic strategies: established approaches for reducing execution time.

By definition, *MI* is a measure of entropy for a pair of feature vectors [4]. The application of entropy has a wide number of use cases, *e.g.*, with respect to the testing of the hypothesis [31,32]. A major challenge of established approaches concerns their high computational complexity [23]. *MIs* are (among others) applied in generic similarity functions, generic functions which are "not limited to specific function types (such as linear, exponential, or periodic), or even to all functional relationships" [5]. Similarly, the work of [33] argues that *averaged MI* [4] may be used to measure the overall independence between two feature-vectors (such as time-series of news stories).

While *MI* is known to have broad applicability, there are not any approaches which explicitly seeks to address the high time-cost of *MI* computation. To investigate established approaches for optimization of entropy-based measures, we review different software approaches, focusing on how they address the requirement for accurate and fast computation of *MI*. In the recent years the *MINE*

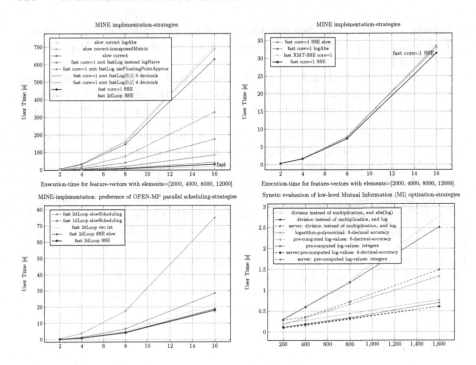

Fig. 1. Time consumption of different *MINE* implementations for increasing vector size: (1.a) top-left: different *MINE* implementation strategies; (1.b) top-right: detailed view of computer hardware close optimization strategies, *e.g.*, with respect to different assembly level SSE [30] optimization strategies; (2.a) bottom-left: effect of different OpenMP parallel scheduling policies; (2.b) bottom-right: effects of logarithm optimization strategies, a subfigure capturing the performance patterns on different computer architectures. (To re-produce the above timing results call the ***performance_comparison.pl*** Perl script located in the *MINE* software repository.)

similarity metric has seen numerous permutations, *e.g.*, with respect to the optimization efforts of [8–10]. Established implementations of the *MINE* similarity metric suffers from high execution time of computations, *e.g.*, as observed from Fig. 1. To optimize the computation of *MINE* [9] has unsuccessfully tried to optimize the software implementation. In contrast, [8,10] present algorithm modifications where the accuracy of their proposed heuristics is widely discussed in the research community [7].

When evaluating alternative approaches to the *MINE* metric of [5,24] asserts that "altogether they do not match the popularity gained by the original *MIC* statistic, also in the computational biology community, *e.g.*, in the analysis and inference of various kinds of biological networks". An example is seen in the work of [6] where the authors use *MI* for parameter estimation. The authors' goal is to "combine concepts from Bayesian inference and information theory in order to identify experiments that maximize the information content of the resulting data" [6]. In their computation, the authors use the slow performing

MI software by [12]. Similarly, the work of [12–14] present software libraries for entropy computations. However, none of the latter software approaches are designed to reduce the tasks computational complexity. The latter results are in contrast to new *MINE* software for fast *MI* computation: Fig. 2 identifies how a pre-computation of entropy scores provides a significant decrease in execution time, a performance difference which outperforms the application of any polynomial approximation strategies for entropy computations, eg, when compared to [34, 35].

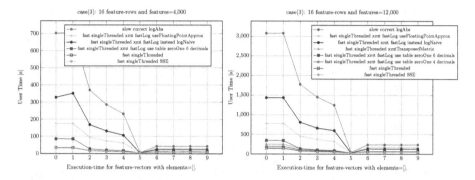

Fig. 2. Influence of data topology on execution time. The above subfigures evaluate the execution time for distinct data topologies and *MINE* implementations: (1.a) top-left: data-set with 4000 features; (1.b) top-right: data-set with 12,000 features. In the above subfigures, the horizontal x-axis describes the 'identity' of the different datasets (which are investigated). The above results identify the benefit of the proposed approach, eg, with respect to the pre-computation of logarithm scores. For each of the above subfigures the x-score maps to the data sets of: "random" at index = 0; "uniform" at index = 1; "binomial p05"; at index = 2; "binomial p010" at index = 3; "binomial p005" at index = 4; "flat" at index = 5; "linear-equal" at index = 6; "linear-different-b" at index = 7; "linear-differentCoeff-a" at index = 8; "sinus" at index = 9. The subfigures reveal how the execution time relates directly to the distribution of values. In the evaluation of data topologies, we observe how the optimization approach, proposed in this paper, is consistent across different feature sizes, hence the proposed strategy may significantly out-perform established *MINE* implementations.

3   Implementation: Design of Our Approach for Evaluation of Entropy-Optimization Techniques

The proposed *MINE* methodology is implemented in C/C++. Through the application of detailed performance analysis, exemplified in Fig. 2, we identify a *MINE* implementation strategy which provides the largest performance-enhancing capabilities. Given the lack of research concerning how to efficiently implement fast and accurate software for *MI*, 30+ different software implementation strategies are explored for the *MINE* metric. The implementation of multiple implementation strategies identifies a method for fast and accurate *MI* computation. The different *MINE* implementations evaluate the cases of:

1. arithmetic: the performance influence of replacing "division" with "multipli-cation", *e.g.*, by re-writing "list[i]/value" into "list[i]*(1/value)";
2. data access: to replace transposed data access with non-transposed data access, *e.g.*, "matrix[*][column]" (where "*" is the incremental variable) into "transposed(matrix)[column][*]";
3. logarithm: respectively evaluate logarithm computation for different approaches: the established strategy used in [9]; implement two different poly-nomial approximations (for logarithm computation), and separately evaluate the performance improvements of different strategies for pre-computation of logarithm scores;
4. parallelism: to address cases of in-efficient parallel computation, *e.g.*, exem-plified in the Minerva software [9]; we implement multiple parallel scheduling approaches;
5. SSE[1] [30]: when exploring different SSE optimization strategies in implemen-tations of the *MINE* metric, we observe that the high logical complexity of *MINE* preempts effective utilization of SSE;

In the above list, we have exemplified multiple approaches to improve the efficiency of *MI* software, *e.g.*, with respect to the computation of the *MINE* metric. The above parametrization of optimization strategies is used to construct multiple software implementations for *MINE* computation. The optimization strategy with the highest performance boost is observed when we pre-compute logarithms, *i.e.*, as exemplified in Fig. 1.

The logarithm optimization which we apply is based on the observation that established entropy software approaches computes "result = log(value/size)". In contrast we compute "result = log(value) − log(size) = table[value] − table[size]", where table hold the pre-computed logarithm scores *table[value] = log(value)*. Section 2 identifies the novelty of the above optimization strategy. While the strategy is simple it is not used, hence its correctness, applicability, an impor-tance. An explanation for the latter may be the unawareness in algorithm con-struction concerning the time cost associated with the computation of loga-rithms.

The speed-ups enabled through above algorithm optimization strategy are higher than for *MINE* approximate approaches. When we compare our speedups to the software of [8,10] we observe how our accurate *MINE* computation provide a lower execution time than those achieved by in-accurate *MINE* approximations (Sect. 2). In contrast to established approaches we have not applied simplified heuristics, hence the proposed *MINE* implementation manages to reduce exe-cution time without reducing prediction-accuracy. Therefore, the proposed soft-ware improvement strategy is not limited by the general assumption of dataset features, hence addressing issues observed by [24].

Given the large data sets to evaluate and availability of multi-core computers, it is of particular importance to maximize the efficiency of approaches for parallel computation. In our work, we have implemented multiple approaches for parallel

---

[1] To use low-level assembly instructions for hardware parallel computations (SSE) to reduce execution time.

computations. To simplify our performance evaluations, we narrow the scope of our micro-benchmarks to only measure the influence of scheduling approaches in the "Open-MP" parallel software API [36]. Given the high utilization of parallel cores observed in Fig. 1, we assert that the choice of the parallel software library for *MINE* computation does not influence the execution time. The high degree of parallelization enabled in the proposed approach is due to the independent computations for each pair of feature-vectors, *i.e.*, where theroe is no need for thread communication. The latter observation is used in our micro-benchmarking of different Open-MP scheduling policies. Figure 1 observe a distinct difference between the default Open-MP scheduling-approach of "#pragma omp parallel" versus optimized selection through "#pragma omp parallel for schedule(static)".

In the construction of the 30+ software implementations the *MINE* Minerva software [9] as a template. Therefore, all of our 30+ different software implementations have functions names and variable names similar to the Minerva implementation. The strength of this approach concerns representativeness and accuracy in evaluation of implementation strategies. The implication is that there is a 1:1 match between the benchmarked software implementation versus the established approach for *MINE* computation, hence the optimization strategies (identified in this paper) are not due to erroneous interpretations of the *MINE* metric. Therefore, the evaluation of software implementations correctly captures established approaches for computation of the *MINE* metric (and similar for *MI-based* metrics).

3.1 Subset of the 30+ Different *MINE* Implementations

This paper evaluates the performance effect of 30+ different *MINE* software implementations. To ease reproducibility, all of the software implementations are compiled into executables. To provide details of the evaluated parameter space, the below paragraphs describe core properties for a subset of the generated *MINE* implementations:

(1) x_mine_fast_singleThreaded: fast software which does not make use of parallelism;

(2) x_mine_fast_singleThreaded_logAbs: similar to "x_mine_fast_singleThreaded" with difference that the absolute-value of log-scores are used, *i.e.*, similar to "x_mine_slow_correct_logAbs";

(3) x_mine_fast_singleThreaded_SSE: fast software which does not make use of parallelism, though makes use of "Intels SSE intrinsics" [30].

(4) x_mine_fast_singleThreaded_vec_int: similar to the "x_mine_fast_single Threade" option, with difference that "integer" (instead of floats) are used to store intermediate elements;

(5) x_mine_fast_2dLoop: fast software where rows are computed in parallel;

(6) x_mine_fast_1dLoop_slowScheduling: medium-speed software where features are computed in parallel;

(7) x_mine_slow_correct: slow software which makes use of the naive *Minerva* *MINE implementation*;

(8) x_mine_slow_correct_logAbs: similar to "x_mine_slow_correct", with difference that we use absolute value for the log-score;
(9) x_mine_slow_errnous: slow software which makes use of the naive *Minerva MINE implementation*;

4   Result

This paper presents a *MINE* software which significantly out-performs established *MINE* based software approaches. Section 3 describes how a conceptually simple optimization approach may significantly reduce the execution time of the popular *MINE* similarity metric [5]. This paper uses a structured approach to identify best-fit optimization strategies. The efficiency of the performance enhancing strategies (*e.g.*, Fig. 2) is verified through the granularity enabled through the 30+ evaluated implementation strategies (Sect. 3). By measuring numerous data distributions we observe how a best-performing implementation strategy consistently provides out-performs the implementation strategies currently in use.

4.1   Empirical Evaluation

This section describes a strategy to evaluate the 30+ *MINE* implementations (Subsect. 3.1). The proposed *MINE* optimization is enabled through a structured evaluation different implementation approaches. To validate broadness and applicability of the identified performance-enhancing strategies, different data topologies are evaluated, *e.g.*, as exemplified in Fig. 2. The empirical evaluation of *MI* based computation strategies may be captured through the parameters of:

1. *MINE* implementations: Fig. 1 compare differences in execution time for 30+ different *MINE* implementations strategies, covering both serial and parallel execution;
2. datasets: Fig. 2 describes the influence of different data topologies, observing how *MINE* computation on a randomized data set is required in order to capture the worst time for computations of *MI-based* measures;
3. feature size: Figs. 1 and 2 compare the performance of the implementation strategies with respect to size of the evaluated data set and data topologies. From the figure, we observe how our approach consistently outperforms established approaches for computation of *MINE*.

The above parameters are used to identify the high-performance *MINE* implementation proposed in this paper. Our *MINE* software is the result of an empirical investigation of different implementation strategies. A comparison of 30+ different *MINE* implementation strategies identify the best performing software approach. What we argue is that our approach manages a 530x+ execution time improvement. The recommended software is the result of this evaluation strategy: to identify the best performing implementation strategy selected from an ensemble of 30+ permutations of software implementations.

4.2   The Performance Benefit of 30+ *MINE* Implementation Strategies

Figure 1 identifies how the proposed *MINE* implementation substantially out-performs established approaches, *e.g.*, when compared to established approaches such as [5,9]. From above Fig. 1 we derive multiple inferences:

1. logarithm optimization: execution time benefit of different logarithm optimization strategies in the *MINE* metric;
2. SSE [30] application: a significant improvement in the application of SSE, hence the need for our broad identification for low-level optimization strategies;
3. parallelism: implication of different parallel optimization strategies.

The performance evaluation measures the execution time implication of 30+ different software implementations. To avoid configurations of the *MINE* comparison to given preference to a certain execution time pattern, *e.g.*, with respect to if-clauses to identify the implementations strategy to use, the applied strategy is to generate distinct programs for each of the 30+ different software implementations. For details of the latter see the *configure_optimization.h* configuration file. The result is a comprehensive evaluation of implementation strategies preference in the computation of the *MINE* similarity metric.

Figure 1 captures how a combination of logarithm optimization, and replacement of "division" with "multiplication", results in significant performance boosts. A different example concerns the configuration of "Open-MP", for which the measurements identify a 4x performance improvement between different parallelization strategies.

4.3   Comparison of Data-Topologies

A comparison of data topologies (Fig. 2) captures the worst case time consumption for different *MINE* implementation strategies. The results demonstrate how the execution time of algorithms such as *MINE* are strongly influenced by the evaluated data topologies (Fig. 2. However, the influence of data topology (when evaluating algorithm heuristics) is often omitted from performance evaluations, *e.g.*, as seen in [9].

Importantly, the strategy proposed in this paper is best the best performer across all evaluated data distributions (Fig. 2). The measurements identify how *MI* computation on feature vectors on randomized datasets gives the worst-case execution time, which is due to how *MI* is computed: detailed micro-benchmarking reveals a 530x+ difference in execution time between [9] versus the proposed approach. The use of the *MINE* similarity metric is of specific importance in unsupervised data mining. The latter unsupervised approach is in contrast to data with known properties, *i.e.*, for with similarity metrics such as Euclidian may be selected to efficiently identify relationships.

5   Conclusion and Future Work

In this work, we have evaluated the applicability of established techniques for optimization of *Mutual Information (MI)* software. The performance measurements presented in this paper identifies a strong relationship between implementation strategy and execution time, an aspect which is often omitted in research papers. The optimization method presented in this paper applies an empirical study of implementation strategies to identify the best performing strategy, hence we assert that our approach may be used to optimize a number of algorithms and software for data mining.

This paper has identified a strategy to decrease the execution time of *MI* by a factor of 530x. Through a detailed evaluation of 30+ different implementation strategies, this paper demonstrates how our approach out-performs established approaches, *i.e.*, without the reduction in prediction quality. Through our optimization strategy, we have exemplified how a novel implementation may significantly improve the performance of established approaches, *i.e.*, without the need to introduce simplifications in computations. Therefore, we argue that the method described in this paper may be used as a template for improving existing approaches for high-quality data mining.

5.1   Future Work

We plan to apply our systematic optimization efforts to related software approaches, such as with respect to *pairwise similarity metrics, network-similarity*, and *measures for cluster-convergence*.

Acknowledgements. The authors would like to thank MD K.I. Ekseth at UIO, Dr. O.V. Solberg at SINTEF, Dr. S.A. Aase at GE Healthcare, MD B.H. Helleberg at NTNU–medical, Dr. Y. Dahl, Dr. T. Aalberg, and K.T. Dragland at NTNU, and Professor P. Sætrom and the High Performance Computing Group at NTNU for their support.

References

1. Ehsani, R., Drabløs, F.: TopoICSim: a new semantic similarity measure based on gene ontology. BMC Bioinform. **17**(1), 296 (2016)
2. Faith, J.J., Hayete, B., Thaden, J.T., Mogno, I., Wierzbowski, J., Cottarel, G., Kasif, S., Collins, J.J., Gardner, T.S.: Large-scale mapping and validation of Escherichia coli transcriptional regulation from a compendium of expression profiles. PLoS Biol. **5**(1), 8 (2007)
3. Leach, S.M., Tipney, H., Feng, W., Baumgartner Jr., W.A., Kasliwal, P., Schuyler, R.P., Williams, T., Spritz, R.A., Hunter, L.: Biomedical discovery acceleration, with applications to craniofacial development. PLoS Comput. Biol. **5**(3), 1000215 (2009)
4. Fraser, A.M., Swinney, H.L.: Independent coordinates for strange attractors from mutual information. Phys. Rev. A **33**(2), 1134 (1986)

5. Reshef, D.N., Reshef, Y.A., Finucane, H.K., Grossman, S.R., McVean, G., Turnbaugh, P.J., Lander, E.S., Mitzenmacher, M., Sabeti, P.C.: Detecting novel associations in large data sets. Science **334**(6062), 1518–1524 (2011)
6. Liepe, J., Filippi, S., Komorowski, M., Stumpf, M.P.: Maximizing the information content of experiments in systems biology. PLoS Comput. Biol. **9**(1), 1002888 (2013)
7. Villaverde, A.F., Ross, J., Morán, F., Banga, J.R.: MIDER: network inference with mutual information distance and entropy reduction. PLoS ONE **9**(5), 96732 (2014)
8. Tang, D., Wang, M., Zheng, W., Wang, H.: RapidMic: rapid computation of the maximal information coefficient. Evol. Bioinform. **10**, 11 (2014)
9. Albanese, D., Filosi, M., Visintainer, R., Riccadonna, S., Jurman, G., Furlanello, C.: Minerva and minepy: a C engine for the MINE suite and its R, Python and MATLAB wrappers. Bioinformatics, 707 (2012)
10. Chen, Y., Zeng, Y., Luo, F., Yuan, Z.: A new algorithm to optimize maximal information coefficient. PLoS ONE **11**(6), 0157567 (2016)
11. Wang, K., Phillips, C.A., Saxton, A.M., Langston, M.A.: EntropyExplorer: an R package for computing and comparing differential Shannon entropy, differential coefficient of variation and differential expression. BMC Res. Notes **8**(1), 832 (2015)
12. Hausser, J., Strimmer, K.: Entropy inference and the James-Stein estimator, with application to nonlinear gene association networks. J. Mach. Learn. Res. **10**(July), 1469–1484 (2009)
13. Marcon, E., Hérault, B.: Entropart: an R package to measure and partition diversity. J. Stat. Softw. **67**(8), 1–26 (2015)
14. Guevara, M.R., Hartmann, D., Mendoza, M.: diverse: an R package to analyze diversity in complex systems. R J. **8**(2), 60–78 (2016)
15. Ince, R.A., Mazzoni, A., Petersen, R.S., Panzeri, S.: Open source tools for the information theoretic analysis of neural data. Front. Neurosci. **3**, 11 (2010)
16. Mazandu, G.K., Mulder, N.J.: Information content-based gene ontology functional similarity measures: which one to use for a given biological data type? PLoS ONE **9**(12), 113859 (2014)
17. Morgan, H.D., Sutherland, H.G., Martin, D.I., Whitelaw, E.: Epigenetic inheritance at the agouti locus in the mouse. Nat. Genet. **23**(3), 314–318 (1999)
18. Lee, H.-S., Chen, Z.J.: Protein-coding genes are epigenetically regulated in Arabidopsis polyploids. Proc. Nat. Acad. Sci. **98**(12), 6753–6758 (2001)
19. Carro, M., Lim, W., Alvarez, M., Bollo, R., Zhao, X., Snyder, E., Sulman, E., Anne, S., Doetsch, F., Colman, H., et al.: The transcriptional network for mesenchymal transformation of brain tumours. Nature **463**(7279), 318 (2010)
20. Yeger-Lotem, E., Sattath, S., Kashtan, N., Itzkovitz, S., Milo, R., Pinter, R.Y., Alon, U., Margalit, H.: Network motifs in integrated cellular networks of transcription-regulation and protein-protein interaction. Proc. Nat. Acad. Sci. U.S.A. **101**(16), 5934–5939 (2004)
21. Kashtan, N., Itzkovitz, S., Milo, R., Alon, U.: Efficient sampling algorithm for estimating subgraph concentrations and detecting network motifs. Bioinformatics **20**(11), 1746–1758 (2004)
22. Sommerfelt, R.M., Feuerherm, A.J., Jones, K., Johansen, B.: Cytosolic phospholipase A2 regulates TNF-induced production of joint destructive effectors in synoviocytes. PLoS ONE **8**(12), 83555 (2013)
23. Lee, W.-P., Tzou, W.-S.: Computational methods for discovering gene networks from expression data. Brief. Bioinform. **10**(4), 408–423 (2009)
24. Riccadonna, S., Jurman, G., Visintainer, R., Filosi, M., Furlanello, C.: DTW-MIC coexpression networks from time-course data. PLoS ONE **11**(3), 0152648 (2016)

25. Ekseth, K., Hvasshovd, S.: hpLysis similarity: a high-performance software-approach for computation of 320+ simliarty-metrics (2017)
26. Cha, S.-H.: Comprehensive survey on distance/similarity measures between probability density functions. City 1(2), 1 (2007)
27. Lord, E., Diallo, A.B., Makarenkov, V.: Classification of bioinformatics workflows using weighted versions of partitioning and hierarchical clustering algorithms. BMC Bioinform. 16(1), 1 (2015)
28. Kanungo, T., Mount, D.M., Netanyahu, N.S., Piatko, C.D., Silverman, R., Wu, A.Y.: An efficient k-means clustering algorithm: analysis and implementation. IEEE Trans. Pattern Anal. Mach. Intell. 24(7), 881–892 (2002)
29. Ekseth, O.K., Hvasshovd, S.-O.: How an optimized DB-SCAN implementation reduce execution-time and memory-requirements for large data-sets (2017)
30. Intel: SSE computer-hardware-low-level parallelism. https://software.intel.com/sites/landingpage/IntrinsicsGuide/. Accessed 06 June 2017
31. Chao, A., Shen, T.-J.: Nonparametric estimation of Shannons index of diversity when there are unseen species in sample. Environ. Ecol. Stat. 10(4), 429–443 (2003)
32. Frery, A.C., Cintra, R.J., Nascimento, A.D.: Entropy-based statistical analysis of PolSAR data. IEEE Trans. Geosci. Remote Sens. 51(6), 3733–3743 (2013)
33. Moon, Y.-I., Rajagopalan, B., Lall, U.: Estimation of mutual information using kernel density estimators. Phys. Rev. E 52(3), 2318 (1995)
34. Jiao, J., Venkat, K., Han, Y., Weissman, T.: Minimax estimation of functionals of discrete distributions. IEEE Trans. Inf. Theory 61(5), 2835–2885 (2015)
35. Jourdan, J.-H.: Vectorizable, approximated, portable implementations of some mathematical functions. https://github.com/jhjourdan/SIMD-math-prims. Accessed 06 June 2017
36. Open-MP: Open-MP: a parallel software-wrapper. http://www.openmp.org/. Accessed 17 Nov 2017

Distributed Nonnegative Matrix Factorization with HALS Algorithm on Apache Spark

Krzysztof Fonał[(✉)] and Rafał Zdunek[(✉)]

Department of Electronics, Wroclaw University of Technology,
Wybrzeze Wyspianskiego 27, 50-370 Wroclaw, Poland
{krzysztof.fonal,rafal.zdunek}@pwr.edu.pl

Abstract. Nonnegative Matrix Factorization (NMF) is a commonly-used unsupervised learning method for extracting parts-based features and dimensionality reduction from nonnegative data. Many computational algorithms exist for updating the latent nonnegative factors in NMF. In this study, we propose an extension of the Hierarchical Alternating Least Squares (HALS) algorithm to a distributed version using the state-of-the-art framework - Apache Spark. Spark gains its popularity among other distributed computational frameworks because of its in-memory approach which works much faster than well-known Apache Hadoop. The scalability and efficiency of the proposed algorithm is confirmed in the numerical experiments, performed on real data as well as synthetic ones.

Keywords: Distributed nonnegative matrix factorization
Large-scale NMF · HALS algorithm · Spark · Recommendation systems

1 Introduction

Many matrix decomposition methods are used for extracting latent factors from an input matrix. The most popular method applied in many fields of science and engineering is Principal Component Analysis (PCA) [1]. However, the problem with the methods, such as PCA, is that its latent factor matrices contain negative values, which, if an input matrix is nonnegative (images, spectrograms, etc.) has no physical representation. Another disadvantage is a holistic representation that is not always desirable, especially in image pattern recognition. Therefore, Nonnegative Matrix Factorization (NMF) [2,3] methods became very popular and successfully applied to many computational problems in image recognition, signal processing, recommender systems, etc. What differs them from other decompositions is that the input for NMF is nonnegative and the produced factors are also nonnegative and often sparse. The consequence of nonnegativity and sparsity is a better physical representation of the latent structure in data as well as parts-based features, e.g. a set of facial images can be decomposed into the

© Springer International Publishing AG, part of Springer Nature 2018
L. Rutkowski et al. (Eds.): ICAISC 2018, LNAI 10842, pp. 333–342, 2018.
https://doi.org/10.1007/978-3-319-91262-2_30

features that contain the parts of faces (hair, nose, eyes, etc.). These aspects of NMF have been presented by Lee and Seung in [3]. They had a significant impact on the popularization of NMF by proposing simple multiplicative algorithms for updating the nonnegative factors. Since then, many researchers were attracted to develop NMF methods. Nowadays, many methods exist for updating the factors in various NMF models.

Most of existing NMF algorithms are designed for synchronous data processing, assuming there are enough RAM and a computational power to get the results in reasonable time. But, nowadays, a fast growth in the amount of collected data and needs to process them makes this assumption cannot always be satisfied. In the era of big data, many researchers pay attention to distributed NMF approaches. As a result, many research papers report a potential of processing massive data in a parallel way. The most popular approach is to partition a computational problem and processing block-wise updates using the MapReduce concept. Liu *et al.* [4] proposed the way of partitioning data and arranging the processing using the MapReduce paradigm to factorize very large matrices with the multiplicative algorithms for NMF. In the paper [5], the MapReduce paradigm was used to scale up convex NMF [6]. Unfortunately, multiplicative approaches are confirmed to have very slow convergence. Yin *et al.* presented another approach to distributed NMF, where the input matrix is split into blocks in the mapping phase, and then partial results are computed in the reduction phase. Regardless the block-wise approach, the updating rules which are based on multiplicative algorithms cannot converge fast.

To face up the slow convergence problem, various numerical algorithms were developed for NMF. The Hierarchical Alternative Least Square (HALS) [7] belongs to a family of block-coordinate algorithms. It was developed by the Cichocki's team from the RIKEN Brain Science Institute. Many independent researches [8–12] confirmed its high efficiency and very fast convergence. Following the success of the HALS, the authors of this study proposed a distributed version of the HALS, called the D-HALS in [13], using the MapReduce paradigm in Matlab.

In this study, we improved the D-HALS by implementing it in the state-of-the-art framework - Apache Spark[1], originally developed at University of California, Berkley [14]. The most time-consuming steps in the Matlab implementation of the D-HALS are the I/O operations that are applied to the map and reduce functions. This is also a commonly known disadvantage of Apache Hadoop (Matlab MapReduce shares the same approach and can be even launched on Apache Hadoop). The large number of I/O operations in the MapReduce paradigm and the MapReduce paradigm itself are not efficient for iterative problems, such as NMF iterative updates. Apache Spark with its in-memory approach is much faster than its predecessor (Apache Hadoop). Moreover, Apache Spark is not based on the MapReduce paradigm, but on the transform and action operations on Resilient Distributed Datasets (RDD). MapReduce can be also implemented using Apache Spark, but Spark provides more flexibility. We leverage from this

---

[1] https://spark.apache.org/.

fact proposing "one map-multi reduce" approach to compute NMF with the HALS.

The remainder of this paper is organized as follows. The first section introduces the topic discussed here and gives the motivation for taking this issue. A short review of mathematical models for NMF with the HALS can be found in Sect. 2. Section 3 discusses our approach to distributed computation of NMF with the HALS using the Apache Spark framework. Section 4 contains the numerical experiments performed for real and synthetic data. The last section summarizes the experimental results.

2 HALS Algorithm

The HALS for solving NMF problems is known from its very fast convergence and monotonicity. The first version of the HALS was proposed in 2007 in [7], and since then, this computational approach was significantly improved by Phan and Cichocki [15], leading to its much faster version (sometimes called the Fast HALS). The idea was to shift the BLAS-3 computations from the inner to the outer loop. In consequence, a current version of the HALS has fast convergence and a low computational complexity, which makes it very efficient in overall.

HALS, like ALS, belongs to a family of alternative algorithms. To approximate the nonnegative input matrix $Y = [y_{it}] \in \mathbb{R}_+^{I \times T}$ by a product of two lower-rank factor matrices $A = [a_{ij}] \in \mathbb{R}_+^{I \times J}$ and $X = [x_{jt}] \in \mathbb{R}_+^{J \times T}$, where usually $J << \min\{I, T\}$, the following alternating optimization scheme is used:
For $s = 1, 2, \ldots,$ do:

$$X^{(s)} = \arg\min_{X \geq 0} \Psi(Y \| A^{(s-1)} X), \tag{1}$$

$$A^{(s)} = \arg\min_{A \geq 0} \Psi(Y \| A X^{(s)}), \tag{2}$$

where $\Psi(Y \| AX)$ is an assumed objective function that represents dissimilarity between the observed data and the model obtained by the product of the factors A and X. The $A^{(0)}$ and $X^{(0)}$ are the initial random matrices. The solution to the above schema leads to the following approximative model: $Y \cong AX \in \mathbb{R}_+^{I \times T}$.

The general concept of the HALS is very simple. Assuming $\Psi(Y \| AX)$ be expressed by the Euclidean distance, it can be written as:

$$\Psi(Y \| AX) = \frac{1}{2} \| Y - AX \|_F^2 = \frac{1}{2} \| Y - \sum_{j=1}^{J} a_j \underline{x}_j \|_F^2$$

$$= \frac{1}{2} \| Y - \sum_{r \neq j}^{J} a_r \underline{x}_r - a_j \underline{x}_j \|_F^2 = \frac{1}{2} \| Y^{(j)} - a_j \underline{x}_j \|_F^2 \tag{3}$$

Performing some straightforward computations on the above equations leads us to the fast HALS updating rule:

$$\underline{x}_j \leftarrow \left[\underline{x}_j + \frac{[A^T Y]_{j,*} - [A^T A]_{j,*} X}{[A^T A]_{jj}} \right]_+, \tag{4}$$

$$a_j \leftarrow \left[a_j + \frac{[\boldsymbol{YX}^T]_{j,*} - [\boldsymbol{XX}^T]_{j,*}\boldsymbol{A}}{[\boldsymbol{XX}^T]_{jj}} \right]_+, \tag{5}$$

It is worth to notice that the products $\boldsymbol{A}^T\boldsymbol{Y}$ and $\boldsymbol{A}^T\boldsymbol{A}$ do not depend on updated variables in $\boldsymbol{x}_j$, and thus, they can be computed before doing the iterative updates of $\boldsymbol{x}_j$ for all $j = 1, 2, \ldots, J$. A similar statement applied to $\boldsymbol{YX}^T$, and $\boldsymbol{XX}^T$ for computation of and a_j. The final form of the HALS for updating the factor $\boldsymbol{X}$ is presented in Algorithm 1. It contains two **for** loops - first, outer, for an iterative improvement of convergence, and the second, internal, for updating blocks of variables. The calculation of $\boldsymbol{C}$ and $\boldsymbol{B}$ are done before the outer loop. The presented algorithm shows only the pseudocode for updating $\boldsymbol{X}$ but updating $\boldsymbol{A}$ is similar is based on the formula 5. The overall computational complexity for updating $\boldsymbol{A}$ or $\boldsymbol{X}$ can be estimated as $O(J^2Tk_{max}) + O(IJ^2) + O(IJT)$. As we mentioned earlier, usually $J << \min\{I, T\}$, and $k_{max} \approx J$, therefore the dominant part is $O(IJT)$, which comes from the calculation of $\boldsymbol{C}$. Thus, this part is processed in a distributed way, which will be presented in the next section.

Algorithm 1. HALS

Input : $\boldsymbol{A} \in \mathbb{R}^{I \times J}$, $\boldsymbol{Y} \in \mathbb{R}^{I \times T}$, $\boldsymbol{X}^{(0)} \in \mathbb{R}^{J \times T}$, k_{max} - maximum number of iterations,

Output: $\boldsymbol{X}$ - estimated factor

1 Initialization: $\boldsymbol{C}^{(X)} = \boldsymbol{A}^T\boldsymbol{Y}$, $\boldsymbol{B}^{(X)} = \boldsymbol{A}^T\boldsymbol{A}$;

2 **for** $k = 0, 1, \ldots, k_{max}$ **do**

3   **for** $j = 1, \ldots, J$ **do**

4          $\boldsymbol{x}_j^{(k+1)} = \left[\boldsymbol{x}_j^{(k)} + \frac{\boldsymbol{c}_j^{(X)} - \boldsymbol{b}_j^{(X)}\boldsymbol{X}^{(k)}}{b_{jj}^{(X)}} \right]_+$;   // Projected updates

---

3   Distributed HALS Algorithm

The updating rules (4) and (5) can be implemented in many computational environments and launched on a single-node machine. However, if I or T is large, a computation of $\boldsymbol{C}$ can be very time-consuming. Moreover, a very large $\boldsymbol{Y}$ might not fit an available memory on a single-node machine. Therefore, there is a demand to distribute the computation of $\boldsymbol{C}$ by splitting the matrix $\boldsymbol{Y}$. Since $\boldsymbol{A}$ and $\boldsymbol{X}$ are much smaller, they can be shared among the nodes.

We assume the matrix $\boldsymbol{Y}$ can be divided into blocks – row-blocks or column-blocks – and then, the blocks might be distributed throughout nodes. The idea of splitting $\boldsymbol{Y}$ and its multiplication with $\boldsymbol{A}$ or $\boldsymbol{X}$ is presented on Fig. 1.Considering such partitioning, the matrices $\boldsymbol{C}$ can be computed as follows:

$$\boldsymbol{C}^{(A)} = [\boldsymbol{C}_1^{(A)}\boldsymbol{C}_2^{(A)} \ldots \boldsymbol{C}_M^{(A)}], \text{where } \boldsymbol{C}_m^{(A)} = \boldsymbol{Y}_m\boldsymbol{X}^T, \tag{6}$$

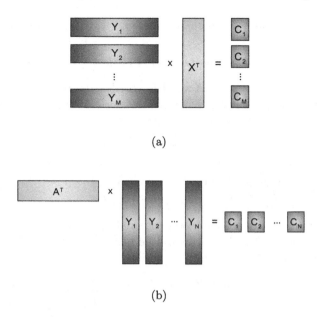

(a)

(b)

Fig. 1. The concept of distributed computing: (a) $C^{(A)} = YX^T$ by partitioning Y into row-blocks; (b) $C^{(X)} = A^T Y$ by partitioning Y into column-blocks.

$$C^{(X)} = [C_1^{(X)} \, C_2^{(X)} \ldots C_N^{(X)}], \text{where } C_n^{(X)} = A^T Y_n, \tag{7}$$

where M is the number of row-blocks, and N is the number of column-blocks. It is good to split computational operations to be as atomic as possible because Spark manages them in the most efficient way. Hence, the above presented blocks can be at the end single rows or single columns, that is $M = I, N = T$. Assuming Y is a sparse matrix, its final sparse representation as a RDD set can be represented by the following key-value pairs:

$$Y_{it} \rightarrow \{key : i, value : \{t, Y_{it}\}\}, \text{ where } i = 1 \ldots I, t = 1 \ldots T \tag{8}$$

$$Y_{it} \rightarrow \{key : t, value : \{i, Y_{it}\}\}, \text{ where } t = 1 \ldots T, i = 1 \ldots I \tag{9}$$

The representation (8) is used for computing (6), and (9) for (7). To achieve such representations, the *map()* operation is needed after reading Y from the storage. Then, the *groupByKey()* operation does shuffling, and the finally partitioned Y is distributed on nodes. Such operations are heavy - because of large data transfer while shuffling - thus, we propose to do it once for (8) and (9) before the iterations start. The data is persisted twice - for row's and column's partitions - but this double-memory usage, in this case, is worthwhile. In [13], the iterative mapping and shuffling is the most time-consuming part of decomposing input data.

As we mentioned earlier, A and X are much smaller than Y, therefore we keep them as non-RDD, regular *Map* data structure consists of following key-value pairs:

$$\boldsymbol{A}_{i*} \rightarrow \{key : i, value : float[J]\}, \text{ where } i = 1 \dots I, J - \text{ factor rank} \quad (10)$$

$$\boldsymbol{X}_{*t} \rightarrow \{key : t, value : float[J]\}, \text{ where } t = 1 \dots T, J - \text{factor rank} \quad (11)$$

Before updating $\boldsymbol{C}^{(A)}$ or $\boldsymbol{C}^{(X)}$, $\boldsymbol{X}$ or $\boldsymbol{A}$ is broadcasted to nodes.

Finally, using such representations, we can formulate the *aggregateByKey()* operation which is responsible for computing $\boldsymbol{C}^{(A)}$ and $\boldsymbol{C}^{(X)}$:

- $\boldsymbol{C}^{(A)}$: Take $< \{i, \{t, \boldsymbol{Y}_{it}\}\}, \boldsymbol{X} >$, where $t = 1 \dots T$
  and emit $\{i, \forall_j^J \sum_t^T \boldsymbol{Y}_{it} \boldsymbol{X}[t]_j\}$, where $j = 1 \dots J$
- $\boldsymbol{C}^{(X)}$: Take $< \{t, \{i, \boldsymbol{Y}_{it}\}\}, \boldsymbol{A} >$, where $i = 1 \dots I$
  and emit $\{t, \forall_j^J \sum_i^I \boldsymbol{Y}_{it} \boldsymbol{A}[i]_j\}$, where $j = 1 \dots J$

Despite, $\boldsymbol{A}$ and $\boldsymbol{X}$ are not RDD sets and are kept as the whole on every node, the calculation of $\boldsymbol{B}^{(A)}$ or $\boldsymbol{B}^{(X)}$ can still be parallelized when being calculated on the master node using *aggregate()* function:

- $\boldsymbol{B}^{(A)}$: Take $< \boldsymbol{A} >$ and emit $\boldsymbol{B} \in \mathbb{R}_+^{J \times J}$, where $\boldsymbol{B}_{j_1 j_2} = \sum_{i=1}^I \boldsymbol{A}[i]_{j_1} \boldsymbol{A}[i]_{j_2}$,
- $\boldsymbol{B}^{(X)}$: Take $< \boldsymbol{X} >$ and emit $\boldsymbol{B} \in \mathbb{R}_+^{J \times J}$, where $\boldsymbol{B}_{j_1 j_2} = \sum_{i=1}^I \boldsymbol{X}[i]_{j_1} \boldsymbol{X}[i]_{j_2}$,

where each case for $i = 1 \dots I$ can be calculated in parallel. The final implementation of the algorithm in the Scala language can be found on the public GitHub repository[2].

4   Experiments

The D-HALS algorithm has been tested on the following datasets:

- **Benchmark I:** The matrix $\boldsymbol{Y}$ is created from the dataset (ml-latest) by MovieLens[3] [16]. It contains 5-star rating and free-text tagging activity from a movie recommendation service. We used the dataset that has 22884377 ratings and 586994 tag applications across 33670 movies, evaluated by 247753 users within the period from January 09, 1995 to January 29, 2016. Thus $\boldsymbol{Y} \in \mathbb{R}_+^{247753 \times 33670}$ is a sparse matrix, containing about 27.43% nonzero entries.
- **Benchmark II:** The matrix $\boldsymbol{Y} \in \mathbb{R}_+^{10212 \times 36771}$ is created from the TDT2 dataset[4]. It consists of six types of data: two newswires (APW, NYT), two radio programs (VOA, PRI), and two TV programs (CNN, ABC). For the tests, 10212 documents are used that contain 36771 distinct words.
- **Benchmark III:** The matrix $\boldsymbol{Y} \in \mathbb{R}_+^{18774 \times 61188}$ is created from the dataset *20-newspapers*[5]. It is a word-document matrix that represents a collection of approximately 20000 newsgroup documents partitioned (nearly) evenly across 20 different newsgroups. We used 18774 post-rainbow-processed documents containing 61188 distinct terms.

---

[2] https://github.com/krzysiekfonal/dhals.
[3] https://grouplens.org/datasets/movielens/.
[4] https://catalog.ldc.upenn.edu/LDC2001T57.
[5] http://qwone.com/jason/20Newsgroups/.

- **Benchmark IV:** The matrix $\boldsymbol{Y} \in \mathbb{R}_+^{I \times T}$ is generated synthetically from the factor matrices: $\boldsymbol{A} = [a_{ij}] \in \mathbb{R}_+^{I \times J}$ and $\boldsymbol{X} = [x_{jt}] \in \mathbb{R}_+^{J \times T}$, where $a_{ij} = \max\{0, \hat{a}_{ij}\}$, $x_{jt} = \max\{0, \hat{x}_{jt}\}$ and $\forall i, j, t : \hat{a}_{ij}, \hat{x}_{jt} \sim \mathcal{N}(0, 1)$. We set: $I = T = 4 \times 10^4$ and $J = 10$.

The aim of the numerical experiments is to show that the Spark D-HALS is scalable when a number of nodes grows and that the D-HALS keeps the characteristics of the HALS, i.e. its convergence is monotonic and fast. Moreover, we compared it with the Spark's MLlib[6] ALS[7] implementation and also with our D-HALS implementation in Matlab [13].

The Spark D-HALS is coded in Scala 2.11[8]. All the tests but one compare the runtime of the Spark D-HALS versus the Matlab's D-HALS and the ALS, and they are conducted on the Amazon Elastic Compute Cloud (Amazon EC2)[9]. The tests, which use Benchmarks I-III, are launched on the *c4.xlarge* instances equipped with CPU Intel Xeon E5-2666 v3 (4 cores) and 7.5 GB RAM. The test, based on Benchmark IV, is launched on *c4.4xlarge* equipped with CPU Intel Xeon E5-2666 v3 (16 cores) and 30 GB RAM. To compare the runtime of Matlab's D-HALS with Spark's D-HALS and ALS implementations, the workstation equipped with CPU Intel Core i7-7700 (4 cores, 8 threads), 16 GB RAM, and 512 GB SSD, under Linux Ubuntu was used. Due to the non-convexity of NMF algorithms, each analyzed case is repeated 10 times for a random initialization.

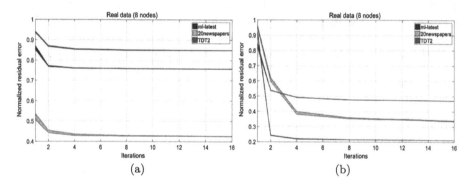

(a)   (b)

Fig. 2. Residual error versus iterations for (a) Spark D-HALS; (b) ALS. Benchmarks I-III were used. The color patch shows the area of STD. (Color figure online)

Figure 2 depicts the residual errors versus the number of iterations obtained with the Spark D-HALS and ALS for Benchmarks I-III. The residual error for

---

[6] https://spark.apache.org/mllib/.
[7] Whenever we mention the ALS in this study, we refer to the distributed ALS implementation from MLlib in the ML package. This is an important note because there is also an older implementation in the mllib package.
[8] https://www.scala-lang.org/.
[9] https://aws.amazon.com/ec2.

Benchmark IV is presented in Fig. 3(a). The color patches determine the area of the Standard Deviation (STD). The experiments showing scalability of the proposed D-HALS are illustrated in Fig. 3(b). The range of STD is marked with the whiskers.

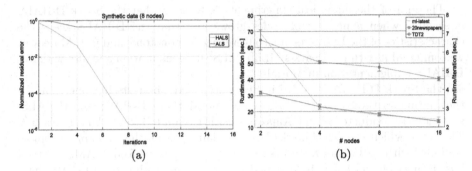

(a)   (b)

Fig. 3. (a) Residual error versus iterations, obtained for the synthetic data (Benchmark IV) using the Spark D-HALS and ALS; (b) Runtime/iteration ratio versus the number of nodes obtained with the Spark D-HALS for the tested datasets.

The runtime of processing 10 iterations of the Matlab's D-HALS, Spark's D-HALS and Spark's ALS, applied to Benchmark I and launched on the standalone workstation, is listed in Table 1. The Matlab's implementation is run with four workers using the *mapreduce* function from the Parallel Computing Toolbox in Matlab 2016a. In this experiment, the computations in Spark are distributed across the cores. All the experiments were launched with the rank of factorization equal to 10.

Table 1. The runtime of performing 10 iterations on Benchmark I with: the Matlab's D-HALS, Spark's D-HALS and Spark's ALS

Algorithm	Matlab's D-HALS	Spark's D-HALS	Spark's ALS
Runtime [sec.]	6960	227.8	47.8

5 Conclusions

In this study, we have proposed a first distributed HALS implementation in the Spark framework. This distribution approach differs from the one proposed in [13] because it does not use the MapReduce paradigm. The results show that the proposed solution is scalable (Fig. 3b), which was the main purpose of this study. The runtime/iteration curve versus the number of nodes is obviously nonlinear because the benefits of distribution are decreased by data traffic load when a number of nodes increases. Figure 3(b) demonstrates how the benefits change together with the size of datasets and the number of worker nodes. We can see also how much faster is the new approach with respect to the previous Matlab's MapReduce implementation - see Table 1. This is mainly due to the *single map - multiple reduce* approach instead of the classic MapReduce as well as the Spark's in-memory computation instead of numerous I/O operations.

We have compared our solution also with the already existing MLlib's ALS implementation. The computational complexity of the ALS is lower than of the HALS, but this is a widely-known fact. The advantage of the HALS over the ALS is faster and monotonic convergence [2]. However, the results presented in Fig. 2 demonstrate that the ALS has better convergence behavior but only for Benchmarks I-III. Despite, the following issues make our proposed Spark's solution promising:

- It is worth to mention the current Spark's ALS implementation is far from the classic ALS algorithm, and seems it gives very good results for the problems similar as in recommender systems (ALS is embedded in the 'recommendation' package). However, for Benchmark IV our D-HALS overcomes the ALS tremendously of a few orders. This shows that the proposed solution might work much better in a different kind of problems (in future we will consider to put this implementation into a new 'factorization' package of MLlib).
- The current ALS implementation has been developed for several years and even it was reimplemented once completely. Our proposed solution is the very first attempt to distribute the HALS algorithm in the Spark's distribution model, and many improvements can be added later on.

Acknowledgment. This work was supported by the grant 2015/17/B/ST6/01865 funded by National Science Center (NCN) in Poland.

References

1. Jolliffe, I.T.: Principal Component Analysis. Springer Series in Statistics, 2nd edn. Springer, New York (2002). https://doi.org/10.1007/978-1-4757-1904-8
2. Cichocki, A., Zdunek, R., Phan, A.H., Amari, S.I.: Nonnegative Matrix and Tensor Factorizations: Applications to Exploratory Multi-way Data Analysis and Blind Source Separation. Wiley, Hoboken (2009)
3. Lee, D.D., Seung, H.S.: Learning the parts of objects by non-negative matrix factorization. Nature **401**, 788–791 (1999)
4. Liu, C., Yang, H.C., Fan, J., He, L.W., Wang, Y.M.: Distributed nonnegative matrix factorization for web-scale dyadic data analysis on MapReduce. In: Proceedings of 19th International Conference on World Wide Web. WWW 2010, pp. 681–690. ACM, New York (2010)
5. Sun, Z., Li, T., Rishe, N.: Large-scale matrix factorization using MapReduce. In: ICDM Workshops, pp. 1242–1248. IEEE Computer Society (2010)
6. Ding, C., Li, T., Jordan, M.I.: Convex and semi-nonnegative matrix factorizations. IEEE Trans. Pattern Anal. Mach. Intell. **32**(1), 45–55 (2010)

7. Cichocki, A., Zdunek, R., Amari, S.: Hierarchical ALS algorithms for nonnegative matrix and 3D tensor factorization. In: Davies, M.E., James, C.J., Abdallah, S.A., Plumbley, M.D. (eds.) ICA 2007. LNCS, vol. 4666, pp. 169–176. Springer, Heidelberg (2007). https://doi.org/10.1007/978-3-540-74494-8_22

8. Han, L., Neumann, M., Prasad, U.: Alternating projected Barzilai-Borwein methods for nonnegative matrix factorization. Electron. Trans. Numer. Anal. **36**, 54–82 (2009–2010)

9. Kim, J., Park, H.: Fast nonnegative matrix factorization: an active-set-like method and comparisons. SIAM J. Sci. Comput. **33**(6), 3261–3281 (2011)

10. Gillis, N., Glineur, F.: Accelerated multiplicative updates and hierarchical ALS algorithms for nonnegative matrix factorization. Neural Comput. **24**(4), 1085–1105 (2012)

11. Chen, W., Guillaume, M.: HALS-based NMF with flexible constraints for hyperspectral unmixing. EURASIP J. Adv. Signal Process. **54**, 1–14 (2012)

12. Laudadio, T., Croitor Sava, A.R., Sima, D.M., Wright, A.J., Heerschap, A., Mastronardi, N., Van Huffel, S.: Hierarchical non-negative matrix factorization applied to three-dimensional 3T MRSI data for automatic tissue characterization of the prostate. NMR Biomed. **29**(6), 751–758 (2016)

13. Zdunek, R., Fonal, K.: Distributed nonnegative matrix factorization with HALS algorithm on MapReduce. In: Ibrahim, S., Choo, K.-K.R., Yan, Z., Pedrycz, W. (eds.) ICA3PP 2017. LNCS, vol. 10393, pp. 211–222. Springer, Cham (2017). https://doi.org/10.1007/978-3-319-65482-9_14

14. Zaharia, M., Chowdhury, M., Das, T., Dave, A., Ma, J., McCauly, M., Franklin, M.J., Shenker, S., Stoica, I.: Resilient distributed datasets: a fault-tolerant abstraction for in-memory cluster computing. In: Gribble, S.D., Katabi, D. (eds.) Proceedings of the 9th USENIX Symposium on Networked Systems Design and Implementation, NSDI 2012, San Jose, CA, USA, 25–27 April 2012, pp. 15–28. USENIX Association (2012)

15. Cichocki, A., Phan, A.H.: Fast local algorithms for large scale nonnegative matrix and tensor factorizations. IEICE Trans. Fundam. Electron. Commun. Comput. Sci. **E92-A**(3), 708–721 (2009)

16. Harper, F.M., Konstan, J.A.: The movielens datasets: history and context. ACM Trans. Interact. Intell. Syst. **5**(4), 19:1–19:19 (2015)

Dimensionally Distributed Density Estimation

Pasi Fränti$^{(\boxtimes)}$ and Sami Sieranoja

School of Computing, University of Eastern Finland, Joensuu, Finland
{pasi.franti,sami.sieranoja}@uef.fi

Abstract. Estimating density is needed in several clustering algorithms and other data analysis methods. Straightforward calculation takes O(N^2) because of the calculation of all pairwise distances. This is the main bottleneck for making the algorithms scalable. We propose a faster O(N log$N$) time algorithm that calculates the density estimates in each dimension separately, and then simply cumulates the individual estimates into the final density values.

Keywords: Clustering · Density estimation · Density peaks · K-means

1 Introduction

The goal of clustering is to partition a set of N data points of D dimensions into K clusters. Using density in the clustering process is appealing as the cluster centroids are typically high density points. Several density-based methods have already been proposed in literature [1, 4, 6, 12, 13, 21, 26].

Most common approach is to estimate the density of every data point individually, and select the K points of highest density as the cluster centroids. In [1] the k centroids are selected in a decreasing order, with the condition that they are not closer than a given distance threshold to an already chosen centroid. The average pairwise distance (*pd*) of all data points was used in [28] to calculate the threshold.

DBSCAN is probably the most cited density-based algorithm [6]. It uses a simple threshold for selecting core points as the points having more than *minPts* other points within their *R*-radius neighborhood. All points within *R*-radius of a core point are then considered *density reachable* and merged to the same cluster. Other points are marked as outliers. An alternative approach is to calculate the density between the points [17]. However, the correct choice of the parameters is the main challenge in both of these approaches.

Density peaks algorithm [26] calculates not only the density but it also finds the nearest neighbor point with higher density. It then applies sorting heuristic based on the density and the distance to this neighbor. The k highest ranked points are chosen as the cluster seeds. The rest of the points are assigned to the clusters by following the neighbor pointers.

Some algorithms use the density-based methods merely as initialization for k-means, which is used to obtain the final clustering result. For instance, the maximum density point is selected as the first centroid and the rest are selected as the points furthest from previously chosen centroids [3, 14, 24]. The distance was also weighted by the density of the points in order to reduce the effect of outliers [14, 24].

L. Rutkowski et al. (Eds.): ICAISC 2018, LNAI 10842, pp. 343–353, 2018.
https://doi.org/10.1007/978-3-319-91262-2_31

Density has also been used to detect outliers. For example, a data point is considered as an outlier if there are less than k points within a given distance d [15]. Another method [23] calculates the k nearest neighbors (k-NN), and uses the distance to the kth neighbor as the outlier detector; points with the largest distance are labeled as outliers.

A bottleneck of using density is that it requires $O(N^2)$ distance calculations. It is possible to speed-up by taking a smaller sub-sample of the data at the cost of compromising the accuracy of the density estimations. Alternative to sub-sampling is to pre-cluster the data [2] so that the neighbors are first taken from the same cluster. Additional check-outs of the neighbor clusters are also performed.

In this paper, we propose a significantly faster algorithm called *dimension distributed density estimation* algorithm (DDDE). The idea is as follows. We sort the data, once per dimension. In each dimension, we use sliding window to find k points ($k/2$ before and $k/2$ after) and calculate their average. With the sorted data, this can be trivially obtained in linear time. We then sum up the cumulated average distances in each dimension. Their sum represents the density estimation of the data points. The time complexity is $O(D{\cdot}N \log N)$ due to sorting D times.

We show by experiments that the proposed density estimation drops the median processing time significantly by a factor of 160:1 in comparison to the brute force density calculation using k-NN, and 50:1 in comparison to the sub-sampling (2%) strategy. We test the effect of this speed-up technique with two clustering algorithms: density-initialized k-means, and the density peaks algorithm [26]. The clustering accuracy had a slight decrease comparable to that of sub-sampling strategy. Considering the remarkable speed-up, such a small degradation in quality might be tolerated.

2 Density Estimation

In general, density is defined as *mass* divided by *volume*. There are two common practices to realize this:

- Distance-based (R-radius)
- Neighbor-based (k-NN)

The *distance-based* approach calculates the number of points (mass) within a fixed neighborhood (volume). The neighborhood is given by a distance threshold (R), which defines R-radius hyper ball in D–dimensional space, see Fig. 1. The algorithm then counts how many data points are within this ball. The approach is also referred to as *cut-off kernel*. A variant called *Gaussian kernel* [12, 20] gives higher weight for nearby points.

The *neighbor-based* approach calculates the distance (volume) within a fixed neighborhood (mass). The neighborhood is defined by the k-nearest neighbors, (k-NN) where k is the input parameter defining the mass. Then average distance to the neighbors is calculated, which indirectly defines the volume, see Fig. 1. Distance to the kth nearest neighbor was also used in [19, 22, 23] but the average distance was found more robust in [11]. This variant is also referred to as *density kernel* in [12].

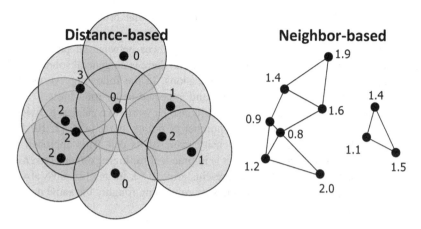

Fig. 1. Two ways to calculate density estimates: distance-based (cutoff kernel) and neighbor-based (density kernel).

In other words, the distance-based approach estimates the mass by counting the number of points for a given volume (distance). The neighbor-based approach estimates the volume by measuring the average distances for a given mass (number of points k). In both cases, the bottleneck is to find the neighbor points and there is no shortcut in this: $O(N^2)$ distance calculations are needed. From computational point of view, neither approach has an obvious benefit over the other.

However, their parameterization is different. Distance-based approach has the distance parameter (R), which depends on the distances in the data. Neighbor-based approach requires the number of neighbors (k), which depends only on the size of data. At least the following parameter choices have been used in literature for estimating density, divergence, or used in other applications of k-NN:

- $R = 10–100\%$ * average distance to data center [2]
- $R =$ Average pairwise distance of all data points [28]
- $R = 90\%$ * first peak in the pairwise distance histogram [17]
- $R = 0.07$ [26]
- $k = 10$ [18]
- $k = 30$ [12]
- $k = 10–100$ [27]
- $k = 30–200$ [5]
- $k = \sqrt{N}$ [19]
- $k = \min\{50, N/(2K)\}$ where K is the number of clusters [this paper]

The optimal choice of the parameter depends on the data. The number of neighbors (k) is simpler to determine and expected to be more robust than the radius (R) although contradicting recommendations have also been reported in estimating divergence [30]. According to [19], k should have sub-linear dependency on N. They recommended $\sqrt{N}$. In general, automatic choice of the parameter may appear simple in the eye of theoretician but is hardly so in the eye of practitioners [29]. As a consequence, some methods leave the choice to the user [17], or assume that brute force manual optimization is performed [22].

Both of the approaches require calculating distances between all pairs of points. Brute force implementation takes quadratic time and there is no general solution to do it faster except some special cases in low-dimensions. In the following, we use the k-nearest neighbor due to its wide popularity and expected better robustness.

There is also a third alternative which might also be worth to consider. It divides the space via a *regular grid*, and counts the number of points in each cell [2, 10, 14, 24, 34]. The individual points inherit the density value of its cell. This approach might work well in low-dimensional space but it is impractical for higher dimensions. *Kd-tree* [14, 24], *space-filling curve* [10], and pre-clustering by k-means [2] have also been used aiming to partition the space into buckets containing roughly the same number of points.

3 Dimensionally Distributed Density Estimation

We present next our algorithm DDDE. It was inspired by the method in [3] used for categorical data. They estimate the density based on the popularity of the individual attributes of the objects. For example, consider an imaginary dataset of 7.6 billion points, representing the name, occupation and nationality of people in the world, and take two samples from it: A = [Zhang, Farmer, Mandarin] and B = [Sieranoja, Scientist, Finnish]. The frequencies of Zhang, Farmer and Mandarin are significantly higher than their counterparts Sieranoja, Scientist and Finnish. The attributes of the first data point are much more common, and thus, it is density estimate is much higher.

The implementation of algorithm consists of two internal loops. The outer loop iterates through all the dimensions, and the inner loop through all the points. In each dimension, we first sort the data according to the values of this dimension. This takes O $(N \cdot \log N)$ time. We then calculate the density for point x by using a sliding window of size $k + 1$ with x at the centre of the window. The density is calculated as the (one-dimensional) mean distance from x to the other points inside the window.

We optimize this process by dividing the window into two halves and maintaining two cumulative sums: one for the values before x (s^-) and another for values after x (s^+). The corresponding mean values are:

$$m^+ = \frac{2s^+}{k} \quad m^- = \frac{2s^-}{k}$$

Since all the values before x all are smaller than x, and all the values after x are greater than x, we can calculate the average distance from x to all points inside the window as follows:

$$Dens(i) = \frac{(x - m^-) + (m^+ - x)}{2} = \frac{m^+ - m^-}{2} = \frac{2s^+ - 2s^-}{2k} = \frac{s^+ - s^-}{k}$$

This average distance serves as our density estimate in this dimension. When sliding the window to the next point, we only need two additions and two subtractions to update the cumulative sums, see Fig. 2. This reduces the time complexity of the sliding window from $O(kN)$ to $O(N)$.

Algorithm DDDE:

```
DDDE (X[1,N]: dataset, k: neighbors) → Dens[1,N]

{
FOR dim=1 to D DO
      Z = Project(X, dim);      // Take dimᵗʰ values
      Y = Sort(Z);
      FOR i=1 TO N DO
            Dens[i] += 2*(m⁺ - m⁻)/k;
            m⁻ = m⁻ + (Y[i] − Y[i-k/2]);
            m⁺ = m⁺ - Y[i+1] + Y[i+k/2+1];
}
```

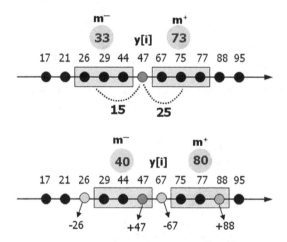

Fig. 2. Maintaining cumulative sums (s^- and s^+) of the two halves of the sliding window allows efficient $O(N)$ implementation of the density calculations in a given dimension.

The overall process is illustrated in Figs. 3 and 4 for two-dimensional toy data. The drawback of the proposed approach is that it does not consider joint influence of the dimensions. This may cause that, in some dimensions, a point can have high density because of having similar value than points in a far away dense cluster. It is therefore possible to detect false density peaks especially with low dimensional data, see Fig. 4. However, errors in one dimension tend to diminish when there are more dimensions.

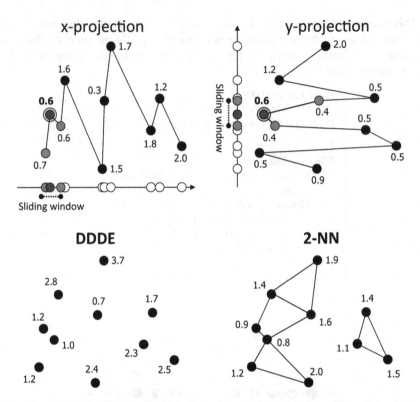

Fig. 3. Example of the DDDE algorithm for a dataset of $N = 10$ points. The dimension-wise density estimates are shown above; the cumulative sums and the 2-NN results below.

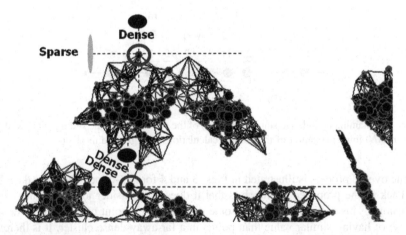

Fig. 4. The potential problems of the fast density estimation in distance-based clustering. The k-neighbors in some dimensions can be located far away, and remote density peaks can influence the density estimation giving false impression of high density.

4 Experiments

We test the proposed density estimation within two clustering algorithms:

- Density-based sorting + k-means
- Density peaks [26]

Both algorithms require the number of clusters given as input. If it is unknown and needs to be solved, the following strategy can be applied. First cluster the data several times with different number of clusters, and then select the one that minimizes the WB-index [33]. The choice of the suitable index is an open problem. In case of density peaks, one can use the ranking scores directly and apply some heuristic to decide how many of the highest ranked centroids are used. Knee point detection heuristic was considered in [31].

For density estimation we consider three alternatives:

- Full search $O(N^2)$
- Using subsample (s=2%) $O(sN^2)$
- Using DDDE $O(N \cdot \log N)$

For sub-sampling, we vary the sample size between 0.1%–10%. Since the goal is to have as small sample size as possible without completely destroying the clustering quality, we select 2% as the default value.

We use the following datasets and parameters. Since our purpose is to evaluate the density estimation rather than the clustering performance, we select only datasets that both algorithms are expected to cluster (at least with reasonable accuracy). The datasets and their properties are shown below (Table 1):

Table 1.

Dataset:	Size:	Clusters:
A1–A3 [16];	$N = 3000$–7500	$K = 20$–50
S1–S4 [8];	$N = 5000$	$K = 15$
Dim32 [9];	$N = 1024$	$K = 16$
Birch1, Birch2 [32];	$N = 100,000$	$K = 100$
Unbalance [25];	$N = 6500$	$K = 8$

For measuring the success of the clustering we use *Centroid Index* (CI) [7], which indicates how many centroids are wrong. In specific, CI = 0 indicates that the clustering structure is correct with respect to the ground truth. The results of the density-based methods appear in Table 2. Reference results are given for k-means, and repeated k-means (restarted 100 times with different random initialization).

The first observation is that the density-based initialization of k-means is not very good as such. Sometimes the faster density estimations (sub-sampling and DDDE) provide even better result. Density peaks, however, is a good algorithm and it finds the correct clustering with all of these sets (CI = 0 in all cases).

Density peaks was implemented as follows. We first calculate density using the three alternative methods (full search, sub-sampling, DDDE). The nearest neighbor with higher density and k-means are then performed for the full dataset (no

Table 2. Clustering results (CI) of the algorithms with various speed-up techniques.

Method	S1	S2	S3	S4	A1	A2	A3	Unb	B1	B2	D32	Av.
Density-based sorting + k-means												
Full search	3.0	5.0	1.0	1.0	2.0	8.0	12.0	5.0	13.0	44.0	7.0	**9.2**
Sub-sample	2.0	2.7	1.3	1.3	5.6	7.4	14.9	4.0	7.4	17.2	12.0	**6.9**
DDDE	3.0	2.0	1.0	1.0	4.0	7.0	14.0	4.0	9.0	33.0	4.0	**7.5**
Density peaks + k-means [26]												
Full search	*0.0*	*0.0*	*0.0*	*0.0*	*0.0*	*0.0*	*0.0*	*0.0*	*0.0*	*0.0*	*0.0*	**0.0**
Sub-sample	0.3	0.7	0.9	0.7	1.5	3.0	4.0	*0.0*	*0.0*	*0.0*	*0.0*	**1.0**
DDDE	*0.0*	*0.0*	*0.0*	1.0	1.0	1.0	3.0	*0.0*	2.0	1.0	*0.0*	**0.8**
Random + k-means												
Single	1.8	1.4	1.3	0.9	2.5	4.5	6.6	3.9	6.6	16.6	3.6	**4.5**
Repeated	0.1	*0.0*	*0.0*	*0.0*	0.3	1.8	2.9	2.9	2.8	10.9	1.1	**2.1**

sub-sampling). The selection is made using the delta-criterion: selecting the K centroids with biggest distances (delta-values) to its nearest neighbor having higher density.

The results show, that sub-sampling by 2% increases error from CI = 0 to CI = 1, on average, whereas DDDE increases it to CI = 0.8. We therefore compare next the processing times. Density peaks algorithm has two bottlenecks: calculating the densities, and finding the nearest higher density neighbor. Since we study the density calculation, we report only processing times of this part. To realize the benefit in the full algorithm, similar speed-up technique should be developed also on the nearest neighbor search. The processing time results are summarized in Table 3.

Table 3. Processing times (s) of the density estimation and k-means.

Method	S1	S2	S3	S4	A1	A2	A3	Unb	B1	B2	D32
Density estimation											
Full search	0.36	0.36	0.35	0.40	0.19	0.45	0.85	0.59	193	552	0.04
Sub-sample	0.10	0.10	0.11	0.11	0.04	0.11	0.21	0.16	55	66	0.01
DDDE	0.00	0.00	0.00	0.00	0.00	0.00	0.00	0.00	0.04	0.04	0.01
K-means											
Single	0.03	0.03	0.04	0.08	0.02	0.04	0.08	0.05	8.8	17.2	0.04
10 repeats	2.6	3.2	4.1	8.2	2.0	4.4	7.5	4.8	882	172	0.40

Our first observation is that the DDDE takes only fraction of the processing required by sub-sampling and by the full search. For the smaller datasets the $O(N^2)$ full search is probably fast enough, but the difference with larger datasets becomes significant. In case of Birch1, the full search of both variants takes about 3.5 min, and the subsample variant about 1 min. DDDE takes only a fraction of a second. The difference is huge. The median speed-up factor of all datasets is about 160:1 compared to full search (both algorithms), and 50:1 compared to sub-sample. These are in line with the time complexities (S1-S4: $N = 5000$, $\log N = 12$; Birch: $N = 100,000$; $\log N = 16$).

The effect of the sub-sampling size is also shown in Fig. 5. The results show that with such large dataset ($N = 100,000$) sub-sampling is effective; 2% sample is enough

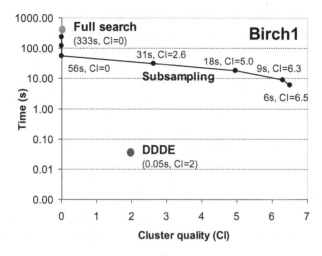

Fig. 5. Effect of the sub-sample size to the clustering result.

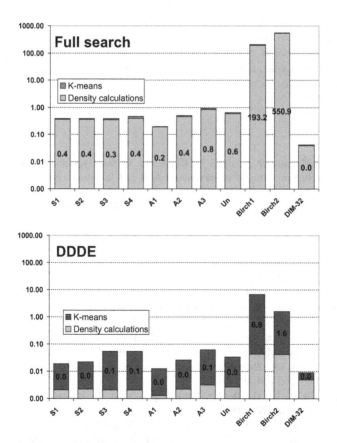

Fig. 6. Processing times of the density initialization (gray) and the k-means (blue). (Color figure online)

to get CI = 0 result. However, further speed-up causes the error to increase soon to about CI = 6 long before reaching real-time (1 s) processing times. We conclude that sub-sampling, although useful, is not as effective as the proposed algorithm. It also seems to lose the benefit of the cache that the full search exploits, making it even less efficient that it otherwise could be.

Figure 6 summarizes the processing times in case of the density-based sorting variant. We observe that the density estimation is the bottleneck in with the full search, but the k-means becomes now the bottleneck if DDDE algorithm is used. The median speed-up factor still remains remarkably high, 18:1. In case of Birch1, it is even 346:1.

5   Conclusion

Rapid $O(D \cdot N \log N)$ density estimation algorithm is proposed. Its median speed-up is remarkable 160:1 compared to the full search for typical data, at the cost of minor degradation of the accuracy. When used in density-based clustering algorithms, the accuracy of the density estimator may not be critical. The faster density estimator can therefore play important role to speed-up density-based clustering methods.

As a result, the density estimation is no longer the bottleneck. In the density-sorted initialization, k-means becomes the bottleneck. In case of the Birch2 dataset, the median speed-up factor is still 18:1 when the time taken by k-means is also taken into account. In density peaks, the nearest neighbor search is still the main bottleneck that should also be solved.

References

1. Astrahan, M.M.: Speech Analysis by Clustering, or the Hyperphome Method, Stanford Artificial Intelligence Project Memorandum AIM-124, Stanford University, Stanford, CA (1970)
2. Bai, L., Cheng, X., Liang, J., Shen, H., Guo, Y.: Fast density clustering strategies based on the k-means algorithm. Pattern Recognit. **71**, 375–386 (2017)
3. Cao, F., Liang, J., Bai, L.: A new initialization method for categorical data clustering. Expert Syst. App. **36**(7), 10223–10228 (2009)
4. Cao, F., Liang, J., Jiang, G.: An initialization method for the k-means algorithm using neighborhood model. Comput. Math. App. **58**, 474–483 (2009)
5. Denoeux, T., Kanhanatarakul, O., Sriboonchitta, S.: EK-NNclus: A clustering procedure based on the evidential K-nearest neighbor rule. Knowl.-Based Syst. **88**, 57–69 (2015)
6. Ester, M., Kriegel, H.P., Sander, J., Xu, X.: A density-based algorithm for discovering clusters in large spatial databases with noise. In: International Conference on Knowledge Discovery and Data Mining (KDD), pp. 226–231 (1996)
7. Fränti, P., Rezaei, M., Zhao, Q.: Centroid index: cluster level similarity measure. Pattern Recognit. **47**(9), 3034–3045 (2014)
8. Fränti, P., Virmajoki, O.: Iterative shrinking method for clustering problems. Pattern Recognit. **39**(5), 761–765 (2006)
9. Fränti, P., Virmajoki, O., Hautamäki, V.: Fast agglomerative clustering using a k-nearest neighbor graph. IEEE Trans. Pattern Anal. Mach. Intell. **28**(11), 1875–1881 (2006)

10. Gourgaris, P., Makris, C.: A density based k-means initialization scheme. In: EANN Workshops, Rhodes Island, Greece (2015)
11. Hautamäki, V., Kärkkäinen, I., Fränti, P.: Outlier detection using k-nearest neighbour graph. In: International Conference on Pattern Recognition (ICPR'2004), Cambridge, UK, pp. 430–433, August 2004
12. Hou, J., Pellilo, M.: A new density kernel in density peak based clustering. In: International Conference on Pattern Recognition, Cancun, Mexico, pp. 468–473, December 2014
13. Jain, A.K., Dubes, R.C.: Algorithms for clustering data. Prentice-Hall, Upper Saddle River (1988)
14. Katsavounidis, I., Kuo, C.C.J., Zhang, Z.: A new initialization technique for generalized Lloyd iteration. IEEE Sig. Process. Lett. **1**(10), 144–146 (1994)
15. Knorr, E.M., Ng, R.T.: Algorithms for mining distance-based outliers in large datasets. In: International Conference on Very Large Data Bases, New York, USA, pp. 392–403 (1998)
16. Kärkkäinen, I., Fränti, P.: Dynamic local search algorithm for the clustering problem, Research Report A-2002-6
17. Lemke, O., Keller, B.: Common nearest neighbor clustering: why core sets matter. Algorithms (2018)
18. Lulli, A., Dell'Amico, M., Michiardi, P., Ricci, L.: NGDBSCAN: scalable density-based clustering for arbitrary data. VLDB Endow. **10**(3), 157–168 (2016)
19. Loftsgaarden, D.O., Quesenberry, C.P.: A nonparametric estimate of a multivariate density function. Ann. Math. Stat. **36**(3), 1049–1051 (1965)
20. Mak, K.F., He, K., Shan, J., Heinz, T.F.: Nat. Nanotechnol. **7**, 494–498 (2012)
21. Melnykov, I., Melnykov, V.: On k-means algorithm with the use of Mahalanobis distances. Stat. Probab. Lett. **84**, 88–95 (2014)
22. Mitra, P., Murthy, C.A., Pal, S.K.: Density-based multiscale data condensation. IEEE Trans. Pattern Anal. Mach. Intell. **24**(6), 734–747 (2002)
23. Ramaswamy, S., Rastogi, R., Shim, K.: Efficient algorithms for mining outliers from large data sets. ACM SIGMOD Rec. **29**(2), 427–438 (2000)
24. Redmond, S.J., Heneghan, C.: A method for initialising the K-means clustering algorithm using kd-trees. Pattern Recognit. Lett. **28**(8), 965–973 (2007)
25. Rezaei, M., Fränti, P.: Set-matching methods for external cluster validity. IEEE Trans. Knowl. Data Eng. **28**(8), 2173–2186 (2016)
26. Rodriquez, A., Laio, A.: Clustering by fast search and find of density peaks. Science **344** (6191), 1492–1496 (2014)
27. Sieranoja, S., Fränti, P.: High-dimensional kNN-graph construction using z-order curve. ACM J. Exp. Algorithmics (submitted)
28. Steinley, D.: Initializing k-means batch clustering: a critical evaluation of several techniques. J. Classif. **24**, 99–121 (2007)
29. Steinwart, I.: Fully adaptive density-based clustering. Ann. Stat. **43**(5), 2132–2167 (2015)
30. Wang, Q., Kulkarni, R., Verdu, S.: Divergence estimation for multidimensional densities via k–nearest-neighbor distances. IEEE Trans. Inf. Theory **55**(5), 2392–2405 (2009)
31. Wang, J., Zhang, Y., Lan, X.: Automatic cluster number selection by finding density peaks. In: IEEE International Conference on Computers and Communications, Chengdu, China, October 2016
32. Zhang, T., Ramakrishnan, R., Livny, M.: BIRCH: a new data clustering algorithm and its applications. Data Min. Knowl. Discov. **1**(2), 141–182 (1997)
33. Zhao, Q., Fränti, P.: WB-index: a sum-of-squares based index for cluster validity. Data Knowl. Eng. **92**, 77–89 (2014)
34. Zhao, Q., Shi, Y., Liu, Q., Fränti, P.: A grid-growing clustering algorithm for geo-spatial data. Pattern Recogn. Lett. **53**(1), 77–84 (2015)

Outliers Detection in Regressions by Nonparametric Parzen Kernel Estimation

Tomasz Galkowski[1]([✉]) and Andrzej Cader[2,3]

[1] Institute of Computational Intelligence, Czestochowa University of Technology,
Czestochowa, Poland
`tomasz.galkowski@iisi.pcz.pl`
[2] Information Technology Institute, University of Social Sciences, 90-113 Lodz,
Poland
[3] Clark University, Worcester, MA 01610, USA

Abstract. A certain observation which is unusual or different from all other ones is called the outlier or anomaly. Appropriate evaluation of data is a crucial problem in modelling of the real objects or phenomena. Actually investigated problems often are based on data mass-produced by computer systems, without careful inspection or screening. The great amount of generated and processed information (e.g. so-called Big-Data) cause that possible outliers often go unnoticed and the result is that they can be masked. However, in regression, this situation can be more complicated. The identification and evaluation of the extremely atypical measurements in observations, for instance in some areas of medicine, geology, particularly in seismology (earthquakes), is precisely the outliers that are the subjects of interest. In this paper, a nonparametric procedure based on Parzen kernel for estimation of unknown function is applied. Evaluation of which measurements in input data-set could be recognized as outliers and possibly should be removed has been performed using the Cook's Distance formula. Anomaly detection is still an important problem to be researched within diverse areas and application domains.

Keywords: Outlier detection · Regression · Nonparametric estimation

1 Introduction and Short Review

This article is not aimed to be a wide up-to-date survey on outlier detection and evaluation methodology. But some key notions should be enumerated. The approach presented in this article concerns the problem of finding the patterns in observations that do not conform to expected behavior. They are often referred to as anomalies, outliers, discordant observations, exceptions, aberrations, surprises, peculiarities, or contaminants in different application domains (see [6]). An unusual data may be a result of keypunch errors, misplaced decimal points, recording or transmission errors, exceptional population slipping into the sample, also intended action of criminals, hackers, and many other situations.

© Springer International Publishing AG, part of Springer Nature 2018
L. Rutkowski et al. (Eds.): ICAISC 2018, LNAI 10842, pp. 354–363, 2018.
https://doi.org/10.1007/978-3-319-91262-2_32

In industry, for instance, the problem of abnormalities identification is known as fault detection and aims to identify defective states of industrial systems, subsystems and/or its components. Early detection of such unwelcome states can help to rectify system behaviour and to prevent unplanned breakdowns and ensure system safety (see [24]).

In medicine, the aberrations or peculiarities recorded in e.g. ECG or EEG signals, or in characteristic reagents in blood, or in other human organic liquids can decide on correct diagnosis results. These examples show instances of a particular situation when outliers help to improve or make safe something.

Many areas of human activity, like the economy, social sciences and other, are analyzed using tools of mathematical statistics through building and further application of the models of researched processes. The model of the object or system is commonly constructed basing on measurement data, often within the additive noise, and then used in equations which depend on a finite number of unknown parameters to be estimated. Well known are e.g. Bayes methods of density estimation or linear regressions. A very important problem is the presence of anomalies which can affect the model parameters. Then the accuracy of the model is corrupted and its application in practice seems to be uncertain. Note that data containing noise tends to be similar to the anomalies and hence difficult to identify and remove. Robust regression methods are designed to be not overly affected by violations of assumptions by the underlying data-generating process [1]. Some outlying points will have more influence on the regression model than other.

Many different approaches in research are applied for detection and identification of outliers. The most significant techniques are: classification based, clustering based methods - including neural networks and other artificial intelligence algorithms, nearest neighbor and statistical - based on distance assessment, spectral analysis, image processing domain, etc. They find a wide range of application areas such as: medical anomaly detection, industrial damage detection, fraud detection, cyber-intrusion detection, image processing, sensor networks, textual anomaly detection [6].

Hence, it is necessary to properly assess what effect on the model has an outlying observation, each one separately or grouped (e.g. collective anomalies [19]).

The main idea of this article is a new proposition of the algorithm helping the detection and evaluation of the outliers in measurement data with additive white noise. We are going to apply the nonparametric methodology previously used in many tasks concerning modelling of unknown objects and systems in the presence of noise, see [10–17], or classification and pattern recognition [20, 29–32]. Similar research problems have been investigated using neuro-fuzzy approach, see e.g. [2, 7–9, 21–23, 25, 27, 33–38].

A nonparametric procedure based on Parzen kernel [26] for estimation of unknown function is applied. Note that nonparametric methodology has no assumption on the mathematical model of function to be estimated. Evaluation of which measurements in input data set could be recognized as outliers and possibly should be removed is performed using the Cook's Distance [4]. Non-parametric methodology has this advantage that can be used either in the linear or nonlinear environment. The results of the simulation analysis are presented.

2   Algorithm of Nonparametric Regression Estimation

We investigate the model of type

$$y_i = R(x_i) + \epsilon_i, \ i = 1, 2, \ldots, n \tag{1}$$

where x_i is assumed to be the set of deterministic inputs, $x_i \in S$, y_i is the set of probabilistic outputs, and ϵ_i is a measurement noise with zero mean and bounded variance. $R(.)$ is an unknown function. In the nonparametric methodology we have completely no assumption neither on its shape (like e.g. in the spline methods or linear regression) nor on any mathematical formula with a certain set of parameters to be found (so-called: parametric approach).

We consider a nonparametric estimator of unknown function $R(.)$ in the form

$$\hat{R}_n(x) = \frac{1}{b_n} \sum_{i=1}^{n} y_i \int\limits_{S_i} K\left(\frac{x-u}{b_n}\right) du \tag{2}$$

where $K(.)$ is the kernel function described by (3), b_n is a smoothing parameter depending on the number of observations n. Interval S is partitioned into n disjunctive segments S_i such that $\cup S_i = [0,1], S_i \cap S_j = \emptyset$ for $i \neq j$. The measurement points x_i are chosen from S_i, i.e.: $x_i \in S_i$. The kernel function is defined by Eq. (3):

$$\left.\begin{array}{ll} \text{(i)} & K(t) = 0, \text{ for } t \notin (-\tau, \tau), \tau > 0, \\ \text{(ii)} & \int_{-\tau}^{\tau} K(t)\, dt = 1 \\ \text{(iii)} & |K(t)| < \infty \end{array}\right\} \tag{3}$$

The set of input values x_i (independent variable in the model (1) are chosen in the process of collecting data, e.g. sampled equally distributed values of ECG signal in time domain, or stock exchange information, or internet activity on specified TCP/IP port of the web or ftp server logs recorded in time, for instance. This data points should provide a balanced representation of function R in the domain S. The standard assumption in theorems on convergence is that the $max\,|S_i|$ maximum size in some measure of S_i) tends to zero if n tends to infinity (see e.g. [10, 11, 18]). We may suppose that in the set of pairs (x_i, y_i) there is present (in some way *inscribed*) the information on essential properties of function R, like its smoothness. Assume that there are some values of y_i in the set of data pairs (x_i, y_i) which are more deviated from others than we expect. Our aim is to detect them and identify as distinctly possible outliers.

3   Detection and Identification of Outliers

Existing in data outliers do not come from the same data-generating process as the rest of the data. They cause in models using least squares the predictions pull out to the outliers. The variance of the estimate is artificially increased and the result is that outliers can be masked. Then estimation is inefficient and can be biased. The usual measure which helps to identify outliers is a distance between the data point and its predicted estimate.

$$d_i = y_i - \hat{R}_i \tag{4}$$

Such distance d_i is defined for each observation, $i = 1, 2, \ldots, n$ and is known as ordinary residual. Standardized residual (or studentized residual) is defined as an ordinary residual divided by an estimate of its standard deviation using as a normalizing factor the mean square error.

$$Resid_i = \frac{d_i}{\sqrt{MSE}} \ where \ MSE = \frac{1}{n} \sum_{i=1}^{n} \left(y_i - \hat{R}_i \right)^2 \tag{5}$$

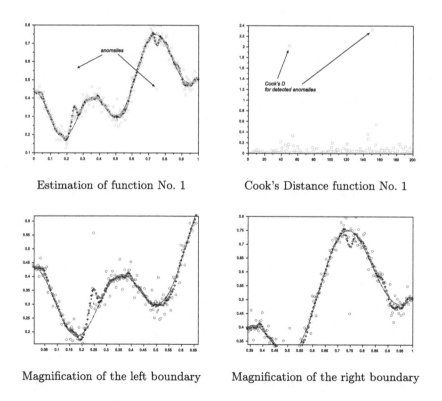

Estimation of function No. 1   Cook's Distance function No. 1

Magnification of the left boundary   Magnification of the right boundary

Fig. 1. Simulation example - function No. 1

The idea is to inspect large values of the residuals with respect to standard deviation and therefore identify outliers. In literature, the standardized residual larger than 3 is generally considered as an outlier score (see [3]).

Classical statistical methods, when the model of the system depends on the finite number of unknown parameters to be estimated, often adapt the least square error algorithm. When the anomalies are present in initial data then the regression model is influenced by them and might not produce the accurate results. Although it is declared that least squares are the robust methods. Robust regression methods are designed to be not overly affected by wrong assumptions by the primary data-generating process. Still while fitting regression the existing anomalies can remain hidden. But authors [28] argue that the robust techniques can help detect anomalies because of their larger residuals.

Let us formulate a question in another way: how to measure an influence on model accuracy by particular point, potentially the outlier? We propose the use of the formula derived from Cook's Distance (shortly *Cook's D*) defined by

$$D_i = \frac{\sum\limits_{j=1 \wedge j \neq i}^{n} \left(\hat{R}_j - \hat{R}_{j\{i\}} \right)^2}{MSE} \ , i = 1, ..., n. \tag{6}$$

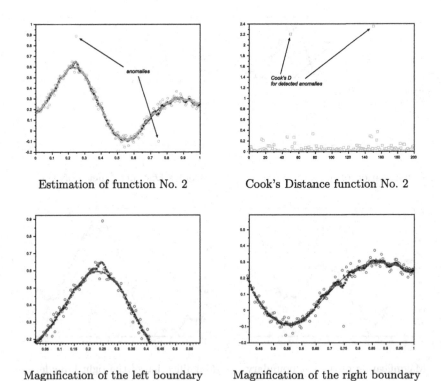

Estimation of function No. 2   Cook's Distance function No. 2

Magnification of the left boundary   Magnification of the right boundary

Fig. 2. Simulation example - function No. 2

In classical statistics, D_i is used in the least square regression analysis to find influential outliers in a set of predictor variables in parametric modelling. It helps to identify points that negatively affect parametric regression model. In our work it is applied in nonparametric approach with slight modification (for comparison see e.g. [4,5]). The sense of the idea is still preserved and the implication is: the higher the residuals, the higher the Cook's Distance.

In Eq. (6) the notation $\hat{R}_{j\{i\}}$ means: estimate of the output $\hat{R}$ obtained using data with removed i-th point from initial data-set. Therefore Cook's D is calculated by removing the i-th data point from the model and recalculating the regression estimate. It shows how much the values of the outputs in the regression model change when the i-th observation is removed.

In literature one may find usually proposed *Rule of Thumb* to treat i-th point as an outlier:

$$D_i \geq \frac{4}{n} \qquad (7)$$

But generally, it is suggested to deeper investigate all the points with D_i conspicuously larger relative to the others. The threshold level depends on user decision.

Let us mention that detection of the existing anomalies in data, even if it negatively affect our regression model, do not imply that it should be automatically removed. The decision depends on the general description of the phenomena studied and requires a deeper analysis. Elimination of outliers always improves the accuracy of regressions but may end up destroying the most important information in data, like for instance in the medical diagnostics. The researchers always should perform very carefully their analysis on possible special circumstances or properties of unusual data.

4   Simulation Example

To show how the proposed algorithm works we made series of simulation experiments. The original functions taken into comparison are in the forms

1. $R_1(x) = 0.3 + 0.3x + \exp(-2x + 05) \cdot sin(4x + 8)) \cdot \cos(12x - 1.2)) \cdot log(x + 1.1)$
2. $R_2(x) = 0.2 + 0.5\exp(-2x) \cdot \sin(7x + \pi/5) - 0.75cos(12x - \pi/6)$
3. $R_3(x) = 0.3 + exp(-x + 0.2) \cdot \sin(2x + 0.4)$

$$(8)$$

The figures present the charts of original function to be estimated (thin dot-line), observed outputs with additive noise (little rings), estimates using classic Parzen procedure (2) without any modifications (black pluses "+"), and estimates using Parzen method, but excluding from data the point in which estimate is currently calculated (little triangles). Note that the NMS algorithm and method of reducing of boundary effect, originally introduced by the author in [13], has been incorporated in the main simulation program. In the right-side picture in every figure, the Cook's Distance chart is presented.

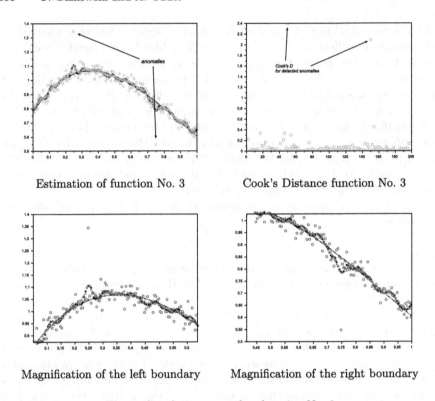

Estimation of function No. 3   Cook's Distance function No. 3

Magnification of the left boundary   Magnification of the right boundary

Fig. 3. Simulation example - function No. 3

The simulations were performed on data set generated as follows: output values of each function (8) are biased by additive white noise with variance limited to $\sigma^2 = 0.3$. The smoothing parameter in the Parzen algorithm was $b_n = 0.02$. We applied the Parzen kernel of the parabolic type - fulfilling the assumptions (3). The set of input data contained $n = 200$ measurements in the interval $S = [0, 1]$. The outputs y_{50} and y_{150} have been artificially changed by increasing (and decreasing) its values that they would be potential anomalies to be detected. The presented graphs show that the proposed procedure is effective.

5   Remarks and Extensions

In this paper, we have proposed the new algorithm helping the detection and evaluation of the outliers in measurement data with the additive white noise. It is based on nonparametric methodology. This approach is applicable to tasks when there is the lack of any information on function defining the object, and it can be used in both linear and non-linear systems. The series of simulations on the different choice of functions confirmed the effectiveness of the algorithm. The multivariate case for resolving this problem will be studied in the future works.

References

1. Andersen, R.: Modern Methods for Robust Regression. Quantitative Applications in the Social Sciences, vol. 152. Sage, Thousand Oaks (2008)
2. Beg, I., Rashid, T.: Modelling uncertainties in multi-criteria decision making using distance measure and topsis for hesitant fuzzy sets. J. Artif. Intell. Soft Comput. Res. **7**(2), 103–109 (2017)
3. Bollen K.A., Jackman R.W.: Regression diagnostics: an expository treatment of outliers and influential cases. In: Fox, J., Scott, L.J. (eds.) Modern Methods of Data Analysis, pp. 257–291. Sage, Newbury Park (1990). ISBN 0-8039-3366-5
4. Cook, R.D.: Detection of influential observations in linear regression. Technometrics **19**, 15–18 (1977). American Statistical Association
5. Cook, R.D.: Residuals and Influence in Regression. Weisberg, Sanford, New York (1982)
6. Chandola, V., Banerjee A., Kumar, V.: Anomaly detection: a survey. ACM Comput. Surv. **41**(3), Article 15, 58 p. Chapman and Hall (2009). https://doi.org/10.1145/1541880.1541882 ISBN 0-412-24280-X
7. Cpalka, K., Rebrova, O., Nowicki, R., et al.: On design of flexible neuro-fuzzy systems for nonlinear modelling. Int. J. Gen. Syst. **42**(6), 706–720 (2013)
8. Cpałka, K., Łapa, K., Przybył, A.: A new approach to design of control systems using genetic programming. Inf. Technol. Control **44**(4), 433–442 (2015)
9. Duch, W., Korbicz, J., Rutkowski, L., Tadeusiewicz, R. (eds.): Biocybernetics and Biomedical Engineering 2000. Neural Networks, vol. 6. Akademicka Oficyna Wydawnicza, EXIT, Warsaw (2000). (in Polish)
10. Galkowski, T., Rutkowski, L.: Nonparametric recovery of multivariate functions with applications to system identification. In: Proceedings of the IEEE, vol. 73, pp. 942–943, New York (1985)
11. Galkowski, T., Rutkowski, L.: Nonparametric fitting of multivariable functions. IEEE Trans. Autom. Control **AC–31**, 785–787 (1986)
12. Galkowski, T.: Nonparametric estimation of boundary values of functions. Arch. Control Sci. **3**(1–2), 85–93 (1994)
13. Gałkowski, T.: Kernel estimation of regression functions in the boundary regions. In: Rutkowski, L., Korytkowski, M., Scherer, R., Tadeusiewicz, R., Zadeh, L.A., Zurada, J.M. (eds.) ICAISC 2013. LNCS (LNAI), vol. 7895, pp. 158–166. Springer, Heidelberg (2013). https://doi.org/10.1007/978-3-642-38610-7_15
14. Galkowski, T., Pawlak, M.: Nonparametric extension of regression functions outside domain. In: Rutkowski, L., Korytkowski, M., Scherer, R., Tadeusiewicz, R., Zadeh, L.A., Zurada, J.M. (eds.) ICAISC 2014. LNCS (LNAI), vol. 8467, pp. 518–530. Springer, Cham (2014). https://doi.org/10.1007/978-3-319-07173-2_44
15. Galkowski, T., Pawlak, M.: Orthogonal series estimation of regression functions in nonstationary conditions. In: Rutkowski, L., Korytkowski, M., Scherer, R., Tadeusiewicz, R., Zadeh, L.A., Zurada, J.M. (eds.) ICAISC 2015. LNCS (LNAI), vol. 9119, pp. 427–435. Springer, Cham (2015). https://doi.org/10.1007/978-3-319-19324-3_39
16. Galkowski, T., Pawlak, M.: Nonparametric estimation of edge values of regression functions. In: Rutkowski, L., Korytkowski, M., Scherer, R., Tadeusiewicz, R., Zadeh, L.A., Zurada, J.M. (eds.) ICAISC 2016. LNCS (LNAI), vol. 9693, pp. 49–59. Springer, Cham (2016). https://doi.org/10.1007/978-3-319-39384-1_5

17. Galkowski, T., Pawlak, M.: The novel method of the estimation of the Fourier transform based on noisy measurements. In: Rutkowski, L., Korytkowski, M., Scherer, R., Tadeusiewicz, R., Zadeh, L.A., Zurada, J.M. (eds.) ICAISC 2017. LNCS (LNAI), vol. 10246, pp. 52–61. Springer, Cham (2017). https://doi.org/10.1007/978-3-319-59060-8_6

18. Gasser, T., Müller, H.-G.: Kernel estimation of regression functions. In: Gasser, T., Rosenblatt, M. (eds.) Smoothing Techniques for Curve Estimation. LNM, vol. 757, pp. 23–68. Springer, Heidelberg (1979). https://doi.org/10.1007/BFb0098489

19. Goldberger, A.L., Amaral, L.A.N., Glass, L., Hausdorff, J.M., Ivanov, P.C., Mark, R.G., Mietus, J.E., Moody, G.B., Peng, C.-K., Stanley, H.E.: Components of a new research resource for complex physiologic signals, PhysioBank, PhysioToolkit, and PhysioNet. Circulation 101(23), 215–220 (2000)

20. Greblicki, W., Rutkowski, L.: Density-free Bayes risk consistency of nonparametric pattern recognition procedures. Proc. IEEE 69(4), 482–483 (1981)

21. Grycuk, R., Gabryel, M., Nowicki, R., Scherer, R.: Content-based image retrieval optimization by differential evolution. In: 2016 IEEE Congress on Evolutionary Computation (CEC), pp. 86–93 (2016)

22. Grycuk, R., Scherer, R., Gabryel, M.: New image descriptor from edge detector and blob extractor. J. Appl. Math. Comput. Mech. 14(4), 31–39 (2015)

23. Korytkowski, M., Rutkowski, L., Scherer, R.: On combining backpropagation with boosting. In: International Joint Conference on Neural Networks, pp. 1274–1277 (2006)

24. Zhang, L., Lin, J., Karim, R.: Adaptive kernel density-based anomaly detection for nonlinear systems. Knowl.-Based Syst. 139, 50–63 (2018)

25. Liu, H., Gegov, A., Cocea, M.: Rule based networks: an efficient and interpretable representation of computational models. J. Artif. Intell. Soft Comput. Res. 7(2), 111–123 (2017)

26. Parzen, E.: On estimation of a probability density function and mode. Anal. Math. Stat. 33(3), 1065–1076 (1962)

27. Rotar, C., Iantovics, L.B.: Directed evolution - a new metaheuristc for optimization. J. Artif. Intell. Soft Comput. Res. 7(3), 183–200 (2017)

28. Rousseeuw, P.J., Leroy, A.M.: Robust Regression and Outlier Detection. Wiley, Hoboken (2003)

29. Rutkowski, L.: A general approach for nonparametric fitting of functions and their derivatives with applications to linear circuits identification. IEEE Trans. Circuits Syst. 33(8), 812–818 (1986)

30. Rutkowski, L.: Sequential pattern recognition procedures derived from multiple Fourier series. Pattern Recognit. Lett. 8, 213–216 (1988)

31. Rutkowski, L.: Non-parametric learning algorithms in the time-varying environments. Sig. Process. 18(2), 129–137 (1989)

32. Rutkowski, L.: Multiple Fourier series procedures for extraction of nonlinear regressions from noisy data. IEEE Trans. Sig. Process. 41(10), 3062–3065 (1993)

33. Rutkowski, L., Cpalka, K.: Compromise approach to neuro-fuzzy systems. In: Intelligent Technologies-Theory and Applications, 2nd Euro-International Symposium on Computation Intelligence, Kosice, Slovakia. Frontiers in Artificial Intelligence and Applications, vol. 76, pp. 85–90 (2002)

34. Starczewski, A.: A new validity index for crisp clusters. Pattern Anal. App. 20(3), 687–700 (2017)

35. Starczewski, A., Krzyżak, A.: Improvement of the validity index for determination of an appropriate data partitioning. In: Rutkowski, L., Korytkowski, M., Scherer, R., Tadeusiewicz, R., Zadeh, L.A., Zurada, J.M. (eds.) ICAISC 2017. LNCS (LNAI), vol. 10246, pp. 159–170. Springer, Cham (2017). https://doi.org/10.1007/978-3-319-59060-8_16
36. Tezuka, T., Claramunt, C.: Kernel analysis for estimating the connectivity of a network with event sequences. J. Artif. Intell. Soft Comput. Res. **7**(1), 17–31 (2017)
37. Yan, P.: Mapreduce and semantics enabled event detection using social media. J. Artif. Intell. Soft Comput. Res. **7**(3), 201–213 (2017)
38. Łapa, K., Cpałka, K., Wang, L.: New method for design of fuzzy systems for nonlinear modelling using different criteria of interpretability. In: Rutkowski, L., Korytkowski, M., Scherer, R., Tadeusiewicz, R., Zadeh, L.A., Zurada, J.M. (eds.) ICAISC 2014. LNCS (LNAI), vol. 8467, pp. 217–232. Springer, Cham (2014). https://doi.org/10.1007/978-3-319-07173-2_20

Application of Perspective-Based Observational Tunnels Method to Visualization of Multidimensional Fractals

Dariusz Jamroz[(⊠)]

Department of Applied Computer Science, AGH University of Science
and Technology, al. A. Mickiewicza 30, 30-059 Krakow, Poland
jamroz@agh.edu.pl

Abstract. Methods of multidimensional data visualization are frequently applied in the qualitative analysis allowing to state some properties of this data. They are based only on using the transformation of
the multidimensional space into a two-dimensional one which represents
the screen in a way ensuring not to lose important properties of the data.
Thanks to this it is possible to observe some searched data properties
in the most natural way for human beings–through the sense of sight.
In this way, the whole analysis is conducted excluding applications of
complex algorithms serving to get information about these properties.
The example of a multidimensional data visualization method is a relatively new method of perspective-based observational tunnels. It was
proved earlier that this method is efficient in the analysis of real data
located in a multidimensional space of features obtained by characters
recognition. Its efficiency was also shown by the analysis of multidimensional real data describing coal samples. In this paper, another aspect
of using this method was shown–to visualize artificially generated five-
dimensional fractals located in a five-dimensional space. The purpose
of such a visualization can be to obtain views of such multidimensional
objects as well as to adapt and teach our mind to percept, recognize
and perhaps understand objects of a higher number of dimensions than
3. Our understanding of such multidimensional data could significantly
influence the way of perceiving complex multidimensional relations in
data and the surrounding world. The examples of obtained views of five-
dimensional fractals were shown. Such a fractal looks like a completely
different object from different perspectives. Also, views of the same fractal obtained using the PCA, MDS and autoassociative neural networks
methods are presented for comparison.

Keywords: Multidimensional data analysis · Data mining
Multidimensional visualization · Observational tunnels method
Multidimensional perspective · Fractals

1   Introduction

The perspective-based observational tunnels method is a new method of qualitative analysis of multidimensional data through its visualization. It was presented

© Springer International Publishing AG, part of Springer Nature 2018
L. Rutkowski et al. (Eds.): ICAISC 2018, LNAI 10842, pp. 364–375, 2018.
https://doi.org/10.1007/978-3-319-91262-2_33

for the first time in the paper [1] in which its efficiency was proved by the construction of pattern recognition systems (characters recognition) and by the analysis of seven-dimensional data containing samples representing three types of coal. It was shown in the example of real data located in a five-dimensional space of features obtained as a result of reception of printed characters that, by construction of image recognition systems, it allows to indicate the possibility to separate individual fractions in the multidimensional space of features even in the case when other methods fail. Furthermore, the perspective-based observational tunnels method occurred to be the best one in the created ranking by the analysis of seven-dimensional data containing samples representing three coal fractions in the context of readability of the results. This paper shows another way of using this method – to visualize artificially generated five-dimensional fractals located in a five-dimensional space.

The main purpose of this paper is to present the effectiveness of the perspective-based observational tunnels method in presenting multidimensional fractals as representatives of artificially generated data. It is the new approach because this method has never been used before on this kind of artificially generated and at the same time such complicated data. Fractals described in this paper comprise points in the multidimensional space. Thus, the purpose of this paper became to demonstrate that the perspective-based observational tunnel method can serve as a tool allowing to explore such a space by obtaining its views.

The purpose of such a visualization can be both to obtain views of such multidimensional objects and to adapt and teach our mind to percept, recognize and also understand objects of a higher number of dimensions than 3. Also other methods are used to analyze the quality of multidimensional data. For example, the PCA method [2–4] which uses the orthogonal projection on two eigenvectors corresponding to two highest absolute values of eigenvalues of the data set covariance matrix. Another method is multidimensional scaling [5,6] which uses transformation of the multidimensional space into a two-dimensional one in such a way that mutual distances between two images of points are as close as possible to the distance between the corresponding points in the input space. The next method is relevance maps [7,8], in which for a n-dimensional space n special points $P_1, P_2, \ldots, P_n$ are additionally used. The transformation of the multidimensional space into a two-dimensional image occurs in such a way that the distance between the image of a specific data point and special point P_i is possibly as close as possible to the ith coordinate of the point which the specific image concerns. Furthermore, the method of parallel coordinates is also used to visualize multidimensional data [9] in which n parallel axes are located next to each other on a plane. The similar method is star graphs [10] where n axes spread radially outwards from one point. Also, autoassociative neural networks [11,12] and Kohonen maps [13,14] are applied in order to visualize multidimensional data.

In the following paper, the fractals obtained by means of Iterated Functions Systems (IFS) were used to generate five-dimensional fractals. There are many papers concerning fractals. Among others, they are used in image compression [15], face recognition [16], character recognition [17] and shape recognition [18]. Usually, fractals are created in a two- or three-dimensional space. However, it is possible to create fractals located in the space of more than three dimensions. Such an approach is presented in this paper.

2   Perspective-Based Observational Tunnels Method

The perspective-based observational tunnels method is the new method. It was first presented in the paper [1]. It intuitively consists in the prospective parallel projection with the local orthogonal projection. In order to understand its idea, the following terms must be introduced [1]:

Definition 1. *The observed space X is defined as any vector space, over an F field of real numbers, n-dimensional, $n \geq 3$, with a scalar product.*

Definition 2. *Let $p_1, p_2 \in X$ - be linearly independent, $w \in X$. An observational plane $P \subset X$ is defined as:*

$$P = \delta(w, \{p_1, p_2\}) \tag{1}$$

where:

$$\delta(w, \{p_1, p_2\}) \stackrel{def}{=} \{x \in X : \exists \beta_1, \beta_2 \in F, \tag{2}$$

such that $x = w + \beta_1 p_1 + \beta_2 p_2\}$.

The two-dimensional computer screen will be represented by vectors p_1, p_2 in accordance with the above definition.

Definition 3. *The direction of projection r onto the observational plane $P = \delta(w, \{p_1, p_2\})$ is defined as any vector $r \in X$ if vectors $\{p_1, p_2, r\}$ are an orthogonal system.*

Definition 4. *The following set is called the hypersurface $S_{(s,d)}$, anchored in $s \in X$ and directed towards $d \in X$:*

$$S_{(s,d)} \stackrel{def}{=} \{x \in X : (x - s, d) = 0\} \tag{3}$$

Definition 5. *A tunnel radius of point $a \in X$ against observational plane $P = \delta(w, \{p_1, p_2\})$ is defined as:*

$$b_a = \psi \xi r + a - w - (1 + \psi)(\beta_1 p_1 + \beta_2 p_2) \tag{4}$$

where:

$$\psi = \frac{(w - a, r)}{\xi(r, r)} \tag{5}$$

$$\beta_1 = \frac{(\psi \xi r + a - w, p_1)}{(1 + \psi)(p_1, p_1)} \tag{6}$$

$$\beta_2 = \frac{(\psi \xi r + a - w, p_2)}{(1 + \psi)(p_2, p_2)} \tag{7}$$

$r \in X$ - *direction of projection onto observational plane P*,
$\xi \in (0, \infty)$ - *coefficient of perspective.*

Figure 1 presents three observational tunnels corresponding to three points belonging to observational plane P. For the readability of the figure, observational plane P is 1-dimensional. The direction in which each tunnel spreads is deviated in relation to the direction of projection r in order to obtain the effect of perspective. The degree of such a tunnel deviation is directly affected by the distance of point e corresponding to it from the zero point of observational plane P and the perspective coefficient.

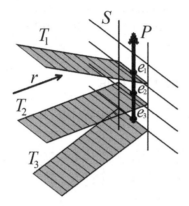

Fig. 1. Three observational tunnels T_1, T_2, T_3 correspond to three different points e_1, e_2, e_3, belonging to observational plane P. The degree of the tunnel deviation in relation to the direction of projection r depends on the distance of point e corresponding to it from the zero point of observational plane P.

3   The Drawing Procedure

Applying the presented theory, the procedure of drawing each point a in accordance with the direction of projection r onto observational plane $P = \delta(w, \{p_1, p_2\})$ consists in executing several steps [1]:

1. we calculate the distance of projection of observed point a: $\psi = \frac{(w-a, r)}{\xi(r, r)}$
2. we calculate the position of the projection (i.e. the pair $\beta_1\beta_2 \in F$) of observed point a: $\beta_1 = \frac{(\psi\xi r + a - w, p_1)}{(1+\psi)(p_1, p_1)}$, $\beta_2 = \frac{(\psi\xi r + a - w, p_2)}{(1+\psi)(p_2, p_2)}$
3. we calculate tunnel radius b_a of point a: $b_a = \psi\xi r + a - w - (1+\psi)(\beta_1 p_1 + \beta_2 p_2)$
4. we verify whether scalar product (b_a, b_a) is smaller than the assumed maximum value of $b_a max$ and whether the distance of projection of observed point a is smaller than the assumed value ψmax. If the condition is met, then we draw the point on observational plane $P = \delta(w, \{p_1, p_2\})$ in the position with coordinates $(\beta_1\beta_2)$, otherwise we do not draw the point.

In this way, we can obtain the image of multidimensional sets of points on the two-dimensional screen. It should be noted that the described theory is true for any scalar product, while the results presented in the further part were obtained with the scalar product given by the formula:

$$(x, y) = \sum_{i=1}^{n} x_i y_i \tag{8}$$

where: $x = (x_1, x_2, ..., x_n)$, $y = (y_1, y_2, ..., y_n)$, n-number of dimensions, $n \geq 3$. It should be noted that the presented method works for any, finite $n \geq 3$. It follows from the algorithm above that the perspective-based observational tunnels method has a linear time complexity in relation to the number of multidimensional points. The time complexity in relation to the number of dimensions is also linear.

4   Two Dimensional Fractals

The IFS method was proposed by Barnsley [19]. It is based on the introduction of k affine mappings on a plane:

$$W_i \left(\begin{bmatrix} x \\ y \end{bmatrix} \right) = \begin{bmatrix} a_{11} & a_{12} \\ a_{21} & a_{22} \end{bmatrix} \begin{bmatrix} x \\ y \end{bmatrix} + \begin{bmatrix} b_1 \\ b_2 \end{bmatrix}, i = 1, ..., k \tag{9}$$

Each mapping is associated with probability p_i and obviously $\sum_{i=1}^{k} p_i = 1$. Mappings W_i must be approaching, so: $d(W_i(x), Wi(y)) < sd(x, y)$, where d is metrics and $0 < s < 1$. The creation of a fractal occurs in the following way:

1. Any starting point (x_0, y_0) is taken.
2. On the basis of probabilities p_i one of the mappings W_i is selected randomly.
3. The location of the next point is calculated as:

$$\begin{bmatrix} x_{k+1} \\ y_{k+1} \end{bmatrix} = W_i \left(\begin{bmatrix} x_k \\ y_k \end{bmatrix} \right) \tag{10}$$

Steps 2 and 3 are performed as many times as many fractal points we want to obtain. Such a two-dimensional fractal can be presented on a computer screen. The difficulties occur only in the case of multidimensional fractals.

5   Multidimensional Fractals

We will refer here to fractals whose points are in the multidimensional space as multidimensional fractals. The presented multidimensionality should not be confused with the notion of the fractal dimension, which describes a completely different property of fractals, that is the Hausdorff dimension. Let us expand the method described above for the case with n dimensions. In such a case the affine mappings take the following form:

$$
W_i\left(\begin{bmatrix} x_1 \\ x_2 \\ . \\ x_n \end{bmatrix}\right) = \begin{bmatrix} a_{11} & a_{12} & ... & a_{1n} \\ a_{21} & a_{22} & ... & a_{2n} \\ . & . & ... & . \\ a_{n1} & a_{n2} & ... & a_{nn} \end{bmatrix} \begin{bmatrix} x_1 \\ x_2 \\ . \\ x_n \end{bmatrix} + \begin{bmatrix} b_1 \\ b_2 \\ . \\ b_n \end{bmatrix} , i = 1, ..., k \qquad (11)
$$

For each mapping, probability p_i is associated and obviously $\sum_{i=1}^{k} p_i = 1$. Mappings W_i must be approaching, so: $d(W_i(x), W_i(y)) < sd(x, y)$, where d is n-dimensional metrics and $0 < s < 1$. The creation of a fractal occurs in the following way:

1. Any starting point is taken: $(x_1^0, x_2^0, ..., x_n^0)$
2. On the basis of probabilities p_i, one of the mappings W_i is selected randomly.
3. The location of the next point is calculated as:

$$
\begin{bmatrix} x_1^{k+1} \\ x_2^{k+1} \\ . \\ x_n^{k+1} \end{bmatrix} = W_i\left(\begin{bmatrix} x_1^k \\ x_2^k \\ . \\ x_n^k \end{bmatrix}\right) \qquad (12)
$$

Steps 2 and 3 are performed as many times as many fractal points we want to obtain. The presentation of the shape of such a multidimensional fractal is much more difficult.

6   Obtained Results

Multidimensional fractals obtained by means of Iterated Functions Systems (IFS) were presented on a computer screen applying the perspective-based observational tunnels method. A special system written in the C++ programming language was created to generate multidimensional fractals and present them on a computer screen. During the investigations, many fractals were visualized in four-dimensional and five-dimensional spaces. Figures 2, 3, 4, 5 and 6 present one of the obtained five-dimensional fractal obtained by means of the perspective-based observational tunnels method. It was created from four following affine mappings:

$$
W_1\left(\begin{bmatrix} x_1 \\ x_2 \\ x_3 \\ x_4 \\ x_5 \end{bmatrix}\right) = \begin{bmatrix} 0.7 & 0.03 & 0.01 & -0.02 & 0.04 \\ -0.03 & 0.7 & 0.02 & 0.1 & -0.04 \\ 0.5 & 0.5 & 0.1 & 0.2 & 0.1 \\ -0.03 & 0.7 & 0.02 & 0.1 & -0.04 \\ -0.03 & 0.7 & 0.02 & 0.1 & -0.04 \end{bmatrix} \begin{bmatrix} x_1 \\ x_2 \\ x_3 \\ x_4 \\ x_5 \end{bmatrix} + \begin{bmatrix} 0 \\ 0.5 \\ 0.1 \\ -0.1 \\ -0.25 \end{bmatrix}
$$

$$
W_2\left(\begin{bmatrix} x_1 \\ x_2 \\ x_3 \\ x_4 \\ x_5 \end{bmatrix}\right) = \begin{bmatrix} -0.1 & 0.4 & 0.1 & 0.05 & 0.01 \\ 0.05 & 0.2 & 0.2 & 0.02 & 0.2 \\ 0.01 & 0.2 & 0.3 & 0.6 & -0.02 \\ -0.05 & 0.01 & 0.03 & 0.3 & 0.2 \\ 0.1 & 0.02 & 0.2 & 0.01 & 0 \end{bmatrix} \begin{bmatrix} x_1 \\ x_2 \\ x_3 \\ x_4 \\ x_5 \end{bmatrix} + \begin{bmatrix} 0 \\ 0.2 \\ 0.3 \\ 0.1 \\ -0.25 \end{bmatrix}
$$

$$W_3\left(\begin{bmatrix} x_1 \\ x_2 \\ x_3 \\ x_4 \\ x_5 \end{bmatrix}\right) = \begin{bmatrix} 0.3 & -0.2 & 0 & -0.15 & 0 \\ 0 & 0.3 & 0.2 & 0.1 & 0.2 \\ 0.2 & 0.1 & 0.2 & 0.1 & 0 \\ 0.04 & 0 & 0 & 0.2 & 0.2 \\ 0.02 & 0.087 & 0.022 & 0 & 0 \end{bmatrix}\begin{bmatrix} x_1 \\ x_2 \\ x_3 \\ x_4 \\ x_5 \end{bmatrix} + \begin{bmatrix} 0 \\ 0.9 \\ 0.5 \\ -0.1 \\ -0.25 \end{bmatrix}$$

$$W_4\left(\begin{bmatrix} x_1 \\ x_2 \\ x_3 \\ x_4 \\ x_5 \end{bmatrix}\right) = \begin{bmatrix} 0 & 0 & 0 & 0.01 & 0.03 \\ 0.04 & 0 & 0.1 & 0 & 0 \\ 0 & 0 & 0 & 0.1 & 0.01 \\ 0.2 & 0.02 & 0.3 & 0 & 0 \\ 0 & 0 & 0 & 0 & 0 \end{bmatrix}\begin{bmatrix} x_1 \\ x_2 \\ x_3 \\ x_4 \\ x_5 \end{bmatrix} + \begin{bmatrix} 0 \\ 0 \\ 0.35 \\ -0.1 \\ -0.25 \end{bmatrix}$$

The probabilities associated with individual mappings are equal to: $p_1 = 0.7$, $p_2 = 0.2$, $p_3 = 0.08$, $p_4 = 0.02$. The obtained figures seem to present completely different objects – while in fact the same fractal is presented, only from a different perspective. Along with the movement of the location of observer in space, settings of observational plane, direction of projection, tunnel radius and perspective coefficient were changed. By changing these parameters, the obtained views fluently were transformed into completely different views. It even seemed that the transformation of one shape into another one occurred. Only after a longer period

Fig. 2. The view of the analyzed five-dimensional fractal obtained by means of the perspective-based observational tunnels method. The view seems like a tree with a narrowing treetop.

Fig. 3. The view of the analyzed five-dimensional fractal obtained by means of the perspective-based observational tunnels method. The view seems like a bulrush.

of watching such a fractal from various perspectives, some sort of adaptation to the obtained changes occurred. But it was only an adaptation, not understanding yet. Probably our brains need these new types of views and their changes to be transmitted to them for longer periods of time. Then there is a chance that we will at least partially understand multidimensional objects being watched. The purpose of this sort of visualization can then serve to obtain views of such multidimensional objects as well to adapt and teach our mind to percept, recognize and perhaps understand objects of a higher number of dimensions than 3. From the presented figures it follows that many different kinds of information can be contained inside such a five-dimensional fractal. The same fractal from a different perspective can look like a tree with a narrowing treetop (Fig. 2), bulrush (Fig. 3), flying dragonfly (Fig. 4), striped shrimp (Fig. 5) and human shape (Fig. 6). As a result of visualization, many other images of the analyzed fractal were obtained. Figures 7, 8 and 9 present views of the same five-dimensional fractal for comparison using other frequently used methods to visualize multidimensional data. Programmes written in the C++ language were created to obtain each of them. Figure 7 presents the view obtained using the PCA method. This method consists in the orthogonal projection of each point onto two specially selected axes. These axes are two eigenvectors corresponding to two highest eigenvalues of the analyzed set covariance matrix as for the module. In the case of the described fractal, the vectors are: $v_1 = (-0.0604, 0.4169, 0.2928, 0.5811, 0.6318)$, $v_2 = (0.7751, -0.2109, 0.5933, -0.0104, -0.0521)$. It follows from the above, that the method can generate only one view for a given data set. For this reason, the analysis of such a fractal using this method is very limited. Autoassociative neural networks are also used to visualize multidimensional data. Such a network during the analysis of n-dimensional data consists of n inputs, one of interlayers comprising 2 neurons and n outputs. Such a network is taught that the value as close as possible to the value which is at the i-th input appeared at the i-th output. The interlayer consisting of 2 neurons represents the screen in such a way that the value of outputs of these neurons denote directly 2 coordinates of the screen. Such a taught network performs compression of the n-dimensional space to the 2-dimensional one, and then decompression back to the n-dimensional space. Figure 8 presents the view obtained using the autoassociative neural network. In this method, as the network learns, we obtain one view. Sometimes, a different view can be obtained by drawing different initial weight values. However, the number of views obtained in this way is very limited. The MDS method consists in such a transformation of the multidimensional input space into the two-dimensional target space representing a screen that for each pair of points their mutual distance in the input space is as close as possible to the distance of their images in the target space. This can be done by the random generation of the initial location of each point image in the target space and iterative change of location of each of these images. Such a change of location should proceed in such a way that the criterion presented above is met to the highest extent. Figure 9 presents the view obtained using the MDS method. In this method, as we obtain better match to the criterion, we obtain one view. Sometimes, a differ-

Fig. 4. The view of the analyzed five-dimensional fractal obtained by means of the perspective-based observational tunnels method. The view seems like a flying dragonfly.

Fig. 5. The view of the analyzed five-dimensional fractal obtained by means of the perspective-based observational tunnels method. The view seems like a striped shrimp.

Fig. 6. The view of the analyzed five-dimensional fractal obtained by means of the perspective-based observational tunnels method. The view seems like a human shape.

ent view can be obtained by the generation of different random values specifying initial locations of point images. However, the number of views obtained as a result of matching to the criterion is very limited.

Fig. 7. The view of the analyzed five-dimensional fractal obtained by means of the PCA method.

Fig. 8. The view of the analyzed five-dimensional fractal obtained by means of the autoassociative neural networks method.

Fig. 9. The view of the analyzed five-dimensional fractal obtained by means of the MDS method.

7   Conclusions

During the conducted investigations, it was found that the perspective-based observational tunnels method serving to visualize multidimensional data allows to obtain significantly different views of a five-dimensional fractal. During the observation, the point of view was interactively changed through a gradual change in observational parameters, the so-called observational plane, direction of projection, tunnel radius and perspective coefficient. Therefore, each next view was obtained from the previous one gradually through slight changes. It can be said that it seemed that the transformation of one shape into another one occurred. Initially, it was difficult to accept that this is still the same object. Only after a longer period of observation of such a fractal from various perspectives some kind of adaptation to the changes occurred. It seems that our brains need such a new type of views and their changes to be transmitted for a longer time to get a possibility of at least partial understanding of the observed multidimensional objects. The purpose of this sort of visualization can also be to adapt and teach our mind to percept, recognize and even understand objects of a higher number of dimensions than 3.

According to the conducted investigations, the perspective-based observational tunnels method allows to present views of multidimensional fractals as representatives of artificially generated data on the screen. It is the new approach because this method has never been used before on this kind of artificially generated and at the same time such complicated data. It follows from this that the perspective-based observational tunnels method can constitute the tool allowing to explore such a space filled with artificially generated data through obtaining its views. Therefore, this method can be used to exercise the perception of multidimensional data through a gradual increase in the degree of complexity of artificially generated data on which this method will be used.

During the conducted investigations, it was found that the multidimensional fractal observed from various perspectives has such different views that can replace views of several two-dimensional fractals. Thus, it can be applied in various areas where various two-dimensional fractals are used, e.g. in image compression. However, the creation of such a fractal involves difficulties in finding such a perspective which would be appropriate for a certain purpose. The paper shows that searching for views of multidimensional fractals can be conducted by means of methods of multidimensional data visualization. It is possible to recognize shapes of created fractals from various perspectives. The paper shows the example of views of a five-dimensional fractal, but it is possible to conduct a similar analysis for fractals of a higher number of dimensions.

The number of views of multidimensional fractals possible to obtain using the PCA method is limited to one and in the case of the MDS method and autoassociative neural networks, it is very limited. The perspective-based observational tunnels method does not have such a limitation.

The perspective-based observational tunnels method has a linear time complexity in relation to the number of multidimensional points and in relation to the number of dimensions. This is the best result for methods presented in the paper.

References

1. Jamroz, D.: The perspective-based observational tunnels method: a new method of multidimensional data visualization. Inf. Vis. 16(4), 346–360 (2017)
2. Jamroz, D., Niedoba, T.: Comparison of selected methods of multi-parameter data visualization used for classification of coals. Physicochem. Probl. Mineral Process. 51(2), 769–784 (2015)
3. Hotelling, H.: Analysis of a complex of statistical variables into principal components. J. Educ. Psychol. 24, 417–441, 498–520 (1933)
4. Jolliffe, I.T.: Principal Component Analysis, Series. Springer Series in Statistics, 2nd edn. Springer, New York (2002). https://doi.org/10.1007/b98835

5. Kruskal, J.B.: Multidimensional scaling by optimizing goodness of fit to a non-metric hypothesis. Psychometrika **29**, 1–27 (1964)
6. Kim, S.S., Kwon, S., Cook, D.: Interactive visualization of hierarchical clusters using MDS and MST. Metrika **51**, 39–51 (2000)
7. Assa, J., Cohen-Or, D., Milo, T.: RMAP: a system for visualizing data in multidimensional relevance space. Vis. Comput. **15**(5), 217–234 (1999)
8. Niedoba, T.: Application of relevance maps in multidimensional classification of coal types. Arch. Min. Sci. **60**(1), 93–107 (2015)
9. Inselberg, A.: Parallel Coordinates: VISUAL Multidimensional Geometry and its Applications. Springer, New York (2009). https://doi.org/10.1007/978-0-387-68628-8
10. Akers, S.B., Horel, D., Krisnamurthy, B.: The star graph: an attractive alternative to the n-cube. In: Proceedings of International Conference On Parallel Processing, pp. 393–400. Pensylvania State University Press (1987)
11. Aldrich, C.: Visualization of transformed multivariate data sets with autoassociative neural networks. Pattern Recogn. Lett. **19**(8), 749–764 (1998)
12. Jamroz, D.: Application of multi-parameter data visualization by means of autoassociative neural networks to evaluate classification possibilities of various coal types. Physicochem. Probl. Mineral Process. **50**(2), 719–734 (2014)
13. Kohonen, T.: Self Organization and Associative Memory. Springer, Heidelberg (1989). https://doi.org/10.1007/978-3-642-88163-3
14. Jamroz, D., Niedoba, T.: Application of multidimensional data visualization by means of self-organizing Kohonen maps to evaluate classification possibilities of various coal types. Arch. Min. Sci. **60**(1), 39–51 (2015)
15. Fisher, Y.: Fractal Image Compression. Springer, New York (1995). https://doi.org/10.1007/978-1-4612-2472-3
16. Kouzani, A.Z.: Classification of face images using local iterated function systems. Mach. Vis. Appl. **19**, 223–248 (2008)
17. Mozaffari, S., Facz, K., Faradji, F.: One dimensional fractal coder for online signature recognition. In: International Conference on Pattern Recognition, pp. 857–860 (2008)
18. Gdawiec, K.: Shape recognition using partitioned iterated function systems. In: Cyran, K.A., Kozielski, S., Peters, J.F., Stańczyk, U., Wakulicz-Deja, A. (eds.) Man-Machine Interactions. AISC, vol. 59, pp. 451–458. Springer, Heidelberg (2009). https://doi.org/10.1007/978-3-642-00563-3_48
19. Barnsley, M.: Fractals Everywhere. Academic Press, Boston (1988)

Estimation of Probability Density Function, Differential Entropy and Other Relative Quantities for Data Streams with Concept Drift

Maciej Jaworski[1]([✉]), Patryk Najgebauer[1], and Piotr Goetzen[2,3]

[1] Institute of Computational Intelligence, Czestochowa University of Technology,
Armii Krajowej 36, 42-200 Czestochowa, Poland
{maciej.jaworski,patryk.najgebauer}@iisi.pcz.pl
[2] Information Technology Institute, University of Social Sciences,
90-113 Łódź, Poland
goetzen@swspiz.pl
[3] Clark University, Worcester, MA 01610, USA

Abstract. In this paper estimators of nonstationary probability density function are proposed. Additionally, applying the trapezoidal method of numerical integration, the estimators of two information-theoretic measures are presented: the differential entropy and the Renyi's quadratic differential entropy. Finally, using an analogous methodology, estimators of the Cauchy-Schwarz divergence and the probability density function divergence are proposed, which are used to measure the differences between two probability density functions. All estimators are proposed in two variants: one with the sliding window and one with the forgetting factor. Performance of all the estimators is verified using numerical simulations.

Keywords: Data stream · Concept drift · Density estimation
Differential entropy · Kernel function

1 Introduction

Development of machine learning algorithms designed for time-changing data streams is a very challenging task of data mining [1,3,12,15,22]. The reason is that data streams are of potentially infinite size. Therefore, each data element has to be processed at most once because of the limited amounts of memory. Moreover, the distribution of data can often change over time, which is known in the literature under the name 'concept drift' [2,5,6,9,11]. The development of algorithms able to deal with nonstationary data is very important from the practical point of view, since they are applicable to many areas of data science [17,24–26].

In this paper, we analyze the problem of probability density function estimation for data streams with concept drift. Having a good estimator of the density, one can also estimate the differential entropy of the distribution as well as other relative quantities, e.g. the Renyi's quadratic entropy. Moreover, using estimators of two different probability density functions it is also possible to estimate some divergence measures between them, e.g. the Cauchy-Schwarz divergence or simply the probability density function (PDF) divergence.

We assume that the data elements X_i, $i = 1, 2, \ldots$, are derived from the probability distribution for which the probability density function is given by $f(x)$. The commonly known estimator of function $f(x)$ is based on Parzen kernel functions [4,18]. This estimator is also known in the literature under the name of probabilistic neural networks [16,23] and it is given by

$$\hat{f}_n(x) = \frac{1}{n} \sum_{i=1}^{n} \frac{1}{h_i} K\left(\frac{x - X_i}{h_i}\right),\tag{1}$$

where $K(u)$ is a kernel function. There are many possible kernel functions. In this paper we use the Epanechnikov's kernel [7], the uniform kernel and the triangular kernel, given below in formulas (2), (3) and (4), respectively

$$K(u) = \begin{cases} 0.75\left(1 - u^2\right), & |u| \le 1, \\ 0, & |u| > 1. \end{cases}\tag{2}$$

$$K(u) = \begin{cases} 0.5, & |u| \le 1, \\ 0, & |u| > 1, \end{cases}\tag{3}$$

$$K(u) = \begin{cases} 1 - |u|, & |u| \le 1, \\ 0, & |u| > 1. \end{cases}\tag{4}$$

The elements of the sequence h_i are called bandwidths and their values determine the shape (width) of the kernel. The commonly used formula for this sequence is given as follows

$$h_i = D i^{-H},\tag{5}$$

where D and H are positive real numbers. We will also apply this formula for h_i in this paper. Estimator (1) cannot be directly applied to concept-drifting data streams. However, it will stand as a starting point for estimators proposed in this paper.

It should be noted that the idea of nonparametric estimation using kernel functions (also in the form of orthogonal series) was also used in the regression task [8,10,14,20]. Applying the stochastic approximation to the nonparametric regression function estimators makes them able to work also in nonstationary environments [19,21]. The sliding window technique and the forgetting mechanism were also successfully applied to the task of nonparametric nonstationary regression function tracking [13].

The rest of the paper is organized as follows. In Sect. 2 two probability density function estimators for time-changing densities are proposed. In one of them the sliding window approach is used, whereas the second one applies the forgetting factor. In Sect. 3 it is shown how to use the estimators of density function to estimate the differential entropy and the Renyi's quadratic differential entropy of the data distribution. The estimators for the Cauchy-Schwarz divergence and the PDF divergence are presented in Sect. 4. In Sect. 5 the performance of the proposed estimators is demonstrated in a series of numerical experiments on synthetic datasets. Finally, Sect. 6 concludes the paper.

2   Density Estimation for Data Streams with Concept Drift

The main advantage of estimator (1) is that it can be expressed in the recurrent manner, which is a very desired feature in data stream mining algorithms

$$\hat{f}_n(x) = \frac{n-1}{n}\hat{f}_{n-1}(x) + \frac{1}{nh_n}K\left(\frac{x-X_n}{h_n}\right). \tag{6}$$

In this form, all data elements X_i are equally important. In data streams, however, the distribution of data can change over time. Therefore, it is desired to treat the most recent data as more important than the data from the past. It means that recent data should be included in the estimator with higher weights. We propose to do it in two manners: by using the sliding window and by applying the forgetting factor.

2.1   The Sliding Window Approach

By applying the sliding window with size W only the last W data elements are taken into account (if the current index of data element n is lower than W, then the estimator is equivalent to estimator (1)). The appropriate estimator can be formulated as follows

$$\overline{f}_n(x;W) = \frac{\sum_{i=\max\{n-W+1,1\}}^{n}\frac{1}{\overline{h}_i(W)}K\left(\frac{x-X_i}{\overline{h}_i(W)}\right)}{\min\{n,W\}} \tag{7}$$

The number of data elements, which affect the estimator after n data elements, is given by $\min\{n,W\}$. Then, by analogy to (5), the formula for the sequence of bandwidth sequence can be given as follows

$$\overline{h}_i(W) = D\left(\min\{i,W\}\right)^{-H}. \tag{8}$$

Obviously, estimator (7) can be reformulated in the recurrent way

$$\overline{f}_n(x;W) = \begin{cases} \frac{n-1}{n}\overline{f}_{n-1}(x;W) + \frac{K\left(\frac{x-X_n}{\overline{h}_n(W)}\right)}{n\overline{h}_n(W)}, & n \leq W, \\[3mm] \overline{f}_{n-1}(x;W) - \frac{K\left(\frac{x-X_{n-W}}{\overline{h}_{n-W}(W)}\right)}{W\overline{h}_{n-W}(W)} + \frac{K\left(\frac{x-X_n}{\overline{h}_n(W)}\right)}{W\overline{h}_n(W)}, & n > W. \end{cases} \tag{9}$$

2.2   The Forgetting Factor Approach

In the estimator with forgetting factor, all data elements are taken into account. However, the recent data elements receive higher weights than the older ones. The estimator is given as follows

$$\widetilde{f}_n(x;\lambda) = \frac{\sum_{i=1}^{n} \lambda^{(n-i)} \frac{1}{h_i(\lambda)} K\left(\frac{x-X_i}{h_i(\lambda)}\right)}{\sum_{i=1}^{n} \lambda^{(n-i)}}, \tag{10}$$

where $0 < \lambda < 1$ is a forgetting factor. For estimators (1) and (7) it was easy to point out the number of elements which participate in the value of estimator. In the forgetting factor approach, it is slightly more complicated. Since the i-th data element takes part in the estimator with weight λ^{n-i} (after processing n data elements), then the number of data elements can be defined as a sum of subsequent weights

$$M(n) = \sum_{i=1}^{n} \lambda^{n-i} = \frac{1-\lambda^n}{1-\lambda}. \tag{11}$$

Then, by analogy to (5) and (8), the sequence of bandwidths can be proposed as follows

$$\widetilde{h}_n(\lambda) = D\left(M(n)\right)^{-H} = D\left(\frac{1-\lambda^n}{1-\lambda}\right)^{-H}. \tag{12}$$

As previously, estimator (10) can be easily formulated in the recurrent manner

$$\widetilde{f}_n(x;\lambda) = \frac{\lambda - \lambda^{(n-1)}}{1-\lambda^n} \widetilde{f}_{n-1}(x;\lambda) + \frac{1-\lambda}{1-\lambda^n} \frac{1}{\widetilde{h}_n(\lambda)} K\left(\frac{x-X_n}{\widetilde{h}_n(\lambda)}\right) \tag{13}$$

3   Estimation of Information-Theoretic Quantities

In this section estimators for the differential entropy and Renyi's quadratic differential entropy will be proposed. To calculate numerically any integral $\int_{-\infty}^{\infty} g(x)dx$ a simple trapezoidal rule will be applied. Since it is not possible to deal with infinities, the integral bounds will be reduced from $(-\infty;\infty)$ to $[x_0;x_N]$ such that $g(x) \approx 0$ for $x < x_0$ and for $x > x_N$. The interval $[x_0;x_N]$ is divided into N equidistant points

$$x_i = x_0 + i\Delta x, \tag{14}$$

where

$$\Delta x = \frac{x_N - x_0}{N-1}. \tag{15}$$

Taking into account the reduction of integral bounds and the trapezoidal rule, the integral is approximated as follows

$$\int_{-\infty}^{\infty} g(x)dx \approx \int_{x_0}^{x_N} g(x)dx \approx \left(\left(\sum_{i=1}^{N-1} g(x_i)\right) + \frac{g(x_0)}{2} + \frac{g(x_N)}{2}\right)\Delta x. \tag{16}$$

The points x_0 and x_N take part in the sum with weight $\frac{1}{2}$, whereas the middle points have weight 1. To simplify the further notations, integral (16) will be expressed as follows

$$\int_{-\infty}^{\infty} g(x)dx \approx \int_{x_0}^{x_N} g(x)dx \approx \sum_{i=0}^{N} a_i g(x_i) \Delta x, \tag{17}$$

where

$$a_i = \begin{cases} 1, & i \in \{1, \ldots, N-1\}, \\ \frac{1}{2}, & i \in \{0, N\}. \end{cases} \tag{18}$$

3.1   Differential Entropy Estimation

The differential entropy of probability density function is defined as follows

$$H(f) = -\int_{-\infty}^{\infty} f(x) \log f(x)dx. \tag{19}$$

To estimate the differential entropy, estimators (7) and (13) will be applied to scheme (17). For the sliding window approach, the following estimator is obtained

$$\overline{H}_n(f; W) = \sum_{i=0}^{N} a_i \overline{f}_n(x_i; W) \log \left(\overline{f}_n(x_i; W) \right) \Delta x. \tag{20}$$

In the case of forgetting factor the differential entropy estimator is given as follows

$$\widetilde{H}_n(f; \lambda) = \sum_{i=0}^{N} a_i \widetilde{f}_n(x_i; \lambda) \log \left(\widetilde{f}_n(x_i; \lambda) \right) \Delta x. \tag{21}$$

3.2   Renyi's Quadratic Differential Entropy Estimation

Another information-theoretic quantity related to the differential entropy is the Renyi's quadratic differential entropy. It is defined as follows

$$H_2(f) = -\log \left(\int_{-\infty}^{\infty} f^2(x)dx \right). \tag{22}$$

As previously, taking into account estimators (7) and (13) and applying them to integration scheme (17) one obtains appropriate estimators based on the sliding window and the forgetting factor, respectively

$$\overline{H}_{2,n}(f; W) = -\log \left(\sum_{i=0}^{N} a_i \left(\overline{f}_n(x_i; W) \right)^2 \Delta x \right). \tag{23}$$

$$\widetilde{H}_{2,n}(f; \lambda) = -\log \left(\sum_{i=0}^{N} a_i \left(\widetilde{f}_n(x_i; \lambda) \right)^2 \Delta x \right). \tag{24}$$

4   Estimation of Probability Density Divergence Measures

In this section, we assume that there are two probability density functions $f_1(x)$ and $f_2(x)$ from which two separate substreams of data elements are derived (each probability function can change over time). The difference between the two distribution can be calculated using some divergence measures. In this paper, we consider the Cauchy-Schwarz divergence and the probability density function divergence. Both measures are defined as appropriate integrals, hence the corresponding estimators can be based on scheme (17).

4.1   Estimation of the Cauchy-Schwarz Divergene

The Cauchy-Schwarz divergence between two density functions $f_1(x)$ and $f_2(x)$ is defined as follows

$$CS(f_1, f_2) = \frac{\int_{-\infty}^{\infty} f_1(x) f_2(x) dx}{\sqrt{\int_{-\infty}^{\infty} f_1^2(x) dx} \sqrt{\int_{-\infty}^{\infty} f_2^2(x) dx}}. \tag{25}$$

Let $\overline{f}_{1,n}(x_i; W)$ denote estimator (7) of $f_1(x)$ and let $\overline{f}_{2,n}(x_i; W)$ be the analogous estimator of $f_2(x)$. Then the estimator of Cauchy-Schwarz divergence based on the sliding window can be formulated as follows (after applying approximation (17))

$$\overline{CS}_n(f_1, f_2; W) = \frac{\sum_{i=0}^{N} a_i \overline{f}_{1,n}(x_i; W) \overline{f}_{2,n}(x_i; W) \Delta x}{\sqrt{\sum_{i=0}^{N} a_i \left(\overline{f}_{1,n}(x_i; W)\right)^2} \sqrt{\sum_{i=0}^{N} a_i \left(\overline{f}_{2,n}(x_i; W)\right)^2}}. \tag{26}$$

If $\tilde{f}_{1,n}(x_i; \lambda)$ and $\tilde{f}_{2,n}(x_i; \lambda)$ denote estimators (10) for $f_1(x)$ and $f_2(x)$, respectively, then the Cauchy-Schwarz divergence estimator with the forgetting factor is given by

$$\widetilde{CS}_n(f_1, f_2; \lambda) = \frac{\sum_{i=0}^{N} a_i \tilde{f}_{1,n}(x_i; \lambda) \tilde{f}_{2,n}(x_i; \lambda) \Delta x}{\sqrt{\sum_{i=0}^{N} a_i \left(\tilde{f}_{1,n}(x_i; \lambda)\right)^2} \sqrt{\sum_{i=0}^{N} a_i \left(\tilde{f}_{2,n}(x_i; \lambda)\right)^2}}. \tag{27}$$

4.2   Estimation of the Probability Density Function Divergence

Similarly to the previous cases, the same procedure can be applied to derive estimators for the probability density function divergence between densities $f_1(x)$ and $f_2(x)$, which is simply given by

$$\Delta PDF(f_1, f_2) = \int_{-\infty}^{\infty} (f_1(x) - f_2(x))^2 \, dx. \tag{28}$$

The corresponding estimators based on the sliding window and the forgetting factor are given by formulas (29) and (30), respectively

$$\overline{\Delta PDF}_n(f_1, f_2; W) = \sum_{i=0}^{N} a_i \left(\overline{f}_{1,n}(x_i; W) - \overline{f}_{2,n}(x_i; W)\right)^2 \Delta x, \tag{29}$$

$$\widetilde{\Delta PDF}_n(f_1, f_2; \lambda) = \sum_{i=0}^{N} a_i \left(\tilde{f}_{1,n}(x_i; \lambda) - \tilde{f}_{2,n}(x_i; \lambda) \right)^2 \Delta x. \tag{30}$$

5   Experimental Results

In this section the proposed estimators were examined experimentally. In all the experiments the size of training dataset was 10000. To measure quantitatively the performance of any density estimator $\hat{f}_n(x)$ the Mean Squared Error (MSE) was computed for the grid of points defined by (14) and (15) with $x_0 = -3$ and $x_N = 3$, where $N = 101$. The MSE for estimator $\hat{f}_n(x)$ is calculated as follows

$$MSE\left(\hat{f}_n\right) = \frac{1}{N} \sum_{i=1}^{N} \left(f_n(x) - \hat{f}_n(x) \right)^2. \tag{31}$$

Parameters D and H of sequences (8) and (12) were set to $D = 2$ and $H = 0.3$. In the first experiment, the performance of probability density function estimators was investigated. Data elements were generated synthetically according to the time-changing Gaussian distribution

$$f_n(x) = \frac{1}{\sqrt{\pi}} \exp\left(-\left(x + 1 - 2\frac{n}{10000} \right)^2 \right). \tag{32}$$

Estimator (7) was run with $W = 510$ whereas λ for estimator (10) was set to $\lambda = 0.9957$ (this values were chosen experimentally and ensure the highest average MSE obtained for values of n from 100 up to 10000). The MSE for estimators $\overline{f}_n(x; 510)$ and $\tilde{f}_n(x; 0.9957)$ are presented in Fig. 1. The comparison between the investigated estimators with the true density function at the end of the simulation (i.e. for $n = 10000$) is shown in Fig. 2.

The results demonstrate that the proposed estimators track satisfactorily well the time-varying density function.

Density function (32), although is time-varying, does not change its shape. Since the entropy of such distribution is constant over time, it seems to be uninteresting to demonstrate the performance of entropy estimators in this case. Therefore, in the second experiment, another nonstationary Gaussian probability density function with time-varying variance was analyzed

$$f_n(x) = \frac{1}{\sqrt{2\pi \left(0.5 + 0.3 \sin\left(\frac{2\pi n}{10000} \right) \right)}} \exp\left(-\frac{x^2}{2 \left(0.5 + 0.3 \sin\left(\frac{2\pi n}{10000} \right) \right)} \right) \tag{33}$$

Differential entropy and Renyi's quadratic differential entropy estimators (20), (23) and (21), (24) were computed with $W = 700$ and $\lambda = 0.997$, respectively. The comparison of the differential entropy estimators with the true entropy of density function (33) is shown in Fig. 3. Analogous results concerning the Renyi's quadratic entropy are presented in Fig. 4.

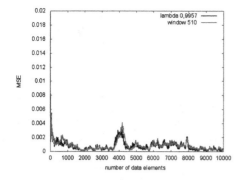

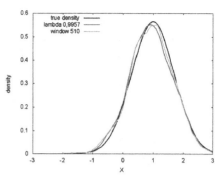

Fig. 1. The MSE values in a function of the number of processed data elements for estimators $\overline{f}_n(x; 510)$ and $\widetilde{f}_n(x; 0.9957)$.

Fig. 2. Comparison of estimators $\overline{f}_n(x; 510)$ and $\widetilde{f}_n(x; 0.9957)$ with the true density function given by (32) for $n = 10000$.

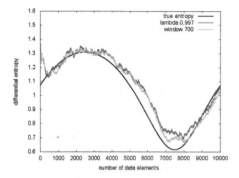

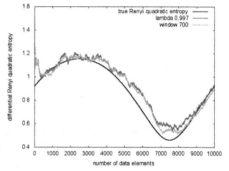

Fig. 3. The differential entropy in a function of the number of processed data elements for estimators $\overline{H}_n(x; 700)$ and $\widetilde{H}_n(x; 0.997)$ compared to the true differential entropy of the density function given by (33).

Fig. 4. Renyi's quadratic differential entropy in a function of the number of processed data elements for estimators $\overline{H}_{2,n}(f; 700)$ and $\widetilde{H}_{2,n}(f; 0.997)$ compared to the true Renyi's quadratic entropy of the density function $f(x)$ given by (33).

At the very beginning, when the density estimators are not learned properly, the entropy of estimators is significantly higher than the true value. However, for $n > 500$ estimators of both types (i.e. with the sliding window and the forgetting factor) follow the true value of considered information-theoretic measures quite well.

In the last experiment, the performance of estimators of divergence measures was investigated. The following two nonstationary Gaussian probability density functions were considered

$$f_{1,n}(x) = \frac{1}{\sqrt{\pi}} \exp\left(-\left(x + 1 - 2\frac{n}{10000}\right)^2\right), \tag{34}$$

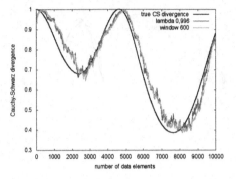

Fig. 5. The Cauchy-Schwarz divergence in a function of the number of processed data elements for estimators $\overline{CS}_n(f_1, f_2; 700)$ and $\widehat{CS}_n(f_1, f_2; 0.997)$ compared to the true Cauchy-Schwarz divergence between density functions (34) and (35).

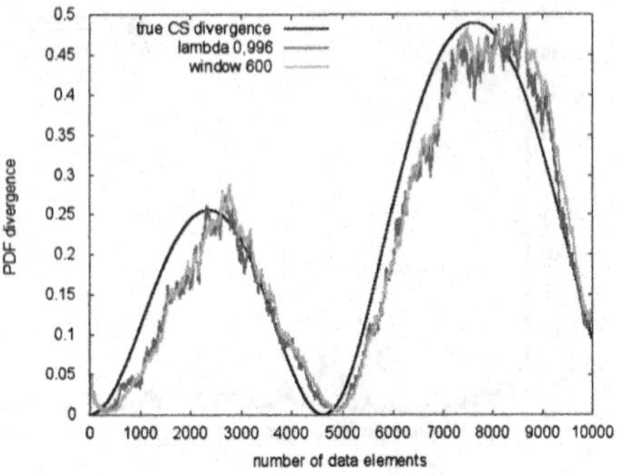

Fig. 6. The PDF divergence in a function of the number of processed data elements for estimators $\overline{\Delta PDF}_n(f_1, f_2; 700)$ and $\widehat{\Delta PDF}_n(f_1, f_2; 0.997)$ compared to the true Cauchy-Schwarz divergence between density functions (34) and (35).

$$f_{2,n}(x) = \frac{1}{\sqrt{\pi}} \exp\left(-\left(x + 1 - 2\sin\left(\frac{2\pi n}{10000}\right)\right)^2\right). \tag{35}$$

In the estimators with sliding window $W = 600$ was used. It the estimators with forgetting factor $\lambda = 0.996$ was applied. The comparison between estimators $\overline{CS}_n(f_1, f_2; 600)$ and $\widehat{CS}_n(f_1, f_2; 0.996)$ with the true Cauchy-Schwarz divergence is demonstrated in Fig. 5. Analogous results obtained for PDF divergence estimators are presented in Fig. 6.

As can be seen, in both cases the tracking properties of estimators with the sliding windows and the forgetting factor are very satisfying.

6   Conclusions

In this paper, the nonstationary probability density function estimators were proposed. To ensure the tracking properties of the estimators two approaches were applied. It the first one, the sliding window are used in which only a number of recent data elements is taken into account. In the second approach the forgetting factor is applied, in which the most recent data affect the estimator value with higher weights than the old, possibly depreciated data elements. Additionally, the estimators of two information-theoretic measures were proposed: the differential entropy and the Renyi's quadratic differential entropy. These estimators are based on the corresponding probability density estimators are on the trapezoidal method of numerical integration. Moreover, using an analogous methodology, estimators of the Cauchy-Schwarz divergence and the probability density function divergence were proposed, which are used to measure the differences between two probability density functions. All estimators were verified experimentally which demonstrated that their performance is satisfactory.

Acknowledgments. This work was supported by the Polish National Science Centre under Grant No. 2014/15/B/ST7/05264.

References

1. Bilski, J., Smolag, J.: Parallel architectures for learning the RTRN and Elman dynamic neural networks. IEEE Trans. Parallel Distrib. Syst. **26**(9), 2561–2570 (2015)
2. Chang, O., Constante, P., Gordon, A., Singana, M.: A novel deep neural network that uses space-time features for tracking and recognizing a moving object. J. Artif. Intell. Soft Comput. Res. **7**(2), 125–136 (2017)
3. Devi, V.S., Meena, L.: Parallel MCNN (PMCNN) with application to prototype selection on large and streaming data. J. Artif. Intell. Soft Comput. Res. **7**(3), 155–169 (2017)
4. Devroye, L.P.: On the pointwise and the integral convergence of recursive kernel estimates of probability densities. Utilitas Math. (Canada) **15**, 113–128 (1979)
5. Ditzler, G., Roveri, M., Alippi, C., Polikar, R.: Learning in nonstationary environments: a survey. IEEE Comput. Intell. Mag. **10**(4), 12–25 (2015)
6. Duda, P., Jaworski, M., Rutkowski, L.: On ensemble components selection in data streams scenario with reoccurring concept-drift. In: 2017 IEEE Symposium Series on Computational Intelligence (SSCI), pp. 1821–1827, November 2017
7. Epanechnikov, V.A.: Non-parametric estimation of a multivariate probability density. Theory Probab. Appl. **14**(1), 153–158 (1969)
8. Galkowski, T., Rutkowski, L.: Nonparametric fitting of multivariate functions. IEEE Trans. Autom. Control **31**(8), 785–787 (1986)
9. Gama, J., Žliobaitė, I., Bifet, A., Pechenizkiy, M., Bouchachia, A.: A survey on concept drift adaptation. ACM Comput. Surv. (CSUR) **46**(4), 44 (2014)
10. Greblicki, W., Pawlak, M.: Nonparametric System Identification. Cambridge University Press, Cambridge (2008)
11. Jaworski, M., Duda, P., Rutkowski, L.: On applying the Restricted Boltzmann Machine to active concept drift detection. In: 2017 IEEE Symposium Series on Computational Intelligence (SSCI), pp. 3512–3519, November 2017
12. Jaworski, M., Duda, P., Rutkowski, L.: New splitting criteria for decision trees in stationary data streams. IEEE Trans. Neural Netw. Learn. Syst. **PP**(99), 1–14 (2018)
13. Jaworski, M., Duda, P., Rutkowski, L., Najgebauer, P., Pawlak, M.: Heuristic regression function estimation methods for data streams with concept drift. In: Rutkowski, L., Korytkowski, M., Scherer, R., Tadeusiewicz, R., Zadeh, L.A., Zurada, J.M. (eds.) ICAISC 2017. LNCS (LNAI), vol. 10246, pp. 726–737. Springer, Cham (2017). https://doi.org/10.1007/978-3-319-59060-8_65
14. Krzyzak, A., Pawlak, M.: The pointwise rate of convergence of the kernel regression estimate. J. Stat. Plan. Inference **16**, 159–166 (1987)
15. Lemaire, V., Salperwyck, C., Bondu, A.: A survey on supervised classification on data streams. In: Zimányi, E., Kutsche, R.-D. (eds.) eBISS 2014. LNBIP, vol. 205, pp. 88–125. Springer, Cham (2015). https://doi.org/10.1007/978-3-319-17551-5_4

16. Napoli, C., Pappalardo, G., Tramontana, E., Nowicki, R.K., Starczewski, J.T., Woźniak, M.: Toward work groups classification based on probabilistic neural network approach. In: Rutkowski, L., Korytkowski, M., Scherer, R., Tadeusiewicz, R., Zadeh, L.A., Zurada, J.M. (eds.) ICAISC 2015. LNCS (LNAI), vol. 9119, pp. 79–89. Springer, Cham (2015). https://doi.org/10.1007/978-3-319-19324-3_8

17. Notomista, G., Botsch, M.: A machine learning approach for the segmentation of driving maneuvers and its application in autonomous parking. J. Artif. Intell. Soft Comput. Res. 7(4), 243–255 (2017)

18. Parzen, E.: On estimation of probability density function and mode. Ann. Math. Stat. 33, 1065–1076 (1962)

19. Pietruczuk, L., Rutkowski, L., Jaworski, M., Duda, P.: The Parzen kernel approach to learning in non-stationary environment. In: 2014 International Joint Conference on Neural Networks (IJCNN), pp. 3319–3323 (2014)

20. Rutkowski, L.: Sequential estimates of a regression function by orthogonal series with applications in discrimination. In: Révész, P., Schatterer, L., Zolotarev, V.M. (eds.) The First Pannonian Symposium on Mathematical Statistics. LNS, vol. 8, pp. 236–244. Springer, New York (1981). https://doi.org/10.1007/978-1-4612-5934-3_21

21. Rutkowski, L.: Generalized regression neural networks in time-varying environment. IEEE Trans. Neural Netw. 15, 576–596 (2004)

22. Rutkowski, L., Jaworski, M., Pietruczuk, L., Duda, P.: A new method for data stream mining based on the misclassification error. IEEE Trans. Neural Netw. Learn. Syst. 26(5), 1048–1059 (2015)

23. Specht, D.F.: Probabilistic neural networks. Neural Netw. 3(1), 109–118 (1990)

24. Yan, P.: Mapreduce and semantics enabled event detection using social media. J. Artif. Intell. Soft Comput. Res. 7(3), 201–213 (2017)

25. Yang, S., Sato, Y.: Swarm intelligence algorithm based on competitive predators with dynamic virtual teams. J. Artif. Intell. Soft Comput. Res. 7(2), 87–101 (2017)

26. Zalasiński, M., Cpałka, K.: New algorithm for on-line signature verification using characteristic hybrid partitions. In: Wilimowska, Z., Borzemski, L., Grzech, A., Świątek, J. (eds.) ISAT 2015. AISC, vol. 432, pp. 147–157. Springer, Cham (2016). https://doi.org/10.1007/978-3-319-28567-2_13

System for Building and Analyzing Preference Models Based on Social Networking Data and SAT Solvers

Radosław Klimek$^{(\boxtimes)}$

AGH University of Science and Technology,
Al. Mickiewicza 30, 30-059 Kraków, Poland
rklimek@agh.edu.pl

Abstract. Discovering and modeling preferences has an important meaning in the modern IT systems, also in the intelligent and multi-agent systems which are context sensitive and should be proactive. The preference modelling enables understanding the needs of objects working within intelligent spaces, in an intelligent city. There was presented a proposal for a system, which, based on logical reasoning and using advanced SAT solvers, is able to analyze data from social networks for preference determination in relation to its own presented offers from different domains. The basic algorithms of the system were presented as well as the validation of practical application.

Keywords: Preference model · SAT solvers · Social networking data
Facebook · Twitter

1 Introduction

Modelling users' preferences enables to support decision making. The choice of decision can be carried on in an interactive process which describes user's aspirations and goals. Those aspects can have a fundamental meaning in intelligent systems which need to work pro-actively, understand their own context and undertake actions which will improve comfort and safety of city residents. Also the companies try to get data to build profiles of their present and potential clients. The building process of such profiles can be automatized which helps to avoid costly and time-consuming market surveys.

The priceless source of information are social networking platforms and data in form of posts, photos, statements, opinions, circles of friends, etc. This type of information is quite easily accessible and hardly removable.

The aim of this work is to propose a system which builds online and in real time preference models. Later on, those models can be adjusted to a particular offer by logical reasoning based on accessible SAT solvers. The project and system can find a wider use in relation to multi agent systems where preference models are built on the basis of results which have already been required. The

© Springer International Publishing AG, part of Springer Nature 2018
L. Rutkowski et al. (Eds.): ICAISC 2018, LNAI 10842, pp. 387–397, 2018.
https://doi.org/10.1007/978-3-319-91262-2_35

next stage is choosing, in logical reasoning process, the agents which meet best certain expectations. All of the processes are carried on fully automatically.

Recommendation system are always at the center of interest, see [12], where a preference model are build to find the user neighbor set. Also, mining online reviews and tweets for predicting [13] have an impact on the future sales. Preference models support sensing and decision-making behavior analysis [1]. However, there is a lack of works with preference models involving social media mining and SAT-based analysis. Last but not least, preferences and their models can be a part of smart cities [5–8,14].

2   Functional Model

At the beginning the functional model of the system will be presented, see Fig. 1. Some used terms will be discussed in the next section. The following actors are taken into consideration:

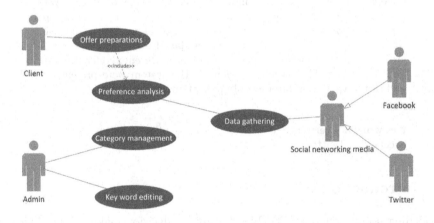

Fig. 1. The use case diagram for the proposed system

- **Client** – a person or another system sending request to the system, the goal is to get an offer for presented preferences. It includes a possibility of giving authorization data which enables logging and getting an access to data on social networks which are the sources of basic (mass) data.
- **Admin** – a person who is empowered to modify knowledge database, categories and particular key words, adding the new and editing the already existing ones, deleting outdated records or incorrect entries.
- **Social networking media** – networking site which gathers and shares user's data. It is a source of basic pieces of data which is furtherly processed and the preferences are formulated on the basis of logical reasoning.

The following use cases are offered:

- **Offer preparations** – preparation of an offer from a particular branch, for example: tourism, real estate market, etc. This data is transformed into a version which enables logical reasoning.
- **Preference analysis** – the process of logical reasoning which aim is to compare data collected from different social networks and data from prepared offers. Among the searched positions are those which have the maximum of compliance in relation to other offers and already collected data.
- **Data gathering** – Data gathering - collecting the basic data from particular social networks on the basis of well-known and widely accessible searching algorithms, mentioned in Sect. 3.
- **Key word editing** – editing single key words, adding the new ones, deleting outdated words and modifying them.
- **Category management** – Category management - the process of managing the categories which enable sorting the key words.

The following scenarios are build, see Tables 1, 2, 3, 4 and 5. Scenarios might form the basis for modelling activity diagrams, see [4,9,10].

Table 1. Scenario: adjusting the offer to the user's preferences

Adjusting the offer to the user's preferences:
Trigger event: a client sends a request to the system
Preconditions: the system should have a necessary amount of offers and key words in its database
Scenario:
1. Client sends a request to the system together with giving a token
2. «include» Use case: searching for user's preferences
3. Creating a logical formula on the basis of user's preferences and the data from an offer
4. Finding formula quantifying
5. The system sends back preferences and a proposed offer
Postconditions for success: in a request there should be sent active tokens of the user
Postconditions for failure: disabled user tokens, too small amount of data in social media, insufficient database of the own offers

3   Data and Preferences Modelling

The draft version of interests and preferences map for a particular networking site is built on the basis of existing and well-known algorithms which will be discussed here very briefly.

One of those algorithms is *Latent Dirichlet Allocation* LDA intended to process natural language [2]. It is a generative statistical model. On the input all text documents are analyzed. On the output we get topics with key words which fit best into the context of analyzed documents. The algorithm treats every document as a collection of different topics. The topics are described by a set of key

Table 2. Scenario: searching for user's preferences

Searching for user's preferences:
Trigger event: a client sends a request to the system
Actors: a client, social media (for example Facebook, Twitter)
Shareholders and their goals: a client wants to get user's preferences. Social media share their data
Preconditions: the system should have a necessary amount of offers and key words in its database
Scenario:
1. Client sends a request to the system together with giving a token
2. «include» Use case: downloading user's data
3. Searching for user's preferences
4. System sends back preferences on the basis of well-known algorithms
Postconditions for success: in a parameter there should be sent an active token of the user
Postconditions for failure: disabled user tokens, too small amount of data in social media, insufficient amount of key words

Table 3. Scenario: downloading user's data

Downloading user's data:
Actors: social media (Facebook, Twitter)
Shareholders and their goals: social media (Facebook and Twitter) share their data
Trigger event: system sends a proper request
Preconditions: possibility of establishing a connection with social media
Scenario:
1. System sends a request with an active user token
2. Social media share their data through their API
Postconditions for success: in a request parameter there should be sent an active token of the user
Postconditions for failure: disabled user token, too many requests in a limited time period

words. LDA algorithm was used to determine the topic of documents from the year 2003 by David Blei, Andrew Ng and Michael I. Jordan.

Another approach is presented by algorithm *Bag-of-words model* BOW – simplified representation of documents in processing natural languages and looking for information. It enables to easily classify documents. In BOW model text is written as a set (bag) of words without paying attention to their grammar and sequence. The important factor is amount repetitions of particular words. The BOW model was described in an article by Harris [3].

This kind of approach in both algorithms is comfortable and proper for our designed system. We assume thematic division on categories and key words.

Table 4. Scenario: adding a category

Adding a category:
Actors: administrator
Shareholders and their goals: the aim of administrator is to add a new preference category
Preconditions: a person has administrator privileges
Scenario:
1. System displays a formula on its website
2. Administrator fills in the form and sends it back
3. System confirms adding a new category
Postconditions for success: the particular category does not exist in a system
Postconditions for failure: a category has already existed in the system

Table 5. Scenario: adding a key word

Adding a key word
Actors: administrator
Shareholders and their goals: the aim of administrator is to add a new key word to a category
Trigger event: administrator sends a formula to the system
Preconditions: a person has administrator privileges
Scenario:
1. System displays a formula on its website
2. Administrator fills in the form and sends it back
3. System confirms adding a new key word
Postconditions for success: the particular key word does not exist in a system. A category is attributed to the key word
Postconditions for failure: a key word has already existed in the system

Categories match topics. Key words match concepts. However, it is possible that one key word can be attributed to more than one category. The example is "swimming" which can be attributed to "sport" category but also to "seaside holiday" category. The exemplary categories, analyzed in this work, were presented in Table 6. The following exemplary offers are presented in Tables 7, 8 and 9.

Data acquisition from social media requires API knowledge. In case of Facebook it is REST endpoint and it is called Graph API. In this way we can get descriptions of websites liked by the user. Authentication purposes require generating a token with the user's consent. The exemplary listing with a request was presented in Listing 1. As the answer we can get tables of objects which represent particular websites. Each of those objects has website id and its description. A collection of descriptions of all pages liked by the user creates input data for algorithm looking for preferences. The exemplary answer was presented in Listing 2.

Table 6. Cattegories and key words

Category	Key word
Sport	Cycling, football, skiing, basketball, volleyball, box, running, hockey, swimming, canoeing, climbing
Drinks	Coffee, tea, juice, beer, wine, water, cocktail
Food	Bread, apple, banana, beef, pork, veal, bacon, borscht, black pudding, French fries, cookies, duck, fish, vegetarian
Nature	Sea, mountains, forest, river, lake
Dwelling	Flat, hotel, hostel, apartments, shelter
Premises	Cafeteria, restaurant, bar, pub, fast food restaurant, inn
Transport	Plane, car, bike, bus, helicopter, motorbike, scooter, ship, walking
Entertainment	Cinema, bowling, billiard, snooker, theater, dance, museum, zoo
Season	Spring, summer, autumn, winter
Travelling	Camping, snorkeling, yacht, climbing, sights
Facilities	Air conditioning, phone, Internet, Wi-Fi, computer

Table 7. Exemplary tourist offer: a trip to Morskie Oko

Category	Parameters
Sport	Climbing
Drinks	Coffee
Nature	Mountains, forest, lake
Dwelling	Shelter
Transport	Car, walking
Season	Spring, summer, autumn, winter
Travelling	Camping, sights

Table 8. Exemplary tourist offer: Beach volleyball tournament

Category	Parameters
Sport	Volleyball
Dwelling	Hotel
Nature	Sea
Transport	Bike, walking
Season	Summer

Table 9. Exemplary tourist offer: Ski station

Category	Parameters
Sport	Skis, hockey
Drinks	Tea, coffee
Nature	Mountains, forest
Dwelling	Hotel, apartments
Premises	Inn, restaurant
Transport	Car, bus, walking
Season	Winter

The similar procedure can be repeated in case of Twitter. According to API documentation a request is sent together with Ouath authentication heading which was presented in Listing 3. The answer contains of table of objects expressed in JSON format. The important fields are tweet contents, see Listing 4. Those pieces of data are an input for programs mentioned above which gather data about preferences.

Listing 1. Request for favorite pages (Facebook)

```
https://graph.facebook.com/v2.8/me?
access_token=EAAKldGZCnym4BAKCPpkSjiDHvv2NSU6jWmOgpZCmPRH
0vQloryvuZBQ8Jrb5uliBZARkheALC8fmHGuCTMclnbbaw82AfLotKFRc
kXZCqv4NKEMKMPEvl33NvyuFwGtoXmZAVIG4LDZBgNuiB9YsZCGxrw5aZ
CjaFw6XZA7Q7eF6TZA8lzpgZCB7rnkaZA9R9KaBRjp8ZD
&callback=FB.__globalCallbacks.f2ced5f8a183a7c
&fields=likes%7Babout%7D &method=get
&pretty=0 &sdk=joey
```

Listing 2. Answer with favorite pages (Facebook)

```json
{"likes": {
 "data": [ {
    "about": "Java and JVM based technologies, dynamic
         languages, RIA, enterprise architectures,
         patterns, distributed computing and much more...",
    "id": "354953985700"
         },
 { "about": "Interesting Engineering is a cutting edge,
         leading community designed for all lovers of
         engineering, technology and science.",
    "id": "139188202817559"
         },
 { "about": "Pasja. Każdy ma swojąc?\n\n Prywatna
         strona o szeroko pojętych nowych technologiach",
    "id": "125612537464918"
         },
 {"about": "Koło Naukowe BIT gromadzi najlepszych
         studentów Informatyki na Akademii Górniczo−Hutniczej
         w Krakowie. Odwiedź naszą stronę aby dowiedzieć się więcej!",
    "id": "719620231484798"
         },
 {"about": "Sekcja Koła Naukowego BIT, która tworzy framework
         do zdobywania wiedzy i realizacji własnych pomysłów. ",
    "id": "1535988496663169"
         },
 {"about": "This page is managed by Marcin Marczewski and
         Jakub Hankiewicz. It follows the IBM Social
         Computing Guidelines.\n ",
    "id": "348362861915681"
         } ] } }
```

Listing 3. Request for favorite tweets (Twitter)

```
https://api.twitter.com/1.1/favorites/list.json?count=2&
screen_name=vapsel21&include_entities=false}
```

Listing 4. Answer with favorite tweets (Twitter)

```
[
{"created_at": "Wed May 17 08:56:03 +0000 2017",
"id": 864766453119152129, "id_str": "864766453119152129",
"text": "@GeeCON starts today in Krakow! Participants were
welcomed by David Moore, SVP of TN Product Development.
https://t.co/ CGTCoEpiJJ", ...},
{"created_at": "Tue Apr 11 22:52:43 +0000 2017",
"id": 851931044789886976, "id_str": "851931044789886976",
"text": "See all sessions from this year's ng-conf 2017
with Angular core team and many many others from around
the world. ... https:// t.co/s43U6O2yir", ...}
]
```

The basic processing algorithm, in relation to data gathering, can be presented in a following way:

1. The user agrees to download data from social networks.
2. Collecting data from social networks.
3. Processing collected data according to BOW model.
4. Filtering out short words, links and emoticons.
5. Looking for key words from glossary.
6. Determining the frequency of words. When they exceed threshold limit, they are marked as preferences.

The users permission is connected with sending authorization data. In case of Facebook, the subject of analysis are liked websites and in case of Twitter – published posts. The data is analyzed on the basis of programs, mentioned above, which look for key words and topics. The texts are divided into single words. Furtherly, the categories and key words are determined. The threshold limit, based on word frequency and necessary to determine preferences, can be changed at any time. In current experiments it was set as number 3.

4   Offer Analysis with the Use of Solvers

In order to perform preference analysis, together with assessment of notified offers, SAT solvers will be used. The SAT satisfiability problem is a classical IT problem which challenge is to find a substitution satisfying certain logical formula. For propositional calculus it will be substitution for sentential variables. Nowadays, the big development in searching the whole space of states was made and it routinely solves tasks consisting of tens of thousands of variables. There exist ready-to-use SAT solvers. The problem similar to SAT is MaxSAT problem based on finding a substitution which will maximally satisfy a formula. Thus, it considers the possibility of nonexistence of (full) satisfying substitution but in such case we look for substitution for the biggest possible number of clauses.

The system will have embedded MaxSAT solver in a form of SAT4J library [11]. The formulas will be saved in CNF form. The exemplary offer, presented in Table 6 written as a logical formula, will be as follows:

$$O = \neg cycling \land \neg football \land \neg skiing \land \neg basketball \land \neg volleyball \land \neg box$$
$$\land \neg running \land \neg hockey \land \neg swimming \land \neg canoeing \land climbing \land coffee$$
$$\land \neg tea \land \neg juice \land \neg beer \land \neg wine \land \neg water \land \neg coctail \land \neg bread$$
$$\land \neg apple \land \neg banana \land \neg beef \land \neg pork \land \neg veal \land \neg bacon \land \neg borscht$$
$$\land \neg blackpudding \land \neg Frenchfries \land \neg cookies \land \neg duck \land \neg fish$$
$$\land \neg vegetarian \land \neg sea \land mountains \land forest \land \neg river \land lake$$
$$\land \neg dwelling \land \neg hotel \land \neg hostel \land \neg apartments \land shelter \land \neg cafeteria$$
$$\land \neg restaurant \land \neg bar \land \neg pub \land \neg fastfoodrestaurant \land \neg inn$$
$$\land \neg plane \land car \land \neg bike \land \neg bus \land \neg helicopter \land \neg motorbike \land \neg scooter$$
$$\land \neg ship \land walking \land \neg cinema \land \neg bowling \land \neg billiard \land \neg snooker$$
$$\land \neg theater \land \neg dance \land \neg museum \land \neg zoo \land spring \land summer$$
$$\land autumn \land winter \land camping$$
$$\land \neg snorkeling \land \neg yacht \land \neg climbing \land sights \land \neg airconditioning$$
$$\land \neg phone \land \neg Internet \land \neg Wi - Fi \land \neg computer \quad (1)$$

Suppose that we have preferences for a user expressed by a formula:

$$P1 = skiing \land coffee \land mountain \land walking \land winter \quad (2)$$

After converting the offer as well as users preferences to logical formula, we can write the final formula which can be used as an input to MaxSAT solver and has a form as follows:

$$O \land P$$

If MaxSAT solver finds a proper assignments to satisfy this formula, it means that it meets all users preferences and can be recommended. This algorithm can be repeated involving all offers from database for every user.

Suppose that in the base we have three offers, as in Tables 7, 8 and 9. For a user described by Formula 2 the system will recommend "Ski station" option because it meets all user's preferences. The offer "Trip to Morskie Oko" does not satisfy "skiing" parameter and the offer "Beach Volleyball Tournament" – skiing, coffee, mountains and winter. For another user described by a formula:

$$swimming \land volleyball \land juice \land sea \land river \land summer \land spring \land diving$$

the system will not be able to propose an offer because none of them satisfy user's preferences.

The proposed system will be REST server receiving data necessary to integrate with external servers and algorithm working results. The input data for

Fig. 2. Class RequestParametersDTO and its parameters for the server. Class PreferencesResponseDTO and its response model for server

server is presented by RequestParametersDTO class (Fig. 2). The system will download users data by FacebookGraphAPIv2.9 and Twitter API1.1. After receiving a request, server downloads user's data from a social network and sends it for preference searching and later on to MaxSAT solver module. A class with server response model is presented in Fig. 2. The first answer field includes a name of proposed offer. The second one is a list of user preferences sorted according to probability of existing of such preference.

The system will consist of a few modules:

- Core – module responsible for handling the operations connected with database. In a project PostgreSQL 9.4.10 is used. Database is installed on the same machine as server with web module. In this module there is also implemented algorithm of searching user's preferences based on ideas of LDA algorithms.
- MaxSAT solver – module used to convert offers and preferences to logical formulas, later on MaxSAT solver (treated as a "black box") is launched.
- Web-module: communicates with clients and social networks by REST service.

5   Conclusions

In this paper we propose some solutions that refer to preference modeling of social network data on the basis of logical reasoning using available solvers. We can imagine a system which processes the flows of data online and in real time.

The further works should concentrate on building a prototype version of the system, another algorithms of building preferences (namely data acquisition form social networks and its analysis), including new logical reasoning schemes (deductive reasoning, modus ponens rule and the others) and another social networks.

Acknowledgments. I would like to thank my students Vadym Perepeliak and Karol Pietruszka (AGH UST, Kraków, Poland) for their valuable cooperation when preparing this work.

References

1. Abdar, M., Yen, N.Y.: Design of a universal user model for dynamic crowd prefer-ence sensing and decision-making behavior analysis. IEEE Access **5**, 24842–24852 (2017)
2. Blei, D.M., Ng, A.Y., Jordan, M.I.: Latent Dirichlet allocation. J. Mach. Learn. Res. **3**, 993–1022 (2003). http://dl.acm.org/citation.cfm?id=944919.944937
3. Harris, Z.: Distributional structure. Word **10**(23), 146–162 (1954)
4. Klimek, R.: Deduction-based formal verification of requirements models with auto-matic generation of logical specifications. In: Maciaszek, L.A., Filipe, J. (eds.) ENASE 2012. CCIS, vol. 410, pp. 157–171. Springer, Heidelberg (2013). https://doi.org/10.1007/978-3-642-45422-6_11
5. Klimek, R.: Behaviour recognition and analysis in smart environments for context-aware applications. In: Proceedings of the IEEE International Conference on Sys-tems, Man, and Cybernetics (SMC 2015), 9–12 October 2015, City University of Hong Kong, Hong Kong, pp. 1949–1955. IEEE Computer Society (2015)
6. Klimek, R., Kotulski, L.: Proposal of a multiagent-based smart environment for the IoT. In: Augusto, J.C., Zhang, T. (eds.) Workshop Proceedings of the 10th International Conference on Intelligent Environments, 30th June–1st July 2014, Shanghai, China. Ambient Intelligence and Smart Environments vol. 18, pp. 37–44. IOS Press (2014)
7. Klimek, R., Kotulski, L.: Towards a better understanding and behavior recogni-tion of inhabitants in smart cities. A public transport case. In: Rutkowski, L., Korytkowski, M., Scherer, R., Tadeusiewicz, R., Zadeh, L.A., Zurada, J.M. (eds.) ICAISC 2015. LNCS (LNAI), vol. 9120, pp. 237–246. Springer, Cham (2015). https://doi.org/10.1007/978-3-319-19369-4_22
8. Klimek, R., Rogus, G.: Proposal of a context-aware smart home ecosystem. In: Rutkowski, L., Korytkowski, M., Scherer, R., Tadeusiewicz, R., Zadeh, L.A., Zurada, J.M. (eds.) ICAISC 2015. LNCS (LNAI), vol. 9120, pp. 412–423. Springer, Cham (2015). https://doi.org/10.1007/978-3-319-19369-4_37
9. Klimek, R., Szwed, P.: Verification of ArchiMate process specifications based on deductive temporal reasoning. In: Proceedings of Federated Conference on Com-puter Science and Information Systems (FedCSIS 2013), 8–11 September 2013, Kraków, Poland, pp. 1131–1138. IEEE Xplore Digital Library (2013)
10. Kluza, K., Jobczyk, K., Wisniewski, P., Ligeza, A.: Overview of time issues with temporal logics for business process models. In: Ganzha, M., Maciaszek, L.A., Paprzycki, M. (eds.) Proceedings of the 2016 Federated Conference on Computer Science and Information Systems, FedCSIS 2016, 11–14 September 2016, Gdańsk, Poland, pp. 1115–1123 (2016). https://doi.org/10.15439/2016F328
11. Le Berre, D., Parrain, A.: Sat4j - the Boolean satisfaction and optimization library in Java (2017). http://www.sat4j.org/. Accessed 8 Jun 2017
12. Liu, Y., Xie, Q., Xiong, F.: Recommendations based on collaborative filtering by tag weights. In: 2017 13th International Conference on Semantics, Knowledge and Grids (SKG), pp. 62–68, August 2017
13. Magdum, S.S., Megha, J.V.: Mining online reviews and tweets for predicting sales performance and success of movies. In: 2017 International Conference on Intelligent Computing and Control Systems (ICICCS), pp. 334–339, June 2017
14. Wisniewski, P., Kluza, K., Ligeza, A.: Decision support system for robust urban transport management. In: FedCSIS, pp. 1069–1074 (2017)

On Asymmetric Problems of Objects' Comparison

Maciej Krawczak[✉] and Grażyna Szkatuła

Systems Research Institute, Polish Academy of Sciences,
Newelska 6, 01-447 Warsaw, Poland
{krawczak,szkatulg}@ibspan.waw.pl

Abstract. In the paper, we describe selected problems which appear during the process of comparison of the objects. The direction of objects' comparison seems to have essential role because such comparison may not be symmetric. Thus, we can say that two objects may be viewed as an attempt to determine the degree to which they are similar or different. Asymmetric phenomena of comparing such objects is emphasized and discussed.

Keywords: Proximity of objects · Directional comparison
Asymmetric proximities · Measures of proximity

1 Introduction

Evaluation of the proximity of the compared objects is an actual problem in many areas. The role of similarity or dissimilarity of two objects is fundamental in many theories of cognitive as well as behavior knowledge, and therefore for comparison of objects there are commonly used different measures of objects' similarity. However, in a famous critique, Goodman [6] dismissed similarity as scientifically useless notion. He claimed, that the concept of similarity of one object to another object is ill-defined, because does not include the concept "under what term". It seems to be obvious that objects are similar with respect to "something".

Cognitive visual illusions are very good examples of complexity of perception of reality [15]. A contrast effect either strengthens or weakens of our perception, for example a simultaneous contrast effect depends on the mutual influence of colors. In Fig. 1 there are two identical images on the left side and the right side, but shown on different backgrounds. Each image consists of a gray house with a red roof and a brown chimney, and the Moon. It must be emphasized that sizes, colors and shades of each image are exactly the same, while colors of the backgrounds are different. The left hand side background is gray while the right hand side is black. It is easy to notice the simultaneous contrast effect, namely the right image of the house seems to be significantly brighter and larger, and the Moon seems to be brighter, compare to the left image. Since we seldom can see

© Springer International Publishing AG, part of Springer Nature 2018
L. Rutkowski et al. (Eds.): ICAISC 2018, LNAI 10842, pp. 398–407, 2018.
https://doi.org/10.1007/978-3-319-91262-2_36

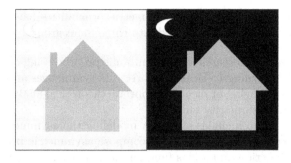

Fig. 1. Mutual affecting of colors (Color figure online)

colors separately, thus the perception of colors and shadows may substantially depend on the background color and our perception not always overlaps the reality.

In the majority of theoretical works of objects' similarity there is an essential assumption about symmetry, i.e., the similarity an object A to another object B equals the similarity from B to A. However, some research (e.g., in psychological literature) does not follow this assumption, and it is believed that the similarity can be asymmetric.

The issue of symmetry was extensively analyzed by Tversky [16,17], who considered objects represented by sets of features, and proposed measuring of similarity via comparison of their common and distinctive features. Such assumptions generate different approach to comparisons of objects. Namely, comparing two objects A and B there are the following fundamental questions: "how similar are A and B?", "how similar is A to B?" and "how similar is B to A?". The first question does not distinguish the directions of comparison and corresponds to symmetric similarity. The next two questions are directional and the similarity of the objects may not be a symmetric relation. For example, comparing a person and his portrait, we say that "the portrait resembles the person" rather than "the person resembles the portrait" [18]. The perceived similarity is strictly associated with data representation. In general, the direction of asymmetry is dependent on "salience of the stimuli". Thus, "the less salient stimulus is more similar to the more salient than the more salient stimulus is similar to the less salient" [16]. If the object B is more salient than the object A, then A is more similar to B. In other words, the variant is more similar to the prototype than the prototype to the variant. A toy train is quite similar to a real train, because most features of the toy train are included in the real train. On the other hand, a real train is not as similar to a toy train, because many of the features of a real train are not included in the toy train.

In many applications the data may be intrinsically asymmetric, e.g., in case of preferences of the people, exchanges (import-export, brand switching), migration data, etc. Possible examples are like telephone calls between cities, e.g. the number of telephone calls from the city A to the city B can be different from the number of telephone calls from the city B to the city A.

Many approaches to modeling asymmetric proximities have been proposed in the literature. They can be divided into three main groups.

(1) In the first group, researchers perform some preprocessing of the data to get symmetry. According to Beals et al. [1], "if asymmetries arise they must be removed by averaging or by an appropriate theoretical analysis that extracts a symmetric dissimilarity index".

(2) The second group includes explicitly modeling the asymmetries in addition to a symmetric component, e.g., decompose asymmetric proximities into a symmetric function and a bias function, multiplicative asymmetric weights.

(3) In the third category of approaches, the asymmetries are represented as the directed distances. In such a point of view, asymmetric proximity data are treated in accordance with the original form of the data, and analyzed in view of the asymmetry (e.g., like in the papers of Krawczak and Szkatuła [9,12,13]).

In Sect. 2, we present a brief discussion about the asymmetric human perception, for instance, the prospect theory, "salient" and "goodness" of the form, and "cost" of objects' transformation. The content of this section is largely is based on the works [5,7,8,18,19].

2   Selected Issues of Data Proximity

There are many types of data proximity which are non-symmetric. It happens that considering two objects one can notice that the object A is more associated with the object B than the other way round. It is important to notice that e.g., in psychological literature, especially related to modeling of human similarity judgments, similarity between objects can be asymmetric (e.g., Tversky [16]). The idea of asymmetries appearing in comparison of objects comes directly from the Tversky and Kahneman prospect theory (e.g., Tversky and Kahneman [19]).

Let us recall some exemplary cases of asymmetries of proximity of data, when the people compare the objects: the different values of losses and winnings in lotteries; a different perception of the so-called stimulus (in particular, the geometric figures) in psychological experiments.

Tversky and Kahneman Prospect Theory
Human perception can be modeled by the prospect theory developed by Tversky and Kahneman [19]. In outline, this theory describes people rationality in taking decisions. The theory states, that people make decisions based on the potential value of losses and gains. The value function is s-shaped and asymmetrical, see Fig. 2, and is the most characteristic for the prospect theory. The graph shows the psychological value of profits and losses. The concept of profit and loss is the basic principle in is the prospect theory. On both sides of the reference point (in the case of Fig. 2 the reference point is 0) the graph has a different shape and, in this way shows, different human sensibility to profits and losses. At the reference point of the graph there is stepwise change of the gradient of the function, and

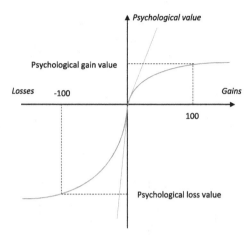

Fig. 2. A hypothetical psychological value function [19]

there is greater sensitivity to losses (the left side curve) than to profits (the right side curve) for the same considered value.

The most evident characteristics of the prospect theory is that the same loss creates greater feeling of pain compared to the joy created by an equivalent gain. For example, see Fig. 2, the feeling of joy due to obtaining $100 is lower than the pain caused by losing $100.

There are many experiments demonstrating people attitude to risk. Another illustrations of the prospect theory, it means the asymmetry between decisions, in which there are involved gains and losses can be found in e.g. [2].

For example, there is possibility to have a choice between getting $1000 for sure and $3000 with 50% chance. Definitely most people take the first choice, it means people prefer to get smaller but certain gain than bigger but with some level of uncertainty. Such behavior is called as risk aversion behavior.

In the counterpart example losses are considered. Let us consider a case when people have to decide what is better: to lose $1000 for sure or $3000 with 50% chance. In general, people chose the second choice, and such behavior is described as risk seeking.

Thus, it means, that considering profits people prefer certain gains while considering losses people are ready to face risk.

"Salient" and "goodness" of Form

The psychological nature of human perception was discussed among others by Tversky and Gati [18]. They hypothesized, that both "goodness of form" and complexity contribute to the salience of geometric figures. Moreover, they expected that the "good figure" to be more salient than the "bad figure". To investigate these hypotheses, they conducted two sets of eight pairs of geometric figures. In the first set, one figure in each pair (denoted p) had "better" form than the other figure (denoted q). In the second set, one figure in each pair

Figure q   Figure p

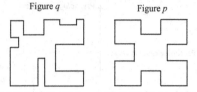

Fig. 3. A pair of figures from first set, used to test the prediction of asymmetry [18]

Figure q   Figure p

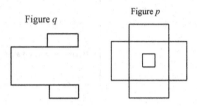

Fig. 4. A pair of figures from second set, used to test the prediction of asymmetry [18]

(denoted p) was "richer or more complex" than the other (denoted q). Two figures from each set are presented in Figs. 3 and 4.

A group of 69 respondents were involved in the experiment whom two elements of each pair were displayed side by side. The respondents were asked to choose one of the following two statements:

(1) "the left figure is similar to the right figure," or
(2) "the right figure is similar to the left figure".

The order of the presented figures were randomized so that figures appeared an equal number of times on the left as well as on the right side. In results, more than 2/3 of the respondents selected the statement 1, i.e. "q is similar to p".

Within the second experiment, the same pairs of figures were used. One group of respondents was asked to estimate (on a 20-point scale) the degree to which the figure on the left was similar to the figure on the right, while the second group was asked to estimate the degree to which the figure on the right was similar to the figure on the left. In results, the hypothesis was confirmed that the average pairs' similarity of the figures q to the figures p, $S(q,p)$, was significantly higher than the average pairs' similarity of the figures p to the figures q, $S(p,q)$.

These experiments confirmed their hypothesis that similarity is asymmetrical, however it does not clarify the concept of "goodness of the form".

"Cost" of Transformation
The objects' distance may be referred as a transformational distance between two objects. Such distance is described by the minimal costs (the smallest number of elementary operations) of transformation by a computer program of the first object's representation to the second object's representation. This concept is known as Levenshtein's distance [14].

According to Tversky [16] as well as Garner and Haun [5], the objects' transformations involve the operations of additions and deletions. It seems that deleting of some feature typically requires less complete specification compare to addition of it. Each comparison of the representations has a "short" and a "long" transformation, the arrows indicate the temporal order of stimulus presentation. Such transformations for the exemplary shapes A and B can be illustrated in Fig. 5. In order to generate the right figure from the left, the bottom line should be deleted. In the opposite case, the process of adding bottom line is more complex because requires specification of "what" and "where" exactly to add.

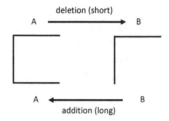

Fig. 5. Example of two shapes A and B [5]

Also can be considered the overall transformation distance between two representations, which is characterized by the number of steps required to change one representation to other [7]. They distinguished three general transformations for comparing shapes:

(1) *create a new feature*, that is unique to the target representation;
(2) *apply feature*, this operation takes a feature created via step 1 and applies it to one or both of the objects in the target representation;
(3) *swap feature between a pair of objects*, e.g. shape or color.

The transformation from the exemplary pair of shapes A to the pair of shapes B, and in the opposite direction, can be illustrated in Fig. 6.

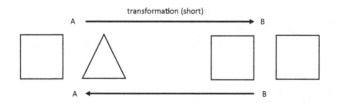

Fig. 6. Example of two pairs of shapes A and B [7,8]

Let us consider first case, in order to calculate the transformation distance from the pair of shapes A to the pair of shapes B. Then, there are required to use only one transformation *apply* for existing *square*, i.e., *apply*(*square*) = 1. In the second case, the transformation distance from the pair of shapes B to the pair of shapes A requires using two transformations, i.e. creation of a new triangle and application of this new triangle, thus *create*(*triangle*) + *apply*(*triangle*) = 2. Thus, the transformation distance in the first case is "short" (requires one operation), whereas the transformation in the second case is "long" (required two operations).

3   Difficulties in Proximity Measure Selection

Much work has been done to determine proximity measure of the objects described by continuous-valued attributes. In general, handling proximity of the objects described by nominal-valued attributes is much more difficult. Therefore, for nominal attributes the comparison of one object to another can be considered whether the objects have the same or different values. In such cases, there are two main approaches, namely:

(1) *simple matching* - the dissimilarity is defined as 0 if they are identical, or 1 otherwise, for two possible values; and then the ratio of numbers of matched and total elements is calculated;
(2) *binary encoding* - the nominal attributes are replaced by binary-valued attributes, and next, some quantitative matching method is used.

The first approach can be used for very specific data, therefore the second one is commonly exploited, and e.g. the simple matching coefficient or Jaccard's coefficient can be employed. The concise overview of the proximity measures can be found in the paper [3].

It is obvious that new binary attributes do not retain semantics as well as dimensionality of the original attributes. Additionally, in general, application of the conventional methods causes neglecting of asymmetry of compared data sets.

Easy to see, that the crucial point in data analysis is the proper selection of the proximity measure.

It seems, that the new proximity measure, called the measure of perturbation of sets, developed by the authors in the papers [9–13], is a challenging approach to determine not only the proximity, but also the asymmetry of the objects described by nominal attributes. In general, let us consider a set V and two subsets $A_i, A_j \subseteq V$ then the measure of the perturbation of the set A_j by the set A_i can be written as follows

the measure of perturbation type 1, $Per^1(A_i \mapsto A_j) = \frac{cardinality(A_i \setminus A_j)}{cardinality(V)}$,

the measure of perturbation type 2, $Per^2(A_i \mapsto A_j) = \frac{cardinality(A_i \setminus A_j)}{cardinality(A_i \cup A_j)}$.

The idea of perturbation was developed by the authors for different kind of sets, namely for ordinary sets in e.g. [9], for fuzzy sets in e.g. [11], and for multisets in e.g. [10,13].

In the subsequent text we provide the numerical analysis of proximity of two binary vectors, partially following [9]. Thus, let us consider two objects o_i and o_j described by two binary vectors A_i and A_j, respectively. In order to analyze these two vectors, the following numbers are introduced:

a - is the number of corresponding elements equal 1 in both vectors,
b - is the number of corresponding elements equal 1 for vector A_i and 0 for vector A_j,
c - as the number of corresponding elements equal 0 for vector A_i and 1 for vector A_j,
d - is the number of corresponding elements equal 0 in both vectors.

Thus, the sum $a + b + c + d$ is always equal to the dimension of the binary vectors; the sum $a + d$ represents the number of matches between A_i and A_j; the sum $b + c$ represents the number of mismatches between A_i and A_j.

Now, let us recall some selected forms of the proximity measures for binary vectors, see e.g. [3,4,9]:

Jaccard extended similarity, $S_J(A_i, A_j) = \frac{a+d}{a+b+c+d}$,

Sokal and Michener similarity, $S_{S-M}(A_i, A_j) = S_J(A_i, A_j)$,

mean-Manhattan distance, $S_M(A_i, A_j) = \frac{b+c}{a+b+c+d}$,

variance distance, $D_V(A_i, A_j) = \frac{b+c}{4(a+b+c+d)}$,

Faith similarity, $S_F(A_i, A_j) = \frac{a+d/2}{a+b+c+d}$,

Russel and Rao similarity, $S_{R-R}(A_i, A_j) = \frac{a}{a+b+c+d}$,

Hamann similarity, $S_H(A_i, A_j) = \frac{(a+d)-(b+c)}{a+b+c+d}$.

It should be mentioned that in the literature there are several cases that the same form of proximity measure is recalled with different names, for example the Jaccard extended similarity and the Sokal and Michener similarity.

Next, the measure of sets' perturbation type 1 introduced in the paper [9] can be written as follows,

$Per^1(A_i \mapsto A_j) = \frac{b}{a+b+c+d}$,

$Per^1(A_j \mapsto A_i) = \frac{c}{a+b+c+d}$.

In order to show interesting relationship between the selected proximity measures compared to the perturbation measures, let us consider the following example [9].

For two exemplary binary vectors $[1, 1, 1, 1, 0, 1, 0, 0, 0]$ and $[1, 1, 0, 1, 1, 1, 1, 0, 0]$ we have calculated the degrees of proximity between these vectors using above recalled measures' definitions, see Fig. 7. It is interesting, that for the considered two vectors the different measures of proximity give different respective values. Therefore, it is impossible to say which measure is better because known proximity measures were developed especially for specific real data. It is obvious, that considering another two objects represented by a pair of different binary vectors we obtain different values of the considered proximity measures. And thus, the illustration picture about relationships between the considered proximity measures will look differently compared to Fig. 7.

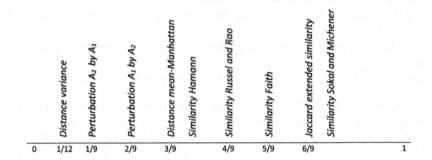

Fig. 7. A graphical illustration of few selected measures [9]

4   Conclusions

In this paper we studied a look at problems appearing during the process of comparison of objects. It seems, that the comparison of two objects should be considered with regard to order of compared objects. Therefore, we called such phenomena as the directional comparisons.

Nowadays, there is a problem of automatic comparison of objects by computers. Therefore, proximity of objects must be modeled and then measured. There are several ways to model asymmetries of proximity of data. The only assumption is, that the measure of similarity or dissimilarity between two objects must be defined. In general, there are two classes of proximity representation of objects. In the first class, each object is represented by a point in the adequate multidimensional Cartesian coordinates, and an appropriate measure of the proximity between two such objects is specified just by the distance between these two corresponding points in that space. In the second class, each object is represented as a collection of some features or attributes. Usually similarity between objects is expressed as a matching function of their common and distinctive features, however similarity can also be expressed as structural compatibility or simple features matching.

In the literature, including the authors' developments, a lot of measures of similarity or dissimilarity between objects and some formulas to calculate them can be found. It is easy to notice, that different measures of sets proximity generate different respective values, in general. It means that there does not exist the best measure for evaluation of proximity between two arbitrary objects and the choice depends on the nature of data under consideration.

References

1. Beals, R., Krantz, D.H., Tversky, A.: The foundations of multidimensional scaling. Psychol. Rev. **75**, 127–142 (1968)
2. Bernstein, P.L.: Against the Gods: The Remarkable Story of Risk. Wiley, New York (1996)
3. Choi, S., Cha, S., Tappert, C.C.: A survey of binary similarity and distance measures. Syst. Cybern. Inform. **8**(1), 43–48 (2010)
4. Cross, V.V., Sudkamp, T.A.: Similarity and Compatibility in Fuzzy Set Theory. Physica, Heidelberg (2002). https://doi.org/10.1007/978-3-7908-1793-5
5. Garner, W.R., Haun, F.: Letter identification as a function of type of perceptual limitation and type of attribute. J. Exp. Psychol. Hum. Percept. Perform. **4**(2), 199–209 (1978)
6. Goodman, N.: Seven strictures on similarity. In: Goodman, N. (ed.) Problems and Projects, pp. 437–450. Bobs-Merril, New York (1972)
7. Hodgetts, C.J., Hahn, U., Chater, N.: Transformation and alignment in similarity. Cognition **113**, 62–79 (2009)
8. Hodgetts, C.J., Hahn, U.: Similarity-based asymmetries in perceptual matching. Acta Psychol. **139**(2), 291–299 (2012)
9. Krawczak, M., Szkatuła, G.: On asymmetric matching between sets. Inf. Sci. **312**, 89–103 (2015)
10. Krawczak, M., Szkatuła, G.: Multiset approach to compare qualitative data. In: Proceedings 6th World Conference on Soft Computing, Berkeley, pp. 264–269 (2016)
11. Kacprzyk, J., Krawczak, M., Szkatuła, G.: On bilateral matching between fuzzy set. Inf. Sci. **402**, 244–266 (2017)
12. Krawczak, M., Szkatuła, G.: Geometrical interpretation of impact of one set on another set. In: Rutkowski, L., Korytkowski, M., Scherer, R., Tadeusiewicz, R., Zadeh, L.A., Zurada, J.M. (eds.) ICAISC 2017. LNCS (LNAI), vol. 10245, pp. 253–262. Springer, Cham (2017). https://doi.org/10.1007/978-3-319-59063-9_23
13. Krawczak, M., Szkatuła, G.: Bidirectional comparison of multi-attribute qualitative objects. Inf. Sci. **436–437**, 367–387 (2018)
14. Levenshtein, V.I.: Binary codes capable of correcting deletions, insertions, and reversals. Sov. Phys. Dokl. **10**, 707–710 (1966)
15. Tanca, M., Grossberg, S., Pinna, B.: Probing perceptual antinomies with the watercolor illusion and explaining how the brain resolves them (PDF). Seeing Perceiving **23**, 295–333 (2010)
16. Tversky, A.: Features of similarity. Psychol. Rev. **84**(4), 327–352 (1977)
17. Tversky, A.: Preference, Belief, and Similarity. Selected Writings by Amos Tversky. Edited by Eldar Shafir. Massachusetts Institute of Technology, MIT Press, Cambridge (2004)
18. Tversky, A., Gati, I.: Studies of similarity. In: Rosch, E., Lloyd, B. (eds.) Cognition and Categorization, vol. 1, pp. 79–98. Lawrence Elbaum Associates, Hillsdale (1978)
19. Tversky, A., Kahneman, D.: The framing of decisions and the psychology of choice. Science **211**, 453–458 (1981)

A Recommendation Algorithm
Considering User Trust and Interest

Chuanmin Mi[1]([☒])[ID], Peng Peng[1], and Rafał Mierzwiak[1,2]

[1] Nanjing University of Aeronautics and Astronautics,
Nanjing 210016, People's Republic of China
michuanmin@nuaa.edu.cn
[2] Poznan University of Technology, Poznan, Poland

Abstract. A traditional collaborative filtering recommendation algorithm has problems with data sparseness, a cold start and new users. With the rapid development of social network and e-commerce, building the trust between users and user interest tags to provide a personalized recommendation is becoming an important research issue. In this study, we propose a probability matrix factorization model (STUIPMF) by integrating social trust and user interest. First, we identified implicit trust relationship between users and potential interest label from the perspective of user rating. Then, we used a probability matrix factorization model to conduct matrix decomposition of user ratings information, user trust relationship, and user interest label information, and further determined the user characteristics to ease data sparseness. Finally, we used an experiment based on the Epinions website's dataset to verify our proposed method. The results show that the proposed method can improve the recommendation's accuracy to some extent, ease a cold start and solve new user problems. Meanwhile, the STUIPMF approach, we propose, also has a good scalability.

Keywords: Data mining · Recommender system
Collaborative filtering · Social trust · Interest tag
Probability matrix factorization

1 Introduction

With the expansion of network and information technology, the amount of data generated from human activities is rapidly growing. "Information Overload" problem is becoming serious [2]. Therefore, a recommender system might be an important tool to help users find the interested items and solve the problem of information overload. More and more e-commerce service providers, such as Amazon, Half. Com, CDNOW, Netflix, and Yahoo!, are using recommendation

This work was supported by the Project of National Social Science Foundation of China (17BGL055).

systems for their own customers with "tailored" buying advice [19]. A recommended algorithm as the most core and key part of these types of systems, determines their performance quality to a great extent [1]. Due to a simple operation, a reliable explanation, easiness of realization in the technical level, and collaborative filtering (CF), a recommendation algorithm is becoming one of the most widely used recommendation algorithm [4], which mainly makes uses of the ratings of the user to calculate the similarity to give recommendation. However, studies have shown that in big e-commerce systems, items rated by users generally will not exceed 1% in total number, so user rating inevitably has problems such as data sparseness, a cold start etc., which affect the precision and quality of the recommendation [21]. On the one hand, introducing a user trust relationship in a recommender system can solve a cold start problem. On the other hand, adding user interest can alleviate the problem of data sparseness.

In recent years, the number of Internet users has exponentially grown and as a result social network also has developed. The 38th China Internet development statistics report issued by China Internet Network Information Center (CNNIC) on August 3rd 2016 in Beijing shows that, the scale of Internet users in China has reached 710 million up to June 2016, whereas the Internet penetration rate reached 51.7%. Nielsen research agency examined factors that influence users' trust in recommendations. The investigation showed that nearly ninety percent of users can believe the recommendation given by their friends [17].

According to the facts presented above, we established an implicit trust relationship between users and potential interest tags from the perspective of user rating. Next, we combined users trust relationship and user interest tag information into a probability matrix factorization (PMF) model. Finally, on the basis of PMF model, we proposed a probability matrix factorization model (STUIPMF) by integrating social trust and user interest.

2   Related Works

In a traditional CF recommendation algorithm there are problems with data sparseness and a cold start, which affect the precision and quality of a recommendation algorithm based on social trust [4,6,10,11,18]. In order to improve the performance of a recommender system, it ought to be roughly divided into two categories. One is a recommended approach, which examines a trust relationship based on a neighborhood model. Here we can distinguish a Mole Trust model, which utilizes a depth-first strategy to search users and predicts the trust value to target user B by considering the passing of the trust on the side of user A's social network [18]. Similarly, Golbeck proposed a Tidal Trust model that improves a breadth-first strategy to forecast a user trust value [4], whereas Jamali proposed a Trust Walker model, which combines a recommender system based on an item with a recommender system based on trust [6]. However, these methods only consider a trust relationship between the neighboring users, but also neglect an implicit trust relationship between users, and the influence user ratings exerted to the result of recommendation.

The second category is a recommended approach, which fuses a trust relationship among users with rating data based on MF model. Take for example a recommendation method that adds a social regularization term to a loss function, which measures the difference between the latent feature vector of a user and those of their friends [16]. On the other hand, referring to Social MF model, it integrates all user trust information, which introduces a concept of trust propagation. Moreover, the model considers information about direct trustable users and "two-steps" users to generate recommendation. However, computing complexity is high, and it does not adopt different trust metrics [7]. Another example is a MF recommendation method that predicts the variation of ratings with time [13]. Next proposal is a stratified stochastic gradient descent (SSGD) algorithm to solve general MF problem. It provides sufficient conditions for convergence [3]. Finally, we can also find an incremental CF recommendation method based on regularized MF [14] and a Social Recommendation method, which connects users rating information and social information for research by sharing user implicit feature vector space [15]. All the methods described above focus on direct trust network, however they ignore mining implicit trust relationships between users.

When analyzing research referring to MF models, it is worth to notice that a recommendation algorithm based on MF uses latent factors. Thus, it is difficult to give an accurate and reasonable explanation to recommended results. Hence, Salakhutdinov described a matrix factorization problem from the perspective of probability, and put forward PMF model, which obtained the prior distribution of the user-recommended item's characteristic matrix, and maximized the posterior probability of the forecast evaluation to make recommendations [20]. This model achieved very good prediction results on Netflix data sets. It is worth to mention that Koenigstein integrated some characteristic information of the items in the process of probability matrix factorization, and carried out experiments on the Xbox movie recommendation system, which verified the effectiveness of the proposed model [9].

Research also has shown that we take into account user interests, such as tags, categories and user profiles, there is a huge opportunity to improve the accuracy of recommendation. Considering the user interest model, it is conducive to make more accurate personalized recommendation. Lee combined user's preference information with a trust propagation of social networking, and improved the quality of recommendation [5], while Tao proposed a CF algorithm based on user interest classification adapting to the user's interests diversity. After that the improved fuzzy clustering algorithm was used to search the nearest neighbor [22]. What is more, Ji put forward a similarity measure method based on user interest degree. This way the combination of a degree of user interest in the different item category and user ratings were utilized to calculate the similarity between them [18]. However, most of these methods focus on the user's rating value of the item. They do not consider user preferences and influence on the relationship between the user ratings and the item properties that affect the accuracy of recommendation. Furthermore, they also ignored the user trust relationships.

Therefore, in this paper, we have a comprehensive consideration of user rating and implicit trust relationship between users. Moreover, users trust relationship and user interest tag information on the basis of PMF model are introduced. And then we identify latent user characteristics hidden behind a trust relationship and user ratings. As a result, a STUIPMF model is proposed. According to the experimental results, this method comprehensively utilizes various information, which can enhance recommended accuracy.

3   Probability Matrix Factorization Recommendation Algorithm Combining User Trust and Interests

3.1   Probability Matrix Factorization Model (PMF)

The principle of PMF model is to predict user ratings for an item from the perspective of probability. To make the notation clearer, the symbols that we will use are shown in Table 1. The calculation process of PMF is as follows.

Assume that latent factors of users and items are subject to Gaussian prior distribution,

$$P\left(U \mid \sigma_U^2\right) = \prod_{i=1}^{M} \mathrm{N}\left(U_i \mid 0, \sigma_U^2 I\right) \tag{1}$$

$$P\left(V \mid \sigma_V^2\right) = \prod_{j=1}^{N} \mathrm{N}\left(V_j \mid 0, \sigma_V^2 I\right) \tag{2}$$

Table 1. Notation

Symbols	Descriptions
M, N, S	Number of users, number of items, number of interest labels respectively
K	Number of latent factors
U_{M*K}	User latent factors
V_{N*K}	Item latent factors
F_{M*K}	Trust latent factors
L_{S*K}	Interest tag latent factors
R_{ij}	Rating matrix
P_{ik}	Tagging times
T_{il}	Trust degree
$\tilde{R}_{ij}$	Predicted rating

Moreover, let's assume that conditional probability of user ratings data obtained are subject to Gaussian prior distribution,

$$P\left(R \mid U, V, \sigma_R^2\right) = \prod_{i=1}^{M} \prod_{j=1}^{N} \left[\mathrm{N}\left(R_{ij} \mid g\left(U_i^T V_j\right), \sigma_R^2\right)\right]^{I_{ij}^R} \tag{3}$$

I_{ij}^R is an indicator function, if user U_i has rated V_j, $I_{ij}^R = 1$, otherwise 0. $g(x)$ maps the value of $U_i^T V_j$ to the interval, in this paper $g(x) = 1/(1 + e^{-x})$.

Through the Bayesian inference, we can gain posterior probability of users and items' implicit characteristics.

$$P\left(U, V \mid R, \sigma_R^2, \sigma_U^2, \sigma_V^2\right) \propto P\left(R \mid U, V, \sigma_R^2\right) \times P\left(U \mid \sigma_U^2\right) \times P\left(V \mid \sigma_V^2\right) \quad (4)$$

This way, we can learn about latent factors of users and items through rating matrix, and then get the most similar user rating by means of inner product formulated as follows:

$$\tilde{R} \approx U_i^T V_j \tag{5}$$

The corresponding probability graph model is presented in Fig. 1.

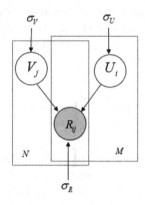

Fig. 1. Probability matrix factorization graph model

3.2   Mining User Implicit Trust Relationships

Most existing algorithms only consider a direct trust network, namely the dominant trust relationship between users [6,10,18]. They have less attention to mining user implicit trust relationship. Therefore, a user behavior coefficient and a user trust function are introduced to improve the measurement of user trust relationships.

After a trust inference based on user rating and calculation of rating accuracy, a user behavior coefficient will be determined. Next, user implicit trust relationships will be established on the basis of user rating similarity. Accuracy of rating is denoted by the difference of the item rating between a target user and all the users. In general, whether user ratings are accurate or not will directly affect the degree of other users' trust. A user behavior coefficient is expressed with φ_u symbol and it is depended on accuracy of rating.

$$\varphi_u = \frac{1}{1 + \sum\limits_{i=1}^{N} \left(R_{ui} - \bar{R}_i\right) \cdot I_{ui}} \tag{6}$$

R_{ui} expresses rating of user u to the item i. $\bar{R}_i$ expresses an average rating of all users to the item i if user u has rated the item i, $I_{ui} = 1$, otherwise $I_{ui} = 0$.

Rating similarity $sim_{i,j}$ is measured by a popular Pearson correlation coefficient. The computational formula is as follows:

$$sim_{i,j} = \frac{\sum\limits_{c \in U} (r_{i,c} - \bar{r}_i) \times (r_{j,c} - \bar{r}_j)}{\sqrt{\sum\limits_{c \in U} (r_{i,c} - \bar{r}_i)^2 \times \sum\limits_{c \in U} (r_{j,c} - \bar{r}_j)^2}} \tag{7}$$

$r_{i,c}$ and $r_{j,c}$ express ratings of user i and user j to the item c respectively, and $\bar{r}_i$ and $\bar{r}_j$ express the average.

User implicit trust relationships are denoted by TI, between user i and user j:

$$TI_{ij} = \varphi_i \cdot sim_{ij} \tag{8}$$

Use t_{ij} to denote the explicit trust relationships between user i and user j, when user i trusts user j, $t_{ij} = 1$, otherwise 0. Due to the asymmetry of trust, t_{ij} cannot reflect a dominant trust relationship between users accurately, which should be related to the number of users' trust and trustable users. For example, when user t_i trusts many users, a trust value t_{ij} between user t_i and user t_j will be reduced. On the contrary, when many users trust user t_i, the trust value between user t_i and user t_j should increase. Therefore, the dominant trust value between users is upgraded on the basis of user influence TE_{ij}, which expresses the improved dominant trust value.

$$TE_{ij} = \sqrt{\frac{d^-(u_i)}{d^+(u_j) + d^-(u_i)}} \cdot t_{ij} \tag{9}$$

$d^-(u_i)$ points out the number of users u_i by a user who is trusted, $d^+(u_j)$ is the number of users by user u_j trust.

A user trust function is denoted by T_{ij}, which is calculated after determining the weight coefficient of a dominant trust and implicit trust combined with a dominant trust relationship stated in trust network. α expresses a weight coefficient.

$$T_{ij} = \alpha \cdot TE_{ij} + (1 - \alpha) \cdot TI_{ij} \tag{10}$$

A user trust relationship matrix is denoted by T. T_{il} expresses the trust degree of user U_i and a friend F_l. A conditional probability distribution function of a user's trust is known as:

$$P\left(T \mid U, F, \sigma_T^2\right) = \prod_{i=1}^{M} \prod_{l=1}^{M} \left[N\left(T_{il} \mid g\left(U_i^T F_l\right), \sigma_T^2\right)\right]^{I_{il}^T} \tag{11}$$

I_{il}^T is an indicator function if user U_i and user F_l are friends, $I_{il}^T = 1$, otherwise 0.

The probability distribution of U and F is as follows:

$$P\left(U\mid\sigma_U^2\right)=\prod_{i=1}^{M}N\left(U_i\mid 0,\sigma_U^2 I\right),P\left(F\mid\sigma_F^2\right)=\prod_{l=1}^{M}N\left(F_l\mid 0,\sigma_F^2 I\right) \quad (12)$$

Through the Bayesian inference, we achieve:

$$P\left(U,F\mid T,\sigma_T^2,\sigma_U^2,\sigma_F^2\right)\propto P\left(T\mid U,F,\sigma_T^2\right)\times P\left(U\mid\sigma_U^2\right)\times P\left(F\mid\sigma_F^2\right) \quad (13)$$

The corresponding probability graph model based on a user trust relationship is demonstrated in Fig. 2.

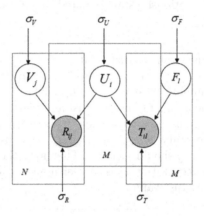

Fig. 2. Probability graph model based on user trust relationship

3.3   Mining the User Interest Similarity Relationship

A current recommendation algorithm based on user interest classification pays less attention to the influence, which user preference and the relationship between user ratings and item properties have, on recommended results [8,9,12,23]. Thus, it is legitimated to combine the item information with user threshold based on the user-item rating matrix, and mining user implicit tag. As a result, a user-interest tag matrix is received and it is useful to fill user information and solve the problem of data sparseness.

The corresponding median rating threshold set to user rating set of all the items is $A = \{A_1, A_2, \cdots, A_m\}$, the attribute set of items is $L = \{L_1, L_2, \cdots, L_k\}$. When $R_{ui} \geq A_i$, we regard that user u likes the item i. The attribute tag L_c of the item i is signed as an interest tag of the user u. We can extract the user's interest tag according to the item's attributes and the user's rating threshold. A user may be signed with the same interest tags repeatedly, and when the times are accumulated, we can get the user interest tag matrix $L_{me} = \{L_{uy}\}$. L_{uy} expresses the times of the interest tag user u signed to the item attributes L. Then, we make the rating which is below the user ratings threshold

0 to get a user-item median rating matrix. Combined with the item-attribute matrix, if the item belongs to some attribute, it is signed 1, otherwise 0. Therefore, when a link is established between the user and the item-attribute, we will get the user-interest tag matrix P.

The user-interest tag matrix is denoted by P, P_{ik} expresses the signed times of user U_i signed on the interest tag L_k. The probability distribution function of a user interest tag is known as follows:

$$P\left(P \mid U, L, \sigma_P^2\right) = \prod_{i=1}^{M} \prod_{k=1}^{Q} \left[N\left(P_{ik} \mid g\left(U_i^T L_k\right), \sigma_P^2\right)\right]^{I_{ik}^P} \tag{14}$$

I_{ik}^P expresses an indicator function, if the user U_i has signed on the interest tag L_k at least one time, otherwise 0.

The probability distribution of U_i and L is the following way:

$$P\left(U \mid \sigma_U^2\right) = \prod_{i=1}^{M} N\left(U_i \mid 0, \sigma_U^2 I\right), \quad P\left(L \mid \sigma_L^2\right) = \prod_{k=1}^{Q} N\left(L_k \mid 0, \sigma_L^2 I\right) \tag{15}$$

According to the Bayesian inference, we achieve:

$$P\left(U, L \mid P, \sigma_P^2, \sigma_U^2, \sigma_L^2\right) \propto P\left(P \mid U, L, \sigma_P^2\right) \times P\left(U \mid \sigma_U^2\right) \times P\left(L \mid \sigma_L^2\right) \tag{16}$$

The corresponding probability graph model based on a user interest tag is shown in Fig. 3.

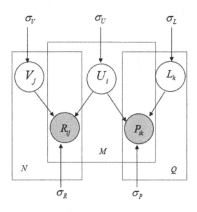

Fig. 3. Probability graph model based on user interest tag

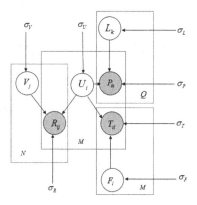

Fig. 4. STUIPMF probability graph model

4   STUIPMF Model Application

PMF algorithm is merely based on a user-item rating matrix and it studies the corresponding feature factor. However, it does not consider the trust relationship between the user and the user's interest on the result of recommendation. In order to reflect the effect, the model was improved by integrating the factorizations of three matrixes, which are a user trust relationship matrix, a user-interest tag matrix, and a user rating matrix respectively, and connected by a user latent feature factor matrix. Therefore, STUIPMF model is put forward, as it is demonstrated in Fig. 4.

The logarithm of posterior probability, after the conjunction, comes down to the Eq. (17).

In this research, a stochastic gradient descent method is used to study a corresponding latent feature factor matrix. Assuming that $\lambda_U = \lambda_V = \lambda_T = \lambda_L = \lambda$, a computational complexity is reduced. The values of λ_P and λ_F will be discussed in the latter part.

$\lambda_P = \sigma_R^2/\sigma_P^2$, $\lambda_T = \sigma_R^2/\sigma_T^2$, $\lambda_U = \sigma_R^2/\sigma_U^2$, $\lambda_V = \sigma_R^2/\sigma_V^2$, $\lambda_L = \sigma_R^2/\sigma_L^2 \lambda_F = \sigma_R^2/\sigma_F^2$ are all fixed regularization parameters, $\| \cdot \|_F$ expresses the Frobernius of the matrix.

$$\ln P\left(U, V, L, F \mid R, P, T, \sigma_R^2, \sigma_P^2, \sigma_T^2, \sigma_U^2, \sigma_V^2, \sigma_L^2, \sigma_F^2\right)$$

$$= -\frac{1}{2\sigma_R^2} \sum_{i=1}^{M} \sum_{j=1}^{N} I_{ij}^R \left(R_{ij} - g\left(U_i^T V_j\right)\right)^2 - \frac{1}{2\sigma_P^2} \sum_{i=1}^{M} \sum_{k=1}^{Q} I_{ik}^P \left(P_{ik} - g\left(U_i^T L_k\right)\right)^2$$

$$- \frac{1}{2\sigma_T^2} \sum_{i=1}^{M} \sum_{l=1}^{M} I_{il}^T \left(T_{il} - g\left(U_i^T F_l\right)\right)^2 - \frac{1}{2\sigma_U^2} \sum_{i=1}^{M} U_i^T U_i - \frac{1}{2\sigma_V^2} \sum_{j=1}^{N} V_j^T V_j$$

$$- \frac{1}{2\sigma_L^2} \sum_{k=1}^{Q} L_k^T L_k - \frac{1}{2\sigma_F^2} \sum_{l=1}^{M} F_l^T F_l - \frac{1}{2} \left(\sum_{i=1}^{M} \sum_{j=1}^{N} I_{ij}^R\right) \ln \sigma_R^2$$

$$- \frac{1}{2} \left(\left(\sum_{i=1}^{M} \sum_{k=1}^{Q} I_{ik}^P\right) \ln \sigma_P^2 + \left(\sum_{i=1}^{M} \sum_{l=1}^{M} I_{il}^T\right) \ln \sigma_T^2\right) - \frac{1}{2}((M \times K) \ln \sigma_U^2$$

$$+ (N \times K) \ln \sigma_V^2) - \frac{1}{2} \left((M \times K) \ln \sigma_F^2 + (Q \times K) \ln \sigma_L^2\right) + C \qquad (17)$$

$$S\left(U, V, L, F, R, P, T\right)$$

$$= \frac{1}{2} \sum_{i=1}^{M} \sum_{j=1}^{N} I_{ij}^R \left(R_{ij} - g\left(U_i^T V_j\right)\right)^2 + \frac{\lambda_P}{2} \sum_{i=1}^{M} \sum_{k=1}^{Q} I_{ik}^P \left(P_{ik} - g\left(U_i^T L_k\right)\right)^2 +$$

$$\frac{\lambda_T}{2} \sum_{i=1}^{M} \sum_{l=1}^{M} I_{il}^T \left(T_{il} - g\left(U_i^T F_l\right)\right)^2 + \frac{\lambda_U}{2} \sum_{i=1}^{M} \|U\|_F^2 + \frac{\lambda_V}{2} \sum_{j=1}^{N} \|V\|_F^2$$

$$+ \frac{\lambda_L}{2} \sum_{k=1}^{Q} \|L\|_F^2 + \frac{\lambda_F}{2} \sum_{l=1}^{M} \|F\|_F^2 \qquad (18)$$

$$\frac{\partial S}{\partial U_i} = \sum_{j=1}^{N} I_{ij}^{P} g' \left(U_i^T V_j\right) \left(g \left(U_i^T V_j\right) - R_{ij}\right) V_j$$

$$+ \lambda_P \sum_{k=1}^{Q} I_{ik}^{P} g' \left(U_i^T L_k\right) \left(g \left(U_i^T L_k\right) - P_{ik}\right) L_k$$

$$+ \lambda_T \sum_{l=1}^{M} I_{il}^{T} g' \left(U_i^T F_l\right) \left(g \left(U_i^T F_l\right) - T_{il}\right) F_l + \lambda_U U_i \tag{19}$$

$$\frac{\partial S}{\partial V_j} = \sum_{i=1}^{M} I_{ij}^{R} g' \left(U_i^T V_j\right) \left(g \left(U_i^T V_j\right) - R_{ij}\right) U_i + \lambda_V V_j \tag{20}$$

$$\frac{\partial S}{\partial L_k} = \lambda_P \sum_{k=1}^{Q} I_{ik}^{P} g' \left(U_i^T L_k\right) \left(g \left(U_i^T L_k\right) - P_{ik}\right) U_i + \lambda_L L_k \tag{21}$$

$$\frac{\partial S}{\partial F_l} = \lambda_T \sum_{l=1}^{M} I_{il}^{T} g' \left(U_i^T F_l\right) \left(g \left(U_i^T F_l\right) - T_{il}\right) U_i + \lambda_F F_l \tag{22}$$

U_i, V_j, L_k, F_l are adjusted in each iteration as follows: $U_i \leftarrow U_i - \gamma \cdot \frac{\partial S}{\partial U_i}$, $V_j \leftarrow V_j - \gamma \cdot \frac{\partial S}{\partial V_j}$, $L_k \leftarrow L_k - \gamma \cdot \frac{\partial S}{\partial L_k}$, $F_l \leftarrow F_l - \gamma \cdot \frac{\partial S}{\partial F_l}$. γ is a predefined step length.

A repeated training process, after each iteration, calculates and validates an average absolute error. When the change of the objective function S value is smaller than a predefined small constant iterative process is terminated. After obtaining a terminated iteration U_i, V_j, L_k, F_l, we can predict the user U_i unknown rating to the item V_j. To each target user, a proposed commodity is sorted from high to low according to a calculated predicting rating, and then Top-N recommended list is produced.

5   Experiment and the Analysis of the Results

Dataset in this research is provided from the studies conducted by Massa and Avesani [18] and "Epinions.com" website, since it is among the most often-used datasets for evaluating trust inference performance. Due to the fact that a trust system was built, it expresses the trust relationship between the users and helps the users determine whether to trust the comments of the item [5,10,11]. Statistics concerning this dataset is presented in Table 2.

Commonly used evaluation indexes, namely MAE (Mean Absolute Error) and RMSE (Root Mean Squared Error) were adopted to evaluate the accuracy of the prediction, and then compare the effect of our proposed algorithm with models proposed in literature i.e. PMF model [20], SocialMF model [7], and SoReg [15].

The assignments of λ_P, λ_F are crucial in the proposed method, which plays the role of balance. When we assign $\lambda_P = 0$, the system only considers the user rating matrix and an implicit interest tag. When it recommends something, it

Table 2. Characteristics of dataset

Dataset	Epinions
Number of user	49290
Number of item	139738
Number of rating	664813
Number of trust relationship	487181
Number of interest tag	154

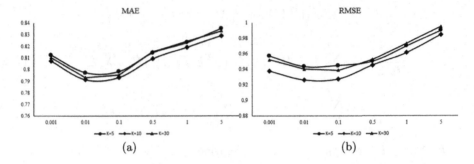

Fig. 5. The influence of parameter λ_P on MAE and RMSE

does not consider the trust relationship between users. If we assign high values to λ_P, the system only recognizes the trust relationships between users, but when recommending, it does not analyze other factors. Similarly, when $\lambda_F = 0$, the system only examines the user rating matrix and the trust relationship between users, however, when recommending, it does not deal with an implicit interest tag of users. When λ_F is enormous, the system only studies the implicit interest tag of users when recommending, not considering other factors.

Figure 5 shows the influence of a parameter λ_P on MAE and RMSE when the number of latent factors are 5, 10 and 30, and other parameters set constant. With the increase of λ_P, MAE and RMSE decrease, namely the accuracy of the prediction is improved. When λ_P reaches a certain threshold with the increase of λ_P, MAE and RMSE increase, namely the accuracy of the prediction is reduced. In conclusion, when $\lambda_P \in [0.01, 0.1]$, the accuracy of recommendation is higher. In latter experiments, we adopt the interval average $\lambda_P = \lambda_F = 0.005$ as the approximate optimal value to conduct an experiment. Figure 6 shows that the influence of a parameter λ_F has similarly more details.

In order to verify the experimental effect, we choose 80% of the whole data as the training set and the remaining 20% of the data constitutes the test set. Recommendations are generated on the basis of known information in the training set. Subsequently, the test set is used to evaluate the performance of recommendation algorithms [1,5,12]. Respectively, 90% of the whole data is the training set, and 10% of the remaining data is the test set for experiments.

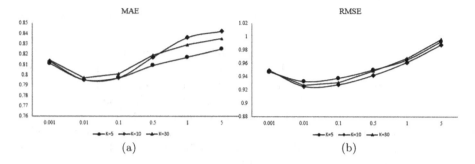

Fig. 6. The influence of parameter λ_F on MAE and RMSE

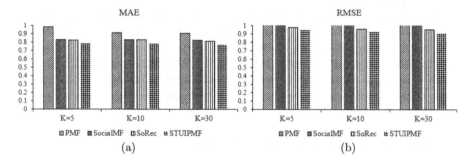

Fig. 7. Comparison of STUIPMF method and other methods under 80% training set

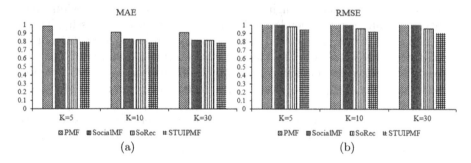

Fig. 8. Comparison of STUIPMF method and other methods under 90% training set

In the experimental process, relevant parameters are selected mainly according to the experimental results for the optimal choice. The parameters' settings in STUIPMF are as follows: $\lambda_U = \lambda_V = \lambda_T = \lambda_L = \lambda = 0.001$, $\lambda_P = \lambda_F = 0.005$. The numbers of latent factors are 5, 10 and 30 respectively.

The parameters' settings in other methods are as follows: in PMF model $\lambda_U = \lambda_V = 0.001$, in Social MF model $\lambda_U = \lambda_V = 0.001$, $\lambda_T = 0.5$, in SoReg model $\lambda_U = \lambda_V = 0.001\alpha = 0.1$. The comparison of experimental results of a STUIPMF method with other methods is presented in Figs. 7 and 8.

According to Figs. 7 and 8, we can come up with the following conclusions, namely

(1) STUIPMF model, we proposed, comprehensively considers the user rating information, user's trust and interest in the case of all experimental parameters chosen optimally. When 80% is the training set, 20% is the test set, then compared with PMF, Social MF, and SoReg, MAE has reduced 17%, 5.8%, 5.3% respectively and RMSE has reduced 21%, 13%, 4% respectively. When 90% is the training set, 10% is the test set, then compared with PMF, Social MF, and SoReg, MAE has reduced 16.2%, 4.1%, 3.7% respectively and RMSE has reduced 20.8%, 13.5%, 4.1% respectively. Therefore, taking into account the analyzed data, the proposed method has improved the recommendation accuracy.
(2) With the increase of latent factors' dimensions, the accuracy of recommendation has improved, but on the other hand, there may be fitting problems. Moreover, computational complexity has increased.
(3) The probability matrix factorization of a user's trust relationship matrix and an interest tag matrix can increase the prior information of user characteristics, so as to solve the problems of a cold start and new users in recommender systems to a large extent.

6   Conclusions and Further Works

With the status and importance of the personalized service in modern economics and social life, it is increasingly prominent to accurately grasp the user's real interests and requirements through user's behavior. What is more, providing high quality personalized recommendation has become the current necessity. Taking into consideration a cold start and data sparseness problems in traditional CF method, we proposed STUIPMF model by integrating a social trust and user interest. We studied an implicit trust relationship between users and potential interest tags from the perspective of user rating. Next, we used PMF model to conduct MF of user ratings information, users trust relationship, and user interest tag information. In result, we analyzed the user characteristics to use data and generate more accurate recommendations. Our proposed method was verified with the use of an experiment based on representative data. The results showed that STUIPMF can improve the recommendation accuracy, make a cold-start easier and solve new user problems to some extent. Meanwhile, it occurred that the STUIPMF approach also has good scalability.

However, our research has revealed many challenges for further study. Take for example, the value λ we used in the model is the approximate optimal value, thus we will determine the optimal value λ and dynamic value changes to improve accuracy of recommendation. In the further research, we are going to verify the effects of the proposed algorithm for new users and for new items in detail. In addition, we will consider adding more information into the proposed model, e.g. text information, location information, time, etc., and pay more attention to the update of the user trust and interest. What is more, we will recognize a conjunction of the distrust relationship between users into the proposed model.

References

1. Bobadilla, J., Ortega, F., Hernando, A., Gutiérrez, A.: Recommender systems survey. Knowl.-Based Syst. **46**, 109–132 (2013)
2. Borchers, A., Herlocker, J., Konstan, J., Reidl, J.: Ganging up on information overload. Computer **31**(4), 106–108 (1998)
3. Gemulla, R., Nijkamp, E., Haas, P.J., Sismanis, Y.: Large-scale matrix factorization with distributed stochastic gradient descent. In: Proceedings of the 17th ACM SIGKDD International Conference on Knowledge Discovery and Data Mining, pp. 69–77. ACM (2011)
4. Golbeck, J.: Personalizing applications through integration of inferred trust values in semantic web-based social networks. In: 2005 Proceedings on Semantic Network Analysis Workshop, Galway, Ireland (2005)
5. Guo, G., Zhang, J., Zhu, F., Wang, X.: Factored similarity models with social trust for top-N item recommendation. Knowl.-Based Syst. **122**, 17–25 (2017)
6. Jamali, M., Ester, M.: Trustwalker: a random walk model for combining trust-based and item-based recommendation. In: Proceedings of the 15th ACM SIGKDD International Conference on Knowledge Discovery and Data Mining, pp. 397–406. ACM (2009)
7. Jamali, M., Ester, M.: A matrix factorization technique with trust propagation for recommendation in social networks. In: Proceedings of the Fourth ACM Conference on Recommender Systems, pp. 135–142. ACM (2010)
8. Kim, H., Kim, H.-J.: A framework for tag-aware recommender systems. Expert Syst. Appl. **41**(8), 4000–4009 (2014)
9. Koenigstein, N., Paquet, U.: Xbox movies recommendations: variational Bayes matrix factorization with embedded feature selection. In: Proceedings of the 7th ACM Conference on Recommender Systems, pp. 129–136. ACM (2013)
10. Lee, W.P., Ma, C.Y.: Enhancing collaborative recommendation performance by combining user preference and trust-distrust propagation in social networks. Knowl.-Based Syst. **106**, 125–134 (2016)
11. Li, J., Chen, C., Chen, H., Tong, C.: Towards context-aware social recommendation via individual trust. Knowl.-Based Syst. **127**, 58–66 (2017)
12. Lim, H., Kim, H.-J.: Item recommendation using tag emotion in social cataloging services. Expert Syst. Appl. **89**, 179–187 (2017)
13. Lu, Z., Agarwal, D., Dhillon, I.S.: A spatio-temporal approach to collaborative filtering. In: Proceedings of the Third ACM Conference on Recommender Systems, pp. 13–20. ACM (2009)
14. Luo, X., Xia, Y., Zhu, Q.: Incremental collaborative filtering recommender based on regularized matrix factorization. Knowl.-Based Syst. **27**, 271–280 (2012)
15. Ma, H., Yang, H., Lyu, M.R., King, I.: SoRec: social recommendation using probabilistic matrix factorization. In: Proceedings of the 17th ACM Conference on Information and Knowledge Management, pp. 931–940. ACM (2008)
16. Ma, H., Zhou, D., Liu, C., Lyu, M.R., King, I.: Recommender systems with social regularization. In: Proceedings of the Fourth ACM International Conference on Web Search and Data Mining, pp. 287–296. ACM (2011)
17. Ma, H., Zhou, T.C., Lyu, M.R., King, I.: Improving recommender systems by incorporating social contextual information. ACM Trans. Inf. Syst. (TOIS) **29**(2), 9 (2011)
18. Massa, P., Avesani, P.: Trust-aware recommender systems. In: Proceedings of the 2007 ACM Conference on Recommender Systems, pp. 17–24. ACM (2007)

19. Mi, C., Shan, X., Qiang, Y., Stephanie, Y., Chen, Y.: A new method for evaluating tour online review based on grey 2-tuple linguistic. Kybernetes **43**(3/4), 601–613 (2014)
20. Mnih, A., Salakhutdinov, R.R.: Probabilistic matrix factorization. In: Advances in Neural Information Processing Systems, pp. 1257–1264 (2008)
21. Sun, X., Kong, F., Ye, S.: A comparison of several algorithms for collaborative filtering in startup stage. In: 2005 IEEE Proceedings of Networking, Sensing and Control, pp. 25–28. IEEE (2005)
22. Tao, J., Zhang, N.: Similarity measurement method based on user's interesting-ness in collaborative filtering. Comput. Syst. Appl. **20**(5), 55–59 (2011)
23. Zuo, Y., Zeng, J., Gong, M., Jiao, L.: Tag-aware recommender systems based on deep neural networks. Neurocomputing **204**, 51–60 (2016)

Automating Feature Extraction and Feature Selection in Big Data Security Analytics

Dimitrios Sisiaridis$^{(\boxtimes)}$ and Olivier Markowitch

Departement d'Informatique, QualSec Group, Université Libre de Bruxelles,
Brussels, Belgium
{dimitrios.sisiaridis,olivier.markowitch}@ulb.ac.be
https://qualsec.ulb.ac.be/

Abstract. Feature extraction and feature selection are the first tasks in pre-processing of input logs in order to detect cybersecurity threats and attacks by utilizing data mining techniques in the field of Artificial Intelligence. When it comes to the analysis of heterogeneous data derived from different sources, these tasks are found to be time-consuming and difficult to be managed efficiently.

In this paper, we present an approach for handling feature extraction and feature selection utilizing machine learning algorithms for security analytics of heterogeneous data derived from different network sensors. The approach is implemented in Apache Spark, using its python API, named pyspark.

Keywords: Machine learning · Feature extraction
Security analytics · Apache Spark

1 Introduction

The augmentation of cyber security attacks during the last years emerges the need for automated traffic log analysis over a long period of time at every level of the enterprise or organisation information system. By utilising Artificial Intelligence (AI) techniques leveraged by machine learning and data mining methods, a *learning engine* would enable the consumption of seemingly unrelated disparate datasets, to discover correlated patterns that result in consistent outcomes with respect to the access behaviour of users, network devices and applications involved in risky abnormal actions, and thus reducing the amount of security noise and false positives. Machine learning algorithms can be used to examine, for example, statistical features or domain and IP reputation [4].

Data acquisition and data mining methods, with respect to different types of attacks such as *targeted* and *indiscriminate* attacks, provide a perspective of the threat landscape. Enhanced log data are then analysed for new attack patterns and the outcome, e.g. in the form of behavioural risk scores and historical baseline

© Springer International Publishing AG, part of Springer Nature 2018
L. Rutkowski et al. (Eds.): ICAISC 2018, LNAI 10842, pp. 423–432, 2018.
https://doi.org/10.1007/978-3-319-91262-2_38

profiles of normal behaviour, is forwarded to update the learning engine. Any unusual or suspected behaviour can then be identified as an anomaly or an outlier in real or near real-time. In this way, the analysis leverages the integration of credible and actionable threat data to other security devices, in order to protect, guarantee and remediate actual threats, to get insight on how the breach occurred, thus to aid forensic investigations and to prevent future attacks [3].

In this paper, we propose an automated approach for feature extraction and feature selection using machine learning methods, as the first stages of a modular approach for the detection and/or prediction of cybersecurity attacks. For the needs of our experiments we employed the Spark framework and more specifically its python API, pyspark. Section 2 deals with the task of extracting data from logs of increased data complexity. In Sect. 3 we propose methods for the task of feature selection, while our conclusions are presented in Sect. 4.

2 Extracting Features from Heterogeneous Data

In our experiments, we examine the case where we have logs of records derived as the result of an integration of logs produced by different network tools and sensors (heterogeneous data from different resources). Each one of them monitors and records a view of the system in the form of records of different attributes and/or of different structure, implying thus an increased level of interoperability problems in a multi-level, multi-dimensional feature space; in the end, each network monitoring tool produces its own schema of attributes.

In such cases, information is usually hidden in multi-level complex structures. It is typical that the number of attributes is not constant across the records, while the number of complex attributes varies as well. On the other hand, there are attributes, e.g., dates, expressed in several formats, or other attributes referred to the same piece of information by using slightly different attribute names (Fig. 1). Most of them are categorical, in a string format while the inner datatype varies from nested dictionaries, linked lists or arrays of further complex structure; each one of them may present its own multi-level structure which increases the level of complexity. In such cases, a clear strategy has to be followed for feature extraction. Therefore, we have to deal with flattening and interoperability solving processes.

2.1 Time Series in Heterogeneous Unlabelled Data

While working with the analysis of heterogeneous data taken from different sources, pre-process procedures, such as feature extraction, feature selection and feature transformation, need to be carefully designed in order not to miss any security-related significant events in the time series. These tasks are usually time-consuming producing thus significant delays to the overall time of the data analysis. That is main motivation in this work: to reduce the time needed for feature extraction and feature selection in data exploration analysis by automating the process. In order to achieve it, we utilise the data model abstractions and we keep to a minimum any access to the actual data.

LOGS FROM DIFFERENT SOURCES

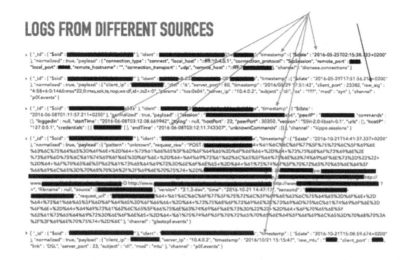

Fig. 1. Logs from different input sources

These time-series are defined in terms of time spaces as the contextual attributes: date attributes will be decomposed to time windows such as *year, month, day of a week, hour* and *minute*, following the approach proposed in [2] and stored in `parquet` files as can be seen in Fig. 2. This particular format has been also chosen for another reason. In parallel with the actual storing of data in the compressed parquet format, metadata, such as the actual attribute labels, are also stored in a separate file (Fig. 2), which can be deployed to extract the abstract schemas in the input data. Statistics then are calculated and stored for batch or online mode; they will be stored in `HIVE` tables, or in temporary views for ad-hoc temporal real-time analysis. Alternatively, they can be calculated for a single time space (e.g. a specific day), or by using a user-defined variable window time space. Thus, the structure in Fig. 2, may be extended to monitor and keep statistics for an *hour* or a *minute*, by defining the relevant time windows utilizing functions available in the `MLlib` library such as the *window()*, *month()*, *hour()* or *minute()*, or by defining *User Defined Functions* (UDFs) for more flexibility.

Experiments at the exploratory data stage revealed that the number of single feature attributes in this log were between a range of 7 (the smallest number of attributes of a distinct feature space) up to 99 attributes (corresponding to the total number of the overall available feature space). A fact, that led us to carry on with feature extraction by focusing on flattening multi-nested records separately for each different structure (under a number of 13 different baseline structures).

2.2 Local Flattening of Input Data

In order to reduce the data complexity while working with security analytics of heterogeneous data, we propose the local flattening of input data, in terms of

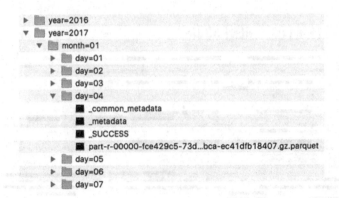

Fig. 2. Time series stored in parquet files following a tree structure

identifying all the different schemas in metadata. First, it is a *bottom-up* analysis by re-synthesing results to answer to either simple of complex questions. In the same time, we can define hypotheses to the full set of our input data (i.e. *top-down* analysis). Thus, it is a complete approach in data analytics, by allowing data to *tell their story*, in a concrete way, following a minimum number of steps. In this way, we are able to:

- keep the number of assumptions to a minimum
- look for misconfigurations and data correlations into the abstract dataframes definitions
- keep access to the actual data to a minimum
- provide solutions in interoperability problems, such as:
  - different representations of date attributes
  - namespace inconsistencies (e.g. attributes with names such as *prot, protocol, connectionProtocol*)
- cope with complex structures of different number of inner levels
- deal with event ordering and time-inconsistencies (as it is described in [5]).

2.3   Feature Extraction in Apache Spark

In `Apache Spark`, data are organised in the form of *dataframes*, which resemble the well-known *relational tables*: there are *columns* (aka *attributes* or *features* or *dimensions*) and *rows* (i.e. events recorded, for example, by a network sensor, or a specific device). The list of columns and their corresponded datatypes define the *schema* of a dataframe. In each dataframe, its columns and rows i.e. its schema is unchangeable. An example of a schema could be the following:

```
DataFrame[id: string, @timestamp: string, honeypot: string,
payloadCommand: string]
```

A sample of recorded events of this dataframe schema is shown in Fig. 3:

▸events

```
+------------------------------+-----------------------------------+---------+--------------+
|id                           |@timestamp                        |honeypot |payloadCommand|
+------------------------------+-----------------------------------+---------+--------------+
|5776664eb3c585471bef1bb5|2016-07-01T14:48:37.839108389+02:00|1█████████|              |
|57b309abb3c585471bf4d920|2016-08-16T14:43:23.40957412+02:00 |1█████████|              |
|57c738b6b3c585471bf60c81|2016-08-31T22:10:00.535912321+02:00|1█████████|              |
|57ddf66bb3c585471bf7b508|2016-09-18T04:09:56.331084531+02:00|1█████████|              |
|57ddf76fb3c585471bf7b514|2016-09-18T04:14:16.533762285+02:00|1█████████|wget          |
|57de0864b3c585471bf7b5be|2016-09-18T04:52:29.215091494+02:00|1█████████|              |
|57de016bb3c585471bf7b5c8|2016-09-18T04:56:52.565757121+02:00|1█████████|wget          |
|57dfe3f2b3c585471bf7e90b|2016-09-19T15:15:42.293763881+02:00|1█████████|curl          |
|57dfe478b3c585471bf7e913|2016-09-19T15:17:56.123639101+02:00|1█████████|              |
|57dfe88ab3c585471bf7e953|2016-09-19T15:35:19.002310154+02:00|1█████████|              |
+------------------------------+-----------------------------------+---------+--------------+
```

Fig. 3. A sample of recorded events

The following steps refer to the case in which logs/datasets are ingested in .json format. Our approach examines the data structures on their top-level, focusing on abstract schemas and re-synthesis of previous and new dataframes, in an automatic way. Access to the actual data is only taken place when there is a need to find schemas in dictionaries and only by *retrieving* just one of the records (thus, even if we have a dataframe of million/billions of events, we only examine the schema of the first record/event).

Steps for feature extraction

A. load the logfile in a spark dataframe
B. find and remove all single-valued attributes
   (this steps applies also to the 'feature selection' section)
C. flatten complex structures
   a. find and flatten all columns of complex structure
      (the steps are run recursively, down to the lowest complex
      attribute of the hierarchy of complex attributes)
      i. e.g. struct, nested dictionaries, nested lists,
         arrays, etc.
         (i.e. currently those which their value is of RowType)
   b. remove all the original columns of complex structure
D. convert all time-fields into timestamps, using distinct
   time fields in the dataframes
E. integrate similar fields in the list of dataframes

. In Fig. 4, in the left-hand schema, attributes _id is of the datatype struct. The actual value is given by the inner-level attribute, $oid. The same stands for the outer attribute timestamp: the actual date value can be searched in the inner-level attribute $date. In both cases, attributes $oid and $date are extracted in the form of two new columns, named _id_ and dateOut; the original attributes _id and timestamp are then deleted, having thus a new schema on the right-side. In this way, we achieved to reduce the complexity of the original input schema to a new one of lower complexity.

Quite often there are attributes which act as *containers* of information. In Fig. 4, the exploratory analysis has revealed that the *payload* attribute, although of a *string* datatype, it represents actually a dictionary in the form of a list of

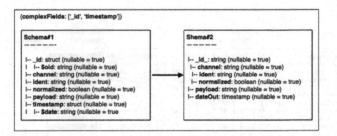

Fig. 4. Transform complex fields: attributes _id and *timestamp*

multi-nested dictionaries; each one of the latter present a complex structure with further levels. These different schemas found in *payload* are presented in Fig. 5.

The new-created dataframes schemas correspond to the different schemas of the payload attribute in Fig. 5. By following this approach, data are easier to be handled: in the next stages, they will be cleaned, transformed from categorical to numerical and then they will be further analyzed in order to detect anomalies in entities bevaviour.

the **Payload** column

```
1:  ['remote_host', 'connection_protocol', 'local_port', 'connection_type',
'remote_hostname', 'remote_port', 'local_host', 'connection_transport']

2:  ['client_ip', 'app', 'timestamp', 'server_ip', 'params', 'raw_sig', 'dist',
'client_port', 'mod', 'server_port', 'subject']

3:  ['client_ip', 'server_ip', 'timestamp', 'uptime', 'subject', 'client_port',
'raw_freq', 'server_port', 'mod']

4:  ['client_ip', 'server_ip', 'timestamp', 'reason', 'raw_hits', 'subject',
'client_port', 'mod', 'server_port']

5:  ['client_ip', 'server_ip', 'timestamp', 'os', 'params', 'raw_sig', 'dist',
'client_port', 'mod', 'server_port', 'subject']

6:  ['client_ip', 'server_ip', 'timestamp', 'link', 'subject', 'client_port', 'mod',
'server_port', 'raw_mtu']

7:  ['hostIP', 'loggedin', 'commands', 'unknownCommands', 'startTime', 'peerPort',
'version', 'urls', 'session', 'ttylog', 'credentials', 'endTime', 'peerIP', 'hostPort']

8:  ['sensorid', 'request_raw', 'request_url', 'filename', 'source', 'pattern',
'version', 'time']

9:  ['tos', 'ttl', 'ethdst', 'ethtype', 'udplength', 'sensor', 'priority',
'destination_ip', 'signature', 'classification', 'id', 'ethlen', 'dgmlen',
'destination_port', 'header', 'source_port', 'proto', 'source_ip', 'iplen', 'ethsrc']

10: ['destination_port', 'timestamp', 'tcpflags', 'tcpwin', 'dgmlen', 'tcpack',
'classification', 'sensor', 'proto', 'tcpseq', 'header', 'source_ip', 'iplen', 'tos',
'ttl', 'ethtype', 'priority', 'destination_ip', 'id', 'tcplen', 'ethlen', 'ethdst',
'source_port', 'signature', 'ethsrc']

11: ['timestamp', 'destination_ip', 'dgmlen', 'classification', 'sensor', 'proto',
'header', 'source_ip', 'iplen', 'tos', 'ttl', 'ethtype', 'priority', 'icmpcode', 'id',
'icmpseq', 'ethlen', 'ethsrc', 'ethdst', 'icmpid', 'signature', 'icmptype']

12: ['daddr', 'md5', 'url', 'dport', 'sport', 'sha512', 'saddr']

13: ['url', '@timestamp', 'honeypot', 'payloadCommand', 'headers', 'method',
'payloadMd5', 'form', 'payloadBinary', 'payload', 'payloadResource', 'type', 'source']
```

Fig. 5. Schemas in the *payload* attribute

In the next step, we proceed with the process of flattening further the *payload* attribute into its inner-level attributes, illustrated in Fig. 6. Here for example, feature *raw_sig* is in the form of an array. By applying consecutive transformations automatically, we manage to extract all inner attributes, which simplifies the process of correlating data in the next stage. Thus, by looking into the *raw_sig* column, we identify inner values separated by ':', which further are decomposed into new features derived by the inner levels, as it is depicted e.g. for column *attsCol5*; the latter could be further split by leading to two new columns (e.g. with values 1024 and 0, respectively), as this process is recursive and automated; special care is given how we name the new columns, in order to follow the different paths of attributes decomposition.

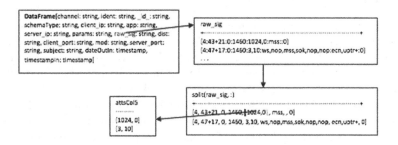

Fig. 6. Transforming array fields

3 Feature Selection

The process of feature selection (FS) is crucial for the next analysis steps. As was explained in Sect. 3.1, our motivation in our approach is to reduce data complexity in parallel with a significant reduction of the time needed for applying security analytics in un-labelled data. As we are aiming ultimately to detect anomalies as strong form of outliers in order to improve quantitative metrics such as, to increase accuracy and detection rates or to decrease security noise to a minimum, we need to select the data that are more related to our questions. Dimensionality reduction can play a significant role in complex event processing, especially when data are coming from different sources and different forms.

We present four methods to achieve this goal:

- Leave-out single-value attributes
- Namespace correlation
- Data correlation using the actual values
- FS in case of having a relative small number of categories.

3.1 Leave Out Single-Value Attributes

The first method is quite simple: all single-valued attributes are removed from the original dataframe. For example, consider the dataframe schema in Fig. 7. Attribute *normalized* of datatype Boolean takes the value *True* for all the events in our integrated log and therefore we drop the relevant column, which leads to a new dataframe schema.

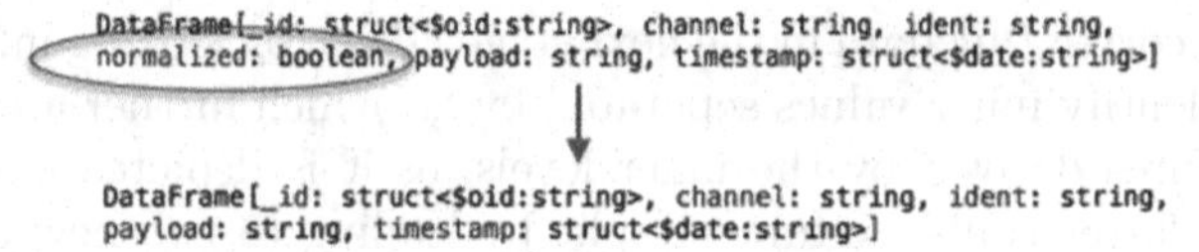

```
DataFrame[_id: struct<$oid:string>, channel: string, ident: string,
normalized: boolean, payload: string, timestamp: struct<$date:string>]
```

↓

```
DataFrame[_id: struct<$oid:string>, channel: string, ident: string,
payload: string, timestamp: struct<$date:string>]
```

Fig. 7. Attribute *normalized* is left out as it presents a value of *True* in all records

3.2 Namespace Correlation

It is quite common when data inputs are coming from different sources to deal with entity attributes which refer to the same piece of information although their names are slightly different. For example, attributes *proto* and *connection_protocol* refer to the actual protocol used in a communication channel. Different tools used by experts to monitor network traffic do not follow a unified namespace scheme. This fact, could lead to misinterpretations, information redundancy and misconfigurations in data modelling, among other obstacles in data exploration stage; all these refer mainly to problems in interoperability, as can be seen in Fig. 1. By solving such inconsistencies, we achieve to further reduce data complexity as well as to reduce the overall time for data analysis. In [5] we have presented an approach to handle such interoperability issues by utilizing means derived by the theory of categories while we extend the work presented in [1].

3.3 Using Pearson Correlation to Reduce the Number of Dimensions

As long as data inputs, in the form of dataframes, are cleaned, transformed, indexed and scaled into their corresponding numerical values, and before the process of forming the actual feature vectors that will be used in clustering, by using data correlation, we are able to achieve a further reduction of the dimensions that will be used for the actual security analytics.

The outcome of applying this technique, using Pearson correlation, is presented in Fig. 8. Attributes highly correlated may be omitted while defining the relevant clusters; the choice of the particular attribute to be left out is strongly related to the actual research of interest. For example, we are interested to monitor the behaviour in local hosts and to detect any anomalies deviate by patterns of normal behaviour. Experiments have shown that this technique can be used effectively for categorical attributes presenting at least five categories.

3.4 Feature Selection in Case of Having a Relative Small Number of Categories

In case where we deal with categorical attributes presenting a relative small number of categories, i.e. numberOfCategories less than 4 we propose the following steps in order to achieve a further feature reduction. We distinguish the cases

Fig. 8. Applying *Pearson correlation* to indexed and scaled data for feature selection

where data are unlabelled (lack of any indication for a security-related event) and the case where some or all the labels are available. We need to mention that in real scenarios, usually we need to cope with either fully un-labelled data or highly-unbalanced data (i.e. where only few instances of the rare/anomalous class are available). While working with un-labelled data, for the set of these features, select each one of them as the feature-label attribute and then either:

- Use a *decision tree* with a *multi-class classification evaluator* to further reduce the number of the dimensions (by following one or more of the aforementioned techniques)
- Create 2^n sub-dataframes with respect to the number of categories
- Calculate features importance using a *Random Forest* classifier
- Use an ensemble technique in the form of a *Combiner* e.g. a *Neural Network* or a *Bayes* classifier, running a combination of the above techniques to optimize results in the next levels of the analysis (e.g. to further optimize detection rates).

While working with labelled data, we select features using the *ChiSquare* test of independencies. In our experiments, with respect to the input data, we have used four different statistical methods, available in **Spark** *MLlib* library, such as the number of top features, a fraction of the top features, p-values below a threshold to control the false positive rate, or p-values with false discovery rate below a threshold.

4 Conclusions

We have presented an approach to handle efficiently the tasks of feature extraction and feature selection while working with security analytics. It is an automated solution to handle interoperability problems. It is based on a continuous transformation of the abstract definitions of the data inputs, as access to the actual data is limited to a minimum read actions of the first record of a dataframe, and only when it is needed to extract the inner schema of a dictionary-based attribute. The latter is especially important for big data security analytics, while analysing vast amount of heterogeneous data from different sources.

In our experiments we used as input data an integrated log of recorded events produced by a number of different network tools, applied on telco system.

It worths to be mentioned that for this pre-processing analysis stage was used a single server of 2xCPUs, 8cores/CPU, 64 GB RAM, running an Apache Hadoop installation v2.7 with Apache Spark v2.1.0.

We are currently working into formalizing the approach by utilizing novel structures derived from the theory of categories as it has been presented in [5], towards an overall optimization.

References

1. Bird, S., Klein, E., Loper, E.: Natural Language Processing with Python. O'Reilly Media Inc., Sebastopol (2009)
2. Veeramachaneni, K., Arnaldo, I., Cuesta-Infante, A., Korrapati, V., Bassias, C., Li, K.: AI2: training a big data machine to defend. In: IEEE International Conference on Big Data Security, New York, NY, USA, June 2016
3. Shyu, M.-L., Huang, Z., Luo, H.: Efficient mining and detection of sequential intrusion patterns for network intrusion detection systems. In: Yu, P.S., Tsai, J.J.P. (eds.) Machine Learning in Cyber Trust, pp. 133–154. Springer, Heidelberg (2009). https://doi.org/10.1007/978-0-387-88735-7_6
4. Sisiaridis, D., Carcillo, F., Markowitch, O.: A framework for threat detection in communication systems. In: Proceedings of the 20th Pan-Hellenic Conference on Informatics, pp. 68:1–68:6. ACM (2016)
5. Sisiaridis, D., Kuchta, V., Markowitch, O.: A categorical approach in handling event-ordering in distributed systems. In: Parallel and Distributed Systems (ICPADS), pp. 1145–1150. IEEE (2016)

Improvement of the Simplified Silhouette Validity Index

Artur Starczewski[1]([✉]) and Krzysztof Przybyszewski[2,3]

[1] Institute of Computational Intelligence, Częstochowa University of Technology,
Al. Armii Krajowej 36, 42-200 Częstochowa, Poland
artur.starczewski@iisi.pcz.pl
[2] Information Technology Institute, University of Social Sciences,
90-113 Łódź, Poland
[3] Clark University, Worcester, MA 01610, USA

Abstract. The fundamental issue of data clustering is an evaluation of results of clustering algorithms. Lots of methods have been proposed for cluster validation. The most popular approach is based on internal cluster validity indices. Among this kind of indices, the *Silhouette* index and its computationally simpled version, i.e. the *Simplified Silhouette*, are frequently used. In this paper modification of the *Simplified Silhouette* index is proposed. The suggested approach is based on using an additional component, which improves clusters validity assessment. The performance of the new cluster validity indices has been demonstrated for artificial and real datasets, where the *PAM* clustering algorithm has been applied as the underlying clustering technique.

Keywords: Clustering · Cluster validity index
PAM clustering technique

1 Introduction

Data clustering aims to discover natural existing structures in a dataset. For this purpose, data are partitioned into groups (clusters) of objects. Objects within a cluster are similar, whereas they are dissimilar in different clusters. Since there is a large variety of datasets different clustering algorithms and their configurations are still created, e.g. [9,11,12,31]. Note that among clustering methods two major categories are distinguished: partitioning and hierarchical clustering. For example, the well-known partitioning algorithms are, e.g. *K-means*, *Partitioning Around Medoids* (*PAM*) [5,24] and *Expectation Maximization* (*EM*) [21]. Whereas the agglomerative hierarchical clustering includes such methods as, e.g. the *Single-linkage*, *Complete-linkage* or *Average-linkage* [16,22,25]. Data clustering is applied in many areas, such as biology, spatial data analysis, business and so on. It can be noted that there is no a clustering algorithm, which creates the right data partition for all datasets. Moreover, the

© Springer International Publishing AG, part of Springer Nature 2018
L. Rutkowski et al. (Eds.): ICAISC 2018, LNAI 10842, pp. 433–444, 2018.
https://doi.org/10.1007/978-3-319-91262-2_39

same algorithm can also give different results depending on the input parameters. Therefore, cluster validation should be used to assess the results of data clustering. Generally, it is a very difficult task and is the most frequently realized by validity indices. Techniques of the cluster validation are usually classified into three groups, i.e. external, internal and relative validation [16,30]. The external validation is based on a comparison of partitions of a dataset obtained by a clustering algorithm with the correct partition of this data. In turn, the internal approach uses only the intrinsic properties of the dataset. On the other hand, the relative validation method compares the data partitions obtained by chaining input parameters of a clustering algorithm. It should be noted that the number of clusters is the key parameter for many clustering algorithms. So far, a number of authors have proposed different validity indices or modifications of existing indices, e.g., [1,10,18,29,32,33,36,38]. Among internal cluster validity indices, the *Silhouette (SIL)* [26] and *Simplified Silhouette (SimSIL)* [15] indices are frequently used to evaluate the efficacy of the clustering algorithms in detecting the right data partitioning. It is important to note that clustering methods in conjunction with cluster validity indices can be used during a process of designing various neural networks [2–4,6,17], neuro-fuzzy structures [7,8,20,27,28] and creating some algorithms for identification of classes [13,14].

In this paper, new cluster validity indices called the $SimSILA$ and the $SimSILAv1$, are presented. These new indices modify the *Simplified Silhouette (SimSIL)* index. The proposed approach is based on an additional component and it is a detailed explanation in Sect. 3. In order to present effectiveness of the validity indices, several experiments were performed for various datasets. This paper is organized as follows: Sect. 2 presents a detailed description of the *Silhouette*, $SILA$ and $SILAv1$ indices. In Sect. 3 the *Simplified Silhouette*, $SimSILA$ and $SimSILAv1$ indices are outlined. Section 4 illustrates experimental results on datasets. Finally, Sect. 5 presents conclusions.

2   Modification of the Silhouette Index

In this section modification of the *Silhouette (SIL)* index is described. This approach was proposed and discussed in papers [34,35]. Let us denote K-partition scheme of a dataset X by $C = \{C_1, C, ..., C_K\}$, where C_k indicates k_{th} cluster, $k = 1, .., K$. The original *SIL* index is presented as follows:

$$SIL = \frac{1}{K} \sum_{k=1}^{K} SIL(C_k) \tag{1}$$

where $SIL(C_k)$ is the *Silhouette width* for the given cluster C_k and is defined as:

$$SIL(C_k) = \frac{1}{n_k} \sum_{\mathbf{x} \in C_k} \frac{b(\mathbf{x}) - a(\mathbf{x})}{max\,(a(\mathbf{x}), b(\mathbf{x}))} \tag{2}$$

n_k is a number of elements in C_k, and $a(\mathbf{x})$ is the within-cluster mean distance, i.e. it is the average distance between $\mathbf{x}$ and the rest of the patterns belonging to the same cluster, $b(\mathbf{x})$ is the smallest of the mean distances of $\mathbf{x}$ to the elements belonging to the other clusters. The values of the index are from the range -1 to 1 and a maximum value (close to 1) provides the best partitioning of the dataset.

Now let us turn to the modification of this index [34]. This approach is based on using an additional component, which improves a performance of the index. The new index is called $SILA$ index and it is defined as follows:

$$SILA = \frac{1}{n} \left(\sum_{\mathbf{x} \in X} \left(\frac{b(\mathbf{x}) - a(\mathbf{x})}{\max{(a(\mathbf{x}), b(\mathbf{x}))}} \cdot \frac{1}{(1 + a(x))^q} \right) \right) \tag{3}$$

where the exponent q is equal 1 and n is the number of elements in a dataset. A maximum value of the new index indicates the right partition scheme. Noted that the choice of the value of the q is very important and $q = 1$ can be too small for the very large differences of distances between data points. Hence, the new concept was proposed in paper [35]. This new index, called $SILAv1$, can be presented by Eq. (3), where q is defined as below:

$$q = 2 + \frac{K^2}{n} \tag{4}$$

Generally, the $SILA$ and $SILAv1$ indices ensure a better performance compared to the original *Silhouette* index. In the next section, a detailed explanation of modification of the *Simplified Silhouette* index is presented.

3   Modification of the Simplified Silhouette Index

It can be noted that the *Silhouette* index depends on of the computation of all the distances between data elements and it can lead to a computational cost $O(mn^2)$ [37], where m is the number of features. On the other hand, the *Simplified Silhouette* index is much less computationally expensive, and the overall complexity of the computation of the index is estimated as $O(kmn)$ [37]. Although the *Simplified Silhouette* index is similar to the *Silhouette* index, there are very significant differences. First, the distance of $\mathbf{x}$ to the cluster is not the average distance between $\mathbf{x}$ and the rest of the elements belonging to the same cluster. It is calculated as the distance between $\mathbf{x}$ and the centroid of the cluster and can be written as follows:

$$\hat{a}(\mathbf{x}) = d\left(\mathbf{x}, \bar{C}_k\right) \tag{5}$$

where $\bar{C}_k$ is the centroid of the cluster C_k and $d\left(\mathbf{x}, \bar{C}_k\right)$ is a function of the distance between $\mathbf{x}$ and $\bar{C}_k$. Next, the distance of $\mathbf{x}$ to the other cluster is defined as follows:

$$\hat{b}(\mathbf{x}) = \min_{\substack{l=1 \\ l \neq k}}^{K} d(\mathbf{x}, \bar{C}_l) \tag{6}$$

where $\bar{C}_l$ is the centroid of the cluster C_l and $l \neq k$. Finally, the *Simplified Silhouette* (*SimSIL*) index is defined as:

$$SimSIL = \frac{1}{n} \sum_{x \in X} \frac{\hat{b}(x) - \hat{a}(x)}{\max(\hat{a}(x), \hat{b}(x))} \tag{7}$$

where n is the number of elements in the dataset X. The value of the index is also from the range -1 to 1 and a maximum value indicates the right partition scheme.

As in the previous index, the modification of the *Simplified Silhouette* index is based on using the additional component, which is expressed as:

$$\hat{A}(x) = \frac{1}{(1 + \hat{a}(x))^q} \tag{8}$$

For the exponent $q = 1$, the newly proposed index is called *SimSILA* and can be written as:

$$SimSILA = \frac{1}{n} \left(\sum_{x \in X} \left(\frac{\hat{b}(x) - \hat{a}(x)}{\max\left(\hat{a}(x), \hat{b}(x)\right)} \cdot \frac{1}{(1 + \hat{a}(x))^q} \right) \right) \tag{9}$$

It can be noted that the additional component $\hat{A}(x)$ corrects the value of the index. When a clustering algorithm greatly increases sizes of clusters, the ratio of $1/(1 + \hat{a}(x))^q$ decreases significantly and the value of the index is also decreased. However, the value $q = 1$ can be too small to appropriately correct the *SimSILA* index. Hence, the issue of the choice of the exponent q for $\hat{A}(x)$ is a very significant problem. As with the previous index, the new index called *SimSILAv1* is proposed and contains a formula of the change of the exponent q depending on the number of clusters. This formula is expressed by (4). Thus, the *SimSILAv1* index can be presented by Eq. (9), where q is calculated by (4). It should be noted that the new indices can take values between $1/(1 + \hat{a}(x))^q$ and $-1/(1 + \hat{a}(x))^q$. A maximum value of the index selects the right data partitioning for a dataset. In the next section, the results of the experimental studies are presented to confirm the effectiveness of these new indices.

4 Experimental Results

In this section, several experiments have been conducted on artificial and real datasets using the *Partitioning Around Medoids* (*PAM*) clustering algorithm. This algorithm is a realisation of *K-medoid* clustering, which is a more robust version of *K-means* method. Both *K-medoids* and the *K-means* algorithm are partitional, but the first method searches K representative data elements (medoids) among all elements of a dataset. After finding K medoids, K clusters are created by assigning each data point to the nearest medoid. In contrast to the *K-means*, the *K-medoids* algorithm chooses data elements as centers (medoids). Moreover, the Manhattan Norm is used to define distances between elements of the dataset. These make that the PAM algorithm is robust to noise and outliers. As mentioned in Sect. 1, the different parameter configurations of clustering algorithms can lead to different results. Thus, the choice of these input parameters is a key issue. Furthermore, one of the essential configuration parameters is a number of clusters. This parameter should be set before the start of the algorithm, but it is not usually known in advance. The common way to resolve this problem is to run the clustering algorithm multiple times with a different number of clusters and select the best result. For the clustering analyze, the range of the different number of clusters should be varied from $K_{min} = 2$ to $K_{max} = \sqrt{n}$, [23]. Whereas, the evaluation of results is usually realized by cluster validity indices. In experiments conducted on artificial and real datasets, the six indices, i.e. the *Silhouette (SIL)*, *SILA*, *SILAv1*, *Simplified Silhouette (SimSil)*, *SimSILA* and *SimSILAv1* are used to determine the right number of clusters. To show the efficacy of the new validity indices, the results are also presented on the plots. It is assumed that the value of the validity indices equals 0 for $K = 1$. Furthermore, the *min-max* normalization of data has been applied to all the datasets used in the experiments. In order to better compare of the new indices, the maximum value of all the indices is modified and it is equal to 1.

4.1   Datasets

In the conducted experiments four artificial and six real datasets are used. The artificial data was called *Data* 1, *Data* 2, *Data* 3 and *Data* 4 and they were 2-dimensional with 3, 5, 8 and 11 clusters, respectively. Note that they consist of various cluster structure and densities. The scatter plot of these data is presented in Fig. 1. As it can be observed on the plot the distances between clusters are very different and some clusters are quite close. Generally, clusters are located in groups and some of the clusters are very close and others quite far. Moreover, the sizes of the clusters are different and they contain the various number of elements. Hence, many clusters validity indices can provide incorrect partitioning schemes. The real datasets are numeric data from the UCE Irvine Machine Learning Repository [19]: *Diabetes*, *Ecoli*, *Glass*, *Iris*, *Spectf*, *Wine*. The *Diabetes* dataset includes results of studies relating to the signs of diabetes in patients. This set includes 768 instances belonging to 2 classes and each item is described by 8 features. The second set is *Ecoli* dataset consisting of 336 instances, and the number of attributes equals 7. It has 8 classes, which represent the protein localization sites. Next comes the *Glass* dataset, which contains information about 6 types of glass defined in terms of their oxide content. The set has 214 instances and each of them is described by 9 attributes. The well-known *Iris* data are extensively used in many comparisons of classifiers. This set has three classes, which contain 50 instances per class. Moreover, each item is represented by four features. The *Spectf* dataset describes diagnosing of cardiac Single Proton Emission Computed Tomography images. This set includes 267 instances and each of them is described by 44 features. It has 2 classes. Finally, the *Wine* dataset shows the results of a chemical analysis of wines. It comprises

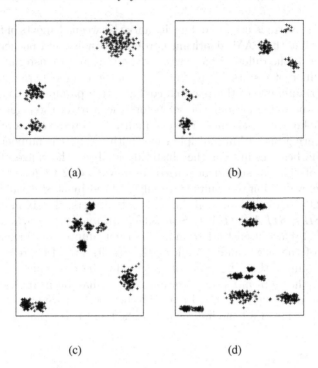

Fig. 1. 2-dimensional artificial datasets: (a) *Data* 1, (b) *Data* 2, (c) *Data* 3, and (d) *Data* 4

Table 1. A detailed description of the artificial datasets

Datasets	No. of elements	Features	Classes
Data 1	300	2	3
Data 2	170	2	5
Data 3	495	2	8
Data 4	665	2	11
Diabetes	768	8	2
Ecoli	336	7	8
Glass	214	9	6
Iris	150	4	3
Spectf	267	44	2
Wine	178	13	3

three classes of wines. Altogether, the dataset contains 178 patterns, where each of them is described by 13 features.

Additionally, Table 1 shows a detailed description of these datasets used in experiments.

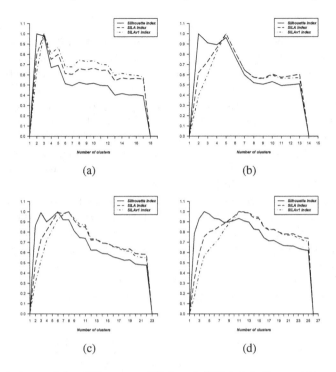

Fig. 2. Variations of the *Silhouette*, *SILA* and *SILAv*1 indices with respect to the number of clusters for 2-dimensional datasets: (a) *Data* 1, (b) *Data* 2, (c) *Data* 3, and (d) *Data* 4 partitioned by the *PAM* method.

4.2   Experiments

The experimental analysis is designed to evaluate the performance of the new indices. In these studies, the partitional *PAM* method as the underlying clustering method was adopted to clustering of the datasets. First of all, the *Silhouette*, *SILA* and *SILAv*1 indices are analyzed. For this purpose, the 2-dimensional *Data* 1, *Data* 2, *Data* 3 and *Data* 4 datasets have been clustered by the *PAM* algorithm. As shown in Fig. 1, these datasets create groups of clusters, which are far away from each other and their sizes are very different. As mentioned above, the number of clusters is the key configuration parameter of clustering methods and it is usually varied from $K_{min} = 2$ to $K_{max} = \sqrt{n}$. It is assumed that the value of the validity indices is equal 0 for $K = 1$. In Fig. 2 the comparison of the variations of the *Silhouette*, *SILA* and *SILAv*1 indices with respect to the number of clusters is presented for the artificial datasets. It is also noticeable that the *SILA* and *SILAv*1 indices provide the correct number of clusters for all the artificial datasets. In addition, the value of the *SILAv*1 index more decreases than the value of *SILA* for the small number of clusters, i.e. when the number $K < c^*$ (where c^* is the right number of clusters). This means that the additional component $A(\mathbf{x})$ used in the *SILAv*1 index more reduces the value of the index than the value of the *SILA* index. On the other hand, when the number of clusters $K > c^*$ the component $A(\mathbf{x})$ can increase the values of these indices slightly (see Fig. 2). On the contrary, the *Silhouette* index incorrectly selects all partitioning schemes and mainly provides the greatest values when the number of clusters $K = 2$. Next, the *Simplified Silhouette*, *SimSILA*, and *SimSILAv*1 indices are analyzed. As with the previous studies, four artificial datasets, i.e. *Data* 1, *Data* 2, *Data* 3 and *Data* 4 have been clustered by the *PAM* algorithm. The comparison of the variations of the *Simplified Silhouette*, *SimSILA* and *SimSILAv*1 indices with respect to the number of clusters is presented in Fig. 3. Despite the fact that the differences of distances between clusters are large, the *SimSILA* and *SimSILAv*1 indices provide the correct partitioning for all these data. It can be noted that the component $\hat{A}(\mathbf{x})$ strongly reduces values of the *SimSILAv*1 index when the number of clusters $K < c^*$. This is due to the fact that the exponent q in $\hat{A}(\mathbf{x})$ is calculated by the formula (4). Generally, the component $\hat{A}(\mathbf{x})$ improvements the results especially when the clustering algorithm combines clusters into larger ones and differences of distances between clusters are large. Then the influence of the separability measure is significant and consequently, it can strongly affect the value the index. On the other hand, when

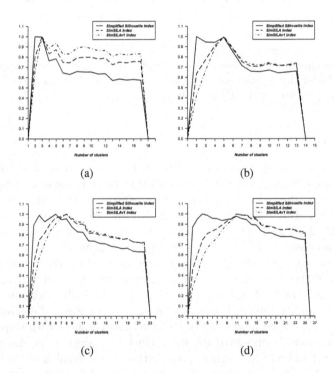

Fig. 3. Variations of the *Simplified Silhouette*, *SimSILA* and *SimSILAv*1 indices with respect to the number of clusters for 2-dimensional datasets: (a) *Data* 1, (b) *Data* 2, (c) *Data* 3, and (d) *Data* 4 partitioned by the *PAM* method.

$K > c^*$ the values of these new indices are increased slightly. It can be noted that the *Simplified Silhouette* and the *Silhouette* indices incorrectly select the number of clusters, whereas the new indices provide the right results for all the artificial datasets. The next experiments are related to the real datasets. As outlined above, the real datasets are numeric data: *Diabetes*, *Ecoli*, *Glass*, *Iris*, *Spectf*, *Wine*. In the experimental process, these datasets have been clustered by the *PAM* algorithm. Moreover, for the evaluation of the clustering validity, the six indices have been used. Table 2 shows the comparison of these indices taking into account the number of clusters, which is the configuration parameter for the clustering algorithm. In addition, the Table also includes results from previous experiments related to the artificial data. From the Table 2, it can be noted that for the real datasets the best results are achieved by the *SILA*, *SILAv1*, *SimSILA* and *SimSILAv1* indices. Moreover, for the *Glass* and *Iris* data, the results of the *SimSILAv1* index are better in comparison with other indices. Based on these results, it can be concluded that for all the experiments carried out on artificial and real data the best clustering results are selected by using these new indices.

Table 2. Comparison of the number of clusters obtained when using the *PAM* algorithm in conjunction with the *SIL*, *SILA*, *SILAv1*, *SimSil*, *SimSILA* and *SimSILAv1* indices. N denotes the actual number of clusters in the datasets.

Datasets	N	Number of clusters					
		SIL	*SILA*	*SILAv1*	*SimSIL*	*SimSILA*	*SimSILAv1*
Data 1	3	2	3	3	2	3	3
Data 2	5	2	5	5	2	5	5
Data 3	8	6	8	8	6	8	8
Data 4	11	4	11	11	4	11	11
Diabetes	2	2	2	2	2	2	2
Ecoli	8	4	4	4	4	4	4
Glass	6	2	2	7	2	7	7
Iris	3	2	2	2	2	2	3
Spectf	2	2	2	2	2	2	2
Wine	3	2	3	3	2	3	3

5   Conclusions

In this paper new indices called *SimSILA* and *SimSILAv1* are proposed, which are the modification of the *Simplified Silhouette* index. As mentioned above, neither the *Simplified Silhouette* index nor the *Silhouette* index performs well when there are large differences of distances between clusters in a dataset. Similarly to the modification of the *Silhouette* index, the change of the *Simplified Silhouette* relies on the application of the additional component, which improves the performance of the index. This additional component contains a measure of cluster compactness and reduces the high values of the index caused by large differences between clusters. In these conducted experiments, several datasets were used, where the number of clusters varied within a wide range. Moreover, the *PAM* clustering algorithm was selected for clustering of all the artificial and the real datasets. It has been noticeable that the *SILA*, *SILAv*1, *SimSILA* and *SimSILAv*1 indices have provided the best results. However, the *Simplified Silhouette* index is much less computationally expensive than the *Silhouette* index. From this perspective, the *SimSILA* and *SimSILAv*1 indices have the competitive performance to the *SILA* and *SILAv*1 indices in the selection of the right clustering results. All the presented results confirm the very high efficiency of the newly proposed indices.

References

1. Arbelaitz, O., Gurrutxaga, I., Muguerza, J., Prez, J.M., Perona, I.: An extensive comparative study of cluster validity indices. Pattern Recogn. **46**, 243–256 (2013)
2. Bilski, J., Smolag, J.: Parallel architectures for learning the RTRN and Elman dynamic neural networks. IEEE Trans. Parallel Distrib. Syst. **26**(9), 2561–2570 (2015)
3. Bilski, J., Wilamowski, B.M.: Parallel learning of feedforward neural networks without error backpropagation. In: Rutkowski, L., Korytkowski, M., Scherer, R., Tadeusiewicz, R., Zadeh, L.A., Zurada, J.M. (eds.) ICAISC 2016. LNCS (LNAI), vol. 9692, pp. 57–69. Springer, Cham (2016). https://doi.org/10.1007/978-3-319-39378-0_6
4. Bologna, G., Hayashi, Y.: Characterization of symbolic rules embedded in deep DIMLP networks: a challenge to transparency of deep learning. J. Artif. Intell. Soft Comput. Res. **7**(4), 265–286 (2017). https://doi.org/10.1515/jaiscr-2017-0019
5. Bradley, P., Fayyad, U.: Refining initial points for k-means clustering. In: Proceedings of the Fifteenth International Conference on Knowledge Discovery and Data Mining, pp. 9–15. AAAI Press, New York (1998)
6. Chang, O., Constante, P., Gordon, A., Singana, M.: A novel deep neural network that uses space-time features for tracking and recognizing a moving object. J. Artif. Intell. Soft Comput. Res. **7**(2), 125–136 (2017). https://doi.org/10.1515/jaiscr-2017-0009
7. Cpałka, K., Rebrova, O., Nowicki, R., Rutkowski, L.: On design of flexible neurofuzzy systems for nonlinear modelling. Int. J. Gen. Syst. **42**(6), 706–720 (2013)
8. Cpałka, K., Rutkowski, L.: Flexible Takagi-Sugeno fuzzy systems. In: Proceedings of the 2005 IEEE International Joint Conference on Neural Networks, IJCNN (2005)
9. Devi, V.S., Meena, L.: Parallel MCNN (PMCNN) with application to prototype selection on large and streaming data. J. Artif. Intell. Soft Comput. Res. **7**(3), 155–169 (2017). https://doi.org/10.1515/jaiscr-2017-0011
10. Fränti, P., Rezaei, M., Zhao, Q.: Centroid index: cluster level similarity measure. Pattern Recogn. **47**(9), 3034–3045 (2014)
11. Gabryel, M.: A bag-of-features algorithm for applications using a NoSQL database. Inf. Softw. Technol. **639**, 332–343 (2016)

12. Gabryel, M., Grycuk, R., Korytkowski, M., Holotyak, T.: Image indexing and retrieval using GSOM algorithm. In: Rutkowski, L., Korytkowski, M., Scherer, R., Tadeusiewicz, R., Zadeh, L.A., Zurada, J.M. (eds.) ICAISC 2015. LNCS (LNAI), vol. 9119, pp. 706–714. Springer, Cham (2015). https://doi.org/10.1007/978-3-319-19324-3_63

13. Gałkowski, T.: Kernel estimation of regression functions in the boundary regions. In: Rutkowski, L., Korytkowski, M., Scherer, R., Tadeusiewicz, R., Zadeh, L.A., Zurada, J.M. (eds.) ICAISC 2013. LNCS (LNAI), vol. 7895, pp. 158–166. Springer, Heidelberg (2013). https://doi.org/10.1007/978-3-642-38610-7_15

14. Galkowski, T., Pawlak, M.: Nonparametric estimation of edge values of regression functions. In: Rutkowski, L., Korytkowski, M., Scherer, R., Tadeusiewicz, R., Zadeh, L.A., Zurada, J.M. (eds.) ICAISC 2016. LNCS (LNAI), vol. 9693, pp. 49–59. Springer, Cham (2016). https://doi.org/10.1007/978-3-319-39384-1_5

15. Hruschka, E.R., de Castro, L.N., Campello, R.J.: Evolutionary algorithms for clustering gene-expression data. In: Fourth IEEE International Conference on Data Mining, ICDM 2004, pp. 403–406. IEEE (2004)

16. Jain, A., Dubes, R.: Algorithms for Clustering Data. Prentice-Hall, Englewood Cliffs (1988)

17. Ke, Y., Hagiwara, M.: An English neural network that learns texts, finds hidden knowledge, and answers questions. J. Artif. Intell. Soft Comput. Res. 7(4), 229–242 (2017). https://doi.org/10.1515/jaiscr-2017-0016

18. Lago-Fernández, L.F., Corbacho, F.: Normality-based validation for crisp clustering. Pattern Recogn. 43(3), 782–795 (2010)

19. Lichman, M.: UCI Machine Learning Repository. University of California, School of Information and Computer Science, Irvine, CA (2013). http://archive.ics.uci.edu/ml

20. Liu, H., Gegov, A., Cocea, M.: Rule based networks: an efficient and interpretable representation of computational models. J. Artif. Intell. Soft Comput. Res. 7(2), 111–123 (2017). https://doi.org/10.1515/jaiscr-2017-0008

21. Meng, X., van Dyk, D.: The EM algorithm - an old folk-song sung to a fast new tune. J. Roy. Stat. Soc. Ser. B (Methodol.) 59(3), 511–567 (1997)

22. Murtagh, F.: A survey of recent advances in hierarchical clustering algorithms. Comput. J. 26(4), 354–359 (1983)

23. Pal, N.R., Bezdek, J.C.: On cluster validity for the fuzzy c-means model. IEEE Trans. Fuzzy Syst. 3(3), 370–379 (1995)

24. Park, H.S., Jun, C.H.: A simple and fast algorithm for K-medoids clustering. Expert Syst. Appl. 36(2), 3336–3341 (2009)

25. Rohlf, F.: Single-link clustering algorithms. In: Krishnaiah, P.R, Kanal, L.N. (eds.) Handbook of Statistics, vol. 2, pp. 267–284 (1982)

26. Rousseeuw, P.J.: Silhouettes: a graphical aid to the interpretation and validation of cluster analysis. J. Comput. Appl. Math. 20, 53–65 (1987)

27. Rutkowski L., Cpałka K.: Compromise approach to neuro-fuzzy systems. In: Sincak, P., Vascak, J., Kvasnicka, V., Pospichal, J. (eds.) Intelligent Technologies - Theory and Applications. New Trends in Intelligent Technologies. Frontiers in Artificial Intelligence and Applications, vol. 76, pp. 85–90 (2002)

28. Rutkowski, L., Cpałka, K.: A neuro-fuzzy controller with a compromise fuzzy reasoning. Control Cybern. 31(2), 297–308 (2002)

29. Saha, S., Bandyopadhyay, S.: Some connectivity based cluster validity indices. Appl. Soft Comput. 12(5), 1555–1565 (2012)

30. Sameh, A.S., Asoke, K.N.: Development of assessment criteria for clustering algorithms. Pattern Anal. Appl. 12(1), 79–98 (2009)

31. Serdah, A.M., Ashour, W.M.: Clustering large-scale data based on modified affinity propagation algorithm. J. Artif. Intell. Soft Comput. Res. **6**(1), 23–33 (2016). https://doi.org/10.1515/jaiscr-2016-0003

32. Shieh, H.-L.: Robust validity index for a modified subtractive clustering algorithm. Appl. Soft Comput. **22**, 47–59 (2014)

33. Starczewski, A.: A new validity index for crisp clusters. Pattern Anal. Appl. **20**(3), 687–700 (2017)

34. Starczewski, A., Krzyżak, A.: A modification of the silhouette index for the improvement of cluster validity assessment. In: Rutkowski, L., Korytkowski, M., Scherer, R., Tadeusiewicz, R., Zadeh, L.A., Zurada, J.M. (eds.) ICAISC 2016. LNCS (LNAI), vol. 9693, pp. 114–124. Springer, Cham (2016). https://doi.org/10.1007/978-3-319-39384-1_10

35. Starczewski, A., Krzyżak, A.: Improvement of the validity index for determination of an appropriate data partitioning. In: Rutkowski, L., Korytkowski, M., Scherer, R., Tadeusiewicz, R., Zadeh, L.A., Zurada, J.M. (eds.) ICAISC 2017. LNCS (LNAI), vol. 10246, pp. 159–170. Springer, Cham (2017). https://doi.org/10.1007/978-3-319-59060-8_16

36. Wu, K.L., Yang, M.S., Hsieh, J.N.: Robust cluster validity indexes. Pattern Recogn. **42**, 2541–2550 (2009)

37. Vendramin, L., Campello, R.J., Hruschka, E.R.: Relative clustering validity criteria: a comparative overview. Stat. Anal. Data Min. **3**(4), 209–235 (2010)

38. Zhao, Q., Fränti, P.: WB-index: a sum-of-squares based index for cluster validity. Data Knowl. Eng. **92**, 77–89 (2014)

Feature Extraction in Subject Classification of Text Documents in Polish

Tomasz Walkowiak[(✉)], Szymon Datko, and Henryk Maciejewski

Faculty of Electronics, Wrocław University of Science and Technology,
Wrocław, Poland
{tomasz.walkowiak,szymon.datko,henryk.maciejewski}@pwr.edu.pl

Abstract. In this work we evaluate two different methods for deriving features for a subject classification of text documents. The first method uses the standard Bag-of-Words (BoW) approach, which represents the documents with vectors of frequencies of selected terms appearing in the documents. This method heavily relies on the natural language processing (NLP) tools to properly preprocess text in the grammar- and inflection-conscious way. The second approach is based on the word-embedding technique recently proposed by Mikolov and does not require any NLP preprocessing. In this method the words are represented as vectors in continuous space and this representation of words is used to construct the feature vectors of the documents. We evaluate these fundamentally different approaches in the task of classification of Polish language Wikipedia articles with 34 subject areas. Our study suggests that the word-embedding based features seem to outperform the standard NLP-based features providing sufficiently large training dataset is available.

Keywords: Text mining · Subject classification · Bag of words
Word embedding · fastText

1 Introduction - Problem Formulation

Automatic classification of text documents in terms of the subject areas is one of the important tasks of text mining. Promising applications of this technology range from classification of articles in the Internet or newspaper repositories to categorization of scientific papers or tech-support requests.

Commonly used methods rely on representing documents with feature vectors and training machine learning models, such as SVM, Naïve Bayes, logistic regression, etc., using a collection of documents with known class labels. The key challenge in this lies in deriving the most informative features while restricting the dimensionality of feature vectors. Most effective methods of feature generation, broadly referred as the bag of words (BoW), are based on frequencies of words occurring in documents [3]. These methods heavily rely on pre-processing text with the language-specific NLP (natural language processing) algorithms/tools

© Springer International Publishing AG, part of Springer Nature 2018
L. Rutkowski et al. (Eds.): ICAISC 2018, LNAI 10842, pp. 445–452, 2018.
https://doi.org/10.1007/978-3-319-91262-2_40

in order to derive base forms of words/terms (lemmatization), as well as to select words/terms into feature vectors using language knowledge, such as POS (part-of-speech) tagging or named-entity identification. In this way the feature vectors can be restricted to specific, presumably most informative for subject classification, parts of speech (e.g. nouns or adjectives) while omitting e.g. adverbs or prepositions. All this leads to significant reduction of dimensionality of otherwise very high-dimensional BoW feature vectors.

It should be noted that the NLP-based pre-processing is especially important in languages with rich inflection, such as in Polish or other Slavic languages, because using raw forms of words boosts dimensionality of feature vectors [1].

In this work we confront this approach with the emerging methodology based on word-embedding techniques, recently proposed by Mikolov [7]. The idea is to represent words in continuous vector spaces in which regularities between vectors reflect semantic of syntactic regularities in the language. Efficient algorithms for learning such representations using (large) corpora of text documents were proposed [6]. Based on word embedding, representation of documents can be constructed for text classification, as proposed in the *fastText* algorithm [4]. It should be noted that this method is entirely data-driven, as it does not rely on any language-specific NLP technology.

We evaluate performance of these two entirely different approaches in the task of subject classification of Polish language Wikipedia articles [8,9]. This publicly available corpus contains ca. 10.000 articles representing 34 classes (subject categories). Our experiments suggest that the data-driven, NLP-free method outperforms the commonly used BoW approach in terms of classification accuracy, additionally generating lower-dimensionality feature vectors. This result seems appealing as it entirely leaves out the laborious NLP step commonly regarded as mandatory in text classification. However, this is possible providing sufficiently large training collection of documents is available.

The paper is organized as follows. In Sect. 2 we provide the technicalities pertaining to the standard BoW and the emerging fastText methods. In Sect. 3 we describe the Wikipedia corpus used in our experiment and compare performance of the linear classifiers fed with BoW and word embedding-based features. We discuss benefits and costs of these approaches in Sect. 4.

2   Methods of Representation of Text for Subject Classification

In this section we describe the Bag-of-Words-based method of text classification and the fastText method based on the concept of word-embedding. In the first part we deal with generation of features from Polish language texts, as we focus on NLP tools specific to this language.

2.1   Bag of Words

The most common vector representation of texts is the bag of words [3]. BoW models are based on the assumption that text can be represented as an unordered collection of words frequencies [11]. The method has a lot of modifications depending on different classification tasks. In subject classification, the BoW dictionary usually consists of words from texts with the most common, the most rare and the stop words filtered out [13]. Furthermore the words may be lemmatized to limit the number of features.

In this study we have followed the BoW schema found as most suitable for Polish in experiments described in [15]. Firstly, all texts were processed by a morphosyntactic tagger for Polish. The WCRFT [10] tagger was used, which joins the Conditional Random Fields (CRF) and tiered tagging of plain text, for POS tagging and lemmatisation. Secondly, lemmas of all found nouns were selected. Third, we selected the most frequent 1000 nouns (lemmas) in training corpora. Finally, each document was represented by counts of particular selected nouns. All processing was performed using Clarin-PL infrastructure [14].

In most of the processing schemes proposed in literature the raw counts are weighted in relation to the document length and also the relative importance of the occurrences of these features for analyzed texts. The most common weighting scheme is tf-idf [12]. Other suggested schemes are Lnu.ltu and OKAPI [5].

Experiments reported in [15], as well as experiments conducted by authors, show that the selection of weighting schema is meaningless for text classification results. This could be justified by the fact that weighting is just a linear modification of feature vectors. Most of supervised classifiers (like logistic regression, SVM or Multilayer Perceptron) do linear modification of the feature vectors in the first step of their algorithms and values of these modification (i.e. weights) are tuned during learning process. Moreover, the tf-idf effectively filters out words (nouns, in our case) that exist in all documents so such information cannot be used during classification.

On the other hand, standardization of feature vectors is a common requirement for many classifiers. Some classifiers assume that features are normally distributed with variance equal to 1. Therefore, feature vector were weighted by removing the mean and scaling to unit variance. The feature vector mean and variance were calculated for the training set and the values are used for weighting training and testing set as well.

Summarizing, a vector of occurrences of 1000 most frequent nouns was calculated for each document in the corpus, forming a noun count matrix. Next, the raw counts were normalized (by linear transformation defined by two vectors: training set mean and variance) forming the feature matrix.

2.2 FastText

The second method analyzed in this paper uses a recent deep learning method for text classification fastText [4]. It is based on representation of documents as an average of word embeddings and uses a linear soft-max classifier [2]. The main idea is to perform word representation and classifier learning in parallel. As a result the (linear) model is very effective to train achieving several orders of magnitude faster solution than other competing methods [4] and in many text mining tasks fastText seems to outperform state of art classifiers with BoW features. FastText builds the word embedding model (a look-up table that maps words to p dimensional vectors of real numbers) on the training corpora. Each document is represented as an average of word embeddings. Words that are not existing in the embedding model (due to not existing in training corpora) are omitted from the averaging. This hidden representation is used by linear classifier for all classes, allowing information about word embedding learned for one class to be used by others. FastText by default ignores word order, much like the BoW method. However, fastTetx allows to use word n-grams, to take into account local word order, but this feature was not used in our experiments.

3 Evaluation

3.1 Data Set and Experiment Organization

To evaluate the competing methods of representation of text documents, we train (i) a linear classifiers based on BoW features and (ii) a fastText classifier based on word-embedding features. We want to predict subject class of articles extracted from the Polish language Wikipedia, with the following 34 subject areas used as class labels: *Airplanes, German military, Football, Diseases, Karkonosze, Comic books, Catholicism, Political propaganda, Culture of China, Plants ecology, Optics, Strength sports, Branches of law, Chess, Skiing, Animated films, Albania, Classical music, Astronautics, Accountancy, Sailing, Healthcare, Drug addiction, Coins, Chemical elements, Computer games, Computers, American prose writers, Armored troops, Egypt, Cars, Jews, Arabs, Cats.*

The training partition [9] includes 6885 articles which translates into ca. 200 articles per class (with the class *Arabs* slightly underrepresented). Performance of classifiers was evaluated based on the test partition [8] of 2952 articles.

3.2 Results

We start with comparing accuracy of the logistic regression based on the BoW features with the fastText classifier - results are given in Table 1.

We observe that the fastText method outperforms the BoW-based linear classifier by ca. 7.0%, with the accuracy of ca. 88.7%. Considering the fact that the total number of class labels in this study is 34 should, this result exceed the baseline random classifier by the factor of ca. 26. It is interesting to note that this performance of fastText is demonstrated with 100-dimensional feature vectors (due to word-embedding realized into 100-dimensional continuous vector space). The 1000-dimensional BoW feature vectors were found as most effective in a meticulous fine-tuning study of this method (results reported in [15]). Lower dimensionality of fastText feature vectors as compared with the BoW method is another appealing characteristic of this approach as it allows to obtain simpler classifiers.

Next we performed a more in-depth analysis of performance of the fastText classifier, with the confusion matrix presented graphically in Fig. 1. Consistently with the high global accuracy of the method, we observe that vast majority of test examples for every class is classified correctly, with relatively few missclassifications. Note however, that the missclassification events may be often accounted for by relatively close relationships between classes being confused (see e.g. that the class 'German military' is most likely confused as 'Political propaganda' or 'Branches of law', which may be considered as somewhat related).

Finally, in Table 2 we report precision, recall and the F-measure calculated for each individual class. Precision is defined as $P = \frac{Tp}{TP+FP}$, recall as $R = \frac{TP}{TP+FN}$, and the F-measure as the harmonic mean of P and R, $F = \frac{2PR}{P+R}$, where TP, FP and FN denote the number of true positive, false positive and false negative recognitions of a given class, respectively.

Table 1. Accuracy for the fastText and BoW with logistic regression classifier.

Method	fastText	BoW
Accuracy	0.887	0.811
Number of features	100	1000
Vocabulary	231 831	1000

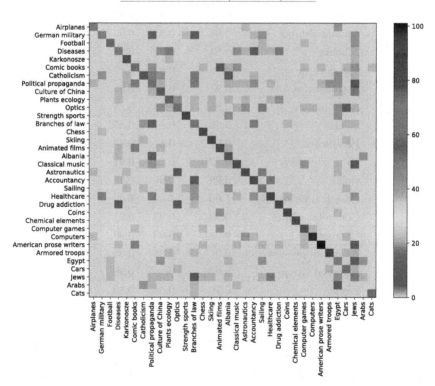

Fig. 1. Confusion matrix for fastText method in the form of a heatmap. Rows represent the actual class, and columns - the predicted class.

Table 2. Precision, recall and the F-measure for individual classes, as obtained with the fastText classifier.

Class	Precision	Recall	F-measure
Airplanes	0.96	0.96	0.96
German military	0.92	0.79	0.85
Football	0.96	0.95	0.96
Diseases	0.86	0.80	0.83
Karkonosze	0.95	0.98	0.96
Comic books	0.88	0.80	0.84
Catholicism	0.92	0.80	0.86
Political propaganda	0.71	0.71	0.71
Culture of China	0.86	0.93	0.89
Plants ecology	0.88	0.88	0.88
Optics	0.83	0.78	0.81
Strength sports	0.98	0.96	0.97
Branches of law	0.57	0.81	0.67
Chess	0.95	0.99	0.97
Skiing	0.97	0.98	0.97
Animated films	0.87	0.93	0.90
Albania	0.84	0.92	0.88
Classical music	0.90	0.83	0.87
Astronautics	0.91	0.89	0.90
Accountancy	0.85	0.86	0.85
Sailing	0.83	0.87	0.85
Healthcare	0.91	0.86	0.88
Drug addiction	0.93	0.88	0.91
Coins	0.98	0.97	0.97
Chemical elements	1.00	1.00	1.00
Computer games	0.93	0.95	0.94
Computers	0.98	0.93	0.96
American prose writers	0.98	0.88	0.93
Armored troops	0.97	0.96	0.96
Egypt	0.76	0.85	0.80
Cars	0.88	0.95	0.91
Jews	0.65	0.79	0.71
Arabs	0.85	0.72	0.78
Cats	0.98	0.98	0.98

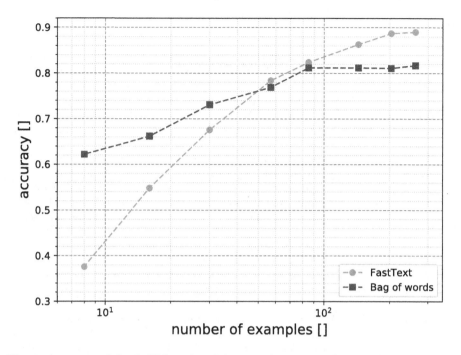

Fig. 2. Accuracy of the BoW-based and fastText linear classifier as a function of the average number of per-class training examples.

4   Conclusion

In this work we compared two conceptually different approaches to represent text documents with feature vectors in the task of subject classification: (i) the bag of words method which is based on frequencies of important words/terms in documents, and (ii) the fastText method which relies on vector space representation of words. The former method heavily relies on the natural language processing technologies which must be available for a specific language of interest, while the latter is entirely data-driven and does not use the NLP technology.

Our study involved the classification of Wikipedia articles in Polish into 34 subject areas. Results of this study prove that the fastText method seems to outperform the commonly used BoW method, and moreover, the higher accuracy of prediction is obtained with lower dimensionality feature vectors which is an appealing characteristic of this approach. This can be achieved providing that sufficiently many training examples per class are available, as illustrated in Fig. 2. With very few training examples per class the BoW still promises higher accuracy of classification.

Acknowledgement. This work was sponsored by National Science Centre, Poland (grant 2016/21/B/ST6/02159).

References

1. Eder, M., Piasecki, M., Walkowiak, T.: An open stylometric system based on multilevel text analysis. Cogn. Stud.—Etudes Cogn. (17) (2017). https://doi.org/10.11649/cs.1430
2. Goodman, J.: Classes for fast maximum entropy training. In: Proceedings of 2001 IEEE International Conference on Acoustics, Speech, and Signal Processing, (Cat. No.01CH37221), vol. 1, pp. 561–564 (2001). https://doi.org/10.1109/ICASSP.2001.940893
3. Harris, Z.: Distributional structure. Word (1954)
4. Joulin, A., Grave, E., Bojanowski, P., Mikolov, T.: Bag of tricks for efficient text classification. In: Proceedings of the 15th Conference of the European Chapter of the Association for Computational Linguistics, Short Papers, vol. 2, pp. 427–431. Association for Computational Linguistics (2017). http://aclweb.org/anthology/E17-2068
5. Manning, C.D., Raghavan, P., Schutze, H.: Introduction to Information Retrieval. Cambridge University Press, Cambridge (2009)
6. Mikolov, T., Chen, K., Corrado, G., Dean, J.: Efficient estimation of word representations in vector space. CoRR abs/1301.3781 (2013). http://arxiv.org/abs/1301.3781
7. Mikolov, T., Yih, W., Zweig, G.: Linguistic regularities in continuous space word representations. In: Proceedings of the 2013 Conference of the North American Chapter of the Association for Computational Linguistics: Human Language Technologies, pp. 746–751. Association for Computational Linguistics, Atlanta, June 2013. http://www.aclweb.org/anthology/N13-1090
8. Młynarczyk, K., Piasecki, M.: Wiki test - 34 categories (2015). http://hdl.handle.net/11321/217. CLARIN-PL digital repository
9. Młynarczyk, K., Piasecki, M.: Wiki train - 34 categories (2015). http://hdl.handle.net/11321/222. CLARIN-PL digital repository
10. Radziszewski, A.: A tiered CRF tagger for Polish. In: Bembenik, R., Skonieczny, L., Rybinski, H., Kryszkiewicz, M., Niezgodka, M. (eds.) Intelligent Tools for Building a Scientific Information Platform. Studies in Computational Intelligence, vol. 467, pp. 215–230. Springer, Heidelberg (2013). https://doi.org/10.1007/978-3-642-35647-6_16
11. Salton, G., Buckley, C.: Term-weighting approaches in automatic text retrieval. Inf. Process. Manag. **24**(5), 513–523 (1988)
12. Salton, G., McGill, M.: Introduction to Modern Information Retrieval. McGraw-Hill, New York (1986)
13. Torkkola, K.: Discriminative features for text document classification. Formal Pattern Anal. Appl. **6**(4), 301–308 (2004). https://doi.org/10.1007/s10044-003-0196-8
14. Walkowiak, T.: Language processing modelling notation - orchestration of NLP microservices. In: Zamojski, W., Mazurkiewicz, J., Sugier, J., Walkowiak, T., Kacprzyk, J. (eds.) DepCoS-RELCOMEX 2017. AISC, pp. 464–473. Springer International Publishing, Cham (2018). https://doi.org/10.1007/978-3-319-59415-6_44
15. Walkowiak, T., Malak, P.: Polish texts topic classification evaluation. In: Proceedings of the 10th International Conference on Agents and Artificial Intelligence, ICAART 2018, vol. 2, pp. 515–522. INSTICC, SciTePress (2018)

Efficiency of Random Decision Forest Technique in Polish Companies' Bankruptcy Prediction

Joanna Wyrobek[1](✉) and Krzysztof Kluza[2]

[1] Corporate Finance Department, Cracow University of Economics,
ul. Rakowicka 27, 31-510 Kraków, Poland
wyrobekj@uek.krakow.pl
[2] AGH University of Science and Technology,
al. A. Mickiewicza 30, 30-059 Kraków, Poland
kluza@agh.edu.pl

Abstract. The purpose of the paper was to compare the accuracy of traditional bankruptcy prediction models with the Random Forest method. In particular, the paper verifies 2 research hypotheses (verification was based on the representative sample of Polish companies): [H1]: The Random Forest algorithm (trained on a representative set of companies) is more accurate than traditional bankruptcy prediction methods: logit and linear discriminant models, and [H2]: The Random Forest algorithm efficiently uses normalized financial statement data (there is no need to calculate financial ratios).

1 Introduction

Machine learning methods nowadays can be found in every aspect of human life. Data processing capabilities which are offered by constantly improved ML algorithms are abundantly used in such areas as: traffic predictions, track control systems, money transfers, fraud prevention, credit analysis, voice, picture and text recognition, video surveillance, or search engines. For commercial companies the most important application of ML methods seems to be bankruptcy prediction [1]. The purpose of the paper is to look into bankruptcy prediction accuracy enhancement introduced by machine learning algorithms. In particular, the paper compares classical bankruptcy analysis tools such as discriminant analysis and logit models with a base ML model which is the Random Forest algorithm (which is often considered as the first algorithm for ML learning and the reference model for other techniques, it also requires very little tuning - compared, for instance to the Boosting of Decision Trees, which is equally popular among data researchers). Comparisons were done on a representative sample ($p = 5\%$) of 1415 bankrupt and 1450 active companies in Poland one year before insolvency proceedings are commenced. Data sample covered years 2008–2017.

The paper is supported by the Cracow University of Economics research grant.

L. Rutkowski et al. (Eds.): ICAISC 2018, LNAI 10842, pp. 453–462, 2018.
https://doi.org/10.1007/978-3-319-91262-2_41

The paper is organized as follows. In Sect. 2, we present previous research on the accuracy of traditional and Random Forest methods in bankruptcy prediction for various years and countries. Section 3 presents research methodology: hypotheses to be tested, a basic outline of discriminant analysis and logit models as well as the Random Forest algorithm and the description of their implementation. Section 4 shows model training results. Finally, Sect. 5 includes hypotheses verification, final conclusions and directions for further research.

2   Literature Survey

Conventional financial understanding of bankruptcy prediction models is based on using either: the linear discriminant analysis models or logit (logistic) models. Such models have numerous advantages: they are easily transferable because the model is described as a linear combination of financial ratios and fixed coefficients, they are also extremely easy to use: a simple substitution of financial ratios allows the calculation of the final score, which informs about the risk of a company's insolvency. Although they assume the normal distribution of independent variables discriminant analysis is quite resistant to the lack of this assumption.

Above mentioned methods have, however, limitations. Many of the limitations are the consequences of underlying assumptions: homoscedasticity (both methods), lack of multicollinearity, independence of scores on variables for different participants. DA models are also quite sensitive to outliers and for logit models one has to determine a cut-off point. Both methods use cross-sectional data and require values of variables to be comparable (in size) between companies. Another limitation is that both methods create a linear combination of independent variables, so they are unable to construct their own ratios. This is why usually for the estimation purposes one uses only financial ratios, not the direct data from the financial statements (because firstly, the methods are unable to create their own ratios and secondly, companies would have to have very similar values of: assets, sales, all other financial statement elements).

Generally, DA and logit models used in bankruptcy prediction are not extremely accurate, but do quite well with time (if the selection of variables is rational and well-though through).

Recently, quite a popular alternative for bankruptcy prediction in finance are Random Forest (RF) models (based on decision trees). They are very simple in construction and do not make assumptions about the behavior of variables. They usually train well on the given sample and give precise predictions for data from the same time period or nearest future. They are not sensitive to outliers. RF also have their disadvantages. They do not deal very well with time (it is believed that one has to provide new, timely data to the model), they require many underlying trees to be precise and the model is very complex – with many variables it is hard to explain relationships between existing data.

All of this does not change the fact that RF models can be very accurate. This can be seen in Table 1 where we presented findings of previous research on the accuracy of DA and logit (logistic) models trained on non-Polish companies data.

As it can be seen in Table 1, cross-validation average testing accuracy for discriminant analysis was in the range between 52.18% [1] and 93.5% [2]. For logit/logistic function models the range was between 69.75% [2] and 97.2%, but cross validation accuracy above 95% was observed only for 1 model [3] which was trained only on 250 companies. For Random Forest model, half of the analyzed papers had accuracy above 95% [2–4]. Minimum accuracy was 73.1% [5] and the maximum was 97.4% [3]. Even though these results are comparable to logit models, distribution of results is better for RF models.

Table 2 shows findings of previous Polish research relating to the application of DA, logit and Random Forest methods in bankruptcy prediction. According to the Polish research, the most accurate method was discriminant analysis because its accuracy was in the range between 86.11% [6] and 96.29% [7]. The second best method was logit/logistic model with the accuracy between 83.33% [6] and 92.59% [7]. The Random Forest model had the lowest accuracy, as it was in the range in between 75.3% (for a balanced panel) [8] and 90.1% (for an unbalanced panel) [8]. Presented results for Polish data is not very promising. First of all, except for Korol DA model no other model achieved testing accuracy above 95%. Secondly, RF models had the lowest accuracy.

If one looks, however, at the number of companies, it can be seen that all Polish models were trained on a very small number of bankrupt firms. The same problem concerned research for non-Polish data. Only Jardin [9] and Min and Jeong [10] used a representative sample of bankrupt firms. In our opinion, this creates the research gap because only the representative sample of companies allows the collected results to be generalized to a larger population. In other words, we believe that models which are trained on the representative sample of companies can do a better job with learning about the true nature of relationships in the economy and consequently, better predict future bankruptcies.

3 Research Method

For the reasons mentioned above, the purpose of the paper was to analyze the accuracy of basic bankruptcy prediction models and RF algorithm using a representative sample of Polish companies.

Data were extracted from Orbis database and missing data was filled with both: data from EMIS database and averages (only in the situations where it could be done in a reliable way). As it was explained earlier, it was necessary for logit and DA models. We did not want to use a different sample for the RF algorithm, so we used the same sample for this algorithm too (in general, the RF does not require this adjustment and it could even hamper its efficiency).

After extraction, we tested whether assets are equal to the sum of equity and liabilities, removed any visible error records and any other suspicious data.

Table 1. Previous foreign research on the accuracy of linear discriminant analysis (DA), Logit and Random Forest (based on decision trees) methods in bankruptcy prediction (cross-validation average accuracy in [%])

Studies	Country	Years	No. of companies	Base classifiers	Accu. [%]
[11] Alfaro et al. (2008)	Spain	2000–2003	590 + 590	DA	79.66
[12] Anandarajan et al. (2001)	USA	1989–1996	265 + 319	DA	52.25
[1] Barboza et al. (2016)	USA+Canada	1985–2013	449 + 449	DA	52.18
[1] Barboza et al. (2016)	USA+Canada	1985–2013	449 + 449	logit	76.29
[1] Barboza et al. (2016)	USA+Canada	1985–2013	449 + 449	Random Forest (DT)	87.06
[13] Cho et al. (2010)	South Korea	2000–2002	500 + 500	Logit	70.58
[14] Cho et al. (2009)	South Korea	2000–2002	500 + 500	DA	78.15
[15] Fedorova et al. (2013)	Russian Federation	2007–2011	444 + 444	DA	82.00
[16] Ghodselahi, Amirmadhi (2011)	Germany	n.a	300 + 700	DA	65.91
[16] Ghodselahi and Amirmadhi (2011)	Germany	n.a	300 + 700	Random Forest (DT)	76.07
[17] Hu and Tseng (2007)	USA	1975–1982	65 + 65	DA	85.42
[17] Hu and Tseng (2007)	USA	1975–1982	65 + 65	Logit	88.73
[5] Huang et al. (2017)	China	2000–2011	156 + 156	Random Forest (DT)	73.1
[5] Huang et al. (2017)	China	2000–2011	156 + 156	DA	74.2
[5] Huang et al. (2017)	China	2000–2011	156 + 156	Logit	74.2
[2] Jabeyr and Fahmi (2017)	France	2006–2009	400 + 400	Random Forest (DT)	96.75
[2] Jabeyr and Fahmi (2017)	France	2006–2009	400 + 400	DA	93.5
[2] Jabeyr and Fahmi (2017)	France	2006–2009	400 + 400	Logit	69.75
[9] Jardin (2016)	France	2003–2012	8010 + 8010	DA	80.05
[9] Jardin (2016)	France	2003–2012	8010 + 8010	DA	82.64
[18] Li and Sun (2009)	China	n.a	135 + 135	DA	83.13
[19] Li and Sun (2010)	China	n.a	135 + 135	DA	88.09
[20] Li and Sun (2011)	China	n.a	135 + 135	DA	88.93
[21] Li et al. (2011)	China	n.a	135 + 135	DA	82.82
[4] Liao et al. (2014)	Taiwan	2005–2011	63 + 2680	DA	92.44
[4] Liao et al. (2014)	Taiwan	2005–2011	63 + 2680	Random Forest (DT)	94.91
[10] Min and Jeong (2009)	South Korea	2001–2004	1271 + 1271	DA	69.1
[22] Min and Lee (2005)	South Korea	2000–2002	944 + 944	DA	78.81
[22] Min and Lee (2005)	South Korea	2000–2002	944 + 944	Logit	79.87
[3] Nagaraj and Sridhar (2015)	India	n.a	107 + 143	Logit	97.2
[3] Nagaraj and Sridhar (2015)	India	n.a	107 + 143	Random Forest (DT)	97.4
[23] Pena et al. (2009)	UK	1989–2002	140 + 140	DA	86.6
[18] Sun and Li (2009)	China	2000–2005	135 + 135	DA	80.68
[18] Sun and Li (2009)	China	2000–2005	135 + 135	Logit	84.72
[24] Tseng and Hu (2010)	UK	1985–1994	32 + 45	Logit	86.25

Table 2. Previous Polish research on the accuracy of linear discriminant analysis (DA), Logit and Random Forest (based on decision trees) methods in bankruptcy prediction (cross-validation average accuracy in (%)

Studies	Country	Years	No. of companies	Base classifiers	Accu. (%)
[8] Pawelek and Grochowina (2017)	Poland	2013–2015	42 + 7181	RF (DT)	90.1
[8] Pawelek and Grochowina (2017)	Poland	2013–2015	42 + 42	RF (DT)	75.3
[6] Pociecha et al. (2014)	Poland	2005–2009	182 + 7147	DA	86.11
[6] Pociecha et al. (2014)	Poland	2005–2009	182 + 182	DA	89.58
[6] Pociecha et al. (2014)	Poland	2005–2009	182 + 7147	Logit	88.89
[6] Pociecha et al. (2014)	Poland	2005–2009	182 + 182	Logit	83.33
[7] Korol (2010)	Poland	2005–2009	50 + 56	DA	96.29
[7] Korol (2010)	Poland	2005–2009	50 + 56	Logit	92.59
[7] Korol (2010)	Poland	2005–2009	50 + 56	RF (DT)	88.88

Companies were classified as bankrupt if they had such a status in the database. As a status change date, we assumed the year when the company had negative equity for the first time, and we also assumed that the insolvency application must have been filed one year before this situation. The model should warn about the forthcoming bankruptcy application 1 year ahead (in other words, two years before the actual bankruptcy announcement). We had financial information from balance sheets and income statements and we calculated several financial ratios.

In total, we collected 1415 useful records for bankrupt companies and 1450 records of active companies. Data covered various types of economic activity. Data sample was then divided into training set which included companies data for years 2008–2013 and the testing (evaluation) set which included companies data for years 2014–2017. The training set included 1376 bankrupt companies and 1411 active companies. The testing (evaluation) set included 39 bankrupt companies and 39 active companies. The selection of bankrupt companies was based on the time order, and the selection of active companies was based on the time order and on the similar type of economic activity (for every bankrupt company we drew at random one active company from the same industry, we also added a small amount of other randomly chosen firms). The sets were almost balanced because without adjustments to the loss function a strongly imbalanced sample (with f.e. twice as many active companies than bankrupt companies) would result with a model which could have high general accuracy, but it could have a high I-st type error (it would have a tendency to classify a firm as active when it does go bankrupt).

After checking and processing, data was normalized and we applied the one-hot encoding to discrete data.

We used Python library called skleran and Anaconda Jupyter IDE. The training set was divided into 10 parts and for each iteration we used 9 parts for training and 1 part for testing (cross-validation). Finally, we tested the model on the test sample. For the discriminant analysis model estimation we used the approach W1 described in [6] only we wrote the code in Python and for the accuracy analysis we used cross-validation and a separate testing set as described previously. We used accuracy as the comparison metric because it is the most popular and mostly used in the literature and we wanted to maintain comparability with previous papers.

4   Model Training Results

Table 3 shows the average accuracy of cross-validation of trained models. As it can be seen, contrary to many other papers, the most efficient algorithm proved to be the Random Forest. Its average accuracy was 98.91% (for the logit model it was 86.31% and for the linear discriminant model 87.10%). Another interesting feature was that the RF used base, "raw" information from financial statements and only to a small extent financial ratios (Table 4). Classical methods of bankruptcy prediction only used financial ratios (we explained the reasons at the beginning of the paper, cross-sectional data had to be comparable).

Estimation of the models was automatic, but the calculation of ratios was done manually (DA and logit models, as explained earlier, required calculation of financial ratios as variables and we used financial ratios used in previous publications). In the case of DA and logit models, one had to test assumptions of the models, which was particularly important for a logit model.

The Random Forest algorithm was based on 100 decision trees which used the Gini impurity criterion every time a split of a node was made.

Table 3. Testing accuracy of DA, logit and RF algorithms

Sample	RF (DT)	DA	Logit
1	100.00	88.82	87.12
2	99.28	87.12	87.47
3	99.64	88.43	87.39
4	99.64	88.31	87.45
5	99.64	88.42	87.51
6	99.27	87.12	87.58
7	99.64	88.81	86.96
8	98.55	88.21	86.12
9	99.64	88.41	87.38
10	99.91	88.1	87.22
av: train set	**99.42**	**88.18**	**87.22**
av. test set	**98.91**	**87.10**	**86.31**

Table 4. Important of independent variables in Random Forest model training

Lp	Name	Importance [%]
1	industry	27.73
2	pozkoszop	6.86
3	emplcost_oprev	3.03
4	currliab	3.02
5	currass	2.13
6	creditors	2.02
7	totalass	1.92
8	debtors	1.83
9	pkd	1.72
10	tangfixass	1.66
11	nace	1.66
12	netcurrass	1.62
13	financialpl	1.53
14	loans	1.32
15	zadlakt	1.32
16	eat	1.24
17	othercurrliab	1.24
18	ebt	1.22
19	stock	1.22
20	profitmargin	1.21

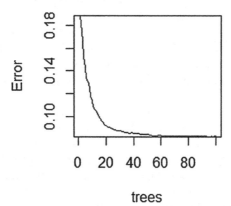

Fig. 1. Dependency between the number of decision trees and general error

5   Concluding Remarks

The goal of the research presented in this paper was to verify two research hypotheses:

[H1]: The Random Forest algorithm (trained on a representative sample of companies) is more accurate than traditional bankruptcy prediction methods: logit and linear discriminant models, and

[H2]: The Random Forest algorithm efficiently uses normalized financial statement data (there is no need to calculate financial ratios).

Based on the presented results, there was no empirical evidence to reject the hypothesis [H1]: the RF for the analyzed representative data set turned to be the most efficient method. Decision trees can learn everything, and they efficiently learned from the provided sample. DA and Logit models also did relatively well, but their predictive accuracy was below 95%. If one had to guess why RF did so well in this case, one could risk several answers.

First of all, we did not remove forcibly too many outliers. We checked their existence and analyzed whether they were caused by some errors of they represent real observations from the economy. The RF is resistant to outliers, it puts them into its leaves and they do not influence the prediction process.

Secondly, the RF deals well with lack of the normal distribution of independent variables. It also handles automatically missing values, so it can learn from the cases which were useless for classical methods [25].

Finally, it is based on majority voting (compare Fig. 1) so it uses a technique quite popular in the financial analysis where one often uses different bankruptcy prediction models and estimates solvency of a company based on multiple models taken together. Summing up, on diversified, representative quite unpredictable set of real-life companies, the RF did very well.

Hypothesis [H2] was more problematic. The Random Forest generally used only variables which were taken directly from the financial statement. There was, however, one exception – it used the relation between employment costs and operating revenue. When we removed it, the model accuracy dropped from 99.42% to 99.38% but when we increased the number of decision trees from 100 to 150, the accuracy of such a model increased to 99.45% (but generally, as it can be seen, accuracy loss was insignificant). In our opinion, since the removal of the only important ratio did not impact significantly the model's accuracy, there was no evidence to reject also the second hypothesis. The RF did well without financial ratios. The RF model algorithm managed to train itself mostly from raw data (only normalized) from the financial statement.

Our research results led to certain conclusions. Classical bankruptcy prediction models (linear discriminant analysis and logit/logistic function models) usually are estimated once (by some researcher – such as Altman), and then they are used in the economy by substituting financial rations to the given, linear formula. Modern bankruptcy prediction methods are based on another assumption – constant learning of the model and commercial access to the server by the interested entrepreneurs. It is far less convenient and more expensive but yet, every bigger company is using some form of AI commercial credit scoring system. The reason is simple – the much higher accuracy of ML methods pays for itself. How accurate ML algorithms can be was shown in this paper. The RF model accuracy was above 99%.

The second conclusion is that the RF model helps to avoid making decisions which observation is and which observation is not an outlier. Since financial data very rarely follows the normal distribution, there is a big discussion how to recognize outliers in data and whether one should remove them (if such observations are not errors).

References

1. Barboza, F., Kimura, H., Altman, E.: Machine learning models and bankruptcy prediction. Expert Syst. Appl. **83**, 405–417 (2017)
2. Jabeur, S., Fahmi, Y.: Forecasting financial distress for french firms: a comparative study. Empir. Econ. **3**, 1–14 (2017)
3. Nagaraj, K., Sridhar, A.: A predictive system for detection of bankruptcy using machine learning techniques. Int. J. Data Min. Knowl. Manag. Process (IJDKP) **5**, 29–40 (2015)
4. Liao, J.J., Shih, C.H., Chen, T.F., Hsu, M.F.: An ensemble-based model for two-class imbalanced financial problem. Econ. Model. **37**, 175–183 (2014)
5. Huang, J., Wang, H., Kochenberger, G.: Distressed chinese firm prediction with discretized data. Manag. Decis. **55**, 786–807 (2017)
6. Pociecha, J., Pawelek, B., Baryla, B.: Statystyczne metody prognozowania bankructwa w zmieniajacej sie koniunkturze gospodarczej. Wydawnictwo UEK (2014)
7. Korol, T.: Systemy ostrzegania przedsiebiorstw przed ryzykiem upadlosci. Oficyna Wolters Kluwer Business (2010)
8. Pawelek, B., Grochowina, D.: Podejscie wielomodelowe w prognozowaniu zagrozenia przedsiebiorstw upadloscia w polsce. Prace Naukowe Uniwersytetu Ekonomicznego we Wroclawiu, pp. 171–179 (2017)
9. Jardin, P.: A two-stage classification technique for bankruptcy prediction. Eur. J. Oper. Res. **254**, 236–252 (2016)
10. Min, J., Jeong, C.: A binary classification method for bankruptcy prediction. Expert Syst. Appl. **36**, 5256–5263 (2009)
11. Alfaro, E., Garcia, N., Games, M., Elizondo, D.: Bankruptcy forecasting: an empirical comparison of ada boost and neural networks. Decis. Support Syst. **45**, 110–122 (2008)
12. Anandarajan, M., Lee, P., Anandarajan, A.: Bankruptcy prediction of financially stressed firms: an examination of the predictive accuracy of artificial neural networks. Int. J. Intell. Syst. Acc. **10**, 69–81 (2001)
13. Cho, S., Hong, H., Ha, B.: A hybrid approach based on the combination of variable selection using decision trees and case-based reasoning using the mahalanobis distance: for bankruptcy prediction. Expert Syst. Appl. **37**, 3482–3488 (2010)
14. Cho, S., Kim, J., Bae, J.K.: An integrative model with subject weight based on neural network learning for bankruptcy prediction. Expert Syst. Appl. **10**, 403–410 (2009)
15. Fedorova, E., Gilenko, E., Dovzhenko, S.: Bankruptcy prediction for russian companies: application of combined classifiers. Expert Syst. Appl. **40**, 7285–7293 (2013)

16. Ghodselahi, A., Amirmadhi, A.: Application of artificial intelligence techniques for credit risk evaluation. Int. J. Model. Optim. **1**, 243–249 (2011)
17. Hu, Y.C., Tseng, F.M.: Functional-link net with fuzzy integral for bankruptcy prediction. Neurocomputing **3**, 2959–2968 (2007)
18. Sun, J., Li, H.: Financial distress prediction based on serial combination of multiple classifiers. Expert Syst. Appl. **18**, 8659–8666 (2009)
19. Li, H., Sun, J.: Business failure prediction using hybrid2 case-based reasoning. Comput. Oper. Res. **37**, 137–151 (2010)
20. Li, H., Sun, J.: Principal component case-based reasoning ensemble for business failure prediction. Inf. Manag. **48**, 220–227 (2009)
21. Li, H., Lee, Y.C., Zhou, Y.C., Sun, J.: The random subspace binary logit (RSBL) model for bankruptcy prediction. Knowl.-Based Syst. **24**, 1380–1388 (2011)
22. Min, J., Lee, Y.: Bankruptcy prediction using support vector machine with optimal choice of kernel function parameters. Expert Syst. Appl. **28**, 603–614 (2005)
23. Pena, T., Martinez, S., B., A.: Bankruptcy prediction: a comparison of some statistical and machine learning techniques. SSRN's eLibrary (18) (2009)
24. Tseng, F., Hu, Y.: Comparing four bankruptcy prediction models: logit, quadratic interval logit, neural and fuzzy neural networks. Expert Syst. Appl. **37**, 1846–1853 (2010)
25. Lewis, N.: Machine Learning Made Easy with R: Intuitive Step by Step Blueprint for Beginners. CreateSpace (2017)

TUP-RS: Temporal User Profile Based Recommender System

Wanling Zeng[1,2(✉)], Yang Du[1,2], Dingqian Zhang[1,2], Zhili Ye[1,2],
and Zhumei Dou[1]

[1] Institute of Software Chinese Academy of Sciences, 4# South Fourth Street,
ZhongGuanCun, Haidian, Beijing 100190, China
`zhumei@iscas.ac.cn`
[2] University of Chinese Academy of Sciences, 80# ZhongGuanCun East Road,
Haidian, Beijing 100190, China
{`zengwanling15,duyang15,zhangdingqian15,yezhili15`}`@mails.ucas.edu.cn`

Abstract. As e-commerce continues to emerge in recent years, online stores compete intensely to improve the quality of recommender systems. However, most existing recommender systems failed to consider both long-term and short-term preferences of users based on purchase behavior patterns, ignoring the fact that requirements of users are dynamic. To this end, we present TUP-RS (Temporal User Profile based Recommender System) in this paper. Specifically, the contributions of this paper are two folds: (i) the long-term and short-term preferences from the topic model are combined to construct the temporal user profiles; (ii) the co-training method which shares the parameters in the same feature space is employed to increase the accuracy. We study a subset of data from Amazon and demonstrate that TUP-RS outperforms state-of-the-art methods. Moreover, our recommendation lists are time-sensitive.

Keywords: Recommender system · Topic model
Temporal user profile

1 Introduction

Recommender system, which is a popular research field in recent years, is applied to e-commerce extensively, e.g., Amazon, Tmall, etc. [1,2]. Recommendation techniques simplify the processing of large-scale products transactions confronting users' massive preferences.

In recent years, more and more online retailers begin to notice the importance of customer loyalty, and purchase pattern analysis techniques are employed to improve the performance of recommender systems [3,4]. As we know, there are rich textual features in most e-commerce applications. As shown in Fig. 1, the sequence of books purchased at different time points are inherently relevant; on the lower side, customer's personal needs tend to differ over time. For example, if a first-year student buys a *C Programming Language* book, it is very likely

© Springer International Publishing AG, part of Springer Nature 2018
L. Rutkowski et al. (Eds.): ICAISC 2018, LNAI 10842, pp. 463–474, 2018.
https://doi.org/10.1007/978-3-319-91262-2_42

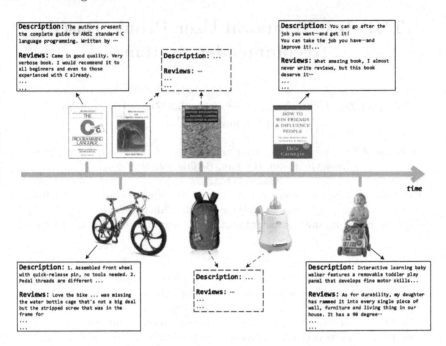

Fig. 1. A vivid illustration of a typical user's purchase trace in Amazon.com: the chances are high that items purchased at different time points by the same user are correlated. TUP-RS is capable of efficiently dealing with such circumstances via extracting valuable information (i.e., topics) from every transaction to portray users' temporal profiles.

that he or she tends to buy a *Pattern Recognition and Machine Learning* book in his or her senior year; while a customer who purchased bags and bikes during his or her schooldays may buy baby walkers after the birth of his or her first child.

Typically, content-based methods utilize text content to embed a user's profile and item to vector space based on keyword matching, TF-IDF or LDA, where proper similarity measures are taken to predict the relevance of a user to a particular item [5]. Most work handles all user profiles of all purchasing records equally regardless of time variance. However, user preferences for products are drifting from time to time. Our proposed system aims to handle this issue.

Therefore, the initiative of our work is to develop a scalable framework to utilize the long-term (intrinsic interests) and short-term (time-sensitive attentions) user preferences to improve the effectiveness of recommendation. Moreover, we focus on the algorithms which can lever the heterogeneous, multivariate and temporal datasets.

In this work, we present TUP-RS, where we first assume that every user's topic feature subjects to a continuous distribution over time and then obtain the users' short-term preferences. To achieve this, TUP-RS combines dynamic topic model [6] with the prediction of the user-item relation and reflects users' long-term preferences as well as the temporal preference variations. What's more, our model automatically arranges the product information across a category hierarchy, and hierarchical topic model is employed to improve the performance of the previously cached topic model.

The main contribution of our work is thus highlighted as follows:

- TUP-RS is a novel recommender system that matches items with users efficiently in a time-sensitive way;
- TUP-RS handles long-term and short-term preferences of users, which significantly improves the performance of recommendation.

2 Related Work

Recommendation technique is an active field of research in both data mining community and commercial applications. Our work relates to several sub-areas including recommender systems, topic models, and time variance issue.

Recommender Systems. Existing methods for recommender systems can be divided into two categories: collaborative-filtering based (CF-based) methods and content-based methods. CF-based methods assume that users with similar rating patterns share more shopping preferences. So these methods cannot work successfully when the rating data is sparse.

Content-based recommendation techniques are methods that recommend items to users based upon descriptions of items and the users' profiles [5]. [7] addresses the item cold-start problem in the content-based recommender system by applying the attribute-to-feature mapping approach to expedite random exploration of new items. [8] employs semantic technologies to recommend health websites from MedlinePlus. [9] proposes a framework that nests feature-based matrix factorization to balance both preferences and price sensitivities. Furthermore, they highlight that their method is capable of feeding economic insights into consumer behavior. Based on the activity data from the Mendeley reference manager, [10] details how implicit user feedback and collaborative filtering is used to generate the recommendations for Mendeley Suggest.

Topic Models. In recent years, topic models, which determine the abstract "topics" that occur in a group of documents, are extensively employed in recommender systems [11–14]. [11] proposes an Online Bayesian Inference algorithm, which is efficient and scalable for learning in data streams. [13] proposes a model based on Poisson factorization models combined with a social factorization model and a topic-based factorization to tackle rich contextual and content information. [12] targets at the long tail phenomena of user behaviors and the sparsity of item features via introducing a compound recommendation framework for online video recommendation. Their framework models the sample level topic proportions as a multinomial item vector. In addition, they utilize a probit classifier for topical clustering on the user part.

Time Variance Issue. Interests of users are shifting from time to time, which introduces new changes in temporal recommendation. [15] builds a framework via matrix factorization (MF) and Markov chains (MC). They analyze sequential basket data to recommend the next item. However, this method is susceptible to contingency. The method is not robust since it fails to deal with interference information introduced by infrequent products that a user randomly buys. Moreover, their model discards those data which may lead to the loss of valuable information. [16] handles user dynamics by learning a transition matrix for each user's latent vectors between consecutive time windows. It summarizes the time-invariant pattern of the evolution for the user.

In summary, the most significant distinction of our proposed method compared with other recommendation is that we consider the user's interest changes of every time interval. To achieve this, we associate each topic of every user with a continuous distribution and construct a temporal user profile vector to predict the relation between users and items.

3   Methodologies

The main idea of TUP-RS is to portray user profile that changes over time by utilizing the long-term (intrinsic interests) and short-term (time-sensitive attentions) user preferences. As illustrated in Fig. 2, TUP-RS consists of two continuous parts: (i) topic models: to extract features of temporal user profiles and item representations; (ii) similarity calculation: to predict relations between items and users. Table 1 describes the notations we use throughout this section.

Table 1. Notations

Symbol	Description
α	A Dirichlet prior
β	A Dirichlet prior
θ_u	K dimensional topic distribution for user u
θ_v	K dimensional topic distribution for item v
t_{uj}	Timestamps associated with jth token in *user document* for user u
w_{uj}	jth token in *user document* for user u
w_{vj}	jth token in *item document* for item v
φ_k	Word distribution for topic k
z	Topic assignments for each word
ψ_{uk}	Beta distribution over time for topic k of user u
γ^1, γ^2	The two parameters in Beta distribution ψ

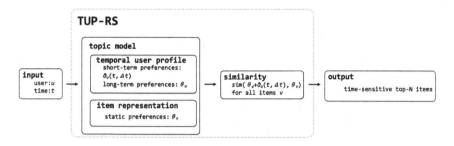

Fig. 2. The overall architecture of TUP-RS, a topic model module proposing temporal user profiles and item representations in the same feature space. Because our user profiles are time-sensitive, our top-N recommendation list also changes by time.

First, we give a formal definition of *user document* and *item document*:

- *user document*, $d_u = \{(review_v, description_v, timestamp_{uv})$ *for all items related to the user u*$\}$, is a time-stamped document which includes descriptions and reviews records of items related to a particular user's shopping trace. Note that, words share the same timestamp in one record.
- *item document*, $d_v = \{description, \ldots, (review_u)$ *for all users related to the item v*$\}$, consists of item information (titles, brands, and descriptions) and a set of reviews corresponding users, regardless of timestamps.

In order to learn temporal user profiles and item representations, we use a dynamic topic model summarized in Algorithm 1. The presented approach has two parts to learn users' and items' representations respectively. Note that, the two parts share the same parameters and can be trained together if we treat *user document* and *item document* as the same object. The only difference is that the *item document* does not consider the effect of time. We elaborate each part in the following subsections.

3.1   Temporal User Profile and Item Representations

Temporal User Profile model. Temporal user profile model quantifies the purchase willingness and adhesion of users to specific goods on sale. Topic models robustly extract the abstract "topics" that occur in a collection of documents.

The proposed Temporal User Profile Model is summarized in part one of Algorithm 1. As is shown, a *user document* is affected by the user topic proportions θ and the user topic trends ψ, which reflect user's long-term preferences and dynamic purchase interest, respectively.

As a consequence, the topic models are utilized to build the temporal user profile by constructing the likelihood of a particular *user document* corpus U in the recommender system. Given topics θ, topic assignments for each word z, word distribution φ and topic trends ψ, we obtain Eq. 1 based on the concept that the higher the likelihood is, the better the user topic proportions θ and topic trends ψ are.

Algorithm 1. Archived model for temporal user profile and item representation

---

Input: topic distribution θ, word distribution φ, beta distribution ψ
Output: a *user document* corpus U and a *item document* corpus V
1: **for** *user document* d_u in U **do**   ▷ Part One: Temporal User Profile
2:      draw topic proportions $\theta_u \sim Dirichlet(\alpha)$ for user u
3:      draw word proportions $\varphi_k \sim Dirichlet(\beta)$ for topic k
4:      **for** word in *user document* d_u **do**
5:          draw a topic z_{uj} from multinomial distribution θ_u
6:          draw a word ω_{uj} from multinomial distribution $\varphi_{z_{uj}}$
7:          draw a timestamp t_{uj} from beta distribution $\psi_{z_{uj}}$
8:      **end for**
9: **end for**
10: **for** *item document* d_v in V **do**   ▷ Part Two: Item Representation
11:      draw topic proportions $\theta_v \sim Dirichlet(\alpha)$ for item v
12:      draw word proportions $\varphi_k \sim Dirichlet(\beta)$ for topic k
13:      **for** word in *item document* d_v **do**
14:          draw a topic z_{vj} from multinomial distribution θ_v
15:          draw a word ω_{vj} from multinomial distribution $\varphi_{z_{vj}}$
16:      **end for**
17: **end for**

$$p(U|\theta,\varphi,z,\psi) = \prod_{u \in U} \prod_{j=1}^{N} \theta_k \cdot \varphi_{k,\omega_{uj}} \cdot \int_{t_{uj}-\Delta t}^{t_{uj}} \psi_{u,k}(t;\gamma_{u,k}^1,\gamma_{u,k}^2)dt \qquad (1)$$

where $\psi_{u,k}(t;\gamma_{u,k}^1,\gamma_{u,k}^2) = \frac{(1-t_{uj})^{\gamma_{u,k}^1} \cdot t_{uj}^{(\gamma_{u,k}^2-1)}}{B(\gamma_{u,k}^1,\gamma_{u,k}^2)}$, $k = z_{u,j}$, and the timestamp t is normalized to range between 0 to 1. $\psi(t;\gamma^1,\gamma^2)$ stands for the Beta probability density function, which allows for versatile shapes. By maximizing the likelihood in Eq. 1, we obtain the user topic trends and long-term preference for every user.

Our next goal is to get temporal user profile $P_u(t,\Delta t)$ during the time interval $[t - \Delta t, t]$. Temporal user profile $P_u(t,\Delta t)$ is a combination of long-term preference θ_u and time-localized user interest $\vartheta_u(t,\Delta t)$. In particular, after the time-localized user interest through user topic trends is obtained, the specific choice of temporal user interest is defined as Eq. 2

$$\vartheta_u(t,\Delta t) = \underbrace{(cul_{u,1}(t,\Delta t), cul_{u,2}(t,\Delta t), \cdots, cul_{u,k}(t,\Delta t))}_{K} \qquad (2)$$

where $cul_{u,k}(t,\Delta t) = (I_t(\gamma_{u,k}^1,\gamma_{u,k}^2)) - (I_{t-\Delta t}(\gamma_{u,k}^1,\gamma_{u,k}^2))$, and $I_x(\gamma^1,\gamma^2)$ is Cumulative Distribution Function of the beta distribution. Therefore, temporal user profile is acquired via Eq. 3.

$$P_{u,k}(t,\Delta t) = \theta_{u,k} + \lambda\vartheta_{u,k}(t,\Delta t) \qquad (3)$$

where λ is a hyperparameter, and we set it by trial and error.

Item Representation. In order to embed user and item into the same feature space, we also apply the topic model to achieve item representations. We assume that the inherent nature of item features are stable over time. Under this assumption, we model user temporal profile and item representation simultaneously in the archived model described in the Algorithm 1. The likelihood of the *user document* and *item document* corpus (given θ, z, φ, ψ) we employed is concluded in Eq. 4.

$$p(U, V | \theta, \varphi, z, \psi) = \prod_{u \in U} \prod_{j=1}^{N} \theta_{z_{uj}} \cdot \varphi_{z_{uj}, \omega_{uj}} \cdot \int_{t_{uj} - \Delta t}^{t_{uj}} \psi_{u, z_{uj}}(t) dt \cdot \prod_{v \in V} \prod_{j=1}^{N} \theta_{z_{vj}} \cdot \varphi_{z_{vj}, \omega_{vj}}$$

$$(4)$$

3.2 Parameter Training

Given a *user document* corpus U and *item document* corpus V, the learning procedure of our model is to estimate the unknown model parameter set $\Omega = \{\theta, \varphi, \psi\}$. The goal of parameter estimation is to maximize the log-likelihood in Eq. 4, which is formulated as follows:

$$argmax \log P(U, V | \theta, \varphi, \psi, z)$$

$$(5)$$

We solve Eq. 5 using the following EM-like procedure. In the E-step of the EM approach, we update z via maximizing $p(z_{uj} = k | \Omega)$ and $p(z_{vj} = k | \Omega)$, which are posterior probabilities of choosing a topic for user and item on the jth word, given $\{\theta, \varphi, \psi\}$. In the M-step, parameters are updated by maximizing the expected log-likelihood in Eq. 5 based on the posterior probability computed in the previous E-step. The entire EM-process is detailed as follows:

E-step: z is updated by maximizing posterior probabilities $p(z_{uj} = k | \Omega)$ and $p(z_{vj} = k | \Omega)$ via Gibbs sampling:

$$p(z_{uj} = k) = \frac{topic_{uk} + \alpha_k}{\sum_{k=1}^{K} topic_{uk} + \alpha_k} \cdot \frac{word_{kj} + \beta_j}{\sum_{j=1}^{N} word_{kj} + \beta_j} \cdot \frac{(1 - t_{uj})^{\gamma_{u,k}^1} \cdot t_{u,j}^{(\gamma_{u,k}^2 - 1)}}{B(\gamma_{u,k}^1, \gamma_{u,k}^2)}$$

$$p(z_{vj} = k) = \frac{topic_{vk} + \alpha_k}{\sum_{k=1}^{K} topic_{vk} + \alpha_k} \cdot \frac{word_{kj} + \beta_j}{\sum_{j=1}^{N} word_{kj} + \beta_j}$$

$$(6)$$

Where $topic_{uk}$ (or $topic_{vk}$) is number of topic k in *user document* u (or *item document* v), and $word_{kj}$ is number of jth word in *user document* u (or *item document* v) assigned to topic k.

M-step: First, we find the estimation of $\{\theta, \varphi\}$ which maximizes the expectation of log-likelihood in Eq. 4 with z updated in the preceding E-step. Then, we update the two parameters $\gamma_{u,k}^1$, $\gamma_{u,k}^2$ in Beta distribution ψ according to $\{\theta, \varphi\}$ through Eq. 7:

$$\gamma_{u,k}^1 = \bar{t}_{uk} \cdot \left(\frac{\bar{t}_{uk}(1 - \bar{t}_{uk})}{s_{uk}^2} - 1 \right)$$

$$\gamma_{u,k}^2 = (1 - \bar{t}_{uk}) \cdot \left(\frac{\bar{t}_{uk}(1 - \bar{t}_{uk})}{s_{uk}^2} - 1 \right)$$

$$(7)$$

where $\bar{t}_{uk}$ and s^2_{uk} indicate the sample mean and biased sample variance of the timestamps belonging to topic k of user u.

With an initial random guess of z, we alternately apply the E-step and M-step until convergence. In practice, due to the massive number of users and items in the scoped datasets, our temporal user profile model maintains hundreds of topics, which results in huge computational cost in the training phase. In order to address this problem, we build an explicit category tree [1] to encode product item via associating each node of the category tree with some topics (1–5), which significantly simplifies the process of training.

3.3   Prediction

In this paper, users and items are projected to the same feature space and described as the topic vector space model. The cosine similarity measure is often used when using a vector space model [17]. Here, we utilize the cosine similarity measure to calculate the relation between users and items, which is defined as follows (Eq. 8):

$$sim(P_u(t, \Delta t), \theta_v) = cos(P_u(t, \Delta t), \theta_v) = \frac{P_u(t, \Delta t) \cdot \theta_v}{|P_u(t, \Delta t)||\theta_v|} \tag{8}$$

Finally, we sort these similarities in descending order and get the top-N list for the user u.

4   Experiments

In this section, we describe the datasets we used as well as the baselines and the experimental results.

4.1   Datasets

We evaluate our proposed TUP-RS on the Amazon dataset[1], one of the biggest and highest-quality publicly-available recommendation field dataset to date. This dataset contains product reviews (i.e., ratings and text) as well as metadata (i.e., description, category information, price, and brand) from Amazon, consisting of 142.8 million reviews ranging from May 1996 to July 2014 in total.

To directly train models with millions of users is not practical. As a consequence, we crawl reviews ranging from August 2010 to July 2014 to reduce the complexity of sample space and remove users who bought less than 5 products during this period. We then draw two random samples from the data, each of size 2000, according to the following principles: the number of reviews of every single user is between 5 and 20 in Amazon-a while between 20 and 80 in Amazon-b.

We employ 5-fold cross-validation, and all recall and NDCG scores reported here are the averaged scores over 5 folds. After the training model is finalized, we apply the model to the test set to construct the relationships between the users and items, and obtain a list of the top-N recommended products.

---

[1] http://jmcauley.ucsd.edu/data/amazon/links.html.

4.2 Evaluation Measures

Generally, we aim to recommend a dynamic list of items for each user in various time periods, denoted by $T_t^{\Delta t}(u)@N$. The set of items interested by the user u between time $[t - \Delta t, t]$ is denoted by $R_t^{\Delta t}(u)$. Here, we set Δt to 180 days, and $t \in \{180d | d \in Z^+\}$. In order to evaluate the recommendation performance in different time periods, we apply the following measures to evaluate the estimated ranking against the actual bought items:

- Recall@N: Recall is a measure of relevance which is widely used in recommendation [18]:

$$Recall@N(\Delta t) = \sum_u \frac{T_t^{\Delta t}(u)@N \cap R_t^{\Delta t}(u)}{|R_t^{\Delta t}(u)|} \qquad (9)$$

- NDCG@N: Normalized Discounted Cumulative Gain (NDCG) is a ranking measure which takes into account the order of recommended items in the list [19].

$$NDCG@N(\Delta t) = \frac{1}{M_n} \sum_{j=1}^{n} \frac{2^{I(T_t^{\Delta t}(u)_j \in R_t^{\Delta t}(u))} - 1}{log_2^{(j+1)}} \qquad (10)$$

where $I(\cdot)$ is an indicator function, M_n is a constant value determined by the maximum value of NDCG@N given $T_t^{\Delta t}(u)@N$, and $T_t^{\Delta t}(u)_j$ stands for the item recommended in the jth position.

4.3 Results and Analyses

We compare the proposed method with several baseline methods:

- **ItemKNN:** ItemKNN is a CF-based method finding K nearest neighbors of items based on Pearson similarity measure. In this paper, we use ItemKNN recommender solver in [20].
- **WCB:** WCB is a typical weighted average content-based model that generates topic vector representation for each item by LDA [21]. WCB constructs a user profile by taking a weighted average of topic vectors of purchased items. Similar to TUP-RS, WCB also relies on product category tree to build a variant of hierarchy topic model and calculates the similarity between user and item based on the cosine similarity measure. However, this method does not consider the time factor or the modeling of user's preferences.
- **MCB:** Like TUP-RS, MCB also uses a variant of hierarchy topic model for user and item, which shares the same parameters and trains *user document* and *item document* together. However, this method doesn't consider the time factor.
- **TUP-RS:** Temporal user profile based recommender system is our proposed model as described above.

The results over the two datasets are shown in Fig. 3 and our analyses are as follows:

(1) Encoding and training users' profiles with items play an important role in our experiments. As shown in Fig. 3(a) and (b), TUP-RS and MCB outperform the other baseline methods on both datasets. Although WCB also uses the topic model, it does not depict users' profiles particularly.

(2) The more items users have ever bought, the better TUP-RS will be. When the number of records is small (Fig. 3(a)), TUP-RS and MCB have similar performance. But when the number of records becomes large (Fig. 3(b)), TUP-RS performs better than MCB. The reason is that the sparse purchase record cannot provide enough information to model temporal user profile minutely. Note that there are more testing items in the Amazon-b dataset for each user are increased, so the specific value in Fig. 3(b) is lower than Fig. 3(a).

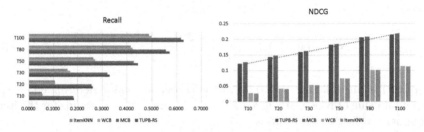

(a) the result for Amazon-a

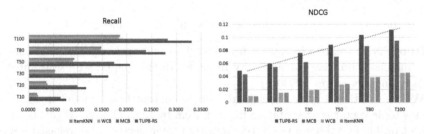

(b) the result for Amazon-b

Fig. 3. Performance comparison of ItemKNN, WCB, MCB, TUP-RS based on Recall@N and NDCG@N for dataset Amazon-a, Amazon-b. Our TUP-RS outperforms other methods on both datasets, and when the number of user's reviews becomes larger, our performance exceeds more.

Furthermore, we give an example to explain the change of a user's interest in TUP-RS. We list out the user's purchase record in the training set (Fig. 4(left)). From the record, we can find that the user preferred items about war and religion during the first half of the time but turned to self-cultivation and life during the second half of the time. So the user's preference (similarity) for the test item named *The Battle of the Neretva* reached the peak in the middle of the period (around 2012.4) and decreased gradually to both sides (the blue curve in Fig. 4(right)). These results show that our recommendations are time-sensitive. Therefore, we can get the up-to-date top-N recommendation list by setting time to present.

Date	Title
05 28, 2011	Bette Davis in the Dark Secret of Harvest Home
01 8, 2012	The 7th Dawn
01 8, 2012	Flight From Ashiya
12 3, 2012	Indian Myth and Legend
03 19, 2013	Lakota Star Knowledge : Studies in Lakota Stellar Theology
03 19, 2013	(12x12) Walkers of the Wind - 2013 Wall Calendar
05 12, 2013	Native Wisdom: Perceptions of the Natural Way
07 28, 2013	Salem witch Trials
08 23, 2013	Radio City Christmas Spectacular

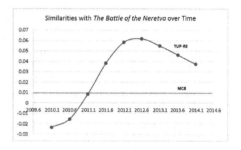

Fig. 4. A user's purchase record and interest curves about test item *The Battle of the Neretva*. We give the user's purchase records in training set and compute the similarities with *The Battle of the Neretva* at different time points. Here we use two methods: TUP-RS and MCB. Note that, the real time to buy this item is March 12, 2012. (Color figure online)

5   Conclusion

In this work, we present a novel time-sensitive recommender system, TUP-RS, which captures not only users' long-term intrinsic interests, but also users' time-sensitive preferences. The experimental results in Amazon dataset demonstrate that the proposed TUP-RS outperforms recent state-of-the-art methods by a significant margin. In the future, we will extend our method to a number of promising application areas such as video recommendation and online recruitment services. Meanwhile, we will resort to more cutting-edge systems for parallelization.

References

1. McAuley, J., Pandey, R., Leskovec, J.: Inferring networks of substitutable and complementary products. In: Proceedings of 21st ACM SIGKDD International Conference on Knowledge Discovery and Data Mining, pp. 785–794. ACM (2015)
2. Mooney, R.J., Roy, L.: Content-based book recommending using learning for text categorization. In: Proceedings of 5th ACM Conference on Digital Libraries, pp. 195–204. ACM (2000)
3. Lu, H.: Recommendations based on purchase patterns, US Patent App. 14/300,248, 10 December 2015
4. Sheehan, N.T., Bruni-Bossio, V.: Strategic value curve analysis: diagnosing and improving customer value propositions. Bus. Horiz. **58**(3), 317–324 (2015)

5. Pazzani, M.J., Billsus, D.: Content-based recommendation systems. In: Brusilovsky, P., Kobsa, A., Nejdl, W. (eds.) The Adaptive Web. LNCS, vol. 4321, pp. 325–341. Springer, Heidelberg (2007). https://doi.org/10.1007/978-3-540-72079-9_10

6. Wang, X., McCallum, A.: Topics over time: a non-Markov continuous-time model of topical trends. In: Proceedings of 12th ACM SIGKDD International Conference on Knowledge Discovery and Data Mining, pp. 424–433. ACM (2006)

7. Cohen, D., Aharon, M., Koren, Y., Somekh, O., Nissim, R.: Expediting exploration by attribute-to-feature mapping for cold-start recommendations. In: Proceedings of 11th ACM Conference on Recommender Systems, pp. 184–192. ACM (2017)

8. Bocanegra, C.L.S., Ramos, J.L.S., Rizo, C., Civit, A., Fernandez-Luque, L.: HealthRecSys: a semantic content-based recommender system to complement health videos. BMC Med. Inform. Decis. Mak. **17**(1), 63 (2017)

9. Wan, M., Wang, D., Goldman, M., Taddy, M., Rao, J., Liu, J., Lymberopoulos, D., McAuley, J.: Modeling consumer preferences and price sensitivities from large-scale grocery shopping transaction logs. In: Proceedings of 26th International Conference on World Wide Web, pp. 1103–1112. International World Wide Web Conferences Steering Committee (2017)

10. Hristakeva, M., Kershaw, D., Rossetti, M., Knoth, P., Pettit, B., Vargas, S., Jack, K.: Building recommender systems for scholarly information. In: Proceedings of 1st Workshop on Scholarly Web Mining, pp. 25–32. ACM (2017)

11. Liu, C., Jin, T., Hoi, S.C., Zhao, P., Sun, J.: Collaborative topic regression for online recommender systems: an online and Bayesian approach. Mach. Learn. **106**(5), 651–670 (2017)

12. Lu, W., Chung, F.L., Lai, K., Zhang, L.: Recommender system based on scarce. Neural Netw. **93**, 256–266 (2017)

13. da Silva, E.d.S.: New probabilistic models for recommender systems with rich contextual and content information. In: Proceedings of 10h ACM International Conference on Web Search and Data Mining, pp. 839–839. ACM (2017)

14. Wang, H., Wang, N., Yeung, D.Y.: Collaborative deep learning for recommender systems. In: Proceedings of 21st ACM SIGKDD International Conference on Knowledge Discovery and Data Mining, pp. 1235–1244. ACM (2015)

15. Rendle, S., Freudenthaler, C., Schmidt-Thieme, L.: Factorizing personalized markov chains for next-basket recommendation. In: Proceedings of 19th International Conference on World Wide Web, pp. 811–820. ACM (2010)

16. Zhang, C., Wang, K., Yu, H., Sun, J., Lim, E.P.: Latent factor transition for dynamic collaborative filtering. In: Proceedings of 2014 SIAM International Conference on Data Mining, pp. 452–460. SIAM (2014)

17. Salton, G.: Automatic Text Processing: The Transformation, Analysis, and Retrieval of. Addison-Wesley, Reading (1989)

18. Steck, H.: Evaluation of recommendations: rating-prediction and ranking. In: Proceedings of 7th ACM conference on Recommender systems, pp. 213–220. ACM (2013)

19. Järvelin, K.: IR evaluation methods for retrieving highly relevant documents. In: International ACM SIGIR Conference on Research and Development in Information Retrieval, pp. 41–48 (2000)

20. Guo, G., Zhang, J., Sun, Z., Yorke-Smith, N.: LibRec: a Java library for recommender systems. In: UMAP Workshops (2015)

21. Blei, D.M., Ng, A.Y., Jordan, M.I.: Latent Dirichlet allocation. J. Mach. Learn. Res. **3**(January), 993–1022 (2003)

Feature Extraction of Surround Sound Recordings for Acoustic Scene Classification

Sławomir K. Zieliński[✉]

Faculty of Computer Science, Białystok University of Technology,
Białystok, Poland
s.zielinski@pb.edu.pl

Abstract. This paper extends the traditional methodology of acoustic scene classification based on machine listening towards a new class of multichannel audio signals. It identifies a set of new features of five-channel surround recordings for classification of the two basic spatial audio scenes. Moreover, it compares the three artificial intelligence-based classification approaches to audio scene classification. The results indicate that the method based on the early fusion of features is superior compared to those involving the late fusion of signal metrics.

Keywords: Machine listening · Acoustic scene classification
Feature extraction · Ensemble-based classifiers

1 Introduction

Machine listening algorithms still have difficulties in attaining the abilities of human listeners in analysis of acoustic scenes [1]. While considerable improvements have recently been made in such areas as speech recognition, speaker verification and music information retrieval, the domain of acoustic scene classification (ASC) seems to remain under-researched. The aim of ASC is typically to identify the environment in which the sound sources were recorded, e.g. "noisy street", "office", "train station" [2], or to characterize its properties ("large", "small", "immersive", etc.). The applications of ASC include surveillance, automatic optimization of audio devices (e.g. hearing aids) and environmental navigation of robots [3].

The state-of-the-art machine listening algorithms for automatic ASC typically employ such artificial intelligence methods as deep neural networks (DNN) [4, 5], convolutional neural networks (CNN) [6], hidden Markov models (HMM) [7], random forests (RF) [6, 8], and Support Vector Machines (SVM) [3, 9]. Most of the work in the area of ASC so far has been limited to the systems with a single audio channel input and, consequently, restricted to monaural sound classification. Little progress has been made towards devising ASC algorithms capable of classifying multichannel audio signals. Recently, Trowitzsch et al. [10] developed a system for detection of environmental sounds in two-channel binaural auditory scenes. Imoto and Ono [11] proposed a method for automatic scene classification using a distributed multichannel microphone array. However, to the best of the author's knowledge, no work has been done towards an automatic classification of five-channel surround sound.

© Springer International Publishing AG, part of Springer Nature 2018
L. Rutkowski et al. (Eds.): ICAISC 2018, LNAI 10842, pp. 475–486, 2018.
https://doi.org/10.1007/978-3-319-91262-2_43

The literature provides many examples of feature extraction methods. The standard features, adapted from such areas as speech recognition or music information, are commonly used for ASC. They predominantly include Mel-frequency cepstral coefficients (MFCC) [2], spectral content indicators (e.g. centroid, brightness, flatness) [12, 13], metrics based on spectro-temporal information [5, 8], and intermediate features obtained by non-negative matrix factorization (NMF) [5]. However, the standard features were designed to represent information derived from single-channel signals and may not fully characterize spatial aspects of multichannel audio recordings. Hence, there is a need to design new features, accounting for spatial characteristics of multichannel audio signals, which is the purpose of this study.

In this work, a new machine listening method for classification of five-channel surround sound recordings according to their basic spatial audio scenes is proposed. The two generic approaches were considered, namely *early fusion* and *late fusion* of features, depending on whether a fusion of information took place at the feature extraction level or at the classification level.

The paper offers the following contributions: (1) it extends the methodologies of ASC towards a new class of spatial audio signals; (2) it identifies a set of signal features, including the original "spatial" metrics, allowing for classification of five-channel surround sound; (3) it provides a comparison of the artificial intelligence-based classification schemes, based on early and late fusion of multichannel audio features. This can help to direct future developments of machine listening algorithms.

2 Approach Overview

The standard layout of loudspeakers allowing for reproduction of five-channel surround sound recordings, adapted for the purposes of this work, is depicted in Fig. 1. It consists of the five loudspeakers positioned on the circle surrounding a listener in the median plane.

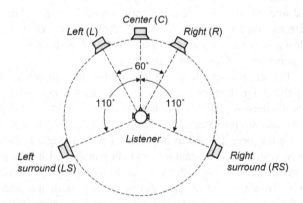

Fig. 1. Loudspeaker layout of the five-channel surround sound system conformant to the ITU-R BS.775 recommendation [14].

2.1   Fusion of Multichannel Audio Information

Majority of the traditional algorithms for audio scene classification exploit a single-channel audio input (Fig. 2a). In this study, new classification schemes were implemented allowing for classification of five-channel surround sound. In the first implemented method, referred to as an early fusion technique, information extracted from individual audio signals was merged prior to undertaking a classification procedure (Fig. 2b). The late fusion technique involved combining inter-channel information during the classification procedure. As can be seen in Fig. 2, the late fusion method could be further subdivided into two variants, depending on the number of classifiers used. In the first variant, channel features were fed to a single classifier (Fig. 2c). In the second case, inspired by the recent work of Sánchez-Hevia et al. [15], channel features were used as input information of the ensemble of five classifiers, whose outputs (meta-features) were fused in the combination classifier (Fig. 2d).

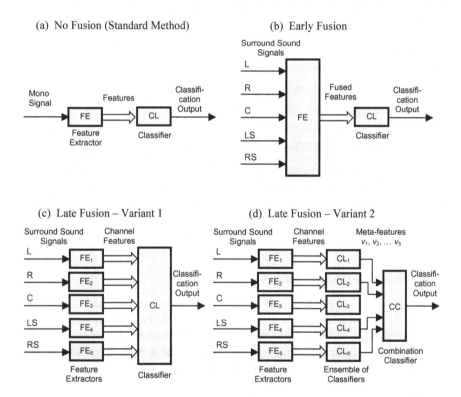

Fig. 2. Classification layouts: (a) no fusion (standard method), (b) early fusion, (c) late fusion – variant 1, (d) late fusion – variant 2.

2.2   Corpus of Surround Sound Recordings

A corpus of 110 five-channel audio recordings was gathered for the purpose of this work. The recordings were extracted from commercially available DVD recordings in the form of short excerpts. The mean duration of the acquired audio samples was equal to 11.6 s with a standard deviation of 7.2 s. The recordings represented a broad range of genres, including classical music, pop music, jazz, and movies. During the selection procedure attention was paid that each excerpt exhibited stationary spatial character-istics and represented a single spatial scene (either *FB* or *FF* – see below). The recordings were sampled at a 48 kHz rate with a 16-bit resolution and stored as uncompressed multichannel audio files.

2.3   Annotation of Audio Recordings

For the purpose of this study, a simple two-category taxonomy of spatial audio scenes was adopted from Rumsey's spatial audio scene paradigm [16], according to a distri-bution of foreground/background audio content across individual channels of the surround sound system. The two basic scenes were distinguished, signified as *FB* and *FF* respectively. They were defined in Table 1. The recordings were annotated man-ually by this author. The corpus of audio recordings was slightly imbalanced, since 57 recordings represented the *FB* scene (52%), whereas the remaining 53 recordings were annotated as exhibiting the *FF* scene (48%).

Table 1. Taxonomy of basic spatial audio scenes (inspired by Rumsey [16])

Acoustic scene	Distribution of audio content across channels
FB	Front loudspeakers reproduce predominantly foreground audio content (identifiable, important and clearly perceived audio sources), whereas the rear loudspeakers reproduce only background audio content (room response, reverberant, unimportant, unclear, ambient, and "foggy" sounds)
FF	Both front and rear loudspeakers reproduce foreground content. This scene may refer to the audio impression where a listener is surrounded by an ensemble of musicians or a group of simultaneously talking speakers

2.4   Approach to Feature Extraction

In total, 19 features were extracted for the early fusion scheme and additional 75 features were acquired for the late fusion methods. While the features selected for the late-fusion schemes could be considered as standard ones, commonly used for ASC, most of the features proposed for the early fusion scheme were designed for this study. The procedure of feature extraction was described in detail in Sect. 3.

2.5   Automatic Classification

The purpose of automatic classification was to categorize the audio recordings according to one of the basic spatial audio scenes (either *FB* or *FF*). Regardless of the fusion technique, all the classifiers, except the combination ones, were based on ran-dom forests. To this end, the *randomForest* algorithm implemented in *R* system by Breiman and Cutler was used [17]. Each forest consisted of 500 trees. A number of signal features randomly sampled as candidates at each split (parameter *mtry*) was optimized during the cross-validation procedure, and set to 2 and 38 features for the early and late fusion (variant 1) schemes respectively. Other classification techniques were also trialed. However, according to the initial results (not presented in the paper), random forests exhibited a reasonable trade-off between the obtained accuracy and a computational load.

An ensemble of classifiers was used in variant 2 of the late fusion algorithm. These classifiers were also based on random forests (500 trees in each forest) and were almost identical in its operation principle to those described above. The only difference was that they operated on limited sets of features (see Fig. 2d). Each of the classifiers was fed with 15 channel specific features. The value of parameter *mtry* was optimized separately for every fold of data during the cross-validation procedure and was set to either 2 or 15 features for each classifier, depending on the cross-validated fold of data.

While many sophisticated methods of combining information from ensembles of classifiers exist [18, 19], in this study the final classification output was obtained using the technique of stochastic gradient boosting [20]. For this purpose, *gbm* algorithm, developed in *R* system by Ridgeway [21], was applied. Its parameters were optimized during the cross-validation procedure, and for most of the cross-validated data folds were set to the following values: number of trees – 50, interaction depth – 1, shrinkage – 0.1, a minimum number of observations in the trees terminal nodes – 10.

Since the database used for the classification was relatively small (110 audio recordings), the performance of the classification methods was tested using a 10-fold cross-validation procedure repeated 10 times. The classification results were compared using the two standard metrics: accuracy and Cohen's kappa coefficient.

3   Feature Extraction

Let matrix $\mathbf{X} = (\mathbf{x}_1\mathbf{x}_2...\mathbf{x}_5)$ represents a set five-channel surround sound signals, where indices $k = 1, 2 ... 5$ refer to the consecutive channels of the standard surround sound system in the following order: left, right, center, left surround, right surround (see Fig. 1), while $\mathbf{x}_k$ represents a column vector containing samples of the k-th channel signal.

3.1   Late-Fusion Features

Mel-Frequency Cepstral Coefficients (mfcc_*kl*). Mel-frequency cepstral coefficients (MFCC) are commonly used as spectral features for acoustic scene classification of monaural signals. In line with the literature [2, 9, 15], 13 coefficients l were calculated for every k-th channel and 12 of them were retained for the classification purposes (omitting the first one as irrelevant), yielding a vector of 60 features (12 features × 5 channels).

Mean Energy (energy_*k*). Mean energy (power) is a commonly used feature in ASC. It was calculated for each audio channel k as

$$E_k = \frac{1}{N}\sum_{n=1}^{N} x_{nk}^2, \quad k = 1, 2, \ldots 5,$$ (1)

where N represents the total number of samples in a given recording.

Crest Factor (crest_k). Crest factor is another standard metric. It was calculated for each k-th channel as follows:

$$s_k = 20 \log_{10}\left(\frac{\max|\mathbf{x}_k|}{\sqrt{E_k}}\right) \quad k = 1, 2, \ldots 5.$$ (2)

Zero Crossing (zcrossing_k). Zero crossing is also a popular metric used for acoustic scene analysis [9, 12]. It was calculated for each channel separately ($k = 1, 2, \ldots 5$) and normalized to the total number of samples N in a given signal.

3.2 Early-Fusion Features

Front-to-Back Energy Ratio (fb_energy). This new spatial feature was introduced due to an informal observation that for some recording exhibiting an FB scene the energy of the rear channels was less than that of the front channels. It was calculated using the following formula:

$$E_{FB} = 10 \log_{10}\left(\frac{E_1 + E_2 + E_3}{E_4 + E_5}\right),$$ (3)

where mean energy of the individual channels E_1, E_2, ... E_5 was calculated using Eq. (1).

Lateral Energy (lateral_energy). This feature was inspired by the early work of Bradley and Soulodre [22] in concert hall acoustics. The lateral energy E_{LE} is an estimation of the lateral acoustic energy E_{lateral} captured by a simulated figure-of-eight microphone normalized to the total energy of the down-mixed signal E_{omni}, according to the following equation:

$$E_{LE} = 10 \log_{10}\left(\frac{E_{\text{lateral}}}{E_{\text{omni}}}\right).$$ (4)

The lateral energy was calculated using a cosine-type directivity pattern as

$$E_{\text{lateral}} = \sum_{k=1}^{5} |\cos(\theta_k)| E_k,$$ (5)

where θ_k denotes azimuth of the k-th loudspeaker (see Fig. 1). E_{omni} was calculated as a mean energy of a mono signal $\mathbf{x}_{\text{mono}} = \sum_{k=1}^{5} \mathbf{x}_k$.

Centroid of PCA Coefficients (pca_centroid). The rationale for selecting this new feature was an observation that for some recordings exhibiting the FF scene the absolute values of the PCA eigenvectors associated with the surround channels were prominent. We denote $\mathbf{R}$ as the covariance matrix of the multichannel audio signals $\mathbf{X}$. All the channel signals $\mathbf{x}_k$ were centered (mean-value equalized) prior to the calculation of the covariance matrix, however, they were deliberately left unstandardized, so that the energy of the individual channels could affect the coefficients of the covariance matrix $\mathbf{R}$. The eigen-decomposition of $\mathbf{R}$ could be expressed as

$$\mathbf{R} = \mathbf{EDE}^T, \tag{6}$$

where $\mathbf{E}$ represents the eigenvector matrix, whereas $\mathbf{D}$ denotes a matrix whose the diagonal elements are the eigenvalues of the decomposed matrix. Both matrices $\mathbf{E}$ and $\mathbf{D}$ have dimensions of 5×5. Let coefficients w_q represent the following sums:

$$w_q = \sum\nolimits_{p=4}^{5} e_{pq}, \quad q = 1, 2, \ldots 5, \tag{7}$$

where e_{pq} are the elements of the eigenvector matrix $\mathbf{E}$. The centroid of the absolute values of the PCA coefficients associated with the rear channels was calculated as

$$c = \sum\nolimits_{q=1}^{5} |qw_q| \Big/ \sum\nolimits_{q=1}^{5} w_q. \tag{8}$$

Variance of PCA Components (pca_var_k). The variance of the PCA components was calculated using the standard formula [23]. This metric indirectly reflects the level of inter-channel correlation between the channels and hence might be useful for distinguishing between the *FB* and *FF* scenes.

Inter-channel Cross Correlation Coefficients (corr_l_ls, corr_r_rs, corr_l_ls, corr_fb). Another way of estimating a magnitude of correlation between audio signals is to calculate a set of cross-correlation coefficients. The standard cross-correlation coefficients were calculated between the following pairs of signals: $\mathbf{x}_1$ and $\mathbf{x}_4$, $\mathbf{x}_2$ and $\mathbf{x}_5$, $\mathbf{x}_4$ and $\mathbf{x}_5$, and $\mathbf{x}_{\text{front}}$ and $\mathbf{x}_{\text{rear}}$; where $\mathbf{x}_{\text{front}} = \mathbf{x}_1 + \mathbf{x}_2 + \mathbf{x}_3$ and $\mathbf{x}_{\text{rear}} = \mathbf{x}_4 + \mathbf{x}_5$.

Inter-aural Cross-Correlation Coefficient (IACC) (iacc, d_iacc). IACC is a popular metric used for evaluation of concert hall acoustics as well as for the objective assessment of spatial audio quality [24, 25]. The binaural signals, necessary for estimation of IACC, were synthesized by convolving the multichannel audio signals with the head-related transfer functions (HRTFs) acquired from the MIT KEMAR database [26]. In addition to the standard IACC metric (iacc), a difference feature (d_iacc) was calculated as IACC − IACC$_{\text{front}}$, where IACC is the feature computed for all the channels, whereas IACC$_{\text{front}}$ represents the feature estimated for only the front channels.

Crest Factor Difference (d_crest). A difference in crest factor was computed as $\Delta s = s_{\text{front}} - s_{\text{rear}}$, where s_{front} and s_{rear} represent crest factors calculated for signals $\mathbf{x}_{\text{front}}$ and $\mathbf{x}_{\text{rear}}$ respectively.

Zero Crossing Difference (d_zcrossing). A difference in zero crossing was computed between signals x_{front} and x_{rear} respectively. The obtained difference was normalized to the value of zero crossing obtained for the front channels x_{front}.

MFCC-Based Coefficients (mfcc_dist, mfcc_corr). The additional two features were included: (1) Euclidean distance between MFCC coefficients calculated for x_{front} and x_{rear} signals respectively, and (2) correlation coefficient between the above coefficients.

Spectral Coherence (coherence). The standard function of spectral coherence was computed between signals x_{front} and x_{rear}. Its mean value was subsequently calculated across the frequency spectrum.

4   Results

Prior to undertaking the classification tests, the extracted features were explored using the PCA method. For the early fusion features, its first PCA dimension was predominantly related to the IACC feature and to the features accounting for the variance of principal components obtained from the analysis of surround sound signals (pca_var1, pca_var2, etc.) (Fig. 3a). The second dimension was related to the features describing inter-channel cross-correlation coefficients (corr_l_ls, corr_fb, corr_r_rs). Due to a large number of overlapping variables, the interpretation of the factor map obtained for the late fusion database was more challenging (Fig. 3b). The first dimension was predominantly related to the MFCCs, whereas the second dimension seemed to be affected by a mixture of MFCCs and zero crossing features.

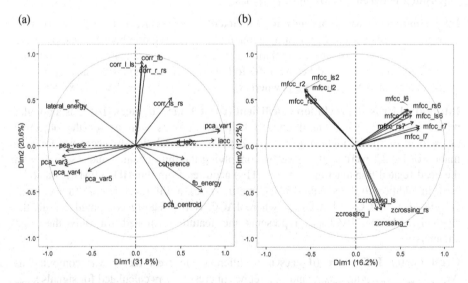

Fig. 3. Variable PCA factor maps: (a) early fusion features, (b) late fusion features. For clarity, the plot was limited to 15 features showing the highest contribution to the model.

PCA factor maps for the individual audio recordings were presented in Fig. 4 for both early and late fusion databases respectively. There was a high degree of overlap between the *FB* and *FF* recordings, which revealed that regardless of the feature fusion scheme, the task of classification of the audio scenes constituted a nontrivial problem.

(a)   (b)

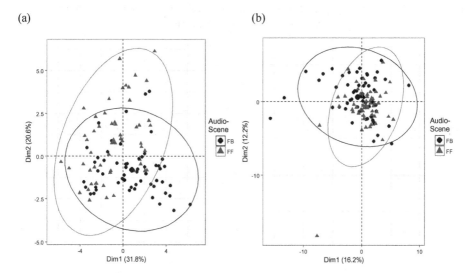

Fig. 4. Individual audio recordings factor maps obtained using: (a) early fusion data, (b) late fusion data. Ellipses represent a concentration of individual recordings for *FB* and *FF* audio scenes respectively with 95% confidence-boundaries around group means.

Regardless of the aforementioned challenge, the obtained classification results were promising for all compared methods (Table 2). Despite the fact that the database used for the early fusion scheme consisted of only 19 features, compared to 75 features used for the late fusion methods, the results obtained for the early fusion scheme were the best. The accuracy obtained for the early fusion scheme was equal to 84%, whereas the accuracy of the two methods based on the late fusion approach was less, reaching a level of 75% and 73% respectively. A similar tendency could be observed when comparing Cohen's kappa coefficients. Hence, despite the mentioned above disproportion of the database sizes, the method involving the early fusion scheme outperformed the other two schemes, which was confirmed statistically in a t-test at $p < 0.05$ level. No inferences concerning the late fusion schemes could be made since the difference between their accuracy was not statistically significant ($p > 0.05$).

Table 2. Classification results. Numbers in brackets represent a standard deviation

Performance metric	Early fusion scheme	Late fusion scheme variant 1	Late fusion scheme variant 2
Accuracy	0.84 (0.12)	0.75 (0.14)	0.73 (0.14)
Kappa	0.67 (0.24)	0.50 (0.28)	0.46 (0.27)

The most important features identified during the classification were overviewed in Fig. 5. For the early fusion scheme, centroid of PCA coefficients was identified as the most prominent feature (pca_centroid). Other significant features included fb_energy, lateral_energy, d_zcrossing, and d_crest. For the late fusion classification scheme, the following features exhibited the highest level of importance: energy_ls, energy_rs, mfcc_rs2 (Mel-frequency 2nd cepstral coefficient of the right surround channel), and crest_rs. They were all related to the surround channels.

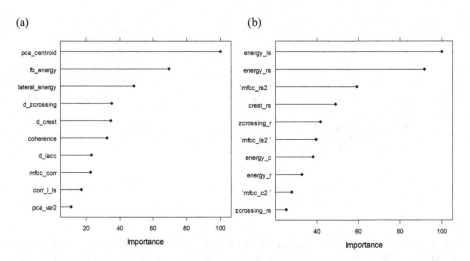

Fig. 5. Top 10 important features: (a) early fusion scheme, (b) late fusion scheme – variant 1.

To verify the above observations, the fourth strategy of fusion of information was also examined (not illustrated in Fig. 2). Both the early and the late fusion features were concatenated forming a vector of 94 metrics (19 + 75), which were subsequently fed to the random forest-based classifier. The results were almost identical to the ones obtained using the early fusion approach (discrepancy less than 0.03%). This outcome supports the observation regarding the superiority of the early fusion topology.

5   Conclusions and Future Work

This work extended the traditional methodologies of acoustic scene classification (ASC) based on machine listening towards five-channel surround sound signals. The three artificial intelligence-based classification topologies were compared. According to the obtained results, the method based on the early fusion of signal features was superior compared to the two algorithms involving the late fusion of metrics and therefore it should be considered for further developments of the spatial ASC algorithms.

The feature derived from the PCA transformation of the surround sound signals proved to be the most prominent in terms of the spatial audio scene classification using the early fusion scheme. This outcome indicates that the features extracted during principal component decomposition or signal matrix factorization might be particularly effective in terms of spatial audio scene classification and, therefore, such methods should be further explored.

This work identified the features as well as the classification topologies suitable for automatic categorization of spatial audio scenes. It is difficult to compare the proposed method against the conventional algorithms since the work presented in this paper can be considered as an extension rather than improvement of the existing ASC techniques. Moreover, the currently presented study was concerned with the automatic classification of the basic spatial audio scenes (*FB* and *FF*); a scenario which has not been incorporated in the traditional ASC algorithms yet.

The conclusions reached in this study should be treated as preliminary, since only random forests were used as the classification algorithms across the experimental conditions. One cannot exclude a possibility that other classification methods, in particular deep neural networks (DNNs), may yield different results. Another limitation of the study is related to a small number of audio samples used. The tasks of a systematic comparison of various types of classification techniques as well as extending the number of audio recordings used in the experiments were left to future work.

Acknowledgements. This work was supported by a grant S/WI/1/2013 from Bialystok University of Technology and funded from the resources for research by Ministry of Science and Higher Education.

References

1. Richard, G., Virtanen, T., Bello, J.P., Ono, N., Glotin, H.: Introduction to the special section on sound scene and event analysis. IEEE/ACM Trans. Audio Speech Lang. Process. **25**(6), 1169–1171 (2017)
2. Stowell, D., Giannoulis, D., Benetos, E., Lagrange, M., Plumbley, M.D.: Detection and classification of acoustic scenes and events. IEEE Trans. Multimedia **17**(10), 1733–1746 (2015)
3. Chu, S., Narayanan, S., Jay Kuo C.-C., Matarić, M.J.: Where am I? Scene recognition for mobile robots using audio features. In: Proceedings of IEEE International Conference on Multimedia and Expo, Toronto, Canada, pp. 885–888. IEEE (2006)
4. Petetin, Y., Laroche, C., Mayoue, A.: Deep neural networks for audio scene recognition. In: Proceedings of 23rd European Signal Processing Conference (EUSIPCO), Nice, France, pp. 125–129. IEEE (2015)
5. Bisot, V., Serizel, R., Essid, S., Richard, G.: Feature learning with matrix factorization applied to acoustic scene classification. IEEE/ACM Trans. Audio Speech Lang. Process. **25**(6), 1216–1229 (2017)
6. Phan, H., Hertel, L., Maass, M., Koch, P., Mazur, R., Mertins, A.: Improved audio scene classification based on label-tree embeddings and convolutional neural networks. IEEE/ACM Trans. Audio Speech Lang. Process. **25**(6), 1278–1290 (2017)
7. Dargie, W.: Adaptive audio-based context recognition. IEEE Trans. Syst. Man Cybern. – Part A: Syst. Hum. **39**(4), 715–725 (2009)

8. Stowell, D., Benetos, E.: On-bird sound recordings: automatic acoustic recognition of activities and contexts. IEEE/ACM Trans. Audio Speech Lang. Process. **25**(6), 1193–1206 (2017)
9. Geiger, J.T., Schuller, B., Rigoll, G.: Large-scale audio feature extraction and SVM for acoustic scene classification. In: Proceedings of IEEE Workshop on Applications of Signal Processing to Audio and Acoustics, New Paltz, NY. IEEE (2013)
10. Trowitzsch, I., Mohr, J., Kashef, Y., Obermayer, K.: Robust detection of environmental sounds in binaural auditory scenes. IEEE/ACM Trans. Audio Speech Lang. Process. **25**(6), 1344–1356 (2017)
11. Imoto, K., Ono, N.: Spatial cepstrum as a spatial feature using a distributed microphone array for acoustic scene analysis. IEEE/ACM Trans. Audio Speech Lang. Process. **25**(6), 1335–1343 (2017)
12. Yang, W., Kirshnan, S.: Combining temporal features by local binary pattern for acoustic scene classification. IEEE/ACM Trans. Audio Speech Lang. Process. **25**(6), 1315–1321 (2017)
13. Peeters, G., Giordano, B.L., Susini, P., Misdariis, N., McAdams, S.: The timbre toolbox: extracting audio descriptors from musical signals. J. Acoust. Soc. Am. **130**(5), 2902–2916 (2011)
14. ITU-R Rec. BS.775: Multichannel stereophonic sound system with and without accompanying picture. International Telecommunication Union, Geneva, Switzerland (2012)
15. Sánchez-Hevia, H.A., Ayllón, D., Gil-Pita, R., Rosa-Zurera, M.: Maximum likelihood decision fusion for weapon classification in wireless acoustic sensor networks. IEEE/ACM Trans. Audio Speech Lang. Process. **25**(6), 1172–1182 (2017)
16. Rumsey, F.: Spatial quality evaluation for reproduced sound: terminology, meaning, and a scene-based paradigm. J. Audio Eng. Soc. **50**(9), 651–666 (2002)
17. Breiman, L., Cutler, A.: Random Forests for Classification and Regression. https://www.stat.berkeley.edu/~breiman/RandomForests. Accessed 18 Nov 2017
18. Woźniak, M., Graña, M., Corchado, E.: A survey of multiple classifier systems as hybrid systems. Inf. Fusion **16**, 3–17 (2014)
19. Trajdos, P., Kurzynski, M.: A dynamic model of classifier competence based on the local fuzzy confusion matrix and the random reference classifier. Int. J. Appl. Math. Comput. Sci. **26**(1), 175–189 (2016)
20. Friedman, J.H.: Stochastic gradient boosting. Comput. Stat. Data Anal. **38**(4), 367–378 (2002)
21. Ridgeway, G.: Generalized Boosted Regression Models. http://code.google.com/p/gradientboostedmodels. Accessed 18 Nov 2017
22. Bradley, J.S., Soulodre, G.A.: Objective measures of listener envelopment. J. Acoust. Soc. Am. **98**(5), 2590–2597 (1995)
23. Jollifee, F.: Principal Component Analysis, 2nd edn. Springer, Berlin (2002). https://doi.org/10.1007/b98835
24. George, S., Zieliński, S., Rumsey, F.: Feature extraction for the prediction of multichannel spatial audio fidelity. IEEE Trans. Audio Speech Lang. Process. **14**(6), 1994–2005 (2006)
25. Conetta, R., Brookes, T., Rumsey, F., Zieliński, S., Dewhirst, M., Jackson, P., Bech, S., Meares, D., George, S.: Spatial audio quality perception (part 2): a linear regression model. J. Audio Eng. Soc. **62**(12), 847–860 (2014)
26. Gardner, B., Martin, K.: HRTF Measurements of a KEMAR Dummy-Head Microphone. http://sound.media.mit.edu/resources/KEMAR.html. Accessed 16 Nov 2017

Artificial Intelligence in Modeling, Simulation and Control

Cascading Probability Distributions in Agent-Based Models: An Application to Behavioural Energy Wastage

Fatima Abdallah, Shadi Basurra[(✉)], and Mohamed Medhat Gaber

School of Computing and Digital Technology, Birmingham City University,
Birmingham, UK
{fatima.abdallah,shadi.basurra,mohamed.gaber}@bcu.ac.uk

Abstract. This paper presents a methodology to cascade probabilistic models and agent-based models for fine-grained data simulation, which improves the accuracy of the results and flexibility to study the effect of detailed parameters. The methodology is applied on residential energy consumption behaviour, where an agent-based model takes advantage of probability distributions used in probabilistic models to generate energy consumption of a house with a focus on energy waste. The implemented model is based on large samples of real data and provides flexibility to study the effect of social parameters on the energy consumption of families. The results of the model highlighted the advantage of the cascading methodology and resulted in two domain-specific conclusions: (1) as the number of occupants increases, the family becomes more efficient, and (2) young, unemployed, and part-time occupants cause less energy waste in small families than full-time and older occupants. General insights on how to target families with energy interventions are included at last.

1 Introduction

The building sector accounts for more than one-third of the total worldwide energy consumption which is also expected to increase with the increase in population [1]. From this high percentage, more than a half is caused by human behavioural energy waste (e.g. leaving appliances ON while not in use) [2]. Beisdes, human behaviour is gaining more interest in the zero carbon design as it is considered one of the barriers against the efficiency of zero carbon buildings [3].

This concern about the effect of human behaviour on energy consumption has been considered in several energy simulation models which are used to analyse buildings energy performance. One approach of simulation models are *Probabilistic Models* (PM) whose aim is to add the human behaviour factor to building simulation tools. These models simulate the activities of occupants, and as a result the energy consumption of the house. Furthermore, PM enable modelling different household characteristics such as occupants' ages, employment types, and household income [3,4]. However, these models do not simulate behavioural

© Springer International Publishing AG, part of Springer Nature 2018
L. Rutkowski et al. (Eds.): ICAISC 2018, LNAI 10842, pp. 489–503, 2018.
https://doi.org/10.1007/978-3-319-91262-2_44

energy waste because they assume ideal and identical behaviour among occupants [5]. While in fact, occupants may have different energy awareness levels thus different energy consumption habits [6]. Another emerging trend of energy simulation models are *Agent-Based Models* (ABM). Several ABM approaches have been used to model behavioural energy waste in both residential [7] and commercial buildings [8]. In these models, occupants/energy consumers are modelled as agents that change their state and make decisions by interacting with their environment (electric appliances) and other occupants [9]. However, most of these models do not capture the low level interaction between occupants and appliances which is important to determine the causes of energy waste in buildings [10], and to produce high level data. These limitations in existing PM and ABM in simulating energy consumption motivates the approach of this paper, where the integration process can overcome their limitations when they work separately. The ABM takes advantage of probability distributions used in PM to produce more detailed data at occupant and appliance level, and simulates various levels of energy awareness of occupants. The same cascading approach can be used in other human behaviour models such as transport modelling and human communications to ensure the accuracy and flexibility of the results. The energy simulation model has been validated in [11] and proved that there is an effect of employment type on the energy efficiency of the house. Therefore, beside proposing the integration approach, detailed results of the effect of varied social parameters are presented to gain insights towards energy efficiency plans.

The paper is organised as follows. The next section presents existing PM and ABM highlighting their limitations and advantages of integrating them. Section 3 presents the proposed cascading approach. Section 4 illustrates how the proposed model can be used to analyse energy consumption based on occupants energy awareness and varied social parameters. Based on the results, the model is compared with existing PM and ABM in Sect. 5 and the results of the experiments are discussed providing general recommendations for policy makers on how to target family members to achieve less energy waste in buildings. Finally, conclusion and future work are presented in Sect. 6.

2   Related Work

2.1   Probabilistic Models

Probabilistic (or stochastic) Models (PM) have been widely proposed to enhance the prediction of energy demand in residential buildings by simulating occupant activities. They utilise time-use surveys, where occupants record the activities they do throughout the day, to calculate the probability that an action occurs. Using large amounts of data from time-use surveys enables generating the data based on different socio-economic factor like income, household size, occupants ages or employment types [3,4]. These models are considered as bottom-up approaches because they use highly detailed data (at activity and appliance level) to build up the energy consumption of the house [12]. Bottom-up approaches make it possible to detect energy waste when having information about what the occupant is doing, what is her/his location, which appliances are turned ON, etc. In addition, this level of granularity is useful to study the changes in occupant behavioural characteristics [13].

Although PM produce detailed data which is useful when modelling energy waste, the existing models only aim to reproduce realistic occupant activities and energy consumption. Therefore, they are not capable of capturing how occupants react to changes in their environment [14]. From the computational view, PM follow a linear modelling process where occupancy and activity data are generated, then the resulting electricity consumption. This linear process cannot be used to model dynamic humna behaviour which is non-linear and can change based on several individual and environmental attributes [9]. Existing PM assume that all occupants are the same and consume energy in an ideal way, i.e. energy is consumed only when occupants are active at home or doing an activity [4,5,12]. However, human behaviour is more complex and is unlikely to be always the same, which can be one of the most influential factors of energy consumption in buildings. For example, more than 50% of energy consumption in commercial buildings is consumed during unoccupied hours, and sometimes even in occupied hours [2]. In addition, occupants can be categorised based on their greenness of behaviour [6]. Assuming that no energy is wasted have caused an underestimation of the real data in some existing models. For example, Aerts [4] realised that the developed model failed to produce high energy consumption levels, and explained that the reason could be behavioural energy waste.

2.2 Agent-Based Models

Besides PM, buildings energy consumption can be generated using Agent-Based Models (ABM). In ABM, agents are defined as autonomous software components that take decisions based on their state and rules of behaviour [9]. ABM are widely used in social sciences to study dynamic human behaviour and its influential factors [8]. Azar and Menassa [15] developed an ABM that represents social network structures in commercial buildings to study the effectiveness of energy interventions. Similarly, Chen et al. [8] studied structural properties of peer networks in residential buildings. These models differentiate between occupants by varying the average daily/yearly consumption. This factor is not only affected by how aware the occupants are, but also how long they spend in the building or what appliances they use. Therefore, no consideration was made whether high energy consumption is a result of occupant behaviour. In another way, Zhang et al. [7] represented energy-consumer agents at household level to study experience development of households when using smart meters. Modelling the household as a whole entity with one energy awareness level makes it difficult to model occupants-appliances interaction and study the effect of occupants behaviour on the consumption of the family. Therefore, the aforementioned models [7,8,15] do not produce detailed data like location and activity of occupants which are important attributes when studying behavioural energy waste.

Among the existing ABM, only a few capture the occupant-appliance interaction and produce detailed data that is useful in energy waste analysis. Zhang et al. [16] tested the effectiveness of automated lighting strategy against manual lighting strategy in a university building. They found that the manual strategy is more effective when occupants have high energy awareness level, and the automatic one is better when occupants have low awareness level. Similarly, Carmenate et al. [10] developed an ABM that models the human-appliance-building interaction to understand determinants of energy waste in an office environment. By including this interaction level they highlighted the effect of both building structure and occupants awareness on energy consumption of the building. The advantage of these models is that they simulate the detailed movement of occupants in the building and study the factors that affect energy consumption within the building environment (physical, social or others). However, the limitation of these models is that they are implemented from hypothetical [10] and small [16] case studies which questions the accuracy of the results, limits the variation of parameters, and offers energy efficiency strategies specific for these environments, while using large samples allows more realistic data, more varied parameters, and more generalised conclusions.

2.3   Cascading Probabilistic and Agent-Based Models

PM utilise large samples of data, therefore, it is guaranteed that the produced data are realistic and possible to study the effect of social parameters on energy consumption of the house. PM also provide detailed data at appliance and occupant level. Therefore, cascading PM with ABM overcomes the limitations that existed in some of the ABM presented above. Besides, ABM overcome the linear approach in PM by enabling dynamic human behaviour modelling where occupant agents take decisions based on their personal characteristics and the state of the environment. Furthermore, various energy awareness levels can be modelled at occupant level in ABM which enables the study of energy awareness in a family setting. Therefore, an approach that combines ABM and PM overcomes limitations of both models when they are separated.

3   The Agent-Based and Probabilistic Model Cascading Methodology

The model proposed in this paper cascades PM and ABM, where the first stage is obtaining probability distributions from realistic data to simulate the occupants daily behaviour, and the second stage is using these distributions in an ABM to simulate the dynamic interaction of occupants and appliances. To get the probability distributions, we take advantage of an existing PM which is developed by *Aerts* [3,4]. Aerts model is one of the recent models which has advantages over other models and satisfies the requirements of modelling energy waste. The model was selected because it includes the following features: (1) Obtains more realistic duration of activities and occupancy states (opposed to [5,12]); (2) enables multitasking where occupants can be doing more than one activity at a time (opposed to [12]); (3) includes nine activities that are linked to energy usage opposed to [13] that includes activities that may not be connected to energy consumption; (4) simulates household dynamics by distinguishing between household tasks and personal activities; and (5) uses 7 patterns of typical occupancy behaviour based on age and employment type, which results in more realistic occupancy data. The main approach followed in Aerts model is generating realistic occupancy and activity data using higher order Markov Process. The process is based on transition probability from one state to another, and the probability distribution for the duration of the state. Probability Distribution Functions (PDF) were extracted from Belgian Time-Use Survey and Household Budget Survey which include 6400 respondents from 3455 households. The PDFs are generated based on several social and environmental parameters such as occupants ages and employment types, household type, and day of week. The model is composed of three stages: (1) occupancy model, (2) activity model and (3) electricity model. The occupancy and activity models with their associated PDFs are used in the ABM to produce realistic human behaviour. However, in order to model behavioural energy waste, modifications were made mainly on the electricity model by adding an energy awareness and location attributes for occupant agents. These attributes control when occupants turn appliances and lights ON or OFF. Thus, behavioural energy waste is modelled by combining data about occupants activity, location, energy awareness, and time of day.

The following subsections explain the components of the agent-based model: 'Occupants Agents', 'Appliances Agents', and the 'Environment' that the agents act in. Details about the usage of the probability distributions in the ABM is explained where necessary.

3.1   The Environment and Appliances Agents

Occupant agents live and interact in a house environment composed of a number rooms, each having a set of appliances. The number of rooms affects the mobility and number of locations that the occupants can be in, and consequently the energy consumption. Therefore, the number of rooms was obtained from the Income and Living Conditions Database by Eurostat [17]. The database contains data about the average number of rooms per person by household type and income group. The data were normalised and fitted to the included household types. Every household is assigned a kitchen, a living room, at least one bedroom and at least one bathroom, in addition to dining and laundry rooms when necessary. The size of basic rooms was set to $20 \, m^2$ based on the average room size in Belgium [18] (the room size was used to calculate the amount of lights consumed in every room). In terms of the day and time, occupant agents are aware of the day of the week, time of day in a 10-min time step, and the amount of external daylight. Electric appliances in the house are modelled as dummy agents that react to occupant agents. Occupants change appliances state from ON to OFF or vice versa. The types and number of appliances in the house are obtained from appliances PDFs in the PM. Before initialising the simulation and based on the household type and income, the household is assigned a number and types of appliances which identifies the amount of energy that the appliance consumes.

The simulation environment **E** can be defined using the triplet $<\mathbf{T}, \mathbf{R}, \mathbf{A}>$, where:

- **T** is a one-year simulation time defined by the triplet $<t, d, daylight_{td}>$ where, $t \in [1\text{--}144]$ is a 10-min time step in 24 h, d is the day of the year, and $daylight_{td}$ is the amount of external daylight at every time step and day.
- **R** is the set of rooms in the house: For every room $r \in \mathbf{R}$, r is defined by the triplet $<size, A_r, O_r>$, where $size$ is the size of the room, A_r is the set of appliances in the room, and O_r is the set of occupants that are in the room.
- **A** is the set of appliances in the house: For every appliance $a \in \mathbf{A}$, a is defined by the set $<inUseConsumption, r, O_a, C_{td}>$, where $inUseConsumption$ is the amount of energy used when the device is ON, r is the room that the appliance is in, O_a is the set of occupants using the appliance, and C_{td} is the consumption array of the appliance over a whole year, where every $c_{td} \in C_{td}$, $c_{td} = \{0, inUseConsumption\}$ based on its ON-OFF state.

3.2 The Occupant Agent

Initially, occupants' ages and employment types are given as input for the model. Employment types include: full time job, part time job, unemployed, retired and school, where under 18 occupants are school children and above 65 are retired. Another input attribute of the model is the energy awareness which will be explained in this section. Based on the defined household type (occupants' ages and employment types) the income group of the household is assigned using the income PDF in the PM. Next, the appliances and rooms of the house are determined as functions of the household type and income group. At this stage, all occupant agents are initialised and start doing activities in the house. At every time step, the occupants change the state of the environment by changing their location and using the electric appliances.

Occupant Daily and Weekly Behaviour. In order to simulate occupancy of members, work routines and occupancy patterns are needed. Working occupants can belong to one of ten work routines (wr) to decide working days and duration of occupants. Every day, and based on the occupant's age, employment and day type, the occupant chooses one occupancy pattern op_d for the day. The PM includes 7 occupancy patterns which could be referred to in Aerts et al. [3].

At every time step, the occupant either selects a new occupancy state os_{td} based on PDFs in the PM, or decrements the duration of an already running occupancy state. OS is the function to select a new occupancy state.

$$OS : op_d, os_{(t-1)d}, t \to os_{td} \tag{1}$$

$$op_d, os_{td}, t \to dr \tag{2}$$

where, os_{td} is the new occupancy state, $os_{td} \in \{Away, Sleeping, Active\}$ ($Away$: out of home, $Active$: at home and not sleeping, $Sleeping$: at home but sleeping). The agent first selects a new state as function of his/her occupancy pattern op_d, previous state $os_{(t-1)d}$, and time of day t, then decides the duration dr of the state based on his occupancy pattern, current occupancy state, and time of day.

The PM distinguishes between household tasks which are performed by one occupant at a time, and personal activities that can be performed and shared by all occupants. When the occupant is in the *Active* occupancy state, he/she can do several tasks or personal activities. The occupant can either select to start the activity, or decrement the duration of an ongoing activity. The action of selecting new activities is defined by the function AC

$$AC : age, emp, t, d \rightarrow \{0,1\}_{ac/tk}, dr \tag{3}$$

This function is performed by the occupant agent for every personal activity $ac \in \{$*Using the computer, Watching television, Listening to music, Taking shower/bath*$\}$ and task $tk \in \{$*Preparing food, Vacuum cleaning, Ironing, Doing dishes, Doing laundry*$\}$. The function returns a Boolean value $\{0,1\}$ to distinguish if the action will take place or not. This way of modelling enables the occupant to perform more than one activity at a time. The decision of doing an activity is based on the occupant *age*, employment type (*emp*), time of day t, and day type d; and similarly the duration dr of the activity.

Occupant Location. Whenever the occupant is at home, he/she needs to be in one of the rooms. Every activity is assigned to a room or a set of possible rooms. The occupant agent determines his/her location using the function OL

$$OL : os_{td}, AC_{td}, TK_{td} \rightarrow r_{td} \tag{4}$$

The occupant decides his/her location r_{td} based on his occupancy state os_{td}, ongoing personal activities AC_{td}, and ongoing tasks TK_{td}. If the occupant is doing more than one activity at a time, he/she may have a set of possible rooms and his/her location alternates among the rooms of this set at every time step.

Occupant Energy Awareness and Energy Usage. Occupants' energy awareness have been modelled in existing literature in different ways. For example, Carmenate et al. [10] distinguishes between energy literate and energy illiterate occupants. Similarly, Zhang et al. [6] categorises occupants into high and low consumers. Another way is using average yearly/daily consumption as a characteristic of the occupant [8,15]. The most detailed and flexible definition of energy awareness was proposed in Zhang et al. [7] where energy consumers can belong to one of four consumer types: 'Follower Green', 'Concerned Green', 'Regular Waster', and 'Disengaged Waster'. Based on the consumer type, the agent's energy awareness attribute is assigned a value between 0 and 100. This attribute is used to decide the probability that an occupant follows energy saving actions such as turning off devices when they are not in use. The value is calculated based on a normal distribution for every consumer type (Table 1). In the current model, the occupant types and energy awareness attribute defined in Zhang et al. [7] are used to model energy awareness of occupant agents.

Table 1. Mean and standard deviation of consumer types

Consumer types	Mean μ	Standard deviation σ
Follower green	0.74	0.041
Concerned green	0.72	0.043
Regular waster	0.41	0.033
Disengaged waster	0.25	0.057

The action of turning appliances ON/OFF is defined by the function TO_a

$$TO_a : ac_{td} \to turnOn_a \qquad (5)$$

$$ac_{td}, O_a, ea \to \{keepOn, turnOff\}_a \qquad (6)$$

Every activity ac_{td} that the occupant performs is associated to an appliance a. When the occupant starts an activity, he/she turns ON the appliance associated to this activity. When the activity ends and based on the occupants energy awareness attribute, he/she may turn OFF the appliance or keep it ON. The occupant may also communicate with other occupant/s O_a who may be using the same appliance at the same time to decide whether to turn the appliance OFF. The action of turning OFF appliances is also executed every time an occupant visits a room and finds appliances that are ON but unused.

The action of turning lights ON/OFF is different from using appliances, because using lights depends on the amount of daylight and the location of occupants.

$$TO_r : r_{td}, daylight_{td} \to \{turnOn, !turnOn\}_r \qquad (7)$$

$$r_{td}, O_r, ea \to \{turnOff, !turnOff\}_r \qquad (8)$$

Every time the occupant is in a room r_{td} he may decide to turn ON the light in this room based on the amount of daylight ($daylight_{td}$) [4]. When the occupant leaves the room, he/she checks other occupants in the room O_r, and based on his energy awareness (ea) he/she may decide whether to turn off the light.

In summary, the occupant agent **OA** is defined using the set $<age, emp, wr,$ $op_d, os_{td}, ea, AC_{td}, TK_{td}, r_{td}>$ and can perform the actions $<OS, AC, OL, TO_a,$ $TO_r>$ to act in the house environment.

The model was implemented in Repast Simphony (https://repast.github.io), a Java-based agent-based platform. For validation of the model refer to [11]. Three appliances were implemented: Lights, TV, and PC. These appliance are clearly affected by the energy awareness of occupants like leaving lights ON when leaving a room or leaving the TV/PC ON when the activity ends.

4   Simulation Experiments and Results

This section presents a set of experiments that were done to study the effect of social parameters on the energy consumption of the house with various occupants energy awareness. Every simulation run (or scenario) calculates the average energy consumption of 100 simulated households of the same type, but different work routines, income, appliances number and types, and house rooms.

The energy awareness of occupants is reduced to two types: *Follower Green* and *Disengaged Waster* in order to limit the number of scenarios, while achieving the objectives of this study. The validation of the model which includes all occupant types can be found in [11]. A total of 244 scenarios were tested; for every simulated scenario the total amount of energy per day for three appliance types (lights, TVs, and PCs) is calculated using the formula:

$$C_n = \sum_{t,d,a} \overline{c}_{td} \tag{9}$$

where C_n is the total energy consumption of scenario n and $\overline{c}_{td}$ is the average energy consumption at time step t and day d. In order to calculate the energy efficiency of each scenario, the distance to the ideal energy saving behaviour D_n is calculated using the formula:

$$D_n = \frac{C_n}{C_{base}} \tag{10}$$

where, C_{base} is the total energy consumption in the ideal scenario where devices are only ON when they are being used. As much as D_n is closer to 1 means that the household is closer to the ideal scenario, thus more efficient. In the below experiments, family size, employment type, and occupants' ages are the tested social parameters. These parameters were selected, because they are available in the real data provided in the PM. Other social parameters can be included if the corresponding real data are available.

4.1 Experiment 1: Effect of Family Size

This experiment is intended to study the effect of number of occupants in the house. Scenarios of the age group 25–39 in full-time job are presented in Table 2. The table consists of two groups of scenarios, each group has the same age and employment type for adults, same energy awareness type, but different number of occupants. In the first group of scenarios, where all family members are green occupants, it is observed that as the number of occupants increases, D_n decreases. This means that more green occupants in the house makes the family more energy efficient. For the second group of scenarios, where all occupants are energy wasters, it could be expected that when the number of wasters increases, D_n should increase. However, it is observed that as the number of wasters increases, D_n decreases and the family is closer to the ideal scenario. This indicates that more occupants in the house, whether they are green or waster occupants, causes the house to be more efficient. Similar observations were noticed for other age groups and employment types.

Table 2. Scenarios and results for the effect of family size

Adults age group/empl. type/energy awareness	No. of occupants	Household type	D_n
25–39/full-time job/All *Green* occupants	1	One adult	2.31
	2	One adult, one child	1.97
	2	Two adults	1.99
	3	One adult, two children	1.42
	3	Two adults, one child	1.58
	4	Two adults, two children	1.59
25–39/full-time job/all *Waster* occupants	1	One adult	10.49
	2	One adult, one child	6.13
	2	Two adults	6.66
	3	One adult, two children	3.92
	3	Two adults, one child	4.18
	4	Two adults, two children	4.17

4.2   Experiment 2: Effect of Employment Type

The purpose of this experiment is to test the effect of employment type on the energy consumption of the house. In order to do that, it is important to fix occupants ages and number of occupants while varying the employment types. Therefore, based on the household types available in the PM, it is only possible to study the effect of full-time, part-time, and unemployed occupants. Table 3 represents the scenarios for age group 40–45. The Occupant Types column encodes the energy awareness of occupants where G refers to green occupants and W refers to waster occupants. The sequence of letters (G and W) has the same sequence of occupants defined in the previous columns.

For every household type, the first two occupants (which are full-time/part-time or full-time/unemployed) are involved in the energy awareness variation, while the rest are put all green or all waster occupants in order to observe the effect. The difference between every two varied scenarios is calculated in the last column. Among the total number of simulated scenarios, there are cases when two occupants belong to the same age group and have the same employment type. It was observed that swapping the energy awareness between these occupants resulted in similar amounts of energy consumption with very slight differences. This difference is expected to be due to random numbers generation. The average difference between these scenarios was calculated and found to be 0.1. Therefore, whenever the difference between two scenarios is more than 0.1, it is considered a significant difference and further analysis is made to identify the cause of the difference. The first three household types in Table 3 are for comparing full-time and part-time employment types. It is observed in all of these scenarios that whenever the part-time occupant is the green occupant, the energy consumption of the house is closer to the ideal scenario. This means

Table 3. Scenarios and results for the effect of employment type

Occupants age group/employment type				Occ. types	D_n	Difference
Occ. 1	Occ. 2	Occ. 3	Occ. 4			
40–54/full-time	40–54/part-time			GW	3.89	0.23
				WG	3.66	
40–54/full-time	40–54/part-time	12–17/school		GWG	2.33	0.1
				WGG	2.23	
				GWW	3.14	0.07
				WGW	3.07	
40–54/full-time	40–54/part-time	12–17/school	12–17/school	GWGG	1.88	0.05
				WGGG	1.83	
				GWWW	3.10	0.09
				WGWW	3.01	
40–54/full-time	40–54/unemployed			GW	3.05	0.37
				WG	2.68	
40–54/full-time	40–54/unemployed	12–17/school		GWG	2.12	0.24
				WGG	1.88	
				GWW	2.75	0.14
				WGW	2.61	
40–54/full-time	40–54/unemployed	12–17/school	12–17/school	GWGG	1.69	0.03
				WGGG	1.66	
				GWWW	2.62	0.15
				WGWW	2.47	

that green part-time occupants are responsible for improving the house energy consumption when compared to full-time occupants. A similar observation is noticed when comparing full-time and unemployed occupants in the next three household types. This observation was noticed in our previous paper [11] and is further supported in Table 3. Looking at the difference values, part-time occupants efficiency effect is significant (>0.1) in two cases: (1) the two-occupant family and (2) the three-occupant family when the third occupant is a green occupant. This indicates that part-time occupants can make an energy saving effect in small families (a small family is a family less than 4 occupants) and when there are more green occupants in the house, but not in big families where the difference is 0.05 and 0.09. However, for unemployed occupants, the efficiency effect is significant in most of the cases except for the four-occupant family when all of the other occupants are green occupants. It is also observed that unemployed occupants, in general, have higher effect than part-time occupants. These observations show that unemployed occupants are more efficient than part-time occupants, and the latter are more efficient than full-time occupants in small families.

4.3   Experiment 3: Effect of Occupants Ages

In order to study age groups for adults, households that have the same employment type and number of occupants with no children were considered (Table 4).
As for the children effect, households with an equal number of adults and children, with the same employment type for adults were studied (Table 5). Table 4
shows that as the age of adults in small families increase, the household is becoming less efficient (both for waster and green households). And for children, it is
observed in Table 5 that children were more efficient than adults in small families
(0.26 and 0.1), but not in big families where adults were more efficient in some
of the cases (-0.17). These observations imply that younger occupants including
children can make more efficiency effect in small families but not in big families.

Table 4. Scenarios and results for the effect of adults ages in full-time job

Energy awareness	Occupant 1 age group	Occupant 2 age group	D_n
All green occupants	25–39		2.31
	40–54		2.47
	55–64		2.78
All waster occupants	25–39		10.49
	40–54		11.35
	55–64		13.29
All green occupants	25–39	25–39	1.99
	40–54	40–54	1.87
	55–64	55–64	1.85
All waster occupants	25–39	25–39	6.66
	40–54	40–54	6.68
	55–64	55–64	6.75

Table 5. Scenarios and results for studying the effect of children

Adults age group	Household type	Occupant types	D_n	Difference
25–39	One adult, one child	GW	3.94	0.26
		WG	3.68	
40–54	One adult, one child	GW	3.10	0.10
		WG	3.20	
25–39	Two adults, two children	WWGG	2.41	0.04
		GGWW	2.45	
40–54	Two adults, two children	WWGG	2.57	-0.17
		GGWW	2.74	

5   Discussion and Insights

This study proposes a methodology to combine ABM and PM to produce fine grained data. The implemented model simulates the dynamic interaction of occupants with appliances to produce detailed activities and energy consumption of houses. Opposed to exiting PM [3–5,12,13] the cascaded model simulates dynamic occupants behaviour which is affected by occupants personal characteristics and surrounding environment. In addition, an energy awareness level can be assigned at occupant level which varies based on the occupant's greenness level, while PM assume same and ideal energy consumption behaviour of occupants. The proposed model simulates energy waste caused by human behaviour. Existing ABM that simulate the effect of human behaviour [7,8,15] produce the consumption data at household or building level, however, the proposed model generates energy consumption data at appliance level as shown in our previous paper [11]. This is because exiting models either model consumer agents at household level or characterise occupant agents by yearly/monthly consumption. The most similar ABM in terms of output are Carmenate et al. [10] (hypothetical case study) and Zhang et al. [16] (real case study). These models can produce appliance level consumption and model energy awareness at occupant level. The difference is that the proposed model uses PM (embedded Markov Process technique) to get the realistic occupants activities as a preprocessing stage to ABM, while existing models use the real data directly in the ABM to simulate human activity. Using PM ensures that the produced data are realistic and enables the inclusion of data for a whole city (6400 respondent vs. 143 respondent in Zhang et al. [16]) which leads to more varied scenarios and generalised conclusions.

Besides the above discussions, the integration of PM with ABM has given the advantage of studying the effect of social parameters on the energy consumption of families. Experiment 1 showed that as the number of occupants increases, the household becomes more energy efficient even if all of the occupants are unaware of energy consumption. Although the implemented model does not model family pressure, which means that family members do not affect the energy awareness of each other, we have shown that merely having more occupants in the house makes the family more efficient (by more efficient we mean that big families waste less than small ones even though they actually consume more). This is explained by the fact that more occupants in the house means more probability that somebody turns off unneeded appliances/lights (knowing that occupant agents can know if a device is being used or a room is being occupied). For example, if one occupant, who lives alone, leaves the house/room while the lights are ON, the lights will never be OFF until he/she returns back to the location. However, in a four-occupant family, if a member leaves something ON and goes away, there is still a probability that somebody will turn it OFF before he/she returns back. The second experiment proved that unemployed occupants have the most efficiency effect in small families compared to part-time and full-time occupants. Whereas part-time occupants are more efficient than full-time occupants, again in small families. This is mainly explained by the occupancy pattern of each employment type, where unemployed and part-time occupants are available at home more than full-time occupants. This enables unemployed and part-time occupants to reduce the waste in small families. However, in big families, this effect is reduced due to the existence of more occupants in the house who may cancel the effect of the green occupant. A similar conclusion was obtained concerning ages of occupants, where younger occupants made the household more efficient in small families. It is important here to note that this conclusion does not imply that younger occupants are more aware than older occupants, but with the same energy awareness levels younger occupants' longer existence at home or longer active durations causes less energy waste than older occupants.

These conclusions are important as they give insights for policy makers and governments about how to target family members to achieve higher energy efficiency. The developed model shows that it is important to target all members of big families with energy efficiency interventions and technologies – not just because big families consume more energy in general, but also because increasing the energy awareness of all members of big families makes more effect than small families. Concerning small families, it is important to concentrate on younger occupants including children, and adults who are housewives, unemployed, carers, or those who work in part-time jobs, because we have shown that these types of people can make more efficiency effect than older occupants and full-time employees.

6   Conclusion and Future Work

This paper presented a methodology to cascade ABM and PM in order to generate detailed and accurate data. The proposed approach was applied on the energy consumption domain, however, it can be used to simulate other human behaviour applications. The energy consumption model incorporates energy awareness at occupant level and produces fine-grained data to simulate behavioural energy waste. The paper have shown that the cascading approach overcomes limitations of exiting PM and ABM when they work separately. Social parameters were varied to gain insights towards energy efficiency plans for families. It was concluded that bigger families cause less energy waste than small families due to the higher probability of somebody to turn OFF unneeded consumption. Besides, young, unemployed and part-time occupants can make more efficiency effect in small families than full-time and older occupants because they are more active at home. The model can be used in the future to study the effect of intervention technologies (e.g. energy waste notifications) or family pressure when varying social parameters. This will give insights about how to target and customise interventions for different types of occupants/households.

References

1. International Energy Agency: Transition to Sustainable Buildings: Strategies and Opportunities to 2050. Technical report (2013). https://www.iea.org/publications/freepublications/publication/Building2013_free.pdf. Accessed 1 Feb 2018
2. Masoso, O.T., Grobler, L.J.: The dark side of occupants' behaviour on building energy use. Energy Build. **42**, 173–177 (2010)
3. Aerts, D., Minnen, J., Glorieux, I., Wouters, I., Descamps, F.: A method for the identification and modelling of realistic domestic occupancy sequences for building energy demand simulations and peer comparison. Build. Environ. **75**, 67–78 (2014)
4. Aerts, D.: Simulations, occupancy and activity modelling for building energy demand, comparative feedback and residential electricity demand characteristics. Ph.d. thesis, Vrije University Brussel (2015)
5. Richardson, I., Thomson, M., Infield, D., Clifford, C.: Domestic electricity use: a high-resolution energy demand model. Energy Build. **42**, 1878–1887 (2010)
6. Zhang, T., Siebers, P.O., Aickelin, U.: A three-dimensional model of residential energy consumer archetypes for local energy policy design in the UK. Energy Policy **47**, 102–110 (2012)
7. Zhang, T., Siebers, P.O., Aickelin, U.: Simulating user learning in authoritative technology adoption: an agent based model for council-led smart meter deployment planning in the UK. Technol. Forecast. Soc. Chang. **106**, 74–84 (2016)
8. Chen, J., Taylor, J.E., Wei, H.H.: Modeling building occupant network energy consumption decision-making: the interplay between network structure and conservation. Energy Build. **47**, 515–524 (2012)
9. Bonabeau, E.: Agent-based modeling: methods and techniques for simulating human systems. Proc. Nat. Acad. Sci. **99**(Suppl. 3), 7280–7287 (2002)
10. Carmenate, T., Inyim, P., Pachekar, N., Chauhan, G., Bobadillaa, L., Batoulib, M., Mostafavi, A.: Modeling occupant-building-appliance interaction for energy waste analysis. Procedia Eng. **145**, 42–49 (2016)
11. Abdallah, F., Basurra, S., Gaber, M.M.: A hybrid agent-based and probabilistic model for fine-grained behavioural energy waste simulation. In: IEEE 29th International Conference on Tools with Artificial Intelligence (ICTAI), pp. 991–995. IEEE (2017)
12. Widén, J., Andreas, M., Ellegårdc, K.: Models of domestic occupancy, activities and energy use based on time-use data: deterministic and stochastic approaches. J. Build. Perform. Simul. **5**(1), 27–44 (2012)
13. Wilke, U., Haldi, F., Robinson, D.: A bottom-up stochastic model to predict building occupants' time-dependent activities. Build. Environ. **60**, 254–264 (2013)
14. Reynaud, Q., Haradji, Y., Sempé, F., Sabouret, N.: Using time-use surveys in multi agent based simulations of human activity. In: Proceedings of the 9th International Conference on Agents and Artificial Intelligence, ICAART, Porto, Portugal, vol. 1, pp. 67–77 (2017)
15. Azar, E., Menassa, C.C.: Framework to evaluate energy-saving potential from occupancy interventions in typical commercial buildings in the united states. J. Comput. Civil Eng. **28**(1), 63–78 (2014)
16. Zhang, T., Siebers, P.O., Aickelin, U.: Modelling electricity consumption in office buildings: an agent based approach. Energy Build. **43**, 2882–2892 (2011)
17. Eurostat: the statistical office of the European Union: Average number of rooms per-person and group by type of household income from 2003 - EU-SILC survey (2003 onwards). https://data.europa.eu/euodp/en/data/dataset/pYzSXuZuS2yZzD3nQKQcWQ. Accessed 3 July 2017
18. Evans, A.W., Hartwich, O.M.: Unaffordable housing: fables and myths. Technical report, Policy Exchange (2005)

Symbolic Regression
with the AMSTA+GP in a Non-linear
Modelling of Dynamic Objects

Łukasz Bartczuk[1(✉)], Piotr Dziwiński[1], and Andrzej Cader[2,3]

[1] Institute of Computational Intelligence, Częstochowa University of Technology,
Częstochowa, Poland
`lukasz.bartczuk@iisi.pcz.pl`
[2] Information Technology Institute, University of Social Sciences, Łódź, Poland
[3] Clark University, Worchester, USA

Abstract. In this paper, we present a new version of the State Transition Algorithm, which allows to automatically determine the number and range of local models that describe the behaviour of a non-linear dynamic object. We used this data as input for genetic programming algorithm in order to create a simple functional model of the non-linear dynamic object which is not computationally demanded and has high accuracy.

1 Introduction

The task of symbolic regression is to determine the mathematical formula which, based on the provided data, best describes the relationship between independent and dependent variables. This task is more complex than ordinary regression problem. This is due to the fact that when solving linear and non-linear regression problems we assume that the form of the model is known and our task is only to determine its parameters. Symbolic regression, in turn, aims to define both the structure of the formula sought (i.e. appropriate operators and mathematical functions), as well as its parameters and coefficients. Formally, the symbolic regression task can be represented by the following equation:

$$f = \arg \min_{f^* \in F} \sum_i \| f^*(\mathbf{v}_{(i)}) - y_i \| \tag{1}$$

where F is a set of all functions that can be created from the set of terminals (i.e. constants and variables) and non-terminals (i.e. arithmetical operators like $+$, $-$, $\times$, $\div$ or mathematical functions like exp, sin, log), $\mathbf{v}_i$ is a vector of function parameters values, and y_i is the value that the function sought should achieve for value $\mathbf{v}_i$.

In the literature, we can find many methods that allow solving the problems of symbolic regression [6,13,16,18,25]. However, one of the most popular are methods based on genetic programming algorithms [14,15].

© Springer International Publishing AG, part of Springer Nature 2018
L. Rutkowski et al. (Eds.): ICAISC 2018, LNAI 10842, pp. 504–515, 2018.
https://doi.org/10.1007/978-3-319-91262-2_45

Symbolic regression is used to solve problems in many different areas such as bioinformatics [17], signal processing [10], cryptanalysis [29], system identification [2] and others. In this paper, we want to present the new method to create models of non-linear dynamic objects. It should be noted that such models can be created not only by symbolic regression algorithms like genetic programming [2,7] but also by other computational intelligence methods like fuzzy systems [8,19,27,33–35], neural networks [3,5,11,12,23] and evolutionary or PSO algorithms [9,21,22,28,31].

Generally, non-linear dynamic objects can be described by the following equation [24]:

$$\frac{d\mathbf{x}}{dt} = f(\mathbf{x}, \mathbf{u}) \tag{2}$$

where $\mathbf{x}$ is a state variables vector, $\mathbf{u}$ is an input values vector, f is a nonlinear function that represents system changes. In case of weakly nonlinear dynamic objects the Eq. (2) can be also presented in the form [4]:

$$\frac{d\mathbf{x}}{dt} = \mathbf{A}\mathbf{x} + \mathbf{B}\mathbf{u} + \eta g(\mathbf{x}, \mathbf{u}) \tag{3}$$

where $\mathbf{A}$ is a transition matrix, $\mathbf{B}$ is input matrix, $g(\cdot)$ is a function that represents system non-linearity, and η is a coefficient that describes the influence of $g(\cdot)$ on the system. However, because defining the function $g(\cdot)$ in the entire range of the modelled system is a difficult task, in practice non-linear models are often approximated by linear once:

$$\frac{d\mathbf{x}}{dt} = \mathbf{A}\mathbf{x} + \mathbf{B}\mathbf{u} \tag{4}$$

It should be noted that such a model is accurate in some strictly limited range around some typical operating point (x_s, u_s) for with it was defined. In order to increase the accuracy of modelling in paper [1], we add the correction matrix $\mathbf{P_A}(\mathbf{x})$ to the matrix $\mathbf{A}$. The elements of $\mathbf{P_A}(\mathbf{x})$ matrix are functions that are depended on the current state of the system. In this case the Eq. (4) can be presented in the following form:

$$\frac{d\mathbf{x}}{dt} = (\mathbf{A} + \mathbf{P_A}(\mathbf{x}))\mathbf{x} + \mathbf{B}\mathbf{u} \tag{5}$$

In order to determine the elements of the $\mathbf{P_A}(\mathbf{x})$ matrix, we can use symbolic regression methods. However, these techniques assume that we know the values of independent variables and the corresponding values of the function sought. In the case of non-linear modelling, this dependency is embedded in the Eq. (5) and the data describing it directly are not available. This causes that the algorithms of genetic programming can find a solution that satisfies the modelling accuracy, but does not reflect the correct shape of the dependencies sought. In the paper [2], we proposed introducing the preprocessing phase into genetic programming. For this purpose, we used the modified version of the State Transition Algorithm method. This algorithm divides training data into subsets of a fixed size, and determine the local approximated linear model for each subset separately. Then these models are used to determine the approximate shape of the desired $\mathbf{P_A(x)}$ dependencies. In this paper, we propose a new version of this method which does not require dividing the training set into equal-sized subsets, but instead allows automatically determine the number and size of individual subsets of data. With this method, we can obtain better results of modelling in preprocessing phase.

2   Automatic Multiple State Transition Algorithm

The Automatic Multiple State Transition Algorithm (AMSTA) is a modification of well known the State Transition Algorithm (STA) procedure which was proposed in paper [38]. The STA method is an iterative way to solve unconstrained optimization problems in the form:

$$\min_{\mathbf{s}\in\mathbb{R}^n} \mathrm{ff}(\mathbf{s}), \tag{6}$$

where $\mathbf{s}$ is a proposed solution (a state), and ff is a function to evaluate this solution. The initial solution is generated randomly and then is corrected by the state transition operators:

1. Expansion

$$\mathbf{s}_{k+1} = \mathbf{s}_k + \gamma \mathbf{R}_e \mathbf{s}_k \tag{7}$$

   where $\gamma \in \mathbb{Z}^+$ is an expansion factor, $\mathbf{R}_e \in \mathbb{R}^{n \times n}$ is a random diagonal matrix in which elements are randomized according to the Gaussian distribution
2. Rotation

$$\mathbf{s}_{k+1} = \mathbf{s}_k + \alpha \frac{1}{n \|\mathbf{s}_k\|_2} \mathbf{R}_r \mathbf{s}_k \tag{8}$$

   where $\alpha \in \mathbb{Z}^+$ is a rotation factor, $\mathbf{R}_r \in \mathbb{R}^{n \times n}$ is a random matrix in which elements belong to $[-1, 1]$ range, and $\|\cdot\|_2$ is the Euclidean norm of a vector
3. Axesion

$$\mathbf{s}_{k+1} = \mathbf{s}_k + \delta \mathbf{R}_a \mathbf{s}_k \tag{9}$$

   where $\delta \in \mathbb{Z}^+$ is an axesion factor $\mathbf{R}_a \in \mathbb{R}^{n \times n}$ is a random diagonal matrix in which one element is randomized according to the Gaussian distribution and others are equal to zero.
4. Translation

$$\mathbf{s}_{k+1} = \mathbf{s}_k + \beta \mathbf{R}_t \left\| \frac{\mathbf{s}_k - \mathbf{s}_{k+1}}{\mathbf{s}_k - \mathbf{s}_{k+1}} \right\|_2 \tag{10}$$

   where $\beta \in \mathbb{Z}^+$ is a translation factor, $R_t \in \mathbb{R}$ is a random value from $[0,1]$ range.

Each of these operators creates SE of new solutions, of which only the best one is selected for further processing. The evaluation function may be, for example, the root mean square error (RMSE) determining the matching of data obtained from the $\overline{\mathbf{Y}}$ model to the reference data $\mathbf{Y}$:

$$\text{ff}(\overline{\mathbf{Y}}) = \text{RMSE}(\overline{\mathbf{Y}}) = \sqrt{\frac{\sum_{i=1}^{N} \epsilon_i^2}{N}} = \sqrt{\frac{\sum_{i=1}^{N} (|\mathbf{y}_i - \overline{\mathbf{y}}_i|)^2}{N}} \tag{11}$$

In the paper [36], authors used this algorithm in order to determine the values of the matrix $\mathbf{A}$ appearing in the Eq. (4). This method allows obtaining one global model describing the behaviour of a non-linear object in the whole range of its activity. However, the assumption that matrix $\mathbf{A}$ is constant may be too strong for systems in which the change of state depends on time or current value of this state.

From this reason in the paper [2], we proposed the Multiple State Transition Algorithm (MSTA). This is the modified version of the STA method, which instead of one global model, tries to determine many local models. From this reason, the MSTA algorithm works on 5-tuples: $\mathbf{S}_m = \{\mathbf{s}_m, IS_m, \text{ff}(\overline{\mathbf{Y}}_m), \overline{\mathbf{Y}}_m, \overline{\mathbf{X}}_m\}$, where m is an index of local model found by the STA algorithm, $\overline{\mathbf{Y}}_m$ is a set of outputs vectors, $\overline{\mathbf{X}}_m$ is a set of state vectors, $\mathbf{s}_m = \mathbf{A}_m$ is a transition matrix found by the STA algorithm, $\text{ff}(\overline{\mathbf{Y}}_m)$ is an evaluation of the m-th local model, and IS_m is an initial condition for m-th local model. In order to preserve the continuity between those local models, the STA algorithm is initialized according to the formula:

$$\mathbf{S}_m^0 = \left\{ \begin{array}{l} \mathbf{s}_m = \begin{cases} \text{random} & \text{if } m = 1 \\ \mathbf{S}_{m-1}^{last}.\mathbf{s}_{m-1} & \text{if } m > 1 \end{cases} \\ IS_m = \begin{cases} \mathbf{X}_0 & \text{if } m = 1 \\ \mathbf{S}_{m-1}^{last}.\overline{\mathbf{X}}_m^{count} & \text{if } m > 1 \end{cases} \end{array} \right\}$$

where $\mathbf{S}_m^0$ is an initial solution for m-th execution of STA algorithm, $\mathbf{S}_{m-1}^{last}$ is a last solution found by the STA algorithm for $(m - 1)$-th local model, $\mathbf{X}_0$ is a initial condition of the entire non-linear system, $count = |\overline{\mathbf{X}}_m|$ is a number of elements of the set $\overline{\mathbf{X}}_m$. The other parts of the solution are defined during its evaluation.

In case of the method presented in the paper [2], we assume that the whole training set is split into M equal-sized subsets $\mathbf{Y} = \mathbf{Y}_1 \cup \cdots \cup \mathbf{Y}_M$, so as the result of MSTA algorithm we obtained the model composed from M local approximated linear models. This allows increasing the accuracy of modelling. On the other hand, when we split data into equal-sized subsets, we can obtain too many models and the single subset can contain data which cannot be described by local approximated linear model accurately.

In this paper, we propose the Automatic Multiple State Transition Algorithm (AMSTA). This method can automatically determine the number of subsets and the count of data in each subset for which the approximate local linear model is generated. The AMSTA method differs from the MSTA algorithm mostly in the way in which the best solution is chosen. As mentioned above, in the STA and the MSTA methods applied to the modelling of non-linear systems, the only criterion for selecting the best solution was to obtain the highest accuracy (11). The evaluation of solution in the proposed method requires the consideration of two criteria: local solution error (11) (which we want to minimize) and the number of data in the chosen subset of training data for which the generated model is valid (which should be maximized). The latter criterion is defined as the number of training data $|\mathbf{S}_m.\overline{\mathbf{X}}_m|$ for which the error value ϵ_i is lower than the predefined threshold τ. In order to compute the value of fitness function we use the normalized values of both criteria:

$$\text{ff}_{acc}(\mathbf{S}_m^i) = \frac{\text{RMSE}(\mathbf{S}_m^i.\overline{\mathbf{X}}_m)}{\max\limits_{j=1,\ldots,SE} \text{RMSE}(\mathbf{S}_m^j.\overline{\mathbf{X}}_m) - \min\limits_{j=1,\ldots,SE} \text{RMSE}(\mathbf{S}_m^j.\overline{\mathbf{X}}_m)} \tag{12}$$

$$\text{ff}_{size}(\mathbf{S}_m^i) = \frac{\max\limits_{j=1,\ldots,SE} |\mathbf{S}_m^j.\mathbf{X}_m| - |\mathbf{S}_m^i.\mathbf{X}_m|}{\max\limits_{j=1,\ldots,SE} |\mathbf{S}_m^i.\mathbf{X}_m| - \min\limits_{j=1,\ldots,SE} |\mathbf{S}_m^i.\mathbf{X}_m|} \tag{13}$$

$$\text{ff}(\mathbf{S}_m^i) = w \cdot \text{ff}_{acc}(\mathbf{X}_m^i) + (1 - w) \cdot \text{ff}_{size}(\mathbf{X}_m^i) \tag{14}$$

where $\mathbf{S}_m^i$ is a i-th solution generated by the STA operator $i = 1, \ldots, SE$, $m = 1, 2, \ldots$ and w is a weight that allows to determine which component has greater impact on the overall assessment of the solution.

As a result of this algorithm we obtain the triple that determine the model of considered nonlinear dynamic system $\langle \overline{\mathbf{Y}}, \overline{\mathbf{X}}, \overline{\mathbf{P_A}} \rangle$, where $\overline{\mathbf{Y}}$ is the set of outputs of created models $\overline{\mathbf{Y}} = \overline{\mathbf{Y}}_1 \cup \overline{\mathbf{Y}}_2 \cup \ldots$, $\overline{\mathbf{X}}$ is the set of values of state variables $\overline{\mathbf{X}} = \overline{\mathbf{X}}_1 \cup \overline{\mathbf{X}}_2 \cup \ldots$, and $\overline{\mathbf{P_A}}$ is the set of values of $\mathbf{P_A}$ matrices generated for each of M local approximated linear models $\overline{\mathbf{P_A}} = \overline{\mathbf{P_A}}_1 \cup \overline{\mathbf{P_A}}_2 \cup \ldots$.

3   Genetic Programming Algorithm

The genetic programming [14] was designed as an evolutionary base method to generate computer programs. This idea can also be used to solve symbolic regression problems [15]. In this paper, we use this algorithm to find a symbolic form of dependency $\mathbf{P_A}(\mathbf{x})$ that determines how the values of the $\mathbf{P_A}$ matrix should change depending on the current state of the non-linear dynamic object. In the classical version, the genetic programming algorithm [14] can find a solution based on data directly describing the dependence sought. Thanks to the introduction of the AMSTA algorithm as a preprocessing phase, we have data that we can use with genetic programming algorithm. However, because the values of the $\overline{\mathbf{P_A}}$ describe the dependency sought only approximately, and genetic programming tries to find a solution that describes best those data, the obtained functions may not give the good accuracy of modelling.

Better results can be achieved, when we use both the data $\overline{\mathbf{P_A}}$ and reference data $\mathbf{Y}$. For this reason, we use the method that is similar to the cooperative coevolutionary approach [26]. This method uses several populations, each of which tries to find a part of the solution. Moreover, those populations cooperate together in order to find the best total solution.

In our case, each population tries to independently find one element of $\mathbf{P_A}(\mathbf{x})$ matrix based on data obtained from the AMSTA algorithm (local optimization). Next, the best solution from each population is taken and connected with all solutions from other populations. This allows checking how this solution works with others solutions in order to obtain high modelling accuracy (global optimization).

Since the evaluation of each individual must take into account the results achieved in both local and global optimization, the fitness function used in genetic programming algorithm takes the following form:

$$\mathrm{ff}(\mathbf{ch}) = w \cdot \mathrm{ff}_{AccL}(\mathbf{ch}) + (1 - w) \cdot \mathrm{ff}_{AccG}(\mathbf{ch}) \tag{15}$$

where: $\mathrm{ff}_{AccL}(\mathbf{ch})$ is the accuracy of the approximation of the points $\mathbf{P_A}$, $\mathrm{ff}_{AccG}(\mathbf{ch})$ is the accuracy of non-linear modelling and w is a weight that determines the influence of particular components. Such form of fitness function allows obtaining a good compromise between local and global accuracy.

4   Simulation Results

To examine the effectiveness of the proposed method, we considered the problem of the harmonic oscillator. Such oscillator can be defined using the following formula:

$$\frac{d^2x}{dt^2} + 2\zeta\frac{dx}{dt} + \omega^2 x = 0, \tag{16}$$

where ζ and ω are oscillator parameters and $x(t)$ is a reference value of the modelled process as function of time. We used the following state variables: $x_1(t) = dx(t)/dt$ and $x_2(t) = x(t)$. In such a case the system matrix $\mathbf{A}$ and the matrix of corrections coefficients $\mathbf{P_A}$ are described as follows:

$$\mathbf{A} = \begin{bmatrix} 0 & \omega \\ -\omega & 0 \end{bmatrix} \qquad \mathbf{P_A} = \begin{bmatrix} 0 & p_{12}(\mathbf{x}) \\ p_{21}(\mathbf{x}) & 0 \end{bmatrix}.$$

In order to introduce variability in the transition matrix $\mathbf{A}$, we assume that in our experiments the ω parameter depends on the state variable x_1 and is modified in accordance with the formula:

$$\omega(x_1) = 2\pi - \frac{\pi}{(1 + |2 \cdot x_1|^6)}. \tag{17}$$

The parameters of the Automatic Multiple State Transition Algorithm were set to the values presented in Table 1:

Simulations were performed for three different values of τ threshold 0.01, 0.005, 0.001. The results obtained for all three values of the τ parameter are presented in Table 2, and for $\tau = 0.001$ also in Figs. 1 and 2.

Table 1. Parameters of the extended version of the State Transition Algorithm used in simulations

Parameter name	Value
SE	10
A_{min}	0.0001
A_{max}, α, β, γ, δ	1
Fc	2

Table 2. Results obtained by the Automatic Multiple State Transition Algorithm

	$\tau = 0.01$	$\tau = 0.005$	$\tau = 0.001$	Results from [2]
Number of generated local models	13	21	44	50
RMSE	0.0025	0.0013	0.0003	0.0008

a)   b)

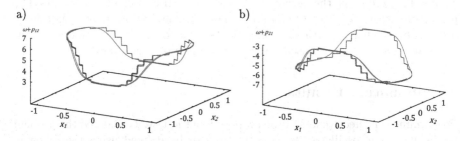

Fig. 1. The graphical representation of dependency between state variables and values of elements of state matrices for reference data and data obtained by the AMSTA algorithm for $\tau = 0.001$.

a)   b)

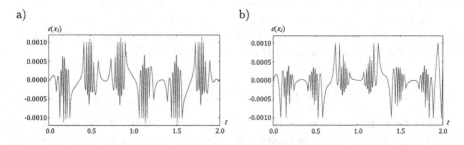

Fig. 2. The errors obtained for signals x_1 and x_2 by the models generated by the AMSTA algorithm for $\tau = 0.001$.

As can be seen for $\tau = 0.01$ and $\tau = 0.005$, a much smaller number of generated local models were obtained, with a slight decrease in modelling accuracy than for the method presented in [2]. On the other hand, for $\tau = 0.001$ the number of generated models and the obtained modelling accuracy were smaller than the results reported in [2].

As stated before the results obtained from the AMSTA method were used as input data for genetic programming algorithm in order to determine the functional dependencies describing changes in the elements of the transition matrix.

The parameters of genetic programming algorithm are presented in the Table 3.

Table 3. Parameters of genetic programming used in simulations

Functions set F	$\{+,-,\cdot,/^{\mathrm{a}}, \mathrm{neg}, \mathrm{pow}, \mathrm{inv}^{\mathrm{a}}$
The number of species	2
Number of constants	61
Constants range	$[1,7]$
Number of epochs	1000
Population size μ	20
Probability of crossover p_c	0.5
Probability of mutation p_m	0.5
Weight w	0.5

[a]In case of division and multiplicative inverse operator we used their safe versions

Genetic programming algorithm was run 30 times for each data set generated by the AMSTA method. The best models obtained are shown below:

$$\tau = 0.01 \begin{cases} p_{12}(\mathbf{x}) = (5x_1^2)^{\left(\frac{2.3}{4x_1}\right)} - 3.1 \\ p_{21}(\mathbf{x}) = x_2^8 + 2x_2^4 \end{cases}$$

$$\tau = 0.005 \begin{cases} p_{12}(\mathbf{x}) = -3.5 \cdot 5.5^{-2x_1^2} \\ p_{21}(\mathbf{x}) = 4.9^{x_1^2 - \frac{1}{1.6 \cdot x_1^2} + 0.375} \end{cases}$$

$$\tau = 0.001 \begin{cases} p_{12}(\mathbf{x}) = \frac{-3.2}{(5.7^{5.8 \cdot x_1^2} + 0.169)^{x_1^2}} \\ p_{21}(\mathbf{x}) = (1.521x_2^2)^{3.048} \end{cases}$$

The results obtained by each of these models are presented in the Table 4 and for data generated by AMSTA method for $\tau = 0.001$ also on Figs. 3 and 4.

Table 4. Results obtained by genetic programming algorithm

	Results obtained by proposed method			Results reported in [2]
	$\tau = 0.01$	$\tau = 0.005$	$\tau = 0.001$	
Best	0.006	0.007	0.004	0.004
Average	0.029	0.018	0.012	0.016
Worse	0.021	0.021	0.016	0.046

a)

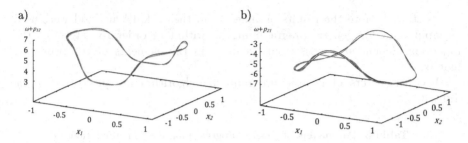

Fig. 3. The graphical representation of dependency between state variables and values of elements of state matrices for reference data and data obtained by the model generated by genetic programming method for $\tau = 0.001$.

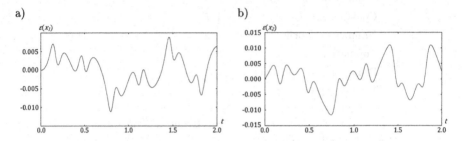

Fig. 4. The errors obtained for signals x_1 and x_2 by the model generated by genetic programming method for $\tau = 0.001$.

As we can see, the best results achieved by all models created on basis of data obtained from the AMSTA algorithm are similar to results reported in [2]. On the other hand, the worst models had the smaller error.

5   Conclusions

In this paper, a new version of the State Transition Algorithm was presented. This method allows generating local models of a non-linear dynamic object that describes the behaviour of this object in strictly defined range. The number of models and the range of data for which local model is valid are determined by the proposed method. This allows obtaining the models that are simpler and more accurate. Next, those models were used to generate functional dependencies describing the changes of parameters of the transition matrix of the model. As shown in Sect. 3 the results obtained by the AMSTA algorithm were better than the results reported in [2], however, the functional models generated by genetic programming method are similar.

References

1. Bartczuk, Ł., Przybył, A., Cpałka, K.: A new approach to nonlinear modelling of dynamic systems based on fuzzy rules. Int. J. Appl. Math. Comput. Sci. **26**(3), 603–621 (2016)
2. Bartczuk, Ł., Dziwiński, P., Red'ko, V.G.: The concept on nonlinear modelling of dynamic objects based on state transition algorithm and genetic programming. In: Rutkowski, L., Korytkowski, M., Scherer, R., Tadeusiewicz, R., Zadeh, L.A., Zurada, J.M. (eds.) ICAISC 2017. LNCS (LNAI), vol. 10246, pp. 209–220. Springer, Cham (2017). https://doi.org/10.1007/978-3-319-59060-8_20
3. Bologna, G., Hayashi, Y.: Characterization of symbolic rules embedded in deep DIMLP networks: a challenge to transparency of deep learning. J. Artif. Intell. Soft Comput. Res. **7**(4), 265–286 (2017)
4. Caughey, T.K.: Equivalent linearization techniques. J. Acoust. Soc. Am. **35**(11), 1706–1711 (1963)
5. Chang, O., Constante, P., Gordon, A., Singana, M.: A novel deep neural network that uses space-time features for tracking and recognizing a moving object. J. Artif. Intell. Soft Comput. Res. **7**(2), 125–136 (2017)
6. Chen, C., Luo, C., Jiang, Z.: Elite bases regression: a real-time algorithm for symbolic regression. arXiv preprint arXiv:1704.07313 (2017)
7. Cpałka, K., Łapa, K., Przybył, A.: A new approach to design of control systems using genetic programming. Inf. Technol. Control **44**(4), 433–442 (2015)
8. Cpałka, K., Rebrova, O., Nowicki, R., Rutkowski, L.: On design of flexible neuro-fuzzy systems for nonlinear modelling. Int. J. Gen. Syst. **42**(6), 706–720 (2013)
9. Dub, M., Stefek, A.: Using PSO method for system identification. In: Březina, T., Jabloński, R. (eds.) Mechatronics 2013, pp. 143–150. Springer, Cham (2014). https://doi.org/10.1007/978-3-319-02294-9_19
10. Gajdoš, P., et al.: A signal strength fluctuation prediction model based on symbolic regression. In: 2015 38th International Conference on Telecommunications and Signal Processing (TSP), Prague, pp. 1–5 (2015)
11. Ke, Y., Hagiwara, M.: An English neural network that learns texts, finds hidden knowledge, and answers questions. J. Artif. Intell. Soft Comput. Res. **7**(4), 229–242 (2017)
12. Khan, N.A., Shaikh, A.: A smart amalgamation of spectral neural algorithm for nonlinear lane-emden equations with simulated annealing. J. Artif. Intell. Soft Comput. Res. **7**(3), 215–224 (2017)
13. Korns, M.F.: A baseline symbolic regression algorithm. In: Riolo, R., Vladislavleva, E., Ritchie, M., Moore, J. (eds.) Genetic Programming Theory and Practice X, pp. 117–137. Springer, New York (2012). https://doi.org/10.1007/978-1-4614-6846-2_9
14. Koza, J.R.: Genetic Programming: On the Programming of Computers by Means of Natural Selection, vol. 1. MIT Press, Cambridge (1992)
15. Krawiec, K.: Behavioral Program Synthesis with Genetic Programming, vol. 618. Springer, Heidelberg (2016). https://doi.org/10.1007/978-3-319-27565-9
16. Kubalík, J., Alibekov, E., Žegklitz, J., Babuška, R.: Hybrid single node genetic programming for symbolic regression. In: Nguyen, N.T., Kowalczyk, R., Filipe, J. (eds.) TCCI XXIV. LNCS, vol. 9770, pp. 61–82. Springer, Heidelberg (2016). https://doi.org/10.1007/978-3-662-53525-7_4
17. La Cava, W., Silva, S., Vanneschi, L., Spector, L., Moore, J.: Genetic programming representations for multi-dimensional feature learning in biomedical classification. In: Squillero, G., Sim, K. (eds.) EvoApplications 2017. LNCS, vol. 10199, pp. 158–173. Springer, Cham (2017). https://doi.org/10.1007/978-3-319-55849-3_11

18. Luo, C., Chen, C., Jiang, Z.: A divide and conquer method for symbolic regression. arXiv preprint arXiv:1705.08061 (2017)
19. Łapa, K., Cpałka, K., Wang, L.: New method for design of fuzzy systems for nonlinear modelling using different criteria of interpretability. In: Rutkowski, L., Korytkowski, M., Scherer, R., Tadeusiewicz, R., Zadeh, L.A., Zurada, J.M. (eds.) ICAISC 2014. LNCS (LNAI), vol. 8467, pp. 217–232. Springer, Cham (2014). https://doi.org/10.1007/978-3-319-07173-2_20
20. Łapa, K., Cpałka, K.: On the application of a hybrid genetic-firework algorithm for controllers structure and parameters selection. In: Borzemski, L., Grzech, A., Świątek, J., Wilimowska, Z. (eds.) Information Systems Architecture and Technology: Proceedings of 36th International Conference on Information Systems Architecture and Technology – ISAT 2015 – Part I. AISC, vol. 429, pp. 111–123. Springer, Cham (2016). https://doi.org/10.1007/978-3-319-28555-9_10
21. Łapa, K., Szczypta, J., Saito, T.: Aspects of evolutionary construction of new flexible PID-fuzzy controller. In: Rutkowski, L., Korytkowski, M., Scherer, R., Tadeusiewicz, R., Zadeh, L.A., Zurada, J.M. (eds.) ICAISC 2016. LNCS (LNAI), vol. 9692, pp. 450–464. Springer, Cham (2016). https://doi.org/10.1007/978-3-319-39378-0_39
22. Szczypta, J., Łapa, K., Shao, Z.: Aspects of the selection of the structure and parameters of controllers using selected population based algorithms. In: Rutkowski, L., Korytkowski, M., Scherer, R., Tadeusiewicz, R., Zadeh, L.A., Zurada, J.M. (eds.) ICAISC 2014. LNCS (LNAI), vol. 8467, pp. 440–454. Springer, Cham (2014). https://doi.org/10.1007/978-3-319-07173-2_38
23. Minemoto, T., Isokawa, T., Nishimura, H., Matsui, N.: Pseudo-orthogonalization of memory patterns for complex-valued and quaternionic associative memories. J. Artif. Intell. Soft Comput. Res. 7(4), 257–264 (2017)
24. Nelles, O.: Nonlinear System Identification: From Classical Approaches to Neural Networks and Fuzzy Models. Springer, Heidelberg (2013). https://doi.org/10.1007/978-3-662-04323-3
25. Pennachin, C.L., Looks, M., de Vasconcelos, J.A.: Robust symbolic regression with affine arithmetic. In: Proceedings of the 12th Annual Conference on Genetic and Evolutionary Computation, pp. 917–924. ACM (2010)
26. Potter, M.A., De Jong, K.A.: A cooperative coevolutionary approach to function optimization. In: Davidor, Y., Schwefel, H.-P., Männer, R. (eds.) PPSN 1994. LNCS, vol. 866, pp. 249–257. Springer, Heidelberg (1994). https://doi.org/10.1007/3-540-58484-6_269
27. Prasad, M., Liu, Y.-T., Li, D.-L., Lin, C.-T., Shah, R.R., Kaiwartya, O.P.: A new mechanism for data visualization with TSK-type preprocessed collaborative fuzzy rule based system. J. Artif. Intell. Soft Comput. Res. 7(1), 33–46 (2017)
28. Rotar, C., Iantovics, L.B.: Directed evolution - a new metaheuristc for optimization. J. Artif. Intell. Soft Comput. Res. 7(3), 183–200 (2017)
29. Smetka, T., Homoliak, I., Hanacek, P.: On the application of symbolic regression and genetic programming for cryptanalysis of symmetric encryption algorithm. In: 2016 IEEE International Carnahan Conference on Security Technology, Orlando, pp. 1–8 (2016)
30. Ugalde, H.M.R., et al.: Computational cost improvement of neural network models in black box nonlinear system identification. Neurocomputing 166, 96–108 (2015)
31. Yang, S., Sato, Y.: Swarm intelligence algorithm based on competitive predators with dynamic virtual teams. J. Artif. Intell. Soft Comput. Res. 7(2), 87–101 (2017)

32. Zalasiński, M., Cpałka, K.: Novel algorithm for the on-line signature verification using selected discretization points groups. In: Rutkowski, L., Korytkowski, M., Scherer, R., Tadeusiewicz, R., Zadeh, L.A., Zurada, J.M. (eds.) ICAISC 2013. LNCS (LNAI), vol. 7894, pp. 493–502. Springer, Heidelberg (2013). https://doi.org/10.1007/978-3-642-38658-9_44

33. Zalasiński, M., Cpałka, K., Hayashi, Y.: New fast algorithm for the dynamic signature verification using global features values. In: Rutkowski, L., Korytkowski, M., Scherer, R., Tadeusiewicz, R., Zadeh, L.A., Zurada, J.M. (eds.) ICAISC 2015. LNCS (LNAI), vol. 9120, pp. 175–188. Springer, Cham (2015). https://doi.org/10.1007/978-3-319-19369-4_17

34. Zalasiński, M., Cpałka, K.: New algorithm for on-line signature verification using characteristic hybrid partitions. In: Wilimowska, Z., Borzemski, L., Grzech, A., Świątek, J. (eds.) Information Systems Architecture and Technology: Proceedings of 36th International Conference on Information Systems Architecture and Technology – ISAT 2015 – Part IV. AISC, vol. 432, pp. 147–157. Springer, Cham (2016). https://doi.org/10.1007/978-3-319-28567-2_13

35. Zalasiński, M., Cpałka, K., Hayashi, Y.: A method for genetic selection of the most characteristic descriptors of the dynamic signature. In: Rutkowski, L., Korytkowski, M., Scherer, R., Tadeusiewicz, R., Zadeh, L.A., Zurada, J.M. (eds.) ICAISC 2017. LNCS (LNAI), vol. 10245, pp. 747–760. Springer, Cham (2017). https://doi.org/10.1007/978-3-319-59063-9_67

36. Zhou, X., Yang, C., Gui, W.: Nonlinear system identification and control using state transition algorithm. Appl. Math. Comput. **226**, 169–179 (2014)

37. Zhou, X., Gao, D.Y., Yang, C., Gui, W.: Discrete state transition algorithm for unconstrained integer optimization problems. Neurocomputing **173**, 864–874 (2016)

38. Zhou, X., Yang, C., Gui, W.: Initial version of state transition algorithm. In: 2011 Second International Conference on Digital Manufacturing and Automation (ICDMA), pp. 644–647. IEEE (2011)

A Population Based Algorithm and Fuzzy Decision Trees for Nonlinear Modeling

Piotr Dziwiński[1]([✉]), Łukasz Bartczuk[1]([✉]), and Krzysztof Przybyszewski[2,3]([✉])

[1] Institute of Computational Intelligence, Częstochowa University of Technology,
Częstochowa, Poland
{piotr.dziwinski,lukasz.bartczuk}@iisi.pcz.pl
[2] Information Technology Institute, University of Social Sciences,
90-113 Łodz, Poland
kprzybyszewski@san.edu.pl
[3] Clark University, Worcester, MA 01610, USA

Abstract. The paper presents a new approach for using the fuzzy decision trees for non-linear modeling based on the capabilities of particle
swarm optimization and evolutionary algorithms. The most nonlinear
dynamic objects have their approximate nonlinear model. Their parameters are known or can be determined by one of the typical identification
procedure. The obtained approximate nonlinear model describes well the
identified dynamic object only in the operating point. In this work, we use
hybrid model composed with of two parts: approximate nonlinear model
and fuzzy decision tree. The fuzzy decision tree contains correction values
of the parameters in terminal nodes. The hybrid model ensures sufficient
accuracy for the practical applications. A particle swarm optimization
and evolutionary algorithm were used for identification of the parameters
of the approximate nonlinear model and fuzzy decision tree. An important benefit of the proposed method is the obtained characteristics of
the unknown parameters of the approximate nonlinear model described
by the terminal nodes of the fuzzy decision tree. They present valuable
and interpretable knowledge for the experts concerning the essence of
the unknown phenomena.

Keywords: Nonlinear modeling · Non-invasive identification
Significant operating point · Particle swarm optimization
Evolutionary strategies · Permanent magnet synchronous motors
Takagi-Sugeno system · Fuzzy decision trees

1 Introduction

The most nonlinear dynamic objects have their Approximate Nonlinear Model
(ANM). Their parameters are known or can be discovered by one of the typical identification procedure. The model obtained in this way represents well the
nonlinear dynamic object only in Operating Point (OP) [7]. Between them, there
are several secondary phenomena, that are not explained precisely enough by the
expert. The observed phenomena must be reproduced in order to obtain the model
precise enough for practical application.

© Springer International Publishing AG, part of Springer Nature 2018
L. Rutkowski et al. (Eds.): ICAISC 2018, LNAI 10842, pp. 516–531, 2018.
https://doi.org/10.1007/978-3-319-91262-2_46

A large number of mathematical models which can describe the linear or nonlinear systems in a universal way had been proposed in the literature, among others, neural networks [9,18,27,29], treated as black box models, fuzzy systems [2,4,30], flexible fuzzy systems [20,31], neuro-fuzzy systems [25,32,34,36], flexible neuro-fuzzy systems [5,37], Takagi-Sugeno systems [19], flexible Takagi-Sugeno systems and rule based networks [15]. Lapa et al. [12] proposed new interpretability criteria for flexible neuro-fuzzy systems for nonlinear classification. They applied an evolutionary algorithm for a new flexible PID-fuzzy controller [13]. Rutkowski [22] proposed a general algorithm for estimation of functions and their derivatives from noisy observations.

The methods mentioned earlier allow modeling in a universal way but do not provide an acceptable precision of the reproduction of the reference values. The much better result can be obtained by using a hybrid model [7] composed of two parts: ANM and Fuzzy Decision Tree (FDT). The ANM allows reproduction of the reference values with sufficient precision in the certain OP, whereas FDT discovers the values of the unknown parameters of the ANM in different OP and between them. This method guarantees to obtain an adequate precision of the identification in all states of the nonlinear dynamic object. In this paper we use the representation of the approximate state and input matrices presented in [7] by including the sparse corrections $\Delta\hat{\mathbf{g}}(\mathbf{x}(t))$ and $\Delta\hat{\mathbf{q}}(\mathbf{x}(t))$ of the estimated parameters $\mathbf{g}$ and $\mathbf{q}$. It allows to obtain characteristics of the parameters of the ANM described by the functions in the terminal nodes of the FDT. This method guarantees to obtain a sufficient precision of the identification in the entire area of the work of the nonlinear deterministic dynamic object.

This article describes a new method for using the FDT [6,17,26], for selection of the important inputs from measurements as criterion of the significant operating points detection. The splitting node is created in the FDT for each selected inputs. The terminal nodes of the tree contain the system matrix values for the detected operating points. Finally, the obtained FDT is converted to Takagi-Sugeno (TS) fuzzy system.

The remainder of this paper is organized as follows. Section 2 describes approximate modeling of nonlinear dynamic objects by the algebraic equations and on the basis of the state variable technique using sparse corrections of the known or estimated parameters in the operating points. Section 3 deals with fuzzy decision tree modeling of the corrections of the parameters in the operating points. Section 4 presents the method for detection of the OP described by FDT interpreted as TS fuzzy system. Finally, Sect. 5 shows simulation results which proves the effectiveness of the proposed method.

2   Approximate Modeling of the Nonlinear Dynamic Object

Let us consider the nonlinear dynamic stationary object described by the algebraic equations and based on the state variable technique [20]

$$\frac{d\mathbf{x}}{dt} = \mathbf{A}(\mathbf{x}(t))\mathbf{x}(t) + \mathbf{B}(\mathbf{x}(t))\mathbf{u}(t), \tag{1}$$

$$\mathbf{y}(t) = \mathbf{C}\mathbf{x(t)}, \tag{2}$$

where $\mathbf{A}(\mathbf{x}(t))$, $\mathbf{B}(\mathbf{x}(t))$ are the system and input matrices respectively, $\mathbf{u}(t)$, $\mathbf{y}(t)$ are the input and output signals respectively, $\mathbf{x}(t)$ is the vector of the state variables. The algebraic equations based on the state variable technique, delivered by the experts, describe the dynamic nonlinear object with a sufficient precision only in some characteristic work state called an operating point. Beyond the OP there are phenomena that are not included in the mathematical model. The overall accuracy of such a model may be too low for many practical applications.

In this work, we propose the hybrid method which increases the effectiveness of the modeling of the nonlinear dynamic object. It is done by the modeling of the system and input matrices parameters which are not described precisely enough by the mathematical model. The entire approximate model can be described by algebraic equations and on the basis of the state variable technique, where unknown linear or nonlinear part can be modeled by the $\hat{\mathbf{A}}(\mathbf{x}(t), \mathbf{g} + \Delta\hat{\mathbf{g}}(\mathbf{x}(t))$ and $\hat{\mathbf{B}}(\mathbf{x}(t), \mathbf{q} + \Delta\hat{\mathbf{q}}(\mathbf{x}(t)))$ approximate matrices [6]. The unknown parameters change and can be described by the correction values $\Delta\hat{\mathbf{g}}(\mathbf{x}(t))$ and $\Delta\hat{\mathbf{q}}(\mathbf{x}(t))$.

For example, consider the specific nonlinear dynamic deterministic system with the element of the system matrix $a_{33} = 1 - T_s(F/J)$ with the unknown or estimated value of the F. The element a_{33} can be written as $a_{33} \approx 1 - T_s((F + \Delta\hat{F})/J)$. The parameter F has constant value in the OP but changes in unknown way between them and can be modeled by the $\Delta\hat{F}$ correction values. So by using the approximate matrices, we obtain the following form of the Eq. (1)

$$f(\mathbf{x}(t), \mathbf{u}(t)) = \hat{\mathbf{A}}\left(\mathbf{x}(t), \mathbf{g} + \Delta\hat{\mathbf{g}}(\mathbf{x}(t))\right)\mathbf{x}(t) \tag{3}$$
$$+ \hat{\mathbf{B}}\left(\mathbf{x}(t), \mathbf{q} + \Delta\hat{\mathbf{q}}(\mathbf{x}(t))\right)\mathbf{u}(t),$$

where $\hat{\mathbf{A}}$, $\hat{\mathbf{B}}$ are approximate state and input matrices respectively, $\mathbf{g}$, $\mathbf{q}$ are known parameters, $\Delta\hat{\mathbf{g}}(\mathbf{x}(t))$, $\Delta\hat{\mathbf{q}}(\mathbf{x}(t))$ are the sparse corrections of parameters $\mathbf{g}$ and $\mathbf{q}$, respectively.

3 Fuzzy Decision Tree Modeling of the System and Input Matrices in the Operating Points

The change of the sparse corrections $\Delta\hat{\mathbf{g}}(\mathbf{x}(t))$, $\Delta\hat{\mathbf{q}}(\mathbf{x}(t))$ of parameters $\mathbf{g}$ and $\mathbf{q}$, respectively takes place between OP, usually does not occur rapidly, but in a smooth manner which is difficult to describe by using the mathematical model. The correction values $\Delta\hat{\mathbf{g}}(\mathbf{x}(t))$ and $\Delta\hat{\mathbf{q}}(\mathbf{x}(t))$ existing in the OP pass fluently among themselves and overlap. So, for activation level of the correction values in the OP, we use FDT. It allows creating the splitting nodes step by step according to detected OP. FDT ensures obtaining the simplest fuzzy rule set simultaneously, simplifies the learning process of the Participle Swarm Optimization (PSO) and Evolutionary Strategy (ES).

The FDT contains two types of nodes: the inner nodes and terminal nodes. The inner nodes of the FDT contain a split fuzzy function for the left (4) and right (5) node.

$$\mu_{n^{left}}(x; a, b) = \begin{cases} 1 & \text{if } (x < a) \\ \frac{x-a}{b-a} & \text{if } (a \leq x \leq b) \\ 0 & \text{if } (x > b) \end{cases}, \tag{4}$$

$$\mu_{n^{right}}(x; a, b) = \begin{cases} 0 & \text{if } (x < a) \\ \frac{b-x}{b-a} & \text{if } (a \leq x \leq b) \\ 1 & \text{if } (x > b) \end{cases}, \tag{5}$$

The terminal nodes contain the correction values $\Delta \hat{\mathbf{g}}(\mathbf{x}(t))$ and $\Delta \hat{\mathbf{q}}(\mathbf{x}(t))$. In the FDT, all terminal nodes can be active, so final values of an element of the input and state matrices are calculated as weighted average of the all terminal nodes. More precise description can be found in the [6]. The FDT can be interpreted as the Takagi-Sugeno (TS) fuzzy system by reading all paths from the root of the tree to all terminal nodes (tree leaves). In the results we obtain the fuzzy rules of TS fuzzy system used finally for modeling correction values $\Delta \hat{\mathbf{g}}(\mathbf{x}(t))$ and $\Delta \hat{\mathbf{q}}(\mathbf{x}(t))$.

The general form of the TS fuzzy system is presented in Eq. (6).

$$R^{(l)} : \text{ IF } \bar{\mathbf{x}} \text{ is } \mathbf{D}^l \text{ THEN } \mathbf{y}^l = \mathbf{f}^{(l)}(\mathbf{x}), \tag{6}$$

where: $\bar{\mathbf{x}} = [\bar{x}_1, \bar{x}_2, \ldots, \bar{x}_N] \in \bar{\mathbf{X}}$, $\mathbf{y}^l \in \mathbf{Y}^l$, $\mathbf{D}^l = D_1^l \times D_2^l \times \ldots \times D_N^l$, $D_1^l, D_2^l, \ldots, D_N^l$, are the fuzzy sets described by the membership functions $\mu_{D_i^l}(\bar{x}_i)$, $i = 1, \ldots, N$, $l = 1, \ldots, n$, L is the number of the rules and N is the number of the inputs of the TS fuzzy system, $\mathbf{f}^{(l)}$ are the functions describing values of the system matrix or input matrix for the l-th fuzzy rule. In case of the a_{33} element of the state matrix, the $f^{(l)}$ function from the consequent takes the form: $f^{(l)}(t) = 1 - T_s \left((F + \Delta \hat{F}^{(l)}(t))/J \right)$.

Assuming the aggregation method as the weighted average, using the Eq. (3) and Euler integration method with a time step T_s, we obtain the discrete approximate hybrid model described by equation [6] (7)

$$f(\mathbf{x}(k), \bar{\mathbf{x}}(k), \mathbf{u}(k+1)) =$$

$$\left(\mathbf{I} + \left(\hat{\mathbf{A}} \left(\mathbf{x}(k), \mathbf{g} + \frac{\sum_{l=1}^{L} \hat{\mathbf{g}}^l \cdot \mu_{\mathbf{D}^l}(\bar{\mathbf{x}}(k))}{\sum_{l=1}^{L} \mu_{\mathbf{D}^l}(\bar{\mathbf{x}}(k))} \right) \right) T_s \right) \mathbf{x}(k)$$

$$+ \left(\hat{\mathbf{B}} \left(\mathbf{x}(k), \mathbf{q} + \frac{\sum_{m=L+1}^{L+M} \hat{\mathbf{q}}^m \cdot \mu_{\mathbf{D}^m}(\bar{\mathbf{x}}(k))}{\sum_{m=L+1}^{L+M} \mu_{\mathbf{D}^m}(\bar{\mathbf{x}}(k))} \right) \right) \mathbf{u}(k+1), \tag{7}$$

where $\bar{\mathbf{x}}(k)$ is the vector of the fuzzy values obtained from the vector $\mathbf{x}(k)$ using singleton fuzzification, $\hat{\mathbf{g}}^l$, $\hat{\mathbf{q}}^m$ are the sparse vectors containing the correction values for the changing parameters in the l and m OP, $l = 1, \ldots, L$, $m = L + 1, \ldots, L + M$, L, M - number of the rules describing the OP for the state and input matrices respectively, $\mu_{\mathbf{D}^m}(\bar{\mathbf{x}}(k))$ and $\mu_{\mathbf{D}^l}(\bar{\mathbf{x}}(k))$ are the membership functions describing activation levels of the operating point and $\mathbf{I}$ is the identity matrix.

The Eq. (7) represents the discrete hybrid model describing the dynamic nonlinear deterministic system. The Euler integration method was selected for simplicity but there should be chosen a better one in the practical application.

So, the local linear or nonlinear model in the operating point l and m is defined through the set of the parameters $\Theta_l = \{\hat{\mathbf{A}}(\mathbf{x}(k)), \mathbf{g}, \hat{\mathbf{g}}^l, \mathbf{D}^l\}$ and $\Theta_m = \{\hat{\mathbf{B}}(\mathbf{x}(k)), \mathbf{q}, \hat{\mathbf{q}}^m, \mathbf{D}^m\}$. The parameters are determined by the fuzzy decision tree method using PSO and ES.

4   Fuzzy Decision Tree Method for the Detection of the Operating Points

The automatic detection of the OP in nonlinear modeling is the very hard and time-consuming task. In the most studies, the authors focus on solutions using grouping and classification algorithms to determine initially the potential areas that stand for the OP. Unfortunately, the most part of them requires complete data set or its random samples to determine initial areas for the OP. Dziwiński et al. [7] proposed a method for non-linear modeling with a new representation of the approximate input and state matrices by including the sparse correction values $\Delta\hat{\mathbf{g}}(\mathbf{x}(t))$ and $\Delta\hat{\mathbf{q}}(\mathbf{x}(t))$ modeled by using of the TS fuzzy system. They use the PSO supported by the Genetic Algorithm (PSO-GA) to determine ANM and the sparse corrections of the parameters $\mathbf{g}$ and $\mathbf{q}$. The PSO algorithm is frequently used by the authors for solving the hard combinatorial problems [1,3,16], as well as GA [8,35]. Korytkowski et al. [10] used combining of backpropagation with boosting to learn neuro-fuzzy systems applied in classification problems. Łapa and Cpałka [11] used hybrid population-based algorithm [12] for solving complex optimization problem composed of the parameters and structure of the system optimization [14]. Rotar and Iantovics [21] proposed novel general algorithm inspired by Directed Evolution. Yang and Sato [28] proposed an improved fitness predator optimizer by increasing population diversity. He applied a new approach to high dimensional well-known benchmark functions. Zalasiński et al. [33] proposed a new Evolutionary Algorithm (EA) for selection of the dynamic signature global features.

Real-world modeling problems usually involve a large number of the candidate inputs for the splitting features. In the case of the nonlinear dynamic objects identification, the selection of the important inputs is sometimes difficult due to nonlinearities. Thus, the input selection is a crucial step to obtain the simple model using only inputs, that are important for the detection of the OP. The methods found in the literature [17,26] can be divided generally into two groups: model-free methods and model-based methods.

Algorithm 1. Pseudocode of the method for identification of the operating points using FDT

---

1 **Algorithm Build-FDT** ($\mathbf{u}(t)$, $\mathbf{x}(t)$, $\mathbf{y}(t)$, T_{max}, T_s, E_{min}, e_{max}, Δe, $\mathbf{Q}$)

   Data: $\mathbf{u}(t)$, $\mathbf{x}(t)$, $\mathbf{y}(t)$ - measurements, $t = 0, (T_s), T_{max}$, T_s - time steep, T_{max} - total time of the measurements, E_{min} - minimal RMSE error, e_{max} - maximum epoch number, Δe - maximum epoch number after adding a new FDT^γ, $\gamma \in \mathbf{Q}$, $\mathbf{Q}$ - the set of available splitting attributes.

   Result: FDT^b - the best fuzzy decision tree, $\mathbf{\Theta}^b$ - set of the operating points corresponding to the terminal nodes of the best FDT^b, $\theta_l^b = \{\hat{\mathbf{g}}^{l,b}, \hat{\mathbf{q}}^{l,b}, \mathbf{D}^{l,b}\}$, $\theta_l^b \in \mathbf{\Theta}^b$, $l = 0, \ldots, L$, L - number of the detected operating points;

2 Set initially: $\mathbf{\Theta}^b = \emptyset$, $L = 0$, $e = 0$, $\Delta e = 0$, $\mathbf{Q}' = \emptyset$, $\theta_0^b = \{\hat{\mathbf{g}}^{1,b}, \hat{\mathbf{q}}^{1,b}, \emptyset\}$;

3 Add the root node to the FDT^b: $L \leftarrow L + 1$, $\mathbf{\Theta}^b \leftarrow \mathbf{\Theta}^b \cup \theta_l^b$;

4 Determine the initial time interval:
   $t_{max}^{(e+1),b} = \text{Extend-Time-Interval}(\mathbf{\Theta}^b, t_{max}^e, x(0), d_{start})$;

5 Initialize the algorithm: $\mathbf{S}^{e,\gamma} = \text{Init-PSO-ES}(\mathbf{\Theta}^b)$;

6 **repeat**

7   **repeat**

8       $e \leftarrow e + 1$, $e_{sn} \leftarrow e_{sn} + 1$;

9   **for** $\gamma \in \mathbf{Q}'$ **do**

10   Run the hybrid algorithm: for FDT^γ

11         $\mathbf{S}^{e,\gamma} = \text{PSO-ES}(\mathbf{S}^{e-1,\gamma})$;

12         $\mathbf{E}^{e,\gamma} = \text{Evaluate}(\mathbf{S}^{e,\gamma}, \mathbf{u}(t), \mathbf{x}(t), \mathbf{y}(t), t_{max}^{e,\gamma})$;

13         $\mathbf{\Theta}_{best}^{e,\gamma} = \text{Get-Best}(\mathbf{S}^{e,\gamma}, \mathbf{E}^{e,\gamma})$;

14   **if** $E_{best}^{e,\gamma} < E_{best}^{e-1,\gamma}$ **then**

15            $t_{max}^{(e+1),\gamma} = \text{Extend-Time-Interval}(\mathbf{\Theta}_{best}^{e,\gamma}, t_{max}^{e,\gamma}, \epsilon_{min})$;

16            $E_{best}^{e,\gamma} = \text{Evaluate}(\mathbf{\Theta}_{best}^{e,\gamma}, \mathbf{x}(t), \mathbf{u}(t), \mathbf{y}(t), t_{max}^{(e+1),\gamma})$;

17   **if** $(e_{sn} > \Delta e)$ & $(|\mathbf{Q}'| > 1)$ **then**

18          $\gamma_{best} = \min_\gamma(E_{best}^{e,\gamma})$, $FDT^b = FDT^{\gamma_{best}}$, $\mathbf{Q}' \leftarrow \gamma_{best}$;

19       $t_{max}^e = \max_\gamma(t_{max}^{e,\gamma})$, $E_{best}^e = \min_\gamma(E_{best}^{e,\gamma})$;

20   **until** $\left(t_{max}^{(e)} > t_{max}^{(e-\Delta z)}\right)$ || $\left(E_{best}^{(e)} - E_{best}^{(e-\Delta z)}\right) < \Delta E$;

21   Set $\mathbf{Q}'$ to all available splitting attributes: $\mathbf{Q}' \leftarrow \mathbf{Q}$, $e_{sn} = 0$;

22   **for** $\gamma \in \mathbf{Q}'$ **do**

23       $FDT^\gamma = \text{Clone-FDT-And-Add-Split-Node}(FDT^b, t_{max}^{e,b}, \gamma)$;

24       $t_{max}^{(e+1),\gamma} \leftarrow \text{Extend-Time-Interval}(\mathbf{\Theta}^{e,\gamma}, t_{max}^{e,\gamma}, x(t_{max}^{e,b}), d_{add})$;

25 **until** $\left(t_{max}^{(e)} < T_{max}\right)$ & $\left(E_{best}^{(e)} > E_{min}\right) | (e < e_{max})$;

Mendonca [17] has used two approaches: top-down and bottom-up. In the top-down approach, he selects all of the input variables and removes the one input variable with the worst results at each stage. In the bottom-up approach, he starts with only one input. At each stage, he builds the fuzzy model for each of the n considered inputs. Next, he evaluates models using different quality criterions. Finally, he selects the best one. The mentioned early approach has the following drawback – it requires estimating the $2 * N$ fuzzy models at each stage of splitting, so it is very computationally expensive and uses all measurement data to estimate quality criterion.

Algorithm 2. The pseudocode of the function for clone DFT and add a split node for the splitting attribute γ

---

1 **Function Clone-FDT-And-Add-Split-Node**(FDT^b, $t_{max}^{e,b}$, γ)

    Data: FDT^b - the best fuzzy decision tree, γ - the split attribute

    Result: FDT$^\gamma$ - copy of the FDT^b extended by the γ split attribute

2   Clone the FDT for the γ: FDT$^\gamma$ = Clone(FDT^b);

3   Chose the best terminal node n_t' from the FDT$^\gamma$ corresponding to the fuzzy rule $R^{(l)}$ on the basis of the activity at the end of the measurements $l = \arg \max\limits_{i=1,\dots,L} (\mu_{\mathbf{D}^i}(\bar{\mathbf{x}}(t_{max}^{e,b})))$;

4   Estimate a new initial value for

    $t_{max}^{(e+1),\gamma} \leftarrow$ Extend-Time-Interval($\mathbf{\Theta}^{e,\gamma}, t_{max}^{e,b}, x(t_{max}^{e,b}), d_{add}$);

5   Create split node n_γ and add two terminal nodes n_t^{left} and n_t^{right};

6   **if** n_t' *is left node* **then**

7   Copy the parameters of the membership function activating n_t' node to the n_γ node activating the left node n_t^{left};

8   Add new operating point θ_L for the new terminal node n_t^{right};

9         $L \leftarrow L + 1, \mathbf{\Theta}^{e,\gamma} \leftarrow \mathbf{\Theta}^{e,\gamma} \cup \theta_L$;

10   **else**

11   Copy the parameters of the membership function activating n_t' node to the n_γ node activating the right node n_t^{right};

12   Add new operating point θ_L for the new terminal node n_t^{left};

13         $L \leftarrow L + 1, \mathbf{\Theta}^{e,\gamma} \leftarrow \mathbf{\Theta}^{e,\gamma} \cup \theta_L$

14   Replace the terminal node n_t' with the split node n_γ;

15   Copy the estimated values of the parameters from the node n_t' to the new terminal nodes n_t^{left} and n_t^{right};

16   Estimate the initial parameters of the membership functions for n_t^{left} and the n_t^{right} terminal nodes based on the values of the split attribute in the t_{max}^e and $t_{max}^{e,\gamma}$ time of the simulation according to the following equations

    $\alpha = \min\limits_{t \in <t_{max}^{e,b}, t_{max}^{e,\gamma}>} \mathbf{x}(t, \gamma), \quad \beta = \max\limits_{t \in <t_{max}^{e,b}, t_{max}^{e,\gamma}>} \mathbf{x}(t, \gamma),$

    $a_{left} = b_{left} = \alpha - (\beta - \alpha) \cdot \rho_1, \quad a_{right} = b_{right} = \beta + (\beta - \alpha) \cdot \rho_2,$

    where ρ_1, ρ_2 are the left and right fuzzy factor.

In this paper, we propose a new method for using the FDT for selecting the important inputs used for detection of the OP. The algorithm presented in the Algorithm 1, starts work with a small amount of the reference data. Next, during the optimization process, new reference data are added according to the error criterion. So the algorithm works in a time-varying environment, just like in the works [23,24] and works in the more effective way, than in case [17].

When the algorithm is adding a new OP, then starts with all available inputs and builds in a parallel way, the different FDT for each available splitting attribute - one for each FDT. In the process of the identification, the method uses only small part of the measured data for learning FDT according to error criterion. After a predefined number of the epochs, the algorithm selects the best splitting attribute. Next, it continues the identification process with the selected splitting attribute.

The new method start for the root node containing only const corrections of the parameters used in the entire work area $\theta_0^b = \{\hat{\mathbf{g}}^{0,b}, \hat{\mathbf{q}}^{0,b}, \emptyset\}$ (Algorithm 1, line 2). Initially, the method determines the initial time of the measurements based on a distance criterion $d\left(x(0), x(t_{max}^{e+1,b})\right) < d_{start}$, where d_{start} is maximum distance between measurements used for the identification of the OP determined by the expert or from the experiments for the best (current) FDT^b (line 4), where Θ^b is set of the terminal nodes for FDT^b. Next, the method initialize the set of all solutions $\mathbf{S}^{e,\gamma}$ (line 5) of the PSO and ES algorithm and runs algorithm for the root node (line 10). Initially, the algorithm determines corrections of the parameters $\theta_0^b = \{\hat{\mathbf{g}}^{0,b}, \hat{\mathbf{q}}^{0,b}, \emptyset\}$. Parameters identification process is performed until the better results are obtained $(E_{best}^{(e)} - E_{best}^{(e-\Delta z)}) < \Delta E$ or method was proceeded to acquire new data samples $t_{max}^{(e)} > t_{max}^{(e-\Delta z)}$ (line 20). For the first root node, the method works with a one FDT^b. The FDT^γ is described by the set Θ^γ. For the root node we do not use splitting attributes, so the $\gamma = \emptyset$. The acquisition of new data samples is done, if RMSE error decreases $E_{best}^{e,\gamma} < E_{best}^{e-1,\gamma}$ and the error criterion presented in the Eq. (8) is meet (lines 14–16).

$$d\left(\mathbf{y}'(\text{FDT}^\gamma, t_{max}^{e,\gamma}) - \mathbf{y}(t_{max}^{e,\gamma})\right) < \epsilon_{min}, \tag{8}$$

where: $\mathbf{y}'(\text{FDT}^\gamma, t_{max}^{e,\gamma})$ is the output obtained for the best created model for the splitting attribute γ in the time $t_{max}^{e,\gamma}$ of the simulation for the FDT^γ, $\mathbf{y}(t_{max}^{e,\gamma})$ is the measured reference value and d is selected distance measure.

If the process of the identification is ineffective by the Δz epoch number (line 20), then the method performs cloning of the FDT^b for each splitting attribute $\gamma \in \mathbf{Q}$ and adds a new splitting node using the Algorithm 2 for each a new FDT^γ. Next, the method extends the time interval according to error criterion (8). The new splitting node is equivalent to a new OP and a new fuzzy rule of the TS fuzzy system. So, at this point, the method learns independent parallel the $|\mathbf{Q}|$ fuzzy decision trees for the predefined number of epoch Δe (Algorithm 1, line 17). For all FDT$^\gamma$, method extends the time of the measurements $t_{max}^{e,\gamma}$. In the next steep (Algorithm 1, line 18), method leaves only FDT^γ with the best value of the $E_{best}^{(e),\gamma}$ and continue identification process with the best $FDT^{(e),\gamma}$.

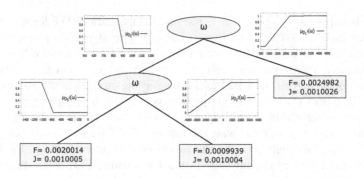

Fig. 1. The obtained FDT with fuzzy membership functions $\mu_{D_2^1}(\omega)$ and $\mu_{D_1^1}(\omega)$ for the root node, $\mu_{D_3^1}(\omega)$ and $\mu_{D_2^2}(\omega)$ for the left splitting node.

The algorithm finishes the works, when all measurement data $t_{max}^{e,b} = T_{max}$ were used, and error criterion $E_{best}^e \leq E_{min}$ has been meet or method works for predefined number of epoch e_{max}. As a criterion of the error, we use Root Mean Square Error measure (RMSE).

5   Experimental Results

The experiments were performed for the PMSM described in the work [7] with unknown values of the friction coefficient F and moment of inertia J. Other parameters are known and were not identified in the experiments. The learning dataset was prepared using the mathematical model of the PMSM with auto-regulation, a constant value of the moment of inertia $J_{ref} = 0.01$, changed value of friction coefficient $F_{ref}(t)$ presented in the Fig. 4-a and given control angular speed $\omega_{set}(t)$ presented in the Fig. 2-a. In the results were obtained the control voltages $U_d(t)$ and $U_q(t)$ showed in the Fig. 2-b, the reference values of the currents $I_d^{ref}(t)$ and $I_q^{ref}(t)$ are illustrated in the Fig. 2-c and the reference value for the angular speed $\omega_{ref}(t)$ is presented in the Fig. 2-d. The goal of the experiments are reproduction of the reference values $\omega_{ref}(t)$, $I_d^{ref}(t)$ and $I_q^{ref}(t)$ presented in the Fig. 2 with the smallest RMSE. It is done by discovering changes of the moment of inertia J (constant) and the friction coefficient F, which were changed in three operating points. Experiments were done using PSO and ES with the parameters presented in the Table 1, where: w is an inertia weight, ψ_1, ψ_2 are acceleration coefficients responsible for global and local search respectively, P_e is the probability of using ES by the PSO algorithm, P_m is the probability of mutation in the ES, N_m is the number of mutated positions in the solutions vectors.

As a result of the experiments were obtained three operating points described by the FDT presented in the Fig. 1. In the Fig. 5 we show the obtained membership functions for the root node (a) and for the left splitting node (b). The

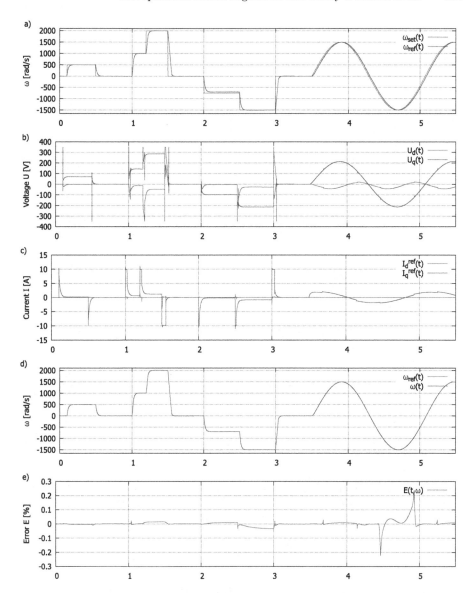

Fig. 2. The results of the experiments: (a) - an angular speed set $\omega_{set}(t)$ of the PMSM and obtained reference angular speed $\omega_{ref}(t)$, (b) - the control voltages $U_d(t)$ and $U_q(t)$ of the PMSM with automatic regulation system, (c) - the reference current response $I_d^{ref}(t)$ and $I_q^{ref}(t)$ of the PMSM, (d) - the obtained angular speed $\omega(t)$ and reference value $\omega_{ref}(t)$, (e) - the obtained relative error $E(t,\omega) = (\omega_{ref}(t) - \omega(t))/(\omega_{max} - \omega_{min}) * 100$

obtained fuzzy rules from the FDT are described by the Eqs. (9–11) containing corrections of the moment of inertia $J = 1e - 3$ and the friction coefficient $F = 1.5e - 3$.

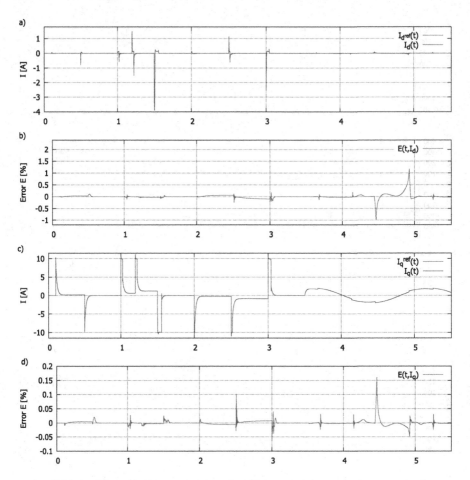

Fig. 3. The results of the experiments: (a) - the obtained current $I_d(t)$ and reference value $I_d^{ref}(t)$, (b) - the obtained relative error $E(t, I_d)$, (c) - the obtained current $I_q(t)$ and reference falue $I_q^{ref}(t)$, (d) - the obtained error $E(t, I_q)$.

Table 1. The parameters of the method for identification of the operating points using FDT and the parameters of the PSO and Evolutionary Strategy

w	ψ_1	ψ_2	P_e	P_m	N_m	d_{start}	d_{add}	ϵ_{min}	Δz	ΔE	Δe	ρ_1	ρ_2
0.7	1.6	1.6	0.25	0.75	3	50	20	0.35	250	0.01	120	0.2	1.2

$$R^{(1)} : \text{ IF } \bar{\omega} \text{ is } D_1^1 \text{ THEN} \begin{cases} a_{32} = & 1.5 T_s P^2 \frac{\lambda_m}{\{1+0.0026\}e{-}3} \\ a_{33} = & 1 - \frac{\{1.5+0.949\}e{-}3}{\{1+0.0026\}e{-}3} T_s \\ b_{33} = & -T_s \frac{P}{\{1+0.0026\}e{-}3} \end{cases} , \qquad (9)$$

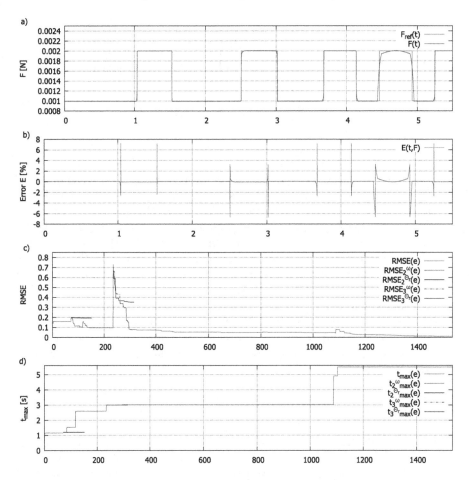

Fig. 4. The results of the experiments: (a) - the obtained value of the unknown friction coefficient F and reference value $F_{ref}(t)$, (b) - the obtained error $E(t, F)$, (c) - RMSE error in the function of epoch number, (d) - time of the simulation t_{max} used for learning in the function of epoch number.

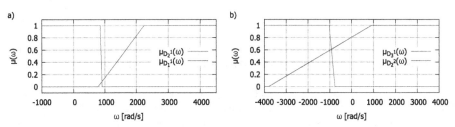

Fig. 5. The obtained fuzzy membership functions $\mu_{D_2^1}(\omega)$ and $\mu_{D_1^1}(\omega)$ for the root node, $\mu_{D_3^1}(\omega)$ and $\mu_{D_2^2}(\omega)$ for the left splitting node.

$$R^{(2)} : \text{ IF } \bar{\omega} \text{ is } D_2^1 \text{ and } \bar{\omega} \text{ is } D_2^2 \text{ THEN } \begin{cases} a_{32} = & 1.5 T_s P^2 \frac{\lambda_m}{\{1+0.0004\}e-3} \\ a_{33} = & 1 - \frac{\{1.5-0.5006\}e-3}{\{1+0.0004\}e-3} T_s \\ b_{33} = & -T_s \frac{P}{\{1+0.0004\}e-3} \end{cases} , (10)$$

$$R^{(3)} : \text{ IF } \bar{\omega} \text{ is } D_3^1 \text{ THEN } \begin{cases} a_{32} = & 1.5 T_s P^2 \frac{\lambda_m}{\{1+0.0005\}e-3} \\ a_{33} = & 1 - \frac{\{1.5+0.50014\}e-3}{\{1+0.0005\}e-3} T_s \\ b_{33} = & -T_s \frac{P}{\{1+0.0005\}e-3} \end{cases} . \qquad (11)$$

In the results of the performed experiments were obtained a very small relative error, what proves the rightness of the proposed algorithm. The obtained relative error E for the angular speed is presented in the Fig. 2-e. The Fig. 3-b,d shows relative errors for the currents $I_d(t)$ and $I_q(t)$ respectively. The value of unknown friction coefficient $F(t)$ is presented in the Fig. 4-a. The Fig. 4-b illustrate the obtained relative error E for the $F(t)$. Finally, the progress of the algorithm in the function of the epoch number for the PSO and ES algorithm is presented in the Fig. 4-c and d. The learning process of the different FDT^γ for available splitting attributes $\gamma \in \{\omega, \Theta_r\}$ is presented in the Fig. 4-c. The observed root square measure error $RMSE_2^\omega$ and $RMSE_2^{\Theta_r}$ correspond to the two different FDT building after adding the second OP for splitting attributes ω and Θ_r, $RMSE_3^\omega$ and $RMSE_3^{\Theta_r}$ – after adding third OP. In the each case, the different FDT are learned in a parallel way by the Δe epochs for available splitting attributes.

References

1. Aghdam, M.H., Heidari, S.: Feature selection using particle swarm optimization in text categorization. J. Artif. Intell. Soft Comput. Res. **5**(4), 231–238 (2015)
2. Bartczuk, Ł., Łapa, K., Koprinkova-Hristova, P.: A new method for generating of fuzzy rules for the nonlinear modelling based on semantic genetic programming. In: Rutkowski, L., Korytkowski, M., Scherer, R., Tadeusiewicz, R., Zadeh, L.A., Zurada, J.M. (eds.) ICAISC 2016. LNCS (LNAI), vol. 9693, pp. 262–278. Springer, Cham (2016). https://doi.org/10.1007/978-3-319-39384-1_23
3. Chen, M., Ludwig, S.A.: Particle swarm optimization based fuzzy clustering approach to identify optimal number of clusters. J. Artif. Intell. Soft Comput. Res. **4**(1), 43–56 (2014)
4. Cpałka, K., Łapa, K., Przybył, A.: A new approach to design of control systems using genetic programming. Inf. Technol. Control **44**(4), 433–442 (2015)
5. Cpałka, K., Rebrova, O., Nowicki, R., Rutkowski, L.: On design of flexible neuro-fuzzy systems for nonlinear modelling. Int. J. Gen. Syst. **42**(6), 706–720 (2013)
6. Dziwiński, P., Avedyan, E.D.: A new method of the intelligent modeling of the nonlinear dynamic objects with fuzzy detection of the operating points. In: Rutkowski, L., Korytkowski, M., Scherer, R., Tadeusiewicz, R., Zadeh, L.A., Zurada, J.M. (eds.) ICAISC 2016. LNCS (LNAI), vol. 9693, pp. 293–305. Springer, Cham (2016). https://doi.org/10.1007/978-3-319-39384-1_25
7. Dziwiński, P., Bartczuk, Ł., Tingwen, H.: A method for non-linear modelling based on the capabilities of PSO and GA algorithms. In: Rutkowski, L., Korytkowski,

M., Scherer, R., Tadeusiewicz, R., Zadeh, L.A., Zurada, J.M. (eds.) ICAISC 2017. LNCS (LNAI), vol. 10246, pp. 221–232. Springer, Cham (2017). https://doi.org/10.1007/978-3-319-59060-8_21

8. El-Samak, A.F., Ashour, W.: Optimization of traveling salesman problem using affinity propagation clustering and genetic algorithm. J. Artif. Intell. Soft Comput. Res. **5**(4), 239–245 (2015)

9. Khan, N.A., Shaikh, A.: A smart amalgamation of spectral neural algorithm for nonlinear Lane-Emden equations with simulated annealing. J. Artif. Intelli. Soft Comput. Res. **7**(3), 215–224 (2017)

10. Korytkowski, M., Scherer, R., Rutkowski, L.: On combining backpropagation with boosting. In: Proceedings of the 2006 International Joint Conference on Neural Networks, IEEE World Congress on Computational Intelligence, pp. 1274–1277 (2006)

11. Łapa, K., Cpałka, K.: On the application of a hybrid genetic-firework algorithm for controllers structure and parameters selection. In: Borzemski, L., Grzech, A., Świątek, J., Wilimowska, Z. (eds.) Information Systems Architecture and Technology: Proceedings of 36th International Conference on Information Systems Architecture and Technology – ISAT 2015 – Part I. AISC, vol. 429, pp. 111–123. Springer, Cham (2016). https://doi.org/10.1007/978-3-319-28555-9_10

12. Łapa, K., Cpałka, K., Galushkin, A.I.: A new interpretability criteria for neuro-fuzzy systems for nonlinear classification. In: Rutkowski, L., Korytkowski, M., Scherer, R., Tadeusiewicz, R., Zadeh, L.A., Zurada, J.M. (eds.) ICAISC 2015. LNCS (LNAI), vol. 9119, pp. 448–468. Springer, Cham (2015). https://doi.org/10.1007/978-3-319-19324-3_41

13. Łapa, K., Szczypta, J., Saito, T.: Aspects of evolutionary construction of new flexible PID-fuzzy controller. In: Rutkowski, L., Korytkowski, M., Scherer, R., Tadeusiewicz, R., Zadeh, L.A., Zurada, J.M. (eds.) ICAISC 2016. LNCS (LNAI), vol. 9692, pp. 450–464. Springer, Cham (2016). https://doi.org/10.1007/978-3-319-39378-0_39

14. Szczypta, J., Łapa, K., Shao, Z.: Aspects of the selection of the structure and parameters of controllers using selected population based algorithms. In: Rutkowski, L., Korytkowski, M., Scherer, R., Tadeusiewicz, R., Zadeh, L.A., Zurada, J.M. (eds.) ICAISC 2014. LNCS (LNAI), vol. 8467, pp. 440–454. Springer, Cham (2014). https://doi.org/10.1007/978-3-319-07173-2_38

15. Liu, H., Gegov, A., Cocea, M.: Rule based networks: an efficient and interpretable representation of computational models. J. Artif. Intell. Soft Comput. Res. **7**(2), 111–123 (2017)

16. Ludwig, S.A.: Repulsive self-adaptive acceleration particle swarm optimization approach. J. Artif. Intell. Soft Comput. Res. **4**(3), 189–204 (2014)

17. Mendonca, L.F.: Decision tree search methods in fuzzy modeling and classification. Int. J. Approx. Reason. **44**, 106–123 (2007)

18. Arain, M.A., Hultmann Ayala, H.V., Ansari, M.A.: Nonlinear system identification using neural network. In: Chowdhry, B.S., Shaikh, F.K., Hussain, D.M.A., Uqaili, M.A. (eds.) IMTIC 2012. CCIS, vol. 281, pp. 122–131. Springer, Heidelberg (2012). https://doi.org/10.1007/978-3-642-28962-0_13

19. Prasad, M., Liu, Y.-T., Li, D.-L., Lin, C.-T., Shah, R.R., Kaiwartya, O.P.: A new mechanism for data visualization with TSK-type preprocessed collaborative fuzzy rule based system. J. Artif. Intelli. Soft Comput. Res. **7**(1), 33–46 (2017)

20. Przybył, A., Cpałka, K.: A new method to construct of interpretable models of dynamic systems. In: Rutkowski, L., Korytkowski, M., Scherer, R., Tadeusiewicz,

R., Zadeh, L.A., Zurada, J.M. (eds.) ICAISC 2012. LNCS (LNAI), vol. 7268, pp. 697–705. Springer, Heidelberg (2012). https://doi.org/10.1007/978-3-642-29350-4_82

21. Rotar, C., Iantovics, L.B.: Directed evolution - a new metaheuristc for optimization. J. Artif. Intelli. Soft Comput. Res. **7**(3), 183–200 (2017)

22. Rutkowski, L.: A general approach for nonparametric fitting of functions and their derivatives with applications to linear circuits identification. IEEE Trans. Circuits Syst. **33**(8), 812–818 (1986)

23. Rutkowski, L.: Non-parametric learning algorithms in time-varying environments. Sig. Process. **182**, 129–137 (1989)

24. Rutkowski, L.: Adaptive probabilistic neural networks for pattern classification in time-varying environment. IEEE Trans. Neural Netw. **15**(4), 811–827 (2004)

25. Rutkowski, L., Cpałka, K.: Compromise approach to neuro-fuzzy systems. In: Proceedings of the 2nd Euro-International Symposium on Computation Intelligence. Frontiers in Artificial Intelligence and Applications, vol. 76, pp. 85–90 (2002)

26. Tambouratzis, T., Souliou, D., Chalikias, M., Gregoriades, A.: Maximising accuracy and efficiency of traffic accident prediction combining information mining with computational intelligence approaches and decision trees. J. Artif. Intell. Soft Comput. Res. **4**(1), 31–42 (2014)

27. Liu, X., Meng, J., Ge, J.: A method research on nonlinear system identification based on neural network. In: Zhu, R., Ma, Y. (eds.) Information Engineering and Applications. LNEE, vol. 154, pp. 1444–1449. Springer, London (2012). https://doi.org/10.1007/978-1-4471-2386-6_193

28. Yang, S., Sato, Y.: Swarm intelligence algorithm based on competitive predators with dynamic virtual teams. J. Artif. Intell. Soft Comput. Res. **7**(2), 87–101 (2017)

29. Zalasiński, M.: New algorithm for on-line signature verification using characteristic global features. In: Wilimowska, Z., Borzemski, L., Grzech, A., Świątek, J. (eds.) Information Systems Architecture and Technology: Proceedings of 36th International Conference on Information Systems Architecture and Technology – ISAT 2015 – Part IV. AISC, vol. 432, pp. 137–146. Springer, Cham (2016). https://doi.org/10.1007/978-3-319-28567-2_12

30. Zalasiński, M., Cpałka, K., Er, M.J.: A new method for the dynamic signature verification based on the stable partitions of the signature. In: Rutkowski, L., Korytkowski, M., Scherer, R., Tadeusiewicz, R., Zadeh, L.A., Zurada, J.M. (eds.) ICAISC 2015. LNCS (LNAI), vol. 9120, pp. 161–174. Springer, Cham (2015). https://doi.org/10.1007/978-3-319-19369-4_16

31. Zalasiński, M., Cpałka, K., Er, M.J.: New method for dynamic signature verification using hybrid partitioning. In: Rutkowski, L., Korytkowski, M., Scherer, R., Tadeusiewicz, R., Zadeh, L.A., Zurada, J.M. (eds.) ICAISC 2014. LNCS (LNAI), vol. 8468, pp. 216–230. Springer, Cham (2014). https://doi.org/10.1007/978-3-319-07176-3_20

32. Zalasiński, M., Cpałka, K.: New algorithm for on-line signature verification using characteristic hybrid partitions. In: Wilimowska, Z., Borzemski, L., Grzech, A., Świątek, J. (eds.) Information Systems Architecture and Technology: Proceedings of 36th International Conference on Information Systems Architecture and Technology – ISAT 2015 – Part IV. AISC, vol. 432, pp. 147–157. Springer, Cham (2016). https://doi.org/10.1007/978-3-319-28567-2_13

33. Zalasiński, M., Cpałka, K., Hayashi, Y.: New fast algorithm for the dynamic signature verification using global features values. In: Rutkowski, L., Korytkowski, M., Scherer, R., Tadeusiewicz, R., Zadeh, L.A., Zurada, J.M. (eds.) ICAISC 2015.

LNCS (LNAI), vol. 9120, pp. 175–188. Springer, Cham (2015). https://doi.org/10. 1007/978-3-319-19369-4_17

34. Zalasiński, M., Cpałka, K., Hayashi, Y.: A new approach to the dynamic signature verification aimed at minimizing the number of global features. In: Rutkowski, L., Korytkowski, M., Scherer, R., Tadeusiewicz, R., Zadeh, L.A., Zurada, J.M. (eds.) ICAISC 2016. LNCS (LNAI), vol. 9693, pp. 218–231. Springer, Cham (2016). https://doi.org/10.1007/978-3-319-39384-1_20

35. Zalasiński, M., Cpałka, K., Hayashi, Y.: A method for genetic selection of the most characteristic descriptors of the dynamic signature. In: Rutkowski, L., Korytkowski, M., Scherer, R., Tadeusiewicz, R., Zadeh, L.A., Zurada, J.M. (eds.) ICAISC 2017. LNCS (LNAI), vol. 10245, pp. 747–760. Springer, Cham (2017). https://doi.org/10.1007/978-3-319-59063-9_67

36. Zalasiński, M., Cpałka, K., Rakus-Andersson, E.: An idea of the dynamic signature verification based on a hybrid approach. In: Rutkowski, L., Korytkowski, M., Scherer, R., Tadeusiewicz, R., Zadeh, L.A., Zurada, J.M. (eds.) ICAISC 2016. LNCS (LNAI), vol. 9693, pp. 232–246. Springer, Cham (2016). https://doi.org/10. 1007/978-3-319-39384-1_21

37. Łapa, K., Cpałka, K., Wang, L.: New method for design of fuzzy systems for nonlinear modelling using different criteria of interpretability. In: Rutkowski, L., Korytkowski, M., Scherer, R., Tadeusiewicz, R., Zadeh, L.A., Zurada, J.M. (eds.) ICAISC 2014. LNCS (LNAI), vol. 8467, pp. 217–232. Springer, Cham (2014). https://doi.org/10.1007/978-3-319-07173-2_20

The Hybrid Plan Controller Construction for Trajectories in Sobolev Space

Krystian Jobczyk[1,2(✉)] and Antoni Ligęza[2]

[1] University of Caen Normandy, Caen, France
krystian_jobczyk@op.pl
[2] AGH University of Science and Technology, Kraków, Poland
jobczyk@agh.edu.pl

Abstract. This paper proposes a new integrated approach to the hybrid plan controller construction. It forms a synergy of the logic-based approach in terms of LTL-description and automata of Büchi with the integral-based approach. It is shown that the integral-based complementation may be naturally exploited in detection of the robot trajectories by the appropriate control functions.

1 Introduction

The plan controling forms an important complementation of planning and scheduling. Generally speaking, it consists in detecting different discrepancies between initial requirements and their real performing by robots, satellites, etc. The plan controling forms a multi-stage procedure. One of its steps usually consists in the appropriate description of robot behavior or initial requirements in terms of a given formal language. Linear Temporal Logic (LTL) – introduced to computer science in 1977 by Pnueli in [26] – is especially useful in it. Among other description languages one can indicate Motion Description Language in [13] and the so-called planning languages: PDDL and its extensions (commonly denoted by PDDL+) – developed by Fox's school in such papers as: [7–10]. Finally, there is a relatively broad class of action description languages – see: [23–25].

The initial situation description may be encoded later by the appropriate automaton – usually by the so-called Büchi automata. Despite of the fact that Büchi automata were already introduced in 1962 in [1–3] and exploited for translation of LTL-formulae in [11, 27, 28] – the construction of the hybrid plan controller automaton has been made relatively recently in [5, 6]. A new preferential extension of Büchi automon has been proposed in [15]. Meanwhile, a purely logical background for the current analysis (with a brief evaluation of different logical systems) were proposed in [14, 16–19].

1.1 Motivation of Current Analyses

Unfortunately, these original approaches suffer from the following difficulties:

© Springer International Publishing AG, part of Springer Nature 2018
L. Rutkowski et al. (Eds.): ICAISC 2018, LNAI 10842, pp. 532–543, 2018.
https://doi.org/10.1007/978-3-319-91262-2_47

D1 They discuss this issue rather in a form of extended outlines and without (often needed) technical details.

D2 Secondly, these constructions do not refer to metalogical restrictions of LTL and HS – such as the proved by Maximova in [21] and by Montanari in [22] with respect to encoding HS with $A\bar{A}B\bar{B}$-operators by formulas of ω-regular languages.

D3 The next, the mutual relationships between relational semantics for LTL and Büchi automata is still unclear.

D4 Finally, the purely logic-based approach to the controller construction requires some analytic complementation. In particular, it should be explained how to predict the robot trajectories by control functions.

Although it seems that difficulties D1, D2 and D3 have been partially overcome by the proposal from [15], D4 has not ye been discussed and overcome.

2 The Objectives of the Paper and Its Organisation

According to these difficulties, above detected, the main objective of the paper is to propose, at first, an integral-based complementation of the logic-based approach to the hybrid plan controller construction.

This complementation may be materialized if only the robot trajectories and the control functions will be considered as functions in the so-called Sobolev function space. Thus, this paper has an additional goal in a form of the task to redefine the robot environment as a Sobolev space and the robot trajectories and control functions by smooth function in this space.

The rest of the paper is organised as follows. In Sect. 2, a terminological background of the paper analysis is put forward. In Sect. 3, an idea of the logic-based construction for the plan controller is repeated. Section 4 introduces the integral-based approach to the controller construction as a support for the logic-based one. This complementation is presented here in a formal, conceptual way. Section 5 elucidates a computational side of the approach.

3 Preliminaries

The formal definition of a Sobolev space requires the definitions of a normed space and of a Banach space. (The basis definition of Legesque integrability may be easily found, for example, in [12]). Thus, let us assume that X is a given vector (linear) space. Each vector space is defined over a scalar field, say K. This fact is denoted by $X(K)$. The usual scalar fields are: the field $\mathbb{R}$ of real numbers or $\mathbb{C}$, or the field $\mathbb{C}$ of complex numbers. Assume now that $X(\mathbb{R})$ is given and introduce a new function $\| \bullet \| : X(\mathbb{R}) \mapsto [0, \infty)$ that respects the following conditions:

1      $\|x\| = 0 \iff x = 0.$
2      $\|\alpha x\| = |\alpha|\|x\|$, for $\alpha \in \mathbb{R}.$
3      $\|x + y\| \leq \|x\| + \|y\|.$

This function is to be called *a norm* and the whole space $(X(\mathbb{R}), |\bullet|)$ forms a *normed space*. A *Banach space* is such a *normed vector space* X, which is complete with respect to that norm, that is to say, each Cauchy sequence $\{x_n\}$ in X converges to an element x in X, formally:

$$\|x_n - x\|_X \to 0. \tag{1}$$

A distinguished class of Banach spaces is a function space $\mathbf{L}^p(\mathbb{R})$ of Lebesque integrable functions, for a parameter $1 \le p < \infty$. In general, this space is usually equipped by the norm $\|f\| = \left(\int |f|^p dx \right)^{\frac{1}{p}}$, $1 < p < \infty$.

In essence, we intend to consider both trajectories and control functions as functions with 'predictable diagrams', thus we consider them as smooth functions. Each smooth function must contain derivative, which additionally is continuous. These functions are said to be of differentiability class C^1. One can naturally expand this definition up to C^k, for a natural k. (If $f \in C^k$, then it has continuous derivatives up to k). Finally, let C_c^∞ denotes a class of infinitely differentiable functions.

Let us assume that $\phi \in C_c^\infty$ and $\phi : U \to \mathbb{R}$ forms a compact support-function in U^1. If a function $u \in C^1(U)$ and $\phi \in C_c^\infty$, then the integration by parts yields that:

$$\int_U u\phi_{x_i} dx = - \int_U u_{x_i}\phi dx, (i = 1, 2 \ldots, n), \tag{2}$$

because integrals disappears on a boundary ∂U – as ϕ has a compact support in U. Generally, if $u \in C^k(U)$, for a natural number k, $\phi \in C_c^\infty$ is as above and $\alpha = (\alpha_1, \ldots, \alpha^n)$ is such a multi-index that $|\alpha| = k$, then

$$\int_U u D^\alpha \phi dx = (-1)^\alpha \int_U D^\alpha u\phi dx, \text{ where } D^\alpha \phi := \frac{\partial^\alpha \phi}{\partial x_1^{\alpha_1}, \ldots \partial x_n^{\alpha_n}}. \tag{3}$$

In essence, (3) introduces the so-called *weak α-derivative* of u. Indeed, we say that a function $v = D^\alpha u$ is *a weak α-derivative of u*, if only (3) holds for $D^\alpha \phi$ as defined by (4).

Example 1. *Consider the following two functions:*

$$u(x) = \begin{cases} 2 + x, & \text{for } -2 < x < 0, \\ 100, & \text{for } x = 0, \\ 2 - x, & \text{for } 0 < x < 2, \\ 0, & \text{else} \end{cases} \quad \text{and } v(x) = \begin{cases} 1, & \text{for } -2 < x < 0, \\ -1, & \text{for } 0 < x < 2, \\ 0, & \text{else.} \end{cases}$$

Obviously, $u(x)$ is not continuous in 0, so it cannot be differentiable in this point, but $v(x)$ is the weak derivative for $u(x)$.

The above definitions of the weak derivative elucidate a sens to consider Sobolev spaces. In fact, this general function space contain functions, which are

---

[1] The role of compact support consists in a fact that integrals (coefficients) disappears in a neighborhood of its boundary, what simplifies the computations.

not so ideal as we wish, but they are not so wrong to avoid them. In other words, the functions of Sobolev spaces are more realistic, what also motivates us to represent robot trajectories and control function by functions from this space. It allows us to define the Sobolev space in a formal way.

Definition 1. *Let* $1 \leq p \leq \infty$ *andf* k *be a natural number (or equal to 0) and* $U \subset \mathcal{R}^\backslash$ *is an open set. The Sobolev space* $W^{k,p}(U)$ *consists of all such locally integrable functions* $u : U \to \mathcal{R}s$, *that for each multi-index* α *with of a length* $\leq k$ *the following conditions hold:*

- *a weak derivative* $D^\alpha u$ *exists and*
- $D^\alpha u \in L^p(U)$ *(is Lebesque integrable in p).*

Example 2. *If* $k = 0$, *then* $W^{0,2}(U) = L^2(U)$. *Generally,* $W^{k,2}$ *form Hilbert spaces.*

The norm in Sobolev space forms a slight modification of the norm definition for usual Banach spaces– as we should consider the sums (over $\alpha \leq k$) from the appropriate integrals for partial derivative functions.

Definition 2. $\|u\|_{W^{k,p}(U)} = \begin{cases} \left(\sum_{|\alpha| \leq k} \int |D^\alpha u|^p dx \right)^{1/p}, & (1 \leq p \leq \infty), \\ \sum_{|\alpha| \leq k} \sup_U |D^\alpha u|, & (p = \infty). \end{cases}$

4 The Logic-Based Side of the Hybrid Plan Controller Contruction

Before we exploit the terminological background in next part of the aper, let us repeat the main steps of to the logic-based approach to the hybrid plan controller construction. Let us begin with a formal specification of robot environment.

4.1 The Stages of the Construction

Therefore, let us assume that E is a polygonal environment of robot motion operations. All possible admitted holes of E have to be enclosed by a single polygonal chain. The motion of robot may be rendered by the clauses:

$$x(t) \in E \subseteq R^2, u(t) \in U \subseteq R^2, u(t) \cap x(t) \neq \emptyset \tag{4}$$

where $x(t)$ is a trajectory of robot's motion (position of a robot in a time t) in E and $u(t)$ is a control time-dependent function (called also a control input). Non-emptiness of intersection $u(t) \cap x(t)$ ensures that a controller detects the robot's trajectory. In such a framework, an exact chronology of the construction steps in a desired controller construction looks as follows.

step1 At first, the environment E is triangulated.
step2 Secondly, we consider some transition system FTS to describe a basic dynamism of E.

step3 The next, we specify E in terms of LTL (ϕ-formula) and of some subsystem of HS logic.

step4 In this step, we transform FTS to the appropriate Büchi automaton $\mathcal{A}_{FTS}$ for it. The similar automaton $\mathcal{A}_{LTL,HS}$ is constructed for representation of a specification of E (with a chosen point x_0) in terms of the considered temporal logic.

step5 Finally, having these automaton, we construct some product automaton $\mathcal{A}$ to 'reconcile' the activity of both automata.

This idea may be represented by the appropriate algorithm – as a specified version of algorithm from [5] (results are on the left side):

Algorithm. The Hybrid Controller Construction

===

Procedure: CONTROLLER(E, ϕ)

1. $\triangle \leftarrow Triangulate(E)$
2. $FTS \leftarrow TriangulationToFTS(\triangle)$
3. $\mathcal{A}_{FTS} \leftarrow FTS\ to\ Buchi\ Automaton$
4. $\mathcal{A}_{LTL,HS^D} \leftarrow LTL \cup HS^D to\ Buchi\ Automaton$
5. $\mathcal{A} \leftarrow Product\mathcal{A}_{FTS}, \mathcal{A}_{LTL,HS^D}$
6. **return:** $Controller(\mathcal{A}, \triangle, \phi)$

End procedure

$\triangle$ represents an effect of E-trangulation, FTS is a finite transition system, modeled o a base of E as the robot motion environment. (The possible moves of robot in E determine transitions in FTS). LTL$\cup$ HS^D denotes a formal description if LTL enriched by Halpern-Shoham logic (HS). $\mathcal{A}_{FTS}$ denotes a Büchi automaton for FTS and $\mathcal{A}_{LTL,HS}$ – the corresponding Büchi automaton for LTL$\cup$HS^D-description.

4.2   An Exemplification

To exemplify this construction – let us consider the following example.

Example 3. *Consider a robot, say R, in some polygonal environment with 4 rooms as depicted on a picture below. Assume that R performs a task to dislocate a black block A from a room 1 to the room 4 and put in on a block B there and the planned (preferred) move trajectory leads from the room 1 by a neighborhood of the room 3 to the room 4 (the blue line on a picture). Let also assume that our robot exchanged this trajectory for another one (marked by a red line).*

According to the appropriate stage of construction of the hybrid path plan controller, we should materialize the following two steps:

1. representing the real situation of the robot task performing by the appropriate FTS.
2. encoding the required situation of the robot task performing in terms of $\mathcal{L}$(LTL$\cup$HS).

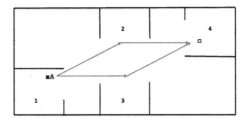

Fig. 1. The polygonal environment of the robot motion with 4 rooms. The blue broken line illustrates the planned trajectory of the robot move from a room no. 1 to the room no. 4. The red one illustrates the deviated trajectory of the robot move. (Color figure online)

These two steps forms a basis for the construction of the appropriate automata of Büchi and the unified product automaton. Let us assume, however, that it is more convenient to represent FTS in terms of $\mathcal{L}(\text{LTL}\cup\text{HS})$. In result, we may obtain, for example, the following juxtaposition.

plan of the robot	the real plan performing
$Take(A)$	$Take(A)$
$Move(R_1^A, R_3) \vee Move(R_1^A, R_4)$	$Move(R_1^A, R_2)$
$HOLDS(R_3^R) \rightarrow Move(R_3, R_4)$	$Move(R_2^A, R_4)$
$Put(A)$	$Put(A)$
$HOLDS(R_4^A)$	$HOLDS(R_4^A)$
behavioral rule	**behavioral rule**
$G(take(A) \rightarrow \langle L \rangle go(R_3))$	?

According to the next stages of the controller construction algorithm, one should built up the corresponding fragment of Büchi automaton (for each LTL∪HS-formula from the table). For example, it leads to the following automaton:

Finally – according to the last step of the construction algorithm – these two incoherent situations may be finally may be rendered in PROLOG as follows:

```
arc(0, stateNotR1). arc(0, stateR1).arc(0, stateR2). arc(0, stateNotR2).
arc(stateNotR1, stateMoveR1R2). arc(stateR1,stateMoveR1R2).
arc(state R1, stateNotMoveR1R2). arc(stateR2, stateMoveR1R2).
arc(stateNotR2P, stateNotMoveR1R2) etc.
```

And corresponding part of $\mathcal{A}^{Pref}$-automaton:
```
arc(0, stateNotR1). arc(0, stateR1). arc(0, stateR3). arc(0,stateNotR3).
arc(stateNotR1, stateMoveR1R3). arc(stateR1, stateMoveR1R3).
arc(state R1, NotMoveR1R3). arc(stateR2, stateMoveR1R3).
arc(stateNotR3P,stateNotMoveR1R3) etc.
```

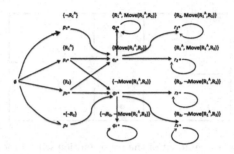

Fig. 2. Fragment of the Büchi automaton for the real task performing for the closure of LTL-formula $Move(R_1^A, R_2)$. The incoherences are marked by blue color. (Color figure online)

The detected differences on a level of PROLOG-description are marked by red color.

5   Integral-Based Approach as a Support of the Logical Approach – A Formal Depiction

In essence, an idea to represent both the robot trajectories and control functions for their detection might be not so clear at the first appearance. In order to approximate it – let us observe the following facts.

1. The robot environment E may be naturally considered as a *metric vector space* – as E is measurable in a metric sense. In fact, each potential move of the robot in E forms a vector in this space and E alone may be assumed to be a subspace of metric space $(\mathbb{R}, d)$, where d is Euclidean, etc.
2. Even more – E may be seen as a unique *Banach space* if we agree to interpret E as a field of all possible trajectories and control functions. In fact – without losing of generality – one can assume that each such a robot trajectory $x(t)$ and each control function $u(t)$ is a Lebesgue integrable function defined over an interval $[a, b]$. All such Lebesgue-integrable functions form a Banach space $\left(\mathbf{L}^p[a, b], \| \bullet \| \right)$ with norms $\| \bullet \|$ defined as follows:

$$\|x\| = |x(t)| + \left(\int_{[a,b]} |x(t)| dt \right)^{\frac{1}{p}}, \|u\| = |u(t)| + \left(\int_{[a,b]} |u(t)| dt \right)^{\frac{1}{p}}, 0 < p \leq 1, \tag{5}$$

Briefly and formally:

$$E = \left\{ x(t)_i, u(t)_j \in \left(\mathbf{L}^p[a, b], \| \bullet \| \right) : x(t)_i \cap u(t)_j, \text{ for } i = j, i, j \in I, 0 < p \leq 1 \right\}. \tag{6}$$

Finally, this reasoning may be generalised for a multi-dimensional case of $\mathbb{R}^n$. Therefore, let us assume that $u(t) \in \mathbb{R}^n$, so $u(t) = u(t_1, \ldots, t_k)$. Since a robot move equations are usually involved in partial derivatives it is comfortable to postulate their differentiability up to α-degree. In addition, one may require the following condition:

C1 both partial derivatives $D^\alpha u(t)$ and $D^\alpha x(t)$ should belong to the same space.

Formally:

$$D^\alpha u(t) := \frac{\partial^\alpha u(t)}{\partial t_1, ... \partial t_n}, \; D^\alpha x(t) := \frac{\partial^\alpha x(t)}{\partial t_1 ... \partial t_n} \in \left(L^p[a,b], \| \bullet \| \right), \qquad (7)$$

In this case, E might be identified with the so-called *Sobolev's space*. For $1 \leq p < \infty$ and natural k we briefly and formally define E as follows.

$$E = \left\{ x(t)_i, u(t)_j \in \left(\mathbf{W}^{k,p}, \| \bullet \| \right) : x(t)_i \cap u(t)_j, \text{ for } i = j, i, j \in I \right\}. \qquad (8)$$

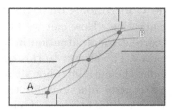

Fig. 3. Robot polygonal environment (with rooms A and B) as a Banach space of Lebesgue integrable trajectories (green lines) and control functions (red line). (Color figure online)

It seems that such a depiction of E has at least two advantages:

1. it delivers new *geometric features* of this environment and
2. it delivers new criteria of robot *controllability*.

We put aside a presentation of geometric features of the robot environment as they may be easily imaginable as a slight modification of Schwartz's and Minkowski's inequalities. We focus our attention on a concept of controllability in new terms.

Controllability in New Terms. Considering the robot environment E in a more abstract way – as a Sobolev space of trajectories and control inputs – has an additional advantage. Namely, it allows us to adopt different concepts of controllability. Some of them have recently been elaborated in [20]. They are as follows.

1. The system (1) is said to be *locally controlable* at z if for each $\epsilon > 0$ there exists $\rho > 0$ such that for all $v \in B(0, \rho)$ there exists a trajectory x for F, with $\|x - z\| \leq \epsilon$, satisfying $(x(0), x(1) + v) \in S$.

2. It is said to be *strongly locally controlable* [1] at z if there exist $a > 0$ and $\rho > 0$ such that for all u and v in $B(0, \rho)$ there exists a trajectory x for F, with $\|x - z\| \leq a(|u| + |v|)$, satisfying $(x(0) + u, x(1) + v) \in S$.

Here $B(0, \rho)$ denotes the closed ball in H centered at 0 and of radius ρ.

Independently of their precision, they seem to lose an intuition regarding to the concept of Controllability as based on some convergence in a metric or a norm-based sense. Therefore, we are willing to adopt another definition of this concept. It follows from a general property of global approximation by smooth functions in Sobolev spaces.

Definition 3 *(Controllability). Assume that a set $U \subset \mathbb{R}^n$ is bounded and it has a boundary ∂U of a class C^1 (functions on ∂U belong to C^1). Let us also assume that $u \in W^{k,p}(U)$ for some $1 \leq p < \infty$. Then there exist such functions $u_m \in C^\infty(U)$ (they are infinitely differentiable) that*

$$u_m \to u \text{ in } W^{k,p}(U). \tag{9}$$

A New Algorithm of the Controller Construction. It remains to answer to the following intriguing question: 'How can we know about the robot plan performing and about the current robot situation in order to encode it in LTL and the appropriate automata'? The answer to this question seems to be clear. A carrier of information about the current robot situations and a real state of its plan performing are just control functions for the robot trajectories – as differentiable functions of the ideal Sobolev's or Hoelder's space.

Each LTL- or automata-based encoding the current situation must be proceeded by an earlier extraction of knowledge about the current robot situation. In a consequence, the general algorithm of the hybrid controller construction from paragraph 5.4 (pp. 127–128) may be enlarged to the following one:

Algorithm. The Hybrid Controller Construction
===

Procedure: CONTROLLER(E, ϕ)

1. $\triangle \leftarrow Triangulate(E)$
2. $FTS \leftarrow TriangulationToFTS(\triangle)$
3. $\mathcal{A}_{FTS} \leftarrow FTS$ to Buchi Automaton
4. $\mathcal{L}TL \cup HS \leftarrow$ Detect Trajectories by Controll Functions
5. $\mathcal{A}_{LTL,HS^D} \leftarrow LTL \cup HS^D$ to Buchi Automaton
6. $\mathcal{A} \leftarrow Product\mathcal{A}_{FTS}, \mathcal{A}_{LTL,HS^D}$
7. **return:** $Controller(\mathcal{A}, \triangle, \phi)$

End procedure

The new 'key' point of this newly extended algorithm is the point 4. It encodes the procedural move from detecting the robot situation (trajectories) to its representing in terms of $LTL \cup HS$.

6 Integral-Based Approach as a Support of the Logical Approach – in a Computational Depiction

In last section, a formal (conceptual) approach to the integral-based approach to the plan controller construction (as a support for the purely logical one) has been proposed. However, it still remains a question, how the computational side of this approach looks in details. In order to describe how this controllability criterion may be exploited in a concrete computational situation, let us consider the following problem.

Problem 1. *Assume that the robot move trajectory is given by*

$$u(t) = |t|^{-\alpha}, \tag{10}$$

and considered in the unit open ball $B(0,1) \subset \mathbb{R}^5$, for $|\alpha| = 3$ and $p = 2$. Is the trajectory $u(t)$ controlable?

Solution. In order to verify whether a controllability in sense of (17) holds, one should check whether $u(t)$ belongs to $W^{1,p}(B(0,1))$. To make it, we find the class of parameters, for which $u(t) \in W^{1,p}$ holds, because it implies (17)-property. Let us generalise that case by the assumption that $B(0,1) \subset \mathbb{R}^n$. Meanwhile, a condition $u(t) \in W^{1,p}$ would mean that the the following condition for weak derivatives would hold for $B(0,1)$:

$$\int_{B(0,1)} u\phi_{t_i} dx = -\int_{B(0,1)} u_{t_i} \phi dx, (i = 1, 2, \ldots, n), \tag{11}$$

for a support function ϕ, which disappears on a boundary $\partial B(0,1)$. In order to show this, not that $u(t)$ is a smooth function, for each $t \neq 0$ and

$$u_{t_i} = \frac{-\alpha t_i}{|t|^{\alpha+2}}, \tag{12}$$

thus

$$|Du(t)| = \frac{-\alpha t_i}{|t|^{\alpha+1}}, \tag{13}$$

Let $\psi \in C_c^\infty$ be a support function and establish $\epsilon > 0$. Then we have

$$\int_{U(B(0,\epsilon))} u\psi_{t_i} dt = -\int_{U(B(0,\epsilon))} u_{t_i} \psi dx + \int u\phi v^i dS, \tag{14}$$

where $\mathbf{v} = (v^1, \ldots, v^n)$ denotes an inner normal vector on $\partial B(0, \epsilon)$. If $\alpha + 1 < n$ then $|Du(t)| \in L^1(B(0,1))$ (is integrable). Let us check a value of ψ on a boundary $\partial B(0,1)$. Since

$$\left| \int_{\partial B(0,\epsilon)} u\psi v^i dS \right| \leq \|\psi\|_{L^\infty} \int_{\partial B(0,1)} \epsilon^{-\alpha} dS \leq C\epsilon^{n-1-\alpha} \to 0, \tag{15}$$

thus

$$\int_{B(0,1)} u\phi_{t_i} dx = -\int_{B(0,1)} u_{t_i}\phi dx, \tag{16}$$

for all $\psi \in C_c^\infty(B(0,1))$, if only $0 \leq \alpha < n-1$. In addition, $|Du(t)| = \frac{\alpha}{|t|^{\alpha+1}} \in L^p$ if and only if $(\alpha + 1)p < n$. Hence, $u \in W^{1,p} \iff \alpha < \frac{n-p}{p}$. Let us return to our case. Since $\frac{n-p}{p} = \frac{5-2}{2} = \frac{3}{2}$ in our case, thus $\frac{n-p}{p} < \alpha = 3$, so the robot motion is not controlable in our case.

7   Conclusion

It has already been shown how the logic-based approach to the hybrid plan controller construction may be complemented by the integral-based one. It appears that the analytic (integral-based) approach may be naturally exploited to detect the trajectories of the robot motion in its polygonal environment. Obviously, this synergy proposal requires deeper analysis and extensions (for example: for other types of of function spaces, such as Orlicz or Hölder ones). It may be a promising subject for further research.

References

1. Antonniotti, M., Mishra, B.: Discrete event models+ temporal logic = supervisory controller: automatic synthesis of locomotion controllers. In: Proceedings of IEEE International Conference on Robotics and Automation (1999)
2. Bacchus, F., Kabanza, F.: Using temporal logic to express search control knowledge for planning. Artif. Intell. **116**, 123–191 (2000)
3. Buchi, R.: On a Decision Method in Restricted Second-order Arithmetic. Stanford University Press, Stanford (1962)
4. Evans, L.: Partial Differential Equations. American Mathematical Society, Providence (1998)
5. Fainekos, G., Kress-gazit, H., Pappas, G.: Hybrid controllers for path planning: a temporal logic approach. In: Proceeding of the IEEE International Conference on Decision and Control, Sevilla, pp. 4885–4890, December 2005
6. Fainekos, G., Kress-gazit, H., Pappas, G.: Hybrid controllers for path planning: a temporal logic approach. In: Proceedings of the IEEE International Conference on Decision and Control, Sevilla, pp. 4885–4890 (2005)
7. Fox, M., Long, D.: PDDL+: planning with time and metric sources. Technical report, University of Durham (2001a)
8. Fox, M., Long, D.: PDDL2.1: an extension to PDDL for expressing temporal planning domains. Technical report, University of Durham (2001b)
9. Fox, M., Long, D.: The third international planning competition: temporal and metric planning. In: Preprints of The Sixth International Conference on Artificial Intelligence Planning and Scheduling, vol. 20, pp. 115–118 (2002)
10. Fox, M., Long, D.: An extension to PDDL for expressing temporal planning domains. J. Artif. Intell. Res. **20**, 61–124 (2003)

11. De Giacomo, G., Vardi, M.Y.: Automata-theoretic approach to planning for temporally extended goals. In: Biundo, S., Fox, M. (eds.) ECP 1999. LNCS (LNAI), vol. 1809, pp. 226–238. Springer, Heidelberg (2000). https://doi.org/10.1007/10720246_18
12. Hewitt, E., Stromberg, K.: Real and Abstract Analysis. Springer, Heidelberg (1965). https://doi.org/10.1007/978-3-662-29794-0
13. Hristu-Varsakelis, D., Egersted, M., Krishnaprasad, S.: On the complexity of the motion description language MDLe. In: Proceedings of the 42 IEEE Conference on Decision and Control, pp. 3360–3365, December 2003
14. Jobczyk, K., Ligeza, A.: Fuzzy-temporal approach to the handling of temporal interval relations and preferences. In: Proceeding of INISTA, pp. 1–8 (2015)
15. Jobczyk, K., Ligeza, A.: A general method of the hybrid controller construction for temporal planing with preferences. In: Proceeding of FedCSIS, pp. 61–70 (2016)
16. Jobczyk, K., Ligeza, A.: Multi-valued halpern-shoham logic for temporal allen's relations and preferences. In: Proceedings of the Annual International Conference of Fuzzy Systems (FuzzIEEE) (2016, to appear)
17. Jobczyk, K., Ligeza, A.: Systems of temporal logic for a use of engineering. toward a more practical approach. In: Stýskala, V., Kolosov, D., Snášel, V., Karakeyev, T., Abraham, A. (eds.) Intelligent Systems for Computer Modelling. AISC, vol. 423, pp. 147–157. Springer, Cham (2016). https://doi.org/10.1007/978-3-319-27644-1_14
18. Jobczyk, K., Ligeza, A., Kluza, K.: Selected temporal logic systems: an attempt at engineering evaluation. In: Rutkowski, L., Korytkowski, M., Scherer, R., Tadeusiewicz, R., Zadeh, L.A., Zurada, J.M. (eds.) ICAISC 2016. LNCS (LNAI), vol. 9692, pp. 219–229. Springer, Cham (2016). https://doi.org/10.1007/978-3-319-39378-0_20
19. Jobczyk, K., Ligęza, A., Bouzid, M., Karczmarczuk, J.: Comparative approach to the multi-valued logic construction for preferences. In: Rutkowski, L., Korytkowski, M., Scherer, R., Tadeusiewicz, R., Zadeh, L.A., Zurada, J.M. (eds.) ICAISC 2015. LNCS (LNAI), vol. 9119, pp. 172–183. Springer, Cham (2015). https://doi.org/10.1007/978-3-319-19324-3_16
20. Jourani, A.: Controllability and strong controllability of differential inclusions. Nonlinear Anal. Theory Methods Appl. **75**, 1374–1384 (2012)
21. Maximova, L.: Temporal logics with operator 'the next' do not have interpolation or beth property. Sibirskii Matematicheskii Zhurnal **32**(6), 109–113 (1991)
22. Montanari, A., Sala, P.: Interval logics and ωB-regular languages. In: Dediu, A.-H., Martín-Vide, C., Truthe, B. (eds.) LATA 2013. LNCS, vol. 7810, pp. 431–443. Springer, Heidelberg (2013). https://doi.org/10.1007/978-3-642-37064-9_38
23. Pednault, F.: Synthetizing plans that contain actions with contex-dependent effects. Comput. Intell. **4**(4), 356–372 (1988)
24. Pednault, F.: Exploring the middle ground between STRIPS and the situation calculus. In: Proceedings of the International Conference on Knowledge Representation and Reasoning (KR), vol. 4, no. 5, pp. 324–332 (1989)
25. Pednault, F.: ADL-and the state-transition model of action. J. Log. Comput. **4**(5), 467–512 (1994)
26. Pnueli, A.: The temporal logic of program focs. pp. 46–57 (1977)
27. Vardi, M., Wolper, P.: An automata-theoretic approach to automatic program verification. In: Proceedings of the 1st Symposium on Logic in Computer Science, pp. 322–331, June 1986
28. Vardi, M., Wolper, P.: Reasoning about infinite computations. Inf. Comput. **115**(1), 1–37 (1994)

Temporal Traveling Salesman Problem – in a Logic- and Graph Theory-Based Depiction

Krystian Jobczyk[1,2(✉)], Piotr Wiśniewski[2], and Antoni Ligęza[2]

[1] University of Caen Normandy, Caen, France
krystian.jobczyk@unicaen.fr
[2] AGH University of Science and Technology, Kraków, Poland
krystian_jobczyk@op.pl, jobczyk@agh.edu.pl

Abstract. In this paper, a new temporal extension of Traveling Sales-man Problem (TSP) – as an old optimization problem – is proposed. This proposal stems from a need to elucidate TSP not only as an optimization problem, but also as a potentially paradigmatic problem for the subject specification of temporal planning. This new Temporal Traveling Salesman Problem is described in two ways – in the graph-based depiction and in terms of logic to be interpreted later by the so-called fibred semantics.

1 Introduction

Traveling Salesman Problem (TSP) belongs to the class of the basis optimization problems in computer science and may be formulated in different ways and contexts. One of the most popular depictions of this problem is the following one:

> *Given a list of cities and the distances between each pair of cities, what is the shortest possible route that visits each city exactly once and returns to the origin city?.*

TSP has a relatively long chronology. It is commonly assumed that it was primary formulated c.a. 1800 year by the British mathematician W.R. Hamilton in terms of Hamiltonian cycles. Anyhow, in a mathematical depiction TSP was formulated in 1932 by Menger in [14] and developed by Heller in [7] and solved by Flood in [5] and in many other places. Since this early period of research on TSP, many researchers have been putting forward different solutions and approximations of this problems. In [12], an approximation algorithm for asymetric TSP was proposed, but Hassin and Rubinstein discussed the most optimal algorithm for some maximal version of TSP in [16]. Finally, new bounds for TSP have been recently found in [17].

From the perspective of the theory of computational complexity, TSP is known to be a paradigmatic example of NP-complete problem (nondeterministic

© Springer International Publishing AG, part of Springer Nature 2018
L. Rutkowski et al. (Eds.): ICAISC 2018, LNAI 10842, pp. 544–556, 2018.
https://doi.org/10.1007/978-3-319-91262-2_48

polynomial problem). It was shown by Papadimitriou in [15] that even if a distances between cities in a basis formulation of TSP are determined by euclidean metrics, the TSP remains NP-complete. In addition, even though this problem is computationally difficult, a large number of heuristics and exact methods to solve it are known. A distinguished role of TSP in computer science reflects itself in the fact that this problem was a subject of the separate monograph (see: [1]). It also is noteworthy that TSP was a foundation for further more advanced problems, such as the *Vehicle Routing Problem* – introduced in [3] and broadly discussed, for example in [2,18]. Some different aspects of TSP have recently been discussed in [8–11].

1.1 The Motivation

Unfortunately, majority of the approaches to TSP refers to a combinatorial, optimization-based side. The objectives of these all approaches is to elaborate a new, more efficient algorithm or a more ergonomic solution for TSP in a general depiction or for TSP in a restricted context. In addition, it seems that the main motivation factor to extend and modify TSP – such as in a case of Vehicle Routing Problem in [2,3,18] – was a rather practical and application-based character only.

Meanwhile, TSP may be also exploited as a practical motivation to develop different logical systems towards their capabilities to represent knowledge, temporal constraints and other determinants of temporal reasoning and acting, as preferences, obligations, etc. This task requires, however, a temporal version of TSP (TTSP) as a basis for further extensions.

The next, this TTSP, newly defined, should be represented in a chosen logical system (of temporal or modal logic) to be interpreted in the appropriate semantics associated to these logical systems. The so-called Halpern-Shoham logic, invented in [6] seems to be a convenient and rich system as it is capable of representing all Allen's relations between temporal intervals. Unfortunately, an ordinary Kripke frame seems to be insensitive to a complexity of the situation of an operating agent (Salesman). The so-called behavioral semantics of Fagin from [4] is more suitable for this task, but rather suitable for relatively simple capabilities of agents (such as recognizing an identity between intervals) – as recently shown in [13].

1.2 The Objectives and Organisation of the Paper

Againts this background the main objective of this paper is to propose:

1 At first – a temporal Traveling Salesman Problem (TTSP) in a graph theory-based depiction,
2 The next – a logical representation for (some extract of) TTSP with a distinguished deontic component (for Salesman's obligations) in terms of deontic Halpern-Shoham logic,

3 Finally – a fibred semantics for combined formulae of deontic Halpern-Shoham
   logic for representation of TTSP.

All these objectives determine a novelty of the paper in a comparison to
earlier approaches. This new TTSP is considered as an extension of the standard
Traveling Salesman Problem (TSP) – a classical optimization problem – over
time component. It is in fact composed as a specific mixture of *travel planning*
and *task performing* subject to temporal constraints.

The rest of the paper is organised as follows. In Sect. 2, a terminological background of analysis is put forward. Section 3 is devoted to the intuitive and the
formal graph theory-based depiction of TTSP. Section 4 contains a logical representation of some extract of TTSP with a distinguished deontic component in
terms of deontic Halpern-Shoham logic. In the same Sect. 4 a new interval-based
fibred semantics for a logical representation of TTSP is introduced. Section 5
contains closing remarks and describes a perspective of future research in this
area.

2   Terminological Background of the Analysis

Before we move to the proper body of the paper, we put forward a terminological
background of further analyses. More precisely, Halpern-Shoham logic will be
briefly characterized and its deontic Halpern-Shoham logic will be proposed in
a form of an outline.

Halpern-Shoham logic. Halpern-Shoham logic forms a modal representation
of Allen's temporal relations between intervals: "after" (or "meets"), ("later"),
"begins" (or "start"), ("during"), "end" and "overlap"; such that each of them
corresponds to the modal HS operators: $\langle A \rangle$ for "after", $\langle B \rangle$ for "begins", $\langle L \rangle$
for "later", etc. The syntax of HS entities ϕ is given by the grammar:

$$\phi := p \,|\, \neg\phi \,|\, \phi \wedge \phi \,|\, \langle X \rangle \phi \,|\, \langle \bar{X} \rangle \phi,$$

where p is a propositional variable and $\langle \bar{X} \rangle$ denotes a modal operator for the
inverse relation w. r. t. $X \in \{A, B, D, E, O, L\}$ being a set of Allen's relations.

The interval-based semantics for HS is based either on continuous or on
discrete intervals. If $\phi \in \mathcal{L}(\text{HS})$, M is an interval Kripke frame-based model, and
$I \in W$, then the satisfaction for the HS-operators is given as follows:

$$M, I \models \langle X \rangle \, \phi \text{ iff } \exists I' \text{ that } I \, X \, I' \text{ and } M, I' \models \phi.$$

Deontic Halpern-Shoham Logic. Let us begin with an initial observation
that Salesman's tasks (to deliver some goods to different cities) may be rendered
as his obligations or permissions. Meanwhile, these entities form a subject of
deontic logic. Hence, we should enrich HS by a deontic component to render
these obligations to a new system DeoHS. Let us specify, at first, a language of

the deontic component of DeoHS. For that reason, we consider two usual modal operators: $O_i(\phi)$ as a box-type operator for 'and agant i is obliged to (do, what is described by ϕ)' and the co-definable operator: $P_i(\phi)$ of a diamond-type for: 'it is permitted for agent i to (do, what is described by ϕ)'. Thus, it allows us to define.

Definition 1 *(Language of MVDeo). The language of MVDeo, $\mathcal{L}(MVDeo)$, is given by a grammar:*

$$\phi := p \,|\, \neg\phi \,|\, \phi \wedge \psi \,|\, O_i\phi \,|\, P_i\phi,$$

where $i \in \mathcal{A}$, and $\mathcal{A}$ is a non-empty set of agents.

One definition should be also adopted in the MVDeo-syntax as representing the co-definability of both operators. As usual, we should adopt (at least) axioms of propositional calculus and the appropriate form of K-axiom in syntax of our system. (Since we are willing to propose a basis system, we do not adopt any additional axioms.)

Def. 1: $O_i\phi \iff \neg P_i \neg\phi$ for each $\alpha \in G$.

1 axioms of Boolean propositional calculus,

2 $O_i(\phi \to \chi) \to (O_i\phi \to O_i\chi)$,   (axiom K)

As *inference rules* we adopt *Modus Ponens*, substitution and the necessitation rule for the O_i-operator: $\frac{\phi}{O_i^\alpha\phi}$. It allows us to introduce syntax of DeoHS as follows.

Definition 2 *(Language of DeoHS). Since we also intend to consider combined formulae of both the deontic and the temporal nature, the language of DeoHS is given by the following grammar.*

$$\phi := p \,|\, \neg\phi \ |\, \phi \wedge \psi | O_i\phi \ | P_i\phi \ | \langle X \rangle\phi \ | \langle \bar{X} \rangle\phi \,|\, [X]\phi \,|\, [\bar{X}]\phi,$$

$$O_i[X]\phi \,|O_i[\bar{X}]\phi \,|P_i[X]\phi \,|P_i[\bar{X}]\phi, |O_i\langle X\rangle\phi \,|O_i\langle \bar{X}\rangle\phi \,|P_i\langle X\rangle\phi \,|P_i\langle \bar{X}\rangle\phi,$$

$$[X]O_i\phi \,|[\bar{X}]O_i\phi \,|[X]P_i\phi \,|[\bar{X}]P_i\phi, |\langle X\rangle O_i\phi \,|\langle \bar{X}\rangle O_i\phi \,|\langle X\rangle P_i\phi \,|\langle \bar{X}\rangle P_i\phi,$$

where p is a propositional variable, $i \in \mathcal{A}$ and X is one of relations of Allen.

Simple formulae of both $\mathcal{L}$(DeoHS) are defined in ordinary Kripke semantics. Since satisfaction for formulae of $\mathcal{L}$(HS) has already been given, we only describe conditions of deontic modal operators.

Definition 3 *(Satisfaction). Let us assume that $\mathcal{M} = \langle S, \precsim_i, Lab \rangle$ is a Kripke frame-based model and the intervals $I, I^{'} \in S$ and Lab is a labeling function. Given a formula $\phi \in \mathcal{L}(DeoHS)$ with a set of propositions Prop we inductively define the fact that ϕ is satisfied in $\mathcal{M}$ and in an interval I (symb.$I \models \phi$) as follows:*

1 *for all $p \in$ Prop, we have $(\mathcal{M}, I) \models p$ iff $p \in Lab(I)$,*

2 $(\mathcal{M}, I) \models \neg\phi$ *iff it is not such that $(\mathcal{M}, I) \models \phi$,*

3 $(\mathcal{M}, I) \models \phi \wedge \psi$ *iff $(\mathcal{M}, I) \models \phi$ and $(\mathcal{M}, I) \models \psi$,*

4 $(\mathcal{M}, I) \models O_i\phi$, *where $i \in A$, iff for all $I \precsim_i I'$ we have $(\mathcal{M}, I') \models \phi$,*

5 $(\mathcal{M}, I) \models P_i\phi$, *where $i \in A$, iff there is I' such that $I \precsim_i I'$ and $(\mathcal{M}, I') \models \phi$.*

The key clauses in this definition are that one referring to the modal operators $O_i\phi$ and $P_i\phi$. These conditions assert that these modal formulae are satisfied in I of IBIS iff the atomic formula ϕ holds in all intervals (at least one interval for $P_i\phi$) accessible from this I via $\precsim_i$-relation as an accessibility relation in $\mathcal{M}$ (*resp.*).

The satisfaction condition for combined formulae will be put forward after introducing the fibred semantics in Sect. 4.

3   Temporal Traveling Selesman Problem in the Graph Theory-Based Depiction

In this section a basic, generic temporal planning problem to be referred to as *The Temporal Traveling Salesman Problem* (or TTSP, for short) will be introduced and described in detail. TTSP is an extension of the standard TSP – a classical optimization problem – over time component. It is in fact composed as a specific mixture of *travel planning* and *task performing* subject to temporal constraints. The formal definition of TTSP and its solution will be prefaced by some informal definition of this problem.

3.1   Informal Defining of TTSP

The *Traveling Salesman* is an agent, capable of traveling (e.g. from town to town) and performing tasks (e.g. deliver goods). In the basic statement there is exactly one such agent and many delivery points located at different nodes of a graph. Each delivery task is assumed to have some temporal extent – i.e. has some assigned length in time. For example, to unload a truck takes a certain interval of time. Moreover, there is given a network of nodes (locations, towns) represented by a graph. Each vertex is assigned a delivery task to be accomplished at the node. Each such task (e.g. a course) is identified by a unique name, and a start time and an end time. Both the start time and the end time are referring to the calendar/clock time and are rigid. For example, they can refer to the working hours of a delivery point. In order to accomplish a task, the agent must:

* arrive at the node *at or after* the start time, and
* stay at the node for a period necessary to complete the delivery task,
* leave it *before or at* the end time.

In a generalized version, there may be other requirements, both of hard and soft nature (e.g. Allen's relations, Fuzzy Temporal Relations).

Further, any travel from a node to another, directly connected node takes some amount of time. In the basic formulation, this is just an interval (a *floating* one). This means that the travel can be started at any time (e.g. the agent uses its own car). Now, a semi-informal depiction of the *The Temporal Traveling Salesman Problem* is as follows:

> Consider an agent – called later the *Salesman* – which moves over a given graph G such the following conditions are satisfied.
>
> 1. The agent starts from some predefined initial node.
> 2. He has to travel through all specified nodes, finishing at some arbitrary node (this may be the initial node, as in the classical version, or any other arbitrarily defined one).
> 3. At any node, the agent must accomplish the delivery task assigned to that node.
> 4. He must act according to the temporal constraints - the task must be accomplished within the operation interval.

The *solution* of the problem is given by a sequence of nodes to be visited such that:

A the sequence start with the predefined initial node,
B it covers all the required nodes (tasks),
C it ends at the required node (e.g. the same as the start node),
D at each node the temporal constraints are satisfied (Node Temporal Constraints),
E there is enough time to travel between any successive nodes (Travel Temporal Constraints).

Usually, it is assumed that the solution of TSP must be optimal (time or costs should be possibly minimal). It follows from the fact that TSP originally forms an optimization problem, as earlier mentioned. For the same reason, the solution of TTSP as a temporal extension of TSP should be also optimal in the same sense if consider TTSP as an optimization problem. However, we avoid this restriction as we are not interested in the optimization aspects of TTSP, but in a modeling and formal representation of this problem.

3.2 Formal Definition of TTSP

In order to formally define the TTSP one needs the following components:

1 a graph representing the delivery points (nodes) and links among them,
2 definition of temporal constraints at each node (Node Temporal Constraints or Task Temporal Constraints),
3 definition of global constraints w.r.t. the travel,
4 definition of global temporal constraints or performance index (e.g. the total time to cover or delivery requirements).

The definition of TTSP is presented below.

Definition 4 *The   Temporal   Traveling   Salesman   Problem (TTSP) is defined as n-tuple*

$$TTSP = (G, \gamma, s, e, \delta) \tag{1}$$

where:

- *G is the graph representing the problem, $G = (V, E)$, V is the set of vertices and E is the set of edges,*
- *γ is the function assigning time to edges, $\gamma : E \to \Delta$, where Δ is the set of admissible (floating) intervals of time (for intuition, any edge is assigned an interval of time necessary to go along it),*
- *s and e are the functions defining start and end of operation of any node $v \in V$ (e.g. $s(v)$, $e(v)$), or the agent a at the node (e.g. $s(a, v)$, $e(a, v)$),*
- *δ is such a mapping defining admissible or necessary time for operation at any node $v \in V$ that: $\delta : v \mapsto \Delta t \subset [s(v), e(v)]$*

The last condition for δ expresses the fact that a task is performed in v in some internal time Δt *within* the interval $[s(v), e(v)]$ associated to v.

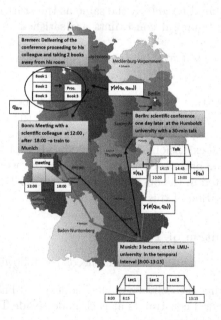

Fig. 1. A visualization of TTSP for visiting professor.

Example 1. *Consider a professor X involved in some temporally restricted activities in German cities. Taking into account his weekly activity timetable, we can distinguish the following activities:*

1. *One day Prof. X carries out 3 lectures at LMU in Munich between 8:00 and 13:15.*
2. *The next day he participates in the scientific conference (between 10:00 and 15:00) at the Humbold's University of Berlin with a 30-min talk (14:15–14:45).*
3. *During the third day of his activity, he meets his colleague in Bremen, delivering him a conference proceeding from Berlin and he visits his office at the university taking 2 books away from his room,*
4. *In the 4th day, he has a scientific appointment at 12:00 in order to manage to take a night train to Munich.*

This activity is also involved in some preferences with respect to the choice of the appropriate communication way between cities – as depicted on in Fig. 1.

Assume that D denotes a day $D, D+1$ – a one day later, $D+2$ – two day later. Adopt also the convention that: X:00D denotes X:00h on D-day, etc. Thus, TTSP may be formally given as follows:

$$TTSP^{Prof} = (G^{Prof}, \gamma^{Prof}, s^{Prof}, e^{Prof}, \delta^{Prof}), \tag{2}$$

where:

- $G^{Prof} = \Big(V^{Prof} = \text{Munich, Berlin, Bremen, Bonn}, \ E = \{\text{Munich} \to$ Berlin, Berlin $\to$ Bremen, Bremen $\to$ Bonn, Bonn $\to$ Munich$\}\Big)$,
- γ^{Prof}(Munich $\to$ Berlin), γ^{Prof}(Berlin $\to$ Bremen), γ^{Prof}(Bremen $\to$ Bonn), γ^{Prof}(Bonn $\to$ Munich),
- δ^{Prof}(Munich) $= (8:15D, 13:00D)$, δ^{Prof}(Berlin) $= (14:15(D+1), 14:45(D+1))$, δ^{Prof}(Bremen) $= (14:15(D+2), 14:45(D+2))$, δ^{Prof}(Bonn) $= (12:00(D+3), 12:00+\epsilon)(D+3))$,
- $(s^{Prof}$(Munich), e^{Prof}(Munich)) $= (8:00, 13:15)$, $(s^{Prof}$(Berlin), e^{Prof}(Berlin)) $= (10:00, 15:00)$, $(s^{Prof}$(Bremen), e^{Prof}(Bremen)) $= (00:00, 24:00)$, $(s^{Prof}$(Bonn), e^{Prof}(Bonn)) $= (00:00, 18:00+\epsilon)$.

4   Temporal Traveling Salesman Problem in Terms of Deontic Halpern-Shoham Logic

TTSP has just been introduced in terms of the graph-based depiction. In this section, the same TTSP (pedantically: some of its extract) will be represented by the appropriate logical formula. This formula (4) will be of a quasi-modal type. The next, we move to the proper modal combined formulae of $\mathcal{L}$(DeoHS) (somehow corresponding to the quasi-modal formula for TTSP) in order to show how fibred semantics works for them. Finally, we will return to TTSP and the quasi-modal formula representing TTSP. As earlier mentioned, the Salesman's tasks imposed on his activity may be rendered as his obligations, what determines a deontic nature of this formulation.

Deontic Formulation of TTSP 1. *Consider a salesman K and a list of n cities (with temporal distances between them). Assume, as usual, that K is visiting all the cities in such a way to find the shortest possible route that visits each city exactly once and leads to the origin city. Assume also that K – being in a city C_1 in some temporal interval I_1 – is obliged to deliver a package A from C_1 to C_2, a temporal distance between C_1 and C_2 amounts 3 h, but A must be delivered in C_2 in the interval I_2. Thus, a situation of K may be rendered by the following expression with a combined modal prefix.*

$$[\text{K is obliged}_{C_1}]\langle\text{Later in 3 h}\rangle\text{Deliver}_{C_2}^A. \tag{3}$$

We adopt the following convention. The outer expression: $[\text{K is obliged}]\phi$ plays a role of a box-type operator that renders an obligation of K and $\phi = \langle\text{Later in 3 h}\rangle\psi$ plays a role of a unique $\langle L\rangle$-operator of HS logic. Finally, $\psi = \text{Deliver}_{C_2}^A$ is already an atomic formula. In these terms, the paper objective is to propose a semantic interpretation of (3) in terms of fibred semantics.

4.1   Fibring Semantics for Combined Formulae of DeoHS

In this section, we demonstrate how the mechanism of fibred semantics works for combined modalities, which cannot be modeled by single models. In fact, one can imagine that a model, say $\mathcal{M}_1$ is not capable of recognizing modal operators interpreted in another model, say $\mathcal{M}_2$.

Example 2. *One can imagine that a model $\mathcal{M}_1$ for MVDeo does not recognize modal operators of HS interpreted in a model $\mathcal{M}_2$ and conversely.*

Therefore, let us consider a formula $\psi = P_i^\alpha\langle X\rangle\phi$ and a model $(\mathcal{M}_1, I_1) \models P_i^\alpha p$, where p is atomic and X denotes an Allen's relation. It may arise a question of satisfiability of the atomic p in $(\mathcal{M}_1, I_1)$, if $p = \langle X\rangle\phi$, or what about $(\mathcal{M}_1, I_1) \models p$, if only $p = \langle X\rangle\phi$? Generally, as it has already been said, it may hold the following case:

– the model $\mathcal{M}_1$ (with I_1) may not 'recognize' $p = \langle X\rangle\phi$.

Fibring Mapping. This difficulty generates the question: "How to deal with this last fact?". In order to evaluate $\langle X\rangle\phi$ at $\mathcal{M}_1$ a unique *fibring mapping* between $\mathcal{M}_1$ and $\mathcal{M}_2$ is needed to 'transfer' the validity checking from $\mathcal{M}_1$ to the validity checking within the second one.

An idea of this mapping construction may be materialized as follows. Having a distinguished interval, say I_1, from a given model $\mathcal{M}_1$, we associate I_1 to a new pair $(\mathcal{M}_2^{I_1}, I_2)$, for $I_2 \in \mathcal{M}_2^{I_1}$ ($\mathcal{M}_2^{I_1}$ is parametrized by I_1). Generalizing, whole class of such pairs may be associated to I_1. Formally, for a given interval I_1 (from a model $\mathcal{M}_1$), we define the *fibring mapping* $\mathbf{F}: I_1 \to \bigcup_{i=2}^{k}(\mathcal{M}_i, I_i)$, for some k, such that the following condition holds:

$$(\mathcal{M}_1, I_1) \models [X]\psi \iff \forall i \leq k(\mathcal{M}_i^{I_1}, I_i \models [X]\psi) \tag{4}$$

for the fixed $I_i \in \mathcal{M}_i$ and X – as earlier. In particular, it may hold $\mathbf{F}(I_1) = (\mathcal{M}_2^{I_1}, I_2)$. Then (4) takes the simplified form:

$$(\mathcal{M}_1, I_1) \models [X]\psi \iff (\mathcal{M}_2^{I_1}, I_2) \models [X]\psi \tag{5}$$

Since the pair $(\mathcal{M}_2, I_2)$ is unambiguously determined by the interval I_2 alone, we can identify it with $\mathbf{F}(I_1)$ alone. It allows us to reformulate a satisfaction condition (5) as follows:

$$\mathcal{M}_1, I_1 \models [X]\psi \iff \mathcal{M}_2^{I_1}, \mathbf{F}(I_1) \models [X]\psi \tag{6}$$

Example 3. *If $X = L$ ('Later'-relation), then we can consider the corresponding operator $[L]\phi$ and (6) may be given as:*

$$\mathcal{M}_1, I_1 \models [L]\psi \iff \mathcal{M}_2^{I_1}, \mathbf{F}(I_1) \models [L]\psi$$

It is convenient to adopt the "switching condition" for $\mathbf{F}$: for each $I \in \mathcal{M}_1$, it holds $\mathbf{F}(I) \in \mathcal{M}_2$ and for each $I \in \mathcal{M}_2$: $\mathbf{F}(I) \in \mathcal{M}_1$. Finally, if $I_1 \neq I_2$, than also $\mathbf{F}(I_1) \neq \mathbf{F}(I_2)$. Obviously, the same procedure may be repeated for all types of combined formulas of $\mathcal{L}(\text{DeoHS})$.

Fibred Models and Fibred Satisfaction. We built up the fibred models from fusions of the corresponding components of the initial models (for simplicity: we consider two models only), but only fibred mapping $\mathbf{F}$ encodes a combination of these models. Formally, it looks as follows.

Let us assume that $\mathcal{M}_1 = \langle \mathcal{S}_1, R_1, h_1, \mathbf{F} \rangle$ and $\mathcal{M}_2 = \langle \mathcal{S}_2, L, h_2, \mathbf{F} \rangle$ are the interval-based Kripke models, where $\mathcal{S}_1 = \{J_1, J_2 \ldots, J_{2k}\}$, $\mathcal{S}_2 = \{I_1 \ldots I_{2k}\}$, for fixed $2k$, $J_j R_1 J_l \iff J_j \precsim_1^\alpha J_l$ in IBIS (for $j, l \in \{1, \ldots, 2k\}$), L is "Later"-relation between intervals from $\mathcal{S}_2$, h_1, h_2 are assignment functions in $\mathcal{M}_1$ and $\mathcal{M}_2$ (resp.). Then a *fibred models* $\mathcal{M}$ (for one agent denoted by 1) is the tuple:

$$\mathcal{M} = \langle \mathcal{S}_1 \otimes \mathcal{S}_2, R_1 \otimes L, h_1 \otimes h_2, \mathbf{F} \rangle, \tag{7}$$

where $\mathbf{F}$ is fibring mapping and $\otimes$ denotes a fusion of the appropriate corresponding components of $\mathcal{M}_1$ and $\mathcal{M}_2$.

Definition 5. *Let $\mathcal{M}$ be a fibred model and a formula $\phi = \langle P_i^\alpha \rangle \langle X \rangle \psi \in \mathcal{L}(DeoHS)$. The satisfaction condition for it in the fibred model $\mathcal{M}$ is put forward as follows:*

$$(\mathcal{M}, I) \models P_i^\alpha \langle X \rangle \psi \iff \exists I_1(I \precsim_i^\alpha I_1 \text{and } \mathcal{M}_1, I_1 \models \langle X \rangle \phi)$$
$$\iff \exists I_1(I \precsim_i^\alpha I_1 \text{and } \mathcal{M}_2, \mathbf{F}(I_1) \models \langle X \rangle \phi)$$
$$\iff \exists I_1(I \precsim_i^\alpha I_1 \text{and } \exists I_2(\mathbf{F}(I_1)XI_2 \text{ and } \mathcal{M}_2, I_2 \models \phi).$$

The satisfaction for all mixed formulae is defined similarly.

4.2   Modeling of the Deontic TSP

Return now to our deontic Traveling Salesman Problem with Salesman K delivering a packet A from a city C_1 to C_2 (Salesman's obligation). Because of a temporal distance between C_1 and C_2 its obligation can be satisfied not earlier than 3 h in some time in C_2. As mentioned, this obligation of Salesman may be rendered by the quasi-modal formula:

$$[\text{K is obliged}]\langle\text{Later in 3 h}\rangle\text{Deliver}_{C_2}^A. \tag{8}$$

(read: "K is obliged to deliver later (in 3 h) a packet A to a city C_2".) The outer operator $[\text{K is obliged}]\phi$ is a kind of a box operator and $\phi = \langle\text{Later in 3 h}\rangle\psi$ plays a role of $\langle L\rangle$-operator of HS and $\psi = \text{Deliver}_{C_2}^A$ is atomic.

Model for a Deontic Component of DeoHS. To find a model for the deontic (outer) component of the formula (8) – let us assume that $\mathcal{I}$ and $\mathcal{I}^{Deliver}$ are two discrete intervals interpreting the obligation for K such that:

- $\mathcal{I}$ is an interval where the obligation is "expressed" and
- $\mathcal{I}^{Deliver}$ is the interval, in which the obligation of K is materialized.

Formally: $\mathcal{I}^{Deliver} \models \text{Deliver}_{C_2}^A$, or a fact of delivering of the packet A to C_2 holds in this interval. Assume also that $\precsim_K$ is an accessibility relation between them, i.e. it holds $\mathcal{I} \precsim_K \mathcal{I}^{Deliver}$. Thus, a model for the deontic component is given as follows:

$$\mathcal{M}_1 = \langle\{\mathcal{I}, \mathcal{I}^{Deliver}\}, \precsim_K, h_1\rangle \tag{9}$$

for some valuation h_1.

Model for the Temporal Component of DeoHS. Similarly, we find a model for a temporal component. For that reason consider two *temporal discrete* intervals: I_1 for representation of "now" and I_2 for representation of "sometimes in a future". Thus, the appropriate model is given now as follows:

$$\mathcal{M}_2 = \langle\{\text{now}, \text{sometimes in a future}\}, \text{Later in 3 h}, h_2\rangle \tag{10}$$

for some valuation h_2. The *fibred model* for the whole formula (8) is determined by the tuple:

$$\mathcal{M} = \langle\mathcal{S}, R^*, h, \mathbf{F}\rangle, \tag{11}$$

where $\mathcal{S} = \{\mathcal{I}, \mathcal{I}^{Deliver}\} \otimes \{I_1 = \{\text{now}\}, I_2 = \text{sometimes in a future}\}$, $R^* = \{\precsim_K\} \otimes \{\text{Later in 3 h}\}$, $h = h_1 \otimes h_2$, $\mathbf{F}(P^{Deliver}) = I_1(= \{\text{now}\})$.

5   Conclusion

A temporal extensions of Traveling Salesman Problem TTSP has already been introduced and described in detail in this paper. This temporal extension was elucidated in two ways: from the graph theory-based depiction and from the logical point of view. In both cases, we were interested in proposing a new conceptual depiction of this problem and in a new form of a semantic modeling of it. The so-called fibred semantics appears to be a convenient type of the interval-based semantics for this task. It seems that research on a practical side of fibred semantics may be developped with respect to other systems of modal logic. This path seems to be a promising direction for future research – perhaps, in the context of Traveling Salesman Problem, too.

Obviously, different optimization problems – associated to the original TSP – may be considered in the context of its temporal extension. Although they seem to be relevant from the modeling perspective of this paper, they may be naturally associated to our TTSP. It is possible that teh so-called spacio-temporal graphs would be a convenient formal tool for elucidating these optimization aspects of TTSP.

References

1. Applegate, D., Bixby, M., Chvatal, V., Cook, W.: The traveling salesman problem, Edinburgh (2006). ISBN 0-691-12993-2
2. Christofides, N., Mingozzi, A., Toth, P.: The vehicle routing problem. In: Christofides, N., Mingozzi, A., Toth, P., Sandi, C. (eds.) Combinatorial Optimization, pp. 315–338. Wiley, Chichester (1979)
3. Dantzig, G., Ramser, J.: The truck dispatching problem. Manag. Sci. **6**(1), 80–91 (1959)
4. Fagin, R., Halpern, J., Moses, Y., Vardi, M.: Reasoning About Knowledge. MIT Press, Cambridge (1995)
5. Flood, M.M.: The travelling-salesman's problem. Oper. Res. **5**, 61–75 (1957)
6. Halpern, J., Shoham, Y.: A propositional modal logic of time intervals. J. ACM **38**, 935–962 (1991)
7. Heller, I.: The travelling salesman's problem. In: Proceedings of the Second Symposium in Linear Programming, vol. 1, pp. 27–29 (1955)
8. Jobczyk, K., Ligeza, A.: Fuzzy-temporal approach to the handling of temporal interval relations and preferences. In: Proceeding of INISTA, pp. 1–8 (2015)
9. Jobczyk, K., Ligeza, A.: A general method of the hybrid controller construction for temporal planing with preferences. In: Proceeding of FedCSIS, pp. 61–70 (2016)
10. Jobczyk, K., Ligeza, A.: Multi-valued halpern-shoham logic for temporal allen's relations and preferences. In: Proceedings of the Annual International Conference of Fuzzy Systems (FuzzIEEE) (2016, page to appear)
11. Jobczyk, K., Ligeza, A.: Dynamic epistemic preferential logic of action. In: Rutkowski, L., Korytkowski, M., Scherer, R., Tadeusiewicz, R., Zadeh, L.A., Zurada, J.M. (eds.) ICAISC 2017. LNCS (LNAI), vol. 10246, pp. 243–254. Springer, Cham (2017). https://doi.org/10.1007/978-3-319-59060-8_23

12. Kaplan, H., Lewenstein, L., Shafrir, N., Sviridenko, M.: Approximation algorithms for assymetric TSP by decomposing directed regular multigraphs. In: Proceedings of 44th IEEE Symposium on Foundation of Computer Science, pp. 56–65 (2004)
13. Lomuscio, A., Michaliszyn, J.: An epistemic halpern-shoham logic. In: Proceedings of IJCAI-2013, pp. 1010–1016 (2013)
14. Menger, K.: Das botenproblem. Ergebnisse eines Mathematisches Kolloquiums **2**, 11–12 (1932)
15. Papadimitriou, C.: The euclidean traveling salesman problem is NP-complete. Theor. Comput. Sci. **4**(3), 237–244 (1977)
16. Rubinstein, S., Hassin, R.: Better approximations for max TSP. Inf. Process. Lett. **75**(4), 181–186 (2000)
17. Steinerberger, S.: New new bounds for the traveling salesman constant. Adv. Appl. Probab. **47**, 27–36 (2015)
18. Vidal, T., Craining, T.G., Gendreau, M., Prins, C.: A unified solution framework for multi-attrobute vehicle routing problem. Eur. J. Oper. Res. **234**, 658–673 (2014)

Modelling the Affective Power of Locutions in a Persuasive Dialogue Game

Magdalena Kacprzak[1(✉)], Anna Sawicka[2], and Andrzej Zbrzezny[3]

[1] Bialystok University of Technology, Bialystok, Poland
m.kacprzak@pb.edu.pl
[2] Polish-Japanese Academy of Information Technology, Warsaw, Poland
asawicka@pja.edu.pl
[3] IMCS, Jan Długosz University in Częstochowa, Częstochowa, Poland
a.zbrzezny@ajd.czest.pl

Abstract. One of the most important contemporary directions of development in the field of artificial intelligence is to equip AI systems with emotional intelligence. This work is part of this trend. The paper presents a mathematical model that allows us to describe changes in players' emotional states as a response to dialogue actions. To this end, we use the paradigm of dialogue games and propose a method of rating locutions. The method is inspired by the affective rating system SAM which uses Mehrabian's PAD space which distinguishes emotions because of three attributes: Pleasantness (valence) (P), Arousal (A), and Dominance (D). Emotions that are analyzed are taken from Ekman's model with five universally accepted labels: fear, disgust, anger, sadness, and joy. In addition, we describe the emerging tool for the realization of dialogue games with emotional reasoning. This tool is the basis for designing a system for verifying the properties of dialog protocols.

Keywords: Dialogue game · Emotions · Locutions · Protocol

1 Introduction

Emotions are becoming an increasingly important factor in designing AI, cyborgs, chatbots, virtual characters, or improving human-computer communication. More and more attention is devoted to adding emotions and emotional reasoning to agents in multi-agent systems [1]. In recent years, we have seen a lot of progress in the field of affective computing [4], which is *"computing that relates to, arises from, or deliberately influences emotion or other affective phenomena"* [21]. The main research directions of affective computing focus on designing new ways for people to communicate affective-cognitive states and showing how computers can be more emotionally intelligent inventing personal technologies for improving self-awareness of affective states. Achieving these goals requires developing techniques for recognizing, analyzing, and simulating the emotional

© Springer International Publishing AG, part of Springer Nature 2018
L. Rutkowski et al. (Eds.): ICAISC 2018, LNAI 10842, pp. 557–569, 2018.
https://doi.org/10.1007/978-3-319-91262-2_49

states of computer users. Unfortunately, more and more often we forget to train emotional skills and also emotional intelligence in human beings, and here modern technologies and methods of artificial intelligence can definitely help. In all these approaches, the mathematical model of representation and dynamic change of emotions plays a crucial role.

In our research, as a formal base we choose the paradigm of dialogue games, which can be used as a means of communication in both multi-agent systems and in user-computer communication In such games, some of the players assume the role of proponents and argue for some thesis, and others are opponents [7]. The specification of these systems is typically given by defining the set of locution rules, protocol and the set of effect rules. Emotions play an essential role in communication. Therefore, it is necessary to incorporate emotional reasoning into this model as well. If we want to model effective communication, including effective persuasion and argumentation, we cannot ignore the emotional factors that impact upon this effectiveness.

Various definitions of persuasion have historically been proposed, starting with those by Perelman and Albrecht-Tyteca. Most of them have a common core, addressing methodologies aiming at changing, by means of communication, the mental state of the receiver. Persuasion mechanisms typically include four aspects: the cognitive state of the participants (beliefs, goals, commitments), their social relations (social power, shared goals), their emotional states and the context in which the interaction takes place [20]. Argumentation is often distinguished from persuasion as a process in which rational arguments are constructed, whereas in persuasion this is not necessarily the case. In particular, arguments referring to emotions are classified as irrational. Thus argumentation focuses more on the correctness of arguments whereas persuasion focuses on their effectiveness.

Many of the works devoted to formal modeling of emotions in multi-agent systems appeal to appraisal theory of emotions [18] where the main cause of change in the intensity of emotions is the belief and intention of the agent. The same emotions depend mainly on some events and their consequences, that is how they impact beliefs, especially those concerning the possibility of achieving the intended goal [6]. Therefore in formal systems, emotions are often determined by the mental state of the agent [5]. The BDI-like formal model of emotions, which merge both empirical and theoretical approaches, is given in [17]. The authors introduce the semantically grounded formal representation of rational dialogue agents and implement agents which can express empathy and recognize situations where it should be shown. BDI description is also used in [10] for deducing and understanding user's emotions during interaction with a pedagogical virtual agent. In [15] the communication theory of emotion [16] is applied. In this model emotional behavior is based on selected mental attitudes expressed in modal logic and emotions which indicate which actions should be performed by the agent. In many works, to generate emotions the Ortony, Clore and Collins (OCC) model is used [18]. It states that the strength of an agent's emotion depends on the events, other agents, and objects found in the multi-agent environment.

One of the important trends in the study of human emotions is to analyze the emotional reactions of the recipients of spoken expressions. It is particularly important in the field of psycholinguistics. It turns out that words have a certain emotional charge. How recipients receive the way that words are spoken is often analyzed. Less often, however, attention is paid to the fact that even when we read words and do not hear the person speaking them, the words themselves can arouse certain emotions in us. In [3] the results of experimental studies with a list of about 1000 words subjected to such evaluation are presented. Using the affective rating system SAM, subjects have rated the words on the basis of pleasure, arousal, and dominance (PAD).

This paper aims to contribute on evaluation of expressions within dialog systems. The entire sentences are evaluated and then used to change the mood and emotional state of the players. Consequently, we propose a formal, mathematical model that allows us to describe changes in players' moods and their emotions as a response to dialogue actions.

2   Emotionally Loaded Dialogue

The fact that words affect our emotional state is widely known. We agree that words can cheer us up, entertain us, but also cause harm. However, we often forget about this when having a conversation. Equally, often we do not realize how our words affect the other person and how she/he receives them. To illustrate this, consider the following dialogue between mother and son.

Ten-year-old Steve returns home and screams furiously from the doorway:
Steve: *I hate John, I could have killed him! I'm never going training again.*
Mother: *What are you so angry about?*
Steve: *Every time I've got the football John says, "Give me it, I'm better than you." That would make anyone angry, wouldn't it!*

Steve insists on giving up sports activities. How should his mother react? The mother would probably start by arguing that sports activities are important for health and by persuading him not to give up, to fight and pursue his goals with determination. These are all appropriate arguments, but they will be ineffective if we do not precede them with a phase of empathy. This is the part of the dialogue where we show acceptance and understanding of both the situation and the emotions of the interlocutor. The answer should be full of empathy:

M: *Oh dear, I know how you feel, it must have made you really angry.*
S: *It did! No one understands me the way you do!*

In a similar situation, many adults would react differently. Let's take a look at two possible dialogues and mother's attitude. In the first dialogue, the parent tries to calm down the situation saying that nothing serious happened. At the same time, she/he does not realize how harmful it is to deny the child's feelings.

M: *I can see you're annoyed with your training again! Why do you get so worked up every time? You are making a mountain out of a molehill! Stop exaggerating.*

In the second dialogue, the parent tries to explain the behavior of Steve's colleague, failing to notice that by doing this she does not respect her child and underestimates his skills.

M: *John's right, if you gave him the ball, he would score.*

In the next sections, we will show how to encode the affective power of locutions and examine its impact on the course of dialogue.

3   A Game-Theoretic Model

To give the theoretical background for parent-child affective dialogues, we use the terminology of persuasive dialogue games where dialogues are viewed as communicative games between two or more agents. In this section, the formal model is presented in the game-theoretic terminology [11–13,23]. When defining this model we use the following standard notation. Given a set Σ, the set of all finite sequences over Σ is denoted by Σ^* and the set of all infinite sequence over Σ is denoted by Σ^ω. The empty sequence is denoted by ε and the operation of concatenation is denoted by $\cdot$. First, we need to define the following parameters of the game: the set of statements, and the set of locutions.

Let S_0 be a non-empty and countable set called the *set of atomic statements* over which we define the set $FORM[S_0]$ of complex expressions composed of negations, conjunctions, alternatives and conditionals of atomic statements. The *set of locutions*, $L[S_0]$, is then defined as follows:

$$L[S_0] = \{\varepsilon,\ \varphi\ since\ \{\psi_1,\dots,\psi_n\},\ claim\ \varphi,\ concede\ \varphi,\ why\ \varphi,\ scold\ \varphi, nod\ \varphi,$$
$$retract\ \varphi,\ retract\ \varphi,\ question\ \varphi : \varphi, \psi_1,\dots,\psi_n \in FORM[S_0]\}.$$

All the expression from the set $FORM[S_0]$, which have been spoken are treated as public declarations of players and are called *commitments*. The commitments of player i are stored in the *commitment set* C_i. To model the change of players' moods we consider five emotions: *fear*, *disgust*, *joy*, *sadness*, and *anger*. These emotions are recognized by Ekman [9] as emotions which are universal despite the cultural context. They are universal for all human beings and are experienced and recognized in the same way all around the world. Other emotions are mixed and built from those basic emotions. The strength (intensity) of emotions is represented by real numbers from the set $[0, 10]$. Thus, the emotion vector E_i is a 5-tuple consisting of five values which refer to *fear*, *disgust*, *joy*, *sadness*, and *anger*, respectively. The change in the intensity of the emotions is dependent on the type of the performed locution as well as on its content. Given a set of atomic statements, S_0, the *parent-child persuasive game* is a tuple

$$\Gamma_{[S_0]} = \langle Pl, \pi, H, T, (\precsim_i)_{i\in Pl}, (A_i)_{i\in Pl}, (AAF_i)_{i\in Pl}, (C_i)_{i\in Pl}, (E_i)_{i\in Pl}, (Init_i)_{i\in Pl}\rangle$$

- $Pl = \{P, C\}$ is the set of players.
- $H \subseteq L[S_0]^* \cup L[S_0]^\omega$ is the *set of histories*, i.e. a sequence of locutions from $L[S_0]$. The set of finite histories in H is denoted by $\bar{H}$.

- $\pi : \bar{H} \to Pl \cup \{\varnothing\}$ is the *player function* assigning to each finite history the player who moves after it, or $\varnothing$, if no player is to move. The set of histories at which player $i \in Pl$ is to move is $H_i = \overrightarrow{\pi}^{-1}(i)$.
- $T = \overrightarrow{\pi}^{-1}(\varnothing) \cup (H \cap L[S_0]^\omega)$ is the set of *terminal histories*, i.e. it consists of the set of finite histories mapped to $\varnothing$ by the player function and the set of all infinite histories.
- $\precsim_i \subseteq T \times T$ is the *preference relation* of player i defined on the set of terminal histories. The preference relation is total and transitive.
- $A_i = L[S_0]$ is the *set of actions* of player $i \in Pl$.
- $AAF_i : H_i \to 2^{A_i}$ is the *admissible actions function* of player $i \in Pl$, determining the set of actions that i can choose from after history $h \in H_i$.
- $C_i : L[S_0]^* \to 2^{FORM[S_0]}$ is the *commitment set function* of player $i \in Pl$, designating the change of commitments.
- $E_i : L[S_0]^* \to Emotion_i$ is the *emotion intensity function* of player $i \in Pl$, designating the change of emotions where $Emotion_i$ is the set of possible emotional states of i.
- $Init_i$ determines the initial commitments IC_i and emotions IE_i.

In what follows we will assume that the set of atomic statements S_0 is fixed and omit it, writing $FORM$ and L rather than $FORM[S_0]$ and $L[S_0]$. In the case of the parent-child persuasive dialogue system, the parent starts the dialogue, and then actions are performed alternately: $\pi(h) \in \{P, \varnothing\}$ if $|h|$ is odd and $\pi(h) \in \{C, \varnothing\}$ if $|h|$ is even. The rules of dialogue are defined using the notion of players' commitment sets and emotion levels. Commitments are public declarations of players and come from a fixed set of expressions $FORM$. The *commitment set function* of player i is a function

$$C_i : L^* \to 2^{FORM},$$

assigning to each finite sequence of locutions $h \in L^*$ the *commitment set* $C_i(h)$ of i at h. For details about the function see [13]. The *emotion intensity function* of player i is a function

$$E_i : L^* \to Emotion_i,$$

assigning to each finite sequence of locutions $h \in L^*$ the *emotion vector* $E_i(h)$ of i at h. The set $Emotion_i$ consists of all possible 5-tuples for levels of emotions:

$$Emotion_i = \{(n_1, \ldots, n_5) : n_k \in [0; 10] \wedge k \in \{1, \ldots, 5\}\}.$$

The emotion intensity function of $i \in Pl$ determines the change of intensity of emotions and is defined inductively as follows.

ER_0 $E_i(\varepsilon) = IE_i$.
    At the beginning of the dialogue, the intensity of emotions is fixed by the vector IE_i.
ER_1 If $h \in L^*$, $a \in L$ and $j = \pi(h)$, then

$$E_i(h \cdot a) = EMOT_i(E_i(h), j, a),$$

where $EMOT_i : Emotion_i \times Pl \times L \rightarrow Emotion_i$ is a function which shows how emotions of player i can change if player j performs action a after the history h. This function is defined for each specific application and depends on player's profile and character.

The rules of the dialogue game are described by the *function of admissible actions*. It actually defines the game protocol. Assume that h is a dialogue where the last action is a, i.e., $h = h' \cdot a$ for some dialogue h'. Then the function defines which actions can be performed next and under what additional conditions. These conditions mean that the player can use *scold* when he is agitated and can use *nod* when he is in a calm mood. Subsequent actions are usually direct responses to actions preceding them, for example after *why* φ can occur φ *since* $\{\psi_1, \ldots, \psi_n\}$. Sometimes a new thread is allowed, e.g. after *claim* φ a player can perform *claim* ψ, where ψ is not related to φ. The function AAF_i of player $i \in Pl$ is defined below, where, for $i \in Pl$, $-i \in Pl \setminus \{i\}$ denotes the opponent for i. Given $h \in H_i$, $AAF_i(h)$ is a maximal set of locutions satisfying the following:

R_0 $AAF_i(\varepsilon) = InitActions$,
  where *InitActions* are locutions that can begin a dialogue. It is mostly a collection of actions of the type *claim, question, since*. Therefore $InitActions \subseteq \{claim\ \varphi,\ question\ \varphi,\ \varphi\ since\ \{\psi_1, \ldots, \psi_n\} : \varphi, \psi_1, \ldots, \psi_n \in FORM\}$.

R_1 If $h = h' \cdot claim\ \varphi$, $i \in \pi(h)$, $\psi \notin C_i(h)$ then $AAF_i(h) = \{why\ \varphi, concede\ \varphi, claim\ \psi, \neg\varphi\ since\ \{\psi_1, \ldots, \psi_n\}\}$ for $\psi, \psi_1, \ldots, \psi_n \in FORM$. Moreover, the set is extended to the following actions, if the following conditions are met:
  – if $E_i(h)[k] > 5$ for $k \in \{1, 5\}$, then $scold\ \psi \in AAF_i(h)$ for $\psi \in FORM$,
  – if $E_i(h)[k] < 5$ for $k \in \{2, 3, 4\}$, then $nod\ \psi \in AAF_i(h)$ for $\psi \in FORM$.

R_2 If $h = h' \cdot scold\ \varphi$, $i \in \pi(h)$, $\psi \notin C_i(h)$, then $AAF_i(h) = \{why\ \varphi, concede\ \varphi, claim\ \psi, scold\ \psi, \neg\varphi\ since\ \{\psi_1, \ldots, \psi_n\}\}$ for some $\psi, \psi_1, \ldots, \psi_n \in FORM$.

R_3 If $h = h' \cdot \varphi\ since\ \{\psi_1, \ldots, \psi_n\}$, $i \in \pi(h)$, $\psi \notin C_i(h)$ then $AAF_i(h) = \{why\ \varphi, concede\ \varphi, claim\ \psi, \neg\varphi\ since\ \{\psi_1, \ldots, \psi_n\}\}$ for $\psi, \psi_1, \ldots, \psi_n \in FORM$. Moreover, the set is extended to the following actions, if the following conditions are met:
  – if $E_i(h)[k] > 5$ for $k \in \{1, 5\}$, then $scold\ \psi \in AAF_i(h)$ for $\psi \in FORM$,
  – if $E_i(h)[k] < 5$ for $k \in \{2, 3, 4\}$, then $nod\ \psi \in AAF_i(h)$ for $\psi \in FORM$,
  – if $\psi \in C_{-i}(h)$ and $E_i(h)[k] > 5$ for $k \in \{1, 5\}$, then $concede\ \psi \in AAF_i(h)$ for some $\psi \in FORM$.

R_4 If $h = h' \cdot why\ \varphi$, $i \in \pi(h)$, then $AAF_i(h) = \{retract\ \varphi,\ \varphi\ since\ \{\psi_1, \ldots, \psi_n\}\}$ for some $\psi, \psi_1, \ldots, \psi_n \in FORM$.

R_5 If $h = h' \cdot question\ \varphi$, $i \in \pi(h)$, then $AAF_i(h) = \{retract\ \varphi,\ claim\ \varphi, claim\ \neg\varphi\}$.

R_6 If $h = h' \cdot a$, $a \in \{concede\ \varphi,\ nod\ \varphi,\ retract\ \varphi\}$, $i \in \pi(h)$, $\psi \notin C_i(h)$, then $AAF_i(h) = \{claim\ \psi,\ nod\ \psi,\ scold\ \psi,\ \psi\ since\ \{\psi_1, \ldots, \psi_n\}\}$ for some $\psi, \psi_1, \ldots, \psi_n \in FORM$.

4   Speech Acts and Their Affective Power

In this paper we focus on modeling and formalizing persuasive communication. According to Walton and Krabbe's theory of dialogue [25], the goal of persuasion is to resolve or clarify some issue. This means that at the beginning of a dialogue participants express their conflicting positions on towards this issue and then try to persuade the other party. This process involves the alternating execution of actions, which are called speech acts [8]. Speech acts are basic language communication units. Austin and Searle's most popular taxonomy of speech acts [2, 24] identifies: *assertives*, committing to the truth of a proposition, e.g., stating; *directives*, which get the listener to do something, e.g., asking; *commissives*, committing the speaker to some future action, e.g., promising; *expressives*, expressing a psychological state, e.g., thanking; *declaratives*, changing reality according to the proposition e.g., baptising.

Now consider only assertives where the speaker is committed to the truth of the expressed proposition. They include: stating, concluding, reporting, asserting, claiming etc. Note that depending on the content and intention, they can trigger a different emotion. For example:

- *A* : "*You always bring flowers to me. Thank you.*" - evokes joy;
- *B* : "*You always bring me flowers! Stop doing this!*" - causes anger;
- *C* : "*You only bring flowers to me. Always.*" - can cause sadness.

Emotional reactions to verbal stimuli were studied in [3]. The authors presented the list of words rated by evaluators along the components elaborated by Osgood et al. in their theory of emotions [19] and then expanded by Mehrabian and Russell [22]. This list sets Affective Norms for English Words. It shows the reaction of people to spoken or read words, and what emotions they wake up in. Several models and theories of emotions and moods are described in the literature. The research from [3] is based on PAD model which distinguishes emotions because of three attributes: Valence (P) (sometimes marked also with V), pleasantness, Arousal (A), the intensity of the emotion, and Dominance (D), the degree of control exerted by the perceiver over the stimulus. Valence indicates whether the emotion is pleasant or not and is ranging from unpleasant to pleasant. Arousal describes a state of being awoken and is ranging from calm to excited. Dominance shows whether an emotion makes the recipient withdrawn or controlling (dominant). Assuming a scale from 1 to 9, the values for the two selected words are given below (see [3]):

holiday: P = 7.55; A = 6.59; D = 6.30; jealousy: P = 2.51; A = 6.36; D = 3.80.

In a similar way, it is possible to measure and compare affective features of locutions and then assign to them some affective power. Following this, we can determine three values to each locution. However, we assume a scale from -1 to 1. Consequently, PAD function Ω_{PAD} is defined as follows:

$$\Omega_{PAD} : L[S_0] \longrightarrow [-1, 1]^3$$

The exact values of the affect function result from experimental tests carried out by a team of psychologists cooperating with us in this project. Examples of the values for the above sentences, marked with symbols A, B, C are as follows:

$$\Omega_{PAD}(A) = (0.4; 0.3; 0.1), \Omega_{PAD}(B) = (-0.3; 0.6; 0.5),$$
$$\Omega_{PAD}(C) = (-0.2; -0.3; -0.4)$$

Similarly, values for parent's statements from Sect. 2 may be as follows:
$\Omega_{PAD}(Oh\ dear,\ I\ know\ how\ you\ feel,\ it\ must\ have\ made\ you\ angry.)=$ (0.4;0.2;0.2)
$\Omega_{PAD}(John's\ right,\ if\ you\ gave\ him\ the\ ball,\ he\ would\ score.)=(-0.6;0.7;0.3)$
$\Omega_{PAD}(You\ are\ making\ a\ mountain\ out\ of\ a\ molehill!)=(-0.2; -0.1; -0.6)$.

The PAD function will be used to describe the change in the emotional state of a participant in the dialogue. In our approach, we consider 5 emotions. They can be placed on three-dimensional PAD model. The values for joy, fear, and anger are as follows (see e.g. [14]):

$$PAD(joy) = (0.4, 0.2, 0.1), PAD(fear) = (-0.64, 0.60, -0.43),$$
$$PAD(anger) = (-0.51, 0.59, 0.25).$$

Given this values the distance between $PAD(e)$ for an emotion e and $\Omega_{PAD}(a)$ for a locution a can be calculated. Let $PAD(e) = (v_1, v_2, v_3)$ and $\Omega_{PAD}(a) = (w_1, w_2, w_3)$ then:

$$DIST(a, e) = \sqrt{((v_1 - w_1)^2 + (v_2 - w_2)^2 + (v_3 - w_3)^2)}.$$

This function allows specifying in what area of the PAD space the localization of a relative to a given emotion e is placed. Thus, it specifies the emotions caused by a. If the locution is close to the position of, for example, *joy* (i.e. the distance is less than or equal to 0.5), then we can conclude that it evokes *joy*. If the distance is greater than 0.5, but less than 1, then a has little effect on *joy*. If the distance is greater than or equal to 1, then it evokes an opposite emotion, i.e. the intensity of *joy* should decrease.

Assume that emotion intensity function of player i after history h is

$$E_i(h) = (t_1, t_2, t_3, t_4, t_5).$$

Then, the update function defines the influence of affective locution on the change of emotions: $UPD(E_i(h), a) = (u_1, u_2, u_3, u_4, u_5)$ where

$$u_i = \begin{cases} 10, \text{ if } t_i + h_i \geq 10 \\ t_i + h_i, \text{ if } 0 < t_i + h_i < 10 \\ 0, \text{ if } t_i + h_i \leq 0. \end{cases}$$

and

$$h_i = \begin{cases} 1, \text{ if } DIST(a, e_i) \leq 0.5; \\ 0, \text{ if } DIST(a, e_i) \in (0.5; 1); \\ -1, \text{ if } DIST(a, e_i) \geq 1. \end{cases}$$

and e_1 is fear, e_2 is disgust, e_3 is joy, e_4 is sadness, e_5 is anger.

In other words, if a evokes the emotion e, the intensity of e increases, if a does not affect e, then the intensity remains at the same level. Otherwise, the intensity of e falls. However, the intensity range of emotions cannot go beyond the fixed range of values from 0 to 10, i.e. the intensity level can not fall below 0 and increase above 10.

The emotional-state update function for a speaker j and receiver i is now defined as follows:

$$EMOT_i(E_i(h), j, a) = UPD(E_i(h), a).$$

Intuitively this means that after completing the dialogue h and then executing the action a, the change of the emotion vector of agent i is consistent with the change determined by UPD function.

5   Playing the Persuasive Dialogue Game

In this section, we show the emerging tool for the realization of dialogue games. This framework focuses on argumentative dialogues and is based on Game with Emotional Reasoning Description Language (GERDL) [23]. It is a description language for a persuasive dialogue game in which speech acts have an emotional undertow. This framework is based on the interpreted system designed for a dialogue protocol in which participants have emotional skills.

GERDL is used on the input of our system to specify required aspects of the dialogue game and to describe players' strategies and preferences. The general input file structure in this language was presented in [23]. We proposed this language because we have found none directly suitable for our domain, even though we have a lot of inspirations (e.g. MCMAS, DGDL). GERDL will be used in our system for semantic verification of properties of dialogue games with emotional reasoning. We can use it also to specify dialogue game for the tool, which allows to play it. To describe a dialogue game, the user should specify a set of available commitments, initial states, and players, each with possible locutions (speech acts), the protocol and the evolution function. The *protocol* is an important element of the model since it gives strict rules which formally describes who, when and which action can perform. To determine results of actions, to express how locutions and their contents affect players' commitments and emotions during the dialogue we need *evolution function*. All these required elements of the dialogue game should be specified in the GERDL file.

The general scheme of our system is shown in the Fig. 1. The base of our system is a specific dialogue game. There are many dialogue games presented in the literature (e.g. Lorenzen DG, MacKenzie DG), but our goal is to allow the user to define his own game and describe it in GERDL. We have also a mathematical formalism (transitional system), which allows modeling such dialogue games. Having the specification of the game, we can use it during protocol realization (play the game). A dialogue game can have many interesting properties which can be formulated in many languages. In our system, we choose to

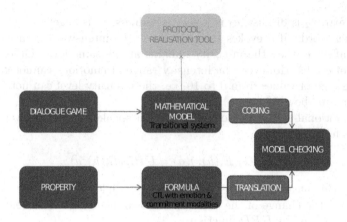

Fig. 1. General project scheme

specify them as formulae of CTL with emotion & commitment modalities. We can check, whether the property is true by the means of model checking, e.g. bounded model checking.

Currently, our work focuses on the tool for the realization of the specified dialogue game protocol (Fig. 2). We can enable one of two running modes: supervised and unsupervised. Within the supervised mode, at each stage of the dialogue, the user has to choose one move from all listed possible ones. Within the unsupervised one, the application will randomly choose the move. The input consists only of GERDL file with the dialogue game specification and the result is a dialogue conducted according to the rules of the game.

After parsing the GERDL file, we have to analyze all the rules described in the GERDL file (sections *Protocol* and *Evolution*). During the dialogue, after each move, the rules are checked whether the condition is fulfilled (the rule is *enabled*). If we enable a rule of the protocol (which describes possible interlocutors' moves), we have to bind some rule's elements e.g. in the rule:

Black.lastAction=scold.X : { scold.Y, why.X, concede.X, claim.X, claim.(!X)} content X is bound and content Y is not. The content of the locution which is not bound can take the form of any of the commitments considered in the system. This rule describes which moves are allowed if the last move of the player "Black" was *scold* with content X (rule construction details are available in [23]). As a consequence of the analysis of the protocol rules, we create or extend set of possible moves with locutions allowed by the rules.

We also check which evolution function rules (describing consequences of actions e.g. changes in the emotion) are enabled. If a condition is satisfied, then we make specified changes in player's variables, sets of commitments or emotions. After analyzing protocol and evolution rules, the opponent takes his turn. We can continue the game as long as the rules of the game allow for some move.

The next stage of our implementation will be related to the verification of the properties which are specified in the separate file (Fig. 3). The verification

Fig. 2. Scheme for realization of the specified dialogue game

will be provided using bounded model checking techniques. On the input, we give the GERDL file with the game protocol specification and the properties file, where the verified properties are described using the extension of CTL logic with commitment and emotion modalities (formulated in [23]). The example of such properties are shown in Fig. 3.

$AG(COM_p(\alpha) \Rightarrow E(true \ U \ \neg \ COM_p(\alpha)))$ expresses that even if a player p has committed to α at some point, then during the dialogue he can change his mind and retract this commitment. Second formula $A(true \ U \ \neg EMO_W(fear))$ claims that every computation contains a state at which mentioned player does not feel a strong fear and we can assume it is the end of the dialogue. On the output, we get the decision, whether the property is true or false. If the property is not true, we get a counterexample dialogue.

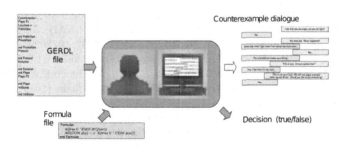

Fig. 3. Scheme for verification of the specified dialogue game properties

6 Conclusions

In this work we presented the theoretical framework that allows for the modelling of changes in emotional states resulting from the realization of dialogue actions. This approach uses the paradigm of dialogue games in which dialogue is treated as a game. We focus mainly on persuasive dialogues between children and parents. We show how one of the more frequently used PAD mood models can be used to evaluate locutions. It is an indispensable element the function of changing emotional states. In this way, we can examine how emotions affect the achievement of success in a dialogue. Our model can be used as a basis for modelling persuasive dialogues as well as verification of the effectiveness of persuasive strategies. The potential for applications of persuasive systems is huge in fields like health, business or safety. However, our main goal is to use it for educational purposes as an aid to learning and in developing the ability of emotional intelligence.

Acknowledgment. The research by Kacprzak have been carried out within the framework of the work S/W/1/2014 and funded by Ministry of Science and Higher Education. The images use icons from Project Icons by Mihaiciuc Bogdan (http://bogo-d.deviantart.com).

References

1. Adam, C., Gaudou, B., Herzig, A., Longin, D.: OCC's emotions: a formalization in a BDI logic. In: Euzenat, J., Domingue, J. (eds.) AIMSA 2006. LNCS (LNAI), vol. 4183, pp. 24–32. Springer, Heidelberg (2006). https://doi.org/10.1007/11861461_5
2. Austin, J.: How to Do Things with Words. Clarendon, Oxford (1962)
3. Bradley, M.M., Lang, P.J.: Affective norms for english words (ANEW): instruction manual and affective ratings. Technical report C-1, The Center for Research in Psychophysiology, University of Florida (1999)
4. Calvo, R., D'Mello, S.K., Gratch, J., Kappas, A. (eds.): The Oxford Handbook of Affective Computing. Oxford University Press, Oxford (2015)
5. Carofiglio, V., De Rosis, F.: In favour of cognitive models of emotions. In: Virtual Social Agents, p. 171 (2005)
6. Castelfranchi, C.: Affective appraisal *versus* cognitive evaluation in social emotions and interactions. In: Paiva, A. (ed.) IWAI 1999. LNCS (LNAI), vol. 1814, pp. 76–106. Springer, Heidelberg (2000). https://doi.org/10.1007/10720296_7
7. Dunin-Kęplicz, B., Strachocka, A.: Paraconsistent multi-party persuasion in TalkLOG. In: Chen, Q., Torroni, P., Villata, S., Hsu, J., Omicini, A. (eds.) PRIMA 2015. LNCS (LNAI), vol. 9387, pp. 265–283. Springer, Cham (2015). https://doi.org/10.1007/978-3-319-25524-8_17
8. Dunin-Keplicz, B., Strachocka, A., Szalas, A., Verbrugge, R.: Paraconsistent semantics of speech acts. Neurocomputing **151**, 943–952 (2015)
9. Ekman, P.: An argument for basic emotions. Cognit. Emot. **6**, 169–200 (1992)
10. Jaques, P.A., Viccari, R.M.: A BDI approach to infer student's emotions. In: Lemaître, C., Reyes, C.A., González, J.A. (eds.) IBERAMIA 2004. LNCS (LNAI), vol. 3315, pp. 901–911. Springer, Heidelberg (2004). https://doi.org/10.1007/978-3-540-30498-2_90
11. Kacprzak, M., Sawicka, A., Zbrzezny, A.: Dialogue systems: modeling and prediction of their dynamics. In: Abraham, A., Wegrzyn-Wolska, K., Hassanien, A.E., Snasel, V., Alimi, A.M. (eds.) Proceedings of the Second International Afro-European Conference for Industrial Advancement AECIA 2015. AISC, vol. 427, pp. 421–431. Springer, Cham (2016). https://doi.org/10.1007/978-3-319-29504-6_40
12. Kacprzak, M., Sawicka, A., Zbrzezny, A.: Towards verification of dialogue protocols: a mathematical model. In: Rutkowski, L., Korytkowski, M., Scherer, R., Tadeusiewicz, R., Zadeh, L.A., Zurada, J.M. (eds.) ICAISC 2016. LNCS (LNAI), vol. 9693, pp. 329–339. Springer, Cham (2016). https://doi.org/10.1007/978-3-319-39384-1_28

13. Kacprzak, M.: Persuasive strategies in dialogue games with emotional reasoning. In: Polkowski, L., Yao, Y., Artiemjew, P., Ciucci, D., Liu, D., Ślęzak, D., Zielosko, B. (eds.) IJCRS 2017. LNCS (LNAI), vol. 10314, pp. 435–453. Springer, Cham (2017). https://doi.org/10.1007/978-3-319-60840-2_32

14. Kasap, Z., Moussa, M.B., Chaudhuri, P., Magnenat-Thalmann, N.: Making them remember—emotional virtual characters with memory. IEEE Comput. Graph. Appl. **29**(2), 20–29 (2009)

15. Meyer, J.J.C.: Reasoning about emotional agents. In: Proceedings of the 16th European Conference on Artificial Intelligence, pp. 129–133. IOS Press (2004)

16. Oatley, K.: Best Laid Schemes: The Psychology of the Emotions. Cambridge University Press, Cambridge (1992)

17. Ochs, M., Sadek, D., Pelachaud, C.: A formal model of emotions for an empathic rational dialog agent. Auton. Agents MAS **24**(3), 410–440 (2012)

18. Ortony, A., Clore, G.L., Collins, A.: The Cognitive Structure of Emotions. Cambridge University Press, Cambridge (1998)

19. Osgood, C., Suci, G., Tannenbaum, P.: The Measurement of Meaning. University of Illinois Press, Champaign (1957)

20. Petta, P., Pelachaud, C., Cowie, R. (eds.): Emotion-Oriented Systems. The Humaine Handbook. Springer, Heidelberg (2011). https://doi.org/10.1007/978-3-642-15184-2

21. Picard, R.W.: Affective Computing. MIT Press, Cambridge (2000)

22. Russell, J.A., Mehrabian, A.: Evidence for a three-factor theory of emotions. J. Res. Personal. **11**, 273–294 (1977)

23. Sawicka, A., Kacprzak, M., Zbrzezny, A.: A novel description language for two-agent dialogue games. In: Polkowski, L., Yao, Y., Artiemjew, P., Ciucci, D., Liu, D., Ślęzak, D., Zielosko, B. (eds.) IJCRS 2017. LNCS (LNAI), vol. 10314, pp. 466–486. Springer, Cham (2017). https://doi.org/10.1007/978-3-319-60840-2_34

24. Searle, J., Vanderveken, D.: Foundations of Illocutionary Logic. Cambridge University Press, Cambridge (1985)

25. Walton, D., Krabbe, E.: Commitment in Dialogue: Basic Concepts of Interpersonal Reasoning. SUNY Series in Logic and Language. SUNY Press, New York (1995)

Determination of a Matrix
of the Dependencies Between Features
Based on the Expert Knowledge

Adam Kiersztyn[1(✉)], Paweł Karczmarek[1], Khrystyna Zhadkovska[1],
and Witold Pedrycz[2,3,4]

[1] Institute of Mathematics and Computer Science,
The John Paul II Catholic University of Lublin,
ul. Konstantynów 1H, 20-708 Lublin, Poland
{adam.kiersztyn,pawelk,khrystyna.zhadkovska}@kul.pl
[2] Department of Electrical and Computer Engineering, University of Alberta,
Edmonton, AB T6R 2V4, Canada
wpedrycz@ualberta.ca
[3] Department of Electrical and Computer Engineering, Faculty of Engineering,
King Abdulaziz University, Jeddah 21589, Saudi Arabia
[4] Systems Research Institute, Polish Academy of Sciences, Warsaw, Poland

Abstract. In the paper, we investigate the problem of replacing long-lasting and expensive research by expert knowledge. The proposed innovative method is a far-reaching improvement of the AHP method. Through the use of a slider, the proposed approach can be used by experts who have not yet met the AHP method or do not feel comfortable when using classic approach related to words and numbers. In the series of experiments, we confirm the efficiency of the method in a modeling of electricity consumption in teleinformatics and in an application of biodiversity to urban planning.

Keywords: Expert system · Analytic Hierarchy Process (AHP)
Decision-making · Electricity consumption · Biodiversity

1 Introduction

One of the key stages of model building is the selection of variables describing this model [2,12,35]. In the vast majority of methods aiming in a selection of the variables for a model, it is necessary to obtain a matrix of the dependencies between the considered variables [8]. In many cases, determining the matrix of the dependencies between the variables requires a lot of work and it involves labour-intensive and cost-intensive research. Of particular note is the research effort required by the natural sciences, where tremendous laboratory tests or field tests are usually necessary. In the method presented here, we propose the solution based on the replacing these tests with expert knowledge. The method is a far-reaching modification of the well-known Analytic Hierarchy Process (AHP)

© Springer International Publishing AG, part of Springer Nature 2018
L. Rutkowski et al. (Eds.): ICAISC 2018, LNAI 10842, pp. 570–578, 2018.
https://doi.org/10.1007/978-3-319-91262-2_50

[27], in which experts determine the hierarchy of traits. It is worth noting that AHP has been widely applied in many problems of prioritization, ordering, finding saliency and choice of the features, see, for instance [28]. There are known its various interval, fuzzy, linguistic, graphical or Granular Computing-based enhancements [16,19,24,25,31]. The problem of an aggregation of assessments obtained by the group of experts was thoroughly discussed in [5]. A survey of AHP applications can be found in [5,24]. A comparison of AHP approaches and applications was thoroughly presented in, among others, [10,14,33]. Here, our predominant goal is to use a comparison of pairs of factors to determine the matrix of the dependencies between the features. This is a very intuitive approach that has not been used so far in the literature of a topic. Its main advantage, in addition to its easy interpretation, is a reduction of testing costs. By replacing the tremendous field and laboratory tests with expert knowledge, one can significantly save both time and money. Thanks to the pairwise comparisons, the experts are not overloaded and an indicator which is a consistency index (CI) determining their credibility is obtained. Moreover, the method enables the experts to not focus on the numeric or linguistic values but on the graphics tool (slider or track bar) which is one of the shortcomings of classic AHP approach.

The work is divided into the following parts. The Sect. 2 describes the assumptions of the proposed method. Section 3 presents the results of experiments on the ground of modelling the energy consumption by teleinformatic objects. Section 4 contains the results of experiments presenting an application of the proposal in the field of natural sciences. The Sect. 5 is devoted to the conclusions and future works.

2   Method Description

Suppose that we have K variables (features) describing the modelled variable. In addition, let us assume that we have the knowledge coming from N experts. The starting point is the determination of the dependencies between the analysed K variables or features by the group of N experts.

2.1   A Case of One Expert

During the research, each of the experts, independently, for every possible pair of features indicates which of the features have a greater impact on the modelled variable. In order to facilitate the research, the characteristics for AHP values (9; 7; 5; 3; 1; 1/3, 1/5; 1/7; 1/9) have been replaced by a graphical interface in which the experts indicate their opinions. An example window of the program with questions to an expert can be seen in Fig. 1 (see the experimental section). It should be noted that the program user does not know the above-mentioned scale. This is to divert his/her attention away from the numerical values and to place greater emphasis on the individual opinions of the expert. For each pair of features, the program saves the indicated number in the range $[-Max; Max]$ and creates an expert response matrix of the form

$$
\begin{bmatrix}
0 & x_{1,2} & x_{1,3} & \cdots & x_{1,k-1} & x_{1,k} \\
x_{2,1} & 0 & x_{2,3} & \cdots & x_{2,k-1} & x_{2,k} \\
x_{3,1} & x_{3,2} & 0 & \cdots & x_{3,k-1} & x_{3,k} \\
\vdots & \vdots & \vdots & \ddots & \vdots & \vdots \\
x_{k-1,1} & x_{k-1,2} & x_{k-1,3} & \cdots & 0 & x_{k-1,k} \\
x_{k,1} & x_{k,2} & x_{k,3} & \cdots & x_{k,k-1} & 0
\end{bmatrix},
$$

where $x_{i,j} \in [-Max; Max]$ for $i, j \in \{1, \ldots, k\}$.

In the next step, the elements of the expert's response matrix are the input to the appropriate analysis algorithm. The values are converted into linear correlation coefficients. This process allows an objective examination of the credibility of the expert and determines the impact that each of the describing variables affects the explained variable. In the first case, the expert responses in the range, say $[-Max; Max]$, are converted into values in the range $[0; 1]$ according to the formula

$$
r_{i,j} = 1 - \frac{|x_{i,j}|}{Max} \tag{1}
$$

Due to transformations such as this, we gain the knowledge about the assumed effect: If it is indicated that both features have the same effect on the explained variable, then the dependence between them is equal to one. In the cases where one feature significantly outweighs the other, the dependence is close to zero.

The AHP method can be used to determine the degree of credibility of an expert and the impact of individual characteristics on an explanatory variable. At the beginning of the process, the expert's responses should be transformed to the values used in the AHP. It is proposed to use the following transformation function

$$
f(x) = \begin{cases}
1, & \text{if } \frac{9}{2}\log(c \cdot x + 1) \in [0, 2] \\
3, & \text{if } \frac{9}{2}\log(c \cdot x + 1) \in (2, 4] \\
5, & \text{if } \frac{9}{2}\log(c \cdot x + 1) \in (4, 6] \\
7, & \text{if } \frac{9}{2}\log(c \cdot x + 1) \in (6, 8] \\
9, & \text{if } \frac{9}{2}\log(c \cdot x + 1) \in (8, 9] \\
\frac{1}{f(-x)}, & \text{if } x < 0
\end{cases} \tag{2}
$$

where $c = \frac{100}{Max}$.

In this way we obtain a typical AHP reciprocal matrix and may determine the credibility of the expert's response by calculating the consistency index (CI), see [27], which can be obtained by the following formula: $CI = (\lambda_{max} - n)/(n - 1)$, where λ_{max} stands for the reciprocal matrix maximal eigenvalue or by getting the consistency ratio (CR) which is given by $CR = CI/RI$, where RI is a so-called random inconsistency index. Its values were obtained in a series of experiments in [26].

In addition, we may determine the degree to which, according to the experts, each of the variables affects the modelled variable. These quantities may be identified by the vector of the dependencies of explanatory variables with the explained variable. The values of the vector are weights determined using the AHP method.

2.2 Aggregation of Expert Responses

It is clear that the answers of one expert may be subjective according to the experts preferences. Therefore it is advisable to calculate the average of the results obtained with the presence of several experts. Suppose that for each of the N experts we have a specific matrix of the dependencies between the describing variables $R[i, j]$, for $i, j = 1, 2, \ldots, k$. Suppose also that we have vectors of the dependencies between dependent variable and predictors R_i^0, for $i = 1, 2, \ldots, N$ and the vector $C = [c_1, c_2, \ldots, c_N]$ containing the inverse of the experts consistency indices CIs. With these assumptions, the weighted dependencies vector R^0 may be determined using the formula

$$R^0 = \sum_{i=1}^{N} \frac{c_i \cdot R_i^0}{\sum_{j=1}^{N} c_i}. \tag{3}$$

The weighted matrix of the dependencies can be yielded using an analogous formula. Each element of the aggregated matrix R reads as

$$R[i, j] = \sum_{n=1}^{N} \frac{c_n \cdot R_n[i, j]}{\sum_{m=1}^{N} c_m}, \quad i, j = 1, 2, \ldots, k. \tag{4}$$

3 Modeling of Energy Consumption

Potential applications of the method will be introduced using the examples of the analysis of energy consumption and biodiversity. The problem of energy consumption in teleinformatic facilities and, more generally, in the industry is widely discussed [7,11,20,23]. Similarly, scientists devote a lot of attention to the issue of biodiversity in urban planning [3,13,29,30,32,34]. Analogous considerations can also be made in other fields of science where classic methods of selecting variables for the model are used. They occur in finances and economy topics [1,36], medicine [9,17], psychology [4], natural sciences [6,21,22], and image analysis [15,18], etc.

In the series of experiments, we describe an application of the proposed method in the case of the selection of variables describing the model of the predictive consumption of electricity in ICT facilities and also an application to the selection of green areas having the greatest impact on biodiversity in the city area. During the preliminary analysis of the data used to determine the model of energy consumption in teleinformatic objects, a list of the most important features influencing the model's construction was proposed. They are:

1. The type of an object.
2. The number of devices permanently consuming energy.
3. Number of devices that periodically consume energy.
4. Type of cooling/heating.
5. The presence of devices in the facility for which fluctuations in energy consumption occur.
6. Type of data available (15 min/h readings, etc.).
7. Number of available data sources.
8. The period for which the data are available.

Based on the responses of five experts $(A, B, C, D,$ and $E)$ who have used an application presented in Fig. 1 the estimated values of the matrix of the dependencies were obtained using the method described above. The surveyed experts are employees of an ICT company working in the facility management department and research workers engaged in R&D projects. For each expert, the responses were transformed to the classical AHP values constituting a reciprocal matrix and the weight vector as well as consistency index CI and consistency ratio CR were calculated. The values of the weight vectors and the values of the CI and CR coefficients for individual experts are presented in the Table 1.

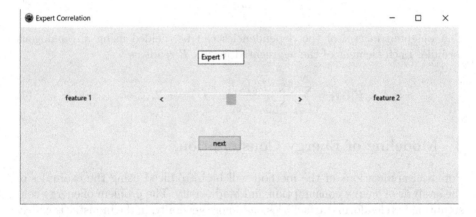

Fig. 1. Conception of collecting the experts opinions (a screen of an application)

It may be seen that, according to the experts, the most important feature is "type of available data (15 min/h readings, etc.)". It describes the details of the available data from the readings of electricity meters. It may also be noted that there is no unanimity among experts for the final classification of the remaining features. In connection with the above, it is reasonable to use the formula for the aggregation of expert responses. After applying the formula (4), the aggregated values shown in the Table 2 were obtained. The method lets us easily find the experts who were not consistent in their judgements (B and C).

The obtained final aggregated results are intuitively appealing and confirm that in the opinion of experts the most important features are "type of data

Table 1. The values of the weight vectors, consistency index, and consistency ratio

Expert	Feature 1	Feature 2	Feature 3	Feature 4	Feature 5	Feature 6	Feature 7	Feature 8	CI/CR
A	0.10	0.04	0.03	0.05	0.06	0.37	0.20	0.15	0.62/0.44
B	0.10	0.10	0.24	0.01	0.03	0.37	0.10	0.05	0.31/0.22
C	0.13	0.13	0.24	0.01	0.02	0.22	0.15	0.10	0.80/0.57
D	0.10	0.10	0.31	0.02	0.03	0.30	0.10	0.04	0.23/0.16
E	0.19	0.06	0.18	0.02	0.06	0.40	0.06	0.03	0.22/0.15

Table 2. Aggregation of experts opinion using the formula (4)

Feature 1	Feature 2	Feature 3	Feature 4	Feature 5	Feature 6	Feature 7	Feature 8
0.13	0.08	0.23	0.02	0.04	0.35	0.10	0.05

available (15 min/h readings, etc.)" and "number of devices that periodically consume energy." The aggregation of expert responses obtained in this way may be a starting point for determining the variables that will be taken into account when building the model of predictive consumption of electricity in ICT facilities.

4 Determination of the Matrix of Dependencies in the Study of Biodiversity

Urban planning needs are constantly increasing, therefore, planning authorities are exerting increasing political pressure for the allocation of urban green areas for development. In this case, it is advisable to develop a method, which allows for the determination of the areas necessary to preserve biodiversity. The appropriate solution was presented in [21]. An important element of the problem discussed here is the disposition of a matrix of the dependencies between particular areas. In the classic approach, it is necessary to conduct long-term field tests. Let us try to replace these considerations with the approach presented in Sect. 2. The research covered 21 green areas located in the city of Lublin. The analysed areas are marked in details on the city map presented in [21]. Five experts in the field of ecology and statistics were asked a question: Which of the analysed areas have the greatest impact on the biodiversity of the city? All respondents are authors of scientific publications in the field of ecology or economics. The answers they gave were subject to the procedure described above and the appropriate values were calculated. The expert matrix of the dependencies was compared with the results of field studies. The function given by the following formula was used as a measure of similarity

$$d(R, \hat{R}) = \sum_{i=1}^{k} \sum_{j=1}^{k} \frac{\left| |R[i,j]| - |\hat{R}[i,j]| \right|}{|R[i,j]| + |\hat{R}[i,j]|} \tag{5}$$

where $\hat{R}$ is the matrix of the dependencies obtained on the basis of field studies. The values of the differences between individual experts are presented in the Table 3.

Table 3. Differences between experts answers and empirical matrix of dependencies

Expert	A	B	C	D	E
$d(R, \hat{R})$	0.02	0.03	0.02	0.03	0.15

Analysing the results obtained above, it may be noticed that the answers of the experts largely coincide with the results of the fieldwork. It is, therefore, justifiable in some cases to replace field studies with expert knowledge. The distance between the matrix obtained through labor-intensive experiments and the matrices obtained from experts is small, which indicates their high similarity. However, it should also be noted that in order to obtain relatively objective data, the series with the presence of much more number of experts should be carried and then the data should be aggregated.

5 Conclusions and Future Studies

In the study, we have proposed an application of a graphical approach to the AHP to obtain the matrix of dependencies between the variables describing the problems of ICT facilities and biodiversity in the urban areas. The proposed approach allows, while maintaining the reliability of results, to significantly accelerate research and to significantly reduce research costs of the field studies. It is advisable to examine other methods of converting the value of experts' responses to values used in AHP. In addition, it is reasonable to use graphic components that allow for the introduction of fuzzy answers by experts in the future.

Acknowledgements. The authors are supported by National Science Centre, Poland [grant no. 2014/13/D/ST6/03244]. Support from the Canada Research Chair (CRC) program and Natural Sciences and Engineering Research Council is gratefully acknowledged (W. Pedrycz).

References

1. Altman, E.I.: Financial ratios, discriminant analysis and the prediction of corporate bankruptcy. J. Finance **23**(4), 589–609 (1968)
2. Bogdan, M., Van Den Berg, E., Sabatti, C., Su, W., Cands, E.J.: SLOPEadaptive variable selection via convex optimization. Ann. Appl. Stat. **9**(3), 1103–1140 (2015)
3. Brown, K.: Integrating conservation and development: a case of institutional misfit. Front. Ecol. Environ. **1**(9), 479–487 (2003)
4. Cohen, S.G., Ledford Jr., G.E., Spreitzer, G.M.: A predictive model of self-managing work team effectiveness. Hum. Relat. **49**(5), 643–676 (1996)
5. Forman, E., Peniwati, K.: Aggregating individual judgments and priorities with the analytic hierarchy process. Eur. J. Oper. Res. **108**, 165–169 (1998)
6. Geijzendorffer, I.R., Regan, E.C., Pereira, H.M., Brotons, L., et al.: Bridging the gap between biodiversity data and policy reporting needs: an Essential Biodiversity Variables perspective. J. Appl. Ecol. **53**(5), 1341–1350 (2016)

7. Gungor, V.C., Hancke, G.P.: Industrial wireless sensor networks: challenges, design principles, and technical approaches. IEEE Trans. Ind. Electron. **56**(10), 4258–4265 (2009)
8. Guyon, I., Elisseeff, A.: An introduction to variable and feature selection. J. Mach. Learn. Res. **3**(Mar), 1157–1182 (2003)
9. Hewett, T.E., Webster, K.E., Hurd, W.J.: Systematic selection of key logistic regression variables for risk prediction analyses: a five-factor maximum model. Clin. J. Sport Med.: off. J. Can. Acad. Sport Med. (2017). https://doi.org/10.1097/JSM.0000000000000486
10. Ho, W.: Integrated analytic hierarchy process and its applications-A literature review. Eur. J. Oper. Res. **186**, 211–228 (2008)
11. Holmberg, K., Kivikyt-Reponen, P., Hrkisaari, P., Valtonen, K., Erdemir, A.: Global energy consumption due to friction and wear in the mining industry. Tribol. Int. **115**, 116–139 (2017)
12. Hooten, M.B., Hobbs, N.T.: A guide to Bayesian model selection for ecologists. Ecol. Monogr. **85**(1), 3–28 (2015)
13. Hoyle, H., Hitchmough, J., Jorgensen, A.: All about the wow factor? The relationships between aesthetics, restorative effect and perceived biodiversity in designed urban planting. Landsc. Urban Plann. **164**, 109–123 (2017)
14. Ishizaka, A., Labib, A.: Review of the main developments in the analytic hierarchy process. Expert Syst. Appl. **38**, 14336–14345 (2011)
15. Karczmarek, P., Pedrycz, W., Kiersztyn, A., Rutka, P.: A study in facial features saliency in face recognition: an analytic hierarchy process approach. Soft Comput. **21**(24), 7503–7517 (2017)
16. Karczmarek, P., Pedrycz, W., Kiersztyn, A.: Graphic interface to analytic hierarchy process and its optimization. IEEE Trans. Fuzzy Syst. (submitted)
17. Khorana, A.A., Kuderer, N.M., Culakova, E., Lyman, G.H., Francis, C.W.: Development and validation of a predictive model for chemotherapy-associated thrombosis. Blood **111**(10), 4902–4907 (2008)
18. Kuo, B.C., Ho, H.H., Li, C.H., Hung, C.C., Taur, J.S.: A kernel-based feature selection method for SVM with RBF kernel for hyperspectral image classification. IEEE J. Sel. Top. Appl. Earth Obs. Remote Sens. **7**(1), 317–326 (2014)
19. van Laarhoven, P.J.M., Pedrycz, W.: A fuzzy extension of Saaty's priority theory. Fuzzy Sets Syst. **11**, 199–227 (1983)
20. Lange, C., Kosiankowski, D., Weidmann, R., Gladisch, A.: Energy consumption of telecommunication networks and related improvement options. IEEE J. Sel. Top. Quantum Electron. **17**(2), 285–295 (2011)
21. Łopucki, R., Kiersztyn, A.: Urban green space conservation and management based on biodiversity of terrestrial faunaa decision support tool. Urban For. Urban Green. **14**(3), 508–518 (2015)
22. Mac Nally, R.: Regression and model-building in conservation biology, biogeography and ecology: the distinction between – and reconciliation of – 'predictive' and 'explanatory' models. Biodivers. Conserv. **9**(5), 655–671 (2000)
23. Palensky, P., Dietrich, D.: Demand side management: demand response, intelligent energy systems, and smart loads. IEEE Trans. Ind. Inform. **7**(3), 381–388 (2011)
24. Pedrycz, W., Song, M.: Analytic hierarchy process (AHP) in group decision making and its optimization with an allocation of information granularity. IEEE Trans. Fuzzy Syst. **19**, 527–539 (2011)
25. Pedrycz, W.: Granular Computing. Analysis and Design of Intelligent Systems. CRC Press, Boca Raton (2013)

26. Saaty, T.L., Mariano, R.S.: Rationing energy to industries: priorities and input-output dependence. Energy Syst. Policy **3**, 85–111 (1979)
27. Saaty, T.L.: Decision-making with the AHP: why is the principal eigenvector necessary. Eur. J. Oper. Res. **145**(1), 85–91 (2003)
28. Saaty, T.L., Vargas, L.G.: Models, Methods, Concepts & Applications of the Analytic Hierarchy Process. Springer, New York (2012). https://doi.org/10.1007/978-1-4614-3597-6
29. Savard, J.P.L., Clergeau, P., Mennechez, G.: Biodiversity concepts and urban ecosystems. Landsc. Urban Plann. **48**(3–4), 131–142 (2000)
30. Standish, R.J., Hobbs, R.J., Miller, J.R.: Improving city life: options for ecological restoration in urban landscapes and how these might influence interactions between people and nature. Landsc. Ecol. **28**(6), 1213–1221 (2013)
31. Sugihara, K., Tanaka, H.: Interval evaluations in the analytic hierarchy process by possibility analysis. Comput. Intell. **17**, 567–579 (2001)
32. Threlfall, C.G., Mata, L., Mackie, J.A., Hahs, A.K., Stork, N.E., Williams, N.S., Livesley, S.J.: Increasing biodiversity in urban green spaces through simple vegetation interventions. J. Appl. Ecol. **54**(6), 1874–1883 (2017)
33. Vaidya, O.S., Kumar, S.: Analytic hierarchy process: an overview of applications. Eur. J. Oper. Res. **169**, 1–29 (2006)
34. Yu, D., Xun, B., Shi, P., Shao, H., Liu, Y.: Ecological restoration planning based on connectivity in an urban area. Ecol. Eng. **46**, 24–33 (2012)
35. Yuan, M., Lin, Y.: Model selection and estimation in regression with grouped variables. J. R. Stat. Soc.: Ser. B (Stat. Methodol.) **68**(1), 49–67 (2006)
36. Zhong, Y.: Analysis of incentive effects of government R&D investment on technology transaction. Mod. Econ. **8**, 78–89 (2017)

Dynamic Trust Scoring of Railway Sensor Information

Marcin Lenart[1,2,3]([⊠]), Andrzej Bielecki[3], Marie-Jeanne Lesot[2],
Teodora Petrisor[1], and Adrien Revault d'Allonnes[2,4]

[1] Campus Polytechnique, Thales, Palaiseau, France
marcin.lenart@thalesgroup.com
[2] Laboratoire d'Informatique de Paris 6, LIP6, CNRS,
Sorbonne Université, 75005 Paris, France
[3] Chair of Applied Computer Science, Faculty of EAIIB,
AGH University of Science and Technology, Cracow, Poland
[4] LIASD EA 4383, Université Paris 8, Saint-Denis, France

Abstract. A sensor can encounter many situations where its readings can be untrustworthy and the ability to recognise this is an important and challenging task. It opens the possibility to assess sensors for forensic or maintenance purposes, compare them or fuse their information. We present a proposition to score a piece of information produced by a sensor as an aggregation of three dimensions called reliability, likelihood and credibility into a trust value that take into account a temporal component. The approach is validated on data from the railway domain.

Keywords: Information scoring · Sensor · Trust · Reliability
Likelihood · Credibility

1 Introduction

Information scoring (see e.g. [1]) aims at assessing the quality of available pieces of information and, in particular, the trust that can be put in them. It plays a crucial role in any decision-aid system. For instance, in an information fusion system, equally considering reliable and unreliable sources may severely cripple the results. Sensors are not an exception, the information they produce is often used to get enhanced knowledge about a given situation and the ability to differentiate between them in terms of quality is a much needed feature.

Indeed, sensors do not always produce correct information. There are many situations in which a sensor can fail, e.g. producing out of range values, when encountering unfavourable operating conditions, communication problems or other interferences. Knowing whether the information produced by sensors is trustworthy can be key in many aspects, for instance, to choose the ones with the highest quality level for a given time interval. It can also be used to predict maintenance operations for sensors with decreased quality of information.

© Springer International Publishing AG, part of Springer Nature 2018
L. Rutkowski et al. (Eds.): ICAISC 2018, LNAI 10842, pp. 579–591, 2018.
https://doi.org/10.1007/978-3-319-91262-2_51

As detailed in Sect. 2, current quality measurements for sensors are mainly based on scoring *reliability* either from meta-data [5,7] or ground truth evaluation [3,8]. Other systems include *credibility* to further improve scoring by comparing information with other sources [8]. These solutions can suffer from lack of external knowledge (meta-data or ground truth) which can make them unusable.

This paper aims to address these limitations by decreasing the dependence on meta-data or ground truth and incorporating statistical analysis into the computation. It proposes new definitions for three dimensions chosen such that different aspects of the source and the information can be captured.

The paper is organised as follows: Sect. 2 presents some of the current approaches to score information quality for sensors, in Sect. 3 the proposed process of information scoring is explained and in Sect. 4 the approach is illustrated on real-world data. Section 5 concludes the paper and discusses future research directions.

2   Literature Review

This section briefly discusses general Information Quality scoring and describes approaches dedicated to the special case where the considered information is provided by sensors.

General Information Quality Assessment. The task of information scoring is mainly addressed through the decomposition of its quality into components, assessed on different dimensions whose list and definitions vary depending on the author. Some examples are relevance and truthfulness [13], reliability and certainty [10], source-trustworthiness and information-credibility [2], sincerity, competence, intention of the source and plausibility [11] or trust [4,8,15,18], see [17] for a complete list. One recent approach, introduced in [15], considers trust evaluation in a multivalued logic framework based on four dimensions: reliability and competence, which evaluate the source, and plausibility and credibility, which relate to the information content, spanning the range from source to information, from general to contextual and from subjective to objective.

Information Quality Dimensions in the Context of Sensor Measurement. Many papers [5,7,9,14] focus on the case of information provided by sensors. They often consider three dimensions, called reliability, contextual reliability and credibility, but vary in the way they are scored, as detailed below.

Reliability is generally understood as the ability of a system to perform its required functions under stated conditions for a specified time. It is an *a priori* assessment of the source.

Different approaches are considered to score reliability. In [5,7] meta-information on the source are considered, e.g. its specification, protocol or environment. The gathered knowledge is then combined to propose a final reliability score. This approach is limited to the case where valuable meta-data are available.

A second approach to define reliability consists in viewing it as accuracy [3,8] in the case where ground truth is available, i.e. knowledge about the expected results. This suffers the same limitation as the previous approach since ground truth is not always available. Blasch [3] views it as a compound notion that aggregates several sub-dimensions. He enriches the previously mentioned approach by considering that reliability requires accurate, confident and timely results. However these three are not always achievable simultaneously, e.g. sometimes having more accurate or confident data leads to longer collecting time, which induces a choice between accuracy and timeliness.

Such an approach also leaves open the question of the aggregation operator to be used to combine the selected components of reliability. Blasch presents a user-driven approach, where these three dimensions are weighted based on a desired utility.

Contextual Reliability aims at changing reliability depending on the task the device is used for and thus the context of each piece of information.

Mercier et al. [12] propose to score reliability in a way where it better reflects the reality of a sensor and its working environment by enriching it with its context. Then, different situations can result in different output qualities for a given sensor. For instance in the case of target recognition [12], the performances of a data acquisition system may depend on weather conditions and on background and target properties, making the reliability of the decision system dependent on the target at hand. A sensor that recognises between three objects (helicopter, aeroplane and rocket) can have different accuracies for each one, effectively creating a vector of three reliabilities with different contexts.

Credibility can be defined as the level of confirmation of a given piece of information by other, independent, sources and constitutes another component of information quality. There are situations where assessing reliability is difficult or even impossible. This is where scoring credibility can provide an alternative or a complement to scoring reliability.

Using a "majority vote" strategy, it is possible to either improve the quality of the acquired piece of information [8] or combine multiple similar and dissimilar sensors to improve the overall quality of calculations by aggregating all outputs into one [8,9,14,16].

Credibility for a piece of information is a relation to other pieces of information provided by independent sources, which ends with two cases, information is either concurring or conflicting [8]. The more pieces of information confirming the given piece of information, the more credible it is. This presents two possibilities of usage: i/ calculating ground-truth-type-of-reference by taking as output the majority of the sensors and then comparing it to evaluated sensor's output [9,14] or ii/ combining all outputs to determine information quality by grouping sensors according to the feature they measure and evaluating the degree of consensus between them [8]. This approach can suffer limitations if sources are lacking or if their information is not comparable.

3   Proposed Process for Information Scoring for Sensors

The information scoring model we propose is inspired by the multidimensional approach introduced in [15] which considers source evaluation, using reliability and competence, as well as content evaluation, using plausibility and credibility. We adapt, evaluate and aggregate three among these dimensions for the case of sensors, more precisely railway monitoring sensors. The three dimensions are also aggregated into a single *trust* value, which, in our case, is attributed to a sensor's reading at a given time. The presented approach has, in addition, a dynamic character: to score dimensions for the current log entry, the previous log entries are considered as well as their computed trust values.

This section gives a high level description of the considered data then details the process of dynamically scoring multi-dimensional trust.

3.1   Data Structure

The data structure we use for information scoring has the following characteristics: it is in the form of a log file whose entries contain a date, a time, a sensor id and a value, as shown in Table 1 for a real data example. The entries are event-triggered, i.e. they occur only when an event happens. The possible values represent the different messages given by the sensor that describe the sensor state. In the real data we consider, these messages can be occupied, clear or some type of disturbance. We aim to give a trust evaluation for each log entry, as illustrated in the last column of Table 1. We exhibit a part of data where a deficiency in quality is encountered which corresponds to a decreased trust value (see the second entry in Table 1).

For the computations, some notions need to be specified. We denote $\mathcal{L}$ the complete log set and $\mathcal{L}_s$ the set of log entries produced by sensor s. The notation l corresponds to one log entry defined as a vector containing three values: $l.fullDate$ corresponding to date and time, $l.sensor$ to the sensor id and $l.message$ contains the provided piece of information describing the sensors state. The set of all sensors is denoted $\mathcal{S}$, and the set of all times $\mathcal{T}$.

3.2   Scoring Trust

As explained at the beginning of this section our proposition is similar to [15], adapting and implementing this theoretical proposition to the specific case of sensors. We also consider reliability and credibility, presenting our view of scoring them in Sects. 3.3 and 3.5 respectively. Regarding competence, its definition and scoring in the case of sensors appear to require knowledge about the system

Table 1. Example of input data structure and output trust scores for a sensor.

Date	Time	Sensor ID	Message	Trust
11.03.2015	07:24:53	AC1	occupied	0.9
11.03.2015	07:25:40	AC1	**occupied**	**0.3**
11.03.2015	08:23:18	AC1	occupied	0.7
11.03.2015	08:24:08	AC1	clear	0.7
11.03.2015	09:15:23	AC1	occupied	0.8
11.03.2015	09:16:08	AC1	clear	0.8
11.03.2015	09:39:45	AC1	occupied	0.8
11.03.2015	09:40:29	AC1	clear	0.8
11.03.2015	10:22:14	AC1	occupied	0.8
11.03.2015	10:23:03	AC1	clear	0.9

that is difficult to acquire e.g. external conditions or range of measurements. For instance, if competence is defined as the capacity of the sensor to provide the measurements it was designed for, this value is high when the sensor is working in its optimal conditions. The lack of that knowledge about the sensor and its surroundings makes it marginally useful in our trust calculation. Finally, we propose to replace plausibility with likelihood: whereas plausibility takes into account user background knowledge, likelihood depends only on the log file and takes into account the entry history.

The rest of this section presents our propositions for scoring each dimension, which takes into account meta-information and statistical analysis. *Reliability* is considered as a function of a sensor and time: $r(s, t)$, *likelihood* is related to the log entry: $lkh(l)$ and *credibility* applies to a log entry as well: $cr(l)$. They are finally aggregated into a trust value: $trust(l)$.

3.3 Reliability

Reliability, as a source metric, focuses on the specifics of the sensor, not the measures it provides. It is an *a priori* assessment of the source. This section first discusses the various approaches that can be proposed, organising them as constant vs. dynamic and meta-data-based vs. history-based; it then formalises the proposed definition.

Discussion. A basic approach could consist in making reliability a constant value, e.g. depending on the sensor type or brand: it might be known that specific sensors are of better quality than others and thus *a priori* provide more trustworthy information.

A way to enrich this basic definition is to take into account time and to define a dynamic reliability, for instance considering that this initial reliability value decreases when the sensor becomes older. This approach requires acquiring the knowledge about the obsolescence speed of the sensors, which might be difficult to know.

Note that it is possible to enrich further such a dynamic definition of reliability by taking into account maintenance operations, if their dates and types are known, although their interpretation can be debatable: they can be seen either as increasing reliability by slowing down the sensor ageing, or can be considered as the sign of the sensor needing repairs, casting doubts on the quality of the information it provides.

These approaches rely on the availability of very rich meta-information about the sensors, among which the type, brand, age, obsolescence speed and dates of maintenance operations. Another source of information that can be exploited to define sensor reliability is offered by the history of its previous outputs, which is available in the log file. Indeed, reliability can be related to the question whether the device is working properly or not, which can be derived from its downtime or from its error messages. However sensor log files usually are event-triggered, which means that the downtime is not reported as such. The next paragraph describes in more detail a reliability definition based on error messages.

Proposed Sensor Reliability Definition. The measure we propose is based on the interpretation according to which the more errors a sensor reports, the less reliable it is: error messages indicate it encounters problems. We propose a dynamical measure that automatically adapts to the current state of the sensor, depending on what happened in its recent history; formally, it is defined as:

$$r : \mathcal{S} \times \mathcal{T} \longrightarrow [0,1]$$

$$(s,t) \longmapsto 1 - \frac{|error(recent(\mathcal{L}_s, t))|}{|recent(\mathcal{L}_s, t)|} \tag{1}$$

where $recent : \mathcal{L} \times \mathcal{T} \to \mathcal{P}(\mathcal{L})$ provides the set of log entries produced by the sensor s in the considered time window t and $error : \mathcal{L} \to \mathcal{P}(\mathcal{L})$ is the function which extracts the set of error entries in this time window.

The definition of the considered time window, which determines the notion of "recent history" and the value of the reference $recent$ can take several forms: it can be directly defined as an entry number, indicating the number of previous messages one may want to take into account; it can also be a temporal window, from which the log entries to be considered must be retrieved.

3.4   Likelihood

Likelihood measures how likely a piece of information is, independently of its source, but usually depending on available external information. Its expression varies according to the type of this considered external information.

For instance, in the case where the considered piece of information describes the position of a train on a track, it might be confronted to a train schedule, so as to check the compatibility with this external knowledge.

In the case considered in this paper, as described in Sect. 3.1, the pieces of information indicate the sensor states. We propose to measure their likelihood according to their compatibility with a model stating the allowed state evolution. Indeed, it can for instance be known that a sensor cannot remain in the 'occupied' state at two consecutive time stamps. A more general state evolution model for our considered sensors is illustrated in Fig. 1: the two main states are *occupied* and *clear* and the several error states are distinguished. It can be seen that this sensor type cannot successively report *clear* and *disturbed* but an intermediary message *section disturbed* is used.

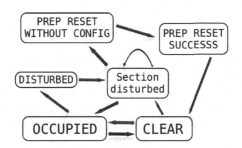

Fig. 1. Example of a state evolution model.

The proposed approach considers two cases: the message flow is compatible with the model or it is not. In the first case, the trust value of the previous log entry is considered. If it is strong, the likelihood will be high; if the log entry was untrustworthy, the likelihood will be lowered accordingly, indicating the fact that it could have been faulty. In the second case, the trust value of the previous log entry is also used to decrease likelihood: when that information was trustworthy, the likelihood will be low, otherwise the information is not considered enough to fully lower the likelihood score. The formal definition of $lkh : \mathcal{L} \rightarrow [0,1]$ is:

$$lkh(l) = \begin{cases} trust(prv(l)), & \text{if } l.message \text{ compatible with } prv(l).message \\ 1 - trust(prv(l)), & \text{otherwise} \end{cases}$$

(2)

where $prv : \mathcal{L} \rightarrow \mathcal{L}$ returns the single log entry l' which is the entry provided by the same sensor just before the current entry l and $l.message$ is compatible with $l'.message$ when that state evolution is allowed by the model.

3.5   Credibility

Credibility aims to confirm or deny a piece of information, independently of its source, by comparing it with information from other sources. Its expression depends on the type of information provided by other sources.

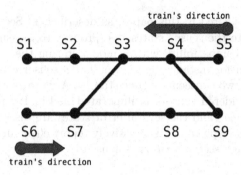

Fig. 2. Representation of the sensor locations on a portion of the railway structure

Discussion. In the case considered in this paper where the piece of information describes the position of a train on tracks, it might be confirmed by its neighbouring source which should have reported the passing train shortly before. To implement this principle, the relative positions of the sensors are required, for instance in the form of sensor network. Figure 2 illustrates such a network: the nodes represent sensors and the lines between them indicate that two sensors are neighbours. For instance, when sensor S2 reports an activity, it means that the train had to pass through sensor S3 and it should have reported that fact with a log entry.

Formalization. The proposed approach considers scoring credibility of the sensor's state by looking through the recent log entries to find the ones which confirm the event and the ones which contradict it. The previously computed trust values for the considered entries are aggregated, ending with the final fusion of two values representing confirmation and contradiction. Formally, the credibility function cr is thus defined as:

$$cr : \mathcal{L} \longrightarrow [0,1]$$

$$l \longmapsto agg_1(agg_2(confirm(l)), agg_3(infirm(l))) \tag{3}$$

where $confirm : \mathcal{L} \to \mathcal{P}(\mathcal{L})$ returns a set of entries that confirm l; $infirm(l) : \mathcal{L} \to \mathcal{P}(\mathcal{L})$ returns a set of entries that contradict l; agg_1, agg_2 and agg_3 are three aggregation operators applied to the trust scores of their set of logs.

Selection of Aggregation. An aggregation operator in general is a function which reduces a set of numbers into a single, meaningful, number. The selection of an operator opens a wide discussion due to the diversity and variety of existing aggregation operators, each with its characteristics and properties, see e.g. [6].

The purpose of $agg_{2,3}$ is to combine the trust values of multiple entries. We propose to discard conjunctive and disjunctive operators, which can be considered as too extreme and to favour compromise operators that allow a compensation effect. As all log entries have the same impact, we consider the average. The

agg_1 operator aims at combining the global confirmation trust (c) and the global information trust (i). We require the following behaviour at the boundaries: if $c = 1$ and $i = 0$, the aggregated result must be 1; if $c = 0$ and $i = 1$, it must be 0.

They ensure that a fully confirmed piece of information has the highest credibility score and a fully contradicted information has the lowest credibility score. Therefore agg_1 needs to be asymmetrical. To meet these requirements we propose to define $agg_1 : [0,1] \times [0,1] \to [0,1]$ as:

$$agg_1(c,i) = \frac{1 + c - i}{2} \qquad (4)$$

For the sake of simplicity, Eq. (3) omits its temporal dependence: confirmations and informations are looked for in recent entries. The notion of *recent* is equivalent to the one presented in Sect. 3.3. A too small window can result in "false negative", if the confirmation is earlier and outside the window However a too large window can result in "false positives": the confirmation does not exist but the previous train passage is included.

3.6 Trust

The final step then consists in aggregating the three dimensions: reliability, likelihood and credibility into the trust score.

We propose an approach to divide the overall trust scoring into two phases. First reliability and plausibility are aggregated. Indeed these dimensions both have an abating effect, leading to decrease trust, therefore, we propose to aggregate them using a t-norm, offering a conjunctive behaviour. The implementation described in Sect. 4 more specifically considers the probabilistic t-norm. Credibility can either increase or decrease trust due to its external factor as an opposite to the first two dimensions. We propose to consider a compromise operator, the weighted average. Trust is thus formally defined as

$$trust : \mathcal{L} \longrightarrow [0,1]$$
$$l \longmapsto \alpha \cdot r(l.sensor, l.fullDate) \cdot lkh(l) + (1 - \alpha) \cdot cr(l) \qquad (5)$$

where the constant $\alpha \in [0,1]$ is set *a priori* to manipulate the influence of both sides.

4   Illustration on Real Data

This section describes the implementation of the proposed approach for real-world data from the railway domain. Among different sensors, the axle counter (AC) was chosen for its crucial role in maintaining safe and efficient train traffic. The aim is thus to verify the information it produces e.g. "the train is on this part of track", "the train left this part of track", "the sensor is not working properly". The dataset contains 60 axle counters to provide information on the train presence in the different part of tracks. The example of messages produced by AC is presented in Table 1 and all types of messages are included in the graph shown in Fig. 1.

This section first describes the experimental protocol and then presents an illustrative example.

4.1   Experimental Protocol

The testing process is challenging due to the lack of a ground truth for the available dataset. We choose to illustrate our scoring by considering the original data as a reference and building a synthetic dataset from it with added random noise. We change 5% of the AC states randomly, where the changes mean replacing the message of the log entry with a different one. We constrain the noise injection to preserve the initial distribution of the sensor states, i.e. if the *disturbed* message appears 1% in the dataset and the *clear* message appears around 49%, the same proportions hold in the noisy data.

The initial values for the 3 dimensions are set to 1.0; the window defining recent entries for reliability and credibility is set to consider entries from the previous 10 min; for Trust, we set $\alpha = 0.75$ in Eq. (5).

4.2   Illustrative Example

Two approaches are considered when testing: applying noise to only one sensor or to multiple sensors.

Single Sensor Subject to Noise. In Fig. 3, the trust values (y-axis) for the modified sensor are plotted over the log-entry number (x-axis). The noise is applied only to this device, its positions are highlighted by the vertical lines. It is noticeable that the corrupted entries are recognised, the trust being lowered for these log entries. Also, the trust level does not recover immediately after the decrease but takes time to do so, which is reflected by the introduction of previous logs trust into the computation. The part of the chart around entry number 220 presents one of the cases where the trust value was not able to fully recover, due to encountering another invalid entry which ended with another decrease. This example shows the ability of this tool to properly handle this scenario as well.

Multiple Sensors Subject to Noise. In this case, noise is applied to all sensors to observe how different sensors affect each other's trust. Figure 4 shows the evolution of trust values for a reference sensor, vertical line show its modified entries, the ones of the other sensors are omitted. The interesting part is the smaller decrease in trust for the entries that were not modified. The explanation for it lies in other sensors and their low trust scores. Due to the correlation between sensors, one of them can influence the other's trust. The level of the

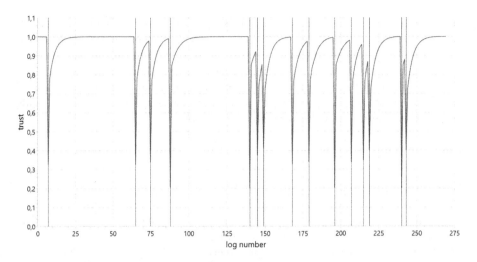

Fig. 3. Trust evolution for a single sensor affected by noise.

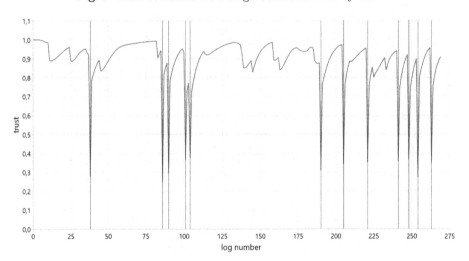

Fig. 4. Trust evolution for one sensor, when noise affects all sensors.

decrease depends on the trust value of the correlated sensor. Even though the trust value can decrease for the entries that were not modified, the level of that decrease is noticeably weaker compared to that of modified logs.

5   Conclusion

The variations in information produced by sensors bring out the need for an information quality scoring system taking into account both sensors and their output characteristics. Our approach proposes a modified version of dynamical trust scoring with three dimensions: reliability, likelihood and credibility. The temporal nature of the sensor's signal is considered in the aggregated trust score. We illustrated the proposed approach on a real-world railway dataset. Future works will include performing an experimental validation with statistical study generalising the illustrative example. Another perspective lies in proposing enriched scoring methods for presented dimensions.

Acknowledgements. This work was supported in part by Thales Polska.

References

1. Batini, C., Scannapieco, M.: Data and Information Quality. DSA. Springer, Cham (2016). https://doi.org/10.1007/978-3-319-24106-7
2. Besombes, J., d'Allonnes, A.R.: An extension of STANAG2022 for information scoring. In: International Conference on Information Fusion, FUSION 2008, pp. 1–7 (2008)
3. Blasch, E.P.: Derivation of a reliability metric for fused data decision making. In: IEEE National Aerospace and Electronics Conference, pp. 273–280 (2008)
4. Demolombe, R.: Reasoning about trust: a formal logical framework. In: Jensen, C., Poslad, S., Dimitrakos, T. (eds.) iTrust 2004. LNCS, vol. 2995, pp. 291–303. Springer, Heidelberg (2004). https://doi.org/10.1007/978-3-540-24747-0_22
5. Destercke, S., Buche, P., Charnomordic, B.: Evaluating data reliability: an evidential answer with application to a web-enabled data warehouse. IEEE Trans. Knowl. Data Eng. **25**(1), 92–105 (2013)
6. Detyniecki, M.: Fundamentals on aggregation operators. Technical report, University of California Berkeley. Ph.D. thesis (2001)
7. Florea, M.C., Bossé, É.: Dempster-Shafer theory: combination of information using contextual knowledge. In: International Conference on Information Fusion, FUSION 2009, pp. 522–528. IEEE (2009)
8. Florea, M.C., Jousselme, A.L., Bossé, É.: Dynamic estimation of evidence discounting rates based on information credibility. RAIRO-Oper. Res. **44**(4), 285–306 (2010)
9. Guo, H., Shi, W., Deng, Y.: Evaluating sensor reliability in classification problems based on evidence theory. IEEE Trans. Syst. Man Cybern. Part B (Cybern.) **36**(5), 970–981 (2006)
10. Lesot, M.J., Delavallade, T., Pichon, F., Akdag, H., Bouchon-Meunier, B., Capet, P.: Proposition of a semi-automatic possibilistic information scoring process. In: Proceedings of the 7th Conference of the European Society for Fuzzy Logic and Technology (EUSFLAT-2011) and LFA-2011, pp. 949–956. Atlantis Press (2011)
11. Lesot, M.-J., Revault d'Allonnes, A.: Information quality and uncertainty. In: Kreinovich, V. (ed.) Uncertainty Modeling. SCI, vol. 683, pp. 135–146. Springer, Cham (2017). https://doi.org/10.1007/978-3-319-51052-1_9
12. Mercier, D., Quost, B., Denœux, T.: Refined modeling of sensor reliability in the belief function framework using contextual discounting. Inf. Fusion **9**(2), 246–258 (2008)
13. Pichon, F., Dubois, D., Denoeux, T.: Relevance and truthfulness in information correction and fusion. Int. J. Approx. Reason. **53**(2), 159–175 (2012)

14. Pon, R.K., Cárdenas, A.F.: Data quality inference. In: Proceedings of the 2nd International Workshop on Information Quality in Information Systems, pp. 105–111. ACM (2005)

15. d'Allonnes, A.R., Lesot, M.-J.: Formalising information scoring in a multivalued logic framework. In: Laurent, A., Strauss, O., Bouchon-Meunier, B., Yager, R.R. (eds.) IPMU 2014. CCIS, vol. 442, pp. 314–324. Springer, Cham (2014). https://doi.org/10.1007/978-3-319-08795-5_33

16. Rogova, G., Hadzagic, M., St-Hilaire, M.O., Florea, M.C., Valin, P.: Context-based information quality for sequential decision making. In: 2013 IEEE International Multi-Disciplinary Conference on Cognitive Methods in Situation Awareness and Decision Support (CogSIMA), pp. 16–21 (2013)

17. Sidi, F., Panahy, P.H.S., Affendey, L.S., Jabar, M.A., Ibrahim, H., Mustapha, A.: Data quality: a survey of data quality dimensions. In: Proceedings of International Conference on Information Retrieval Knowledge Management, pp. 300–304 (2012)

18. Young, S., Palmer, J.: Pedigree and confidence: issues in data credibility and reliability. In: International Conference on Information Fusion, FUSION 2007, pp. 1–8 (2007)

Linear Parameter-Varying Two Rotor Aero-Dynamical System Modelling with State-Space Neural Network

Marcel Luzar[(✉)] and Józef Korbicz

Institute of Control and Computation Engineering,
University of Zielona Góra, ul. Pogórna 50, 65-246 Zielona Góra, Poland
{m.luzar,j.korbicz}@issi.uz.zgora.pl

Abstract. In every model-based approaches, i.e., fault diagnosis, fuzzy control, robust fault-tolerant control, the exact model is crucial. This paper presents a methodology which allows to obtain an exact model of high-order, non-linear cross-coupled system, namely Two Rotor Aero-dynamical System (TRAS), using a state-space neural network. Moreover, the resulting model is presented in a linear parameter-varying (LPV) form making it easier to analyze (i.e., its stability and controllability) and control. Such a form is obtained by direct transformation of the neural network structure into quasi-LPV model. For the neural network modelling, a *SSNN Toolbox* is utilized.

Keywords: Non-linear modelling · Neural networks
Linear parameter-varying system

1 Introduction

During the paste decade, there has been an increasing interest in developing techniques for analyzing and designing linear parameter-varying (LPV) control systems for non-linear plants. Numerous methods involves obtaining linearized dynamic plant model at some operating points, calculating a control procedure to fulfil local performance goals for each point and then, in real-time, regulating ("scheduling") the controller gains as the operating conditions vary. This method has been utilized successfully for many years, especially for process control and aircraft problems. Examples of relatively recent research (some of which involve modern fault-tolerant control design methods) include sewer networks [3], missile autopilots [10], aero-engines [13], wind turbines [7] and tank systems [12]. In spite of the past achievements of gain scheduling in practice, little is known about it theoretically as a non-linear and/or time-varying control technique. Moreover, obtaining the exact scheduling routine is more of an art than a science; while ad-hoc methods like curve fitting or linear interpolation can be good enough for simple static-gain controllers, giving the same results for dynamic multi-variable controllers can be a rather arduous process.

© Springer International Publishing AG, part of Springer Nature 2018
L. Rutkowski et al. (Eds.): ICAISC 2018, LNAI 10842, pp. 592–602, 2018.
https://doi.org/10.1007/978-3-319-91262-2_52

In general, the classical approaches to the control of LPV systems linearize the non-linear system model in some set of operating points and design one or more linear controllers for the system in these points [2]. Modern control paradigms such robust $\mathcal{H}_\infty$ control synthesis methods typically deal with this by requiring a linear nominal (state-space) model. However, a suitable, physical model may not always be available.

In the case when the system is simple, then it is easy to calculate a mathematical model using differential equations. Sometimes the producer of the system provides quite exact analytical model. However, in the case of high-order, non-linear cross-coupled systems classical modelling methods are usually very complicated. In order to solve such a problem, a novel methodology of designing LPV system model on the basis of the neural network is proposed in this paper. It has been decided to use artificial neural networks (ANNs) because they have some interesting properties especially attractive for modelling complex non-linear dynamic systems for which efficient analytic modelling methods do not exist. Among these properties there are the ability of approximating any non-linear functions, modelling system dynamics, parallel processing, generalization and adaptivity features [8]. However, the main disadvantage of ANNs is that disturbances decoupling and convergence to the origin are not guaranteed. Thus, the concept of this approach relies on the combination of neural network modelling abilities with an LPV technique. Thus, the proposed approach combines the positive features of the analytical and soft-computing methods. In order to do it, the state-space representation of the neural model is required. The above property is fulfilled by the recurrent neural network (RNN) [4], and such a neural model has been chosen for system modelling. The RNN has a state-space description, which can be converted into an LPV form [1]. Such a representation, especially attractive in LPV gain-scheduled control schemes [1,6], allows applying e.g., the observer-based methodology to design robust actuators as well as sensor fault detection and estimation schemes.

2   State-Space Neural Network

There are many different dynamic ANNs structures in the literature. Among them, there is a special class of network, called the State Space Neural Network (SSNN), which scheme is presented in Fig. 1. In such a network structure, the hidden layer outputs are propagated back to the input layer through a bank of unit delays. System order is defined by the number of such delays. Usually, it is possible to decide how many neurons are used to produce feedback.

Let $u(k) \in \mathbb{R}^n$ be the input vector, $\bar{x}(k) \in \mathbb{R}^q$ - the output of the hidden layer at time k, and $\bar{y}(k) \in \mathbb{R}^m$ - the output vector. Then the state-space form of the neural model given in Fig. 1 is defined as follows:

$$\bar{x}_{k+1} = \bar{g}(\bar{x}_k, u_k), \tag{1}$$
$$\bar{y}_k = C\bar{x}_k, \tag{2}$$

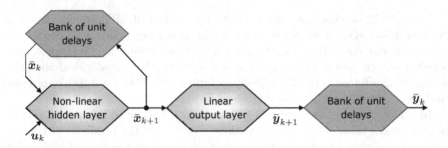

Fig. 1. The state-space neural network with one hidden layer.

where $\bar{g}(\cdot)$ is a non-linear function characterizing the hidden layer, and C represents synaptic weights between hidden and output neurons.

Introducing the weight matrix between input and hidden layers $\boldsymbol{W}^u$ and the matrix of recurrent links $\boldsymbol{W}^x$, the previous Eqs. (1–2) can be rewritten as follows:

$$\bar{\boldsymbol{x}}_{k+1} = h(\boldsymbol{W}^x \bar{\boldsymbol{x}}_k + \boldsymbol{W}^u \boldsymbol{u}_k) \tag{3}$$

$$\bar{\boldsymbol{y}}_k = C\bar{\boldsymbol{x}}_k \tag{4}$$

where $h(\cdot)$ denotes the activation function of the hidden neurons. In most examples, when the hyperbolic tangent activation function is chosen, the modelling results are satisfactory. Generally for the state-space models the outputs of hidden neurons, which constitute feedbacks, are unknown during training. Thus, the only way to train the state-space neural model is to minimize the simulation error. The training process can be done easier when the state measurement are available, using e.g., series-parallel identification scheme, like it is carried out in the external dynamic approach (the feedforward network with tapped delay lines). Despite of this disadvantage, the SSNNs popularity was build due to its number of positive features, contrary to fully and partially recurrent networks [9], and they are as follows:

- The states number (model order) can be determined independently from the number of hidden neurons.
  The main result of this fact is that the responsibility for defining the network state is on the neurons which propagate, through delays, their outputs back to the input layer. As a consequence, the output neurons are eliminated from the state. [9]. Contrary, in the recurrent networks, e.g. Williams-Zipser, Elman, locally recurrent networks, the model order is directly influenced by the number of the neurons making the modelling task more difficult.
- Model states feed the network input (which makes them easily accessible).
  The network input is fed by the model states (which allows direct access to them). In case when state values are available at some time instants, this feature can be very useful.
- State-space neural models are useful in a model-based fault diagnosis and a fault-tolerant control frameworks. State-space form allows to approximate size

and localization of a fault and to deal with different types of faults including multiplicative and additive ones.

Mentioned above advantages of SSNN make models of this kind a very interesting and promising tool used to solve different engineering issues i.e. the fault diagnosis problem. Also the class of non-linear state space models is strongly evaluated in different scientific approaches as a nominal model.

3   LPV Modelling with State-Space Neural Network

As was mentioned in introduction, obviously it is much easier to analyze the linear models instead of non-linear. Thus it is necessary to find a way to present a non-linear neural model in quasi-linear form. To fulfil this objective, this section provides a general methodology to transform a neural state-space model that can represent a general class of non-linear state-space models into a LPV model.

Let us consider the following discrete-time non-linear model:

$$x_{k+1} = g\left(x_k, u_k\right), \tag{5}$$
$$y_k = Cx_k, \tag{6}$$

where $x_k \in \mathbb{R}^n$ denotes the state vector, $y_k \in \mathbb{R}^{n_y}$ the output, $u_k \in \mathbb{R}^{n_u}$ the input vector, and $g\left(\cdot\right)$ is a non-linear function.

The main objective of this section is to represent such a model in the form of a polytopic discrete-time LPV model, described by the following equations:

$$x_{k+1} = A(h_k)x_k + B(h_k)u_k, \tag{7}$$
$$y_k = C(h_k)x_k, \tag{8}$$

where $A(h_k)$, $B(h_k)$, $C(h_k)$ are state-space matrices and $h_k \in \mathbb{R}^l$ is a time-varying parameter vector which ranges over a fixed polytope. The dependence of A, B and C on h_k symbolize a general discrete-time LPV model. To obtain such a model, it is proposed to use the RNN. For this purpose, a general structure of the RNN proposed by Lachhab et al. [4] is used, with some appropriate modifications.

The general structure of the discrete-time non-linear model represented by the proposed RNN is shown in Fig. 2 and described by following equations:

$$x_{k+1} = Ax_k + Bu_k + A_1\sigma(E_1x_k) + B_1\sigma(E_2u_k), \tag{9}$$
$$y_{k+1} = Cx_k + C_1\sigma(E_3x_k). \tag{10}$$

Matrices A, A_1, B, B_1, C, C_1, E_1, E_2 and E_3 are real valued, have appropriate dimensions and represent the weights which will be tuned during the RNN training process. The non-linear activation function $\sigma(\cdot)$, which is applied element-wise in (7)–(8), is considered to be continuous, differentiable and bounded. For that purpose, let us write (9) as

$$x_{k+1} = Ax_k + Bu_k + g\left(x_k\right), \tag{11}$$

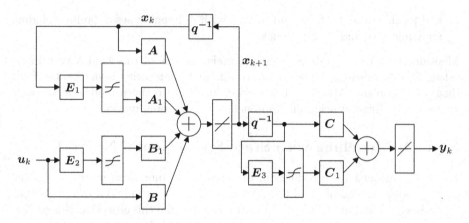

Fig. 2. Structure of the recurrent neural network.

where

$$g\left(x_k\right) = A_1\sigma(E_1 x_k) + B_1\sigma(E_2 u_k). \tag{12}$$

Such a RNN form leads to a general structure of the neural state-space model in the sense that, if it is transformed into a LPV model in the form of (5)–(6), the matrices A, B and C will be parameter dependent. In order to obtain a parameter dependence only in matrices A and B, it is necessary to remove the sigmoidal layers from the input and output paths.

After such a simplification, the modified RNN is implemented and presented in Fig. 3. Note that in such a structure the outputs are taken as the input to the sigmoidal layer, instead of the states. The modified RNN is described as follows:

$$x_{k+1} = Ax_k + Bu_k + A_1\sigma(E_1 C x_k) + B_1\sigma(E_2 u_k), \tag{13}$$
$$y_{k+1} = Cx_k. \tag{14}$$

It can be shown that the stability condition of this custom RNN is the same as that given in Theorem II.1 of [4] with the Lipschitz constant $L = \|E_1 C\|$. Such a structure can be easily implemented using, e.g., *SSNN Toolbox* [5] and trained with the Levenberg–Marquardt algorithm.

The main goal of further discussion in this section is to obtain the LPV model from the presented RNN. Let us assume that vector h_k depends on the vector of measurable signals $\rho_k \in \mathbb{R}^r$, referred to as scheduling signals, according to

$$h_k = s(\rho_k), \tag{15}$$

where $s \in \mathbb{R}^r \rightarrow \mathbb{R}^l$ is a continuous mapping. Generally in LPV systems scheduling parameters (all or some of them) are determined by input and output. A polytope can be represented in a matrix form and described as a convex hull composed of matrices N_i with the same dimension:

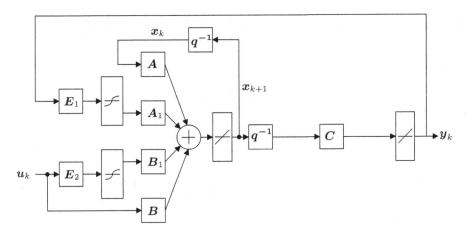

Fig. 3. Structure of a simplified recurrent neural network.

$$\mathrm{Co}\{N_i, \, i = 1, \ldots, l\} := \left\{ \sum_{i=1}^{l} h_k^i N_i, \, \sum_{i=1}^{l} h_k^i = 1, \, h_k^i \geq 0 \right\}. \qquad (16)$$

The time-varying parameter h_k ranges in a polytope $\boldsymbol{\Theta}$, which is considered as a convex set with vertices $v_1, v_2, \ldots, v_r$, i.e.,

$$h_k \in \boldsymbol{\Theta} := \mathrm{Co}\{v_1, v_2, \ldots, v_r\}. \qquad (17)$$

Finally, the objective is to transform the state-space neural model (9)–(10) into a polytopic LPV one (7)–(8) which has the properties presented above with

$$[\boldsymbol{A}(h_k) \, \boldsymbol{B}(h_k)] \in \tilde{\boldsymbol{P}}_h := \mathrm{Co}\{[\boldsymbol{A}_i \, \boldsymbol{B}_i], \quad i = 1, \ldots, l\}, \qquad (18)$$

where $\tilde{\boldsymbol{P}}_h \subset \mathbb{R}^l$.

First, let us define the time-varying parameters

$$h_k^i = \begin{cases} \frac{\sigma(\boldsymbol{E}_1^{(i)} \boldsymbol{x}_k + \boldsymbol{E}_2^{(i)} \boldsymbol{u}_k)}{\boldsymbol{E}_1^{(i)} \boldsymbol{x}_k + \boldsymbol{E}_2^{(i)} \boldsymbol{u}_k \boldsymbol{x}_k}, & \left(\boldsymbol{E}_1^{j} \boldsymbol{x}_k + \boldsymbol{E}_2^{(i)} \boldsymbol{u}_k \boldsymbol{x}_k \right) \neq 0, \\ 1, & \left(\boldsymbol{E}_1^{(i)} \boldsymbol{x}_k + \boldsymbol{E}_2^{(i)} \boldsymbol{u}_k \boldsymbol{x}_k \right) = 0, \end{cases} \qquad (19)$$

for $1 \leq i \leq l$, where $\boldsymbol{E}_1^{(i)}$ and $\boldsymbol{E}_2^{(i)}$ denote the i-th row in the respective hidden layer weight matrices which contains the sigmoid activation functions, and l denotes the number of the neurons in this layer. Then, (9) can be rewritten as

$$\boldsymbol{x}_{k+1} = \boldsymbol{A} \boldsymbol{x}_k + \boldsymbol{B} \boldsymbol{u}_k + \boldsymbol{A}_1 \boldsymbol{\Theta}_k \boldsymbol{E}_1 \boldsymbol{x}_k + \boldsymbol{B}_1 \boldsymbol{\Theta}_k \boldsymbol{E}_2 \boldsymbol{u}_k, \qquad (20)$$

where $\boldsymbol{\Theta}_k \in \mathbb{R}^{l \times l}$ is a diagonal matrix:

$$\boldsymbol{\Theta}_k = \begin{bmatrix} h_k^1 & 0 & \ldots & 0 \\ 0 & h_k^2 & \ldots & 0 \\ \vdots & \vdots & \ddots & \vdots \\ 0 & 0 & \ldots & h_k^l \end{bmatrix} \qquad (21)$$

that contains the variable parameters of the LPV model. In this way the main objective is achieved, i.e., the neural network model is transformed into an LPV model in the form of (7), where

$$A(h_k) = A + \sum_{i=1}^{l} h_k^i A_1^i E_1^{(i)}, \tag{22}$$

$$B(h_k) = B + \sum_{i=1}^{l} h_k^i B_1^i E_2^{(i)}, \tag{23}$$

with X^i being the i-th column of the matrix X, while $X^{(i)}$ stands for its i-th row. The time-varying parameter can be collected into a vector $h_k \in \mathbb{R}^l$ whose elements are contained in $\left[\underline{h_k^i}\ \overline{h_k^i} \right]$, where

$$\underline{h_k^i} = \min_{0 \le k \le T} h_k^i, \quad \overline{h_k^i} = \max_{0 \le k \le T} h_k^i, \tag{24}$$

for $1 \le i \le l$, with $k \in [0 \quad T]$ being the time interval in which the training data have been acquired. Note that $0 \le \underline{h^i}, \overline{h^i} \le 1$, so no further scaling is required.

Summarizing, the neural state-space model (9)–(10) is transformed into an LPV model in a polytope representation (7)–(8) that satisfies (18), where $A(h_k)$ and $B(h_k)$ are given by (22)–(23), respectively. The time-varying parameter vector is defined by (19) and its bounds are given by (24). Further on, an LPV model obtained in such a way will be denoted as an *NN-LPV* model.

4   Two-Rotor Aero-Dynamical System Modelling

4.1   Description of the TRAS

The aim of this section is to implement the proposed methodology and verify its usefulness using real system. For this task, the Two-Rotor Aero-dynamical System (TRAS) is chosen. The TRAS is a laboratory set-up designed for control experiments. In certain aspects its behaviour resembles that of a helicopter. From the control and modelling point of view it exemplifies a high order non-linear system with significant cross-couplings. The system is controlled from a PC. A schematic diagram of the laboratory set-up is shown in Fig. 4. The TRMS consists of a beam pivoted on its base in such a way that it can rotate freely both in thehorizontal and vertical planes. At both ends of the beam there are rotors (the main and tail rotors) driven by DC motors. A counterbalance arm with a weight at its end is fixed to the beam at the pivot. The state of the beam is described by four process variables: horizontal and vertical angles measured by position sensors fitted at the pivot, and two corresponding angular velocities. Two additional state variables are the angular velocities of the rotors, measured by tacho-generators coupled with the driving DC motors. In a casual helicopter the aerodynamic force is controlled by changing the angle of attack. The laboratory set-up from Fig. 1.1 is so constructed that the angle of attack is fixed. The

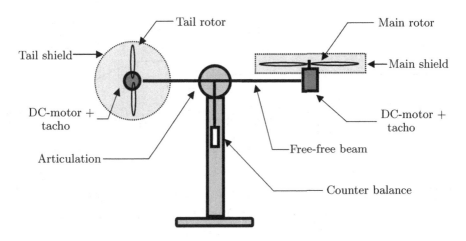

Fig. 4. Two-Rotor Aero-dynamical System.

aerodynamic force is controlled by varying the speed of rotors. Therefore, the control inputs are the supply voltages of the DC motors. A change in the voltage value results in a change of the rotation speed of the propeller which results in a change of the corresponding position of the beam. Significant cross-couplings are observed between the actions of the rotors: each rotor influences both position angles.

4.2   Experimental Results

The objective of this experiment is to provide an illustrative example which emphasize quality of the LPV model obtained with SSNN. The TRAS is chosen because its analytical model is known and given by the producer, thus it is easy to compare it with the NN-LPV model, derived in previous sections. The analytical model details can be found in the system user manual [11]. In order to model the system, the *SSNN Toolbox* [5] is chosen, which is designed by the author of this paper.

To obtain the NN-LPV model, 20000 samples of the training and validation data from the real system were collected during real-time simulation. The TRAS were controlled by tunned PID controller in a closed-loop. The reference signal was composed of square, sine wave and sawtooth wave signals, mixed randomly. The input signal was taken direct from the PID controller. Both input and output training data are presented in Fig. 5.

70% of the data set gathered from the system was taken as a training set, 15% as a validation set and 15% as a testing set. The TRAS can be perceived as a second order system, thus there were 2 neurons in output layer and 15 in hidden layer. The training process stops after 263 iterations, when the prescribed mean squared error (MSE) level was reached.

The result of azimuth angle modelling is presented in Fig. 6 with the associated modelling errors.

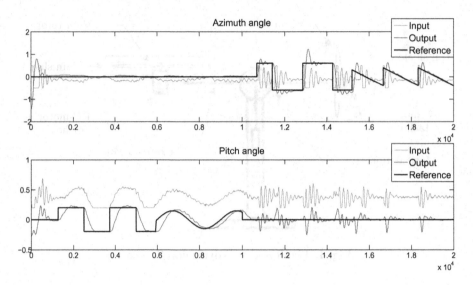

Fig. 5. Training data gathered from TRMS.

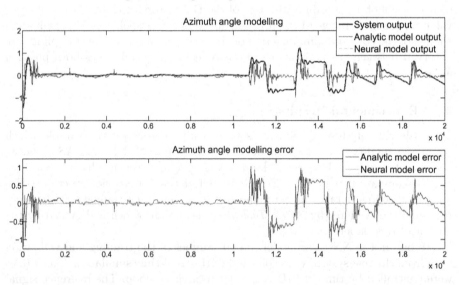

Fig. 6. Comparison of the real system output (azimuth angle) with analytical and NN-LPV model output.

It is clear that when the reference signal is almost constant, both models reflect real system behaviour with a good quality. However, when the reference signal is changed into a square and sawtooth wave, the difference between models quality is huge. Based on the modelling errors presented in the lower graph in Fig. 6, it is easy to see that the NN-LPV model reflects the real system much better than the analytical one. The same conclusion can be drawn by analysing

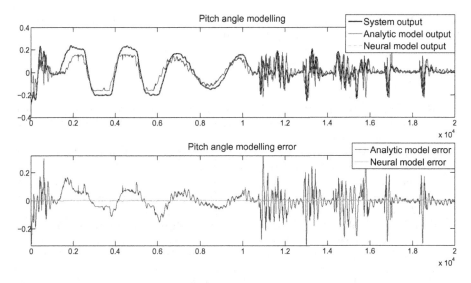

Fig. 7. Comparison of the real system output (pitch angle) with analytical and NN-LPV model output.

Fig. 7, in which the result of pitch angle modelling in the second is depicted. The difference between the NN-LPV model error and the analytical model error seems small in the first phase (when the reference signal is a sine wave). When the reference signal is a constant one, the disturbances caused by azimuth angle change are not modelled properly by the analytical model in contrast to the NN-LPV model.

In conclusion, the NN-LPV model is better than the analytical one and can be used successfully in applications, where the exact LPV model is crucial, e.g., model-based fault diagnosis and fault-tolerant control.

5   Conclusions

The main objective of the paper was to propose a new methodology in neural network modelling, which allows to present a modelled system in the LPV form. Nowadays, most of researchers involved in the non-linear systems analysis and control uses the state-space form for describing the system model. Therefore, utilize the state-space neural networks for system modelling seems a good idea. However, in literature it is difficult to find examples how to build such a neural structure. From this point of view, this paper attempts to fill this gap. Moreover, the SSNN model is transformed into LPV form, which makes it easier to analyze and control using well-known linear techniques. Thus, the aim of future research is to use the developed model in a robust actuator/sensor fault diagnosis and fault-tolerant control schemes.

602 M. Luzar and J. Korbicz

Acknowledgements. The work was supported by the National Science Centre of Poland under grant: UMO-2014/15/N/ST7/00749.

References

1. Abbas, H., Werner, H.: Polytopic quasi-LPV models based on neural state-space models and application to air charge control of a SI engine. In: 17th IFAC World Congress, pp. 6466–6471 (2008)
2. Bendtsen, J., Trangbæk, K.: Robust quasi-LPV control based on neural state-space models. IEEE Trans. Neural Netw. **13**(2), 355–368 (2002)
3. Hassanabadi, A., Shafiee, M., Puig, V.: Robust fault detection of singular LPV systems with multiple time-varying delays. Int. J. Appl. Math. Comput. Sci. **26**(1), 45–61 (2016)
4. Lachhab, N., Abbas, H., Werner, H.: A neural-network based technique for modelling and LPV control of an arm-driven inverted pendulum. In: Proceedings of the 47th IEEE Conference on Decision and Control, Cancun, Mexico, pp. 3860–3865 (2008)
5. Luzar, M., Czajkowski, A.: LPV system modeling with SSNN toolbox. In: American Control Conference, pp. 3952–3957 (2016)
6. Luzar, M., Witczak, M., Witczak, P., Auburn, C.: Neural-network based robust predictive fault-tolerant control for multi-tank system. In: 13th European Control Conference (ECC), pp. 276–281 (2014)
7. Martin, D.P., Johnson, K.E., Zalkind, D.S., Pao, L.Y.: LPV-based torque control for an extreme-scale morphing wind turbine rotor. In: American Control Conference, pp. 1383–1388 (2017)
8. Nørgaard, M., Ravn, O., Poulsen, N.K., Hansen, L.K.: Neural Networks for Modelling and Control of Dynamic Systems. Springer, Heidelberg (2014)
9. Patan, K.: Artificial neural networks for the modelling and fault diagnosis of technical processes. Springer, Heidelberg (2008). https://doi.org/10.1007/978-3-540-79872-9
10. Shen, Y., Yu, J., Luo, G., Mei, Y.: Missile autopilot design based on robust LPV control. J. Syst. Eng. Electron. **28**(3), 536–545 (2017)
11. Two Rotor Aero-Dynamical System–User's Manual (2013). www.inteco.com.pl
12. Xu, F., Puig, V., Ocampo-Martinez, C., Olaru, S., Niculescu, S.: Robust MPC for actuator-fault tolerance using set-based passive fault detection and active fault isolation. Int. J. Appl. Math. Comput. Sci. **27**(1), 43–61 (2017)
13. Yang, D., Zhao, J.: $\mathcal{H}_\infty$ output tracking control for a class of switched LPV systems and its application to an aero-engine model. Int. J. Robust Nonlinear Control **27**(12), 2102–2120 (2017)

Evolutionary Quick Artificial Bee Colony for Constrained Engineering Design Problems

Otavio Noura Teixeira[1(✉)], Mario Tasso Ribeiro Serra Neto[2],
Demison Rolins de Souza Alves[2], Marco Antonio Florenzano Mollinetti[3],
Fabio dos Santos Ferreira[2], Daniel Leal Souza[4], and Rodrigo Lisboa Pereira[5]

[1] Federal University of Para (UFPA), Tucurui, Brazil
`onoura@gmail.com`
[2] University Centre of the State of Para (CESUPA), Belém, Brazil
`marioserra2602@gmail.com`, `demisonalves9@gmail.com`, `ferreira06@gmail.com`
[3] Tsukuba University, Tsukuba, Japan
`marco.mollinetti@gmail.com`
[4] Federal University of Para (UFPA), Belém, Para, Brazil
`daniel.leal@gmail.com`
[5] Federal Rural University of Amazonia (UFRA), Paragominas, Brazil
`rod.lisboa@gmail.com`

Abstract. The Artificial Bee Colony (ABC) is a well-known simple and efficient bee inspired metaheuristic that has been showed to achieve good performance on real valued optimization problems. Inspired by such, a Quick Artificial Bee Colony (QABC) was proposed by Karaboga to enhance the global search and bring better analogy to the dynamic of bees. To improve its local search capabilities, a modified version of it, called Evolutionary Quick Artificial Bee Colony (EQABC), is proposed. The novel algorithm employs the mutation operators found in Evolutionary Strategies (ES) that was applied in ABC from Evolutionary Particle Swarm Optimization (EPSO). In order to test the performance of the new algorithm, it was applied in four large-scale constrained optimization structural engineering problems. The results obtained by EQABC are compared to original ABC, QABC, and ABC + ES, one of the algorithms inspired for the development of EQABC.

Keywords: Metaheuristics · Artificial Bee Colony
Quick Artificial Bee Colony · Optimization
Constrained optimization · Structural Engineering Design

1 Introduction

The Artificial Bee Colony (ABC) is a Swarm Intelligence algorithm proposed by [1] that is based on the foraging behavior of honey bees, which follows the guidelines of a minimal behavioral model of bees established by [2]. The algorithm simulates the search process of bees for foraging food sources around their

© Springer International Publishing AG, part of Springer Nature 2018
L. Rutkowski et al. (Eds.): ICAISC 2018, LNAI 10842, pp. 603–615, 2018.
https://doi.org/10.1007/978-3-319-91262-2_53

hive - its exploration and exploitation process in order to solve numerical optimization problems. The algorithm increasing popularity over the years can be associated to its simplicity, lightness and versatility in solving unconstrained and constrained, single objective and multi-objective, continuous and combinatorial optimization problems present in various fields of study [3,4].

Based on a previous hybrid metaheuristic developed by Mollinetti et al. [5] for real valued optimization problems, together with an efficient local search ABC approach by Karaboga and Gorkmeli [6], a hybrid metaheuristic has been devised by the authors to account for the necessity of increasing the algorithm robustness while producing high quality solutions to prevent the algorithm from getting stuck in suboptimal local optima. The novel metaheuristic combines the Evolutionary Strategies found in Artificial Bee Colony + Evolutionary Strategies proposed by Mollinetti et al. in [5] with QABC, resulting in a novel algorithm named Evolutionary Quick Artificial Bee Colony. The premise of the EQABC is to cover both algorithms deficiencies by performing steps from the ABC+ES and QABC.

To validate the proposed algorithm, the algorithm is tested on several benchmarks present in the optimization literature in order to measure the performance and robustness of the new approach. The results are compared against other adaptations of the ABC.

The paper is organized as it follows: Sect. 2 describes the features of the original ABC, while Sect. 3 explains the QABC. Section 4 describes the proposed method (Evolutionary Quick Artificial Bee Colony). Section 5 discusses correlated work conducted to solve engineering design problems. Section 6 details the experiment, while Sect. 7 presents the results of the method and compares with the results of other techniques. Lastly, Sect. 8 outlines the conclusion of the paper.

2   Artificial Bee Colony

Based on the mathematical model of honey bees foraging behavior of Tereshko and Loengarov [2], Karaboga [1] proposed a metaheuristic (Algorithm 1) named Artificial Bee Colony (ABC) simulating the minimum process of honey bee's behavior for the solution of diversified problems [1,7]. This metaheuristic has three distinct characterizations of candidate solutions, where they can be classified as: (1) Employed Bees; (2) Onlooker Bees; (3) Scout Bees. Each group has different amounts of bee, where each bee in each group contains the food source and a value suitable for such a source in its memory, i.e. the solution for the problem and its fitness value [3].

According to Karaboga [1], ABC is controlled by 4 steps, that will be explained in next subsections: initialization of candidate solutions (Bees); Employed bees phase; Onlooker bees phase; Scout bees phase. And by three parameters: which represent the number of bees in each group; a threshold value whose purpose is to remove a solution from a local minimum; and finally, the stopping criterion that can be defined as a maximum number of generations or until the best overall result is found, i.e., the solution converges to a desired accumulation point [1,3,4].

2.1   Initialization

In the initialization phase, food sources are generated by Eq. (1), using the bounds of each decision variable,

$$X_{ij} = L_i + rand(0,1) * (U_i - L_i), \tag{1}$$

where X_{ij} is the value of the j-th variable of the dimension problem of the i-th bee and L_i and U_i are the respective bounds for each dimension of the problem. Then, obtained solutions are evaluated and their objective function values are calculated.

2.2   Employed Bees Phase

During this phase, the bees decide go out from the hive to perform a search in the neighborhood to obtain better sources of food (solutions) than their current one [4]. This step is modeled mathematically by means of Eq. (2),

$$NX_{ij} = X_{ij} + rand(0,1) * (X_{ij} - X_{kj}), \tag{2}$$

where, X_{ij} is the j-th value of the i-th bee and X_{kj} the j-th value of another bee inside the hive, such as $i \neq k$. After producing a new candidate (NX_{ij}), the new fitness value is calculated. Then, a greedy selection is applied between X and NX. The i-th bee decides to keep X, it's threshold value is increase by one [4,8].

2.3   Onlooker Bees Phase

After the return of the employed bees, information is shared between onlooker bees via waggle dance. The onlooker bees select the bee with the most agitated dance, i.e. the one that stores the best solution to trade information [1]. Computationally, this analogy is equivalent to selecting a candidate solution to perform another local search based on a selection method. The selection mechanism is usually similar to the ones seen in Evolutionary Algorithm, such as fitness proportionate roulette or tournament.

According to Karaboga et al. [8], in the original ABC, to calculate the probability (P_i), of a bee is selected, fitness values are used in Eq. (3),

$$p_i = \frac{Fitness_i}{\sum_{i=1}^{n} Fitness_i}, \tag{3}$$

After a bee is selected, as in the employed bees phase a neighbor source NX_{ij} is generated by Eq. (2), and its fitness value is calculated. Then, a greedy selection is applied between the new and old values. If the old value determines a better fitness value then the new one, the threshold variable is increased by one [4,8,9].

2.4   Scout Bees Phases

After the other steps, if a bee reached a predetermined number of iterations without any improvement, this bee becomes a Scout. For each scout bee, a random value is replaced by a new one generated by Eq. (1). Clearly, the purpose of this phase is to try to move stagnated solutions out of local optima.

Algorithm 1. Pseudo code of the ABC algorithm [10]
1: Objective function: f(x), x = (x1, x2, ..., xN);
2: Generate an initial solutions (bees) for each employed bee (SN) with Eq. (1).
3: Evaluate fitness value for each bee in the hive.
4: Initialize cycle = 1;
5: For each employed bee
5.1: Produce new food sources NX_{ij} with Eq. (2).
5.2: Evaluate new solution
5.3: If new food sources (NX) are better than previous (X), memorize new position.
6: End for.
7: Calculate the probability values of the solution.
8: For each onlooker bee:
8.1: Produce a new population from the selected populations.
8.2: Produce new food sources NX_{ij} with Eq. (2).
8.3: If new food sources (NX) are better than previous (X), memorize new position.
9: End for.
10: If there is any abandoned solution i.e. employed becomes scout, then replace its position with new random source positions
11: Memorizes the best solution achieved so far
12: cycle = cycle + 1
13: if stopping criterion is satisfied, then stop otherwise go to step 5.

3   Quick Artificial Bee Colony

According to Karaboga and Gorkmeli [6], during the steps of employed and onlooker bees, the original ABC generates new solutions through Eq. (2), that is, the onlooker bees are developing the same action as the employed bees. The same author suggests that for a better analogy to the hive of real world bees, the equation needs to be modified to the onlooker phase.

With this information, the QABC proposed in [11], reinforces the idea that the bee selects another bee from the dance of the employed bees and exchanges information about the food sources. However, this information exchange is now modeled by Eq. (4).

$$NX_j^{best} = X_j^{best} + rand(0,1) * (X_j^{best} - X_{ij}), \qquad (4)$$

For the modified equation, NX_j^{best} the new variable generated for the best solution, X_j^{best} corresponds to the best j-th variable of the best solution achieved so far and X_{ij} represents the value of j-th variable of i-th bee, i.e. a neighbor bee.

In [6], the author suggests that to determine a neighboring bee, multiple methods can be used for this purpose, such as the average of Euclidean distance - which had been used by the author -. However, according to [10], restrictive functions tend to improve with small advances until an efficient result is obtained, so for this article, all bees present in the hive were considered as neighbors.

Since the QABC matches the other steps of the ABC, Algorithm 2 presents the modifications inserted in the onlooker phase by [6,11].

Algorithm 2. Pseudo code of the new onlooker phase in QABC algorithm [6,11]

8: For each onlooker bee:

    8.1: Select a solution (bee) depending on its probability.

    8.2: Find the best solution among the neighbors of selected solution.

    8.3: Generate a new candidate solution using Eq. (4)

    8.4: Apply a greedy selection between the new and old solution, memorize the best solution found so far.

9: End for.

---

4 The Proposed Method

The novel metaheuristic mixes the Evolutionary Strategies found in Artificial Bee Colony + Evolutionary (ABC + ES) Strategies proposed by Mollinetti et al. in [5] with QABC described by Karaboga and Gorkmeli [6], resulting in a novel algorithm named Evolutionary Quick Artificial Bee Colony.

4.1 Evolutionary Strategies Phase of Artificial Bee Colony + ES

As said before, ABC + ES is a hybrid metaheuristic that combines ES with ABC using some elements from the Evolutionary Particle Swarm Optimization (EPSO). Equation (5) illustrates the new movement formula applied to improve solutions that came from employed and onlooker bees,

$$X_m^{(t+1)} = X_m^{(t)} - \phi * (X_m^{(t)} - X_k^{(t)}) + \gamma_m^* (P_g^* - X_m^{(t)}), \tag{5}$$

where $X_m^{(t)}$ is the current value of the food source, $X_m^{(t+1)}$ is the clone for the next generation, $\gamma^m *$ stands for the social coefficient and P_g^* for the global best reference position. This concept was borrowed from the Evolutionary Particle Swarm Optimization (EPSO) [12], a well-known modification of the PSO that has been adapted to solve electrical engineering applications.

Analogous to the EPSO, P_g^* acts as a reference point for the candidate solution towards the best know local optimum so far, while $\gamma^m *$ is a weight that dampens the social component and prevents the candidate solution from greatly distancing itself from the global optimum reference by forcing the solution to converge towards P_g^* [5]. According to [5,12], the mutation operator is responsible for mutating the social weight and global P_g^* by a slight perturbation of their values. Doing a Gaussian perturbation to the value of the coefficients may lead solutions to more fruitful regions. The mutation process is demonstrated by (6) and (7).

$$\gamma_m^* = \gamma_m + (1 + \sigma N(0,1)), \tag{6}$$

$$P_g^* = P_g + (1 + \sigma_g N(0,1)), \tag{7}$$

where $\sigma N(0,1)$ stands as a random number generated by a Gaussian distribution of mean 0 and standard deviation 1.

4.2   Evolutionary Quick Artificial Bee Colony

At the QABC, after the search for new food sources in the neighborhood, the employed bees return to the hive and the best global solution exchanges information with their neighbors for a better global search. However, when analyzing the context ABC vs QABC, ABC promotes an extra movement for the scout bees, whereas the QABC removes this movement to apply the global search, that is, the ABC loses one of its local mechanisms.

The new equation for the clones of employed bees of QABC stems from QABC onlooker phase with the insertion of a dump factor and modifications in the random number generate in Eq. (2), as described below in Eq. (8),

$$CX_{ijk} = CX_{ijk} + (\theta * (\gamma_m^* * (CX_{ijk} - X_k^{best}))), \tag{8}$$

where, CX_{ijk} represents the value of k-th variable of i-th clone of j-th bee, θ is dump a factor subject to $\theta \in [0,1]$ and X_k^{best} is the k-th value of the best solution achieved so far. After modifying every clone for every variable in the dimension of the problem, the clone fitness is evaluated with the objective function and a greedy selection is applied to found the best clone generated.

Like the ABC + ES, γ_m^* acts as a weight that dampens the exchange of information between the clone and the best solution and prevents the candidate solution from greatly distancing itself from the global optimum reference by forcing the solutions to converge towards X_k^{best}. Acting in tandem with the global best reference, the reproduction and mutation operators of the ES influence (5) by other means. The reproduction operator replicates the original candidate solution CN times to apply (9), and replace it with the best of the replicas if one of them proves to be better than the original. It is important to state that only the replicas will carry out the updated movement formula (8) while the original candidate solution will maintain the original movement formula (2). This is because if the clones perform poorer than the original solution, the ABC+ES will try to perform as good as the ABC to prevent any further loss.

Overall algorithm steps: (1) Employed bees search for new food sources in the neighborhood; (2) Onlooker bees selects the best food sources based on the employed dance, i.e. the best solution found will learn with every bee for a better global result; (3) Employed bees trade information with the best solution found, i.e. bees learn about a new food source that came from their information with global best (Step introduced by EQABC). With that, the bees that teach the way to the best solution learn a possible best path discovered by such, thus, the local search performed is also used to obtain a possible global best solution in the next iteration of the algorithm.

Algorithm 3, in the next page, describes the application of EQBC for solving constrained engineering design problems, steps with marked in grey (2, 4, 7.1, 9-10) are the steps inserted in the QABC algorithm.

Algorithm 3. Pseudo code of the EQABC for Engineering Design Problems

1: Objective function: f(x), x = (x1, x2, ... , xN);
2: While solutions have violations, do
3: Generate an initial solutions (bees) for each employed bee (SN) with Eq. (1).
4: end while
5, 6: Steps 5 and 6 from Algorithm 1.
7: For each onlooker bee:
    7.1: Select a solution (bee) depending using tournament.
    7.2: Generate a new candidate solution for the selected bee using Eq. (4)
    7.3: Apply a greedy selection between the new and old solution, memorize the best solution found so far.
8: End for.
9: For each employed bee:
    9.1: Generate NC clones.
    9.2: For each clone
        9.2.1: Produce new food sources for the clone using Eq. (8)
    9.3: end for
    9.4: Evaluate every clone and select the best.
    9.5: If the best clone is better than the current bee, replace bee with clone, else add +1 to bee threshold.
10: end for.
11: If there is any abandoned solution i.e. employed becomes scout, then replace its position with new random source positions
12: Memorizes the best solution achieved so far, if it has 0 violations.
13: cycle = cycle + 1
14: if stopping criterion is satisfied, then stop otherwise go to step 5.

---

5   Correlated Work

Work related to hybridizations of the ABC is extensive due to the increasing popularity of the algorithm. A complete survey describing prominent approaches and modifications of the ABC can be found in [13]. Several authors integrated elements that range from other population based heuristics to mathematical global optimization algorithms. Examples of such work is the ABC + ES proposed by [5], that was used as base for EQABC. Yildiz [14], proposed and ABC + Taguchi method that produced good results for the same applications. Jatoth and Rajasekhar [15], proposed and hybrid ABC with Differential Evolution algorithm for designing the fractional order of PI controller. Sundar and Singh [16] presents a hybrid approach combining ABC with a novel local search to solve non-unicost set covering problem.

For engineering design problems, a considerable amount of results obtained by different authors is found at [5, 17–21]. Taking these into consideration, results obtained by [10] indicates that applying ABC with adjustments to bounds and threshold rules may yield even better results than the aforementioned. Therefore, the results obtained by EQBC are mainly compared to the ones from the work of Garg [10], since they are the best obtained so far.

6 Experiment

To validate the proposed algorithm, the algorithm is tested on several benchmarks present in the optimization literature [17–21] in order to measure the performance and robustness of the new approach. These examples have linear and nonlinear constraints, and have been previously solved using variety of other techniques, which is useful to determine the quality of the solutions produced by the proposed approach.

6.1 Settings

To ensure that no results were influenced by the machine load, a total of 30 different runs for each test case were used to obtain any statistical results presented. About the algorithms, Table 1, in the next page, presents the configurations used by ABC [10], which will be used in this article. For comparison purposes, the results obtained by the QABC will be compared to several other methods: the original QABC proposed by [6] and implemented by the authors; the ABC proposed by Garg [10]; and the ABC + ES presented by Mollinetti et al. [5].

Table 1. Settings for every algorithm.

Employed Bees	20 x *Problem Dimension*
Onlooker Bees	Employed Bees ÷ 2
Threshold	(Employed Bees * *Problem Dimension*) ÷ 2
Max number of cycles	500
Clones (ABC+ES, EQABC)	10% of Employed Bees
Θ (EQABC)	0.09

6.2 Design of Pressure Vessel

According to [10, 17–19], Design of Pressure Vessel (DPV) is an engineering problem that aims to minimize the material cost, shaping and welding of a cylindrical container that is limited at both ends by hemispherical lids. The DPV consists of four design variables, which are shell thickness (Ts, x1); Head thickness (Th, x2); Inner radius (R, x3); Length of cylindrical section of the container (G, x4). It is emphasized that the values for the variables are multiples of 0.0625 in., because they relate to the thickness of laminated steel sheets, which are continuous variables. Eq. (9) displays the cost function of the DPV

$$Minimize f(x) = 0.6224x^1x^3x^4 + 1.7781x_2x_3^2 + 3.1661x_1^2x_4 + 19.84x_1^2x_3 \quad (9)$$

This structural problem has been solved by many researchers within the following bounds: $1 \times 0.0065 \le$ x1, x2 $\le 99 \times 0.0625$; $10 \le$ x3 ≤ 200; $10 \le$ x4 ≤ 240.

6.3 Design of Welded Beam Design Problem

The Welded Beam Design (WBD) is a simplified example of structural engineering that deals with issues of complex designs. The objective of this problem is the minimization of Eq. (10) referring to the cost of manufacturing a steel beam that is subject to some constraints such as shear stress, bending stress in the beam, load buckling on the bar, deflection of the end beam and lateral restraints [10]. The variables of the problem are: weld thickness (h or x1); beam width (x2); beam thickness (t or x3); length of the beam (x4).

$$Minimize f(x) = 1.1047x_1^2x_2 + 0.04811x_3x_4(14 + x_2) \quad (10)$$

This structural problem has been solved within the following bounds: $0.1 \le$ x1 ≤ 2; $0.1 \le$ x2 ≤ 10; $0.1 \le$ x3 ≤ 10; $0.1 \le$ x4 ≤ 2.

6.4 Design of Tension/Compression String Problem

The MWTCS problem consists of the minimization of the weight of the tension/compression of a spring, which is subject to some constraints: minimum deflection; shear strain; wave frequency; outside diameter limit and project variables. Th variables of the MWTCS are: wire diameter (d or x1), Average diameter coil (D or x2) and the number of active, coils (P or x3). Eq. (11) is the cost function of the MWTCS [10].

$$Minimize f(x) = (x_3 + 2)x_2x_1^2 \quad (11)$$

This problem has been solved within the following bounds: $0.05 \le$ x1 ≤ 2; $0.25 \le$ x2 ≤ 1.3; $2 \le$ x3 ≤ 15.

6.5 Speed Reducer with 11 Restrictions

The main objective of the Speed Reducer Design with 11 Restrictions (SDR11) is to find the minimum of Eq. (12) with respect to the volume of the gearbox (and therefore its minimum weight), subject to various constraints such as the bending stress of gear teeth, surface tension, transverse rods deviations and the shaft tension [7]. The problem variables are: face width (X1 or b); teeth module

(X2 or m); quantity of the pinion teeth (z or X3); length of the first shaft between the bearings (L1 or X4); length of the second shaft between the bearings (X5 or L2); diameter of the first axis (d1 or X6); diameter of the second axis (X7 ord2). Equation (12) presents the cost function of the SRD11 [5].

$$
\begin{aligned}
Minimize f(x) &= 0.7854 x_1 x_2^2 (3.333 x_3^2 + 14.993 x_3 - 43.0934) \\
&\quad -1.508 x_1 (x_6^2 + x_7^2) + 7.477 (x_4 x_6^2 + x_5 x_7^2) \\
&\quad +0.7054 (x_4 x_6^2 + x_5 x_7^2).
\end{aligned}
\tag{12}
$$

This problem has the following bounds: $2.6 \leq x1 \leq 3.6$; $0.7 \leq x2 \leq 0.8$; $17 \leq x3 \leq 28$; $7.3 \leq x4 \leq 8.3$; $7.8 \leq x5 \leq 8.3$; $2.9 \leq x6 \leq 3.9$; $5.0 \leq x7 \leq 5.5$.

7 Results

For comparative purposes, the statistical metrics, mean, standard deviation and presentation of the best result of each algorithm will be displayed through Table 2, while Table 3 presents the variables found by the best EQABC result.

Table 2. Statistical results for each algorithm

Problem	Algorithm	Best	Mean	Std. Dev
DPV	ABC [10]	5804.44867	**5805.47391**	1.41146
	ABC + ES [5]	5933.91933	6951.80318	7.656265E+06
	QABC	5805.56291	5821.53035	15.65631
	EQABC	**5804.40957**	5805.57158	1.74142
WBD	ABC [10]	1.69526	1.69531	2.83623E-5
	ABC + ES [5]	**1.468497**	**1.60069**	9.81824E-03
	QABC	1.77072	1.75825	0.024323
	EQABC	1.73419	1.73481	0.02012
MWTCS	ABC [10]	0.01266	0.012668	9.42943E+06
	ABC + ES [5]	**0.00282**	**0.00282**	0
	QABC	0.01287	0.01493	0.00131
	EQABC	0.012665	0.012688	2.18691E-5
SDR11	ABC [2]	2996.34816	-	-
	ABC + ES [5]	2894.90134	2894.90134	0
	QABC	2900.64100	2942.33536	31.93720
	EQABC	**2894.38218**	**2894.38219**	7.86238E-6

In Table 2, it is possible to verify that EQABC produced the best solutions when compared to the other algorithms for DPV and SDR11 problems. However, in DPV, ABC proposed by Garg [10] had overall performance better than

EQABC. In SDR11, EQABC was slightly better than ABC + ES at best value and mean. For MWTCS, the best solution found was produced by ABC + ES in [5] and EQABC was slightly worse than this method. For WBD, EQABC was worse than ABC + ES, having medium statistical significance. For every design problem, EQABC was statistically better than original QABC.

Results corroborate the fact that the proposed method is as good as ABC + ES and better than QABC for this class of problems. This suggests that these new steps inserted in QABC algorithm can be a good rival for other metaheuristics in optimization problems, because [10] shows that his method, addressing more exact parameters for initialization, were more efficient than a diversified number of algorithms proposed in [17–21] and EQABC algorithm produced solutions better than ABC.

Table 3. Bests solutions (variables) found by Evolutionary Quick ABC

Design variables	DPV	WBD	MWTCS	SDR11
x1	0.7275975277524102	0.20343968792199735	0.05183332748973103	3.5000000001430376
x2	0.3596506012726204	3.53113706604078	0.360198314412783	0.7000000000071295
x3	37.699192556105054	9.009607336656245	11.08782401247815	17.00000000028451
x4	239.99713824278132	0.20696947416771136	-	7.300000001432028
x5	-	-	-	7.715319922547744
x6	-	-	-	2.9000000009645936
x7	-	-	-	5.286654466654735
Cost	5804.409569077998	1.7341910024535336	0.01266564473383818	2894.382186837499

8   Conclusion

Comparison of the results suggests that the ES operators enhance the quality of the candidate solutions leading to a better outcome when comparing QABC and EQABC. In overall performance, EQABC performed better than ABC in test cases that QABC was not better than ABC. This may indicate that the approach could produce interesting results on more complex domains, for example training of neural networks in supervised learning paradigm and unconstrained functions with a larger amount of parameters.

Despite the fact that increasing the amount of employed bees and maximum number of cycles of the algorithm could be a practical way of increasing the diversity of solution, other alternatives could prove more interesting. One of such, a rather common procedure among metaheuristics to solve different types of problem, is by parallelizing either the candidate or the entire population. The first speeds up the algorithm to allow more iterations on a shorter time, while the second cleverly refines the solutions up to a desired point by creating a specific topology of clusters of individuals.

Further tests on feed forward neural networks, unconstrained problems and mechanical design will be made to further investigate the performance of EQABC against other techniques for different applications and a parameter analysis for the number of clones and θ value found in the main equation of the proposed method.

References

1. Karaboga, D.: An idea based on honey bee swarm for numerical optimization. Technical report-tr06, Erciyes university, engineering faculty, Computer Engineering Department (2005)
2. Tereshko, V., Loengarov, A.: Collective decision making in honey-bee foraging dynamics. Comput. Inf. Syst. **9**(3), 1 (2005)
3. Karaboga, D., Akay, B.: A comparative study of artificial bee colony algorithm. Appl. Math. Comput. **214**(1), 108–132 (2009)
4. Karaboga, D., Basturk, B.: On the performance of artificial bee colony (ABC) algorithm. Appl. Soft Comput. **8**(1), 687–697 (2008)
5. Mollinetti, M.A.F., Souza, D.L., Pereira, R.L., Yasojima, E.K.K., Teixeira, O.N.: ABC+ES: combining artificial bee colony algorithm and evolution strategies on engineering design problems and benchmark functions. In: Abraham, A., Han, S.Y., Al-Sharhan, S.A., Liu, H. (eds.) Hybrid Intelligent Systems. AISC, vol. 420, pp. 53–66. Springer, Cham (2016). https://doi.org/10.1007/978-3-319-27221-4_5
6. Karaboga, D., Gorkemli, B.: A quick artificial bee colony (qABC) algorithm and its performance on optimization problems. Appl. Soft Comput. **23**, 227–238 (2014)
7. Binitha, S., et al.: A survey of bio inspired optimization algorithms. Int. J. Soft Comput. Eng. **2**(2), 137–151 (2012)
8. Karaboga, D., Akay, B., Ozturk, C.: Artificial bee colony (ABC) optimization algorithm for training feed-forward neural networks. MDAI **7**, 318–319 (2007)
9. Karaboga, D., Basturk, B.: A powerful and efficient algorithm for numerical function optimization: artificial bee colony (ABC) algorithm. J. Glob. Optim. **39**(3), 459–471 (2007)
10. Garg, H.: Solving structural engineering design optimization problems using an artificial bee colony algorithm. J. Ind. Manag. Optim. **10**(3), 777–794 (2014)
11. Karaboga, D., Gorkemli, B.: A quick artificial bee colony-qABC-algorithm for optimization problems. In: 2012 International Symposium on Innovations in Intelligent Systems and Applications (INISTA), pp. 1–5. IEEE (2012)
12. Miranda, V., Fonseca, N.: EPSO-evolutionary particle swarm optimization, a new algorithm with applications in power systems. In: Transmission and Distribution Conference and Exhibition 2002: Asia Pacific. IEEE/PES, pp. 745–750. IEEE (2002)
13. Karaboga, D., et al.: A comprehensive survey: artificial bee colony (ABC) algorithm and applications. Artif. Intell. Rev. **42**(1), 21–57 (2014)
14. Yildiz, A.R.: A new hybrid artificial bee colony algorithm for robust optimal design and manufacturing. Appl. Soft Comput. **13**(5), 2906–2912 (2013)
15. Jatoth, R.K., Rajasekhar, A.: Speed control of pmsm by hybrid genetic artificial bee colony algorithm. In: 2010 IEEE International Conference on Communication Control and Computing Technologies (ICCCCT), pp. 241–246. IEEE (2010)
16. Sundar, S., Singh, A.: A hybrid heuristic for the set covering problem. Oper. Res. **12**(3), 345–365 (2012)
17. Gandomi, A.H., Yang, X., Alavi, A.H.: Mixed variable structural optimization using firefly algorithm. Comput. Struct. **89**(23), 2325–2336 (2011)
18. Akay, B., Karaboga, D.: Artificial bee colony algorithm for large-scale problems and engineering design optimization. J. Intell. Manuf. **23**(4), 1001–1014 (2012)

19. Gandomi, A.H., Yang, X.-S., Alavi, A.H.: Cuckoo search algorithm: a metaheuristic approach to solve structural optimization problems. Eng. Comput. **29**(1), 17–35 (2013)
20. Hedar, A., Fukushima, M.: Derivative-free filter simulated annealing method for constrained continuous global optimization. J. Glob. Optim. **35**(4), 521–549 (2006)
21. Mahdavi, M., Fesanghary, M., Damangir, E.: An improved harmony search algorithm for solving optimization problems. Appl. Math. Comput. **188**(2), 1567–1579 (2007)

Various Problems of Artificial Intelligence

Patterns in Video Games Analysis – Application of Eye-Tracker and Electrodermal Activity (EDA) Sensor

Iwona Grabska-Gradzińska$^{(\boxtimes)}$ and Jan K. Argasiński

Department of Games Technology, Faculty of Physics, Astronomy and Applied
Computer Science, Jagiellonian University, Krakow, Poland
{iwona.grabska,jan.argasinski}@uj.edu.pl

Abstract. The aim of the article is to propose a method for evaluating
player's experience during gameplay using an eye-tracker and galvanic
skin response sensor. The method is based on using data obtained from
the game, in the light of patterns in game design. The article presents
a preliminary, qualitative study, along with an exemplary interpretation
of the gameplay of the Hidden Object Puzzle Adventure (HOPA) game.

Keywords: Game metrics · User experience
Patterns in game design · Oculography · Psychogalvanometry

1 Motivation

One of the basic motivations for qualitative analysis of games is to indicate
elements that are crucial to game reception by the player [1]. Specifically understood User Experience (UX) in games relates not only to the interface layer
and the measurements of effectiveness in performing desired activities. It is also
grounded in the relation to the issues regarding the course of the gameplay, such
as challenges or mechanics. We understand mechanics here after Miguel Sicart,
as "methods invoked by agents for interacting with game world" [2]. This means
that games put some problems in front of the player to solve and—at the same
time – provide ways to solve these problems. The art of game design is based on
the ability to stack obstacles and provide the means to solve them, so that the
given arrangement is interesting, challenging and demanding – but not beyond
the capabilities of the player. When we design serious games and simulations,
this aspect is one of the key factors that have to be taken into consideration.
The thing is relatively easy to implement at the level of system consistency.
Laying the plot, mechanics, locations and interfaces so that player always has
the ability to complete the game (while avoiding the state of permanent block)
is relatively easy to achieve. However, the assessment of the game in terms of
optimal cognitive and emotional involvement is almost always based solely on
the experience of the designers.

© Springer International Publishing AG, part of Springer Nature 2018
L. Rutkowski et al. (Eds.): ICAISC 2018, LNAI 10842, pp. 619–629, 2018.
https://doi.org/10.1007/978-3-319-91262-2_54

In order to enhance the process of game evaluation, as well as individual player's experience, we must use methods of probing and depicting the course of the gameplay and be able to generalize them. Abstraction comes in the form of design patterns. The proposed approach is another step towards the creation of comprehensive, pattern based models [3].

2   Methods

2.1   Patterns in Game Design

The subject of design patterns in computer science has been known and explored for a very long time. In the field of video game design, the most popular is the concept by Björk and Holopainen, described in the book "Patterns in Game Design" [4]. The idea proposed by the authors relies on creating special game description language, where individual mechanics are assigned to specific, individual "patterns". The sample pattern consists of the name (e.g. "Alarms"), description ("Alarms are abstract game elements that provide information about particular game state changes") and examples of use. In addition, each pattern contains information on what other patterns it *instantiates, modulates, is instantiated by*, is modulated by and *is potentially conflicting with*. Thanks to this, a network of interrelated elements is created, which facilitates both the design process and the game analysis. Björk and Holopainen distinguished nearly 300 elementary patterns. Their method is somewhat popular among game developers and researchers.

2.2   Psychogalvanometry (Electrodermal Activity)

According to one of the most interesting and popular theories of emotions, initiated by William James and developed in modern version by Prinz [5], emotions can be understood as responses to changes in the bodily state. That is opposite to the intuitive beliefs of most people – predominately, we are convinced that the body's reactions (sweating, accelerated heart rate) are the result of the affects we experience. If, however, it is rather "I strain my muscles so I get nervous" than "I get nervous so I strain my muscles" – it means that we can detect potential or actual affective arousal by monitoring the biophysiological states of the human.

Our emotional arousal is the result of the activity of the Autonomic Nervous System (ANS). For us, the most interesting part of it is the sympathetic system responsible for "handling" violent situations requiring intensive mobilization of the organism ("fight or flight"). The activation of this system is associated with bodily signals activated autonomously, such as elevated heart rate, sweating or increased blood pressure.

One of the easiest ways to record parameters, which proves the stimulation of the sympathetic ANS system, is electrical conductivity of the skin (Galvanic Skin Response – GSR, also known as Electrodermal Activity—EDA or Skin Conductance – SC). In case of stimulation, a special type of sweat glands located mainly in the area of the interior of hands and feet increase the hydration of the skin surface, rising electrical conductivity in these areas. With special sensors, it is possible to detect even small changes of this parameter.

2.3 Oculography

In the HOPA game scenes, the items connected with many patterns are available at the same time. To recognize which of the presented patterns attracted player's attention in the exact moment, eye-tracking is required.

The eye-tracking methods operate on the pupil position and calculate coordinates of the point of visual interest in the environment. For the head-mounted eye-trackers, the world camera is mounted under players eyes and, after calibration of the device, it allows to correlate pupils position with the element of the environment.

The eyes move constantly. The most interesting from the cognitive point of view are movements called fixations. Fixation points correlate with the focusing of attention, and show regions of player's particular attention [6], which is often used in experiments, e.g. [7].

The process of segmentation of the player's activity and conjunction with the proper game pattern is based on the fixations sequences. During the gameplay, the player wears eye-tracker with three cameras. Two are pointed at his pupils and one is directed towards the screen in front of her. Every fixation is noticed. The sequence of the fixations is connected with the pattern if all fixations in the sequence are located on the items associated with that pattern. The longitude of the sequence is calculated as a period between the first frame of the first fixation and the last frame of the last fixation noticed in the region of interest.

3 Experiments

3.1 HOPA Games

Hidden Object Puzzle Adventure is a genre of adventure games. They belong to the rarely studied or analyzed types of games because of their casual nature—the HOPA players most often do not identify themselves with the gaming fandom, and rarely consider this kind of entertainment as something very important in their lives. This does not change the fact that puzzle adventure games are very popular and have loyal fan base.

The gameplay in a typical HOPA consists of three basic elements:

- a story (adventure)—most often the driving force is the plot of some kind – criminal, fantasy, etc. The adventure aspect also means that players must follow the logic of the narrative – if we are dealing with police procedural, we should collect clues, interrogate witnesses etc.;
- finding objects—this is the most characteristic kind of mechanics for this type of games—the player is presented with graphics (often a very high quality digital painting), on which she must find specific objects, and click on them as soon as possible;

– puzzles – in selected moments of the game, players are confronted with puzzles that have to be solved – they most often belong to the game's setting but their logic is not necessarily related to the logic of the real world (e.g. to open a safe player needs to align the sequence of colorful stones; to repair the broken electrical wiring he has to move block through the maze by visiting each intersection only once etc.).

HOPA games perfect as testing grounds in the development of methods for examining patterns because their dynamics are slow paced (the player rarely operates under real time pressure, in principle he never has to perform actions based solely on dexterity). At the same time, the plot of the games in question is characterized by a certain narrative tension, while graphics and sensational cut-scenes make for emotional immersion. An additional advantage of HOPA games is – what is typical for casual games – an intuitive interface; these games do not require training special skills – the interactions are based on pointing and clicking on objects on the screen.

In our experiment we have used the popular 2016 HOPA game by Artifex Mundi – "Crime Secrets: Crimson Lily".

3.2 Hardware Setup

In our experiment we have used Pupil Lab head-mounted 120 Hz eye-tracker, with word camera resolution 1280×720 px. For the fixation detection the Dispersion-Based Algorithm [8] and Pupil Capture software v. 0.8.6.1 were used [9].

For the measurement of EDA the e-Health Sensor Platform V2.0 with Arduino UNO microcontroler was used [10]. The GSR electrodes were placed in the inner side of non-dominant hand. Data was registered using custom Arduino scripts utilizing e-Health libraries.

Game was run on Apple MacBook Pro laptop and MacOS X operating system. Players used wired professional optical Razer gaming mouse. External display was 42″ Sony Bravia flatscreen TV. Data was registered using separate Ubuntu based PC (Sony Vaio) laptop.

Setup schema and photography is presented on Fig. 1.

3.3 Subjects and the Course of Experiment

The four subjects proficient in computers use, differing in experience with computer games, were invited to participate in the experiment. The subjects were asked only to play the first sequence of the game – without any other suggestions. After installing the GSR sensor and calibrating the oculograph they immediately proceeded to the game.

The four gameplays were conducted (one for every subject), then the shortest (Player A, 13 min 57 s, 1805 fixations) and longest (Player B, 43 min 2 s, 2908 fixations) were taken into further consideration.

Subject A – it occurred, have had great experience with HOPA games, subject B on the other hand have had little experience with any games.

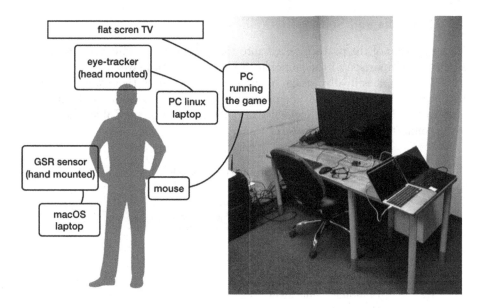

Fig. 1. Setup used in experiment

4   Results

4.1   Patterns Occurred

First sequence of the game "Crime Secrets: Crimson Lily"consists of 12 arbitrary named main scenes: (1) *Cinematic*; (2) *the Car Accident*; (3) *the Frozen Path*; (4) *Cinematic*; (5) *Gate with Frozen Sheriff*; (6) *the Courtyard*; (7) *the Hotel's Hall*; (8) *Cinematic*; (9) *the Hotel's Hall 2 – Blackout*; (10) *the Hotel's Hall 3 – After Blackout*; (11) *the Corridor*; (12) *the Hotel Room*. Scenes 2, 5, 6, 10 and 11 have multiple "sub-scenes" that require some action to be performed by player. Almost every scene (except cinematics) has some activity related.

The whole game is build around meta-patterns such as: "First Person Views", "Single Player Game", "Narrative Structures", "Varied Gameplay", "Consistent Reality Logic" and contains such elementary patterns as: "Cut Scenes", "Emotional Immersion", "Ultra Powerful Events", "Goal Indicators", "Collecting", "Exploration", "Clues", "Tools", "Obstacles", "Inaccessible Areas", "Illusionary Rewards" "Storytelling", "Puzzle Solving", "Tension", "Cognitive Immersion", "Helpers", "Movable Tiles" and "Closure Points". For detailed description of particular patterns see [4].

The main goal of experiment was to correlate occurrence of the patterns and changes in EDA and observational behavior (using oculography), so only the patterns linked with items presented in the scene where taken into further consideration. An interesting issue was also the comparison of experienced and inexperienced player game styles.

4.2 Patterns and Sensors

High precision of the pattern assays into the gameplay give the researchers possibility of comparison of the gameplay style of the both players.

The differences involve usage of the patterns (some of them were used only by player B), summed longitude of the fixation sequences connected with each pattern (see Figs. 4 and 5), incidence of pattern changes (see Table 1) and EDA value connected with each pattern.

Table 1. Incidence of episodes

Pattern	Incidence rate Player A	Incidence rate Player B
Clues	1	2
Cognitive immersion	2	6
Collecting	5	17
Emotional immersion	1	4
Exploration	11	30
Goal indicators	2	9
Helpers	0	6
Illusionary rewards	0	3
Moveable tiles	2	3
Narrative structure	1	2
Obstacles	2	6
OUT	1	3
Puzzle solving	2	5
Storytelling	5	4
Tools	3	4

Difference Between the Players. The main difference between the players is longitude of the whole gameplay; we can see that the distribution of the longitudes of the particular patterns are different. Some patterns are used only by player B: "Illusionary Rewards" were ignored by the experienced user – the related artifacts were shown on the screen just as long, but novice user made few fixation on them every time and the experienced user made no fixation at all. Similar situation can be noticed with the pattern "Helpers". This time the user had to intentionally launch this pattern. The difference can be seen in the Figs. 5 and 4, column 6–7.

The difference in attitude to challenges during the game is shown in EDA values for the following patterns: "Goal Indicators", "Exploring" and "Collecting". The first of them, binded with the reading of the game journal, caused the high level of EDA only for the player B. The player A demonstrated the growing level of the skin conductance at the beginning of every new scene and after the phase of exploring started to collect artifacts (Fig. 2). The player B very often interlaced usage of "Exploring" and "Collecting" patterns, what can be seen in fixation sequences and which was very time consuming strategy.

The significant elements of the HOPA games are "games within the game". In the game "Crime Secrets: Crimson Lily" these games can be connected with two different patterns: "Puzzle Solving" and "Movable Tiles". What is surprising, the level of EDA depends on pattern, not on type of the game at all. EDA value of player A were higher while using pattern "Movable Tiles" than the player's B, who prefers "Puzzle Solving".

Interesting observations can be made regarding plots on Figs. 2 and 3.

On the first plot we can observe Galvanic Skin Response of the player A. Particularly arousing moments are marked with asterisks. For more experienced (are more interested in game) player A, every scene that allows for activity seems to be source of excitement (on the plot changes in color denote changes of scene). *1 is the task of assembling fireworks to alarm the guard (same scene marked *1 on player's B plot – similar excitement) [patterns: "Collection", "Cognitive Immersion", "Tools"]; *2 is dialogue scene when NPC (Non Player Character) faints [patterns: "Storytelling", "Emotional Immersion"]; *3 is unexpected blackout [patterns: "Tension", "Closure Points", "Emotional Immersion", "Ultra Powerful Events", "Surprises"]; *4 is puzzle game (find the sequence to get the key) [patterns: "Puzzle Solving", "Movable Ties"]; *5 is scene in the corridor (which is on fire) [patterns: "Goal Indicators", "Tension"]; *6 is "crime scene" in the hotel room (escaping burglar) [patterns: "Storytelling", "Exploration", "Clues"];

On the second plot we can observe that unexperienced player B was less emotionally affected by gameplay but particular patterns/events had their influence: *1 the very first puzzle game (first time the player saw this kind of gameplay); *2 assembling fireworks; *3 and *4 another puzzle game (get the key).

4.3   Discussion of Results

The conducted study was a part of bigger project that aims into creating system/architecture for creating and evaluating affective serious games and simulation. Qualitative study of two players needs to be expanded to quantitative analysis of various gamers and different game styles.

Ultimately, the authors of the study are convinced that on this small sample they managed to show that constructing the language of patterns and correlating them with the various types of sensors (whether it is an eye-tracker and EDA, or ECG and EMG) may result in interesting findings.

The ultimate goal seems to be the introduction of affective feedback to the main loop of the game so as to be able to dynamically "control" affective responses at the level of game's engine. Such a solutions already have their first applications in simulations and serious games. This article is development of previously presented research [3] and is an introduction to the next stage of the project.

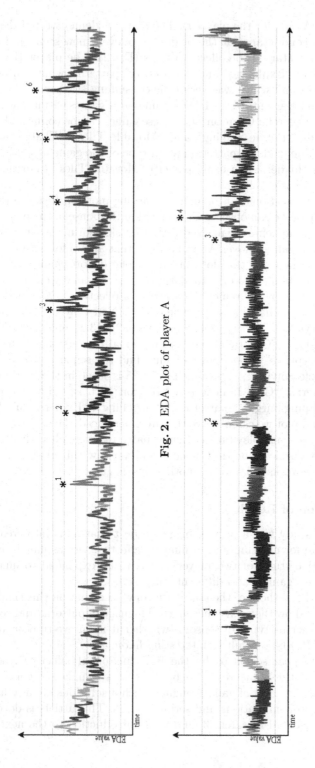

Fig. 2. EDA plot of player A

Fig. 3. EDA plot of player B

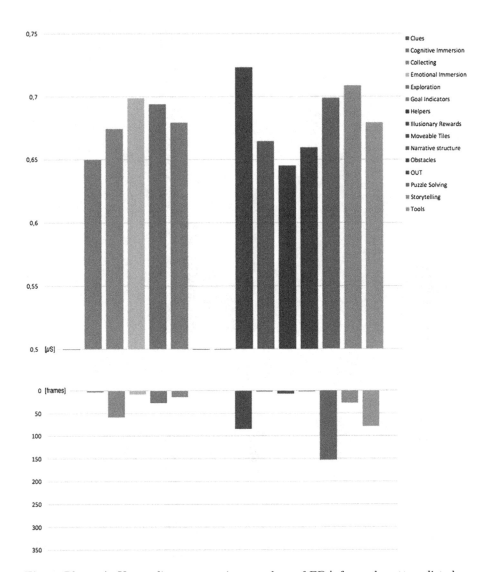

Fig. 4. Player A. Upper diagram: maximum values of EDA for each pattern listed on the right side. Lower diagram: longitude of fixation sequences on each pattern.

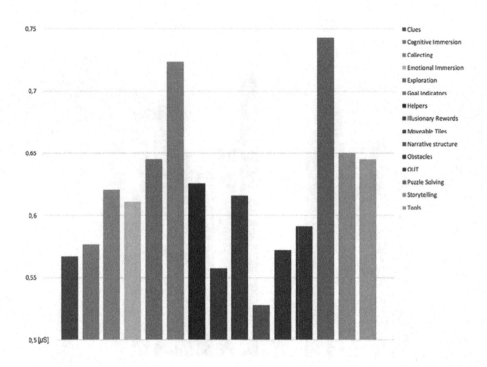

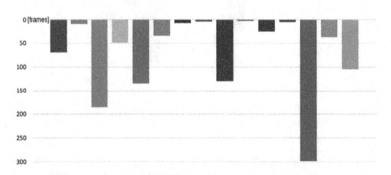

Fig. 5. Player B. Upper diagram: maximum values od EDA for each pattern listed on the right side. Lower diagram: longitude of fixation sequences on each pattern.

References

1. Lankoski, P., Björk, S. (eds.): Game Research Methods. ETC Press, Halifax (2015)
2. Sicart, M.: Defining Game Mechanics. Game Stud. **8**(2), 1–14 (2008). http://gamestudies.org/0802/articles/sicart
3. Argasiński, J.K., Grabska-Gradzińska, I.: Patterns in serious game design and evaluation application of eye-tracker and biosensors. In: Rutkowski, L., Korytkowski, M., Scherer, R., Tadeusiewicz, R., Zadeh, L.A., Zurada, J.M. (eds.) ICAISC 2017. LNCS (LNAI), vol. 10246, pp. 367–377. Springer, Cham (2017). https://doi.org/10.1007/978-3-319-59060-8_33
4. Björk, S., Holopainen, J.: Patterns in Game Design. Charles River Media, Boston (2004)
5. Prinz, J.: Gut Reactions. A Perceptual Theory of Emotion. Oxford University Press, Oxford (2004)
6. Duchowski, A.T.: Eye Tracking Methodology: Theory and Practice. Springer, New York (2003). https://doi.org/10.1007/978-1-84628-609-4
7. Almeida, S., Mealha, O., Veloso, A.: Interaction behavior of hardcore and inexperienced players: "Call of Duty: Modern Warfare" context. In: Proceedings of SBGames 2010 - IX Brazilian Symposium on Computer Games and Digital Entertainment (2010)
8. Salvucci, D.D., Goldberg, J.H.: Identifying fixations and saccades in eye-tracking protocols. In: Proceedings of the 2000 Symposium on Eye Tracking Research & Applications (2000)
9. Platform for eye tracking and egocentric vision research. https://pupil-labs.com/pupil. Accessed 10 Mar 2018
10. e-Health Sensor Platform V2.0 for Arduino and Raspberry Pi [Biometric / Medical Applications]. https://www.cooking-hacks.com/documentation/tutorials/ehealth-biometric-sensor-platform-arduino-raspberry-pi-medical. Accessed 10 Mar 2018

Improved Behavioral Analysis of Fuzzy Cognitive Map Models

Miklós F. Hatwagner[1]([⊠]), Gyula Vastag[2], Vesa A. Niskanen[3],
and László T. Kóczy[4,5]

[1] Department of Information Technology, Széchenyi István University,
Győr, Hungary
miklos.hatwagner@sze.hu
[2] Department of Leadership and Organizational Communication,
Széchenyi István University, Győr, Hungary
vastag.gyula@sze.hu
[3] Department of Economics and Management,
University of Helsinki, Helsinki, Finland
vesa.a.niskanen@helsinki.fi
[4] Department of Information Technology, Széchenyi István University,
Győr, Hungary
koczy@sze.hu
[5] Department of Telecommunications and Media Informatics,
Budapest University of Technology and Economics, Budapest, Hungary
koczy@tmit.bme.hu

Abstract. Fuzzy Cognitive Maps (FCMs) are widely applied for describing the major components of complex systems and their interconnections. The popularity of FCMs is mostly based on their simple system representation, easy model creation and usage, and its decision support capabilities.

The preferable way of model construction is based on historical, measured data of the investigated system and a suitable learning technique. Such data are not always available, however. In these cases experts have to define the strength and direction of causal connections among the components of the system, and their decisions are unavoidably affected by more or less subjective elements. Unfortunately, even a small change in the estimated strength may lead to significantly different simulation outcome, which could pose significant decision risks. Therefore, the preliminary exploration of model 'sensitivity' to subtle weight modifications is very important to decision makers. This way their attention can be attracted to possible problems.

This paper deals with the advanced version of a behavioral analysis. Based on the experiences of the authors, their method is further improved to generate more life-like, slightly modified model versions based on the original one suggested by experts. The details of the method is described, its application and the results are presented by an example of a banking application. The combination of Pareto-fronts and Bacterial Evolutionary Algorithm is a novelty of the approach.

Keywords: Banking · Fuzzy Cognitive Maps · Model uncertainty
Multi-objective optimization · Bacterial Evolutionary Algorithm

© Springer International Publishing AG, part of Springer Nature 2018
L. Rutkowski et al. (Eds.): ICAISC 2018, LNAI 10842, pp. 630–641, 2018.
https://doi.org/10.1007/978-3-319-91262-2_55

1   Introduction

The task of well considered decision making may be really hard, and the consequences of a wrong intervention are often serious, especially in an environment where several important, interrelated factors have to be taken into account. According to this, decision support is in the focus of researchers for a long time, and various methods were suggested [1].

This paper deals with Fuzzy Cognitive Maps (FCMs) [2]. FCM is a bipolar fuzzy graph: its nodes represent the major components of the modeled system and the arcs among them express the direction and strength of relationships. It describes the operation of a system qualitatively [3] and can be used for decision support [4,5]. The main advantages of applying FCM are e.g. transparency, ease of use, can be used to model even complex systems.

The FCM model of a system can be created in two main ways [6]: the first is based on the knowledge, experiences and competence of one or more experts. The cooperation of multiple experts help to decrease the influence of personal beliefs and subjectivity, but even if the developed model is free from these effects it can be inaccurate. For example, if a model contains only 10 nodes, the number of relationships can be up to 90, and it is often hard to define the strength of so many relations with the required accuracy. That is why the recommended, second way of model creation is based on objective, historical, measured data and a suitable machine learning technique. These data are sometimes not available, however, and only the expert-based method can be applied. Unfortunately, even a subtle change in connection strengths may change the behavior of the model, e.g. the final, stable states of two slightly different systems can be entirely different, despite the same initial state, or the number of possible final states may change. It is worth to analyze the effect of uncertainty on model behavior before decision making. This work has already begun [7,8], but the authors improved the method based on their experiences.

The analysis is based on the systematic and automated modification of the strength of relationships. Every modified model version is tested with a predefined huge set of initial states, and the result of simulations are collected and analyzed. The goal of the investigation is to find a slightly modified model that has different or more final stable states, repeats a series of states or behaves chaotically more often. The differentiation of the last two cases is one of the new features of the improved method. The behavioral properties are very interesting for the decision makers. The search was performed by a multi-objective optimization, and the fitness of modified models was defined by a weighted sum. This approach has its disadvantages [9], thus the fitness of models are now expressed on the basis of their Pareto-optimality. The method was already able to find model versions with significantly different behavior, but in order to achieve its goal, it usually had to drastically modify the internal relationships of the model. The improved method strives for more similar original and modified models, because similarity has become one of the optimization targets. Furthermore, the effect of the user-defined λ parameter of FCM's threshold function is also investigated. It is already known that its value has an effect on the behavior of the model [10].

The structure of the paper is the following. Section 2 describes briefly the theoretical basics of the applied methods, including FCM and Bacterial Evolutionary Algorithm (BEA) [11,12]. BEA is an optimization method, used here to find slightly modified models that have significantly different behavior. The reason why this metaheuristic was used is that in comparison with other popular approaches it shows much better convergence and speed [13]. Section 2.3 specifies some specific details of the implemented program. In order to demonstrate the capabilities of the improved method, a case study is provided in Sect. 3. Section 4 concludes the results and states the possible ways of further research.

2   The Applied Methods of Behavioral Analysis

2.1   Fuzzy Cognitive Maps

Cognitive Maps were suggested by Axelrod [14] to describe the cause-effect relations of political groups and their possible acts. His technique was further improved by Kosko [2]: the edges of the graph are weighted to express the strength of relations, and also the nodes have numerically defined status. Formally, an FCM can be defined by a 4-tuple: (C, W, A, f), where $C = C_1, C_2, \ldots, C_N$ is the set of nodes, called *concepts* in FCM terminology. N is the number of concepts. Concepts represent the main factors, components of a system or a variable. The status of concept i at time t $(t = 1, 2, \ldots, T)$ is expressed by the *activation value* $A_i \in \mathbb{R}$. The function $A : (C_i) \rightarrow A_i$ associates the activation value to the node. The function $W : (C_i, C_j) \rightarrow w_{ij}$ defines the *weight* (causal value) of the directed arc between concepts C_i and C_j. The weight values are represented with the *connection matrix*. In our paper FCM of Type I [15] is used, where concepts never influence themselves (the main diagonal contains zeros). The weight must fall in the $w_{ij} \in [-1, +1]$ interval. An example FCM is provided in Fig. 1 together with its corresponding connection matrix (Table 1).

Table 1. Connection matrix of the example FCM model.

	C1	C2	C3	C4	C5
C1	0	0	0	1	0.5
C2	1	0	0.5	0.5	1
C3	0	0.5	0	0	0
C4	0	0	−0.5	0	0
C5	0.5	0	0	0	0

The last component of the tuple is the *transformation* or *threshold* function $f : \mathbb{R} \rightarrow [0, 1]$. This function guarantees that the activation values will remain in their allowed interval during simulations. (In some rare cases, the $A_i \in [-1, +1]$ can also be used with a matching threshold function.) Several threshold functions

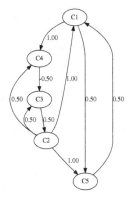

Fig. 1. Graph representation of the example FCM model.

were suggested in the literature [16], but only the most common sigmoid function (1) is used in this paper.

The $\lambda > 0$ parameter defines the steepness of the function, and it is not directly connected to any physically observable properties of the modeled system. Its usual value is 5. With lower λ values the function approximates a linear function, with higher values the sign function.

The activation values are updated by (2) in our case during the consecutive time steps.

$$f(x) = \frac{1}{1 + e^{-\lambda x}} \tag{1}$$

$$A_i^{t+1} = f\left(\sum_{j=1}^{N} w_{ij} A_j^t\right) \tag{2}$$

A model using continues activation values can behave three different ways [16] during simulation: (i) in most cases it converges quickly to an equilibrium point, often called *fixed point attractor* (FP). (ii) Sometimes a series of activation vectors appears repeatedly always in the same order. This infinite transition among states is called *limit cycle* (LC). (iii) If the model behaves *chaotically*, the state of the model never stabilizes.

2.2 Bacterial Evolutionary Algorithm

Bacterial Evolutionary Algorithm (BEA) [11,12] is a member of evolutionary algorithms, capable to solve even non-continuous, non-linear, multi-modal, high dimensional, global optimization problems, and provides the near-optimal solutions of them. Nawa and Furuhashi suggested this straightforward and robust method for the optimization of fuzzy systems' parameters, but it can be successfully applied to other problems as well.

The algorithm works with a collection of possible solutions, called population. The elements of the population often called bacteria as well, because the method

634 M. F. Hatwagner et al.

imitates the evolution of bacteria in nature. Several generations of the population are generated using the two main operators, *bacterial mutation* and *gene transfer*, until one of the stop conditions (e.g. stopped convergence, limit on time or number of generations) are fulfilled. The best bacteria of the final population are considered as result.

Bacterial mutation (c.f. [17]) explores the search space by random modification of bacteria. The bacteria are mutated individually and independently. First, the copies of an original bacterium, the so-called clones are created. Then the operator iterates over every genes of the bacterium in random order. In every iterative step, the current gene is randomly modified in the clones, and they are evaluated. If the modification leads to better objective value, the new allele is kept and copied to both original and clone bacteria. This technique preserves the old alleles if they serves the goals of optimization better, and explicit elitism is not needed.

Gene transfer (c.f. [18]) exploits and combines the genetic information coded in the bacteria of the current population in order to find even better solutions. At first, it sorts the population based on the objective values of bacteria. Then it divides the population to two halves: the sub-population containing better bacteria is called the superior half, while the other is the inferior half. The operator chooses a bacterium randomly from the superior half, and an other from the inferior half. Next, at least one allele is copied from the better bacterium into the other. The modified bacterium have to be re-evaluated, and if it became better, it has the chance to migrate into the superior half, and scatter its genetic code among other bacteria during the consecutive gene transfers.

2.3 Specific Details of the Program Developed for FCM Analysis

The goal of this study was to find a slightly modified model, that behaves radically different than the original model. This way the connections effecting the strongest influence on the behavior of the model (e.g. the simulations lead to more FPs, LCs or chaotic behavior) can be discovered and their values can be further analyzed before using the model for decision support.

The weights in the connection matrix (see Table 2) are given by real values in the allowed interval, thus the original search problem defines an infinitely large search space. Because only some of the possible values are used in practical applications (according to the applied linguistic variables), only 9 different weight values are used in our program (-1, -0.75, -0.5, ..., $+1$). The lack of causal relation between two concepts can be identified by experts with high confidence, therefore the program never changes the zero weight connections of the original model.

The search for modified models is directed by BEA. A bacterium encodes a possible λ value ($0.1 < \lambda < 10.0$) and the new weights of the originally non-zero weight connections. In our case study, which is based on a real life bank management problem, the model contains 13 concepts, thus the number of connections is $12 \times 13 = 156$. Luckily many possible connections do not exist. Thus, there is a constant zero in the respective positions of the connection matrix. There

Table 2. Connection matrix of the FCM model

	C1	C2	C3	C4	C5	C6	C7	C8	C9	C10	C11	C12	C13
C1	0	0	0.5	0	0	0.5	1	0.5	0	0.5	1	0.5	0
C2	1	0	0.5	1	0	0	1	1	0.5	0	1	1	0
C3	1	0.5	0	0	0	−1	0	−1	1	0	1	1	1
C4	0	0	0	0	0	0	0	0	0	0	0	0	0
C5	0	0	1	−0.5	0	0	0	−1	0	0	0	1	0
C6	0	0	0	0	−0.5	0	0	0	0	0	0	0	0
C7	0.5	0	0.5	1	0	0.5	0	0	0	−0.5	0	0	0
C8	0	0	0	0	0	−0.5	0	0	0	0.5	0	0	−0.5
C9	0	0	0	1	0	0.5	0.5	0.5	0	−0.5	0	0.5	0
C10	0.5	0	0	0	0	0	0.5	0.5	0.5	0	1	0	0
C11	0	0	0.5	0.5	0	0	0	0	0.5	0.5	0	0	1
C12	0	0	0.5	0.5	0	0	1	0	0.5	0	0.5	0	−0.5
C13	0	0	1	0	0	0.5	0	0	0	0	1	0	0

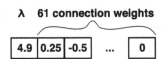

Fig. 2. An example bacterium.

are 61 positions with non-zero connection weights. Including the λ parameter, the model still leads to a 62 variable optimization problem. Figure 2 shows an example bacterium of the given bank model.

The behavior of modified models are tested by simulations. Similarly to connections weights, the set of possible activation values are limited to $0, 0.25, 0.5, 0.75$ and 1. The program starts with the generation of 1000 random initial state vectors (scenarios) and all modified models are tested by using the same set of scenarios. The program automatically detects FPs and counts the initial state vectors leading to the same FP. Simulations consist of at most 100 time steps. (State vectors usually converge quickly to an equilibrium state.) If the state vector of concept values stabilizes earlier, it is considered a FP, otherwise the program starts to find a LC. If a repeated sequence of state vectors is not found, it is considered a chaotic behavior.

The state of the system is considered stable only if the values of all concepts has changed by at most 0.001 during the last five consecutive time steps. Unfortunately, the resulting stable states are often not exactly the same, even if they can be considered identical in practice, because e.g. rounding errors of floating point arithmetic may slightly distort the results. In order to overcome this difficulty, the program creates clusters of final state vectors using k-means clustering [19], and finally these clusters are considered the 'real' FPs. The number of clusters is estimated by gap statistics [20].

The goals of the optimization are the following: (i) maximize the number of FPs, (ii) maximize the number of LCs, (iii) maximize the number of chaotic behavior, (iv) minimize the d difference of modified and original matrix calculated by (3), where N is the number of concepts, o is the connection matrix of the original, and m is the connection matrix of the modified model.

$$d = \sum_{i=1}^{N} \sum_{j=1}^{N} (o_{ij} - m_{ij})^2 \qquad (3)$$

This multi-objective optimization problem is solved by BEA in a Pareto-optimal manner. The bacteria of a population is classified into several sets: the bacteria on the Pareto-front are collected in the set of the 'first' Pareto-front. The Pareto-front of the other, remaining bacteria can also be determined in the same way, and these bacteria will be in the 'second' Pareto-front, etc. The mutation operator is slightly modified in our program: the Pareto-fronts of the sub-population of a bacterium and its clones are detected, and if the original bacterium is not an element of the 'first' Pareto-front, its gene is modified to the allele of the first bacterium of the 'first' Pareto-front. The gene transfer operator is also modified in a similar way. The population is sorted on the basis of the Pareto-front number of bacteria.

The population of our example contained 10 bacteria, 3 clones were created for each bacteria during mutation, 3 infections were made in every generation and the optimization stopped after the 10th generation.

The most important parameters used by the proposed method are collected in Table 3.

Table 3. Parameters and their applied values

Parameter	Values
FCM connection weights	$-1, -0.75, -0.5, \ldots, +1$
Threshold parameter (λ)	Arbitrary value in the [0.1, 10.0] interval
Initial concept values	0, 0.25, 0.5, 0.75 and 1
Number of bacteria	10
Number of clones	3
Number of infections	3
Number of BEA iterations	10

3   Case Study: A Banking Application

The application of the proposed method is demonstrated with a real-life problem. Table 4 contains the description of major concepts of a specific bank, including their unique identifiers and categories. The signs and the weight values were given by bank experts. The connection matrix of the model is given in Table 2. This model was also used in [8], but now it is analyzed with the newer, improved version of the method. This way the results of the earlier and the improved methods are comparable, and the advantages of improvements become visible.

Table 4. Concept IDs, names and categories of the investigated model

Concept ID	Concept name	Category
C1	Clients	Assets
C2	Rules & regulations	
C3	New IT solutions	
C4	Funding	Money
C5	Cost reduction	
C6	Profit/loss	Financials
C7	Investments	
C8	Staff	Human resources
C9	New services	Product and process development
C10	Quality	
C11	Client development	Output measures
C12	Service development	
C13	Productivity	

3.1 Properties of the Original Model

The properties of the original model were analyzed by simulations. The value of the λ parameter of FCM's threshold function was set to 5 (which is the most wide spread used value in the literature). After completing the simulation, two FPs were found: 23.1% of the 1000 investigated scenarios led to the first, and the remaining 76.9% to the second FP. Regardless of the investigated scenarios, most concepts had the same final values. Only the final values of C6 and C8 made a real difference between the two FPs (see Table 5).

While in C8 the fixed points are not entirely different C6 converges in the first case to a 'low' concept value, while in the second case to a 'rather high' state. Depending on the real meaning of C6 this difference may be critical in the steady state operation of the bank.

Table 5. Fixed-point attractors of the model

Concepts	C1–C3, C5, C7, C9–C13	C6	C8
FP#1	1.000	0.150	0.990
FP#2	1.000	0.855	0.922

3.2 Results of the Analysis

Tables 6 and 7 show the connection matrices of the two best bacteria of the last generation. Both of them were located on the first Pareto-front. Other bacteria are members of the other three Pareto-fronts. The elements of Pareto-fronts in the last generation are shown in Fig. 3. The most important properties of these model variants are collected in Table 8.

Table 6. Connection matrix of the 1st model variant

	C1	C2	C3	C4	C5	C6	C7	C8	C9	C10	C11	C12	C13
C1	0.0	0.0	0.75	0.0	0.0	0.5	1.0	0.25	0.0	0.25	0.75	0.5	0.0
C2	1.0	0.0	0.5	1.0	0.0	0.0	0.5	0.75	0.75	0.0	0.75	1.0	0.0
C3	0.25	0.0	0.0	0.0	0.0	−1.0	0.0	−0.75	0.5	0.0	0.5	0.75	1.0
C4	0.0	0.0	0.0	0.0	0.0	0.0	0.0	0.0	0.0	0.0	0.0	0.0	0.0
C5	0.0	0.0	1.0	−0.5	0.0	0.0	0.0	0.0	0.0	0.0	0.0	0.5	0.0
C6	0.0	0.0	0.0	0.0	−0.5	0.0	0.0	0.0	0.0	0.0	0.0	0.0	0.0
C7	0.75	0.0	−0.25	1.0	0.0	0.25	0.0	0.0	0.0	1.0	0.0	0.0	0.0
C8	0.0	0.0	0.0	0.0	0.0	0.25	0.0	0.0	0.0	0.75	0.0	0.0	−0.5
C9	0.0	0.0	0.0	1.0	0.0	0.25	−0.75	0.25	0.0	0.25	0.0	0.75	0.0
C10	−0.75	0.0	0.0	0.0	0.0	0.0	0.25	−0.5	0.25	0.0	−0.25	0.0	0.0
C11	0.0	0.0	0.75	0.5	0.0	0.0	0.0	0.0	0.5	0.5	0.0	0.0	1.0
C12	0.0	0.0	0.5	0.5	0.0	0.0	0.25	0.0	0.25	0.0	−0.75	0.0	−0.5
C13	0.0	0.0	0.75	0.0	0.0	0.75	0.0	0.0	0.0	0.0	0.75	0.0	0.0

Table 7. Connection matrix of the 2nd model variant

	C1	C2	C3	C4	C5	C6	C7	C8	C9	C10	C11	C12	C13
C1	0.0	0.0	0.5	0.0	0.0	0.5	1.0	1.0	0.0	1.0	0.75	0.5	0.0
C2	−0.75	0.0	0.0	1.0	0.0	0.0	0.5	0.0	0.5	0.0	0.0	0.5	0.0
C3	0.25	0.25	0.0	0.0	0.0	−0.75	0.0	−1.0	0.0	0.0	0.5	−0.5	0.0
C4	0.0	0.0	0.0	0.0	0.0	0.0	0.0	0.0	0.0	0.0	0.0	0.0	0.0
C5	0.0	0.0	−0.5	−0.5	0.0	0.0	0.0	0.25	0.0	0.0	0.0	0.5	0.0
C6	0.0	0.0	0.0	0.0	0.0	0.0	0.0	0.0	0.0	0.0	0.0	0.0	0.0
C7	0.5	0.0	0.0	1.0	0.0	0.75	0.0	0.0	0.0	−0.5	0.0	0.0	0.0
C8	0.0	0.0	0.0	0.0	0.0	−1.0	0.0	0.0	0.0	1.0	0.0	0.0	0.5
C9	0.0	0.0	0.0	1.0	0.0	0.0	0.75	1.0	0.0	−0.75	0.0	0.75	0.0
C10	0.5	0.0	0.0	0.0	0.0	0.0	0.0	0.5	−0.25	0.0	−0.75	0.0	0.0
C11	0.0	0.0	−1.0	0.5	0.0	0.0	0.0	0.0	0.25	1.0	0.0	0.0	−1.0
C12	0.0	0.0	0.0	0.5	0.0	0.0	1.0	0.0	0.5	0.0	−0.25	0.0	0.25
C13	0.0	0.0	1.0	0.0	0.0	0.75	0.0	0.0	0.0	0.0	0.5	0.0	0.0

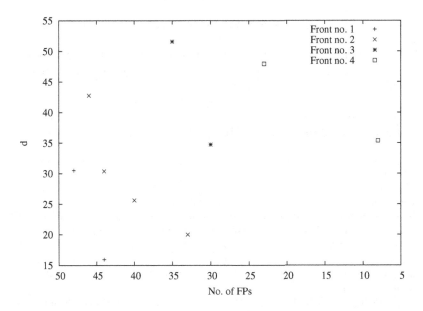

Fig. 3. Bacteria of Pareto-fronts in the last generation

Table 8. Main properties of the modified model variants

Property	1st variant	2nd variant
λ value	2.366	2.070
Number of FPs	44	48
Number of LCs	0	0
Number of chaotic behavior	0	0
Difference from orig. model (d)	15.938	30.500

4 Conclusions

The improved method has reached its goal: it finds interesting model versions with smaller modifications than its preceding version while the modified models still have much more FPs than the original. There are several ways of possible further improvements, however.

The biggest obstacle to the application of the method is its performance: due to the high number of executed simulations, the process is extremely time consuming. BEA could be obviously accelerated: the parallel execution of mutations could be done trivially, but even the parallel version of gene transfer is worked out [18]. The implementation of these techniques are the next tasks.

The analysis could be further accelerated by the selection of some interesting connections, and modify only these connections while preserve the weight of others. It also looks useful to limit the range of new weight values to a specified interval.

BEA is slightly modified in our program to find Pareto-optimal solutions. This aim could be achieved several ways, the different possible implementations should be thoroughly investigated.

Acknowledgement. This research was supported BY the ÚNKP-17-4 New National Excellence Program of the Ministry of Human Capacities.

References

1. Busemeyer, J.R.: Dynamic decision making (1999)
2. Kosko, B.: Fuzzy cognitive maps. Int. J. Man-Mach. Stud. **24**(1), 65–75 (1986)
3. Salmeron, J.L.: Supporting decision makers with fuzzy cognitive maps. Res.-Technol. Manag. **52**(3), 53–59 (2009)
4. Papageorgiou, E.I. (ed.): Fuzzy Cognitive Maps for Applied Sciences and Engineering. ISRL, vol. 54. Springer, Heidelberg (2014). https://doi.org/10.1007/978-3-642-39739-4
5. Baykasoğlu, A., Gölcük, İ.: Development of a novel multiple-attribute decision making model via fuzzy cognitive maps and hierarchical fuzzy topsis. Inf. Sci. **301**, 75–98 (2015)
6. Papageorgiou, E.I.: Learning algorithms for fuzzy cognitive maps—a review study. IEEE Trans. Syst. Man Cybern. Part C (Appl. Rev.) **42**(2), 150–163 (2012)
7. Hatwágner, M.F., Niskanen, V.A., Kóczy, L.T.: Behavioral analysis of fuzzy cognitive map models by simulation. In: 2017 Joint 17th World Congress of International Fuzzy Systems Association and 9th International Conference on Soft Computing and Intelligent Systems (IFSA-SCIS), pp. 1–6. IEEE (2017)
8. Hatwágner, M.F., Vastag, G., van Kóczy, L.T.: Banking applications of FCM models. In: 9th European Symposium on Computational Intelligence and Mathematics, pp. 60–68 (2017). http://escim2017.uca.es/wp-content/uploads/2015/02/OralCommunications.pdf
9. Deb, K.: Multi-Objective Optimization Using Evolutionary Algorithms, vol. 16. Wiley, Hoboken (2001)
10. Hatwágner, M.F., Kóczy, L.T.: Parameterization and concept optimization of FCM models. In: 2015 IEEE International Conference on Fuzzy Systems (FUZZ-IEEE), pp. 1–8. IEEE (2015)
11. Nawa, N.E., Furuhashi, T.: Fuzzy system parameters discovery by bacterial evolutionary algorithm. IEEE Trans. Fuzzy Syst. **7**(5), 608–616 (1999)
12. Nawa, N.E., Furuhashi, T.: A study on the effect of transfer of genes for the bacterial evolutionary algorithm. In: 1998 Second International Conference on Knowledge-Based Intelligent Electronic Systems, Proceedings of KES'98, vol. 3, pp. 585–590. IEEE (1998)
13. Balázs, K., Botzheim, J., Kóczy, L.T.: Comparative investigation of various evolutionary and memetic algorithms. In: Rudas, I.J., Fodor, J., Kacprzyk, J. (eds.) Computational Intelligence in Engineering. Studies in Computational Intelligence, vol. 313, pp. 129–140. Springer, Heidelberg (2010). https://doi.org/10.1007/978-3-642-15220-7_11
14. Axelrod, R.: Structure of Decision: The Cognitive Maps of Political Elites. Princeton University Press, Princeton (1976)

15. Stylios, C.D., Groumpos, P.P.: Mathematical formulation of fuzzy cognitive maps. In: Proceedings of the 7th Mediterranean Conference on Control and Automation, pp. 2251–2261 (1999)

16. Tsadiras, A.K.: Comparing the inference capabilities of binary, trivalent and sigmoid fuzzy cognitive maps. Inf. Sci. **178**(20), 3880–3894 (2008)

17. Nawa, N.E., Hashiyama, T., Furuhashi, T., Uchikawa, Y.: A study on fuzzy rules discovery using pseudo-bacterial genetic algorithm with adaptive operator. In: 1997 IEEE International Conference on Evolutionary Computation, pp. 589–593. IEEE (1997)

18. Hatwagner, M., Horvath, A.: Parallel gene transfer operations for the bacterial evolutionary algorithm. Acta Tech. Jaurinensis **4**(1), 89–111 (2011)

19. Hartigan, J.A., Wong, M.A.: Algorithm as 136: a k-means clustering algorithm. J. Roy. Stat. Soc.: Ser. C (Appl. Stat.) **28**(1), 100–108 (1979)

20. Tibshirani, R., Walther, G., Hastie, T.: Estimating the number of clusters in a data set via the gap statistic. J. Roy. Stat. Soc. Ser. B (Stat. Methodol.) **63**(2), 411–423 (2001)

On Fuzzy Sheffer Stroke Operation

Piotr Helbin[1], Wanda Niemyska[2], Pedro Berruezo[3,4],
Sebastia Massanet[3,4], Daniel Ruiz-Aguilera[3,4], and Michał Baczyński[1(✉)]

[1] Institute of Mathematics, University of Silesia in Katowice,
Bankowa 14, 40-007 Katowice, Poland
{piotr.helbin,michal.baczynski}@us.edu.pl
[2] Institute of Informatics, University of Warsaw,
Banacha 2, 02-097 Warsaw, Poland
wanda@mimuw.edu.pl
[3] Soft Computing, Image Processing and Aggregation (SCOPIA) Research Group,
Department of Mathematics and Computer Science,
University of the Balearic Islands, 07122 Palma, Spain
p.berruezo@uib.cat, {s.massanet,daniel.ruiz}@uib.es
[4] Balearic Islands Health Research Institute (IdISBa), 07010 Palma, Spain

Abstract. The generalization of the classical logical connectives to the
fuzzy logic framework has been one of the main research lines since the
introduction of fuzzy logic. Although many classical logical connectives
have been already generalized, the Sheffer stroke operation has received
scant attention. This operator can be used by itself, without any other
logical operator, to define a logical formal system in classical logic. Therefore, the goal of this article is to present some initial ideas on the fuzzy
Sheffer stroke operation in fuzzy logic. A definition of this operation in
the fuzzy logic framework is proposed. Then, a characterization theorem
in terms of a fuzzy conjunction and a fuzzy negation is presented. Finally,
we show when we can obtain other fuzzy connectives from fuzzy Sheffer
stroke operation.

Keywords: Sheffer stroke · Fuzzy implication · Fuzzy negation
t-norm · t-conorm

1 Introduction

Fuzzy operations such as t-norms, t-conorms, fuzzy implications and fuzzy negations generalize the classical logical connectives, which take values in the set
$\{0,1\}$, to the unit interval $[0,1]$. These functions are not only essential for fuzzy
logic systems and fuzzy control, but they also play a significant role in solving
fuzzy relational equations, in fuzzy mathematical morphology and image processing, and in defining fuzzy subsethood. For the overview of some classes of
such functions see the monographs [1,3].

In classical logic, Sheffer stroke, also called NAND or alternative denial, is
one of the two operations that can be used by itself, without any other logical

© Springer International Publishing AG, part of Springer Nature 2018
L. Rutkowski et al. (Eds.): ICAISC 2018, LNAI 10842, pp. 642–651, 2018.
https://doi.org/10.1007/978-3-319-91262-2_56

operations, to constitute a logical formal system. In this paper, we propose a definition of this operation in the fuzzy logic framework, which generalizes the classical Sheffer stroke when restricted to $\{0, 1\}^2$. We also show how to construct all other main fuzzy connectives when using only a fuzzy Sheffer stroke operator.

The paper is organized as follows. In Sect. 2 we recall basic concepts and definitions used in the paper. Section 3 is devoted to the characterization of fuzzy Sheffer stroke in terms of a fuzzy negation and a fuzzy conjunction. In Sect. 4 we present some basic examples of fuzzy Sheffer strokes. In Sect. 5, we show how using fuzzy Sheffer stroke we can obtain the other fuzzy connectives. The last section contains conclusions and it postulates an open problem.

2   Preliminaries

Fuzzy concepts have to generalize adequately the corresponding crisp objects. In this section first we present the most commonly accepted definitions of fuzzy generalizations of classical connections like conjunction, disjunction, negation and implication, and then we propose a definition of a new fuzzy operation - fuzzy Sheffer stroke.

2.1   Fuzzy Logical Connectives

Let us start recalling the definitions and some immediate facts about the most well-known fuzzy logical connectives.

Definition 2.1 (cf. [3, Definition 11.3]). *A function $C: [0,1]^2 \to [0,1]$ is called a **fuzzy conjunction** if it satisfies, for all $x, y, z \in [0,1]$, the following conditions:*

(C1) $C(x,y) \leq C(z,y)$ *for* $x \leq z$*, i.e., $C(\cdot, y)$ is non-decreasing,*
(C2) $C(x,y) \leq C(x,z)$ *for* $y \leq z$*, i.e., $C(x, \cdot)$ is non-decreasing,*
(C3) $C(0,1) = C(1,0) = 0$ *and* $C(1,1) = 1$.

Definition 2.2 (see [3]).

*(i) An associative, commutative and increasing operation $T: [0,1]^2 \to [0,1]$ is called a **t-norm** if it has the neutral element 1.*
*(ii) An associative, commutative and increasing operation $S: [0,1]^2 \to [0,1]$ is called a **t-conorm** if it has the neutral element 0.*

Definition 2.3 (see [2], [3, Definition 11.3]). *A non-increasing function $N: [0,1] \to [0,1]$ is called a **fuzzy negation** if $N(0) = 1$, $N(1) = 0$. Moreover, a fuzzy negation N is called*

*(i) **strict** if it is strictly decreasing and continuous;*
*(ii) **strong** if it is an involution, i.e., $N(N(x)) = x$ for all $x \in [0,1]$.*

It is important to note that every strong fuzzy negation is strict (see [1, Corollary 1.4.6]). Thus it is an injective and surjective function.

Definition 2.4 (see [1, Definition 1.1.1], [2]**).** *A function* $I: [0,1]^2 \to [0,1]$ *is called a* ***fuzzy implication function*** *if it satisfies, for all* $x, y, z \in [0,1]$, *the following conditions:*

(I1) $I(x,z) \geq I(y,z)$ *for* $x \leq y$, *i.e.,* $I(\cdot, z)$ *is non-increasing,*
(I2) $I(x,y) \leq I(x,z)$ *for* $y \leq z$, *i.e.,* $I(x, \cdot)$ *is non-decreasing,*
(I3) $I(0,0) = I(1,1) = 1$ *and* $I(1,0) = 0$.

2.2 Definition of Fuzzy Sheffer Stroke Operation

In the classical logic, Sheffer stroke operation is denoted by $(\uparrow)$. It is a logical connective whose truth table is presented in Table 1. As it can be seen, $(\uparrow)$ indicates whether one of the inputs is false.

Table 1. Truth table for the classical Sheffer stroke.

p	q	$p \uparrow q$
0	0	1
0	1	1
1	0	1
1	1	0

As any fuzzy logical operation has to coincide with the corresponding classical operation when the inputs are in the set $\{0,1\}$, any potential definition of fuzzy Sheffer stroke should satisfy the previous truth table. Moreover, it is reasonable to impose monotonicity in each variable in the sense that as greater is the truth value of one input, smaller is the output of the operation. This is the key point of our main definition.

Definition 2.5. *A function* $D: [0,1]^2 \to [0,1]$ *is called a* ***fuzzy Sheffer stroke operation*** *(or fuzzy Sheffer stroke) if it satisfies, for all* $x, y, z \in [0,1]$, *the following conditions:*

(D1) $D(x,z) \geq D(y,z)$ *for* $x \leq y$, *i.e.,* $D(\cdot, z)$ *is non-increasing,*
(D2) $D(x,y) \geq D(x,z)$ *for* $y \leq z$, *i.e.,* $D(x, \cdot)$ *is non-increasing,*
(D3) $D(0,1) = D(1,0) = 1$ *and* $D(1,1) = 0$.

On the one hand, it can be easily derived from the above definition that $D(0,x) = 1$ and $D(x,0) = 1$ for all $x \in [0,1]$. On the other hand, the values $D(x,1)$ and $D(1,x)$ are not predetermined from the definition.

Given a Sheffer stroke operation, three natural negations can be defined.

Definition 2.6. *Let* D *be a Sheffer stroke operation.*

(i) *The function* N_D^l *defined by* $N_D^l(x) = D(x,1)$ *for all* $x \in [0,1]$ *is called the left natural negation of* D.
(ii) *The function* N_D^r *defined by* $N_D^r(x) = D(1,x)$ *for all* $x \in [0,1]$ *is called the right natural negation of* D.

(iii) The function N_D^d defined by $N_D^d(x) = D(x,x)$ for all $x \in [0,1]$ is called the diagonal natural negation of D.

It is trivial to check that all of the above functions are fuzzy negations in the sense of Definition 2.3.

3   Characterization of Fuzzy Sheffer Stroke

In classical logic, Sheffer stroke is the negation of the conjunction (NAND), that is, $p \uparrow q = \neg(p \wedge q)$. This result is also valid in the fuzzy logic framework taking into account a fuzzy conjunction and a fuzzy negation.

Theorem 3.1. *Let $D: [0,1]^2 \to [0,1]$ be a binary operation. Then the following statements are equivalent:*

(i) D is a fuzzy Sheffer stroke.
(ii) There exist a fuzzy conjunction C and a strict fuzzy negation N such that $D(x,y) = N(C(x,y))$ for all $x,y \in [0,1]$.

Moreover, in this case, $C(x,y) = N^{-1}(D(x,y))$ for all $x,y \in [0,1]$.

Proof. Let us show that if there exist a fuzzy conjunction C and a strict fuzzy negation N such that $D(x,y) = N(C(x,y))$ for all $x,y \in [0,1]$, then D is a fuzzy Sheffer stroke. Due to the monotonicity of C and N, we have that for all $x_1, x_2, y \in [0,1]$, $x_1 \leq x_2$,

$$D(x_1, y) = N\left(C(x_1, y)\right) \geq N\left(C(x_2, y)\right) = D(x_2, y)$$

and therefore, D is non-increasing in the first variable. It can be shown analogously that D is non-increasing in the second variable. The border conditions are also satisfied:

$$D(0,1) = N(C(0,1)) = N(0) = 1,$$
$$D(1,0) = N(C(1,0)) = N(0) = 1,$$
$$D(1,1) = N(C(1,1)) = N(1) = 0.$$

Thus, $D(x,y) = N(C(x,y))$ is a Sheffer stroke for any fuzzy conjunction C and fuzzy negation N (not necessarily strict one).

Conversely, let us consider now a Sheffer stroke operation D. Let us consider any strict fuzzy negation N and let us define C as the binary function given by

$$C(x,y) = N^{-1}(D(x,y)), \qquad x,y \in [0,1].$$

We will prove that C is a fuzzy conjunction. Due to the monotonicity of D and N^{-1}, we have that for all $x_1, x_2, y \in [0,1]$, $x_1 \leq x_2$,

$$C(x_1, y) = N^{-1}\left(D(x_1, y)\right) \leq N^{-1}\left(D(x_2, y)\right) = C(x_2, y)$$

and therefore, C is increasing in the first variable. It can be shown analogously that C is increasing in the second variable. The border conditions are also satisfied:

$$C(0,0) = N^{-1}(D(0,0)) = N^{-1}(1) = 0,$$
$$C(1,1) = N^{-1}(D(1,1)) = N^{-1}(0) = 1,$$
$$C(0,1) = N^{-1}(D(0,1)) = N^{-1}(1) = 0,$$
$$C(1,0) = N^{-1}(D(1,0)) = N^{-1}(1) = 0.$$

Finally, the result follows since

$$N(C(x,y)) = N(N^{-1}(D(x,y))) = D(x,y),$$

for all $x, y \in [0,1]$.   □

Remark 3.2. Some remarks on the previous theorem are worthy to mention:

(i) The representation of a fuzzy Sheffer stroke is not unique. Indeed, any strict fuzzy negation N can be chosen. However, fixed a strict fuzzy negation N, the fuzzy conjunction C is unique.

(ii) Whenever one of the natural negations of the fuzzy Sheffer stroke is strict, it can be considered to represent the fuzzy Sheffer stroke. In this case, both the fuzzy negation and the fuzzy conjunction are defined from the expression of D.

4   Basic Examples

Using Theorem 3.1, we can obtain fuzzy Sheffer strokes by considering some fuzzy conjunctions and strict fuzzy negations. Let us consider fuzzy Sheffer strokes generated from basic t-norms and fuzzy conjunctions, and the classical negation $N_C(x) = 1 - x$ for all $x \in [0,1]$.

Example 4.1. (i) If we consider the minimum t-norm $T_M(x,y) = \min\{x,y\}$ and the classical negation N_C, we obtain

$$D_M(x,y) = 1 - T_M(x,y) = \max\{1 - x, 1 - y\}.$$

(ii) If we consider the product t-norm $T_P(x,y) = xy$ and the classical negation N_C, we obtain

$$D_P(x,y) = 1 - T_P(x,y) = 1 - xy.$$

We can consider as well the more general fuzzy conjunction $C_P^k(x,y) = (xy)^k$, for any $k > 0$, and the classical negation N_C, and we obtain then

$$D_P^k(x,y) = 1 - C_P^k(x,y) = 1 - (xy)^k.$$

(iii) If we consider the Łukasiewicz t-norm $T_{\mathbf{LK}}(x, y) = \max\{x + y - 1, 0\}$ and the classical negation $N_{\mathbf{C}}$, we obtain

$$D_{\mathbf{LK}}(x, y) = 1 - T_{\mathbf{LK}}(x, y) = \min\{2 - x - y, 1\}.$$

(iv) If we consider the drastic t-norm given by

$$T_{\mathbf{D}}(x, y) = \begin{cases} 0, & \text{if } (x, y) \in [0, 1)^2, \\ \min\{x, y\}, & \text{otherwise,} \end{cases}$$

and the classical negation $N_{\mathbf{C}}$, we obtain

$$D_{\mathbf{D}}(x, y) = 1 - T_{\mathbf{D}}(x, y) = \begin{cases} 1, & \text{if } (x, y) \in [0, 1)^2, \\ \max\{1 - x, 1 - y\}, & \text{otherwise.} \end{cases}$$

Some of these fuzzy Sheffer strokes are displayed in Fig. 1. Table 2 provides the three natural negations of fuzzy Sheffer strokes given in Example 4.1.

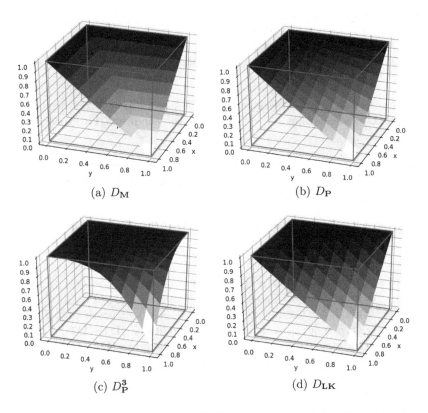

(a) $D_{\mathbf{M}}$

(b) $D_{\mathbf{P}}$

(c) $D_{\mathbf{P}}^3$

(d) $D_{\mathbf{LK}}$

Fig. 1. Plots of some of the fuzzy Sheffer strokes presented in Example 4.1.

Table 2. Natural negations of fuzzy Sheffer strokes given in Example 4.1

	N_D^l	N_D^r	N_D^d
D_M	N_C	N_C	N_C
D_P	N_C	N_C	$1 - x^2$
D_P^k	$1 - x^k$	$1 - x^k$	$1 - x^{2k}$
D_{LK}	N_C	N_C	$\begin{cases} 1, & \text{if } x \leq 0.5, \\ 2 - 2x, & \text{otherwise.} \end{cases}$
D_D	N_C	N_C	greatest fuzzy negation N_{D_2}

5   Construction of Other Fuzzy Connectives from Fuzzy Sheffer Stroke

We need to introduce the following additional properties to obtain other fuzzy connectives:

(D4) $D(D(x,x), D(x,x)) = x$, for all $x \in [0,1]$,
(D5) $D(1,x) = D(x,x)$, for all $x \in [0,1]$,
(D6) $D(x,y) = D(y,x)$, for all $x,y \in [0,1]$,
(D7) $D(x, D(D(y,z), D(y,z))) = D(D(D(x,y), D(x,y)), z)$, for all $x,y,z \in [0,1]$.

Note that the condition **(D4)** means that N_D^d is a strong fuzzy negation, the condition **(D5)** means that $N_D^r = N_D^d$, while the condition **(D6)** means that D is symmetric.

Proposition 5.1. *Let T be a t-norm and N be a strong negation. The function $D(x,y) = N(T(x,y))$ for all $x,y \in [0,1]$ satisfies all the conditions (D1)-(D7) if and only if $T = T_M = \min$.*

Proof. First let us notice that if D satisfies **(D5)** then for every $x \in [0,1]$

$$N(x) = N(T(1,x)) = D(1,x) = D(x,x) = N(T(x,x)),$$

and since strong negation N is a bijection, we obtain that $x = T(x,x)$, for every $x \in [0,1]$. We know (see [3]) that the only idempotent t-norm is T_M.

On the other hand, assume that $T = T_M$. We will prove that $D(x,y) = N(\min(x,y))$, $x,y \in [0,1]$, satisfies all the conditions **(D1)-(D7)**. First three **(D1)-(D3)** arise immediately from Theorem 3.1. Next two **(D4)**, **(D5)** are simple calculations. For every $x \in [0,1]$ we obtain

(D4)

$$\begin{aligned} D(D(x,x), D(x,x)) &= N(\min(N(\min(x,x)), N(\min(x,x)))) \\ &= N(\min(N(x), N(x))) = N(N(x)) = x, \end{aligned}$$

(D5) $D(1, x) = N(\min(1, x)) = N(x) = N(\min(x, x)) = D(x, x).$

Obviously D is commutative, so it satisfies **(D6)**. Finally, for all $x, y, z \in [0, 1]$ we have

$$D(x, D(D(y, z), D(y, z))) = D(x, N(\min(N(\min(y, z)), N(\min(y, z)))))$$
$$= D(x, \min(y, z)) = N(\min(x, y, z)) = D(\min(x, y), z)$$
$$= D(N(\min(N(\min(x, y)), N(\min(x, y)))), z)$$
$$= D(D(D(x, y), D(x, y)), z),$$

thus D satisfies **(D7)**, also.   □

Note that if a binary function D is given by $D(x, y) = N(T(x, y))$ for some t-norm T and some strong fuzzy negation N and it satisfies all the conditions **(D1)**-**(D7)**, then using the well-known representation of strong negations (cf. [1, Theorem 1.4.13]) we obtain

$$D(x, y) = \varphi^{-1}(1 - \varphi(\min(x, y))), \qquad x, y \in [0, 1],$$

where $\varphi \colon [0, 1] \to [0, 1]$ is an increasing bijection.

Example 5.2. From Proposition 5.1 we conclude that the only one fuzzy Sheffer stroke satisfying **(D4)**-**(D7)** given in the Example 4.1 is D_M. Other two operations, generated from t-norm T_M and strong negations from Sugeno and Yager classes are displayed in Fig. 2.

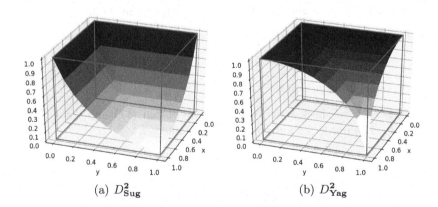

(a) $D^2_{\mathbf{Sug}}$ (b) $D^2_{\mathbf{Yag}}$

Fig. 2. Plots of the fuzzy Sheffer strokes - satisfying **(D4)**-**(D7)** - generated from T_M and strong negations from Sugeno and Yager classes, with constants $\lambda = 2, w = 2$, respectively.

Now we will define families of t-norms, t-conorms and fuzzy implication functions that can be generated from fuzzy Sheffer strokes. In order to generate t-norms and t-conorms, we will apply the following tautologies from classical logic:

$$p \wedge q \equiv ((p \uparrow q) \uparrow (p \uparrow q)),$$
$$p \vee q \equiv ((p \uparrow p) \uparrow (q \uparrow q)).$$

The next two results are not difficult to prove but due to the lack of enough space, they have been omitted.

Theorem 5.3. *Let D be a fuzzy Sheffer stroke that satisfy (D4). Then, the following function*

$$T(x,y) = D(D(x,y), D(x,y)), \qquad x, y \in [0,1] \tag{1}$$

is a t-norm if and only if D satisfies additionally (D5), (D6) and (D7).

Theorem 5.4. *Let D be a fuzzy Sheffer stroke that satisfy (D4). Then, the following function*

$$S(x,y) = D(D(x,x), D(y,y)), \qquad x, y \in [0,1] \tag{2}$$

is a t-conorm if and only if D satisfies additionally (D5), (D6) and (D7).

In classical logic, the two-valued implication can be presented using only Sheffer stroke operation in two ways:

$$p \rightarrow q \equiv p \uparrow (q \uparrow q) \equiv \neg(p \wedge \neg(q \wedge q)), \tag{QQ}$$
$$p \rightarrow q \equiv p \uparrow (p \uparrow q) \equiv \neg(p \wedge \neg(p \wedge q)). \tag{PQ}$$

In the fuzzy logic framework, while the first construction method generates always fuzzy implication functions in the sense of Definition 2.4, the second one cannot guarantee in general the non-increasingness in the first variable.

Theorem 5.5. *Let D be a fuzzy Sheffer stroke. Then the function I defined by*

$$I(x,y) = D(x, D(y,y)), \qquad x, y \in [0,1], \tag{3}$$

is a fuzzy implication function.

Proof. It is easy to check that from **(D1)** we obtain **(I1)** and from **(D2)** we obtain **(I2)**. The border conditions are also satisfied:

$$I(0,0) = D(0, D(0,0)) = D(0,1) = 1,$$
$$I(1,0) = D(1, D(0,0)) = D(1,1) = 0,$$
$$I(1,1) = D(1, D(1,1)) = D(1,0) = 1.$$

$$\square$$

From the equation (PQ) we obtain the following result.

Theorem 5.6. *Let D be a fuzzy Sheffer stroke. Then the function I defined by*

$$I(x,y) = D(x, D(x,y)), \qquad x, y \in [0,1], \tag{4}$$

satisfies (I2) and (I3).

Proof. From (**D2**) we obtain (**I2**) straightforwardly. The border conditions are also satisfied:

$$I(0,0) = D(0, D(0,0)) = D(0,1) = 1,$$
$$I(1,0) = D(1, D(1,0)) = D(1,1) = 0,$$
$$I(1,1) = D(1, D(1,1)) = D(1,0) = 1.$$

$\square$

Theorems 5.5 and 5.6 are analogous to some results obtained in [4] for two new families of fuzzy implication functions denoted as SS_{pq} and SS_{qq}.

6   Conclusions

In this paper we have introduced a novel fuzzy logical connective, called fuzzy Sheffer stroke. We have examined some properties of this logical operator and some basic examples have been provided. In particular, we have given a characterization theorem in terms of a fuzzy conjunction and a fuzzy negation for fuzzy Sheffer strokes. Next, we have shown some construction methods of other fuzzy logical connectives from fuzzy Sheffer stroke. From the results proved in this paper, one open problem immediately arises. Namely, if for every t-norm or t-conorm there exists a fuzzy Sheffer stroke such that this t-norm or t-conorm can be represented by Equations (1) and (2), respectively?

Acknowledgment. M. Baczyński and W. Niemyska were supported by the National Science Centre, Poland, under Grant No. 2015/19/B/ST6/03259. P. Helbin has been supported from statutory activity of the Institute of Mathematics, University of Silesia in Katowice. P. Berruezo, S. Massanet and D. Ruiz-Aguilera acknowledge the partial support by the Spanish Grant TIN2016-75404-P, AEI/FEDER, UE.

References

1. Baczyński, M., Jayaram, B.: Fuzzy Implications. Studies in Fuzziness and Soft Computing, vol. 231. Springer, Heidelberg (2008). https://doi.org/10.1007/978-3-540-69082-5
2. Fodor, J., Roubens, M.: Fuzzy Preference Modelling and Multicriteria Decision Support. Kluwer Academic Publishers, Dordrecht (1994)
3. Klement, E., Mesiar, R., Pap, E.: Triangular Norms. Kluwer Academic Publishers, Dordrecht (2000)
4. Niemyska, W., Baczyński, M., Wąsowicz, S.: Sheffer Stroke Fuzzy Implications. In: Kacprzyk, J., Szmidt, E., Zadrożny, S., Atanassov, K.T., Krawczak, M. (eds.) IWIFSGN/EUSFLAT-2017. AISC, vol. 643, pp. 13–24. Springer, Cham (2018). https://doi.org/10.1007/978-3-319-66827-7_2

Building Knowledge Extraction
from BIM/IFC Data for Analysis
in Graph Databases

Ali Ismail[1], Barbara Strug[2], and Grażyna Ślusarczyk[2(✉)]

[1] Institute of Construction Informatics, TU Dresden, 01062 Dresden, Germany
ali.ismail@tu.dresden
[2] Department of Physics, Astronomy and Applied Computer Science,
Jagiellonian University, Lojasiewicza 11, 30-348 Krakow, Poland
{barbara.strug,gslusarc}@uj.edu.pl

Abstract. This paper deals with the problem of knowledge extraction and processing building related data. Information is retrieved from the IFC files, which are an industry standard for storing building information models (BIM). The IfcWebServer is used as a tool for transforming building information into the graph model. This model is stored in a graph database which allows for obtaining knowledge by defining specific graph queries. The process is illustrated by examples of extracting information needed to find different types of routes in an office building.

Keywords: Knowledge extraction · Graph databases
Building Information Modelling (BIM)
Industry Foundation Classes (IFC)

1 Introduction

In this paper the problem of extracting complex building-related knowledge which is necessary in the process of searching for different types of routes is considered. The information about the building topological structure and attributes of its components is extracted from IFC (Industry Foundation Classes) models and stored in a property graph database. The analysis of the topology of spatial layouts of buildings and semantics of the component elements, like widths of corridors and types of doors, is needed to assess the accessibility of routes in various situations and for different types of users.

In order to process the knowledge obtained from the IFC models some form of representation is required. Graphs offer the possibility to homogeneously encode spatial and non-spatial information of different types, and therefore they constitute an adequate representation for complex relationships among building elements and data within Building Information Models (BIMs) [12]. Graph nodes correspond to building elements or their properties, while edges represent relations between these elements. Attributes assigned to element nodes describe the basic properties of building elements.

© Springer International Publishing AG, part of Springer Nature 2018
L. Rutkowski et al. (Eds.): ICAISC 2018, LNAI 10842, pp. 652–664, 2018.
https://doi.org/10.1007/978-3-319-91262-2_57

Converting BIM models based on the IFC standard into a graph-based effective information retrievable model can significantly facilitate exploring and analysing BIM highly connected data. A graph data model (GDM), which can be used to represent, extract and analyse topological relationships among 3D objects in 3D space, and to perform topological queries is presented in [21]. Another approach towards information retrieval using the IFC object model, where directed graphs serve as semantic data pools, is described in [35].

An automatic workflow for transformation of IFC models into a graph-based model using the property graph database Neo4J [27] as a graph database framework is recently developed [14]. Based on the graph structure of the building information several queries supporting search for different routes can be specified. Moreover functional graph algorithms for searching building data and finding the shortest paths can be implemented. The presented graph model is useful for data management as it allows to explore, check and analyse the complex relationships inside BIM models, and run complex queries for information retrieval.

The problem of searching for routes in buildings has been widely researched. The majority of the research is devoted to finding escape routes in emergency situations. This problem has been dealt with by a number of researchers [1–3,7,26,28,31,32]. The design and construction of optimum escape routes is discussed in [37], while multilevel analysis of fire escape routes, where virtual robots are used to simulate human movement during escape, is presented in [22]. Different evacuation models (BGRAF, DONEGAN'S ENTROPY MODEL, EXIT89, EGRESS, E-ESCAPE, EVACSIM, EXITT, EXODUS, MAGNETODEL, PAXPORT, SIMULEX, VEGAS, STEPS, PATHFINDER, BTS, ELEVATE), which allow to assess the potential evacuation efficiency of a building, are described in [6,23]. In [29] an approach to model emergency situations in buildings based on BIM is also described. However in this approach the process of generating graph networks out of BIM, on which route calculation can be performed, is difficult as it consists of merging several separately generated graphs.

Apart from finding the emergency egress, the problem of searching for routes has also been addressed in different contexts. One of them is supporting the navigation of self-sufficient mobile robots or other automated devices [34,36]. Another context in which this problem plays an important role is evaluating the accessibility of the building for disabled persons. This problem has been researched in two different aspects. The first one deals with the problem of testing if the building satisfies the legal norms of accessibility and has been addressed by a number of works [19,30]. Another aspect in this context deals with searching for best routes for disabled persons and verifying their costs [33].

In this paper a different approach is used. The information are extracted directly from the IFC file and stored in a graph database and then the required knowledge is extracted by querying this database. The IFCWebServer [13] is used as a tool for transforming building information into the graph model. The presented approach is illustrated by examples of extracting knowledge of an office building using graph queries. While computing the length of indoor routes, or the quickest evacuation paths, the information about the existence of large open spaces (like halls, lobbies and corridors), fire-proof doors and width of doors is of vital importance.

2   Building Information Modelling and IFC

Nowadays architectural building designs are often created with the use of CAD tools. BIM technology used for CAD applications enables to represent syntactic and semantic building information with respect to the entire life cycle of designed objects. The 3D object model is created using such elements as parameterized walls, ceilings, roofs, windows or doors [4]. The file format IFC [15] being an interoperable BIM standard for CAD applications, provides an object-oriented and semantic data model for storing and exchanging building information. It supports data exchange among different disciplines and heterogeneous applications. Information retrieved from IFC files is used in applications estimating construction cost for tendering in China [25] managing construction sites [8] or evaluating design solutions [20].

IFC specifies virtual representations of building objects as well as their attributes and relationships. It includes most types of geometry, supports many classes of attributes [5,9–11]. An IFC model is composed of IFC entities arranged in a hierarchical way. Each IFC entity includes a fixed number of IFC attributes and any number of IFC properties. The names of the attributes are defined as part of the IFC standard code and are the main identifiers of the entities. Three fundamental entity types, IfcObjectDefinition, IfcPropertyDefinition and IfcRelationship, of the IFC data schema are child nodes of the IfcRoot entity [14]. IfcPropertyDefinition describes all characteristics that may be attached to objects. IfcObjectDefinition stands for all physical objects or processes. IfcRelationship specifies all relationships among objects, where to each relationship several properties can be attached. The main subtypes of IfcRelationship are IfcRelConnects, IfcRelAssociates, IfcRelDecomposes, IfcRelDefines and IfcRelAssigns. IFC classes have attached attributes describing basic entity properties and referenced attributes which are connected through relationships with other objects.

Information about the building, which is needed from the point of view of the problem considered in this paper, includes information on topology of floor layouts, accessibility between spaces, stairs and doors types and sizes, if available. IFC entities, which store the data required by the proposed system, are of the types IfcSpace, IfcDoor, IfcWall and IfcStair. According to the IFC 2x Edition3 Model Implementation Guide and the IFC specification [15] the above mentioned classes can be described as follows:

- IfcSpace is the instance used to represent a space as the area or volume of a functional region. It is often associated with the class IfcBuildingStorey representing one floor (the building itself is an aggregation of several storeys) or with IfcSite, which represents the construction site. A space in the building is usually associated with certain functions (e.g., hall, bathroom). These functions are specified by attributes of the class IfcSpace (Name, LongName, Description).

- IfcWall is the instance used to represent a vertical element, which is to merge or split the space. In IFC files two representations of a wall can be distinguished. The subclass IfcWallStandardCase of IfcWall is used for all walls that do not change their thickness (the thickness of a wall is the sum of the materials used). IfcWall is used for all other walls, in particular for the walls with non-rectangular cross-sections.
- IfcStair represents a vertical passage allowing for moving from one floor to the other. It can contain an intermediate landing. Instances of IfcStairs are treated as containers, by which we refer to component elements as IfcStairFlight using IfcRelAggregates.
- IfcDoor represents a building element used to provide access to a specific area or room. Parameters of IfcDoors specify dimensions, an opening direction and a style of the door (IfcDoorStyle). IfcDoor is a subclass of IfcBuildingElement. Door instances are usually located in a space IfcOpeningElement to which we refer by IfcRelFillsElement.

The above mentioned instances inherit from the base class IfcProduct, which allows for determining their positions on the basis of some attributes like IfcLocalPlacement and PlacementRelTo. The coordinates obtained in this way specify the relative position of the object against other instances of the class IfcProduct. Obtaining the actual position of the considered IfcProduct instance is possible by tracking all references IfcLocalPlacement and PlacementRelTo.

In Fig. 1 an example of an office building and its IFC file is presented. In Fig. 1a the retrieved tree structure of the IFC file is depicted, while in Fig. 1c a sub-tree with the expanded part of the tree starting from IfcRoot element and showing all the IfcProduct entities is presented. Figure 1b depicts the visualization of the office building.

3   Graph Representation of Buildings

In our approach the spatial configuration of the building structure is obtained from the IFC model and stored in a graph database. Graph databases are based on the concept of so called Property Graph Model. The property graph is build of connected entities (the nodes) which can hold any number of attributes (key-value-pairs). Nodes can also be tagged with labels representing their different roles (classes) in a given domain. In addition to contextualizing node and relationship properties, labels may also be used to attach metadata, index or constraint information to certain nodes. Relationships provide directed, named semantically meaningful connections between two nodes. A relationship in a property graph model always has a direction, a type, a start node, and an end node. Relationships can also have properties similar to those that can be attached to nodes. In most cases, relationships have quantitative properties, such as weights, costs, distances, sizes, positions, among others. Neo4j implements the

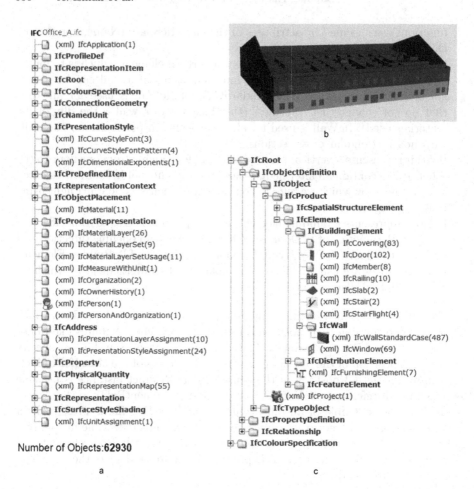

Fig. 1. An example of Building Information Model data for an office building (a) the structure of an IFC file (b) a visualization of this building (c) a subtree of the IFC showing IfcProduct data

Property Graph Model and it also provides full database characteristics including ACID transaction compliance [27].

The relations between IFC entities required to compute the topological relationships of spaces are searched for. Two rooms are adjacent if two IfcSpace entities refer to the same IfcWall or to the same IfcWindowStandardCase using IfcRelSpaceBoundary relation [24]. Two rooms are accessible if the wall between them has an opening or door. Therefore IfcWall or IfcWindowStandardCase entity should refer to IfcOpeningElement by IfcRelVoidsElement relation, or additionally IfcDoor entity should refer to IfcOpeningElement by IfcRelFillsElement relation (the opening is filled with a door). The extracted information is then saved in the graph structure.

We use attributed, labelled and edge-directed graphs. Graph nodes represent building spaces, while edges correspond to accessibility relations between these spaces and therefore represent doors, openings and accessibility between storeys through stairs/lifts. Labels assigned to graph nodes store names of spaces, while node attributes store other properties of spaces, for example their sizes or types.

A labelled directed graph shown in Fig. 4 represents spaces and their surrounding walls on the first floor of the building depicted in Fig. 1b.

4   Accessing IFC Models Through IFCWebServer

IFCWebServer [13] is a BIM data model server and online viewer based on IFC standard (ISO 17639). It aims to simplify sharing and exchanging of information from BIM models using open and standard formats and check the quality of BIM models (Level of Details, Level of Development). IFCWebServer enables full access to all information and relations inside IFC models and it supports all IFC official release through a dynamic EXPRESS parser. It can be used to query, filter and generate reports about any information inside IFC models easily. The online BIMViewer provides a handy way to view, share BIM models and visualize the results of data queries online inside the web browser.

4.1   Converting IFC Models into Graph Models

A workflow for automatic transformation of IFC models into a property graph database has been developed [14]. In this workflow, a special server script has been developed in order to convert IFC models into Neo4j graph database (https://github.com/ifcwebserver/IFC-to-Neoj4). It generates all data import and relationships in Cypher language.

The scope of conversion includes all model elements and the following relationships:

IfcRelAggregates, IfcRelAssignsTasks, IfcRelAssignsToGroup, IfcRelAssignsToActor, IfcRelAssignsToProcess, IfcRelAssociatesClassification, IfcRelAssociatesMaterial, IfcRelCoversSpaces, IfcRelConnectsElements, IfcRelConnectsPorts, IfcRelCoversBldgElements, IfcRelConnectsPathElements, IfcRelContainedInSpatialStructure, IfcRelDefinesByProperties, IfcRelDefinesByType, IfcRelFillsElement, IfcRelSpaceBoundary, IfcRelVoidsElement, IfcRelNests, IfcRelSequence.

The listing in Fig. 2 presents Cypher commands specified to generate (1) BoundedBy relationships between spaces of the building shown in Fig. 1b and their surrounding elements, and (2) hasProperties relationships between IfcPropertySets and IfcProperty objects of the same building.

```
// create the : BoundedBy relationships between spaces and
    building elements
// Graph pattern : ( IfcBuildingElement ) – [ :BoundedBy ] – ( IfcSpace )
MATCH ( n : IfcRelSpaceBoundary { model : " Office_A " } )
UNWIND split ( replace ( replace ( n.   relatedBuildingElement
        ," ( " , " " ) , " ) " , " " ) , " , " ) as o
MERGE ( relatedBuildingElement   : IfcElement {model_id : " Office_A _
    " + o } )
MERGE ( s : IfcSpace { model_id : " Office_A _ " + n. relatingSpace } )
MERGE ( relatedBuildingElement ) – [ BoundedBy ] –> ( s ) ;

// create the : hasProperties relationships between
    IfcPropertySets and IfcProperty objects
// Graph pattern : ( IfcPropertySet ) – [ : hasProperties ] – ( IfcProperty )
MATCH ( n : IfcPropertySet { model : " Office_A " } )
UNWIND split ( replace ( replace ( n.   hasProperties , " ( " , " " ) , " ) " , " " )
        , " , " ) as o
MERGE ( p : IfcProperty { model_id : " Office_A _ " + o } )
MERGE ( n ) – [ : hasProperties ] –> ( p ) ;
```

Fig. 2. An example of Cypher commands specified to generate relationships

5   Case Study

The case study is based on two story office building model, which is one of the
BIM projects provided by the National Institute of Building Sciences as part of
the common building information models [16]. The IFC model (exported from

```
MATCH ( space : IfcSpace { model : ' Office_A ' } ) –[] –( storey :
    IfcBuildingStorey { ifcid : ' 1116 ' } )
MATCH p = ( door : IfcDoor {ifcid : ' 807 ' , model : ' Office_A ' } ) –[ :
    BoundedBy ] –> ( space )
RETURN p UNION
MATCH ( sp1 : IfcSpace { model : ' Office_A ' } ) –[] –( storey :
    IfcBuildingStorey { ifcid : '1116 ' } ) –[] –( sp2 : IfcSpace )
MATCH p = ( space1 { model : ' Office_A ' } ) <–[ :BoundedBy ] –( door :
    IfcDoor ) –[ :BoundedBy ] –> ( sp2 )
WHERE sp1 . ifcid > sp2 . ifcid RETURN p UNION
MATCH ( sp1 : IfcSpace { model : ' Office_A ' } ) –[] –( storey :
    IfcBuildingStorey { ifcid : ' 1116 ' } ) –[] –( sp2 : IfcSpace )
MATCH
p = (sp1 { model : ' Office_A ' } ) –[ :RelatingSpace ] –( :
    IfcRelSpaceBoundary ) –[ :RelatingSpace ] –( sp2 )
WHERE space1 . ifcid > sp2 . ifcid RETURN p
```

Fig. 3. The Cypher query to retrieve emergency routes for the first floor

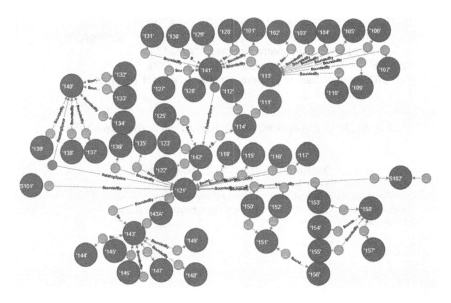

Fig. 4. The graph representation of emergency routes using single exit door

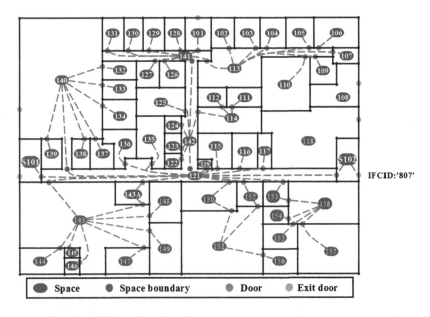

Fig. 5. Emergency routes from Fig. 4 drawn on the floor layout

Revit in IFC2X3-Coordination MVD format) [17] has been uploaded to http://
IFCWebServer.org. The converting into the neo4j graph database is carried out
according to the steps in [18].

```
MATCH (space :  IfcSpace { model : ' Office_A ' } ) –[ :
     IsDefinedByProperties  ] –( :IfcPropertySet ) –[ : hasProperties ]
     –( property :  IfcProperty { name : " ' Area ' " } )
WHERE   toFloat ( property . nominalValue ) > 30
RETURN  space . name, property . name,    toFloat ( property .
     nominalValue  )
```

Fig. 6. Getting spaces which are larger than a given area value

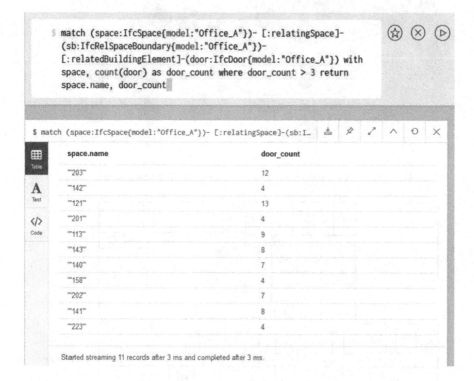

Fig. 7. The query retrieving spaces with more than 3 doors and the listed names of these spaces

In the following some examples of running data retrieval queries, which allow us to analyse the BIM model are presented.

The Cypher query which allows for retrieving emergency routes on the first floor of the considered building is presented in the listing in Fig. 3. The outcome of the graph database, in case when single exit door is used, is shown in Fig. 4. Figure 5 presents the found routes drawn on the floor layout.

In emergency cases the spaces which are large enough to serve as meeting points can be useful. In the listing in Fig. 6 the query allowing us to get spaces with area greater than $30\,m^2$ is shown. While searching for different paths in the building the open spaces and the ones with many entry/leaving points, like halls, lobbies and corridors, require special consideration. As people can cross such spaces in different ways it is important to put up escape signs in the right places to make them visible and to direct people properly. In Fig. 7 the query retrieving all spaces with more than 3 doors and the result in the form of listed space names with numbers of their doors is presented.

In the graph model of the building there are assigned properties for doors specifying whether they constitute fire exit (IsFireExit) and their fire resistance (FireRating). These properties can be used to select fire exit doors or doors with certain FireRating value. In cases of searching for routes accessible for disabled it is important information as such doors placed inside the building can be heavier than standard ones and therefore more difficult to open. In Fig. 8 fire properties of the chosen doors are highlighted.

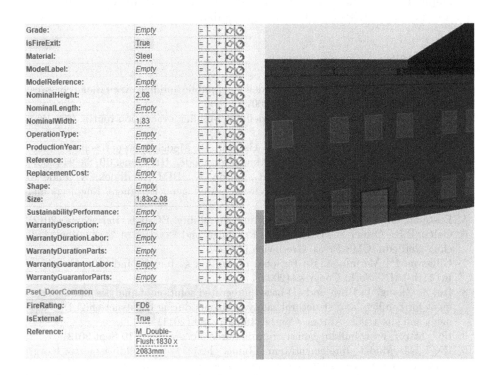

Fig. 8. Fire properties of the chosen doors

6 Conclusions

The paper deals with extraction and analysis of building-specific knowledge coded in BIM models. A workflow for automatic transformation of IFC format for BIM into a graph database is applied. On this database information retrieval queries, which support searching for different routes, are specified. IFCWeb-Server, which enables full access to all information and relations inside IFC models, is used. The results of data queries are visualized online inside the web browser.

The current scope of transformation and queries does not take into account all the geometry information or the process of creating geometry objects based on parameters or Boolean operations. In future an interface between the graph database and an IFC geometry engine will be developed, and thus it will be possible to include the geometry information as a part of the predefined queries. Special procedures, which will allow for accessing the Java API of Neo4j directly and running the retrieval queries much faster than by using Cypher commands, will be worked out. Moreover, IFC-based graph model management will be facilitated by simplifying the way of writing user-defined queries, as up to now advanced queries should be written by IFC and graph database experts.

References

1. AlShboul, A.A., Al-Tahat, M.D.: Modelling of public building evacuation processes. Architectural Sci. Rev. **50**, 37–43 (2007)
2. Cepolina, E.M.: A methodology for defining building evacuation routes. Civ. Eng. Environ. Syst. **22**, 29–47 (2005)
3. Chiu, Y.C., Zheng, H., Villalobos, J., Gautam, B.: Modeling no-notice mass evacuation using a dynamic traffic flow optimization model. IIE Trans. **39**, 83–94 (2007)
4. Eastman, C., Teicholz, P., Sacks, R., Liston, K.: BIM Handbook: A Guide to Building Information Modeling for Owners, Managers, Designers, Engineers and Contractors (2011)
5. Eastman, C.: The Evolution of AEC Interoperability. EG-ICE, Herrsching (2012)
6. Galea, E.R., Owen, M., Gwynne, S.: Principles and Practice of Evacuation Modeling, 2nd edn. CMS Press, Greenwich (1999)
7. Gillieron, P., Merminod, B.: Personal navigation system for indoor applications. In: 11th IAIN World Congress (2004)
8. Hu, Z., Zhang, J.: BIM- and 4D-based integrated solution of analysis and management for conflicts and structural safety problems during construction:2. Development and site trials. Autom. Constr. **20**, 167–180 (2011)
9. IFC. http://www.buildingsmart.org/standards/ifc. Accessed 10 Sept 2013
10. IFC 2x3 Model Implementation Guide. http://www.buildingsmarttech.org/implementation/ifc-implementation/ifc-impl-guide. Accessed 10 Sept 2013
11. IFC2x3 specification. http://www.buildingsmart-tech.org/ifc/IFC2x3/TC1/html/. Accessed 10 Sept 2013
12. Isaac, S., Sadeghpour, F., Navon, R.: Analyzing building information using graph theory. In: International Association for Automation and Robotics in Construction (IAARC)-30th ISARC, Montreal, pp. 1013–1020 (2013)
13. Ismail, A.: IFCWebServer. IFC Data model server and online viewer (2011). http://ifcwebserver.org
14. Ismail, A., Nahar, A., Scherer, R.J.: Application of graph databases and graph theory concepts for advanced analysing of BIM models based on IFC standard. In: Proceedings of EGICE 2017, Nottingham (2017)
15. buildingSMART. http://www.buildingsmart.org. Accessed 12 Sept 2013

16. Common Building Information Model. https://www.nibs.org/?page=bsa_commonbimfiles. Accessed 25 Jan 2018
17. Common Building Information Model: Office building model. http://projects.buildingsmartalliance.org/files/?artifact_id=4284
18. IFC-to-Neoj4. https://github.com/ifcwebserver/IFC-to-Neoj4
19. Iwarsson, S., Stahl, A.: Accessibility, usability and universal design-positioning and definition of concepts describing person-environment relationships. Disabil. Rehabil. **25**, 57–66 (2003)
20. Jeong, S., Ban, Y.: Computational algorithms to evaluate design solutions using Space Syntax. Comput.-Aided Des. **43**, 664–676 (2011)
21. Khalili, A., Chua, D.: IFC-based graph data model for topological queries on building elements. J. Comput. Civ. Eng. **29**(3) (2015). American Society of Civil Engineers. https://doi.org/10.1061/(ASCE)CP.1943-5487.0000331
22. Koutamanis, A.: Multilevel analysis of fire escape routes in a virtual environment. In: Tan, M., Teh, R. (eds.) The Global Design Studio. Centre for Advanced Studies in Architecture, National University of Singapore, Singapore (1995)
23. Kuligowski, E.D., Peacock, R.D., Hoskins, B.L.: A Review of Building Evacuation Models NIST, Fire Research Division. Technical Note 1680, 2nd edn. National Institute of Standards and Technology, Washington, US (2010)
24. Langenhan, C., Weber, M., Liwicki, M., Petzold, F., Dengel, A.: Graph-based retrieval of building information models for supporting the early design stages. Adv. Eng. Inform. **27**, 413–426 (2013)
25. Ma, Z., Wei, Z., Song, W., Lou, Z.: Application and extension of the IFC standard in construction cost estimating for tendering in China. Autom. Constr. **20**, 196–204 (2011)
26. Papinigis, V., Geda, E., Lukošius, K.: Design of people evacuation from rooms and buildings. J. Civ. Eng. Manag. **16**, 131–139 (2010)
27. Robinson, I., Webber, J., Eifrem, E.: Graph Databases: New Opportunities for Connected Data, 2nd edn. O'Reilly Media, Sebastopol (2015)
28. Ronchi, E., Nilsson, D.: Fire evacuation in high-rise buildings: a review of human behaviour and modelling research. Fire Sci. Rev. **2**(7), 1–21 (2013)
29. Rüppel, U., Abolghasemzadeh, P., Stübbe, K.: BIM-based Immersive Indoor Graph Networks for Emergency Situations in Buildings, pp. 65–71. Nottingham University Press, Nottingham (2010)
30. Sakkas, N., Perez, J.: Elaborating metrics for the accessibility of buildings. Comput. Environ. Urban Syst. **30**, 661–685 (2006)
31. Shen, T.S., Chien, S.W.: An evacuation simulation model (ESM) for building evaluation. Int. J. Archit. Sci. **6**, 15–30 (2005)
32. Stringfield, W.H.: Emergency Planning and Management. Government Institutes, Rockville (1996)
33. Strug, B., Ślusarczyk, G.: Reasoning about accessibility for disabled using building graph models based on BIM/IFC. Vis. Eng. **5**, 10 (2017). https://doi.org/10.1186/s40327-017-0048-z
34. Ślusarczyk, G., Łachwa, A., Palacz, W., Strug, B., Paszynska, A., Grabska, E.: An extended hierarchical graph-based building model for design and engineering problems. Autom. Constr. **74**, 95–102 (2017)
35. Tauscher, E., Bargstädt, H.-J., Smarsly, K.: Generic BIM queries based on the IFC object model using graph theory. In: The 16th International Conference on Computing in Civil and Building Engineering, Osaka, Japan (2016)

36. Zender, H., Martinez Mozos, O., Jensfelt, P., Kruijff, G.-J.M., Burgard, W.: Conceptual spatial representations for indoor mobile robots. Robot. Auton. Syst. **56**(6), 493–502 (2008)
37. Yatim, Y.M.: Optimum escape routes designs and specification for high-rise buildings. In: Proceeding of 2012 3rd International Conference in Construction Industry, Indonesia (2012)

A Multi-Agent Problem in a New Depiction

Krystian Jobczyk[1,2(✉)] and Antoni Ligęza[2]

[1] University of Caen Normandy, Caen, France
krystian.jobczyk@unicaen.fr
[2] AGH University of Science and Technology, Kraków, Poland
krystian_jobczyk@op.pl, jobczyk@agh.edu.pl

Abstract. This paper contains a new depiction of the Multi-Agent Problem as motivated by the so-called Nurse Rostering Problem, which forms a workable subcase of this general problem of Artificial Intelligence. Multi-Agent Problem will be presented as a scheduling problem with an additional planning component. The next, the problem will be generalized and different constraints will be put forward. Finally, some workable subcases of Multi-Agent Problem will be implemented in PROLOG-solvers.

1 Introduction

A Multi-Agent Problem may be seen as a relatively far generalization of different problems of scheduling in Artificial Intelligence (see: [22]), although its exact and unified formulation is not known. In essence, it may be seen rather as a reservoir (or a class) of different similar problems that may be commonly specified as problems with:

* an inventing the action sequence in order to perform a goal,
* association actions to agents that could be performed by them due to their skills.

The depiction of Multi-Agent Problem, proposed in this paper, may be viewed as a generalisation of such problems – earlier considered in the specialist literature – as the so-called *Nurse Job Scheduling Problem* (NJSP). This problem is to be also known as *Nurse Rostering Problem* – see for example: [4,17,18]. It appears that NJSP – in formulations known in a subject literature – formed a stimulating problem for operational research, which also supported a broad development of constraints logic programming methodology. This methodology was especially explored by Nottingham's school. All these fact are, somehow, reflected in such works as: [3,4,7,18].

Meanwhile, each expressive formulation of Multi-Agent Problem (MAS) should be involved in some additional concepts, such as temporal constraints and preferences. Meanwhile, these concepts are usually discussed in a variety

© Springer International Publishing AG, part of Springer Nature 2018
L. Rutkowski et al. (Eds.): ICAISC 2018, LNAI 10842, pp. 665–676, 2018.
https://doi.org/10.1007/978-3-319-91262-2_58

of contexts, which are often independent of MAS. In fact, temporal constraints are developped, for example in terms of Simple Temporal Problems (STP) and its extensions in such works as: [6,12,14–16,20,21,25]. In a majority of these works, temporal constraints were rather more associated to graph-based planning than to scheduling as a proper basis for Multi-Agent Problem. Meanwhile, MAS may also find its reflection not only in a construction of different systems of modal and multi-valued logic such as in [9,11,13], but also in a construction of a different types of the plan controllers, such as in [10].

Finally, different approaches have been elaborated with respect to the notion of preferences. This entity still remains a subject of interests of different sciences. The philosophical and psychological provenance of the earlier research on preferences and their nature can be detected in works of Armstrong in 30th and 40th such as [1,2]. Preferences in the contexts of economic analysis was discussed also relatively early by Ramsey in 1928 in [19]. Although they may be seen in different way (for example: as mental states or intentions of rational subject), the main interpretation stream treats preferences as special relations, as in [5,8,23,24].

1.1   The Paper Motivation

Some of difficulties of the current state of art with respect to Multi-Agent Problem have been already described. In fact, it has aready been said that temporal constraints – as crucial for an appropriate depiction of Multi-Agent Problem – are usually described in conceptually inadequate contexts of graph-based planning. Unfortunately, there are (at least three) other difficulties.

A Nurse Job Scheduling Problem – as a basis of Multi-Agent Problem should be rather seen a basis *reservoir* of possible more specified formulations. For example, Ernst's approach in [7] is mathematically general, but refer to simplified situations.
B These works consider this problem as optimization problem of scheduling without planning components and preferences.
C Finally, there is no common consensus with respect to the notion of preference – even if preferences are considered as relations. (For example, they are interpreted as partial orders in [8], but as total orders or linear orders in [23,24].)

1.2   The Objectives of the Paper and Its Organisation

This chapter will be devoted to the Multi-Agent Schedule-Planning Problem (MASPP) and its preferential extension. Their formulation will be motivated by the commonly known formulations of NJSP in a specialist literature. More precisely, we elaborate a new depiction of

A *Multi-Agent Schedule-Planning Problem,*

that will be extended to

B *Preferential Multi-Agent Schedule-Planning Problem.*

The initial depiction of both problems will be later generalised. Finally, some workable cases of Preferential Multi-Agent Schedule-Planning Problem will be solved by means of PROLOG-solvers. By the way, complexity of solutions will be briefly discussed. Novelty of the paper with respect to earlier approaches consists in proposing:

N1 a preferential modification of the Multi-Agent-Problem,
N2 a considering the Multi-Agent-Problem as a synergy synthesis of planning and scheduling components,
N3 a generalization of the Multi-Agent Problem and – finally –
N4 a PROLOG-solvers for some workable fuzzy cubcases of this problem.

Because of different discrepancies between researchers with respect to the notion of preferences, we adopt an intuitive meaning of preferences as wishes or expectations of the operating agents. In this way, we put aside the whole broad discussion on a nature of relations interpreting them semantically.

The rest of the paper is organised as follows. Section 2 contains both the (more) practical depiction of a Multi-Agent Schedule-Planning Problem (MASPP) and its generalization. A Preferential Multi-Agent Schedule-Planning Problem (PMASPP) is also described in this section. Finally, a brief taxonomy of temporal constraints imposed on these problems is put forward here. Section 3 is devoted to the programming-wise aspects of MASPP and PMASPP. Section 4 contains closing remarks.

2   A Multi-Agent Schedule-Planning Problem – a More Practical Depiction

In this chapter, our initial depiction Multi-Agent Schedule-Planning Problem will be put forward in a more practical way. As it has been mentioned – it will be motivated by Nurse Rostering Problem. This solution allows us to grasp many intuitions that form a conceptual foundation of a more general definition of this problem. It will be exploit in the preferential extension of this problem. In order to make it, let us observe that each formulation of Multi-Agent Schedule-Planning Problem must satisfy the following general criteria.

C1 A finite (non-empty) of agents should be given,
C2 Agents should be involved in some activities in some time periods and sub-periods (for examples: days and shifts),
C3 There are some hard constraints imposed on agant activities that must be absolutely satisfied to perform the task,

C4 There are some soft constraints imposed on agent activities that may be satisfied.

C5 One can also admit some preferences imposed on task performing (the external or the internal ones).

Taking into account these general criteria, one can formulate the following basis Multi-Agent Schedule-Planning Problem as follows:

Multi-Agent Schedule-Planning Problem (M-AS-PP). Consider a factory with n-agents working in a rhythm of the day-night shifts: D–the day shift and N–the night shift. Generally – each day at least one person must work at the day shift and at least one – at the night one. Each agent has "working shifts" and "free shifts". These general rules of scheduling is constraints in the following way.

HC1 The charm of the shift organization should be fair: each agent must to have equally: 2 day-shifts and 2 night-shifts.

HC2 Each agent can be associated to at most one shift,

HC3 Some shifts are prohibited for agents,

HC4 Length of the shifts sequences associated to each agent is restricted,

HC5 Quantity of the shifts in a scheduling period is restricted,

HC6 Quantity of the shifts per a day is restricted.

The MASPP consists in a construction of a scheduling diagram, which respects all these constraints.

As one can easily see, a couple of the so-called hard constraints in the above depiction of M-AS-PP was indicated, HC1-HC6.

Generally, *Hard constraints* are described as the constraints, which *should be satisfied* in a scheduling task. They ensure a feasibility of the scheduling task. *The soft constraints* may not be satisfied, but a degree of their satisfaction is a measure how good is a scheduling plan. To make requirements with respect to the hard constraints more liberal, we use the so-called relaxation, i.e. a weakening of the strong constraints. We often use this solution, when satisfaction of all hard constraints leads to an inconsistency.

A further relaxaction of requirements or expectations allows us to consider the next category of preferences. The main nature of preferences as relations was briefly earlier discussed. In this section, we are interested in another sense of this concept. The preferences are wishes or expectations of an agent, for example, with respect to the action execution ora their sequencing.

Both the soft constraints and preferences are admitted in the *Preferential Multi-Agent Schedule-Planning Problem (PMASPP)*. Generally, all hard constraints of *Multi-Agent Schedule-Planning Problem* are preserved in this new problem. In fact, it forms a kind of an extension of the initial MASPP towards soft constraints and preferences. They are also defined in the same context as HC's.

Preferential Multi-Agent Schedule-Planning Problem (PMSPP)
Consider a factory with n-agents working in a rhythm of the day-night
shifts: D–the day shift and N–the night shift. Generally – each day
at least one person must work at the day shift and at least one – at
the night one. Each agent has "working shifts" and "free shifts". These
general rules of scheduling is constraints in the following way.

HC1 The charm of the shift organization should be fair: each agent must
to have equally: 2 day-shifts and 2 night-shifts.
HC2 Each agent can be associated to at most one shift,
HC3 Some shifts are prohibited for agents,
HC4 Length of the shifts sequences associated to each agent is restricted,
HC5 Quantity of the shifts in a scheduling period is restricted,
HC6 Quantity of the shifts per a day is restricted.

Assuming also an agent $n_k \in N$ and the chosen (real) parameters m, M
and α Different soft constraints and preferences of a general form are
also considered in the scheduling procedure.

SC7 A preferential quantity of shifts in a scheduling period is established,
SC8 A preferential scheduling charm's covering by shifts in a scheduling
period is established,
SC9 A preferential lenght of the shifts sequence associated to an agent
is fixed,
Pref1 A number of actions (preferred by an agent n_k) to be associated
to its schedule is greater than m and smaller than M,
Pref2 An agent n_k prefers to perform an action a with a degree α^a.

The M-AS-PP consists in a construction of a scheduling diagram, which
respects all these constraints.

---

[a] The parameters may be chosen arbitrarily, but they are fixed in the whole
M-AS-PP problem. In some particular cases, the choice of them may be
restricted according to the appropriate criteria or other restrictions.

It easy to observe that SC's and preferences have a common denominator:
something is preferred. However, SC's express a global preference of the whole
problem, but preferences render particular preferences of a single agent. Because
of this distinction, we are willing to consider the global external preferences
'more seriously' as soft constraints.

2.1 General Formulation of a Multi-Agent Schedule-Planning Problem

Let us consider a generic temporal multi-agent task scheduling problem. Roughly
speaking, a set of agents, each of them possessing specific skills, is to be assigned

some temporal tasks to be completed. Each agent can accept only tasks consistent with its skills. Execution of tasks should be performed according to predefined partial order relation. Further auxiliary constraints (e.g. Allen's type constraints for execution periods of certain actions) or extensions (e.g. parallel execution of actions by a single agent) are possible. Below, a generic, simple formalization of this problem is put forward.

Consider a set $\mathbb{A} = \{A_1, A_2, \ldots A_n\}$ of n agents. Each agent can possess one or more skills. Let $\mathbb{S} = \{S_1, S_2, \ldots S_k\}$ denote the set of predefined skills. Assume σ is the function defining a two-valued measure for all the skills of any agent; so σ is defined as:

$$\sigma \colon (\mathbb{A}, 2^{\mathbb{S}}) \mapsto \{0, 1\}.$$

For practical reasons, it is convenient to represent this function in a tabular (matrix) form as follows:

	S_1	S_2	...	S_k
A_1	$\sigma_{1,1}$	$\sigma_{1,2}$	...	$\sigma_{1,k}$
A_2	$\sigma_{2,1}$	$\sigma_{2,2}$	...	$\sigma_{2,k}$
$\vdots$	$\vdots$	$\vdots$	$\ddots$	$\vdots$
A_n	$\sigma_{n,1}$	$\sigma_{n,2}$	...	$\sigma_{n,k}$

$$\text{where } \sigma_{i,j} = \begin{cases} 1, & \text{if } S_j \in \sigma(A_i) \text{ (the } i\text{--th agent possess the } j\text{--th skill)} \\ 0, & \text{otherwise.} \end{cases}$$

Similarly, consider a set $\mathbb{T} = \{T_1, T_2, \ldots, T_m\}$ of tasks to be executed. Each tasks, in order to be executable by an agent, requires some specific skills. Assume θ is the function defining all the skills required to execute a specific task; so θ is defined as:

$$\theta \colon (\mathbb{T}, 2^{\mathbb{S}}) \mapsto \{0, 1\}.$$

Again, for practical reasons it is convenient to represent this function in a tabular (matrix) form as follows:

	S_1	S_2	...	S_k
T_1	$\theta_{1,1}$	$\theta_{1,2}$	...	$\theta_{1,k}$
T_2	$\theta_{2,1}$	$\theta_{2,2}$	...	$\theta_{2,k}$
$\vdots$	$\vdots$	$\vdots$	$\ddots$	$\vdots$
T_m	$\theta_{m,1}$	$\theta_{m,2}$	...	$\theta_{m,k}$

$$\text{where } \theta_{i,j} = \begin{cases} 1, & \text{if } S_j \in \theta(T_i) \text{ (the } i\text{--th task requires the } j\text{--th skill)} \\ 0, & \text{otherwise.} \end{cases}$$

For simplicity, it is assumed that a single task can be executed by a single agent, one task at a time. Task T_j can be executed by agent A_i if and only if the agent possesses all the required skills. Formally, skills associated to tasks (obtained by the projection on 2^S in a domain of θ) should be contained in skills (obtained by the projection on 2^S in a domain of σ) associated to agents from $\mathbb{A}$. Symbolically:

$$\pi_{2^S}\Big(dom\{\theta(T_l, S_j)\}\Big) \subseteq \pi_{2^S}\Big(dom\{\sigma(A_i, S_j)\}\Big),$$

and the execution can start whenever the agent is free; this holds for all $i \in \{1, 2, \ldots n\}$, $j \in \{1, 2, \ldots k\}$ and $l \in \{1, 2 \ldots, m\}$.

Now, roughly speaking, the problem consists in efficient assignment of all the tasks to given agents, so that the tasks can be executed, all the constraints are satisfied, and the total execution time will perhaps be minimal. More precisely, the assignment problem is defined as follows.

2.2   Types of Temporal Constraints of PM-AS-PP

As mantioned, temporal constraints in both M-AS-PP and PM-AS-PP might be divided into two groups:

– *Hard constraints* and
– *Soft constraints.*

Hard constraints are specified as these constraints that *should be violated* in a scheduling task. They ensure a feasibility of the scheduling task. *The soft constraints* may not be satisfied, but a degree of their satisfaction is a measure how good is a scheduling plan. To make requirements with respect to the hard constraints more liberal, we use the so-called relaxation, i.e. a weakening of the strong constraints. We often use this solution, when satisfaction of all hard constraints leads to an inconsistency.

For a mathematical representation of temporal constraints imposed on PNJSP we introduce the following set of parameters. Instead of agent skills we will consider agent roles (contracts)[1]:

– $N = \{n_1, n_2 \ldots, n_k\}$ as set of agents (agents),
– $R = \{r_1, r_2, \ldots, r_k\}$ as a set of roles (contracts),
– $D = \{d_1, d_2, \ldots, d_k\}$ as a set of days in a week,
– $Z = \{z_1, z_2\}$ as a set of admissible shifts during days from D,
– $\mathcal{A} = \{a_1, a_2, \ldots, a_k\}$ as a set of actions.

It enables representing now M-ASP-P by its formal instances in the form of the triple

$$(N, D, Z, A, HC), \tag{1}$$

where N, D, Z are given as above and HC denotes a set of hard constraints imposed on actions from A and their performing. Similarly, PM-ASP-P may be given by the n-tuple of the form:

$$(N, D, Z, A, HC, SC, P), \tag{2}$$

---

[1] All of these constraints are typical for scheduling problems of this type to be known as (usually) NP-hard – see: [4].

where N, D, Z and HC are given as above and SC and P denote a set of soft constraints and preferences (*resp.*) Introducing SC to the n-tuple 1.7 follows from the adopted hierarchy of constraints. The hard constraints cannot be violated, the soft ones may be violated, but they should be satisfied before preferences.

This notation allows us to elaborate the following representation of hard and soft constraints. Since their list is not exhaustive[2], it might be relatively naturally extended.

HC 1: The charm of the shift organization should be fair: each agent must to have equally 2–day shifts and 2–night shifts

Assume that Z_{day} denotes a set of day-shifts and Z_{night} denotes a set of night-shifts. Then this strong constraint may be shortly mathematically rendered as follows:

$$\sum_{z \in Z_{day}} X_{n,d,z} = 2 \wedge \sum_{z \in Z_{night}} X_{n,d,z} = 2. \tag{3}$$

HC 2: Each agent can be associated to at most one shift

This strong constraint can be shortly mathematically expressed as follows:

$$\sum_{z \in Z} X_{n,d,z} = 1. \tag{4}$$

HC 3: Some shifts are prohibited for an agent n

This strong constraint renders the following prohibition: some shifts are prohibited for an agent n and a relaxation usually cannot be referred to it. If Z_n denotes a set of shifts prohibited for a agent $n \in \{N_1, N_2 \ldots, N_k\}$, then this constraint can be mathematically depicted as follows:

$$\sum_{z \in Z_n} X_{n,d,z} = 0. \tag{5}$$

HC4: Length of the shifts sequence associated to an agent

It is a strong constraint that defines a restriction for the sequence of shifts associated to a agent. If we denote by a minimal and a maximal number of shifts associated to a agent n by m_z^{min} and by m_z^{max} (resp.), then this constraint can be rendered as follows:

$$m \le \sum_d^{m_z+d} X_{n,d,z} \le M. \tag{6}$$

HC 5: Quantity of shifts in a scheduling period

It defines the minimal and the maximal quantity of shifts during a given scheduling period (day, months, etc. – associated to a single agent.) If we denote – the minimal and the maximal quantity of shifts that can be obtained by agents in a given scheduling period by s and by S (resp.), then this constraint can be rendered mathematically as follows:

---

[2] This fact plays no important role as the main objective of this juxtaposition consists in the *quantitative representation* alone, which will be later combined with qualitative temporal constraints (of Allen's sort) for a use of further investigations.

$$s \leq \sum X_{n,d,z}^{d \in D} \leq S, \tag{7}$$

where $X_{n,d,z}$ is defined as above.

HC6: The Quantity of the shifts per day is restricted

It defines the minimal and the maximal quantity of temporal shifts during a day – associated to a single agent. If we denote the minimal and the maximal number of agents in a role r – which should obtain a shift z during a day d – by r and by $R(resp.)$, then this constraint may be mathematically rendered as follows:

$$r \leq \sum^{n \in N} X_{n,d,z} \leq R, \tag{8}$$

where $X_{n,d,z}$ is defined as the following characteristic function[3]:

$$X_{n,d,z} = \begin{cases} 1 & \text{if an agent } n \text{ works a shift } z \text{ in a day } d, \\ 0 & \text{otherwise.} \end{cases}$$

Obviously, the list of constraints is not exhaustive and it may be naturally enlarged. However, we interrupt their presentation in this point to solve the problem of the appropriate representation of the problems that were indicated.

3 Programming-Wise Aspects of Multi-Agent Schedule-Planning Problem

In this section, we intend to face the *Multi-Agent Schedule-Planning Problem* in order to illustrate how to solve some of its workable subcases with a support of PROLOG-solvers. In order to illustrate a general method of a construction of these solvers, let us assume that a non-empty set $N = \{X, Y \ldots\}$ of agents and a non-empty set $D = \{1, 2, 3, 4, 5\}$ of working days (for simplicity we omit shifts during a day) are given.

In such a framework, the PROLOG-solver task is to give a schedule respecting the temporal constrains imposed on agent task performing and agent activity. The obtained solutions will be returned in a form of lists of the form:

$$X = [X1, X2, X3, \ldots, Xk], \quad Y = [Y1, Y2, Y3, \ldots, Yl], \tag{9}$$

where $X(i), Y(j)$ are characteristic functions representing activity of agents X and Y during i-day and j-day (*resp.*) for $k, l \in \{1, 2 \ldots, 7\}$.

We can consider two types of situations:

– *crisp*-type situations, when $X(i)$ and $Y(j)$ take only two values: 0 or 1 for $i, j \in \{1, 2, \ldots, 7\}$ or

---

[3] This binary representation can be also exchanged by a classical one: $X_{n,d} = z$ as presented in [4].

– $fuzzy^4$-type situations, when $X(i)$ and $Y(j)$ take more than two values for $i, j \in \{1, 2, \ldots, 7\}$, for example: 0, 1, 2, 3, 4. We adopt natural numbers because of restrictions of PROLOG-syntax, which is not capable of representing values from $[0, 1]$. Nevertheless, we intend to think about these values as about normalized values. Namely, we will interpret 1 – taken from a sequence $1, 2 \ldots, k$ – as $\frac{1}{k}$, 2 as $\frac{2}{k}$, k as $\frac{k}{k} = 1$, etc.

We focus our attention on fuzzy cases only as the more interesting ones.

In all these examples, PROLOG-solutions were rendered in the form of the appropriate lists of 0's and 1's. In the current cases we extend a list of admissible values modifying their initial sense. At first, assume the following values:

– 0 – for a representation of the fact that an agent A is absent (on a shift),
– 1 – for a representation of a physical absence of the agent A, but a real disposition to be present.
– 2 – for a representation of a physical presence of A, which is only in a partial disposition to work.
– 3 – for a representation of a full disposition of A to work[5].

Let us begin with an exemplary case of two agents: A1 and A2 working 5 days in a week and having four degrees of disposition denoted by 0, 1, 2 and 3 as above. We denote this problem as MASPP$(2, 5, 4)^{Fuzz}$.

MASPP(2, 5, 4)^{Fuzz}. This situation may be reflected in the following PROLOG-program (As earlier, the sense of lines of the PROLOG-code is explained on the right side of the programm):

```
plan2(A1,A2) :- A1 = [A1D1,A1D2,A1D3,A1D4,A1D5],        /* list of days of agent A1*/
A2 = [A2D1,A2D2,A2D3,A2D4,A2D5],     /* list of days of agant A2 */
A1 ins 0..3,                                 /* Fuzzy degrees of disposition of A1 */
A2 ins 0..3,                                 /* Fuzzy degrees of disposition of A2 */
sum(A1, #<, 12),                        /* Restriction on A1-activity during a week */
sum(A2, #<, 9),                        /* Restriction on A2-activity during a week */
sum([A1D1, A2D1], #<, 4), (A1D1 #> 1) # / (A2D1 #> 2),    /* Restriction on D1 */
sum([A1D2, A2D2], #<, 4), (A1D2 #> 2) # / (A2D2 #> 2),    /* Restriction on D2 */
sum([A1D3, A2D3], #<, 4), (A1D3 #> 2) # / (A2D3 #> 2),    /* Restriction on D3 */
sum([A1D4, A2D4], #<, 4), (A1D4 #> 2) # / (A2D4 #> 2),    /* Restriction on D4 */
sum([A1D5, A2D5], #<, 4), (A1D5 #> 2) # / (A2D5 #> 2),    /* Restriction on D5 */
sum([A1D1, A1D2, A1D3], #<, 6),     /* Restrictions on the next 3 days*/
sum([A1D2, A1D3, A1D4], #<, 7),
sum([A1D3, A1D4, A1D5], #<, 6),
sum([A2D1, A2D2, A2D3], #<, 7),
sum([A2D2, A2D3, A2D4], #<, 5),
```

---

[4] More precisely: multi-valued situations.

[5] As mentioned, we rather prefer to think about these values as normalized to $[0, 1]$ – as $\frac{1}{3}, \frac{2}{3}$ etc. instead of 1, 2, or 3. We make use values 0, 1, 2, 3 because of restrictions imposed on PROLOG-syntax.

```
sum([A2D3, A2D4, A2D5], #<, 7),

label([A1D1,A1D2,A1D3,A1D4,A1D5, A2D1,A2D2,A2D3,A2D4,A2D5]).
```

This PROLOG-solver returns us the following solution-lists:

```
A1 = [2,3,0,3,0]
A2 = [0,0,3,0,3] and
A1 = [2,3,0,3,0]
A2 = [1,0,3,0,3].
```

Further relaxations of temporal constrains usually changes combinatorial explosion of the algorithm relatively quickly. For example, if exchange also a requirement sum([A1D1, A1D2, A1D3], #<, 6) for sum([A1D1, A1D2, A1D3], #<, 13) we get more than 60 solutions. As a nature of things, we omit their exact presentation.

4 Conclusion

It has been already shown how a Multi-Agent Problem may be depicted in some unified version consisting of a planning and a scheduling component. It was also shown how to preferentially expand this formulation. The next, a more general depiction for the Multi-Agent Problem was proposed. Finally, some programming-wise aspects of these problems were discussed in PROLOG-solvers.

It seems that these PROLOG-solvers may be naturally extended. In fact, they were suitable to solve the most workable subcases with a relatively small set of possible fuzzy values representing the agent preferences. It seems that it may constitute a promising research direction for a future.

References

1. Armstrong, W.: Determinates of the utility function. Econ. J. **49**, 453–467 (1939)
2. Armstrong, W.: Uncertainty and the utility function. Econ. J. **58**, 1–10 (1949)
3. Burke, E.K., Curtois, T., Qu, R., Vanden Berghe, G.: A scatter search approach to the nurse rostering problem. J. Oper. Res. Soc. **61**, 1667–1679 (2010)
4. Cheang, B., Li, H., Lim, A., Rodrigues, B.: Nurse rostering problems—a bibliographic survey. Eur. J. Oper. Res. **151**, 447–460 (2003)
5. Davidson, D.: Hempel on explaining action. In: Essays on Actiona and Events, pp. 261–275 (1976)
6. Dechter, R., Meiri, I., Pearl, J.: On fuzzy temporal constraints networks. Temporal Constraints Netw. **49**(1–3), 61–95 (1991)
7. Ernst, A.T., Jiang, H., Krishnamoorthy, M., Sier, D.: Staff scheduling and rostering: a review of applications, methods and models. Eur. J. Oper. Res. **153**(1), 3–27 (2004)
8. Halpern, J.Y.: Defining relative likehood in partially-ordered preferential structure. J. Artif. Intell. Res. **7**, 1–24 (1997)
9. Jobczyk, K., Ligeza, A.: Fuzzy-temporal approach to the handling of temporal interval relations and preferences. In: Proceeding of INISTA, pp. 1–8 (2015)
10. Jobczyk, K., Ligeza, A.: A general method of the hybrid controller construction for temporal planning with preferences. In: Proceeding of FedCSIS, pp. 61–70 (2016)

11. Jobczyk, K., Ligeza, A.: Multi-valued Halpern-Shoham logic for temporal Allen's relations and preferences. In: Proceedings of the Annual International Conference of Fuzzy Systems (FuzzIEEE) (2016, page to appear)
12. Jobczyk, K., Ligeza, A.: Towards a new convolution-based approach to the specification of STPU-solutions. In: FUZZ-IEEE, pp. 782–789 (2016)
13. Jobczyk, K., Ligęza, A., Bouzid, M., Karczmarczuk, J.: Comparative approach to the multi-valued logic construction for preferences. In: Rutkowski, L., Korytkowski, M., Scherer, R., Tadeusiewicz, R., Zadeh, L.A., Zurada, J.M. (eds.) ICAISC 2015. LNCS (LNAI), vol. 9119, pp. 172–183. Springer, Cham (2015). https://doi.org/10.1007/978-3-319-19324-3_16
14. Khatib, L., Morris, P., Morris, R., Rossi, F.: Temporal reasoning about preferences. In: Proceedings of IJCAI-01, pp. 322–327 (2001)
15. Leierson, C.E., Saxe, J.B.: A mixed-integer linear programming problem which is efficiently solvable. In: Proceedings 21st Annual Allerton Conference on Communications, Control, and Computing, pp. 204–213 (1983)
16. Liao, Y.Z., Wong, C.K.: An algorithm to compact a VLSI symbolic layout with mixed constraints. IEEE Trans. Comput. Aided Des. Integr. Circuits Syst. **2**(2), 62–69 (1983)
17. Métivier, J.-P., Boizumault, P., Loudni, S.: Solving nurse rostering problems using soft global constraints. In: Gent, I.P. (ed.) CP 2009. LNCS, vol. 5732, pp. 73–87. Springer, Heidelberg (2009). https://doi.org/10.1007/978-3-642-04244-7_9
18. Qu, R., He, F.: A hybrid constraint programming approach for nurse rostering problems. Technical report, School of Computer Science, Nottingham (2008)
19. Ramsey, F.: A mathematical theory and saving. Econ. J. **38**, 543–59 (1928)
20. Rossi, F., Yorke-Smith, N., Venable, K.: Temporal reasoning with preferences and uncertainty. In: Proceedings of AAAI, vol. 8, pp. 1385–1386 (2003)
21. Shostak, R.: Deciding linear inequalities by computing loop residues. J. ACM **28**(4), 769–779 (1981)
22. Traverso, P., Ghallab, M., Nau, D.: Automated Planning: Theory and Practice. Elsevier, New York City (2004, 1997)
23. van Benthem, J.: Dynamic logic for belief revision. J. Appl. Non-Class. Log. **17**(2), 119–155 (2007)
24. van Benthem, J., Gheerbrant, A.: Game: solution, epistemic dynamics, and fixed-point logics. Fundam. Inform. **100**, 19–41 (2010)
25. Vidal, T., Fargier, H.: Handling contingency in temporal constraints networks: from consistency to controllabilities. J. Exp. Tech. Artif. Intell. **11**(1), 23–45 (1999)

Proposal of a Smart Gun System Supporting Police Interventions

Radosław Klimek[✉], Zuzanna Drwiła, and Patrycja Dzienisik

AGH University of Science and Technology,
Al. Mickiewicza 30, 30-059 Kraków, Poland
rklimek@agh.edu.pl

Abstract. A smart gun idea fits within the smart city concept where the extensive use of information and telecommunication technologies improves working efficiency as well as safety and comfort of residents. There was proposed a system based on Internet of Things (IoT) concept. It detects danger to life and health of policemen and provides them support from other patrols and municipal services. This kind of concept, very different from the previous ones, has already been shown in a prototype version.

Keywords: Smart gun · Use case scenario · Police · IoT
Pervasive computing

1 Introduction

Smart systems have greatly changed over the last years. Smart technologies are more and more popular and are used in order to improve residents' safety and life comfort, efficiently manage resources, lower exploitation costs etc. 'Smart' solutions are usually connected with the use of advanced technologies and lead to the results beneficial for those who are responsible for city management as well as for its residents. Smart city solutions are based on he quickly developing concept known as Internet of Things (IoT). This concept is centered on the idea that almost all kinds of devices can be connected into one network. It is a technology of the future and thanks to that digital devices and physical objects, also everyday objects, will communicate together using special infrastructure. In this way they will provide the user with endless number of new functions. The smart cities idea shifts towards this kind of approach [1–7].

The idea of smart gun seems to be surprising but it is not entirely new. However, the existing solutions are limited to identification of people authorized to carry weapons. Identification process can be carried on in various ways, for example by fingerprints analysis or detection of close presence of a personal device such as smartwatch which guarantees that it is used by a competent person. Authorization can also cover writing a password or a special code. On the other hand, it is noticeable that there is a lack of solutions which would

© Springer International Publishing AG, part of Springer Nature 2018
L. Rutkowski et al. (Eds.): ICAISC 2018, LNAI 10842, pp. 677–688, 2018.
https://doi.org/10.1007/978-3-319-91262-2_59

support everyday work of police officers when they are in dangerous situation. The support could be provided especially in critical situations such as shooting or unplanned fire exchange. One or more police officers who find themselves in a new, dangerous situation, do not have to organize simultaneously two types of actions: fire exchange and help/support notification. This kind of help would be organized automatically by a smart software and it would cover notification the proper services, finding another police patrols in the nearest neighborhood and sending them a support request.

The main purpose of this work is to propose a complex software system for the idea of smart gun which is a part of police equipment. The system is designed to improve officers' safety when operating in the city environment. The further aim is to carry on the initial experiments and working simulation for a few real situations and scenarios.

As it was already mentioned the idea of smart gun is not new [10] but it is discussed rather in the context of identification for the authorized/unauthorized use, as a kind of image preprocessing. The biometric verification system of a smart gun is discussed in paper [8], and experimental results are shown. In paper [9] algorithms for authentication of the police officers with simulation results are presented. However, there is a lack of works with smart scenarios as they are understood in this work.

2 Functional Model of the System

The use case diagram for the proposed system is shown in Fig. 1. The following actors are distinguished:

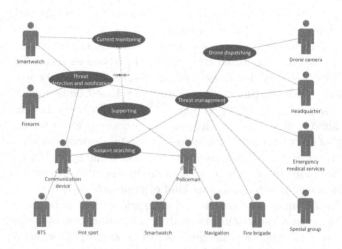

Fig. 1. The use case diagram for the proposed system

- **Firearm** – a physical device and a part of police equipment: the main element of the whole system. Firearm belongs to an officer on duty;
- **Smartwatch** – a physical device and a part of police equipment which belongs to an officer on duty. Smartwatch is directly connected with weapon, has GPS receiver and tracks physical location of a policeman. On the basis of those results, dangerous place is determined and a signal is sent to support groups. Smartwatch also enables communication with Headquarter and quick reaction on signals send by other policemen in need.
- **BTS (Communication device)** – a cell tower which enables communication.
- **Hot spot (Communication device)** – another communication device.
- **Smartwatch (Policeman)** – smartwatch of policeman who is nearby, the device receives a call for help from another policeman.
- **Navigation (Policeman)** – device with navigation which automatically finds coordinates of a policeman in need.
- **Fire brigade** – firefighters team, optionally sent by Headquarter.
- **Special group** – rapid response team, commandos optionally sent on hazardous place.
- **Emergency medical service** – medical help sent on a place of the event by Headquarter.
- **Headquarter** – a place of monitoring and taking decisions, influences the course of events, sends additional support groups.
- **Drone camera** – drone equipped with camera, sent on the place of events to observe dangerous situation. It helps to take further decisions and registers the whole course of events.

The following use cases are taken into consideration:

- **Current monitoring** – systematic recording of location and taking up proper actions by a policeman on duty. If nothing interesting happens, the GPS coordinates are collected and sent to the system. All actions taken up by policeman are registered and they can be stopped by receiving a notification from another policeman who is in a dangerous situation.
- **Threat detection and notification** – detection of situation dangerous for policeman and being a result of unexpected weapon use; sending his/her GPS coordinates to communication system together with help notification.
- **Support searching** – checking possible support options for a policeman in need, looking for another police forces in his/her nearest or further neighborhood, communication with Threat Management use case and taking up decision about further actions.
- **Supporting** – sending help for a policeman in need after receiving help notification.
- **Threat management** – managing threat situations, decisions about sending emergency medical service, special group or fire brigade; influencing searching process parameters (use case "Support searching") by enlargement of analyzed area, finishing searching process after getting to conclusion that there was sent enough support.

– **Drone dispatching** – sending a drone equipped in camera on the place of events for a direct observation. Decision is taken in Headquarter.

The proposed system also combines heterogeneous appliances so that they could adjust themselves correspondingly to various scenarios. Two basic scenarios are shown below in Tables 1 and 2.

Table 1. The use case scenario for Current monitoring

UC name: "Current monitoring"
Precondition: start of patrol
Scenario:
1. Downloading data (on smartwatch) about planned intervention
2. Arriving on intervention place, tracking driving route
3. Performing activities connected with intervention, recording the results in an action logbook
4. Finishing patrol or moving to point 1
Alternative course of events:
After receiving support notification from another policeman, breaking current intervention and immediate going to place of danger, route navigation performed by the policeman's navigation system. ≪extend≫ Supporting
Postcondition: finishing patrol or sending support notification

Table 2. The use case scenario for Threat detection and notification

UC name: "Threat detection and notification"
Precondition: unplanned use of weapon
Scenario:
1. Detecting situation of unplanned weapon use
2. Downloading information about geolocation from smartwatch
3. Sending messages to network devices in order to look for support
4. Sending a message about support necessity to headquarters
Postcondition: support notification

By unplanned weapon use we understand even the situation of taking a gun out of holster It is a rapid situation connected with life threat for a policeman and it is enough to start a special course of events. The ordinary, planned situation takes place when taking a gun out of holster is preceded by sending proper information to smartwatch.

3   Selected Activity Diagrams

Activity diagrams presented for each use case can be found in Figs. 2, 3, 4 and 5. First activity is current monitoring, see Fig. 2. The location is being updated every minute independently whether an incident occurs or not. However, when an incident occurs (in other words, the trigger was pulled) location is being updated again and a message containing gun's location is being sent to headquarters.

The next activity, see Fig. 2, which is called out in case of a gun usage, is threat detection and notification. Whenever a policeman reaches his/her gun, the notification system (built in the smartwatch) is blocked. The main goal of performing such action is that while any shooting, the policeman may not be distracted by incoming notifications from his/her colleagues. If afterwards the gun is used, another activity is performed – the system is looking for support.

The precondition for the activity "Support searching", see Fig. 3, is that a message containing the information that gun's usage and its location was received. Afterwards, the message is delivered to two actors – headquarters and policemen in the nearest area. The headquarters send the drone to monitor state of shooting. Simultaneously, other policemen, who are in the area, are notified about the event. The system recursively searches for policemen in the area. If there was nobody, the search area is being expanded. After receiving such a message, the policeman is obliged to come and support the shooting.

Another important activity is drone dispatching, see Fig. 4. Headquarters have to decide how dangerous the event of using a gun might be. To determine this fact, headquarters are sending drone to monitor the event. Having a live view of a current situation, it is the headquarters' decision which steps should be undertaken: will an ambulance be required? Is this shooting so serious that a special group should be notified?

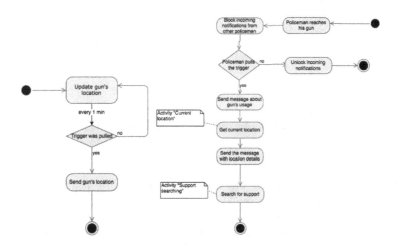

Fig. 2. Activity diagrams, (left) for use case "Current location", (right) for the use case "Threat detection and notification"

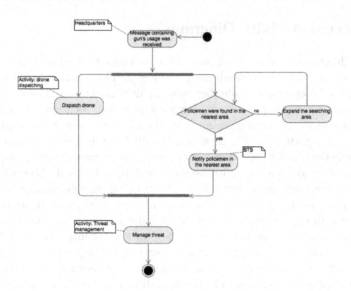

Fig. 3. Activity diagram for use case "Support searching"

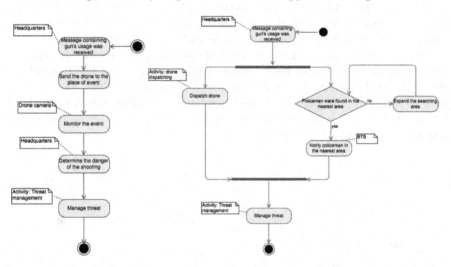

Fig. 4. Activity diagrams, (left) for use case "Drone dispatching" (right) for use case "Support searching"

When considering a dangerous event such as gunfire, the actions must be taken almost instantaneously, see Fig. 4. The most important activity in the system is threat management. The actions must be effective, instant and appropriate. For example, an inappropriate action would be sending fire brigade unless the situation really requires such step. Therefore, proper identification of an event will provide a useful insight to the headquarters about further steps.

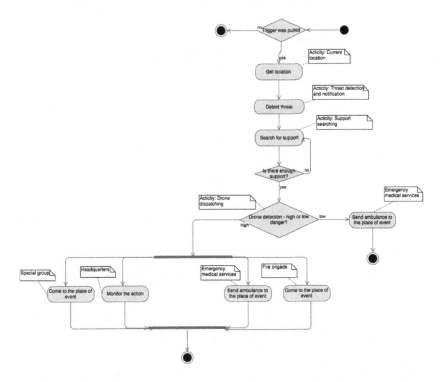

Fig. 5. Activity diagram for use case "Threat management"

After the drone is sent, headquarters determine the level of danger for a specific gun's usage, see Fig. 5. The danger may be high or low. An example of a low danger might be gun usage in case of chasing a thief. Then, the policeman may shoot the thief just to stop or distract him. In that case, the emergency medical services are notified about the event. In case of a mass shooting (high level of danger), the message about gun's usage is being delivered to all actors in the system and actions mentioned in the diagram are performed.

4   System Architecture

Figure 6 presents the system architecture with an indication of particular components.

- **Smartwatch** – application providing interface for system operation by a policeman, it stores the last device location, displays current support notifications and navigates to the place of event.
- **Firearm** – pairs with smartwatch and is controlled by it.
- **Geolocating** – locating the device using GPS, sending data to message broker.
- **Event registering** – registering activities performed by a policeman.

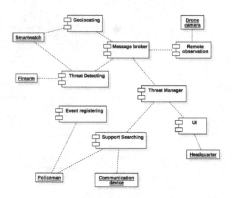

Fig. 6. Component diagram for the proposed system

- **Threat detecting** – detecting the situation of weapon use, downloading the last registered location, generating the request and immediate sending it to headquarters using message broker.
- **Message broker** – component responsible for sending messages in the system.
- **Threat manager** – main component of the system, recognizes the type of message and initiates the proper reaction.
- **Support searching** – finding proper emergency services which are nearby incident place.
- **Remote observation** – Downloading and processing the image from drone camera which was present at the place of event in order to observe and evaluate danger level.
- **Ui (User interface)** – application providing an interface to control the system by headquarters, it shares information about all located devices, enables controlling of notification sending process and its disposition on the place of events.

The most important aspect of a gun system is the way the messages are sent and delivered to actors in the system. In a prototype application message broker was used to communicate between device, BTS stations and headquarters. In a simplest definition message broker is an intermediary program module that translates a message from the formal messaging protocol of the sender to the formal messaging protocol of the receiver. Message brokers are often applied in telecommunication or computer networks, where applications communicate with each other by exchanging formally defined messages. There is a wide variety of message broker software to choose, the most popular including Apache Kafka, Apache ActiveMQ or RabbitMQ. In the prototype application RabbitMQ was chosen as a message broker software. It is a lightweight and reliable messaging system written in Erlang programming language.

There are several components of RabbitMQ application:

- Producer (Publisher) – program which sends the message;
- Consumer – program which waits for message and then receives it;
- Queue – stores messages until they are being read by the consumer;
- Exchange – enables sending messages to multiple queues.

In messaging systems, there are few communication patterns. Having this awareness, publish-subscribe pattern was chosen to be used in smart gun prototype application. This architecture allows sending one message to many consumers and provides message topic, which has a significant meaning in case of a smart gun application. There are three topics appearing in the smart gun application:

- current location;
- trigger was pulled;
- video stream from a drone is being transmitted.

First topic is subscribed by headquarters and updated every minute. By sending this message, headquarters keep track of their policemen and monitor their current location. The second topic informs all actors about the fact that the gun was used and gives information about location of the event. With this message, each authority has to take up actions which are dependent on the severity of case.

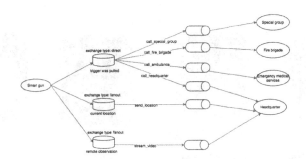

Fig. 7. Structure of the messaging system in smart gun application

In the smart gun application, three exchanges were defined, see Fig. 7. First one is responsible for sending messages about gun usage and location. This exchange is direct, which means that it allows to consume a subset of messages, which key is equal to a given queue key. In a fanout exchange there is no such possibility – the message simply goes to all defined consumers. With this solution we can differentiate, whether a message was sent to fire brigade, emergency medical services or headquarter. Another important topic is gun's current location. Headquarter must be up-to-date with policeman's location, therefore this message becomes a separate topic. Another important aspect of the system is drone video, which must quickly be sent to the headquarter. By streaming the video via message broker, the content can be quickly and reliably delivered to the consumer.

5 Prototype Application

During the research process there was implemented a prototype version of the system, see Fig. 8. We created application for Apple Watch which, first of all, registers weapon use case and displays alerts about support notification. Moreover, we added four additional services: direct contact with headquarter, display of the current emergencies list, preview of current location and settings in which we can pair weapon with watch or pause receiving the notifications. All services are accessible from main menu level.

After choosing the first option, see Fig. 8 (top row, first screen), it is possible to contact headquarters directly and send manually a help notification without necessity of weapon use. In such case, the location of policeman is sent automatically. Moreover, he/she can specify the type of help – ambulance, fire brigade or other services. After choosing the second option, the list of current actions, which are waiting for support, is displayed, see Fig. 8 (top row, second screen). When clicking on one position, the navigation to the target place is activated immediately.

The third option provides information about current position of a policeman. It is constantly updated and sent o headquarters once the weapon is used. Apart

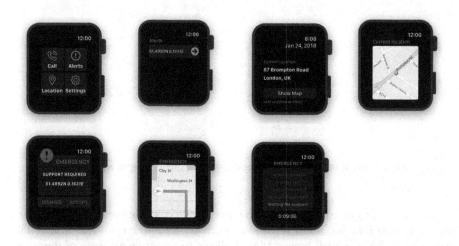

Fig. 8. Screen shots from a prototype application: (top row, first screen) main menu of application, (top row, second screen) list of current actions, (top row, third screen) information about current location of a policeman, (top row, fourth screen) information about current location of a policeman (map), (bottom row, left) emergency notification, (bottom row, middle) navigating to the place of event in emergency mode, and (bottom row, right screen) further stages of notification processing

from text information, route overview is also available, see Fig. 8 (top row, third screen) and Fig. 8 (top row, fourth screen). It supports field orientation of a policeman.

There is also a notification module. In case of alarm, it is immediately displayed on a watch, see Fig. 8 (bottom row, left screen). There appear two options: acceptance or rejection of notification, if a notified policeman cannot take part in action. Choosing the first option, namely clicking Accept button, the navigation process to the target place is initiated, see Fig. 8 (bottom row, middle screen).

In case of detecting weapon use, the watch will show next stages of support notification implementation. After determining current location, the notification is sent to the headquarters. If accepted, the proper services are allocated. The final status of the notification is waiting for support arrival. At the bottom there was placed a timer which counts how much time passed from the moment of the first weapon use, see Fig. 8 (bottom row, right screen).

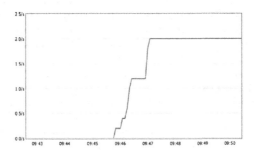

Fig. 9. Message rates of delivered smart gun locations (RabbitMQ Management Console).

Referring to the previous remarks on firearm constant locating issue, we obtained the following results. Figure 9 shows message rates of all smart gun locations delivered to the headquarters. A sample simulation run was conducted for ten units. The messages grow in volume as another gun starts sending current location which means that a police officer begins his shift on patrol.

6   Conclusions

In this paper we propose some solutions that refer to the idea of smart gun supporting police officers. Some basic uses cases, scenarios, as well as activities are proposed. The modern message broker system is used.

This is the first attempt to provide context reasoning for police officers. Future works may include more complex smart scenarios, as well as simulations.

References

1. Bessis, N., Dobre, C. (eds.): Big Data and Internet of Things: A Roadmap for Smart Environments. SCI, vol. 546. Springer, Cham (2014). https://doi.org/10.1007/978-3-319-05029-4

2. Klimek, R.: Behaviour recognition and analysis in smart environments for context-aware applications. In: Proceedings of the IEEE International Conference on Systems, Man, and Cybernetics (SMC 2015), 9–12 October, 2015, City University of Hong Kong, Hong Kong, pp. 1949–1955. IEEE Computer Society (2015)

3. Klimek, R.: Context-aware and pro-active queue management systems in intelligent environments. In: Ganzha, M., Maciaszek, L.A., Paprzycki, M. (eds.) Proceedings of Federated Conference on Computer Science and Information Systems (FedCSIS 2017), 3–6 September 2017, Prague, Czech Republic, pp. 1077–1084. IEEE Xplore Digital Library (2017)

4. Klimek, R.: Proposal of a multi-agent system for a smart outdoor lighting environment. In: Rutkowski, L., Korytkowski, M., Scherer, R., Tadeusiewicz, R., Zadeh, L.A., Zurada, J.M. (eds.) ICAISC 2017. LNCS (LNAI), vol. 10246, pp. 255–266. Springer, Cham (2017). https://doi.org/10.1007/978-3-319-59060-8_24

5. Klimek, R.: Understanding human behavior in intelligent environments: a context-aware system supporting mountain rescuers. In: Rutkowski, L., Korytkowski, M., Scherer, R., Tadeusiewicz, R., Zadeh, L.A., Zurada, J.M. (eds.) ICAISC 2017. LNCS (LNAI), vol. 10246, pp. 267–279. Springer, Cham (2017). https://doi.org/10.1007/978-3-319-59060-8_25

6. Klimek, R., Kotulski, L.: Towards a better understanding and behavior recognition of inhabitants in smart cities. A public transport case. In: Rutkowski, L., Korytkowski, M., Scherer, R., Tadeusiewicz, R., Zadeh, L.A., Zurada, J.M. (eds.) ICAISC 2015. LNCS (LNAI), vol. 9120, pp. 237–246. Springer, Cham (2015). https://doi.org/10.1007/978-3-319-19369-4_22

7. Klimek, R., Rogus, G.: Proposal of a context-aware smart home ecosystem. In: Rutkowski, L., Korytkowski, M., Scherer, R., Tadeusiewicz, R., Zadeh, L.A., Zurada, J.M. (eds.) ICAISC 2015. LNCS (LNAI), vol. 9120, pp. 412–423. Springer, Cham (2015). https://doi.org/10.1007/978-3-319-19369-4_37

8. Shang, X., Veldhuis, R.N.J.: Grip-pattern verification for smart gun based on maximum-pairwise comparison and mean-template comparison. In: IEEE Second International Conference on Biometrics: Theory, Apps and Systems, pp. 1–5 (2008)

9. Shang, X., Veldhuis, R., Bazen, A., Ganzevoort, W.: Algorithm design for grip-pattern verification in smart gun. In: Proceedings of ProRISC 2005, 16th Annual Workshop on Circuits, Systems and Signal Processing, pp. 674–678. STW Technology Foundation, Utrecht (2005)

10. Weiss, D.R.: Smart gun technologies: one method of eliminating unauthorized firearm use. In: 1994 Proceedings of IEEE International Carnahan Conference on Security Technology, pp. 169–172, October 1994

Knowledge Representation in Model Driven Approach in Terms of the Zachman Framework

Krzysztof Kluza[1(✉)], Piotr Wiśniewski[1], Antoni Ligęza[1], Anna Suchenia[2], and Joanna Wyrobek[3]

[1] AGH University of Science and Technology,
al. A. Mickiewicza 30, 30-059 Krakow, Poland
{kluza,wpiotr,ligeza}@agh.edu.pl
[2] Cracow University of Technology, ul. Warszawska 24, 31-155 Kraków, Poland
asuchenia@pk.edu.pl
[3] Corporate Finance Department, Cracow University of Economics,
ul. Rakowicka 27, 31-510 Kraków, Poland
wyrobekj@uek.krakow.pl

Abstract. Model driven approach uses distinct models for representing various kinds of complex knowledge. Use of appropriate models allows for taking advantage of formal checking, testing, and validating possibilities available for the models. Although the notations do not provide any design method or software process, this paper offers a step to integrated modeling using them. We present an overview of the existing OMG solutions for knowledge representation used in software engineering, focusing on UML, BPMN, DMN and CMMN models and the diagrams provided by these notations. We perform an analysis of these approaches in terms of Kruchten's 4+1 view model architecture as well as the Zachman Framework.

1 Introduction

Modern software systems are very complex and so is the process of manufacturing them. Software development consists of several different phases. The sequence of them is called the software life cycle [1]. There are many various life cycle models. The best known is the traditional model, named the *waterfall model*. This model, shown in Fig. 1, assumes a standard sequence of tasks. There are no feedbacks in this model, except feedbacks leading directly to the previous phase.

Other models are usually modifications of the traditional model, e.g., the *prototyping model* [3] builds a throwaway version (or prototype) of the system, to test concepts and requirements. After customer agreement to such a system, the software development usually follows the traditional model. The popular type of life cycle model is the *spiral model* introduced by Boehm [4]. It is a combination of the *prototyping model* and the *waterfall model*. Its aim is to combine

The paper is supported by the AGH UST research grant.

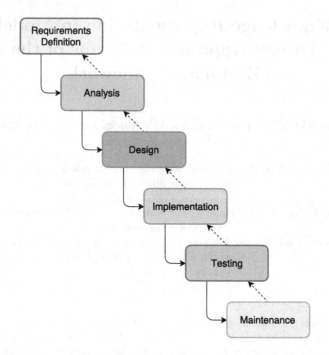

Fig. 1. Waterfall life cycle model – based on [2]

advantages of both concepts. The model is shown in Fig. 2. It is a spiral that starts in the middle and continually revisits the basic phases, including communication with a customer. Agile software development [5] can be seen as the extension of such a spiral model.

In practical software design, the Unified Modeling Language (UML) [6] constitutes the standard for modeling software applications [7]. It provides diagrams to capture requirements, collaboration between parts of the software that realize them, the realization itself and models which show how everything fits together and is executed [8]. Such diagrams can serve as sketches (showing only the key points of the modeling system), as blueprints (providing a detailed system specification), as well as models for model driven approach (requiring the complete system model which can be simulated and used for code generation).

On the other hand, there are more dedicated notations such as Business Process Model and Notation (BPMN) [9] and Decision Model and Notation (DMN) [10] which provide diagrams for modeling processes and decisions. These are the common standards supported and maintained by Object Management Group (OMG).

The rest of this paper is organized as follows: Sect. 2 presents the concept of Model Driven Architecture. In Sect. 3, the UML, BPMN, DMN and CMMN notations are briefly introduced. Section 4 presents our analysis of these approaches in terms of Kruchten's 4+1 view model architecture as well as the Zachman Framework. The paper is summarized in Sect. 5.

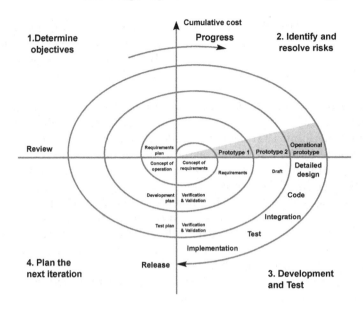

Fig. 2. Spiral life cycle model [4]

2   Model Driven Architecture

The Model Driven Architecture (MDA) [11] is an approach to system development which increases the power of models in this process. In this approach, the software development process is to be driven by modeling the software system. MDA defines an architectural framework that supports detailed specification based on models. According to OMG [11], such specifications should allow producing interoperable, reusable, portable software components and data models. The goal of MDA is to separate the design specification from the specific platform technology. In such a development process a modeler is becoming a programmer, as well as a modeling activity is becoming a programming activity [12].

Models used with MDA can be expressed using the UML. But MDA is not restricted only to UML [13]. However, UML can help to enable the model-driven approach. In the MDA development life cycle one can identify the same phases as in the traditional software life cycle [13]. However, while in the traditional life cycle transformations of models are done manually by a user, in MDA this should be supported by the machine.

Such models can be on different levels of abstraction. Models on certain layer focus on particular viewpoints of the system. At the core of MDA are model abstraction layers [11]:

- *Computation Independent Business Models (CIM)* is on the highest-level of business model [13]. It shows a system from the computation independent viewpoint. This attitude does not show structure details of the system. In [14],

CIM is compared to a domain model specification and an ontology. In practice, CIM should use a business language in which specialized computer knowledge is unnecessary. Process notation, such as BPMN, as well as decision notation like DMN can be useful as such models are well understandable by business users.

– *Platform Independent Models (PIM)* shows a system from the computation-dependent viewpoint [14]. Such a viewpoint hides the details necessary for a certain platform. PIM defines a model that is independent of any implementation technology. It focuses on the operation of a system [11]. That model ignores programming languages, OS, hardware, etc. PIM is mostly described in UML or some UML derivative [8].

– *Platform Specific Models (PSM)* shows a system from the platform specific viewpoint [11]. This viewpoint is based on the platform independent one. It focuses on the usage of a particular platform by a system. PSM specifies a system regarding the certain implementation constructs. PSM can use different UML profiles to define the PSM for each target platform [8]. In PSM some business process models and decision models can be refactored in order to support platform specific environment.

– *Code model* represents the code, usually in a high-level language [8].

Figure 3 shows MDA models on different model abstraction layers. Abstraction increases right to left.

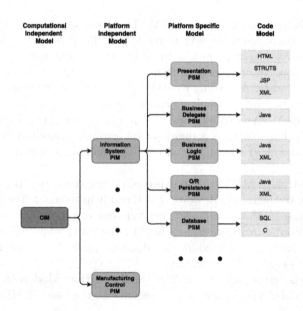

Fig. 3. MDA models and their relationships [8]

The goal of MDA is to be able to transform a high level PIM into a PSM, and then each PSM to code [11], what increases the abstraction level of the design. The MDA transformations with input and output models are shown in Fig. 4 [15].

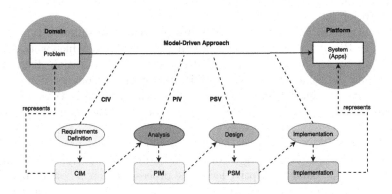

Fig. 4. Foundational concepts of the MDA [15]

3   Software Engineering Notations

3.1   Unified Modeling Language (UML)

In the field of software engineering, Unified Modeling Language (UML) is a general-purpose modeling language. It helps to capture a system in terms of models [8]. The standard provides abstract syntax rules and semantics. However, there are still some unsolved problems when it comes to modeling, e.g., exceptions [16] or rules [17].

Designing the system in UML, one should take care of consistency between different diagrams representing the same model [18]. Such inconsistency can likely occur, especially when different people model different parts of systems and prefer to use different UML diagrams.

The current version of UML defines a variety of UML diagrams, which can be divided into two main categories:

- structure diagrams – representing the structure of a modeled application (Class, Component, Object, Composite Structure, Deployment, Package, and Profile Diagrams),
- behavior diagrams – representing general types of behavior (Activity, State Machine, Use Case and four Interaction Diagrams).

3.2   Business Process Model and Notation (BPMN)

Business Process Model and Notation (BPMN) [9], adopted and maintained by the OMG group, is the most popular notation for business processes representation. BPMN uses a set of models to depict a process and how it is performed. In such a model, a detailed semantics can be specified [19] in terms of workflow representation [20,21].

The current BPMN 2.0 standard [9] distinguishes four different types of diagrams: (1) Process diagrams for describing the ways in which activities are carried out to accomplish the intended goals, (2) Collaboration diagrams for representing the collaborative public B2B processes, (3) Conversation diagrams for specifying the logical relation of message exchanges, and (4) Choreography diagrams for defining the expected behavior between the interacting business participants.

Although BPMN 2.0 defines more than 100 elements, practitioners differentiate them based on the degree of model detail: descriptive, analytical, and executable levels. On a descriptive level, the process reflects a "happy path" scenario by presenting all main activities of the process. The analytical level uses complex structures and elements which allows a modeler to design fully representative processes. The executable level is suitable for technicians in which execution details can be captured in the model, and thus it can be seen as an MDA code layer.

3.3   Decision Model and Notation

Decision Model and Notation (DMN) [10] is an OMG standard for decision modeling. A decision, based on some input data, selects some option or determines the result. The goal of the notation is to support decision modeling so the decision can be easily represented in graphical models understandable by business users [22,23].

According to the specification, there are various purposes of using the notation, e.g., modeling human decision-making, modeling the requirements for automated decision-making, and implementing automated decision-making. Thus, if all decisions and business knowledge models are completely specified using decision logic, the automated decision-making is possible [24].

3.4   Case Management Model and Notation

Case Management Model and Notation (CMMN) [25] provides a declarative way of modeling cases and visualizing the events that may happen in the context of a case, the tasks involved, the milestones, etc. Such various activities may be performed in an unpredictable order in response to evolving situations.

The CMMN notation has the advantage over BPMN, especially in the case of knowledge-intensive processes (e.g., incident management, consulting or sales), where many activities do not form a well-structured business process, but are started and conducted in an ad-hoc way.

4   MDA vs The 4+1 View Model Architecture and The Zachman Framework

To describe software architecture in a more consistent way, a "4+1" model is often used. The model was designed by Philippe Kruchten and used for *"describing the architecture of software-intensive systems, based on the use of multiple, concurrent views"* [26]. The views are used to describe the system from the viewpoint of different users (end-users, developers and project managers) [26,27].

The "4+1" view model supports five main views (see Fig. 5):

1. *Logical View* – an object model of the design.
2. *Process View* – concurrency and synchronization aspects.
3. *Development View* – static organization of the software.
4. *Physical View* – mapping of the software to the hardware.
+1 *Use-cases view* – various usage scenarios.

One of the frameworks which help to manage enterprise architecture modeling is The Zachman Framework [28] (Zachman Framework for Enterprise Architecture and Information Systems Architecture), which provides a structured way of viewing and defining an enterprise. It is often referenced as a standard approach for representing the basic elements of enterprise architecture. This two-dimensional classification schema uses a grid based on six basic questions (what,

Table 1. Overview of diagrams supporting Kruchten's views (O does not support, ◗ partially supports, ● supports)

		Kruchten's 4+1 model architecture views				
		Scenarios	Logical	Process	Development	Physical
UML	Class diagram	O	●	O	◗	O
	Component diagram	O	●	O	●	O
	Composite structure diagram	O	●	O	●	◗
	Deployment diagram	O	O	O	O	●
	Object diagram	◗	●	O	O	O
	Package diagram	O	●	O	●	O
	Profile diagram	O	●	O	O	O
	Activity diagram	◗	O	●	O	O
	Communication diagram	O	●	●	O	O
	Interaction overview diagram	◗	O	●	O	O
	Sequence diagram	◗	◗	●	O	O
	State diagram	O	●	●	O	O
	Timing diagram	O	O	●	O	O
	Use case diagram	●	O	●	O	O
BPMN	Process diagram	◗	O	●	◗	O
	Collaboration diagram	●	●	●	O	O
	Conversation diagram	◗	●	O	O	◗
	Choreography diagram	O	O	●	◗	O
DMN	Decision Requirements diagram	◗	●	O	◗	O
	Decision Requirements graph	●	●	O	O	O
CMMN	Case diagram	●	●	●	O	O

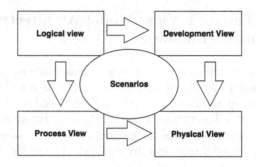

Fig. 5. The "4+1" view model architecture

how, where, who, when, why) and five groups of users (planner, owner, designer, builder, subcontractor). This allows for defining the holistic architecture of corporate systems. The categorization of these elements in terms of MDA approach is presented in Fig. 6.

There were several attempts to mapping UML and RUP phases to The Zachman Framework [29–33]. However, these authors focus only on the main UML diagrams. Based on a discussion of diagrams suitable for representing various Kruchten's model views (see Table 1), we analyzed each of the UML, BPMN, DMN and CMMN diagrams in terms of suitability for modeling of the Zachman Framework cells. The results of our analysis is presented in Fig. 7.

	DATA (what) entities	FUNCTION (how) activities	NETWORK (where) locations	ORGANIZATION (who) people	SCHEDULE (when) time	STRATEGY (why) motivation
SCOPE (contextual) Planner	List of Things Important in the Business	List of Core Business Processes	List of Business Locations	List of Important Organizations	List of Significant Events	List of Business Code
BUSINESS MODEL (conceptual) Owner	Conceptual Data/ Object Model	Business Process Model	Business Logistics System	Work Flow Model	Master Schedule	Business Plan
				Computation Independent Business Models (CIM)		
SYSTEM MODEL (logical) Designer	Logical Data/ Class Model	System Architecture Model	Distributed Systems Architecture	Human Interface Architecture	Processing Structure	Business Role Model
				Platform Independent Models (PIM)		
TECHNOLOGY MODEL (physical) Builder	Physical Data/ Class Model	Technology Design Model	Technology Architecture	Presentation Architecture	Control Structure	Rule Design
				Platform Specific Models (PSM)		
DETAILED REPRESENTATIONS (out of context) Subcontractor	Data Definitions	Program	Network Architecture	Security Architecture	Timing Definition	Rule Specification
				Code Model (CM)		

Fig. 6. The Zachman Framework in the context of MDA

	DATA (what) entities	FUNCTION (how) activities	NETWORK (where) locations	ORGANIZATION (who) people	SCHEDULE (when) time	STRATEGY (why) motivation
SCOPE (contextual) Planner	BPMN (Collaboration, Conversation) UML (Use Case, Package, Class)	BPMN (Process) UML (Use Case, Activity) CMMN	BPMN (Process, Collaboration)	BPMN (Conversation)	BPMN (Choreography)	
BUSINESS MODEL (conceptual) Owner	UML (Use Case, Class, Composite Structure)	BPMN (Process, Collaboration, Choreography) UML (Activity, Interaction) CMMN	BPMN (Process)	BPMN (Conversation, Collaboration) UML (Use Case)	BPMN (Choreography) UML (Timing)	DMN (DRG)
				Computation Independent Business Models (CIM)		
SYSTEM MODEL (logical) Designer	UML (Class, Package, Component)	UML (Activiy, State, Interaction) CMMN	UML (Component, Deployment)	UML (Use Case)	UML (State, Timing)	DMN (DRD)
				Platform Independent Models (PIM)		
TECHNOLOGY MODEL (physical) Builder	UML (Class, Package, Component)	UML (Activiy, State, Interaction)	UML (Component, Deployment)	UML (Deployment)	UML (State, Timing)	DMN (DRD)
				Platform Specific Models (PSM)		
DETAILED REPRESENTATIONS (out of context) Subcontractor						
						Code Model (CM)

Fig. 7. Suitability of UML, BPMN, DMN and CMMN for modeling in terms of the Zachman Framework

5   Concluding Remarks

The goal of the research presented in this paper was to provide an overview of model driven approach for modeling modern information systems. The paper focuses on the standardized OMG knowledge representation notations, what allows for taking advantage of the available techniques for testing, validating or formal checking of such models. We take into account the UML, BPMN, DMN and CMMN models, and provide the analysis of these approaches in terms of Kruchten's 4+1 view model architecture as well as the Zachman Framework, what can help in the integrated modeling of software systems.

References

1. Gustafson, D.A.: Theory and Problems of Software Engineering. McGraw-Hill, Bengaluru (2002)
2. Sommerville, I.: Software Engineering. International Computer Science, 7th edn. Pearson Education Limited, London (2004)

3. Bjorner, D.: Software Engineering 3. Domains, Requirements, and Software Design. Springer, Heidelberg (2006). https://doi.org/10.1007/3-540-33653-2
4. Boehm, B.: A spiral model for software development and enhancement. IEEE Comput. **21**, 61–72 (1988)
5. Strode, D.E.: A dependency taxonomy for agile software development projects. Inf. Syst. Front. **18**(1), 23–46 (2016)
6. Object Management Group: OMG: Unified Modeling Language version 2.5.1 (OMG UML) (2017)
7. Hunt, J.: Guide to the Unified Process featuring UML, Java and Design Patterns. Springer, Heidelberg (2003). https://doi.org/10.1007/b97530
8. Pilone, D., Pitman, N.: UML 2.0 in a Nutshell. O'Reilly, Sebastopol (2005)
9. OMG: business process model and notation (BPMN): version 2.0 specification. Technical report formal/2011-01-03, Object Management Group, January 2011
10. OMG: decision model and notation. version 1.0. Technical report formal/01 Sept 2015, Object Management Group (2015)
11. Object Management Group: OMG: MDA Guide version 1.0.1 (2003)
12. Frankel, D.S.: Model Driven Architecture. Applying MDA to Enterprise Computing. Wiley Publishing, Indianapolis (2003)
13. Kleppe, A., Warmer, J., Bast, W.: MDA Explained: The Model Driven Architecture: Practice and Promise. Addison Wesley, Boston (2003)
14. Gaševic, D., Djuric, D., Devedžic, V.: Model Driven Architecture and Ontology Development. Springer, Heidelberg (2006). https://doi.org/10.1007/3-540-32182-9
15. Alhir, S.S.: Understanding the model driven architecture (MDA). Methods and Tools **11**(3), 17–24 (2003)
16. Klimek, R., Skrzynski, P., Turek, M.: On some problems with modelling of exceptions in UML. In: Software Engineering: Evolution and Emerging Technologies, pp. 87–98. IOS Press (2005)
17. Nalepa, G.J., Kluza, K.: UML representation for rule-based application models with XTT2-based business rules. Int. J. Software Eng. Knowl. Eng. (IJSEKE) **22**(4), 485–524 (2012)
18. Kluza, K., Wiśniewski, P., Jobczyk, K., Ligęza, A., Mroczek, A.S.: Comparison of selected modeling notations for process, decision and system modeling. In: 2017 Federated Conference on Computer Science and Information Systems (FedCSIS), pp. 1095–1098. IEEE (2017)
19. Kluza, K., Nalepa, G.J., Ślażyński, M., Kutt, K., Kucharska, E., Kaczor, K., Łuszpaj, A.: Overview of selected business process semantization techniques. In: Pełech-Pilichowski, T., Mach-Król, M., Olszak, C.M. (eds.) Advances in Business ICT: New Ideas from Ongoing Research. SCI, vol. 658, pp. 45–64. Springer, Cham (2017). https://doi.org/10.1007/978-3-319-47208-9_4
20. Klimek, R.: Towards formal and deduction-based analysis of business models for SOA processes. In: In Filipe, J., Fred, A. (eds.) Proceedings of 4th International Conference on Agents and Artificial Intelligence (ICAART 2012), 6–8 Feb 2012, Vilamoura, Algarve, Portugal, vol. 2, pp. 325–330. SciTePress (2012)
21. Klimek, R.: A system for deduction-based formal verification of workflow-oriented software models. Int. J. Appl. Math. Comput. Sci. **24**(4), 941–956 (2014)
22. Taylor, J., Fish, A., Vanthienen, J., Vincent, P.: Emerging standards in decision modeling - an introduction to decision model and notation. BPM and Workflow Handbook Series. In: iBPMS: Intelligent BPM Systems: Intelligent BPM Systems: Impact and Opportunity, pp. 133–146. Future Strategies, Inc. (2013)

23. Janssens, L., Bazhenova, E., De Smedt, J., Vanthienen, J., Denecker, M.: Consistent integration of decision (DMN) and process (BPMN) models. In: Proceedings of the CAiSE'16 Forum, at the 28th International Conference on Advanced Information Systems Engineering (CAiSE 2016), Vol. 1612, pp. 121–128. CEUR-WS. org (2016)
24. Biard, T., Le Mauff, A., Bigand, M., Bourey, J.-P.: Separation of decision modeling from business process modeling using new "decision model and notation" (DMN) for automating operational decision-making. In: Camarinha-Matos, L.M., Bénaben, F., Picard, W. (eds.) PRO-VE 2015. IAICT, vol. 463, pp. 489–496. Springer, Cham (2015). https://doi.org/10.1007/978-3-319-24141-8_45
25. OMG: Case Model Management and Notation (CMMN): Version 1.1 specification. Technical report formal/16-12-01, Object Management Group, December 2016
26. Kruchten, P.B.: The 4+ 1 view model of architecture. IEEE softw. **12**(6), 42–50 (1995)
27. Wendler, R.: Development of the organizational agility maturity model. In: 2014 Federated Conference on Computer Science and Information Systems (FedCSIS), pp. 1197–1206. IEEE (2014)
28. Lankhorst, M.: Enterprise Architecture at Work: Modelling, Communication and Analysis. Springer, Heidelberg (2009)
29. Ambler, S., Nalbone, J., Vizdos, M.: Extending the RUP with the Zachman Framework. Pearson Education 131914510 (2005)
30. Fatolahi, A., Shams, F.: An investigation into applying UML to the Zachman framework. Inf. Syst. Front. **8**(2), 133–143 (2006)
31. Frankel, D.S., Harmon, P., Mukerji, J., Odell, J., Owen, M., Rivitt, P., Rosen, M., Soley, R.M.: The Zachman framework and the OMG's model driven architecture. Bus. Process Trends **14**(2003), 2003 (2003)
32. Mrdalj, S., Jovanovic, V.: Mapping the UML to the Zachman framework. In: AMCIS 2005 Proceedings, p. 315 (2005)
33. Khoshnevis, S., Aliee, F.S., Jamshidi, P.: Model driven approach to service oriented enterprise architecture. In: Services Computing Conference 2009, APSCC 2009. IEEE Asia-Pacific, pp. 279–286. IEEE (2009)

Rendezvous Consensus Algorithm Applied to the Location of Possible Victims in Disaster Zones

José León[1,5], Gustavo A. Cardona[2], Luis G. Jaimes[3], Juan M. Calderón[1,4(✉)], and Pablo Ospina Rodriguez[1]

[1] Department of Electronic Engineering, Universidad Santo Tomás,
Bogotá, Colombia
{joseleonl,juancalderon,pabloospina}@usantotomas.edu.co
[2] Department of Electrical and Electronics Engineering,
Universidad Nacional de Colombia, Bogotá, Colombia
gacardonac@unal.edu.co
[3] Florida Polytechnic University, Lakeland, FL, USA
ljaimes@floridapoly.edu
[4] Bethun-Cookman University, Daytona Beach, FL, USA
calderonj@cookman.edu
[5] Department of Electronic and Telecommunication Engineering,
Universidad Católica, Bogotá, Colombia
jleon@ucatolica.edu.co

Abstract. In this paper is presented an alternative to performing the analysis of the sensors in the field of applied cooperative robotics for search and location of disaster victims. This work proposes the use of the Rendezvous algorithm to validate the information coming from the sensors of a multi-robot system. The sensors located in each one of the robotic agents provide a measured value according to the existence or not of victims in the surrounding zone to the robot. Since the information coming from the robots is not the same, however, its belong to the same sensed parameters, the Rendezvous algorithm is used to find a consensus of opinion about the existence of victims. In addition, the swarm of robots uses bio-inspired techniques to generate the navigation algorithm. This navigation algorithm was inspired by the foraging technique used by swarms such as bees, birds or bacteria. The results present some Rendezvous algorithm simulation and robot swarm navigation showing the feasibility of the proposed system.

Keywords: Robot swarm · Rendezvous algorithm
Robot navigation · Bio-inspired systems

1 Introduction

The agencies for the attention of emergencies begin their tasks of locating and recovering victims moments after the occurrence of a disaster. These tasks are

© Springer International Publishing AG, part of Springer Nature 2018
L. Rutkowski et al. (Eds.): ICAISC 2018, LNAI 10842, pp. 700–710, 2018.
https://doi.org/10.1007/978-3-319-91262-2_61

stressful and their development takes hours or even days depending on the magnitude of the disaster. Time is a key factor to avoid loss of life since it is extremely urgent to provide the required medical assistance, but fatigue, instability of the land, climatic conditions and lighting problems make these tasks more complicated.

The agencies for the attention of emergencies begin their tasks of locating and recovering victims moments after the occurrence of a disaster. These tasks are stressful and their development takes hours or even days depending on the magnitude of the disaster. Time is a key factor to avoid loss of life since it is extremely urgent to provide the required medical assistance, but fatigue, instability of the land, climatic conditions and lighting problems make these tasks more complicated.

Mobile robotics in search and rescue applications have contributed in different fields such as recognition of terrain, lifting communications networks that allow the organization of rescue teams, and telemedicine among others. Even sometimes the robotic platform is used to provide instructions about medical procedures to victims while Emergency teams arrive to perform rescue actions.

Cooperative robotics is a viable alternative in search and rescue applications since they offer great advantages, such as the robots perform exploration and reconnaissance tasks in less time without risking the lives of the rescue team. If a robot is lost or stuck, others continue to explore the disaster area as depicted by Yanguas-Rojas [1]. The information provided by the robots is validated with sensor fusion techniques since several robots perform the sensing of common areas, allowing to reach a consensus of the information captured individually. In this way, these systems try to avoid failures due to noise or damage of the sensors, allowing these systems to be more reliable, robust, and precise.

2 Related Work

This section presents some applications of consensus algorithms in cooperative control. In particular, swarm robotics in which the problem of communications in wireless sensor networks is still an open issue. In addition, we present some of the most popular consensus algorithms for robust interaction among agents under different operational conditions.

In terms of position-based consensus, the work of Song *et al.* [2] presents a consensus algorithm with double-integrator dynamics and directed topology. Two types of distributed observer algorithms are proposed to solve the consensus problem by utilizing continuous and intermittent position measurements, respectively. Another research in this area corresponds to the work of [3]. Here the authors present a control scheme capable of solving the leader-follower and the leaderless consensus problems in networks composed of multiple Euler-Lagrange systems.

Another communication issue that raises when applying consensus algorithms on multi-point networks is the communication limitation due to interferences such as static and electric noise, and fluctuations in the bandwidth. In that sense, the work of Chen and Ho [4] tackled this problem in the context consensus for Autonomous Underwater Vehicles (AUVs), where transmission faults are common due to the underwater conditions. His work presents two new consensus techniques which work on the presence of transmission faults with respect to leaderless multiple AUV systems and leader-follower multiple AUV systems.

The use of consensus algorithms raises new challenges on wireless sensor networks. Some of these include energy consumption, unification of sensor readings, handling missing sensor reading values, etc. Research in this area includes the work of Chelbi [5] who focus on energy consumption optimization and the unification of sensor readings from multiple sensors. Here, the consensus algorithm allows obtaining a unique value from a set of different sources while minimizing the effect of noise and corrupt measurements. Here, the authors propose a routing algorithm based on clusters that increase the surviving time of the network. In a similar work Nurellari et al. [6] applies consensus algorithm for intrusion detection, are the detection task is distributed through the different nodes of the wireless sensor network. In a first step, the information is disseminated through different channels, the algorithm takes into account the channel capacity. In a second step, the algorithm gathers a global consensus.

3   Performance Evaluation

In order to test the performance of our proposed algorithm, we model an scenario in which a drone swarm is looking for a set of targets on the ground. This could be the case of a rescue mission in which a drone swarm is looking for one or several potential victims on the ground. Thus, we model the target (i.e., victims) and obstacles by using a multi-variable Gaussian distribution. The targets o victims have an attractor effect on the agents which are within a given distance, and obstacles have the opposite effect, namely and a repulsion effect which is used to avoid potential collisions. This surface is defined by $J(x)$, where $x \in \Re^n$, and $J(x)$ is continuous with a finite slope at every point in its domain (i.e., differentiable). Thus, the agents moves toward the negative gradient of $J(x)$ (see Eq. 1) [7,8].

$$- \nabla J(x) = -\frac{\partial J}{\partial x} \tag{1}$$

3.1   Navigation

Even though a swarm of drones seems to be crowded, each member is able to navigate without the interference of any other swarm's member. Thus, avoiding collisions, and keeping the direction and orientation by maintaining always a safe distance from each other. What makes this possible are the constants of attraction and repulsion of the swarm's navigation algorithm. Thus, the attraction constant ensures the swarm stays grouped and moves in the same direction as shown in Eq. 2, while the repulsion constant allows to keep a safe distance among agents in order to avoid collision, see Eq. 3.

$$- k_a(x^i - x^j) \tag{2}$$

Where k_a is the attraction force, this mechanism could be local (e.g., a range in the range of sensing) o global (e.g., an agent can move other group members regardless how far they are each other).

$$k_r exp(\frac{-\frac{1}{2}\|x^i - x^j\|^2}{r_s^2})(x^i - x^j) \tag{3}$$

Where $k_r > 0$, represents the magnitude of repulsion and $r_s > 0$, is the range of repulsion.

3.2 Sub-swarms

Each agent is exposed to different attraction and repulsion forces which help to keep the swarm's cohesion and navigating in the same direction.

The targets generate an attraction force on the closer agents, and thus, this force may modify the normal swarm's behavior by reducing the speed of some agents, namely those closer to that target. Thus, the subset of agents within the range of attraction to the target is separated from the main swarm. This happens because the distance between the main group and those left behind is every time greater, which result in losing communication and breaking the attraction forces that used to keep them together.

Thus, those left behind and with close proximity to the target form a new swarm, this new swarm changes its state to still and send the target's coordinates to the rescue station.

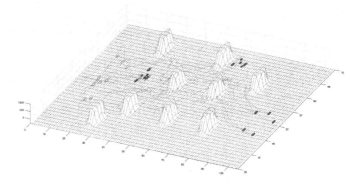

Fig. 1. Sub-swarm [8] (Color figure online)

As can be seen in Fig. 1 the targets are colored in blue, and the swarm agents are colored with red. Around the targets there is a group of agents, those are the ones which were left behind from the main swarm and thus, forming a new small swarm. The members of this small swarm use their sensor readings and the consensus algorithm to determine if there is actually a real target to rescue, and then they proceed to send the target coordinates.

3.3 Sensors

Sensor readings are simulated by using a random normal distribution plus some white noise. These noisy readings are then evaluated by the objective function which determines whether the agent is detecting a potential target. Figure 2 visualize the sensor readings in the target area with five potential targets.

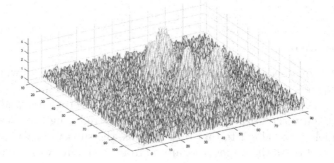

Fig. 2. Sensors

3.4 Consensus

Consensus enables a collective intelligence which allows the swarm to take decisions based on the agents' individual opinions, here, we use consensus among the members of the small swarm, namely the one surrounding a potential target to determine whether or not there is an actual target there.

The algorithm used to reach consensus is [11] which the one that solves the Rendezvous problem which consists in make that several agents meet in given point without knowing its location.

We use Eq. 4 which computes the distance from agent i to the other swarm's members. Thus, the meeting point would be the center of gravity of the swarm [9].

$$\dot{x}_i = \sum_{j \in N_i} (x_j - x_i) \tag{4}$$

For a swarm with many members iterate on Eq. 4 is inefficient, instead we vectorized the computation as show in Eq. 5. This vector is then used as input of Eq. 6 which outputs the swarm's consensus [9].

$$x = \begin{bmatrix} x_1 \\ \cdot \\ \cdot \\ \cdot \\ x_N \end{bmatrix} \tag{5}$$

Here, in Eq. 6 $\dot{x}$ corresponds to rate of change of data, and L to the Laplacian matrix, where is represented the network of the swarm interaction.

$$\dot{x} = -Lx \tag{6}$$

The Laplacian matrix can be found by using Eq. 7 where D is a diagonal matrix as can be seen in Eq. 8 with $d_{in}(k) = \sum_j Adj_{kj}$ and matrix $Adj_{i,j}$ represents the communication network among the swarm agents. Thus, if a member agent i has communication with agent j, the entry (i, j) is one, and zero otherwise [9].

$$L = D - Adj \tag{7}$$

$$D = \begin{bmatrix} d_{in}(1) & & \\ & \ddots & \\ & & d_{in}(k) \end{bmatrix} \tag{8}$$

Thus, the updated new value which results from the consensus algorithm is stored in x as can be seen in Eq. 9.

$$x = k * \dot{x} + x \tag{9}$$

Here, k is convergence factor. If k is high, we might never reach a consensus given hight level of oscillation. On another hand, if k is too small the algorithm may take too much time to converge. Thus, k is one of the most important parameters of the algorithm.

4 Experiment and Results

In this experiment, the sub-swarm (i.e., the one which identified a potential target) is composed by three agents, sensor readings are in the range of 0, 5. Values below a given threshold, in this case 2.5 indicates that there is not target, and values in between 2.5 and 5 will indicate the presence of a target. Thus, the value of sensor 1, 2, and 3 are 2, 2.8, 1 respectively. See Eq. 10.

$$x = \begin{bmatrix} 2 \\ 2.8 \\ 1 \end{bmatrix} \tag{10}$$

Matrix Adj represents the communication network among the agents, because the sub-swarms are in state still (i.e., no movement) after located the potential target, this will allow members of the sub swarm to communicate with other members of the same sub-swarm, but not with itself. This new definition of communication gives place to the matrix Adj which is define in Eq. 11.

$$Adj = \begin{bmatrix} 0 & 1 & 1 \\ 1 & 0 & 1 \\ 1 & 1 & 0 \end{bmatrix} \tag{11}$$

Here, we run the consensus algorithm with 20 iterations, and using a convergence constant $k = 0.1$, the consensus is found after 17 iterations with a final result of 1.93, which indicates that there is not a target in that location. See Fig. 3.

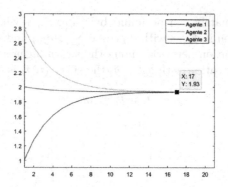

Fig. 3. Test 1

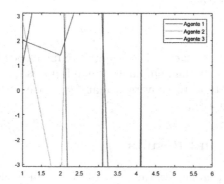

Fig. 4. Test 2

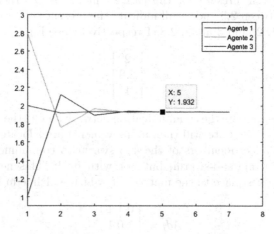

Fig. 5. Test 3

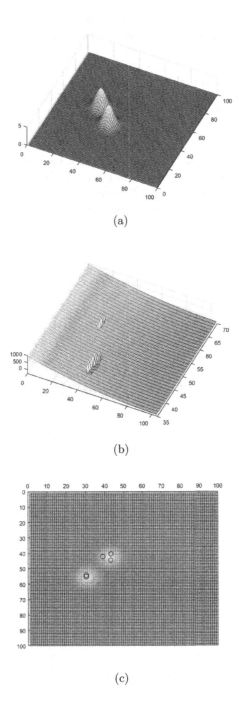

Fig. 6. Test 4

We carried out a second test using the same parameters that the used in first test, but this time using a convergence constant of $k = 3$, using just 5 iteration. Figure 4 shows that the algorithm didn't converge given the value of k and the few number of iterations. A third test was carried out using the same scenario that test 1, and test 2. But this time using a convergence constant of $k = 0.4$, and 7 iterations as shown in Fig. 5. Here, the algorithm found a consensus after iteration 5.

In order to test the performance of the consensus, and navigation algorithms together, we carry out two tests. In the fist test, we simulated an environment with two targets (see Fig. 6b). Here, we did not introduce noise in the sensor readings (Fig. 6a. Around each target we placed a sub-swarm, the first one with three, and two agents respectively. The consensus results in both cases were positive (see Fig. 6c). Here, a positive identification is represented by 0 and a negative with X.

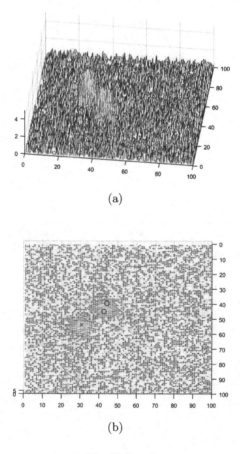

(a)

(b)

Fig. 7. Test 5

In order to evaluate the robustness of system we carried out a second test, this time we added noise to the sensor readings (see Fig. 7a). The consensus in the swarm with less agents resulted negative, while the consensus in sub-swarm with more agents resulted positive (see Fig. 7b).

5   Conclusions and Future Work

This paper presents and consensus algorithm for the identification of targets in the context of swarm robotics. In particular, we present an application of target identification in which a swarm of drones is scanning a target area, looking for potential victims of a disaster, or targets. While the main swarm is moving in a coordinate way scanning the area, a subset of agents whose sensor reading seems to identify a potential target starts to slow down its speed and surrounding that target. This behavior leads to a separation from the main swarm, which results in a loose of communications and a break of the attraction forces that maintain the swarm agents together and navigating in a coordinated way. We carried out a set of experiments to test the ability of the new sub-swarms for the identification of the actual targets. In the first set of the experiment without noise in the sensor readings, all the identification test resulted positive.

We carried out a second set of the experiment using the same set of parameters, but this time adding noise to the sensor reading. Here, parameters such as the number agents in the sub-swarm, the value of convergence constant k, and the number of interactions of the optimization algorithm played a key role in the success of the identification task. In particular, for high values of k the algorithm did not reach a consensus, and for small values of k the algorithm needed a high number of iterations.

Thus, as a future work, we plan to find the range of the parameters that maximize the effectiveness of the combination navigation and target identification. These parameters include the number of agents in the sub-swarm, value of k, the minimum number of iterations before reaching convergence. Another factor might include the relationship between the target area and the minimum number of agents needed per swarm in order to reach a consensus given some level of noise in the agents' sensor readings.

References

1. Yanguas-Rojas, D., Cardona, G.A., Ramirez-Rugeles, J., Mojica-Nava, E.: Victims search, identification, and evacuation with heterogeneous robot networks for search and rescue. In: 2017 IEEE 3rd Colombian Conference on Automatic Control (CCAC), Cartagena, pp. 1–6 (2017). https://doi.org/10.1109/CCAC.2017.8276486
2. Song, Q., Liu, F., Wen, G., Cao, J., Yang, X.: Distributed position-based consensus of second-order multiagent systems with continuous & intermittent communication. IEEE Trans. Cybern. **47**(8), 1860–1871 (2017)
3. Nuno, E.: Consensus of Euler-Lagrange systems using only position measurements. IEEE Trans. Control Netw. Syst. **PP**(99), 1 (2016)

4. Chen, S., Ho, D.W.C.: Consensus control for multiple AUVs under imperfect information caused by communication faults. Inf. Sci. **370**(Suppl. C), 565–577 (2016)
5. Chelbi, S., Duvallet, C., Abdouli, M., Bouaziz, R.: Event-driven wireless sensor networks based on consensus. In: 2016 IEEE/ACS 13th International Conference of Computer Systems and Applications (AICCSA), pp. 1–6 (2016)
6. Nurellari, E., McLernon, D., Ghogho, M.: Distributed two-step quantized fusion rules via consensus algorithm for distributed detection in wireless sensor networks. IEEE Trans. Signal Inf. Process. Netw. **2**(3), 321–335 (2016)
7. Passino, K.M.: Biomimicry for Optimization, Control, and Automation. Springer Science & Business Media, Heidelberg (2004). https://doi.org/10.1007/b138169
8. León, J., Cardona, G.A., Botello, A., Calderón, J.M.: Robot swarms theory applicable to seek and rescue operation. In: Madureira, A., Abraham, A., Gamboa, D., Novais, P. (eds.) ISDA 2016, pp. 1061–1070. Springer, Cham (2016). https://doi.org/10.1007/978-3-319-53480-0_104
9. Mesbahi, M., Egerstedt, M.: Graph Theoretic Methods in Multiagent Networks. Princeton University Press, Princeton (2010)
10. Priolo, A., Gasparri, A., Montijano, E., Sagues, C.: A decentralized algorithm for balancing a strongly connected weighted digraph. In: American Control Conference, pp. 6547–6552 (2013)
11. Lin, J., Morse, A.S., Anderson, B.D.O.: The multi-agent rendezvous problem. Part 2: the asynchronous case. SIAM J. Control Optim. **46**(6), 2120–2147 (2007)

Exploiting OSC Models by Using Neural Networks with an Innovative Pruning Algorithm

Grazia Lo Sciuto[1], Giacomo Capizzi[1,4], Christian Napoli[2], Rafi Shikler[3],
Dawid Połap[1,4], and Marcin Woźniak[1,4(✉)]

[1] Department of Electrical Electronics and Informatics Engineering,
University of Catania, Viale A. Doria 6, 95125 Catania, Italy
`glosciuto@dii.unict.it, gcapizzi@diees.unict.it`
[2] Department of Mathematics and Computer Science, University of Catania,
Viale A. Doria 6, 95125 Catania, Italy
`napoli@dmi.unict.it`
[3] Department of Electrical and Computer Engineering,
Ben-Gurion University of the Negev, 8410501 Beer-Sheva, Israel
`rshikler@ee.bgu.ac.il`
[4] Institute of Mathematics, Silesian University of Technology,
Kaszubska 23, 44-101 Gliwice, Poland
`{Dawid.Polap,Marcin.Wozniak}@polsl.pl`

Abstract. In this paper we have investigated the relationship between the current and the active layer thickness of an organic solar cell (OSC) in order to improve its efficiency by means of a back propagation neural network. In order to preserve the generalization properties of the adopted neural network (NN) in this paper is presented also an innovative pruning technique. The extensive simulations performed show a good agreement between simulated and experimental data with an overall error of about 3%. The obtained results demostrate that the use of an MLP with associated an appropriate pruning algorithm to preserve its generalization capacities permits to accurately reproduce the relationship between the active layer thicknesses and the measured maximum power in an OSC. This neural model can be of great use in manufacturing processes.

1 Introduction

The basic design constraints in an OSC influences significantly its efficiency. In particular exists a strong dependence of the photovoltaic parameters of the film thickness. In this paper we investigate the role of active layer thickness on power conversion efficiency of the OSC. In order to investigate the thickness of layers in OSCs we have prepared and fabricated 80 samples of OSC (Fig. 1). The thicknesses of the organic sandwich structures was obtained using the MFP-3D^{TM} Atomic Force Microscope at Ilse Katz Institute (Fig. 2). To increase the energy conversion efficiency it is important to find the optimal thickness of the

© Springer International Publishing AG, part of Springer Nature 2018
L. Rutkowski et al. (Eds.): ICAISC 2018, LNAI 10842, pp. 711–722, 2018.
https://doi.org/10.1007/978-3-319-91262-2_62

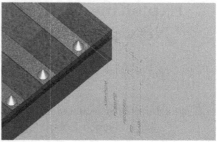

Fig. 1. Geometric representation of the fabricated sample

Fig. 2. Lab in Ilse Katz Institute for Nanoscale Science and Technology at Ben-Gurion University of the Negev

active layer that maximize the effects of near-field light concentration over a broad wavelength range.

In this paper we have investigated the relationship between the current and the active layer thickness of an OSC in order to improve its efficiency by means of a back propagation neural network. Simulations show that the active layer thickness has a great influence on the efficiency. Therefore must be taken into account in manufacturing processes. In order to preserve the generalization properties of the adopted NN in this paper is presented also an innovative pruning technique. The extensive simulations performed show a good agreement between simulated and experimental data with an overall error of about 3%.

Few contributions on the application of neural networks in the field of OSCs modeling are reported in literature [1,14,17]. Furthermore the same authors have investigated the dependence of the conversion efficiency of the active layer in cells manufactured with a different technology [4,5]. Other models oriented on massive data streams can be found in [7,15,19].

2   AFM Measurements

The AFM is a new powerful tool to provide an accurate 3D surface profiles in atomic resolution, measuring the force at nano-newton scale by laser beam deflection system. The measure of the topography can reveal information on the film structure including roughness, texture, abrasion, adhesion, cleaning, corrosion, etching, friction, lubrication, plating, and polishing. The atomic force microscope form the image without lenses, but using a very thin tip trace at the end of a flexible lever (cantilever), performing a scan on the surface of the sample. The cantilever deflections can be detected up to 0.01 nm via an optical detection system consisting of a laser and a photodiode. These signals can be generated from various measurement points in the instrument: the voltage to move the piezo actuator in response to the feed back loop; the closed loop sensor monitoring the movement of the top plate of the optical lever detector; the amplitude, phase or deflection signal from the position sensitive detector; or some kind of error in the feedback loop. All measurements are acquired by the control system software. This software also allows you to connect the AFM tool to a communication network [18].

3   Experimental Procedure

AFM measurements were carried out under ambient conditions in order to define the active layer thickness blend in the fabricated flexible and rigid solar cell structures. In this investigation, to determine the material-specific thickness it was necessary to scratch the materials of the realized samples as shown in Fig. 3. The sample analyzed is rastered in a pattern under the tip, the feed back loop moves the piezo up or down to track the surface. These measurements are digitized in an ADC and sent to the feed back loop [16]. We have analyzed the surface profiles of 80 samples. Each surface profile was sampled at 512 equidistant points with scan rate of 0.50 Hz. A Thermal tune is performed to determine the natural resonant frequency of the cantilever by monitoring the amplitude over a user defined frequency range. During the measurements, the AFM tip has been excited at the resonant frequency to obtain a free oscillating amplitude that corresponds to 800 mV output of the integrated detector (Set Point) with intergral gain of 12.76.

By scans and acquisitions, the 3D pattern is reconstructed providing the roughness measurements with high resolution in order to measure features such as the total height, the top, middle, and bottom width as shown in the Fig. 4.

In addition, the desired performance characteristics of organic cells involve not merely the thickness but the entire design structure.

4   Neural Networks for Multiple Modeling

A neural networks is a directed graph composed by a finite number of nodes, called neurons, and weighted edges, called connections. The neurons are organized in layers: an input layer, an output layer and a variable number of hidden

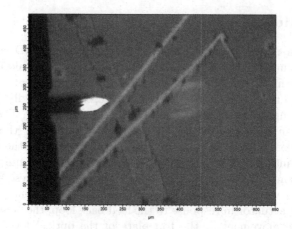

Fig. 3. A scratched sample

Fig. 4. AFM image: row image containing PCBM, PEDOT and ITO layers

layers. The neurons of each hidden layer are called hidden neurons. This latter uses the output of the previous layer's neurons as an input weighted by means of the related connections' weights, and computes on such input a function (called activation o transfer function). In a general formulation, suppose an N-dimensional input $\mathbf{u} \in \mathbb{R}^N$ so that

$$\mathbf{u} = (u_i \in \mathbb{R})_{i=1}^N \tag{1}$$

Then we will define *input layer* the set of neurons $x_i^{(0)}$ so that

$$x_i^{(0)} = u_i \quad \forall\, i \in [1, N] \cap \mathbb{N} : \mathbf{u} \in \mathbb{R}^N \tag{2}$$

Given the l-th hidden layer it can be similarly formalized as (Fig. 5)

$$x_j^{(l)} = \gamma_j^{(l)} \left(b_j^{(l)} + \sum_{i=1}^{N_{l-1}} w_{ij}^{(l)} x_i^{(l-1)} \right) \quad \begin{array}{l} \forall\, l \in [1,\, L] \cap \mathbb{N} \\ \forall\, j \in [1, N_l] \cap \mathbb{N} \end{array} \tag{3}$$

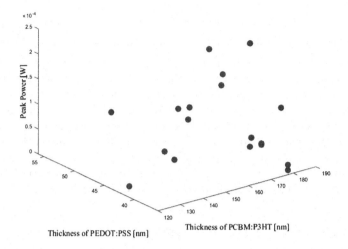

Fig. 5. The maximum power and P3HT:PCBM/PEDOT:PSS thickness dependence of the organic devices

where $x_j^{(l)}$ is the output of the j-th neuron of the l-th layer (composed by N_l neurons), $\gamma_j^{(l)}$ its activation function, and $b_j^{(l)}$ the bias; N_{l-1} is the number of neurons in the previous layer; and where $w_{ij}^{(l)}$ is the connection weight from the i-th neuron (with output $x_i^{(l-1)}$) on the $(L-1)$-th layer (composed by N_l neurons) to the j-th neuron on the L-th layer. Finally the output layer will be composed by M neurons with value y_k given by

$$y_k = \gamma_k^{(L)} \left(b_k^{(L)} + \sum_{j=1}^{N_{L-1}} w_{jk}^{(L)} x_j^{(L-1)} \right) \quad \forall\, k \in [1, M] \cap \mathbb{N} : \mathbf{y} \in \mathbb{R}^M \qquad (4)$$

4.1   MLP Function Approximation

As well known in literature [2,6,11], any Riemann-integrable function, that we will call *signal*, can be arbitrarily approximated by means of feedforward neural network regardless of the activation function or the input space, while signals with a finite support can be exactly approximated also by a single layer of neural units. Although similar to the Cybenko-Hornik theorem, it must be highlighted that the MLP approximation theorem needs not a continuous signal among its hypothesis. A multilayer perceptron network (MLP) [10] can approximate signals using two hidden layers [13]: the first hidden layer approximate the signal by means of a step function σ so that

$$x_j^{(1)} = \sigma \left(b_j^{(1)} - \sum_{i=1}^{N_0} w_{ij}^{(1)} x_i^{(0)} \right) \quad \forall\, j \in [1, N_1] \cap \mathbb{N} \qquad (5)$$

is the output of the first hidden layer (composed by N_1 neurons $x_j^{(1)}$), with respect to the input (composed by N_0 elements $x_i^{(0)}$), where $w_{ij}^{(1)}$ are neural weights and $b_j^{(1)}$ is the bias used as threshold for the step function. The second hidden layer computes the height of the stair step, in which the input value lies in order to return a value to the output layer. The threshold logic units' computation can be interpreted geometrically as they split the input hyperplane in two half-hyperplanes. In this representation the separation line's equation depends by the weights and bias. Such variables are determined by a training algorithm based on the overall approximation error that affects the network. On the other hand the approximation accuracy after training strictly depends by the number of neurons which influence the size of the stair steps.

5   The Proposed Pruning System

In the previous section we introduced the adopted MLPNs swarm methodology and the related pre-training method. On the other hand, due to the nature of the problem at hands, the said MLPNs swarm will be intentionally characterized by slightly over-sized networks, in order to preserve higher generalization properties for the swarm. Although that, after the pre-training phase, a fast and specific training must follow to model each specific component and, consequently, several pruning steps. This latter procedure can be implemented by well known methods [8,9,12], on the other hand, classical pruning algorithms are sometimes low performing [20]. Moreover, in order to obtain device-specific models, more adjustable as well as data-driven pruning algorithms were required [3].

5.1   Adaptive Data-Driven Pruning Algorithm

The developed pruning algorithm takes into account the weight distribution in each layer of a the pre-trained network and devises a variable number of centroids on the weights space using a k-means algorithm. After that the algorithm partitions the space, pruning only the less relevant partition. The algorithm steps are the following (Fig. 6):

1. Randomly distribute centroids
2. Compute weights memberships to centroids-related classes
3. Reposition centroids
4. Recompute until k-means convergence
5. Layer pruning and backward propagation
6. Simulate network and evaluate error
7. Return, or increase centroids' number and iterate

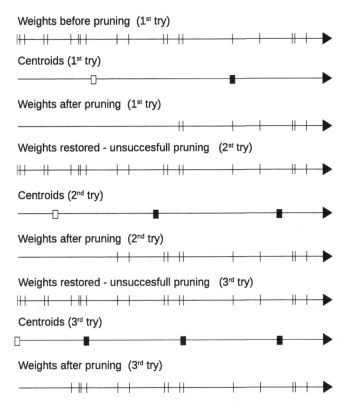

Fig. 6. The proposed pruning

Now follows a detailed description.

Given a network as in (4), and referring to a layer $\bar{l}$, then we will call $W_{\bar{l}}$ the ingoing synapses weights so that:

$$W_{\bar{l}} = \left\{ w_{ij}^{(\bar{l})} \quad \forall\, i \in [1, N_{\bar{l}-1}] \cap \mathbb{N},\ \forall\, j \in [1, N_{\bar{l}}] \cap \mathbb{N}, \right\} \subset \mathbb{R} \tag{6}$$

We can define a metric $\delta : \mathbb{R} \to \mathbb{R}$

$$\delta(x_1, x_2) = ||x_1 - x_2||^2 \quad \forall\, x_1, x_2 \in \mathbb{R} \tag{7}$$

Then, for the n-th iteration of the overall procedure, given a set $C_n^{\bar{l}}$ of centroids in $[\min(W_{\bar{l}}), \max(W_{\bar{l}})]$, randomly distributed at the first, and starting with a user-defined size N_c so that $|C_n^{\bar{l}}| = N_c$, we can define, at each time step τ_n, a membership function $\mu : W_{\bar{l}} \to C_n^{\bar{l}}$ computed for each weight by the following

$$\mu(w_{ij}^{(\bar{l})}) = \bar{c}_n^{\bar{l}}(\tau_n) : \delta\left(w_{ij}^{(\bar{l})}, \bar{c}_n^{\bar{l}}(\tau_n)\right) = \min_{C_n^{\bar{l}}} \left\{ \delta\left(w_{ij}^{(\bar{l})}, c_n^{\bar{l}}(\tau_n)\right) \quad \forall\, c_n^{\bar{l}}(\tau_n) \in C_n^{\bar{l}} \right\} \tag{8}$$

so that, during the time step τ_n, the weight $w_{ij}^{(l)}$ is assigned to the centroid $\bar{c}_n^{\bar{l}}(\tau_n)$. For each centroid, at each time step τ_n, it is possible to define centroid-related classes

$$\mathscr{C}[c_n^{\bar{l}}(\tau_n)] = \left\{ w_{ij}^{(l)} \in W_{\bar{l}} : \mu(w_{ij}^{(l)}) = c_n^{\bar{l}}(\tau_n) \right\} \ \forall \ c_n^{\bar{l}}(\tau_n) \in C_n^{\bar{l}} \qquad (9)$$

Basing on the latter, before each time-step $\tau_n > 1$, the new centroids $c^{\bar{l}}(\tau_n)$ are recomputed as

$$c_n^{\bar{l}}(\tau_n) = \frac{\displaystyle\sum_{\mathscr{C}[c_n^{\bar{l}}(\tau_n-1)]} w_{ij}^{(l)}}{\left| \mathscr{C}[c_n^{\bar{l}}(\tau_n-1)] \right|} \qquad (10)$$

Therefore the new centroids will be positioned to the average weight in the class $\mathscr{C}[c_n^{\bar{l}}(\tau-1)]$, where $c_n^{\bar{l}}(\tau_n-1)$ are the old centroid and $\mathscr{C}[c_n^{\bar{l}}(\tau_n-1)]$ the old classes. Once the new centroids have been computed, for each weight in $W_{\bar{l}}$ a new membership function is evaluated as in (8), and consequently the new classes $\mathscr{C}[c_n^{\bar{l}}(\tau_n)]$ are composed. The process is iterated until convergence of the k-means algorithm. After k-means convergence, we can consider concluded the centroid and classes evaluation for the current iteration. Let call n the iteration index, then, after the k-means algorithm returned, at a time step $\bar{\tau}_n$, we can select the minimum centroid $\chi_n^{\bar{l}}$ at the n-th iteration as

$$\chi_n^{\bar{l}} = \min \left\{ |c_n^{\bar{l}}(\bar{\tau}_n)| \right\} \qquad (11)$$

Finally, basing on (11), a tentative pruning is performed by removing each weight $w_{ij}^{(l)} \in \mathscr{C}[\chi_n^{\bar{l}}]$, and therefore, by pruning the relative network connection. The pruning procedure, is then performed backwardly, for each layer, in a similar fashion. After the n-th iteration of the overall layer-by-layer pruning procedure, the neural network is retrained and tested. The performances variation is computed as

$$\Delta_n = |E_n - E_{n-1}| \qquad (12)$$

where E_n is the network error after the n-th overall pruning iteration, and E_{n-1} the network error at the end of the previous iteration (or before the overall procedure for $n = 1$). Finally the n-th overall iteration is positively evaluated if

$$\begin{cases} \dfrac{\Delta_n}{\Delta_{n-1}} < \vartheta \\[2mm] E_n < \varepsilon \end{cases} \qquad (13)$$

where ϑ and ε are two user-defined thresholds respectively representing the maximum relative performance variation and the maximum network error allowed. If the last iteration is not evaluated positively, then next iteration $(n + 1)$ is computed. The procedure continues until it reaches a positive evaluation or an user-defined iteration limit.

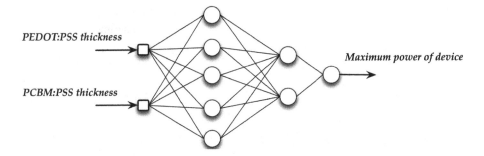

Fig. 7. The resulting network after the pruning phase.

6   The Proposed MLP Based Model

The MLP proposed in this paper is developed in order to obtain an approximate mathematical expression thickness of the active layer and maximum power generated by the OSC. The resulting network after the pruning phase is shown in Fig. 7. It is composed by four layers: an input layer, two hidden layer and an output layer. The input layer has two inputs (thicknesses of the active layer (PEDOT:PSS and PCBM:PSS)), the two hidden layers are composed of 5 and 2 neurons respectively, with hyperbolic tangen transfer function while the output layer is composed of one neuron with linear transfer function.

The network has been trained by using the measured thicknesses of the active layer (PEDOT:PSS and PCBM:PSS) as inputs and the measured maximum power of the device obtained in correspondence with the values of thicknesses present in input as target. The criterion used to train the network has been the minimizion of the MSE.

7   Results

In order to train the network we used the voltage and current data obtained experimentally from 80 devices with different values of thickness of the active layer (PEDOT: PSS and PCBM: PSS) made at the Optoelectronic Organic Semiconductor Devices Laboratory at Ben Gurion University of the Negev. We have evaluated the current e tensione as function of different film thicknesses for active layers of an OSC.

As shown in Fig. 8 the network after the training procedure is able to accurately reproduce the relationship between the active layer thicknesses and the measured maximum power of an OSC. In fact the extensive simulations show a good agreement between simulated and experimental data with an overall error of about 3%. Furthermore, the simulations performed show that the active layer has a great influence on the OSCs efficiency.

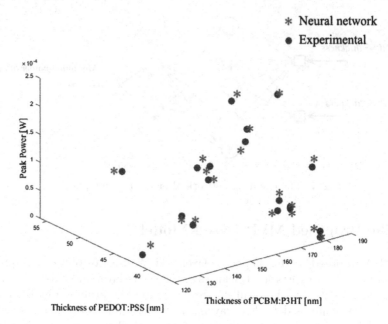

Fig. 8. The maximum power generated of an OSC and the simulated ones by the implemented NN at different thicknesses.

8   Conclusion

In this paper we have investigated the dependence of the maximum power generated by an ultra-thin organic solar cell from the active layer thicknesses in order to improve its efficiency. These investigations were carried out by means of an MLP network. The obtained results show that the use of an MLP with associated an appropriate pruning algorithm to preserve its generalization capacities permits to accurately reproduce the relationship between the active layer thicknesses and the measured maximum power. This neural model can be of great use in manufacturing processes.

Acknowledgment. This work has been supported by the BGU-ENEA joint lab and the ILSE-Joint Italian-Israeli Laboratory on Solar and Alternative Energies. We thank Dr. Jurgen Jopp of Ilse Katz Institute for Nanoscale Science and Technology. The Authors would like to acknowledge contribution to this research from the "Di-amond Grant 2016" No. 0080/DIA/2016/45 funded by the Polish Ministry of Science and Higher Education.

References

1. Barnea, S.N., Lo Sciuto, G., Hai, N., Shikler, R., Capizzi, G., Woźniak, M., Połap, D.: Photo-electro characterization and modeling of organic light-emitting diodes by using a radial basis neural network. In: Rutkowski, L., Korytkowski, M., Scherer, R., Tadeusiewicz, R., Zadeh, L.A., Zurada, J.M. (eds.) ICAISC 2017 Part II. LNCS (LNAI), vol. 10246, pp. 378–389. Springer, Cham (2017). https://doi.org/10.1007/978-3-319-59060-8_34
2. Barron, A.R.: Universal approximation bounds for superpositions of a sigmoidal function. IEEE Trans. Inf. Theory **39**(3), 930–945 (1993)
3. Bonanno, F., Capizzi, G., Lo Sciuto, G.: A neuro wavelet-based approach for short-term load forecasting in integrated generation systems. In: 2013 International Conference on Clean Electrical Power (ICCEP), pp. 772–776, June 2013
4. Bonanno, F., Capizzi, G., Lo Sciuto, G., Napoli, C., Pappalardo, G., Tramontana, E.: A cascade neural network architecture investigating surface plasmon polaritons propagation for thin metals in OpenMP. In: Rutkowski, L., Korytkowski, M., Scherer, R., Tadeusiewicz, R., Zadeh, L.A., Zurada, J.M. (eds.) ICAISC 2014 Part I. LNCS (LNAI), vol. 8467, pp. 22–33. Springer, Cham (2014). https://doi.org/10.1007/978-3-319-07173-2_3
5. Capizzi, G., Lo Sciuto, G., Napoli, C., Tramontana, E.: A multithread nested neural network architecture to model surface plasmon polaritons propagation. Micromachines **7**(7), 110 (2016)
6. Cybenko, G.: Approximation by superpositions of a sigmoidal function. Math. Control Sig. Syst. **2**(4), 303–314 (1989)
7. Duda, P., Jaworski, M., Rutkowski, L.: Convergent time-varying regression models for data streams: tracking concept drift by the recursive parzen-based generalized regression neural networks. Int. J. Neural Syst. **28**(02), 1750048 (2018)
8. Han, S., Pool, J., Tran, J., Dally, W.: Learning both weights and connections for efficient neural network. In: Advances in Neural Information Processing Systems, pp. 1135–1143 (2015)
9. Hassibi, B., Stork, D.G.: Second order derivatives for network pruning: optimal brain surgeon. In: Advances in Neural Information Processing Systems, pp. 164–171 (1993)
10. Haykin, S.S.: Neural Networks: A Comprehensive Foundation, 2nd edn. Prentice-Hall, Englewood Cliffs (1998)
11. Hornik, K., Stinchcombe, M., White, H.: Multilayer feedforward networks are universal approximators. Neural Netw. **2**(5), 359–366 (1989)
12. Huang, G.-B., Saratchandran, P., Sundararajan, N.: A generalized growing and pruning RBF (GGAP-RBF) neural network for function approximation. IEEE Trans. Neural Netw. **16**(1), 57–67 (2005)
13. Kruse, R., Borgelt, C., Braune, C., Mostaghim, S., Steinbrecher, M.: Computational Intelligence: A Methodological Introduction. Springer, Heidelberg (2016). https://doi.org/10.1007/978-1-4471-5013-8
14. Lo Sciuto, G., Capizzi, G., Coco, S., Shikler, R.: Geometric shape optimization of organic solar cells for efficiency enhancement by neural networks. In: Eynard, B., Nigrelli, V., Oliveri, S., Peris-Fajarnes, G., Rizzuti, S. (eds.) Advances on Mechanics, Design Engineering and Manufacturing. LNCS, pp. 789–796. Springer, Cham (2017). https://doi.org/10.1007/978-3-319-45781-9_79
15. Nowicki, R.K., Starczewski, J.T.: A new method for classification of imprecise data using fuzzy rough fuzzification. Inf. Sci. **414**, 33–52 (2017)

16. Ryan, F.: Procedural operation manualette@ONLINE (2009)
17. Lo Sciuto, G., Capizzi, G., Gotleyb, D., Linde, S., Shikler, R., Woźniak, M., Połap, D.: Combining SVD and co-occurrence matrix information to recognize organic solar cells defects with a elliptical basis function network classifier. In: Rutkowski, L., Korytkowski, M., Scherer, R., Tadeusiewicz, R., Zadeh, L.A., Zurada, J.M. (eds.) ICAISC 2017 Part II. LNCS (LNAI), vol. 10246, pp. 518–532. Springer, Cham (2017). https://doi.org/10.1007/978-3-319-59060-8_47
18. Toscano, E., Lo Bello, L.: A topology management protocol with bounded delay for wireless sensor networks. In: IEEE International Conference on Emerging Technologies and Factory Automation. ETFA 2008, pp. 942–951 (2008)
19. Zębik, M., Korytkowski, M., Angryk, R., Scherer, R.: Convolutional Neural networks for time series classification. In: Rutkowski, L., Korytkowski, M., Scherer, R., Tadeusiewicz, R., Zadeh, L.A., Zurada, J.M. (eds.) ICAISC 2017 Part II. LNCS (LNAI), vol. 10246, pp. 635–642. Springer, Cham (2017). https://doi.org/10.1007/978-3-319-59060-8_57
20. Zurada, J.M., Malinowski, A., Cloete, I.: Sensitivity analysis for minimization of input data dimension for feedforward neural network. In: 1994 IEEE International Symposium on Circuits and Systems, ISCAS 1994, vol. 6, pp. 447–450. IEEE (1994)

Critical Analysis of Conversational Agent Technology for Intelligent Customer Support and Proposition of a New Solution

Mateusz Modrzejewski[(✉)] and Przemysław Rokita

Division of Computer Graphics, The Faculty of Electronics and Information Technology, Institute of Computer Science, Warsaw University of Technology, Nowowiejska 15/19, 00-665 Warsaw, Poland
{M.Modrzejewski,P.Rokita}@ii.pw.edu.pl

Abstract. This paper proposes and describes an application of modern, Loebner's Prize-winning conversational agent technologies in the context of intelligent aid for customer service. The paper defines the requirements, design philosophy and main algorithm for a system that would be able to improve customer service efficiency and functionality using innovative artificial intelligence methods, which can be seen as an increasingly common demand in various branches of business. Emerging problems of interface, knowledge engineering and natural language processing nature are discussed along with proposals of technologies suitable for resolving said issues.

Keywords: Conversational agents · Chatbots · NLP
Applied artificial intelligence · AI · Knowledge engineering

1 Introduction

Conversational interfaces, commonly known as chatbots, allow users to engage in conversation using natural language. They are equipped with various artificial intelligence techniques in order to model and simulate human behavior and provide enhanced functionality wherever a text or voice interface is suitable. Chatbots have gone a long way since 1966, when MIT's Joseph Weizenbaum published ELIZA [1], a scripted bot that simulated a psychotherapist's conversation, and this is clearly visible when looking at the milestones achieved in chatbot design at the annual Loebner's Prize contest. [2] displays an evolution of chatbots during the recent years, with techniques ranging from simple pattern matching, through heavy database reliance as in the Jabberwacky bot [3], up to Bruce Wilcox's *ChatScript* [4], a powerful natural language processing and conversation scripting engine.

Practical applications of conversational agents are also gaining wider acceptance and usage in business, entertainment, from mobile applications to everyday

© Springer International Publishing AG, part of Springer Nature 2018
L. Rutkowski et al. (Eds.): ICAISC 2018, LNAI 10842, pp. 723–733, 2018.
https://doi.org/10.1007/978-3-319-91262-2_63

utilities. The Facebook Messenger bot and the Slackbot are prominent example from the social media domain. Amazon's Alexa has been a huge commercial success, selling an estimate of around 10 million smart speakers in 2017. These are only some of the modern examples of applied conversational agent technology, but the list goes a lot further. The market is growing rapidly, with some of the world's biggest companies currently investing in development of such solutions [5].

2   Classic Chatbots

The basic, conceptual structure of a chatbot that has emerged over the years is presented on Fig. 1.

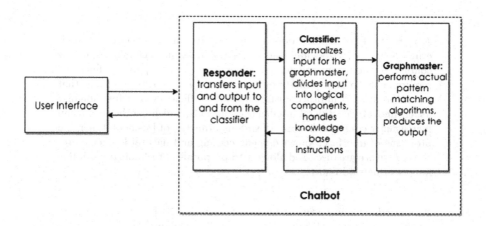

Fig. 1. Basic structure of a Chatbot

Although The Loebner's Prize has been a major catalyst for the development of chatbots, the solutions designed for the Loebner's Prize are far from question-answering machines in terms of functionality. The primary point of interest for those bots is to pass a "chatbot version" of the Turing test: human judges determine whether their conversation partner is human or machine using specific conversation topics that allow lots of small talk and digressions. The chatbots are expected to keep track of the dialogue, show interest (or a lack thereof) and display human-like conversation scenarios with a certain degree of social skill. Most of the Loebner chatbots are therefore equipped with algorithms that imitate human behavior [2], for instance:

– generating occasional typos,
– language tricks,
– non-sequitur sentences,
– canned responses;

Implementing a representation of common knowledge and basic reasoning as the base for the Chatbot is an important issue for the competitive bots, but it's also very easy to exploit by the chatbot designers. It may lead to such situations as Bruce Wilcox, the creator of *ChatScript* and triple Loebner's prize winner, calling the Loebner's judges "foolish and lazy" after figuring out their evaluation methods (*"They spent their time asking unusual questions and making deliberate spelling mistakes or merely walking away after a few minutes - in general being difficult"*, as in [6]).

Chatbots have gone a long way in the recent years and various ideas were applied in the process, including Markov chains, large databases and pattern matching algorithms. Current state of the art tools, like *ChatScript* and AIML, provide simple and elegant mechanisms for developing chatbots: the chatbot logic is usually contained within simple text files or scripts and handled by a provided engine. The focus of these tools, however, is the quality of the conversation with regard to how intelligent the chatbot will seem to the user. This can be achieved in multiple ways, ranging from a "brute-force" implementation utilizing a tremendous amount of scripts to cover different conversation scenarios to clever design and usage of the more advanced techniques provided by the tools. Still, this does not seem to explore the full potential of modern conversational agent technology with regard to the current needs for artificial intelligence support in various applications.

3   Proposition of a New Solution

Conversational interfaces [7] are a strong direction in the development of our basic model of interaction with a machine. The goal is to make the communication as fast and intuitive as possible, and natural language is an obvious idea. Artificial intelligence contained within conversational interfaces can be used in order to improve the data distribution and presentation in sectors like technical support and customer service, in any institution that is likely to provide their services or products to a significant number of people - like banks, medical institutions or huge retailers. The conversational agent may be available directly to the customers or may be used as an internal interface designed specifically for the needs of the staff, thus acting as a natural-language-powered facade for any bigger system.

Specific applications of conversational agent technology usually regard very narrow domains of dispute described by a well-structured data model, which is different from the initial idea behind aforementioned chatbot tools. Additionally, besides the scripts and patterns, the system suited for such needs would have to provide particular question-answering machine and browser features, such as adaptation to operating on granular data sources and various data formats, utilizing advanced answer extraction techniques and highly scalable, modular design, as described in [8].

A basic structure of such a system can be seen as shown on Fig. 2.

The crucial role of the system is providing an answer retrieval mechanism which takes a question written in natural language as input and provides an

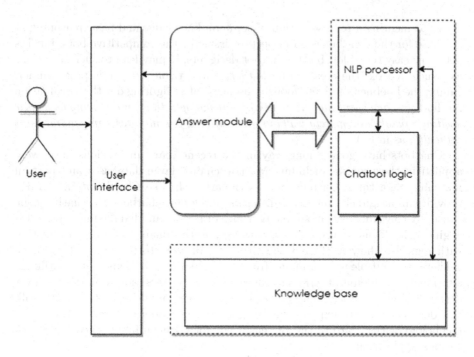

Fig. 2. Basic concept of the agent

answer as an result. This functionality is encapsulated within the answer module. The underlying structure is organized around chatbot technology (including scripts, patterns, topics etc.) with an additional NLP processor for precise handling of domain-specific vocabulary. The knowledge base serves as a data provider and handles underlying knowledge sources. Depending on the desired scope of question recognition capabilities, the answer module can be configured to use different strategies in order to build the answer: for instance, it can resolve directly to scripts or perform domain-specific processing.

The input of the system is a question, which conceptually consists of an intent, corresponding entities and additional attributes. Such questions may be very precise, and represent simple concepts, as *"What is the value of X?"*, *"Who is Y?"*. On the other hand, they may also require reasoning or checking multiple data sources. Retrieving data for the answer may also require attributes that are necessary from the system's point of view, but the askers have not thought of and have not included in their initial question.

The system's reply can therefore consist of any of the following elements:

 – the body of the answer in natural language,
 – a non-empty set of retrieved values,
 – a question (of clarification or continuation),
 – an action;

The answer module is also a point of connection with external systems: integration allows for automatization of certain actions that would have been carried out by the asker as a result of the answer anyway, like copying the answer into an empty e-mail or preparing an archive file.

The following sections elaborate on the specific modules of the system and propose suitable solutions for emerging problems.

4   Scope of Application

4.1   Direct Customer Support

Customer questions are an obvious field of application for conversational agents: providing a friendly, functional chat interface is not only a convenient way of improving the customer service experience, but also an innovative way of aiding the staff's own intelligence with AI.

Many people tend to repeatively use customer service via phone call or e-mail in simple cases that require very little effort to resolve (for instance, all the answers can be found in a supplied manual, warranty card or using an easily available mobile application provided by the company). Answering this type of questions is fairly simple, but many of the problems (and the corresponding solutions) repeat over and over, taking up time and thus generating costs. Modern chatbot technology can be applied in a first-contact interface in order to quickly resolve "mass questions", as the necessary knowledge has no security restrictions, can be easily integrated into a knowledge base and provided to the customers in a natural language interface.

In this case, the agent would be designed as an interactive, intelligent system built upon structured and well-known data contained in frequently-asked-questions lists, as described in [8], but as the customer profiles vary, so will the questions. Some questions may regard complicated issues and therefore cannot be managed by a chatbot, either for complexity or security reasons. In this case, we need to consider a broader concept of such a chatbot.

4.2   Internal Support

As the complexity of the questions rises, the customer service process can no longer be automated and most certainly requires human intervention. [9–11] provide further insight into the challenges of modern customer support. From an artificial intelligence point of view, the customer service process that we want to model is as follows:

1. identify the core and the intent of the client's issue;
2. formulate a solution template that resolves the issue in a satisfying way, for instance:
   - *provide an answer,*
   - *schedule a visit,*
   - *offer a new product,*

– *forward a document to the client,*
– *further the issue to another departament...*

3. if required by the character of the solution template, collect necessary knowledge from the available knowledge sources and formulate the concrete solution;
4. deliver the concrete solution;

If the solution template is insufficient, the agent would ask the user additional questions in order to create an effective query for the knowledge sources. This can be scripted easily using the proposed tools.

A conversational agent able of modelling and implementing the above process would allow not only to increase the efficiency of customer service, but also to focus the intelligence of the staff on resolving the actual issue and providing a precise answer without having to manually search and browse the domain's knowledge base.

5   Chatbot Logic

Although the main logic of the chatbot can be implemented from scratch, several existing tools allow for effective development of chatbot logic. The two most interesting are AIML [12] and *ChatScript* [13]. Both of these technologies provide a functional chatbot scripting environment and have been awarded in the Loebner's competition.

AIML is based on XML and relies on pattern matching. Input rules are matched with scripted output and topics may be used in order to organize the scripts. AIML supports the use of wildcards and is able to call itself recursively, but dictionaries and ontologies have to be supplied externally in the NLP module.

ChatScript, created by Bruce Wilcox, is a full NLP-conversation engine. It operates purely on text files with a comprehensive syntax [13] and introduces some advanced methods, like advanced topic managing, concepts and logical operations. Besides of simple rules, it also uses gambits and wildcards. *ChatScript* is supplied with its own NLP toolkit for English, but it also offers the possibility of expansion to any language when provided with a part-of-speech tagger, dictionaries and ontologies for that language. As stated by the author himself[1], *ChatScript* aims at being a complete, functional and easy to use NLP tool.

ChatScript also supports a mechanism called facts which is used to represent knowledge as triplets of fields called *subject, verb, object*, whether those are actually the contents of those fields or not [14]. These facts may be stored, queried and grouped into tables in order to create a cohesive internal knowledge base. The native JSON support features a mechanism that allows to easily interpret JSON messages as tables of facts. Considering also the built-in PostreSQL

---

[1] Bruce Wilcox is a contributor to the chatbot designer's community board chatbots.org - he is also constantly developing *ChatScript* and updating his website frequently.

support, *ChatScript* allows to greatly simplify integration with external knowledge sources. In this case, the data retrieval mechanism itself could be contained within the text files containing the scripts and fine-tuned for the needs of particular questions.

6   Knowledge Base

Bots submitted to the Loebner's competition usually utilize tens of thousands of rules [6] in order to obtain satisfying conversation skills. In the case of customer service, the knowledge base structure is far more important: the main task of the agent will be answering questions with concrete information and actual knowledge, whenever possible. Furthermore, if we want the agent to recognize and respond not only to questions, but also to requests, we have to equip it with much broader understanding skills. We can therefore say that the knowledge base of such a system consists of:

- dictionaries,
- ontologies,
- databases,
- other systems and their APIs.

Considering the NLP and Chatbot logic modules as a black box, the knowledge base may be understood as a facade for a set of adapters, as presented on Fig. 3.

The facade has to share a common knowledge description with the logic module and delegate the queries to concrete knowledge sources. One source can represent knowledge from one or many domains and it is up to the adapter to collect the data and prepare it for further processing.

Dictionaries and ontologies [16,17], in general, allow the agent to represent its knowledge in a well structured way, using domain-specific vocabulary. As a result, they also allow to recognize whether answering a question or fulfilling a request lies within the agent's scope of recognition - basically, to determine if an action is possible to perform. The agent then retrieves the answer using the concrete knowledge supplied in the databases or produces a request to external systems.

The proposed structure of the knowledge base can be implemented and maintained particularly easily with *ChatScript* as the chatbot logic. *ChatScript* natively supports integration with PostgreSQL databases and JSON webservices, allowing for simple conversion into its inner knowledge representation form. It also has a built-in dictionary and ontology for the English language, providing a powerful template and allowing for recognition and understanding of many basic topics right out of the box.

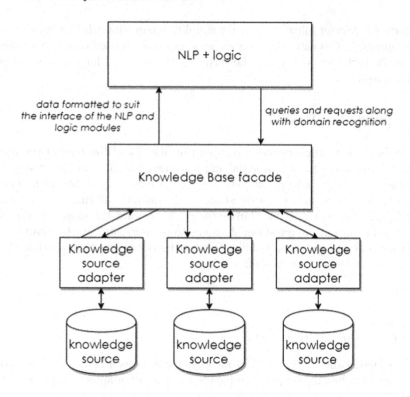

Fig. 3. Structure of the knowledge base

7   Natural Language Processing

As stated in [15], the main part of natural language processing and understanding in question-answering context is recognizing three key parts of a sentence:

– the intent,
– the entities corresponding to the intent,
– the attributes of the entities.

These parts constitute the essence of the sentence and allow the agent to determine the domain of the question. The base of the issue is part-of-speech tagging: for instance, in English and Polish the intent in many cases will be a verb, the entities will be nouns, while parameters are likely to be numbers or adjectives. A POS tagger would be therefore the crucial part of the NLP processor in the proposed solution.

The processor is also responsible for domain recognition. The domain constitutes sets of specific vocabulary that has to be known to the agent in order to recognize and understand the input. We would like to point out that a part of the module should also be used to pre-process the input query, filtering it through the available dictionaries and/or applying ontologies to instantly extract and recognize vocabulary from the domain of dispute (or notify about the lack thereof in the supplied knowledge base). The domain should be represented by an additional tag, as it will be used to specify and query the knowledge sources containing specific data needed to answer the question.

Particular problems may be exceptionally difficult, like multiple negation, amphiboly or polysemy. Whenever the NLP module fails to produce an unequivocal result, the chatbot logic module would take the steering and try to continue with the retrieval algorithm based on pattern matching.

8 Interface Methods

Two obvious approaches suitable for the interface of such a chatbot are speech interfaces [18] and text interfaces [20], both of which of course can be integrated into a GUI (graphical user interface) for convenience of use. Integration with speech interfaces is especially interesting, as it may be used for instance in hot line centers, sending the questions directly into the chatbot logic after recognition in real-time and providing a live prediction of the consultation, whenever security reasons allow to do so.

All of the technologies (AIML, *ChatScript*, OWL/RDF ontologies and so on) mentioned in this paper can be easily integrated with a graphical interface technology of choice in a chosen architecture: for instance, integrating *ChatScript* into a desktop application requires only implementing a single adapter class, which is a very simple task in most modern OOP languages.

Regarding the discussed application, the interface functionality is crucial. The agent's knowledge base can be broadened separately from its other modules, so it is likely that the agent's scope of answering capabilities would expand over more domains in time (for instance, by integration with external systems, knowledge re-engineering etc.). Key factors when designing a flexible graphical interface for such needs are simplicity, functionality of each element of the design and maintaining a clear visual hierarchy.

9 Outline of Proposed Solution

9.1 Advantages of Conversational Agent Interfaces

The application of conversational agents to the context of client questions and staff questions in customer service has very promising advantages in terms of improvement of the efficiency and quality of service. Depending on the branch and size of the enterprise, a customer service employee may have to solve from tens to hundreds of customer cases daily, each of the cases requiring different knowledge sources - a facade system supplied with the company's knowledge sources and procedures is able to encapsulate the browsing process under an user-friendly interface and serve the processed data in a matter of a few intuitive questions. Natural language interfaces are clearly much easier to use than robust, complex and often inconsistent information systems implemented for the needs of big enterprises and allow to reduce time, effort and costs generated by customer service departaments.

Learning algorithms may also be utilized (as described, for instance, in [17]) in order to improve the system's performance with no need to modify the code. Many of the proposed tools operate on text files (like scripts, rules and ontologies) that are read by the software as resources and also may be modified separately form the software's logic. Finally, heuristic methods should be implemented in order to allow the users to evaluate and correct the answers provided by the system, further expanding the system's improvement capabilities.

9.2   Summary of Proposed Approach

Upon the above, the principle idea of the described approach can be summarized as:

- repurposing state-of-the art chatbot technology for narrow domains of dispute,
- building a knowledge base suitable for integration with chatbot technology,
- implementation of a data retrieval and answer building mechanism,
- applying natural language processing w.r.t. to the domain of dispute in order to aid the main algorithm.

The necessary basic input includes scripts, specific dictionaries and knowledge processing mechanisms - once particular, initial issues of domain specification and knowledge representations are solved, the system operates on the framework provided by the logic and is transparent to its associated resources. The structure is highly modular and applicable for a wide selection of implementation technologies.

10   Conclusions

In this paper we have analyzed and discussed the usage of modern conversational agent technology for intelligent aid in customer support. We have also proposed a system for intelligent, automated customer service aid and described the application of algorithms and techniques suitable for resolving emerging issues. Although certain AI tools supporting customer service are commercially available, such use of conversational agents still seems to be in development and research phase. The proposed approach offers high flexibility in terms of functionality, allowing to create precisely tailored solutions, fine-tuned to the needs of particular domains of dispute and data source structure. It is currently under development and will be implemented and tested in the intelligent customer service system of one of the major wireless service providers.

References

1. Wiezenbaum, J.: ELIZA - a computer program for the study of natural language communication between man and machine. Commun. ACM **9**(1), 36–45 (1966)
2. Bradeško, L., Mladenić, D.: A survey of Chabot systems through a Loebner prize competition (2012)
3. Official website of the Jabberwacky Chatbot. http://www.jabberwacky.com
4. Wilcox's, B.: ChatScript - GitHub Repository. https://github.com/bwilcox-1234/
5. Kai Kiat, O.: Business application of ChatBots. Chatbots Magazine, 2 November 2017. https://chatbotsmagazine.com/business-application-of-chatbots-afb952cfdb93
6. Wilcox, B., Wilcox, S.: Winning the Loebner's. Brillig Understanding Inc (2014)
7. Pan, J.: Chatbots Magazine, 25 August 2017. https://chatbotsmagazine.com/conversational-interfaces-the-future-of-chatbots-18975a91fe5a
8. Maybury, M.T. (ed.): New Directions in Question Answering. AAAI/MIT Press, Palo Alto/Cambridge (2004)
9. 2014 Global Customer Service Barometer. Findings in the United States report for Ebiquity (2014)
10. Creating Great Service Experiences - How Modern Customer Service Works. Oracle Cloud (2014)
11. How to Provide Customer Service Excellence. Failte Ireland (2013)
12. A.L.I.C.E. Foundation: AIML language for Chatbots. http://www.alicebot.org/aiml.html
13. Wilcox, B.: Beyond Facade: Pattern Matching for Natural Language Applications. Telltale Games, February 2011
14. Wilcox, B.: ChatScript Fact Manual. Revision 10/22/2017 for ChatScript 7.6
15. Jurafsky, D., Martin, J.H.: Dialog systems and Chatbots, Chap. 29. In: Speech and Language Processing (2017)
16. Al-Zuhaide, H., Issa, A.: OntBot: ontology based Chatbot. In: IEEE Fourth International Symposium on Innovation in Information & Communication Technology, Amman (2011)
17. Lundqvist, K.O., Pursey, G., Williams, S.: Design and implementation of conversational agents for harvesting feedback in elearning systems. In: Hernández-Leo, D., Ley, T., Klamma, R., Harrer, A. (eds.) EC-TEL 2013. LNCS, vol. 8095, pp. 617–618. Springer, Heidelberg (2013). https://doi.org/10.1007/978-3-642-40814-4_79
18. Abdul-Kader, S.A., Wood, J.: Survey on Chatbot design techniques in speech conversation systems. Int. J. Adv. Comput. Sci. Appl. (IJACSA) **6**(7), 76–79 (2015)
19. Setiaji, B., Wibowo, F.W.: Chatbot using a knowledge in database human-to-machine conversation modeling. In: 2016 7th International Conference on Intelligent Systems, Modelling and Simulation (2016)
20. Ask, J.A., Facemire, M., Hogan, A.: The State of Chatbots. Forrester.com report, 20 October 2016
21. Bo, P., Lillian, L., Shivakumar, V.: Thumbs up? Sentiment classification using machine learning techniques. In: Proceedings of the Conference on Empirical Methods in Natural Language Processing (EMNLP), pp. 79–86 (2002)
22. Chatbot Society and Message Board (Bruce Wilcox is a frequent contributor). https://www.chatbots.org/
23. Chatting with LUIS. https://tutorials.botsfloor.com/chatting-with-luis-e7ec94b4e0de

Random Forests for Profiling Computer Network Users

Jakub Nowak[1], Marcin Korytkowski[1,2], Robert Nowicki[1], Rafał Scherer[1,3(✉)], and Agnieszka Siwocha[4,5]

[1] Computer Vision and Data Mining Lab, Institute of Computational Intelligence,
Częstochowa University of Technology,
Al. Armii Krajowej 36, 42-200 Częstochowa, Poland
{jakub.nowak,marcin.korytkowski,robert.nowicki,rafal.scherer}@iisi.pcz.pl
[2] Intigo Sp. z o.o., Łódź, Poland
[3] Passus S.A., Bzowa 21, 02-708 Warsaw, Poland
[4] Information Technology Institute, University of Social Sciences,
90-113 Łódź, Poland
[5] Clark University, Worcester, MA 01610, USA
http://iisi.pcz.pl

Abstract. In this paper, we present a novel system to detect abnormal behaviour of computer network users based on features of web pages which were requested by a user (e.g. URL address, URL category, the day of week or time when the web page was visited). There are many causes of an abnormal behaviour of network users e.g. a computer can be infected by a virus or a Trojan, a stranger can take control of a computer system, etc. Thus, the proposed system can be a very important security mechanism in networks. The system can be also used to make personal user profiles. We use the bag-of-words model to analyse the text data from firewall logs from 63 users collected over a one and half month period. The 500 GB of the network traffic meta-data allowed to achieve satisfactory classification accuracy.

1 Introduction

Efficient computer network intrusion detection and user profiling are substantial for providing computer system security. Each network user leaves traces, some of them are generated directly by the user, e.g. on social networks, others are closely related to the computer network mechanisms. In companies, offices, but also often in home networks, network traffic-filtering devices are used. Network administrators nowadays have an enormous amount of data related to network traffic at their disposal. One of the ways to ensure security is to block traffic based on the categorization of websites. Edge devices (e.g. firewalls or routers with firewall function) verify requested URLs based on the worldwide URL databases and their category ultimately deciding whether a user can access a given page. An example of such devices and reputation databases can be PaloAlto with the Brightcloud database. In order to increase the security, high-end firewalls, simultaneously to filtering, log all the traffic passing through them, storing it e.g. in relational databases, SYSLOG systems, etc. Thus, network operators are able to verify the actions of individual users. Log analysis is a key element of the network security diagnosis. Usually, the log content is analysed after the fact of an attack or a possible error. Registration of logs is also one of the basic requirements of the right to conduct telecommunications activities. It is mandatory for Internet providers to record who and when visited or shared network resources. Depending on the authentication methods used in a given network and the class of security devices, logs contain information from a very general level, e.g. user IP address, time of the event (of page visit), address of the requested page up to the user's name.

In [5] the authors rely on the classification of users with all data stored in network logs. The aim was to identify users for the purposes of forensic applications. An interesting argument about why to identify users using data from network traffic and not using the IP addresses assigned to them is that people use mobile devices more often and identify less and less with one, single network. They do not limit the data, as in our case, to the URLs themselves. They use the meta-data of the traffic, however, that base only includes 46 users. The disadvantage of the system that uses all the data can be its performance. Using only URLs we have fewer data to process, which contributes to the greater efficiency of our system. Events can be also detected by distributed MapReduce approaches [12]. The authors of [1] used a logger on each computer, which additionally logged applications, mouse movements, how were the keys pressed. The error was only 7.1% for 21 users. However, the disadvantage of the method is the interference in the user's system and continuous logging of its behaviour in the system. We expect that artificial neural networks are able to improve the results on the problem presented in the paper [2,7,10]. A promising approach can be using space-time features [4]. As we faced the problem of processing a large amount of data it would be beneficial in the future to utilise some big data processing methods [8].

2   Network Data

This article is based on data collected from a WAN network infrastructure, which is used by residents of four districts in Poland, as well as network users who are employees of the local government offices and their organizational units, e.g. schools, hospitals, etc. Internet access to the analysed network is done with the help of two CISCO ASR edge routers that route packets using RIP version 2. A cluster of PaloAlto devices working in an active-active mode takes care of the network security. The network is routed by the Open Shortest Path First (OSPF) algorithm with virtual routing and forwarding (VRF). In each of the four districts, there is one CISCO core switch. The network is shown in Fig. 1. In the analysed network infrastructure, users are authenticated using accounts in Active Directory services. A RADIUS service is configured for users using the wireless network. PaloAlto devices integrate with the list of accounts contained

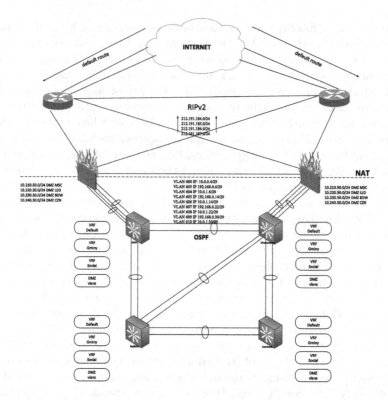

Fig. 1. Schema of the network infrastructure used in the paper to collect traffic data. The Internet is accessed by two routers and the local network is protected by firewalls.

in domain controllers, thanks to which each user's network traffic is logged using its Active Directory name (Fig. 2). The data was acquired in the form of logs from the MS SQL database, then sorted by user names, by the dates when they were recorded, and by the name of the URL address. Sorting by date is designed

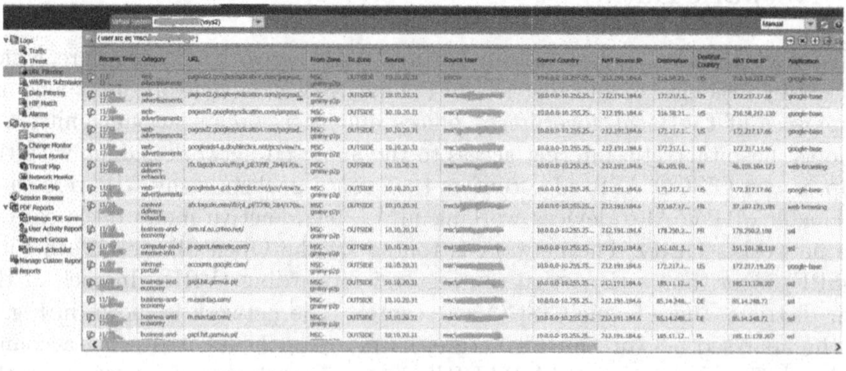

Fig. 2. Part of the network log used to create the training data.

to reflect the order of websites that were visited. We do not operate on exact dates in this case. It is important to note that additional sorting by name is important. It happened that between two identical URLs registered at exactly the same time a foreign address was requested by another program of that user. If the two URLs next to each other were the same, only one was left. This treatment had a big influence on the results and was able to improve the results.

3 Experiments

A simple and popular technique used for text representation in natural-language processing (NLP) is the bag-of-words (BOW) model [9]. It represents words as a numeric vector of features. Thanks to this relatively simple method we can represent text in machine learning. In this representation method, we must create a dictionary based on the collected data and assign an identifier to each word. For this experiment, we created a dictionary with a size of 42,795 for 11 users and 304,946 for 63 users. Then, we count how many times each word from the sequence occurred (in the sentence). A vector is then created that will represent the web pages visited by the user in one session. For URLs, special characters such as [. ,: "" /] are treated as a space between words, so one URL can consist of a few words from the dictionary. In the case of techniques associated with BoW, sorting according to the order of recorded addresses has no effect. The addresses have been grouped into sessions, each of them has no more than 300 addresses and no less than 10. The main criterion for the URL grouping into sessions was time. If there were more than 30 min between the two recorded logs, the URLs belonged to separate sessions. The BoW model despite its simplicity has many disadvantages. In our experiments, it turned out that the BoW model representation alone did not always work perfectly, as each word has the same effect on the classification algorithm. To improve the result, we used the Term Frequency – Inverse Document Frequency (TF-IDF) technique [11]. It is a way to calculate the weight of words, in other words, the effect a word has on the whole text meaning. The weight of each word is calculated based on the number of occurrences in the training text. Using this technique, we to the set of TF-IDF weight values. If we do not add threshold restrictions then the dictionary will grow to a very large size which will have a negative impact on both the results of classification algorithms and the computational demands. Figure 3a presents the word distribution without using threshold values. The number of words has been sorted from the rarest to the most common. In this case, the dictionary had over 300,000 words, most of which were used only a few times. Analysing the graph further, it can be noticed that during the creation of the dictionary we obtained extremely high values for a few words. In our experiments, it is not desirable that a word occurs too many times. In this case, there were words that appeared 75,000 times. During the experiments, the best results were obtained with TF-IDF values ranging from 0.01 to 0.6. The distribution of words in this case is shown in Fig. 3b. This graph shows that the distribution of words is less steep than in the first case (Fig. 3a). The dictionary

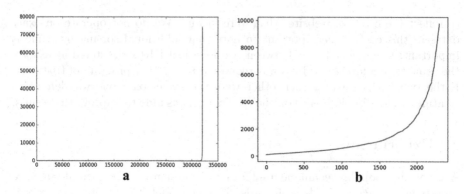

Fig. 3. Word distribution in URL names (a) and after the tf-idf algorithm (b).

in this case, counted 24,001 words. This technique limits words that occur too rarely and too often, that is, those that may have too much or too little impact on the operation of algorithms. In this case, the most often counted word appeared 10,000 times. To classify data we used the random forest classifier [3,6]. Our experiments were mostly implemented with the Scikit-learn (http://scikit-learn.org/) and the NumPy (http://www.numpy.org/) Python packages. The random forest classifier achieved 71.1% accuracy and was trained in 26 s on an Intel i5 machine.

4   Conclusion

In this article, we suggested a method to identify users based on URLs they have visited. It is a relatively novel approach to the subject of security through the observation of how users behave in the network or which websites they visit. Using the method presented in the article, we are able to identify only a part of the events that we can deduce from detailed network logs. However, the ratio of results to the time of the calculations is very promising. It is also important to collect enough data for both the training and the testing. In our case, we collected data for 63 users, however, some of them demonstrated too little activity during the testing period, which negatively affected the results. In our case, testing data was in a different period than training, which is more likely to be used in practice. In the future, but we can modify it on the data available before the action. The system may be also extended with solutions based on artificial neural networks, which should improve the classification results.

Acknowledgments. The research presented in this paper was performed within a project number RPLD.01.02.02-10-0108/17, financed by the Regional Operational Programme for Łódzkie Voivodeship 2014–2020.

References

1. Aupy, A., Clarke, N.: User authentication by service utilisation profiling. In: Proceedings of the ISOneWorld 2005, Las Vegas, USA (2005)
2. Bologna, G., Hayashi, Y.: Characterization of symbolic rules embedded in deep dimlp networks: a challenge to transparency of deep learning. J. Artif. Intell. Soft Comput. Res. **7**(4), 265–286 (2017)
3. Breiman, L.: Random forests. Mach. Learn. **45**(1), 5–32 (2001)
4. Chang, O., Constante, P., Gordon, A., Singana, M.: A novel deep neural network that uses space-time features for tracking and recognizing a moving object. J. Artif. Intell. Soft Comput. Res. **7**(2), 125–136 (2017)
5. Clarke, N., Li, F., Furnell, S.: A novel privacy preserving user identification approach for network traffic. Comput. Secur. **70**, 335–350 (2017)
6. Jordanov, I., Petrov, N., Petrozziello, A.: Classifiers accuracy improvement based on missing data imputation. J. Artif. Intell. Soft Comput. Res. **8**(1), 31–48 (2018)
7. Ke, Y., Hagiwara, M.: An English neural network that learns texts, finds hidden knowledge, and answers questions. J. Artif. Intell. Soft Comput. Res. **7**(4), 229–242 (2017)
8. Marszalek, Z., Wozniak, M., Borowik, G., Wazirali, R., Napoli, C., Pappalardo, G., Tramontana, E.: Benchmark tests on improved merge for big data processing. In: 2015 Asia-Pacific Conference on Computer Aided System Engineering, pp. 96–101, July 2015
9. McTear, M., Callejas, Z., Griol, D.: The Conversational Interface. Springer, Cham (2016). https://doi.org/10.1007/978-3-319-32967-3
10. Minemoto, T., Isokawa, T., Nishimura, H., Matsui, N.: Pseudo-orthogonalization of memory patterns for complex-valued and quaternionic associative memories. J. Artif. Intell. Soft Comput. Res. **7**(4), 257–264 (2017)
11. Salton, G., Buckley, C.: Term-weighting approaches in automatic text retrieval. Inf. Process. Manag. **24**(5), 513–523 (1988)
12. Yan, P.: Mapreduce and semantics enabled event detection using social media. J. Artif. Intell. Soft Comput. Res. **7**(3), 201–213 (2017)

Leader-Follower Formation for UAV Robot Swarm Based on Fuzzy Logic Theory

Wilson O. Quesada[1], Jonathan I. Rodriguez[1], Juan C. Murillo[1],
Gustavo A. Cardona[2], David Yanguas-Rojas[2], Luis G. Jaimes[3],
and Juan M. Calderón[1,4(✉)]

[1] Department of Electronic Engineering, Universidad Santo Tomás,
Bogota, Colombia
{wilsonquesada,jonathan.rodriguez,juanmurillog,
juancalderon}@usantotomas.edu.co
[2] Department of Electrical and Electronics Engineering,
Universidad Nacional de Colombia, Bogota, Colombia
{gacardonac,dryanguasr}@unal.edu.co
[3] Florida Polytechnic University, Lakeland, FL, USA
ljaimes@floridapoly.edu
[4] Bethun-Cookman University, Daytona Beach, FL, USA
calderonj@cookman.edu

Abstract. This paper proposes an algorithm based on a fuzzy logic approach, capable to guide a robot swarm with the aim to keep a leader-follower formation without colliding with other swarm agents. The fuzzy system is programmed and evaluated originally in Matlab, where several experiments were performed. The results depicted a robot swarm showing some bio-inspired behaviors, such as swarm agents surrounding the leader when it is in a static position or when it is traveling from one place to another place. Finally, the proposed fuzzy system was implemented on a drone swarm using V-Rep. The drones simulation shows the swarm navigating together and keeping the leader in the center of the swarm when it is static and following the leader when it is moving. These results could be evaluated in a future work using drone robot swarm in real environments.

Keywords: Swarm robotics · Autonomous mobile robots
Fuzzy logic theory · Drone swarm

1 Introduction

The use of robots' teams for the realization of multiple tasks has become more extended in the recent years, since it is a more reliable system because, in the possibility that one or more robots get to damage the development of the task can take more time due that reduces the number of robots working but the

© Springer International Publishing AG, part of Springer Nature 2018
L. Rutkowski et al. (Eds.): ICAISC 2018, LNAI 10842, pp. 740–751, 2018.
https://doi.org/10.1007/978-3-319-91262-2_65

task will be satisfactory contrary to what happens in the systems of a single sophisticated and expensive robot as it was decades before. Obviously, the arrival on the market of cheaper and improved robotic platforms such as quadcopters opened the possibilities of the implementation of this type of systems at present how is shown in [1], thus improving the performance of the robots in the tasks assigned to them optimizing the time.

There are many applications where multi-robot systems become important because they can perform well in multiple tasks, for example in [2] they show how to make localization of the robots using relative observation between them, this is one of the main challenges to work with multi-robot systems, it is necessary to know all time about the self-localization and the distance with the neighbors. Another approach has been bioinspiration of animals such as ants or the case of swarm bees presented in [3,4], since it may be the case to have a swarm of robots working on the same goal, avoiding collisions and seeking for points of interest how is shown in [5]. Similarly in [6] explore the idea of the implementation of heterogeneous robot network for search and rescue operations. On the other hand, the use of robots' teams have been highly researched and implemented in the case of decentralized navigation and in exploration and mapping tasks, these ideas have been exploited in [7,8] respectively.

The formation control problem has been widely studied in the recent years as a useful tool for the control of multiple homogeneous and heterogeneous robots during the execution of multiple missions such as autonomous formation flight presented in [9]. They consider autonomous aircrafts that can present communication and transmission failure among them and the needed of redundancy to accommodate it. In the same way, other approach was developed in [10] where they used leader and follower formation control on under-actuated autonomous underwater vehicles, it is worth to mention that underwater vehicles have communication range problems taken in to account in this work.

In this work, we present a fuzzy logic based approach oriented to the leader-follower formation control of a swarm of homogeneous robots that allows the swarm to avoid collision between robots and keep the formation based on the leader movements. The other sections that complete this paper are organized as follows: Sect. 3 Swarm Navigation, Sect. 4 Experiments and Results, Sect. 5 Conclusions.

2   Related Work

In the recent years, there have been multiple approaches to the formation control problem from different perspectives. Some of the most representatives are the behavior-based, the artificial potential based and the leader-follower based among others. The behavior-based formation control strategies impose multiple high-level instructions or desired behaviors on the robot's team that allow them to fulfill the mission requirements. In particular, in the work presented in [11] there are presented behavior-based strategies that allow a robot team to navigate through environments with multiple obstacles preserving some basic geometric formations implementing the control signals based on 4 high-level behaviors: avoid static obstacles, avoid robots, maintain formation and move to goal. With those considerations, a robot team is capable of navigating through environments with obstacles without colliding restoring the formation once the obstacle has been sorted.

Alternatively, the artificial potential formation control based strategies employ certain functions called artificial or virtual potentials that in a similar way as the effect of an electromagnetic field or a spring, cause that the robots move according to the desired behaviors. In particular, a repulsive artificial potential may be employed for the robots to avoid collisions. For instance, in the work presented by [12] there is artificial potential based control laws that allow a robots team to maintain formations employing only local information. This distributed approach minimizes the individual artificial potential function of each robot and guarantees the connectivity maintenance and the rigidity of the formation.

On the other hand, regarding the leader-follower formation control based strategies, those employ two types of agents, the leaders who are in charge of determining the formation parameters (position, scale, and rotation) and the followers that maintaining certain distances to their neighbors and to the leaders achieve the desired behaviors. It is worth to mention that there are approaches that do not consider the leader agent as one of the actual robots, but a virtual agent that is tracked by all the roots of the team. For instance, in the work presented in [13] there is presented a leader-follower formation control algorithm where all the robots on the team define a virtual leader that is only a virtual artifact of the optimization process that allows each of the robots to move referenced to it achieving the desired behaviors.

Another example of the leader-follower approach is presented in [14] where it is introduced the idea of combining the Complex-Laplacian matrix with the mutator replicator dynamics. In this work, there are two agents that are the leaders and the rest of them are considered followers. The formation is rigidly defined by the position of the two robot leaders and the position in the graph of each one of the robots is decided through an agreement protocol that employs the mutator-replicator dynamics. Once the positions are decided, the Complex-Laplacian matrix is employed to achieve the appropriate position of each robot on the formation completing the task.

Although most of the formation control algorithms are based on traditional logic principles, there are also approaches that employ the fuzzy logic ideas in order to obtain flexible behaviors with uncertainty and/or with different desired operation regions. In [15] it is presented an exploration and monitoring scheme, where employing fuzzy operators with multiple inputs and outputs, it is possible to control a team of robots to maintain a formation while they are avoiding collisions. In this work, in particular, the robot's model has an important consideration that is that their sensing range does not cover the whole 360 degrees around the robot but only 240 degrees increasing the complexity of the analysis.

3   Swarm Navigation

The principal challenge of the present work is to propose an algorithm to guide a swarm of robots. The robot swarm has to follow a leading robot without colliding with the other robots, but not too far from the nearest robot. To accomplish this requirement, it is proposed a system based on fuzzy logic and this algorithm is simulated using MATLAB and V-Rep for drone robot simulations.

3.1   Initial Variables

Three input variables are taken into account: the number of robots without including the leading robot (Nr), optimal distance to the nearest robot (Lr) and optimal distance to the leader (Ll). The simulation tests were performed with different values for each of input variables.

3.2   Fuzzy System

A fuzzy system is proposed with the aim to accomplish the requirements of the challenge. The robots have to navigate following a leader and at the same time, it has to avoid colliding with other robots of the same swarm. The proposed system has two fuzzy subsystems using the Mamdani approach. The complete system has two outputs, one that determines the velocity towards the closest robot and the second output determines the velocity towards the leader. Also, each fuzzy system has two inputs: the first one is the distance from the robot to the leader and the second one is the distance from the robot to the nearest robot, so two fuzzy subsystems were defined for output 1, output 2 respectively. Figure 1 shows the schema of the complete system, including a vectorial adder of the outputs for determining the final velocity and direction of the reference robot.

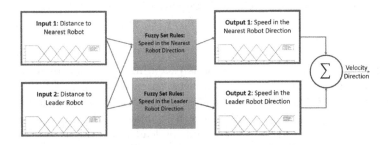

Fig. 1. Outputs of the fuzzy system

Inputs: As was explained previously, the fuzzy system has two inputs, first one depends on the distance to the nearest robot and the second one is the distance to the Leader. Both inputs are composed of five fuzzy set defined as Very Close, Close, Optimal, Far, and Very Far. These fuzzy sets are defined by triangular fuzzy function for the three central sets and two trapezoidal functions for Very Close and Very Far sets as shown in Fig. 2. The universe of discourse for inputs is defined by the optimal distance to the leader robot (Ll) and the nearest robot (Lr). For these, the optimal value is located in 1, the closest distance is zero, and farther distance is 2.

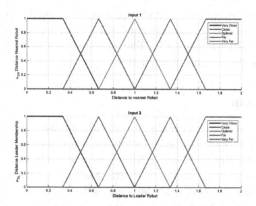

Fig. 2. Inputs of the fuzzy system

Outputs: The fuzzy system has two outputs to determine the displacement velocity toward the nearest robot and the leader respectively. Both outputs have five fuzzy set defined as Moving Far Faster (MFF), Moving Far (MF), Stop (S), Moving Close (MC), and Moving Close Faster (MCF). These set are represented by trapezoidal functions except the "Stop" fuzzy set who uses a triangular function as shown by Fig. 3.

Distance Calculation: The robot that is executing the action of movement is taken as reference robot. All distances between each swarm robot and the reference robot are calculated. The smallest distance is taken as distance to the nearest robot (Lr). This distance is the first input of the fuzzy system. The distance between the reference robot and the lead robot (Ll) is also calculated since this is the second input of the fuzzy system. The aforementioned distances calculation is performed for each of the robots that make up the swarm. The calculation of the distance is made using the formula 1 of distance in the Cartesian plane.

$$Ll, Lr = \sqrt{(x_2 - x_1)^2 + (y_2 - y_1)^2} \tag{1}$$

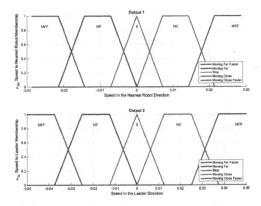

Fig. 3. Outputs of the fuzzy system (Color figure online)

Fuzzy system rules: Given the system has two outputs, there are two sets of decision rules. The rules of each set calculate the minimum product between the membership of two sets, each set belongs to a different system input. The output is calculated by the rule belongs to one of the sets that belong to the system output. The general form of the rule is shown in the Eq. 2

$$\textbf{If}(Input(1) \text{ is } mf(i)) \textbf{ and } (Input(2) \text{ is } mf(j)) \textbf{ then } Output \text{ is } mf(k) \quad (2)$$

where i and j are the subindex for the membership function of the input 1 and 2 respectively, and k is the subindex of the fuzzy set output that is being evaluated.

The fuzzy set rules for output 1 and 2 are depicted in Tables 1 and 2 respectively.

Table 1. Set of Rules: Speed in the nearest robot direction

		Distance to nearest robot				
		Very Close	Close	Optimal	Far	Very Far
Distance to Leader	Very Close	MFF	MF	S	MC	MCF
	Close	MFF	S	S	S	MCF
	Optimal	MFF	S	S	S	MCF
	Far	MFF	S	S	S	MCF
	Very Far	MFF	MF	S	MC	MCF

Finally, to generate the output data, the defuzzification process is required. This method calculates a single scalar quantity from each fuzzy output set. This work used the bisector method for defuzzification process.

Table 2. Set of Rules: Speed in the leader direction

		Distance to nearest robot				
		Very Close	Close	Optimal	Far	Very Far
Distance to Leader	Very Close	MFF	MFF	MFF	MFF	MFF
	Close	MF	S	S	MF	MF
	Optimal	S	S	S	S	S
	Far	MC	MC	MC	MC	MC
	Very Far	MCF	MCF	MCF	MCF	MCF

Final calculation of the speed: The fuzzy system output was described in the previous section. It represents the magnitude of the movement towards the nearest robot and towards the leader since they are two fuzzy systems for two outputs. If the magnitude is negative, it means that the robot must move away from either the nearest robot or the leader respectively. The two outputs are vectorial added, taking into account the angle with respect to the leader and the nearby robot, to generate the vector of movement that points to the direction in which the robot must move. Finally, the new position of the robots is updated according to the vectorial addition of the fuzzy system outputs.

4   Experiments and Results

Several experiments were performed with the aim to test the proposed navigation system based on fuzzy logic approach.

4.1   Initial Robot Position

Since the robots are not initially organized in a specific formation. The robotic agents were initially located randomly within the navigation space. The algorithm does not depend on the initial position of the robots and instead try to organize robots through the navigation process.

4.2   Experiments

Five experiments were performed, where the three first show the behavior of the simulated swarm using MATLAB. These three experiments try to show how the robots are grouped around the leader and how the robots follow the leader when it moves to a different place. In these experiments, the leader robot is a red star and swarm robots are depicted by blue circles. The last two experiments were performed using VRep (Robotics simulation software). These two experiments depicted the self-organization of drones around the leader using the proposed algorithm, and how the drone swarm navigates following the leader when it is moving.

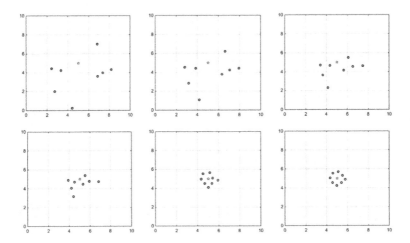

Fig. 4. Experiment 1 (Color figure online)

Experiment 1: This experiment was performed using 8 robots. These robots where localized randomly in a universe of 10X10 units of area. Figure 1 shows the sequence about how robots navigate from the original random position to the nearest point close to the leader. At the end of Fig. 1 is shown how the robot swarm is organized surrounding the leader.

Experiment 2: This experiment uses 40 robots initially localized randomly. The complete sequence shows how the swarm navigates through the field approaching the leader. Once robots have arrived at the nearest point to the leader, it shows how the swarm is organized in layers around the leader and keeping the distance among the robot swarm partners.

Experiment 3: This experiment shows the swarm following the leader robot. The experiment was performed using 15 agents and they were localized randomly at the beginning of the simulation. The sequence shows how the robots navigate close the leader and when the leader starts to displace to another position all the robots navigate following him and keeping the formation established initially. The trace of every robot is shown in green lines in Fig. 3. As was explained initially the leader robot id red and the swarm robots are blue.

Experiment 4: This experiment tests the proposed algorithm using VRep Education Edition. This software allows the simulation of real robots as Drones, humanoids, and several mobile robots. Figure 4 shows the drone swarm navigating towards the drone leader. The initial position of every drone was assigned randomly. At the end of the sequence is shown the final position of the drone swarm, it is showing how the drones are surrounding the leader and keeping the distance to the leader and the distance among other drones.

Experiment 5: This is the last experiment and it uses 15 drones to evaluate the navigation of the drone swarm following the leader. The initial position of every drone was defined randomly as previous experiments. Figure 5 shows the navigation sequence where the drones approach the leader and follow him

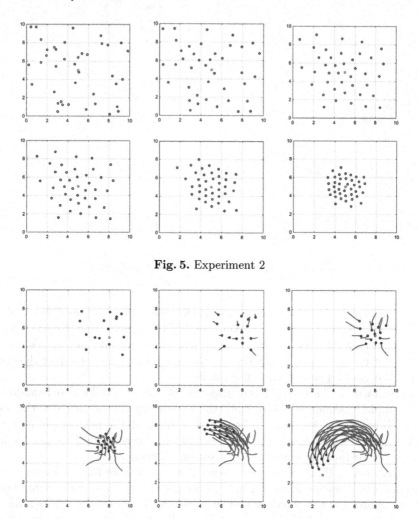

Fig. 5. Experiment 2

Fig. 6. Experiment 3

through the free navigation, just following him. The simulation shows the swarm keeping the formation and the distance among all the swarm robots and avoiding collide with other members of the swarm while they are navigating following the leader (Fig. 6).

5   Conclusions

The present work shows the use of fuzzy logic theory for the generation of leader-follower behavior in a robot swarm (Fig. 7).

The used fuzzy system is based on the Mamdani model. This system is in charge of calculate the velocity and movement direction of every robot in the

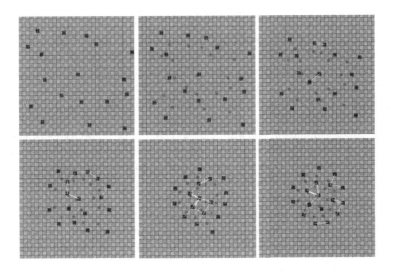

Fig. 7. Experiment 4

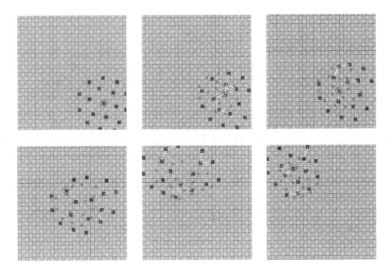

Fig. 8. Experiment 5

swarm. The fuzzy system uses as input two different distance approaches. The first one is the distance between evaluated agent and each other swarm agent.; the second one is the distance between evaluated agent and the leader. Using these two distance approaches, the fuzzy system determines the navigation direction of the evaluated agent (Fig. 8).

The robot swarm presents two very interesting behaviors. The first behavior appears when the leader is static, the agents are grouped around the leader, leaving the leader in the center of the swarm. The second behavior occurs when the leader moves from one place to another, in this behavior the swarm follows the leader towards the new location. Once the leader arrives at the new location, all the swarm agents localize themselves around the leader.

In both behaviors, the swarm agents move according to the leader's location and at the same time, they avoid colliding with the other agents of the swarm. These behaviors were observed through several simulations carried out in Matlab. In addition, the same fuzzy system was programmed in a swarm of 17 drones using the robot simulation software called VRep. The simulations of the drone swarm depicted same behaviors that were observed in the Matlab simulations. In conclusion, this work shows how it is possible to generate navigation systems for swarm robots based on fuzzy logic and at the same time get behaviors from systems inspired by animal swarms such as bees.

Future work is based on the implementation and evaluation of the proposed system on real robot platforms and studies of some navigation features such as generated trajectories, traveled area, and efficiency in the evasion of obstacles.

References

1. Floreano, D., Wood, R.J.: Science, technology and the future of small autonomous drones. Nature **521**(7553), 460–466 (2015)
2. Martinelli, A., Pont, F., Siegwart, R.: Multi-robot localization using relative observations. In: Proceedings of the 2005 IEEE International Conference on Robotics and Automation 2005, ICRA 2005, pp. 2797–2802. IEEE, April 2005
3. Quijano, N., Passino, K.M.: Honey bee social foraging algorithms for resource allocation, part I: algorithm and theory. In: American Control Conference 2007, ACC 2007, pp. 3383–3388. IEEE, July 2007
4. Quijano, N., Passino, K.M.: Honey bee social foraging algorithms for resource allocation, part II: application. In: American Control Conference 2007, ACC 2007, pp. 3389–3394. IEEE, July 2007
5. León, J., Cardona, G.A., Botello, A., Calderón, J.M.: Robot swarms theory applicable to seek and rescue operation. In: Madureira, A.M., Abraham, A., Gamboa, D., Novais, P. (eds.) ISDA 2016. AISC, vol. 557, pp. 1061–1070. Springer, Cham (2017). https://doi.org/10.1007/978-3-319-53480-0_104
6. Yanguas-Rojas, D., Cardona, G.A., Ramirez-Rugeles, J., Mojica-Nava, E.: Victims search, identification, and evacuation with heterogeneous robot networks for search and rescue. In: 2017 IEEE 3rd Colombian Conference on Automatic Control (CCAC), Cartagena, pp. 1–6 (2017). https://doi.org/10.1109/CCAC.2017.8276486
7. Tanner, H.G., Kumar, A.: Towards decentralization of multi-robot navigation functions. In: Proceedings of the 2005 IEEE International Conference on Robotics and Automation 2005, ICRA 2005, pp. 4132–4137. IEEE, April 2005
8. Simmons, R., Apfelbaum, D., Burgard, W., Fox, D., Moors, M., Thrun, S., Younes, H.: Coordination for multi-robot exploration and mapping. In: AAAI/IAAI, pp. 852–858, July 2000
9. Giulietti, F., Pollini, L., Innocenti, M.: Autonomous formation flight. IEEE Control Syst. **20**(6), 34–44 (2000)
10. Cui, R., Ge, S.S., How, B.V.E., Choo, Y.S.: Leaderfollower formation control of underactuated autonomous underwater vehicles. Ocean Eng. **37**(17), 1491–1502 (2010)

11. Balch, T., Arkin, R.C.: Behavior-based formation control for multirobot teams. IEEE Trans. Robot. Autom. **14**(6), 926–939 (1998)
12. Drfler, F., Francis, B.: Formation control of autonomous robots based on cooperative behavior. In: 2009 European Control Conference (ECC), pp. 2432–2437. IEEE, August 2009
13. Droge, G.: Distributed virtual leader moving formation control using behavior-based MPC. In: American Control Conference (ACC), pp. 2323–2328. IEEE, July 2015
14. Guzman-Hernandez, S., Mojica-Nava, E.: Formation control using replicatormutator dynamics for multiple mobile autonomous agents. Neurocomputing **172**, 337–344 (2016)
15. Yasuda, Y., Kubota, N.: Formation behavior of multi-robot for exploration and monitoring. In: 2012 Joint 6th International Conference on Soft Computing and Intelligent Systems (SCIS) and 13th International Symposium on Advanced Intelligent Systems (ISIS), pp. 1232–1237. IEEE, November 2012

Towards Interpretability of the Movie Recommender Based on a Neuro-Fuzzy Approach

Tomasz Rutkowski[1,2]([✉]), Jakub Romanowski[1], Piotr Woldan[1],
Paweł Staszewski[1], and Radosław Nielek[2]

[1] Senfino Technologies, Czestochowa, Poland
tomasz.rutkowski@senfino.com
[2] Polish-Japanese Academy of Information Technology, Warsaw, Poland

Abstract. In the paper, a neuro-fuzzy structure is implemented as a movie recommender. First, a novel method for transforming nominal values of attributes into a numerical form is proposed. This allows representing the nominal values, e.g. movie genres or actors, in a neuro-fuzzy system designed from scratch using the Mendel-Wang algorithm for rules generation. Several experiments illustrate performance of the neuro-fuzzy recommender.

Keywords: Recommendation systems · Neuro-fuzzy systems
MovieLens

1 Introduction

Recommender systems are very useful to suggest various services and products, e.g. movies, books, flights, financial investments or medical doctors. Such systems work based on information about recommended products (items) and/or information about users profiles. The most popular technique in recommendation systems is collaborative filtering (CF), see e.g. [1–5]. The recommender systems based on CF suggest the recommendation for a given user by checking the preferences of other users with respect to a degree of similarity, usually calculated using the Pearson or cosine similarity measures. Various soft computing techniques, see e.g. [6–10], can also be applied to design the CF recommenders. The main problem of the CF approach is a cold start, see e.g. [11,12], and sparsity of the rating matrix, see e.g. [13,14]. In contrast to the CF recommender systems, another approach, called content-based filtering (see e.g. [15–20]) ignores information about preferences of other users. The recommender systems of this type suggest the items characterized by similar features to those ones that the user preferred in the past. As it was mentioned in the literature, various methods have been proposed for designing recommender systems; however, the problem of interpretability of such system was very rarely studied.

© Springer International Publishing AG, part of Springer Nature 2018
L. Rutkowski et al. (Eds.): ICAISC 2018, LNAI 10842, pp. 752–762, 2018.
https://doi.org/10.1007/978-3-319-91262-2_66

In this paper, a neuro-fuzzy structure is implemented as a movie recommender. First, a novel method for transforming nominal values into a numerical form is proposed. This allows representing nominal values, e.g. movie genres or actors, in a neuro-fuzzy system designed from scratch using the Mendel-Wang algorithm for rules generation. We also show several experiments illustrating the performance of the neuro-fuzzy recommender, and discuss the issue of interpretability.

2   Neuro-Fuzzy Recommender System

Recommendation systems require both numerical and nominal values to be considered. It allows achieving much higher efficiency in the recommendation by understanding the context and importance of components describing the situation concerning a person interested in the recommendation. Therefore, we develop a new method for converting nominal values to numerical ones.

For the purposes of our simulations, we have developed a method of converting nominal attributes to numerical form, taking into account the number of occurrences in samples assessed by the user. In the MovieLens database, users ranked their movies on a 1–5 rating scale. We divided all samples rated by one user into two groups: negative for ratings {1; 1.5; 2; 2.5; 3} and positive for ratings {3.5; 4; 4.5; 5}. We determined a smaller range of ratings for positive samples because users are used to movie ranks, which are more positive for them. It is much more difficult to obtain negative movies ranks from users, so this causes a difference in the positive and negative values ranges. Rating range for negative samples is bigger because it allows balancing the number of samples to be learned for both classes. For nominal attributes, we need to determine which item position of the attribute value will be analyzed and what will be the significance of their position (eg. when movie genre is made up of several genres Horror-Drama-Thriller). We analyze the first three values, and their importance is specified in the Table 1.

Table 1. Initialized values of importance relative to position

Attribute	Values of importance on position		
	p_1	p_2	p_3
Genres	1,0	0,6	0,3
Actors	1,0	0,5	0,2
Directors	1,0	0,7	-
Producers	1,0	0,8	0,6
Spoken language	1,0	-	-

In the next step, occurrences of individual values of each attribute relative to the their position should be counted, separately for negative and positive samples as is shown in Table 2 for User 1 and User 2. More detailed sample with values of calculations based on the genre of the movie and on the ratings of one of the users are presented in the Table 3, and the calculation method is shown below:

$$x_{i,(n;p)} = \sum_{j=1}^{n} p_j * c_j \qquad (1)$$

Formula (1) is the initial step of calculations, and it gives the possibility to obtain x_i which is unnormalized form of weight for attribute values. x_i is calculated separately for negative and positive samples, and index "i" represents next attribute values and "j" is index of importance position (in this example from 1 to 3). Value p_j is a value if importance from Table 1. c_j is the occurrence of the attribute value on specific position. The values for c_j are shown in Table 2.

The next step is to calculate normalized weights w_i for positive and negative samples by using Formula (2) where x_i is calculated by Formula (1), k_i represents all occurrences of attributes values. The k_i values are shown in Table 3 just like values of $COUNT$ and ALL. Index "i" in Formula (2) is a representation of next attribute values. Table 3 is an exemplary representation of movie genres calculations for 3 most significant and 3 less significant attribute values.

$$w_{i,(n;p)} = \frac{x_{i,(n;p)}}{COUNT_i} * \frac{k_{i,(n;p)}}{ALL} \qquad (2)$$

To obtain the final value of attribute value weight W_i, we propose Formula (3). Calculated values are obtained in previous steps, and exemplary values are shown in Table 3. The greater the weight, the better the positive recommendation, and the smaller the more negative for this particular feature.

$$W_i = \frac{1 - \frac{wn_i + wp_i}{MAX}}{2} \qquad (3)$$

By applying the attributes analysis algorithm and calculating the composite attribute values, we can analyze the components of the several attribute's values and then reduce it to a single numerical value using the Formula (4). In this formula index "k" is an attribute number (from 1 to 5). Index "z" is an index of the selected value of the selected attribute "k".

$$X_k = \frac{\sum_{j=1}^{n} p_{k,j} * W_{k,z}}{\sum_{j=1}^{n} p_{k,j}} \qquad (4)$$

As we can convert all values to a numerical form, we can build rules in a neuro-fuzzy system. The simulations presented in this article were generated on the basis of fuzzy rules built by the Wang Mendel method [21] and then tuned by the backpropagation algorithm.

3   Data Set

The main source of data for our data set is the MovieLens 20M database. MovieLens is a publicly available database of movies and users who rated these movies.

As a result, it creates perfect conditions for testing recommendation systems. A significant number of movies, large number of users and ratings are the criteria required for testing. The MovieLens database fulfills these conditions. Movies in this database are described by genres, year, tags, 5-star rating and external services movies IDs.

However, we decided to enlarge the number of relevant information about the movies in order to understand the deeper needs of users and provide the possibility of interpreting of the recommendation. We took advantage of external services movies IDs contained in the MovieLens database and we use IMDB and TMDB external services to obtain more information about each of the movies.

Finally, our Data Set consists of 27278 movies, 20000263 ratings and following movies attributes: Genres, Actors, Directors, Producers, Spoken language and the values of the attributes are presented in Table 2. Attributes Genres and Spoken language are presented in Table 2 with a full list of their values. The other attributes are presented in the form of the three most and the three least significant values for the user due to their huge amount.

4   Testing Environment

The tests were performed based on our AI environment which is AI framework developed in C++ programming language using CUDA (Compute Unified Device Architecture) Technology provided by NVidia. Application of the CUDA allows for the high acceleration of calculations by using GPU compared to traditional calculations on CPU.

All of data about users, movies and ranks are stored in database. From the attributes and their values tables stored in our database, we select only those that we use for the experiment. However, for the purposes of other experiments, we store much more information about attributes.

The summary list of applied technologies looks as follows:

- MSSQL Server 2016
- CUDA Toolkit v9.0.176
- CUDNN v7
- BOOST v1.65.1 (vectors and logging features).

5   Experimental Results

5.1   Users Attributes Calculations

As an example of the weight calculation, we presented a list of attributes and their values for two exemplary users which are given in Table 2. Table 2 shows the number of appearances attribute values at three different positions (c_1, c_2, c_3) for both samples evaluated by the user as a negative and a positive. A detailed set of calculations of weights on the *Genres* of the movies is presented in Table 3 and the meaning of the column headers is as follows: k_n, k_p are values of sum

Table 2. Weights calculation and list of attributes and their values for two exemplary users

Attributes	User 1								User 2	
	Attributes values	Negative			Positive			Weight	Attributes values	Weight
		c_1	c_2	c_3	c_1	c_2	c_3			
Genres	Comedy	16	7	5	42	13	10	0,911		
	Drama	8	18	6	19	32	2	0,723	Drama	0,898
	Crime	0	6	5	8	14	6	0,658	Comedy	0,752
	Romance	0	1	8	3	9	12	0,607	Crime	0,614
	Animation	1	0	0	5	2	0	0,572	Romance	0,560
	Family	1	0	0	2	2	6	0,560	History	0,556
	Adventure	7	6	0	11	6	1	0,557	Adventure	0,554
	Documentary	1	0	0	5	0	0	0,554	War	0,551
	History	0	0	0	1	1	1	0,526	Thriller	0,548
	Sci-Fi	0	1	5	1	3	5	0,524	Mystery	0,538
	Mystery	0	2	1	1	2	3	0,521	Animation	0,537
	Fantasy	3	1	1	6	0	0	0,520	Sci-Fi	0,513
	Thriller	1	7	8	2	6	11	0,518	Fantasy	0,508
	War	0	0	0	0	1	2	0,516	Horror	0,500
	Western	0	0	0	0	0	1	0,504	Western	0,500
	Foreign	0	1	0	0	1	0	0,500	Music	0,493
	Music	1	2	1	1	0	0	0,476	Family	0,450
	Horror	2	0	1	0	0	0	0,469	Action	0,317
	Action	19	4	2	9	7	0	0,388		
Actors	Michael Douglas	0	0	0	5	0	0	1,000	S. Connery	0,983
	Harrison Ford	1	0	0	3	3	0	0,871	J. Nicholson	0,873
	Tom Hanks	1	0	0	4	0	1	0,833	M. Broderick	0,850
	...	...	...	...	...	...	...	...	...	...
	Will Smith	1	1	0	0	0	0	0,350	M. J. Fox	0,150
	Ben Stiller	2	0	0	0	0	0	0,300	E. Murphy	0,150
	John Cusack	4	1	0	0	0	0	0,050	B. Willis	0,127
Directors	Nick Park	0	0	–	5	0	–	1,000	J. Coen	1,000
	Joel Coen	0	0	–	3	0	–	0,800	B. Wilder	1,000
	Jay Roach	0	0	–	3	0	–	0,800	R. Zemeckis	1,000
	...	...	...	...	...	...	...	...	...	...
	Tom DiCillo	1	0	0	0	0	–	0,400	J. McTiernan	0,250
	John Dahl	1	0	0	0	0	–	0,400	R. Harlin	0,250
	Michale Bay	2	0	0	0	0	–	0,300	T. Scott	0,125
Producers	Paramount	2	0	0	13	1	0	0,990	Universal	0,919
	20th Cent. Fox	3	0	0	7	5	0	0,840	United Art.	0,888
	PolyGram	0	0	0	3	2	1	0,715	Columbia	0,743
	...	...	...	...	...	...	...	...	...	...
	J. Bruckheimer	3	0	0	0	0	0	0,376	RKO Radio	0,354
	Regency Enterp	4	0	0	0	0	0	0,334	Touchstone	0,303
	Touchstone	3	2	2	0	0	0	0,260	Walt Disney	0,257
Spoken language	en	57	–	–	110	–	–	0,866	en	0,737
	it	0	–	–	3	–	–	0,521	ja	0,508
	fr	0	–	–	1	–	–	0,507	zh	0,508
	cs	0	–	–	1	–	–	0,507	it	0,508

of all negative and positive occurrence of attribute values on each of positions (c_1, c_2, c_3) as shown in Table 2. *COUNT* column is the sum of k_n and k_p. The columns x_n and x_p are calculated based on the Formula (1). Values of sum_1, sum_2, *ALL*, MAX_1, MAX_2 and *MAX* are considered for full list of *Genres* attribute values, not only for six exemplary values.

Formula (2) is used to calculate partial weights w_i for negative and positives samples. In result, final weight W_i of attribute value can be calculated by using Formula (3).

Table 3. A detailed calculations of weights for the Genres attribute values

Attribute values	k_n	k_p	COUNT	x_n	x_p	w_n	w_p	Weight
Comedy	28	65	93	21,7	52,8	0,01597	0,09023	0,911
Drama	32	53	85	20,6	38,8	0,01896	0,05915	0,723
Crime	11	28	39	5,1	18,2	0,00352	0,03195	0,658
...								
Music	4	1	5	2,5	1,0	0,00489	0,00049	0,476
Horror	3	0	3	2,3	0,0	0,00557	0,00000	0,469
Action	25	16	41	22,0	13,2	0,03248	0,01247	0,388

sum_1	sum_2	ALL
134	275	409

MAX_1	MAX_2	MAX
0,03280	0,09023	0,09023

5.2   Measure of Effectiveness

The real measure of the effectiveness of recommendation systems is the user's feeling of the items recommended for him. In fact, the number of items recommended positively or negatively plays a secondary role. The most valuable thing is to recommend even one item, but one that adds the most value to the user of the recommendation system.

In order to be able to determine how much the recommendation is effective, we decide to perform experiments for two separate users, as shown in Table 4.

Here is a shortcut set in the headers of the table columns Tables 4 and 5 and their meaning is described below. *Ranks count* - is total movies ranks count created by user/users, *Train samples* - is part of *Ranks count* and it is used to train the system. *Test samples* - this is the remaining part *Ranks count* ant its amount is always 50 for each tested user. *UNR* - User Negative Ranks, *NR* - Negative Recommendations, *UPR* - User Positive Ranks, *PR*- Positive Recommendations, *Effectiveness* in meaning of ratio of User ranks to recommendations of the system. *UNR* and *UPR* are values of users ranks from our data set. *NR* and *PR* are values calculated by our recommender system. By describing the ratio *UNR* to *NR* and *UPR* to *PR* we show how many realistic movies ratings are equal to the recommendations of our system. It is worth mentioning here about the set of movies with the uncertainty of recommendation. We can not recommend them as positive or negative samples.

Table 4. Table of recommendation results for two separate users (User 1 and User 2)

	Ranks count	Train samples	Test samples	UNR/NR	UPR/PR	Effectiveness
User 1	226	176	50	3/5	15/17	18/22
User 2	251	201	50	4/8	26/34	30/42

Table 5. Table of recommendation results for groups of users (from 100 to 500)

Users count	Ranks count	Train samples	Test samples	UNR/NR	UPR/PR	Effectiveness
100	21731	16731	5000	532/774	2400/3119	2932/3893
200	45772	35772	10000	972/1381	4998/6449	5970/7830
300	71501	56501	15000	1558/2233	7484/9475	9042/11708
400	95263	75263	20000	2188/3155	9765/12315	11953/15470
500	117179	92179	25000	2954/4249	12128/15218	15082/19467

In the next step we made the same experiment as above but for a larger scale of users (from 100 to 500) as is shown in Table 5.

Tables 6 and 7 show the results of Leave-one-out cross-validation. First of them shows results for User 1 and User 2 and the second shows results for 100 users.

Table 6. Leave-one-out cross-validation for User 1 and User 2

	Rank count	Train samples	Test samples	Effectiveness
User 1	226	225	226	148/226
User 2	251	250	251	172/251

Table 7. Leave-one-out cross-validation for 100 users

Users count	Rank count	Train samples	Test samples	Effectiveness
100	16216	16116	10216	12801/16216

5.3   Interpretability of the Recommendation Result

One of the most important values of the recommendation system is to understand the reasons for the recommendation. Interpretability of the result of the recommendation is therefore key value.

For both users User 1 and User 2 we have prepared graphs of linguistic variables and rules between them. There are two forms of graphs - with all of rules and simplified. Simplified one is prepared to clarify the graph and to allow analyzing the interpretability of recommendation results.

Figures 1 and 2 present linguistic variables with full set of rules for User 1 and User 2. Similarly simplified rules for linguistic values for User 1 and User 2 are shown in Figs. 3 and 4.

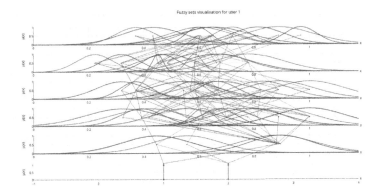

Fig. 1. Linguistic variables and rules for User 1

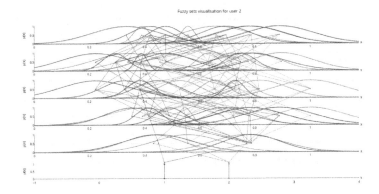

Fig. 2. Linguistic variables and rules for User 2

The order of linguistic variables in the graphs is the same as attributes order in the Table 2 (*Genres, Actors, Directors, Producers, Spoken languages*). The last axis in the graphs corresponds to the output from the system, with two values where value 1 indicating a negative recommendation and value 2 indicating a positive recommendation.

By applying our method of encoding nominal values, it is possible to interpret the recommendation by connecting rules from the systems with the value of the first linguistic variable.

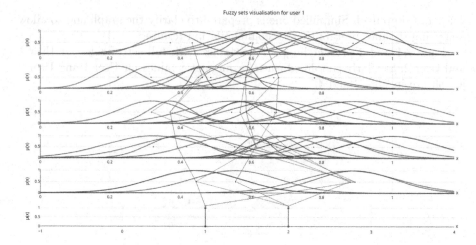

Fig. 3. Linguistic variables and simplified rules for User 1

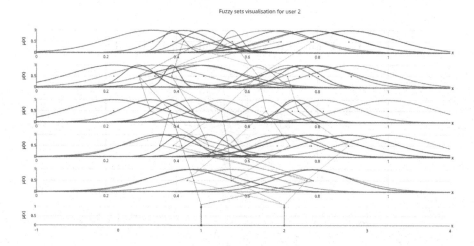

Fig. 4. Linguistic variables and simplified rules for User 2

Based on Fig. 5 and the assumption that the movie consists of the genre Comedy, its output value for User 1 equals 0,911 according to the Formula (4). Analyzing this value with regard to User 1, we see that this value falls almost perfectly into the positive set of the *Comedy* linguistic variable. In order to positively recommend the movie, the rules with the positive conclusion should be activated by input values of other linguistic variables. The rule must contain the set of *Comedy* genre indicated above. By analyzing fuzzy sets along with the rules in this way, we can easily obtain information about the user's preferences with respect to individual attributes and their values. If we distinguish all the values of our attributes with their combinations and extract linguistic variables, we can

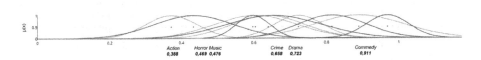

Fig. 5. Example of *Genres* linguistic variable

analyze them based on the fuzzy sets. Values of overlapping fuzzy sets indicate recommendations that are not certain. However, using fuzzy sets allow us to specify a degree of certainty. The rules are formulated by the use of appropriate fuzzy sets presented in Figs. 1, 2, 3, 4 and 5, which allows interpreting the final recommendation by matching the input values with the specific fuzzy sets of those rules.

6   Conclusions

In this paper, we present a way to interpret which attributes mostly influence the movie recommender. Interpretability is the biggest advantage of using a neuro-fuzzy approach. In the future work we plan to apply the rough set theory [22] for rule generation describing the movie recommender. That will allow us to explore different techniques to achieve the goal of building an interpretable and accurate recommender system.

References

1. Nie, F., Wang, H., Huang, H., Ding, C.: Joint schatten lp-norm robust matrix completion for missing value recovery. Knowl. Inf. Syst. **42**(3), 525–544 (2013). https://doi.org/10.1007/s10115-013-0713-z
2. Zhao, K., Pan, L.: A machine learning based trust evaluation framework for online social networks, pp. 69–74 (2015). https://doi.org/10.1109/TrustCom.2014.13
3. Anaissi, M., Goyal, M.: SVM-based association rules for knowledge discovery and classification (2015). https://doi.org/10.1109/APWCCSE.2015.7476236
4. Lu, J., Hoi, S., Wang, J., Zhao, P.: Second order online collaborative filtering. J. Mach. Learn. Res. **29**, 325–340 (2013)
5. Zhao, Q., Zhang, Y., Friedman, D., Tan, F.: E-commerce recommendation with personalized promotion, pp. 219–225 (2015). https://doi.org/10.1145/2792838.2800178
6. Bologna, G., Hayashi, Y.: Characterization of symbolic rules embedded in deep DIMLP networks: a challenge to transparency of deep learning. J. Artif. Intell. Soft Comput. Res. **7**(4), 265–286 (2017)
7. Beg, I., Rashid, T.: Modelling uncertainties in multi-criteria decision making using distance measure and topsis for hesitant fuzzy sets. J. Artif. Intell. Soft Comput. Res. **7**(2), 103–109 (2017)
8. Liu, H., Gegov, A., Cocea, M.: Rule based networks: an efficient and interpretable representation of computational models. J. Artif. Intell. Soft Comput. Res. **7**(2), 111–123 (2017)

9. Riid, A., Preden, J.-S.: Design of fuzzy rule-based classifiers through granulation and consolidation. J. Artif. Intell. Soft Comput. Res. **7**(2), 137–147 (2017)
10. Prasad, M., et al.: A new mechanism for data visualization with TSK-type preprocessed collaborative fuzzy rule based system. J. Artif. Intell. Soft Comput. Res. **7**(1), 33–46 (2017)
11. Wei, J., He, J., Chen, K., Zhou, Y., Tang, Z.: Collaborative filtering and deep learning based recommendation system for cold start items. Expert Syst. Appl. **69**, 29–39 (2017)
12. Park, D.H., Kim, H.K., Choi, I.Y., Kim, J.K.: A literature review and classification of recommender systems research. Expert Syst. Appl. **39**(11), 10059–10072 (2012)
13. Alemeye, F., Getahun, F.: Cloud readiness assessment framework and recommendation system, November 2015. https://doi.org/10.1109/AFRCON.2015.7331995
14. Burke, R.: Hybrid recommender systems: survey and experiments. User Model User-Adap. Interact **12**(4), 331–370 (2002)
15. Baldominos, A., Albacete, E., Saez, Y., Isasi, P.: A scalable machine learning online service for big data real-time analysis (2015). https://doi.org/10.1109/CIBD.2014.7011537
16. Kao, C.-Y., Fahn, C.-S.: A multi-stage learning framework for intelligent system. Expert Syst. Appl. **40**(9), 3378–3388 (2013)
17. Tsuji, K., Yoshikane, F., Sato, S., Itsumura, H.: Book recommendation using machine learning methods based on library loan records and bibliographic information, pp. 76–79 (2014). https://doi.org/10.1109/IIAI-AAI.2014.26
18. Isinkaye, F.O., Folajimi, Y.O., Ojokoh, B.A.: Recommendation systems: principles, methods and evaluation. Egypt. Inform. J. **16**, 261–273 (2015)
19. Portugal, I., Alencar, P., Cowan, D.: The use of machine learning algorithms in recommender systems: a systematic review. Expert Syst. Appl. **97**, 205–227 (2017)
20. Burke, R.: Hybrid web recommender systems. In: Brusilovsky, P., Kobsa, A., Nejdl, W. (eds.) The Adaptive Web. LNCS, vol. 4321, pp. 377–408. Springer, Heidelberg (2007). https://doi.org/10.1007/978-3-540-72079-9_12
21. Wang, L.X., Mendel, J.M.: Generating fuzzy rules by learning from examples. IEEE Trans. Syst. Man Cybern. **22**(6), 1414–1427 (1992)
22. Pawlak, Z.: Rough set theory for intelligent industrial applications. In: Proceedings of the Second International Conference on Intelligent Processing and Manufacturing of Materials, IPMM 1999 (Cat. No.99EX296), vol. 1, pp. 37–44 (1999)

Dual-Heuristic Dynamic Programming in the Three-Wheeled Mobile Transport Robot Control

Marcin Szuster$^{(\boxtimes)}$ (iD)

Rzeszow University of Technology, Rzeszow, Poland
mszuster@prz.edu.pl

Abstract. In this work an intelligent discrete tracking control system of a three-wheeled mobile transport robot is presented. The robot is a model of a forklift truck, with a drive wheel mounted in the rear part of the frame in movable steering module. The dynamics of the mobile transport robot was described using the second order Lagrange's equations. In the tracking control system of the robot the Dual-Heuristic Dynamic Programming algorithm was used, which belongs to the family of Approximate Dynamic Programming algorithms. In the Dual-Heuristic Dynamic Programming algorithm Random Vector Functional Link Neural Networks were used to realize an actor and a critic structure. Numerical tests of robot motion on the desire trajectory were performed. The results of the numerical tests confirmed the correctness of the assumed design assumptions.

Keywords: Actor-critic structure
Approximate dynamic programing
Dual-Heuristic Dynamic Programming · Neural network · Mobile robot

1 Introduction

In recent years, rapid development of mobile robotics has been observed. It is related to the development of many fields of science and industry, eg the development of robots motion control algorithms, artificial intelligence algorithms, microprocessor techniques, drives, energy stores or sensory systems. The combination of knowledge in the field of these areas enables the construction and control of complex mobile robots that perform complex tasks in an autonomous manner. Many examples of mobile robot applications, for example in transport, inspection [4], patrol or exploration of unknown environments, can be found in the literature. The implementation of these tasks is associated with a different degree of autonomy of robots, resulting from the complexity of the control algorithm used. In robot control algorithms, modern artificial intelligence (AI) methods, such as fuzzy logic (FL) algorithms or artificial neural networks (NN) are implemented more and more often [3,9]. The NNs in particular have found a wide application due to their ability to approximate any non-linear waveforms

© Springer International Publishing AG, part of Springer Nature 2018
L. Rutkowski et al. (Eds.): ICAISC 2018, LNAI 10842, pp. 763–776, 2018.
https://doi.org/10.1007/978-3-319-91262-2_67

with assumed accuracy, the ability to learn and adapt [10]. The development of NNs enabled the development of approximate dynamic programming (ADP) algorithms [1, 11–16], derived from the Bellman's theory of optimal control - the dynamic programming (DP) [2]. The application of NNs in ADP algorithms enabled the implementation of this approach in the forward control of dynamic objects, unlike in the Bellman's DP, where the optimal control law is generated backwards from the last step of the process to the first step.

This article discusses the discrete tracking control algorithm of a three-wheeled mobile transport robot (WMTR), in which the ADP algorithm in Dual-Heuristic Dynamic Programming (DHP) configuration was used. This algorithm consists of two structures, an actor generating a suboptimal control law, and a critic evaluating the generated control law by approximating the derivative of the value function with respect to the object's state. In the literature few attempts to implement the DHP algorithm can be found, eg in the task of recognizing targets [7] or controlling dynamic objects such as a turbogenerator [18], a robotic manipulator [5, 15] and a WMR [8, 16]. In the presented approach, the control algorithm with the DHP structure is used to the tracking control of a 3-wheeled mobile robot, with a structure modeled on forklift transport trucks. The aim of performed research is to develop algorithms for controlling the robot's movement in the storage space of high storage with the possibility of transporting large-size facilities and cooperation with automated storage systems [17].

The presented research results are a continuation of the author's earlier work on the application of ADP algorithms in the motion control systems of dynamic objects, such as underactuated systems [6], WMRs [16], or robotic manipulators [5, 15]. The article consists of the following parts: in the second section the construction of the WMTR is presented and the description of the dynamics of the robot is discussed. The third section presents the family of APD algorithms with a detailed discussion of the structure of DHP, used in the tracking control algorithm, described in the section four. The fifth section discusses the results of the numerical test of the tracking movement of the robot, the sixth section summarizes the research work carried out.

2   3-Wheeled Mobile Transport Robot

The control object is the transport WMR. It consists of free wheels 1 and 2 with fixed axes of rotation, which do not change their orientation with respect to the frame, driving wheel 3 with an axis of rotation changing their orientation, fixed to the steering module 4, frame 5 and lift unit 6. The axis of rotation of the drive wheel 3 changes its orientation due to the rotation of the steering module 4 driven by the DC electric motor with the gear. The steering module is mounted in the robot frame and rotates relative to the axis of rotation perpendicular to the running track plane. The driving unit is a wheel 3 driven by an electric motor with a gear. The axis of rotation of the driving wheel 3 lies in the xy plane and its direction results from the angle of rotation of the steering module 4. The lift 6 makes it possible to transport loads. The lift assembly consists of a mast and a

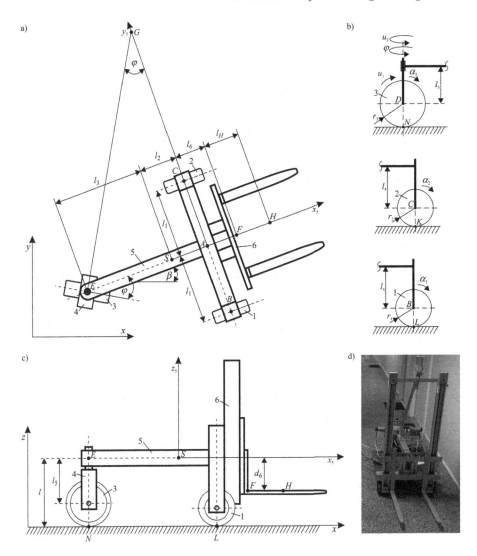

Fig. 1. (a) WMR scheme in the xy plane, (b) diagrams of wheels 1, 2 and 3, (c) WMR scheme in the xz plane, (d) photo of the transport robot

bracket (fork) made of a face plate (carriage) with elements enabling the support of lifting elements. Wheels 1, 2 and 3 are made of plastic discs, in addition wheels 1, 2 are coated with rubber. The WMR scheme is shown in Fig. 1.

The following symbols have been introduced in Fig. 1: α_1, α_2, α_3 - wheel rotation angles 1, 2 and 3 respectively, φ - steering module 4 rotation angle, β - robot frame rotation angle in the assumed fixed coordinate system xy, u_1 - control signal of the drive wheel 1, u_2 - signal controlling the steering module 4 rotation, r_1, r_2, r_3 - radii of the respective wheels, $l_1 - l_6$, l_H, d_6 - appropriate dimensions resulting from the robot's geometry.

WMRs are complex mechanical systems on which nonhomomic constrains were imposed. The WMR motion was described using the second order Lagrange's equations with multipliers, which, after applying the procedure of decoupling multipliers, take the following form

$$M\left(q\right)\ddot{q} + C\left(q,\dot{q}\right)\dot{q} + F\left(q,\dot{q}\right) + \tau_d\left(t\right) = B\left(q\right)u, \tag{1}$$

where $q = [\alpha_3, \varphi]^T$ – vector of generalized coordinates, $\tau_d\left(t\right) = [\tau_{d1}, \tau_{d2}]^T$ – vector of bounded disturbances, $u = [u_1, u_2]^T$ – vector of control signals, $M\left(q\right)$ – the positive defined inertia matrix of the WMR, $C\left(q,\dot{q}\right)\dot{q}$ – the vector of centrifugal and Coriolis momentous, $F\left(q,\dot{q}\right)$ – the vector of resistances to motion, $B\left(q\right)$ – the input matrix.

The continuous model of WMR dynamics (1) was digitized using the Euler method of rectangular approximation in the forward direction. The MRK motion parameters during computer simulation are computed at discrete times $q\left(t_{\{k\}}\right)$, $\dot{q}\left(t_{\{k\}}\right)$, where $t_{\{k\}} = kh$, k – an integer specifying the number of the discrete iteration step, $k = 1, \ldots, N$, N – number of steps, $h = t_{\{k+1\}} - t_{\{k\}}$ – time discretization parameter. Hence $q_{\{k\}} = q\left(t_{\{k\}}\right)$ and $q_{\{k+1\}} = q\left(t_{\{k+1\}}\right)$ – values of the state vector in moments $t_{\{k\}}$ and $t_{\{k+1\}}$, i.e. in steps k and $k + 1$. The discrete model of the WMR dynamics was assumed in the form

$$
\begin{aligned}
z_{1\{k+1\}} &= z_{1\{k\}} + h z_{2\{k\}}, \\
z_{2\{k+1\}} &= z_{2\{k\}} - h M^{-1}\left(z_{1\{k\}}\right)\left[C\left(z_{1\{k\}}, z_{2\{k\}}\right) z_{2\{k\}} + F\left(z_{1\{k\}}, z_{2\{k\}}\right)\right] + \\
&\quad - h M^{-1}\left(z_{1\{k\}}\right)\left[\tau_{d\{k\}} - B\left(z_{1\{k\}}\right) u_{\{k\}}\right],
\end{aligned}
\tag{2}
$$

where $z_{1\{k\}} = \left[z_{11\{k\}}, z_{12\{k\}}\right]^T$ – discrete vector of generalized coordinates, corresponding to vector q in a continuous time domain. The state vector was assumed in the form of $z_{\{k\}} = \left[z_{1\{k\}}, z_{2\{k\}}\right]^T$. Discrete tracking errors of rotation angles $z_{1\{k\}}$ and angular velocities $z_{2\{k\}}$ were defined in the form

$$
\begin{aligned}
e_{1\{k\}} &= z_{1\{k\}} - z_{d1\{k\}}, \\
e_{2\{k\}} &= z_{2\{k\}} - z_{d2\{k\}},
\end{aligned}
\tag{3}
$$

where the desire trajectory $(z_{d\{k\}} = \left[z_{d1\{k\}}, z_{d2\{k\}}\right]^T)$ was generated earlier. On the basis of the Eq. (3), a filtered tracking error was defined in the form

$$s_{\{k\}} = e_{2\{k\}} + \Lambda e_{1\{k\}}, \tag{4}$$

where Λ - a fixed, positive defined diagonal matrix.

Substituting dynamic equations of WMR motion (2) and dependences describing tracking errors (3) to the value of the filtered tracking error in step $k + 1$, $s_{\{k+1\}}$, determined on the basis of Eq. (4), the following dependence was obtained

$$
\begin{aligned}
s_{\{k+1\}} &= Y_d\left(z_{\{k\}}, z_{d\{k\}}, z_{d3\{k\}}\right) - Y_f\left(z_{1\{k\}}, z_{2\{k\}}\right) - Y_{\tau\{k\}}\left(z_{1\{k\}}\right) + \\
&\quad + h M^{-1}\left(z_{1\{k\}}\right) B\left(z_{1\{k\}}\right) u_{\{k\}},
\end{aligned}
\tag{5}
$$

where

$$
\begin{aligned}
Y_d\left(z_{\{k\}}, z_{d\{k\}}, z_{d3\{k\}}\right) &= z_{2\{k\}} - z_{d2\{k+1\}} + \Lambda\left[z_{1\{k\}} + h z_{2\{k\}} - z_{d1\{k+1\}}\right] \\
&= s_{\{k\}} + Y_e\left(z_{2\{k\}}, z_{d2\{k\}}, z_{d3\{k\}}\right), \\
Y_e\left(z_{2\{k\}}, z_{d2\{k\}}, z_{d3\{k\}}\right) &= h\left[\Lambda e_{2\{k\}} - z_{d3\{k\}}\right], \\
Y_f\left(z_{1\{k\}}, z_{2\{k\}}\right) &= h M^{-1}\left(z_{1\{k\}}\right)\left[C\left(z_{1\{k\}}, z_{2\{k\}}\right) z_{2\{k\}} + F\left(z_{1\{k\}}, z_{2\{k\}}\right)\right], \\
Y_{\tau\{k\}} &= h M^{-1}\left(z_{1\{k\}}\right)\tau_{d\{k\}},
\end{aligned}
\tag{6}
$$

where $z_{d3\{k\}}$ – the vector of desire angular accelerations, resulting from writing $z_{d2\{k+1\}}$ using the Euler method of rectangular approximation in the forward direction. The vector $Y_f\left(z_{1\{k\}}, z_{2\{k\}}\right)$ contains all non-linearities of the object.

3   Approximate Dynamic Programming

In recent years, there has been a significant increase in interest in the use of optimal methods in robots control. It results from the technical possibilities of implementing complex computational algorithms in real time and the need to increase the effectiveness of the proposed solutions. The optimal control methods include Bellman's DP, whose idea implementation in on-line control became possible only as a result of the development of AI methods. As a result of the research, a family of ADP algorithms appeared, also called Neural Dynamic Programming (NDP), or Adaptive Critic Designs (ACD) [11–14].

The ADP algorithms family consists of six algorithms, each of which is composed of two parametric structures, an actor and a critic, implemented in the form of algorithms that can map any non-linear waveforms, e.g. NNs. The critic's task is to approximate the value function and/or its derivative, while the actor generates a suboptimal control law. In order to implement five of the six ADP algorithms, it is also necessary to know the mathematical model of the controlled object. The basic structure of the ADP family of algorithms is Heuristic Dynamic Programming (HDP), in which the actor generates control law, while the critic approximates the value function. In the DHP algorithm, the critic's task is to approximate the derivative of the value function with respect to the object's state, while the role of the actor is the same as in the HDP algorithm. In the case of the Globalised Dual-Heuristic Dynamic Programming algorithm (GDHP), the actor is constructed in the same way as in the HDP algorithm, while the critic approximates the value function and its derivative with respect to the object's state. Other algorithms are action dependent versions of the HDP, DHP and GDHP algorithms, in which the actor's NN output signal is also the critic's input. The scheme of the APD algorithms family is shown in Fig. 2.

In the presented control algorithm, the DHP structure generates a control law that minimizes the value function [1,2,11–14] assumed in the form

$$
V\left(s_{\{k\}}\right) = \sum_{k=0}^{N} \gamma^k L_C\left(s_{\{k\}}\right),
\tag{7}
$$

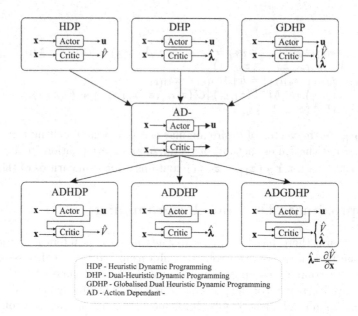

Fig. 2. The scheme of the approximate dynamic programming algorithms family

where $L_C\left(s_{\{k\}}\right)$ – the cost function in the k-th step, N – a number of steps, γ – a discount factor, $0 < \gamma \le 1$. The cost function was assumed in the form

$$L_{C\{k\}}\left(s_{\{k\}}\right) = \frac{1}{2}s_{\{k\}}^T R s_{\{k\}}, \tag{8}$$

where R – a fixed, positive defined diagonal matrix.

The DHP algorithm consists of:

– the predictive model of the control object state described by the equation

$$\begin{aligned} s_{\{k+1\}} = &Y_d\left(z_{\{k\}}, z_{d\{k\}}, z_{d3\{k\}}\right) - Y_f\left(z_{1\{k\}}, z_{2\{k\}}\right) + \\ &+hM^{-1}\left(z_{1\{k\}}\right)B\left(z_{1\{k\}}\right)u_{\{k\}}, \end{aligned} \tag{9}$$

where $u_{\{k\}}$ – the overall tracking control signal, its structure derives from the stability analysis presented in the next section.

– the structure of the actor realized in the form of two Random Vector Functional Link (RVFL) NNs, which generate a suboptimal control law in accordance with the dependence

$$u_{Aj\{k\}}\left(x_{Aj\{k\}}, W_{Aj\{k\}}\right) = W_{Aj\{k\}}^T S\left(D_A^T x_{Aj\{k\}}\right), \tag{10}$$

where $x_{Aj\{k\}}$ – the vector of normalized inputs to the actor's j-th NN, $W_{Aj\{k\}}$ – the output layer weights vector, $S\left(.\right)$ - sigmoidal bipolar neurons activation function vector, D_A – matrix of constant values of input layer weights to the actor's NN, selected randomly in the NN initialization process.

The weights of the output layer of the actor's j-th NN are adapted using a gradient method according to the relation

$$W_{Aj\{k+1\}} = W_{Aj\{k\}} - e_{Aj\{k\}} \Gamma_{Aj} S \left(D_A^T x_{Aj\{k\}} \right), \tag{11}$$

where Γ_{Aj} – a fixed diagonal matrix of positive learning rates. The adaptation process of the actor's NN output layer weights is aimed at minimizing errors $e_{A\{k\}}$ expressed in the form

$$e_{A\{k\}} = \frac{\partial L_C \left(s_{\{k\}} \right)}{\partial u_{\{k\}}} + \left[\frac{\partial s_{\{k+1\}}}{\partial u_{\{k\}}} \right]^T \hat{\lambda} \left(x_{C\{k+1\}}, W_{C\{k\}} \right), \tag{12}$$

where $\hat{\lambda} \left(x_{C\{k+1\}}, W_{C\{k\}} \right)$ – value of the output of the critic NN in step $k + 1$, generated on the basis of the prediction of the object's state.
- the structure of the critic implemented in the form of two RVFL NNs, which approximate the derivative of the value function with respect to the state of the control object in accordance with the dependence

$$\hat{\lambda}_j \left(x_{Cj\{k\}}, W_{Cj\{k\}} \right) = W_{Cj\{k\}}^T S \left(D_C^T x_{Cj\{k\}} \right), \tag{13}$$

where $x_{Cj\{k\}}$ – the vector of normalized inputs to the j-th critic NN, $W_{Cj\{k\}}$ – vector of the output layer weights, $S \left(. \right)$ – sigmoidal bipolar neuron activation functions vector, D_C – matrix of constant values of input layer weights to the critic's NN, selected randomly in the NN initialization process.
The output layer weights of the j-th critic's NN are adapted using a gradient method according to the relation

$$W_{Cj\{k+1\}} = W_{Cj\{k\}} - e_{Cj\{k\}} \Gamma_{Cj} S \left(D_C^T x_{Cj\{k\}} \right), \tag{14}$$

where Γ_{Cj} – constant diagonal matrix of positive gain coefficients. The process of adaptation of NN output-level weights is aimed at minimizing $e_{C\{k\}}$ errors expressed in the form

$$
\begin{aligned}
e_{C\{k\}} = {} & \frac{\partial L_C \left(s_{\{k\}} \right)}{\partial s_{\{k\}}} + \left[\frac{\partial u_{\{k\}}}{\partial s_{\{k\}}} \right]^T \frac{\partial L_C \left(s_{\{k\}} \right)}{\partial u_{\{k\}}} - \hat{\lambda} \left(x_{C\{k\}}, W_{C\{k\}} \right) + \\
& + \gamma \left[\frac{\partial s_{\{k+1\}}}{\partial s_{\{k\}}} + \left[\frac{\partial u_{\{k\}}}{\partial s_{\{k\}}} \right]^T \frac{\partial s_{\{k+1\}}}{\partial u_{\{k\}}} \right]^T \hat{\lambda} \left(x_{C\{k+1\}}, W_{C\{k\}} \right).
\end{aligned} \tag{15}
$$

Schematic structure of DHP algorithm is shown in Fig. 3.

Additional control signals visible in Fig. 3 as the input to the DHP structure are discussed in the next section.

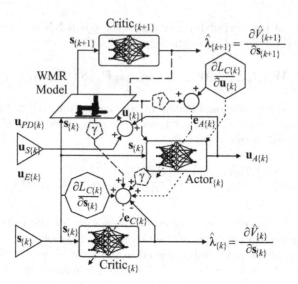

Fig. 3. The scheme of the Dual-Heuristic Dynamic Programming algorithm

4   Tracking Control Algorithm

Motion control systems of objects described by non-linear motion equations usually consist of a stabilizing term and an object's non-linearities compensator, often supplemented by a robust control term. In the presented case, the control system consists of a PD controller ($\boldsymbol{u}_{PD}$) acting as a stabilizing term, a neural compensator of the object's non-linearities in the form of an ADP algorithm in the DHP configuration ($\boldsymbol{u}_A$), a supervisory control term ($\boldsymbol{u}_S$) and an additional control signal ($\boldsymbol{u}_E$) with the structure resulting from discretization process of a continuous control object model, containing known or measured signals. The overall control signal is generated according to the equation

$$\boldsymbol{u}_{\{k\}} = \frac{1}{h}\boldsymbol{B}^{-1}\left(\boldsymbol{z}_{1\{k\}}\right)\boldsymbol{M}\left(\boldsymbol{z}_{1\{k\}}\right)\left[\boldsymbol{u}_{S\{k\}} - \boldsymbol{u}_{A\{k\}} - \boldsymbol{u}_{PD\{k\}} - \boldsymbol{u}_{E\{k\}}\right], \quad (16)$$

where

$$\begin{aligned}
\boldsymbol{u}_{PD\{k\}} &= \boldsymbol{K}_D\boldsymbol{s}_{\{k\}}, \\
\boldsymbol{u}_{S\{k\}} &= \boldsymbol{I}_S\boldsymbol{u}^*_{S\{k\}}, \\
\boldsymbol{u}_{E\{k\}} &= h\left[\boldsymbol{\Lambda}\boldsymbol{e}_{2\{k\}} - \boldsymbol{z}_{d3\{k\}}\right],
\end{aligned} \quad (17)$$

where $\boldsymbol{K}_D$ – a positive defined diagonal matrix of the PD controller gains, $\boldsymbol{I}_S$ – a diagonal matrix, $I_{Sj,j} = 1$ if $|s_{j\{k\}}| \geq \rho_j$, and $I_{Sj,j} = 0$ in the other case, ρ_j – a positive constant, $j = 1, 2$. The scheme of the WMTR tracking control system with DHP structure is shown in Fig. 4.

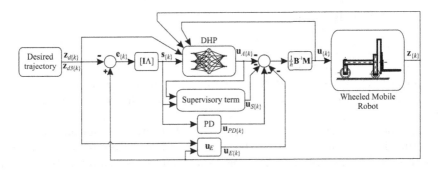

Fig. 4. The scheme of the neural tracking control system with the DHP algorithm

Analysis of the stability of the control system was performed assuming $I_{Sj,j} = 1$. Substituting the control law (16) to the dependence (5), the description of the closed-loop system takes the form

$$s_{\{k+1\}} = Y_d\left(z_{\{k\}}, z_{d\{k\}}, z_{d3\{k\}}\right) - Y_f\left(z_{1\{k\}}, z_{2\{k\}}\right) - Y_{\tau\{k\}} + u_{S\{k\}} + \\ -u_{A\{k\}} - u_{PD\{k\}} - u_{E\{k\}}. \tag{18}$$

Analysis of the stability of the control algorithm was performed assuming a positively defined candidate for the Lyapunov function in the form

$$L_{\{k\}} = \frac{1}{2} s_{\{k\}}^T s_{\{k\}}, \tag{19}$$

whose discreet difference was assumed in the form [15]

$$\Delta L_{\{k\}} = s_{\{k\}}^T \left[s_{\{k+1\}} - s_{\{k\}}\right]. \tag{20}$$

Substituting (18) to (20), $\Delta L_{\{k\}}$ takes the form

$$\Delta L_{\{k\}} = s_{\{k\}}^T \left[-Y_f\left(z_{1\{k\}}, z_{2\{k\}}\right) - Y_\tau\left(z_{1\{k\}}\right) - u_{A\{k\}} - K_D s_{\{k\}} - u_{S\{k\}}^*\right]. \tag{21}$$

Assuming that all elements of the disturbance vector are bounded, $Y_{\tau j}\left(z_{1\{k\}}\right) < b_{dj}$, where b_{dj} – positive constant, the difference of the candidate for the Lyapunov function takes the form

$$\Delta L_{\{k\}} \le -s_{\{k\}}^T K_D s_{\{k\}} + \sum_{j=1}^{2} s_{j\{k\}} u_{Sj\{k\}}^* + \\ + \sum_{j=1}^{2} |s_{j\{k\}}| \left[|Y_{fj}\left(z_{1\{k\}}, z_{2\{k\}}\right)| + |u_{Aj\{k\}}| + |Y_{\tau j}\left(z_{1\{k\}}\right)|\right]. \tag{22}$$

A supervisory control signal was assumed in the form

$$u_{Sj\{k\}}^* = -\text{sgn}\left(s_{j\{k\}}\right)\left[F_j + |u_{Aj\{k\}}| + b_{dj} + \sigma_j\right], \tag{23}$$

where $|Y_{fj}\left(z_{1\{k\}}, z_{2\{k\}}\right)| \le F_j$, F_j – positive constant, σ_j – positive constant. Assuming the above, the Lyapunov function (19) is negatively definite.

5   Research Results

Numerical tests of the control algorithm were carried out using the Matlab/Simulink computational environment. In the current section, in order to simplify the notation, the index k is omitted in the notation of discrete variables, and the waveforms of all variables are shown in diagrams as a function of time t.

The discretization step $h = 0.01$ [s], the PD controller gain matrix $K_D =$ diag $\{0.031, 0.07\}$ and $\Lambda =$ diag $\{0.5, 0.5\}$ have been selected heuristically to ensure adequate quality of tracking motion when, due to zero values of initial weights, the control signal generated by the actor's structure does not provide sufficient control quality. In the DHP algorithm the value of the discount factor for future rewards/penalties $\gamma = 0.5$ and the matrix $R =$ diag $\{1, 1\}$ of the weight of individual elements of the vector $s_{\{k\}}$ in a cost function were assumed. The weight adaptation matrix of the actor's NN $\Gamma_{A1} =$ diag $\{0.25, ..., 0.25\}$, $\Gamma_{A2} =$ diag $\{0.75, ..., 0.75\}$, the weight adaptation matrix of the critic's NN $\Gamma_{C1} =$ diag $\{0.75, ..., 0.75\}$, $\Gamma_{C2} =$ diag $\{1.0, ..., 1.0\}$, were assumed. The values of the weight adaptation matrix elements were selected to achieve the assumed tracking quality. The matrices D_A and D_C of constant values of input layer weights to the actor's and critic's NN were selected randomly in the NN initialization process from the range $D_{Ai,j} = D_{Ci,j} = -0.2$ to $D_{Ai,j} = D_{Ci,j} = 0.2$. The following parameters of the control system were assumed: $\rho_j = 1.5$, $\sigma_j = 0.01$.

The maximum velocity of the A point of the WMR frame $v_A = 0.5$ [m/s] was assumed. The desire value of the velocity of point A is given in Fig. 5(a). Figure 5(b) presents the desire path of the WMR H point, connected with the lift.

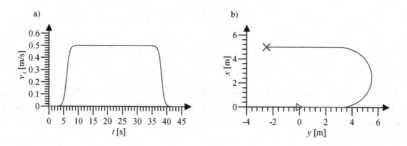

Fig. 5. (a) The desire velocity values of the point A, v_A, (b) the path of the point H

On the basis of the WMR kinematics description, the waveforms of the desire motion parameters of the individual robot elements were generated. The desire rotation angles and angular velocities of rotation of wheels 1, 2 and 3 are shown in Fig. 6(a) and (b) respectively. The desire angles of rotation of the WMR frame β_d and steering module φ_d are shown in Fig. 6(c), whereas the desire angular velocities $\dot{\beta}_d$ and $\dot{\varphi}_d$ are shown in Fig. 6(d).

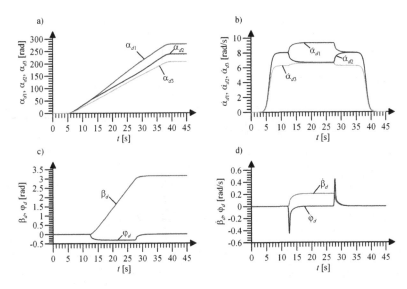

Fig. 6. (a) Desire rotation angles of wheels 1, 2 and 3, α_{d1}, α_{d3}, α_{d3}, (b) desire angular velocity values of wheels 1, 2, and 3, $\dot{\alpha}_{d1}$, $\dot{\alpha}_{d2}$, $\dot{\alpha}_{d3}$, (c) desire angles of rotation β_d and φ_d, (d) desire angular velocity values $\dot{\beta}_d$ and $\dot{\varphi}_d$

The generated motion parameters served as the desire trajectory set in simulation studies of the WMR tracking motion, where the control signals were generated by the proposed control algorithm.

The overall control signals are shown in Fig. 7(a), according to the Eq. (16) they consist of control signals generated by the NNs of the DHP (u_A) algorithm (Fig. 7(b)), PD controller (u_{PD}) control signals (Fig. 7(c)), control signals of the supervisory term (u_S), and additional control signals (u_E) (Fig. 7(d)).

The difference between the desire and executed WMR motion parameters describes tracking errors (3), whose waveforms are shown in Fig. 8(a) for the desire angle of rotation and angular velocity of the wheel 3, and in Fig. 8(b) for the desire angle of rotation and angular velocity of rotation of the steering module. The highest values of tracking errors occur in the initial phase of movement, then they are reduced. This is due to the use of an adaptive structure in the control algorithm, which adjusts control signals to minimize the value function, and this has an effect on minimization of tracking errors.

The weight values of the structure part of the DHP algorithm related to the generation of the u_{A1} signal controlling the motion of the driving wheel 3 are shown in Fig. 9(a) and (c) for the actor's NN ($\boldsymbol{W}_{A1}$) and critic's NN ($\boldsymbol{W}_{C1}$) respectively. The weight values of the part of the DHP structure responsible for controlling the motion of the steering module are shown in Fig. 9(b) and (d) for the actor's NN ($\boldsymbol{W}_{A2}$) and the critic's NN ($\boldsymbol{W}_{C2}$), respectively.

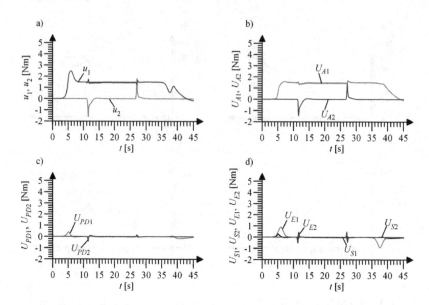

Fig. 7. (a) Overall control signals u_1 and u_2 for wheel 3 and steering module, (b) actor's NNs control signals U_{A1} and U_{A2}, $\boldsymbol{U}_A = -\frac{1}{h}\boldsymbol{B}^{-1}\boldsymbol{M}\boldsymbol{u}_A$, (c) PD control signals U_{PD1} and U_{PD2}, $\boldsymbol{U}_{PD} = -\frac{1}{h}\boldsymbol{B}^{-1}\boldsymbol{M}\boldsymbol{u}_{PD}$, (d) supervisory term's control signals U_{S1} and U_{S2}, $\boldsymbol{U}_S = \frac{1}{h}\boldsymbol{B}^{-1}\boldsymbol{M}\boldsymbol{u}_S$, and additional control signals u_{E1} and u_{E2}, $\boldsymbol{U}_E = -\frac{1}{h}\boldsymbol{B}^{-1}\boldsymbol{M}\boldsymbol{u}_E$

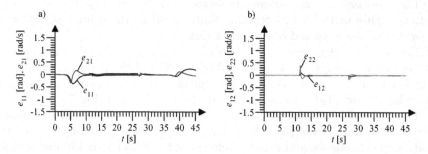

Fig. 8. (a) Tracking errors of the desire angle of rotation (e_{11}) and angular velocity (e_{21}) of the wheel 3, (b) tracking errors of the desire angle of rotation (e_{12}) and angular velocity (e_{22}) of rotation of the steering module

During the numerical tests, the worst case of zero initial values of the NNs output layer weights was assumed. The NN weight values are adapted from the moment when the error values e_A (12) and e_C (15) take values different from zero. It results from the applied trajectory of WMR movement. After the initial volatility period, the weight values are set around certain values and remain bounded during the numerical test. Also the tracking error values remain limited.

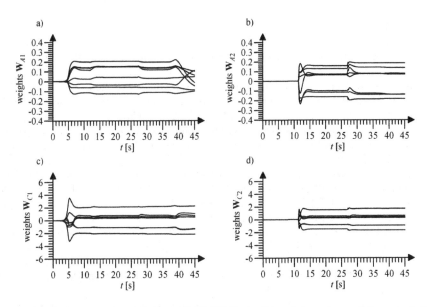

Fig. 9. (a) Weight values of the actor's NN ($\boldsymbol{W}_{A1}$), (b) weight values of the actor's NN ($\boldsymbol{W}_{A2}$), (c) weight values of the critic's NN ($\boldsymbol{W}_{C1}$), (d) weight values of the critic's NN ($\boldsymbol{W}_{C2}$)

6 Summary

The work presents a model of a mobile transport robot construction, adapted for transporting loads and description of its dynamics. Based on the knowledge of the mathematical model of the controlled object, a discrete tracking control algorithm was carried out, in which the ADP algorithm in DHP configuration was used. An analysis of the stability of the control system was carried out using the Lyapunov stability theory. Then, simulation tests of the proposed control algorithm were carried out, which confirmed the high quality of the execution of the tracking motion with the use of the ADP algorithm. The tracking error values obtained are limited and the values of NNs weights stabilize around certain values during the numerical test. The presented robot will be used for research on the possibility of implementing autonomous large-scale transport with the use of a group of mobile robots in industrial conditions. In this type of applications, it is extremely important to ensure high quality of tracking motion in changing operating conditions of the control object, which enables presented control system with the DHP algorithm.

References

1. Barto, A., Sutton, R., Anderson, C.: Neuronlike adaptive elements that can solve difficult learning problems. IEEE Trans. Syst. Man Cybern. **13**, 834–846 (1983)
2. Bellman, R.: Dynamic Programming. Princeton University Press, New York (1957)
3. Giergiel, J., Hendzel, Z., Zylski, W.: Modeling and Control of Wheeled Mobile Robots. Scientific Publishing PWN, Warsaw (2002). (in Polish)
4. Giergiel, J., Kurc, K.: Identification of the mathematical model of an inspection mobile robot with fuzzy logic systems and neural networks. J. Theor. Appl. Mech. **49**, 209–225 (2011)
5. Gierlak, P., Szuster, M., Żylski, W.: Discrete dual–heuristic programming in 3DOF manipulator control. In: Rutkowski, L., Scherer, R., Tadeusiewicz, R., Zadeh, L.A., Zurada, J.M. (eds.) ICAISC 2010. LNCS (LNAI), vol. 6114, pp. 256–263. Springer, Heidelberg (2010). https://doi.org/10.1007/978-3-642-13232-2_31
6. Hendzel, Z., Burghardt, A., Szuster, M.: Reinforcement learning in discrete neural control of the underactuated system. In: Rutkowski, L., Korytkowski, M., Scherer, R., Tadeusiewicz, R., Zadeh, L.A., Zurada, J.M. (eds.) ICAISC 2013. LNCS (LNAI), vol. 7894, pp. 64–75. Springer, Heidelberg (2013). https://doi.org/10.1007/978-3-642-38658-9_6
7. Iftekharuddin, K.M.: Transformation invariant on-line target recognition. IEEE Trans. Neural Netw. **22**, 906–918 (2011)
8. Lendaris, G., Schultz, L., Shannon, T.: Adaptive critic design for intelligent steering and speed control of a 2-axle vehicle. In: Proceedings of the IEEE INNS-ENNS International Joint Conference on Neural Networks, vol. 3, pp. 73–78 (2000)
9. Lewis, F.W., Jagannathan, S., Yesildirak, A.: Neural Network Control of Robot Manipulators and Non-linear Systems. CRC Press, London (1998)
10. Osowski, S.: Neural Networks - An Algorithmic Approach. WNT, Warsaw (1996). (in Polish)
11. Powell, W.: Approximate Dynamic Programming: Solving the Curses of Dimensionality. Wiley, Hoboken (2007)
12. Prokhorov, D., Wunsch, D.: Adaptive critic designs. IEEE Trans. Neural Netw. **8**, 997–1007 (1997)
13. Si, J., Barto, A., Powell, W., Wunsch, D.: Handbook of learning and approximate dynamic programming. Wiley, IEEE Press, Hoboken (2004)
14. Sutton, R., Barto, A.: Introduction to Reinforcement Learning. MIT Press, Cambridge (1998)
15. Szuster, M., Gierlak, P.: Approximate dynamic programming in tracking control of a robotic manipulator. Int. J. Adv. Robot. Syst. **16**, 1–18 (2016)
16. Szuster, M., Hendzel, Z.: Discrete globalised dual heuristic dynamic programming in control of the two-wheeled mobile robot. Math. Probl. Eng. **2014**, 1–16 (2014)
17. Tutak, J., Wiech, J.: Horizontal automated storage and retrival system. Adv. Sci. Technol. - Res. J. **11**, 82–95 (2017)
18. Venayagamoorthy, G.K., Harley, R.G., Wunsch, D.C.: Comparison of heuristic dynamic programming and dual heuristic programming adaptive critics of a turbogenerator. IEEE Trans. Neural Netw. **13**, 764–773 (2002)

Stylometry Analysis of Literary Texts in Polish

Tomasz Walkowiak[1(✉)] and Maciej Piasecki[2]

[1] Faculty of Electronics, Wrocław University of Science and Technology,
Wrocław, Poland
tomasz.walkowiak@pwr.edu.pl
[2] Faculty of Computer Science and Management,
Wrocław University of Science and Technology, Wrocław, Poland
maciej.piasecki@pwr.edu.pl

Abstract. In this work we compare different methods for deriving features for text representation in two stylometric tasks of gender and author recognition. The first group of methods uses the Bag-of-Words (BoW) approach, which represents the documents with vectors of frequencies of selected features occurring in the documents. We analyze features such as the most frequent 1000 lemmas, word forms, all lemmas, selected (content insensitive) lemmas, bigrams of grammatical classes and mixture of bigrams of grammatical classes, selected lemmas and punctuations. Moreover, the approach based on the recently proposed fastText algorithm (for vector based representation of text) is also applied. We evaluate these different approaches on two publicly available collections of Polish literary texts from late 19th- and early 20th-century: one consisting of 99 novels from 33 authors and the second one 888 novels from 58 authors. Our study suggests that depending on the corpora the best are the style features (grammatical bigrams) or semantic features (1000 lemmas extracted from the training set). We also noticed the importance of proper division of corpora into training and testing sets.

Keywords: Stylometric · Natural language processing · Polish
Text analysis · Bag of words · Machine learning

1 Introduction

Stylometry is an interdisciplinary area of science aimed at studying associations between the statistical properties of texts and their meta-properties of texts like: author, author's characteristic features (e.g. gender or native language), place of origin, time of writing etc. Stylometry originated from research on authorship attribution and early attempts to verify authorship of some texts by comparing their statistical descriptions with known examples of somebody's writing. Stylometric techniques are based on identification of textual similarities and dissimilarities between texts and clustering texts according to their linguistic characteristic, cf [3]. They are aimed at detecting signals in texts, e.g. authorship,

© Springer International Publishing AG, part of Springer Nature 2018
L. Rutkowski et al. (Eds.): ICAISC 2018, LNAI 10842, pp. 777–787, 2018.
https://doi.org/10.1007/978-3-319-91262-2_68

genre, gender, origin, style, etc. Typical areas of its applications are authorship attribution (recognition – from a closed set and discovering – from texts or
unlimited set), recognition of period of writing, style, genre place of origin etc.
Stylometry can be also applied in discovery of the features of an author (e.g.
with respect to gender, mother tongue), style and influence of translation (e.g.
source language, native language of the translator).

Stylometry can benefit from Machine Learning techniques [9], beyond
clustering-based analysis in finding more complex, very often latent, dependencies between the statistical features characterizing text documents. Transformation of documents into vectors opens a range of possibilities, but the number of
parameters (e.g. words as features) can be very high, number of potential classes
very large (e.g. the number of authors), dependencies very subtle (e.g. characteristics of a style) while the amount of data limited (e.g. the number of texts of
a given author) and the level of noise present in the data (e.g. text content and
errors in processing) quite substantial.

Most of the stylometric works were concentrated on text in English and the
authorship attribution tasks. Concerning texts in English, it was shown that
frequencies of the most frequent words is the most effective representation for
texts in authorship attribution task [11]. However Polish is a language with a
fairly rich inflection, especially in comparison to English, and weakly constrained
word order, cf e.g. [3]. Moreover, [4] showed that the same stylometric methods
can bring different results for different languages.

The aim of this paper is to compare stylometric methods applied to literary
texts in the Polish language from the point of view of different types of features
used for text representation in two stylometric tasks of recognition of gender and
author. In [1] authors analyzed features based on grammatical classes, statistical
features and on common words, founding all words as the best selection for
Polish literary texts. In this paper, we extend the feature types to bigrams of
grammatical classes, different types of common words and mixture of bi-grams
and common words. Moreover, we apply the emerging fastText [10] algorithm.

The paper is structured as follows. First, we introduce two collections of
Polish literary text and stylometric tasks defined on this basis. Next, we discuss
different features for text representation in the bag-of-words models. Shortly, the
applied classification techniques are presented. We conclude the paper with the
main part discussing the performed evaluation and formulated conclusions.

2 Data Sets

The study was conducted on two collections of novels in Polish. The firsts set [5]
consists of 100 Polish novels written by 34 authors in late 19th- and early 20th-
century. For the analysis we have removed one novel by Magdalena Samozwaniec
since it was the only novel by this author. Each of remaining authors (33) was
represented by three novels. We have classified the whole texts (*full*) as well as
the chunked texts (*divided* into smaller chunks). The chunking into parts of the
same size (c.a. 20,000 bytes) allowed to increase the number of examples for each

class. The algorithm does not break words and tries to keep the same size of all chunks almost constant (smaller texts are included into one chunk only).

For authorship attribution we have divided the corpus into the testing and training sets in two different manners:

- *hand* – two books of each author were placed in the training set and the remaining one in the test set,
- *random* – texts (whole novels or text chunks) were divided randomly into training (66%) and testing (33%) sets.

For gender attribution task, we have divided the data set into the testing and training collections in three different manners:

- *hand* – authors of each gender were divided randomly into two sets (66% training, 33% testing); all novels or parts of novels (text chunks) were assigned to an appropriate set following author membership; as an effect each author is either in training either in testing collection;
- *authorship* – the same as *hand* method in case of authorships;
- *random* – texts (whole novels or chunked texts) were divided randomly into training (66%) and testing (33%) set.

Table 1. Data set statistics

Data set	100 novels				1000 novels	
Classification task	Authorship		Gender		Authorship	
Number of classes	33		2		58	
Method of novel division	Full	Chunk	Full	Chunk	Full	Chunk
Number of examples in training set	66	1927	66	1927	589	4115
Average number of examples per class in training set	2	58	33	963	10	71

The second set [6] consists of 1000 novels written in Polish or translated to Polish by various authors from late 19th- and early 20th-century. We have removed all authors that had less then 3 books. Remaining 888 books of 58 authors were divided into training and testing sets in similar proportion as in case of 100 novel corpora resulting in 589 novels in training collection and 299 in testing one. We have also divided novels into chunks of c.a. 20,000 bytes.

Statistics of the data sets are presented in Table 1.

3 Text Document Representation

3.1 Feature Extraction

Features should reveal properties of text that are characteristic for an author or his style. They should be correlated with the semantic content of the text.

Feature can refer to any level of the language analysis or text form. However, they should be based only on language tools that express relatively small error, as errors produced by language tools increase the level of noise in data. For feature extraction we have used CLARIN-PL infrastructure [13, 20] and features supported by WebSty[1] tool [3]. It can generate several types of features for texts in Polish on the basis of a morpho-syntactic tagger:

- word forms (`orths`), i.e. word types from text,
- punctuations,
- lemmas[2],
- grammatical classes (following the National Corpus of Polish tagset [14]),
- Parts of Speech (as sets of grammatical classes, i.e. generalisation of grammatical classes),
- bigrams and trigrams of grammatical classes (2 and 3 element sequences),

Raw values of the features (i.e. before any transformation) are equal to the frequency of the appropriate occurrences in text (e.g. of a word form, lemma or a tag). In conducted experiments Morfeusz morphological analyser [21] and WCRFT morpho-syntactic tagger [15] were used. WCRFT is based on the idea of tiered tagging (i.e. in performing the decision process on several layers related to the different parts of tags) and uses Conditional Random Fields (CRF) for Machine Learning on the basis of the manually annotated 1 million part of the National Corpus of Polish.

3.2   Bag of Words

The most common vector representation of texts is a *bag-of-words* (BoW) – words together with their frequency – or bag-of-n-grams. BoW models are based on the assumption that text can be represented as an unordered collection of words frequencies [16] and such a representation can characterise the semantics of words in the given text to some extent, e.g. cf [8]. The method has a lot of modifications depending on different classification tasks. For example in stylometry the frequency of letter pairs, letter triplets or Parts of Speech are used [17]. Whereas, usually in subject classification, the BoW dictionary consists of words from texts with the most common, the most rare and stop words filtered out [18]. Furthermore the words may be lemmatized to decrease the number of the descriptive features (by grouping words under lemmas). In general the BoW is based on representing each document by the number of occurrence of the specified features – that forms a feature vector used as an input to a classifier.

In the experiments, we used the following types of features:

- *Base1000* – the most frequent 1000 lemmas from the training corpus,

---

[1] http://ws.clarin-pl.eu/websty.shtml.

[2] Lemmas are simply understood here as basic morphological forms selected to represent sets of word forms that differ only in the values of grammatical categories like number, gender, person etc.

- *BaseAll* – all lemmas from the training corpus,
- *BaseSel* – selected (content insensitive) lemmas,
- *Bigrams* – bigrams of grammatical classes,
- *Bigrams+* – bigrams of grammatical classes, selected lemmas and punctuations,
- *Orth1000* – the most frequent 1000 word forms from the training corpus.

BaseSel – the list of the selected, content insensitive, lemmas – was built on the basis of the initial list of the 500 most frequent lemmas extracted from the National Corpus of Polish [14]. Next, the list was manually filtered on the basis of experiments performed with clustering texts from blogs collected from Internet. All lemmas that appeared to be correlated with some topics and causing thematic clustering of texts were manually eliminated. Finally, several closed classes of lemmas (mostly functional words) located on the top of the reduced list like conjunctions, adverbs and participles were carefully manually expanded. As a result, we obtained the list of 212 lemmas[3].

4   Classification

Most of the statistical classifiers assume that features are normally distributed with variance equal to 1. Therefore, feature vectors generated on the basis of BoW model were weighted by removing the mean and scaling to unit variance. The mean and variance were calculated for the training set and these values were used for weighting the training and testing sets, as well.

We have analysed and tested different statistical classifiers from Scikit-learn package [12], namely:

- ridge regression [9],
- multilayer perceptron (MLP) [9],
- Support Vector Machine,
- random forest [2],
- Elastic net learned by stochastic gradient descent [19],
- Naive Bayes classifier [9],

After initial experiments, finally, we have selected MLP with Broyden–Fletcher–Goldfarb–Shanno (BFGS) nonlinear optimization learning algorithm for the classification of BoW feature vectors. Since MLP gave the best or almost the best results for all analysed cases. It is important to noticed that MLP expresses some randomness built intrinsically into the learning process, so the achieved results differs in each run of learning for the same data. As a consequence, the presented results are the best solution achieved among 20 runs.

We also compared achieved results with a method of purely semantic classification based on the transformation of text into the semantic space of the reduced feature vectors. Namely, we used a recent deep learning method for text classification *fastText* [10]. First, it computes representation of documents as an

---

[3] http://ws.clarin-pl.eu/demo/inc/nkjp360-meaningless-no-prep-freq-above-3500.txt.

average of *word embeddings*. Next, a linear soft-max classifier [7] is trained on the documents represented by word embedding vectors. The main idea is to perform word representation and classifier learning in parallel. As a result, this linear model, which is very effective to be trained, achieves several orders of magnitude faster solution than other competing methods [10]. In many text mining tasks *fastText* seems to outperform state of art classifiers with BoW features. However, as the *fastText* model takes into account all the words in the corpus it is clearly biased towards the semantic content of the documents.

5   Evaluation

5.1   Authorship Attribution – 100 Novels Data Set

Firstly, the *100 Novels* Corpus was analysed in the authorship attribution task. We have used novel chunks manually divided into training and testing set (*divided/hand*) in a way that the same novel (any chunk of it) does not occur in both sets at the same time (i.e. in the same division). Accuracy achieved for the seven different method of feature extraction is presented in Table 2. As it could be noticed, the best results were observed for grammatical bigrams. Features of this type outperformed other methods of representation by at least 5 percent points. The second result was obtained with using the 1000 most frequent lemmas and exactly the same value for the 1000 most frequent word forms (orths).

The usage of all lemmas (*BaseAll*) as features results in a huge dimensionality of feature vector and causes performance problems (due too large memory requirements). Since the accuracy for this feature type is much lower then the best one we have not used it in the following experiments.

Table 2. Authorship attribution accuracy for *100 novels* corpus

Method	Bigrams+	Base1000	BaseAll	**Bigrams**	Orth1000	BaseSel	fastText
Accuracy	0.686	0.75	0.668	**0.799**	0.750	0.700	0.535
Number of features	1226	1000	>100000	1012	1000	209	100

5.2   Authorship Attribution - 1000 Novels Data Set

Secondly, the *1000 Novels* corpus was analysed, again in perspective of the authorship attribution task. In a similar way to the previous experiment, we have used the set of novel chunks manually divided into data and training (*divided/hand*). Results are presented in Table 3. In this case the best results were observed for the 1000 most frequent lemmas (*Base1000*). Moreover, the achieved accuracy is much better (more then 11 percent points) then in the case of *100 novel* set. It looks a little bit strange since we have here more classes (authors) and smaller number of examples per class (average number for training set see Table 1) then in the case of *100 novels* corpora. The results, better accuracy then in case of *100 novels* and semantic features (*Base1000*)[4] outperforming style features (*Bigrams*), could be justified if we analyse the content of the *1000 novels* corpora. Figure 1 presents the histogram of number of books per author in the training set (*full/hand*). As it could be noticed the corpora is unbalanced. On one hand, we have a group of six authors with more then 28 novels (up to 65 for Eliza Orzeszkowa), on the other hand a set of 18 authors with only 2 novels. Moreover, the sizes of literary works in the corpus are very diversified. Some of them are long novels, while many are only short stories. Many long novels can be found among the most prolific authors in the corpus. So, chunks extracted from these works dominate in the data set. As a result, the whole classification task is very difficult or even badly defined. In such a setting, text chunks coming from thematically coherent, larger novels dominate in the data, and the classification process is driven by the semantic features. Features representing personal style of the authors have too small support from imbalanced data in comparison to the semantic 'noise'. Thus, the achieved result can be a little bit misleading for the assessment of the importance of different features for the general task of authorship attribution and idiosyncratic for this particular data set.

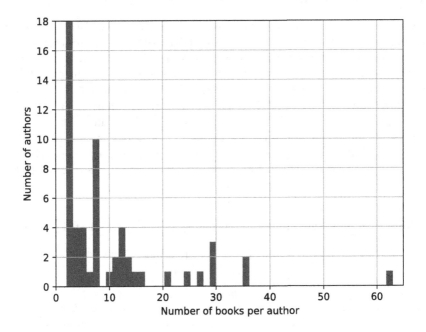

Fig. 1. Histogram of number of books per author in the training set

---

[4] There are many lemmas that express semantic content and are correlated with the semantic content or topics of text among the 1,000 most frequent lemmas.

Table 3. Authorship attribution accuracy for *1000 novels* corpus

Method	Bigrams+	**Base1000**	Bigrams	Orth1000	BaseSel	fastText
Accuracy	0.8152	**0.864**	0.743	0.858	0.801	0.803

5.3   Gender Attribution - 100 Novels Data Set

In the next experiment, we have used the *100 novels* corpora for the gender attribution task. The results for chunked novels manually divided into data and training (*divided/hand*) are presented in Table 4. The best accuracy was achieved for *Base1000* features.

Table 4. Gender attribution accuracy for *100 novels* corpus

Method	Bigrams+	**Base1000**	Bigrams	Orth1000	BaseSel	fastText
Accuracy	0.693	**0.754**	0.673	0.744	0.679	0.712

5.4   Selection Methods of Training and Testing Set

In the previous Sect. 1 we have presented different methods of the corpus division into training and testing set. The Table 5 presents the accuracy for author attribution task, whereas Table 6 for gender attribution. It could be noticed that results for *chunked/random* are the best. However, it is a misleading result, since these data sets are incorrect. The random selection of chunked texts results in placing the same author (a mistake for a gender recognition) or even the same book (a mistake for authorship attribution) in the training and testing set. Such a mistake is quite common due to a popularity of k-fold cross-validation approach [9] and available implementations (i.e. in scikit-learn package [12]). For example in [1] the random selection of training and testing sets resulted in 100% accuracy on a corpora very similar to *100 novels*.

Table 5. Accuracy for authorship attribution for differently selected data sets

Method	Base1000
Full/hand	0.750
Chunked/hand	0.750
Chunked/random	0.982

Table 6. Accuracy for gender attribution for differently selected data sets

Method	Base1000
Chunked/hand	0.754
Chunked/authorship	0.900
Chunked/random	0.956

6 Conclusion

In this work we compared different approaches to represent text documents with feature vectors in the task of gender and author recognition. Our study involved the analysis of two publicly available collections of Polish literary texts from late 19th- and early 20th-century: one consisting of 99 novels from 33 authors and the second one 888 novels from 58 authors. In case of the first set grammatical bigrams representing the style information included in the text outperformed other analysed methods. However, in case of the second data set the semantic features, i.e. 1000 lemmas extracted from the training set, gives the best results. It is caused by the fact, that larger novels dominate in the second collection. So the classification process is driven by the semantic features and personal style features have too small support from imbalanced data in comparison to the semantic 'noise'.

The presented work shows also the importance of carefully selection of training and testing sets. An uninformed random selection sometimes leads to misleading results of gender and authorship attribution.

Future plans include usage analysis of different type of features in authorship and gender attribution of blog texts. Moreover, we plan to investigate open set classification problem in stylometric tasks.

Works funded by the Polish Ministry of Science and Higher Education within CLARIN-PL Research Infrastructure.

References

1. Baj, M., Walkowiak, T.: Computer based stylometric analysis of texts in Polish language. In: Rutkowski, L., Korytkowski, M., Scherer, R., Tadeusiewicz, R., Zadeh, L.A., Zurada, J.M. (eds.) ICAISC 2017. LNCS (LNAI), vol. 10246, pp. 3–12. Springer, Cham (2017). https://doi.org/10.1007/978-3-319-59060-8_1
2. Breiman, L.: Random forests. Mach. Learn. **45**(1), 5–32 (2001). https://doi.org/10.1023/A:1010933404324
3. Eder, M., Piasecki, M., Walkowiak, T.: An open stylometric system based on multilevel text analysis. Cogn. Stud. — Etudes Cogn. **17** (2017). https://doi.org/10.11649/cs.1430
4. Eder, M.: Style-markers in authorship attribution: a cross-language study of the authorial fingerprint. Stud. Polish Linguist. **6**, 99–114 (2011). www.wuj.pl/page,art,artid,1923.html

5. Eder, M., Rybicki, J.: Late 19th- and early 20th-century polish novels. CLARIN-PL Digital Repository (2015). http://hdl.handle.net/11321/57
6. Eder, M., Rybicki, J., Młynarczyk, K., Oleksy, M., Borys, R., Maryl, M., Piasecki, M.: 1000 novels corpus. CLARIN-PL Digital Repository (2016). http://hdl.handle.net/11321/312
7. Goodman, J.: Classes for fast maximum entropy training. In: 2001 IEEE International Conference on Acoustics, Speech, and Signal Processing. Proceedings (Cat. No. 01CH37221), vol. 1, pp. 561–564 (2001)
8. Harris, Z.: Distributional structure. Word 10(2/3), 146–162 (1954)
9. Hastie, T.J., Tibshirani, R.J., Friedman, J.H.: The Elements of Statistical Learning: Data Mining, Inference, and Prediction. Springer Series in Statistics. Springer, New York (2009). https://doi.org/10.1007/978-0-387-84858-7. Autres impressions: 2011 (corr.), 2013 (7e corr.)
10. Joulin, A., Grave, E., Bojanowski, P., Mikolov, T.: Bag of tricks for efficient text classification. In: Proceedings of the 15th Conference of the European Chapter of the Association for Computational Linguistics, vol. 2, Short Papers, pp. 427–431. Association for Computational Linguistics (2017). http://aclweb.org/anthology/E17-2068
11. Koppel, M., Schler, J., Argamon, S.: Computational methods in authorship attribution. J. Am. Soc. Inform. Sci. Technol. 60(1), 9–26 (2009)
12. Pedregosa, F., Varoquaux, G., Gramfort, A., Michel, V., Thirion, B., Grisel, O., Blondel, M., Prettenhofer, P., Weiss, R., Dubourg, V., Vanderplas, J., Passos, A., Cournapeau, D., Brucher, M., Perrot, M., Duchesnay, E.: Scikit-learn: machine learning in python. J. Mach. Learn. Res. 12, 2825–2830 (2011)
13. Piasecki, M.: User-driven language technology infrastructure - the case of CLARIN-PL. In: Proceedings of the Ninth Language Technologies Conference, Ljubljana, Slovenia (2014). http://nl.ijs.si/isjt14/proceedings/isjt2014_01.pdf
14. Przepiórkowski, A., Bańko, M., Górski, R.L., Lewandowska-Tomaszczyk, B. (eds.): Narodowy Korpus Języka Polskiego [Eng.: National Corpus of Polish]. Wydawnictwo Naukowe PWN (2012). http://nkjp.pl/settings/papers/NKJP_ksiazka.pdf
15. Radziszewski, A.: A tiered CRF tagger for Polish. In: Bembenik, R., Skonieczny, L., Rybinski, H., Kryszkiewicz, M., Niezgodka, M. (eds.) Intelligent Tools for Building a Scientific Information Platform. Studies in Computational Intelligence, vol. 467, pp. 215–230. Springer, Heidelberg (2013). https://doi.org/10.1007/978-3-642-35647-6_16
16. Salton, G., McGill, M.J.: Introduction to Modern Information Retrieval. McGraw-Hill Inc., New York (1986)
17. Stamatatos, E.: A survey of modern authorship attribution methods. J. Am. Soc. Inform. Sci. Technol. 60(3), 538–556 (2009)
18. Torkkola, K.: Discriminative features for textdocument classification. Formal Pattern Anal. Appl. 6(4), 301–308 (2004). https://doi.org/10.1007/s10044-003-0196-8
19. Tsuruoka, Y., Tsujii, J., Ananiadou, S.: Stochastic gradient descent training for l1-regularized log-linear models with cumulative penalty. In: Proceedings of the Joint Conference of the 47th Annual Meeting of the ACL and the 4th International Joint Conference on Natural Language Processing of the AFNLP, ACL 2009, pp. 477–485. Association for Computational Linguistics, Stroudsburg (2009)

20. Walkowiak, T.: Language processing modelling notation – orchestration of NLP microservices. In: Zamojski, W., Mazurkiewicz, J., Sugier, J., Walkowiak, T., Kacprzyk, J. (eds.) DepCoS-RELCOMEX 2017. AISC, vol. 582, pp. 464–473. Springer, Cham (2018). https://doi.org/10.1007/978-3-319-59415-6_44

21. Woliński, M.: Morfeusz reloaded. In: Calzolari, N., Choukri, K., Declerck, T., Loftsson, H., Maegaard, B., Mariani, J., Moreno, A., Odijk, J., Piperidis, S. (eds.) Proceedings of the Ninth International Conference on Language Resources and Evaluation, LREC 2014, pp. 1106–1111. ELRA, Reykjavík (2014)

Constraint-Based Identification of Complex Gateway Structures in Business Process Models

Piotr Wiśniewski[✉] and Antoni Ligęza

AGH University of Science and Technology,
al. A. Mickiewicza 30, 30-059 Krakow, Poland
{wpiotr,ligeza}@agh.edu.pl

Abstract. In this paper, we present a method for identifying parallel and alternative gateway structures in BPMN models. It can be applied in the composition of business processes from their declarative specifications. Our approach is based on a directed graph representation of a business process as well as the constraint programming technique. Provided the information about process activities and relations between them, the proposed approach consists in finding a structure of logical data-based gateways that satisfies the set of predefined constraints. A detailed illustration of our method is preceded by a brief description of BPMN and its formal representation.

Keywords: Business process management · Graph theory
Decision support · Structure identification

1 Introduction

Process models are widely used for knowledge representation in business, engineering, administration as well as other domains of life where activities and workflows are present. Real-life business processes include numerous branching flows which represent sets of activities that can be executed in parallel or alternatively, depending on one or more conditions that need to be satisfied. Such flows are controlled by gateways which represent basic logical functions. Manually designed process models with multiple nested branching flows can be vulnerable to various types of anomalies [10,20,22] such as deadlocks, lack of synchronization and infinite loops [28]. Therefore, generating a semantically faultless gateway structure may ensure the correctness of the designed model. In addition, business process models described in non-graphical forms, such as natural language [5] are usually lacking exact information about the form of all the alternative and parallel branches of the process. Identifying an admissible structure of such branches is a significant step towards automation and facilitation of business process modeling.

The identified gateway structures should satisfy the set of constraints. Part of these constraints are predefined according to the process modeling principles,

© Springer International Publishing AG, part of Springer Nature 2018
L. Rutkowski et al. (Eds.): ICAISC 2018, LNAI 10842, pp. 788–798, 2018.
https://doi.org/10.1007/978-3-319-91262-2_69

while the rest is based on the input data (see Fig. 1). The incoming information is then processed by a Constraint Satisfaction Problem (CSP) solver which generates the solution needed to compose a complete BPMN diagram.

This work is related to various types of research in the area of graph theory, causal structure discovery and business process management. One of the approaches consists in identifying causal structures in a given system by verifying the existence of internal connections through reification [19]. Causal dependencies can be also discovered based on Bayesian Networks [25] as well as Answer Set Programming technique [6] including the resolution of conflict between dependency constraints. Another related work aims to generate and execute dynamic workflows based on a product structure, denoted as Product Data Model [31]. An executable process model with parallel gateway structures can be discovered from very simple dependency specification such as Attribute Relationship Diagrams [12]. The proposed approach is also close to the discovery and use of AND/OR graphs [18] as well as temporal structures [1].

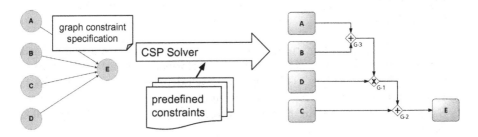

Fig. 1. Schematic presentation of the proposed approach.

Our work includes also the use of formal representations of business process models in Business Process Model and Notation (BPMN) that enables process modelers to design and reconstruct models without a big effort. One of the easily understandable notation is the spreadsheet-based representation of BPMN models [15] where activities are stored in a table including their branches and routing conditions. A semantic model [14] for BPMN can be also created in the CSP (Communicating Sequential Processes) process algebra [35] which facilitates the specification of time constraints between activities in the process. However, the approach which was significantly useful for the purpose of structure identification is the declarative business process modeling [26,36] which focuses on relations between tasks instead of describing the workflow explicitly.

This paper is organized as follows. Section 2 presents an overview of business process management as well as the basics of BPMN including the classification of process flow objects in terms of the number of connected elements. Section 3 describes the principles of creating a directed graph-based representation of a business process model. The detailed information about our approach which includes a illustrative example was included in Sect. 4 while Sect. 5 presents the usage of generated results in business process models. Summary and plans for future work were briefly discussed in Sect. 6.

2 Overview of Business Process Management

Business processes are used to represent workflows in organizations. One of the most significant domains within Business Process Management is the design of process models. It includes manual creation of models performed by process designers or business analysts, as well as model generation from other representations such as: natural text description [5], structured text [11], UML [23], Attribute Relationship Diagrams [12], spreadsheets [15], unordered task list [34] or mining such models from workflow logs generated by existing systems [8,9].

BPMN [24] stands for the leading notation used to represent process models. BPMN provides a uniform standard for process visualization, documentation and communication in form of graphical diagrams whose elements and constructs are easily understandable even for those who are not familiar with process modeling principles [29]. In order to build a BPMN business process model, four main groups of elements are used: *flow objects*, *connecting objects*, *swimlanes* and *artifacts*. However, creating basic models requires only first two groups which designate the flow of activities. The set of flow objects can be divided into three major types of elements:

1. Activities – illustrated by rounded-corner rectangles, describe a well-defined, self-contained piece of work that has to be performed as a part of the process,
2. Events – illustrated by circles, represent well-defined, observable results or outcomes that occur during the execution of the process,
3. Gateways – illustrated by rhombi, control the flow of tokens between flow objects.

This work is mainly related to gateway structures. Therefore, the main focus was put on gateways themselves, as well as activities which are the key flow objects present in a majority of business process models. An activity can either be a task – a single unit of work, or a subprocess – a complex work which can be considered as a separate business process. Gateways control the manner in which a business process can be executed. The most common types of gateways are based on given process data and include the following:

– Exclusive gateways – used to control alternative process flows,
– Inclusive gateways – used to indicate potentially parallel paths based on a condition,
– Parallel gateways – used to indicate parallel flows without the necessity of checking any condition.

Because of the fact that in many cases an inclusive gateway (OR-type) can be replaced by a combination of exclusive and parallel gateways [3] (in BPMN 1.0 it was even possible for every case [2]), the approach proposed in this paper covers the identification of these two basic types.

Flow objects can be also classified into five main groups, regarding their number of inputs and outputs [33]:

1. Sources – elements that can have only output sequence flow.
2. Sinks – elements that can have only inputs.
3. SESE – Single Entry Single Exit.
4. SEME – Single Entry Multiple Exit.
5. MESE – Multiple Entry Single Exit.

All the flow objects in a process need to be connected using the so called connecting objects, which are of three types: sequences, messages, and associations. Between gateways and activities only sequence flows are used. They indicate the order in which particular activities are performed in a process.

In order to demonstrate what business function is responsible for a particular flow objects or by which part of system it is used, the notion of swimlanes is present in BPMN. Process participants are represented by pools, which on the lower level can be divided into lanes that describe objects or roles in a process.

3   Formal Model of a Business Process

A business process model can be represented in the form of a directed graph [32]. Regarding the fact that a mathematical model of a business process is determined by two sets (flow objects and sequence flows) [13,20], it is possible to define a business process graph as a connected, directed graph $G = (V_O, E_F)$ [33], where:

– V_O is a set of vertices corresponding to all flow objects in a process,
– E_F is a set of edges corresponding to all sequence flows in a process.

Vertices of a process graph may contain properties which include additional information such as: pool or lane, object type, object name, as well as optionally a condition or related data entities. They should be all connected with directed edges that may have assigned triggering condition if preceded by a split gateway.

Message flows and artifacts are not included in this notation, as they are optional in business process models and do not influence directly the execution flow of a process. However, process graphs support loops as well as multiple start or end events, which makes them a useful replacement for BPMN diagrams.

4   Constraint-Based Search for Gateway Structures

The term "Gateway Structures" refers to the sets of directly connected gateways with no activities in between. A structure may be considered complex if the number of different gateways exceeds one. Such sets may represent a logical function used to determine which activities may occur before - in case of merge gateways or after a selected activity. Figure 2 presents a sample BPMN subprocess model which depicts product preparation for a sales order. In this diagram two complex structures were identified and grouped.

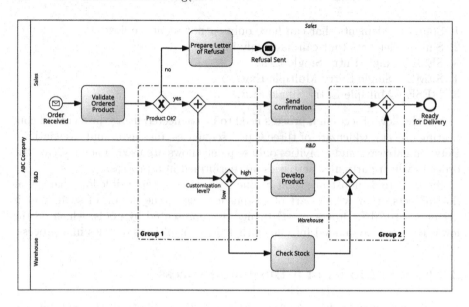

Fig. 2. BPMN diagram representing product preparation for a sales order. Based on [7].

Let us denote the successors of Validate Ordered Product task as follows:

τ_1 – Prepare Letter Of refusal
τ_2 – Send Confirmation
τ_3 – Develop Product
τ_4 – Check Stock

In this case, the set of successors in terms of process execution is determined by a logical formula (Expression 1), given $\oplus$ as the operator for exclusive disjunction.

$$\tau_1 \oplus (\tau_2 \wedge (\tau_3 \oplus \tau_4)) \tag{1}$$

The idea of structure discovery can be applied to activity graphs [17] which can be considered as a simplified version of business process graphs presented in Sect. 3. Such graphs determine admissible chain response and chain precedence [21] relations between activities of a business process. An activity graph G_A can be defined by Expression 2.

$$G_A = (V_A, E, \sigma_I, \sigma_O), \tag{2}$$

where:

- V_A is a finite set of vertices representing process activities,
- E is a set of directed edges,
- $\sigma_I : V_A \longrightarrow \mathbb{N}_0$ determines the number of incoming edges for a vertex and $\mathbb{N}_0$ stands for non-negative integers,
- $\sigma_O : V_A \longrightarrow \mathbb{N}_0$ determines the number of outgoing edges for a vertex.

However, the activity graph lacks the information about allowed concurrency of the process activities. Therefore, in order to identify gateway structures, the user has to indicate which pairs of activities (tasks or subprocesses) can be executed in parallel. Such pairs can be also determined using the process composition technique [34]. Let us PA be the set containing all these pairs.

According to the classification presented in Sect. 2, process activities are in general SESE elements. Thus, in case of every vertex with multiple incident (both incoming and outgoing) edges, a gateway structure should be determined. If vertex v_0 has only two predecessors or successors (v_1 and v_2) then the solution is trivial:

- if $\{v_1, v_2\} \in PA$ then insert a parallel gateway,
- otherwise insert an exclusive gateway.

A more complex method of gateway identification is needed in case when the number of incident edges exceeds two. Let us present in details the model used to discover merge gateways structures, as split gateways are discovered analogically. The input for the search algorithm is a square matrix AR of size $(\sigma_I(v_0) \times \sigma_I(v_0))$, where v_0 is a destination vertex in the activity graph. Provided that vertex v_0 has n predecessors, denoted as $v_1...v_n$ where $n = \sigma_I(v_0)$, elements of matrix AR are defined by Formula 3.

$$AR_{ij} = \begin{cases} 0, \text{if } i = j, \\ 1, \text{if } v_i \text{ and } v_j \text{ can be executed in parallel,} & \forall i, j \in \{1..n\}, \\ -1, \text{otherwise.} \end{cases} \quad (3)$$

According to the BPMN specification, all merge gateways are MESE elements. Therefore, each gateway must have at least two inputs and exactly one output. This implies that the number of gateways cannot exceed $k = n - 1$. The search process consists in finding values for the following decision variables:

- The list of destination gateways for activities $v_1...v_n$:

$$activity\_gateways = [ag_i \in 1..k, \forall i \in \{1..n\}]. \quad (4)$$

- The inner graph structure between all generated gateways represented by an incidence matrix of size $(k \times k)$, denoted as $gateway\_connections$.
- The vector determining the types of all gateways:

$$gateway\_types = [gt_i = \begin{cases} 0, \text{if the gateway is not used,} \\ -1, \text{if the gateway is exclusive,} & \forall i \in \{1..k\}]. \\ 1, \text{if the gateway is parallel.} \end{cases}$$

$$(5)$$

- The index of the output gateway (connected to v_0), denoted as $output\_gateway$.

The identification of the gateway structure is based on the following constraints:

1. Each predecessor has to be connected to exactly one gateway.
2. None of the predecessors can be connected directly to v_0.
3. Each gateway must have at least two inputs and exactly one output.
4. Only one gateway can be linked to v_0.
5. A gateway cannot be connected to itself.

In addition to the decision variables, the algorithm is used to calculate paths leading from each activity to the output gateway. They are saved in form of k-element vectors having gateway indices as values. The paths are used to calculate the first common link, which is the closest gateway which connects two activities in a direct or indirect way. The following constraints determine the shape of such paths:

1. The gateways on each path should be different.
2. Each gateway in a path should be connected to its successor.
3. When the output gateway is reached, all the following elements in a path should be equal to zero.
4. If two activities can be executed in parallel then its first common link should be a parallel gateway. Otherwise, the gateway should be exclusive.

In order to illustrate the proposed approach, let us analyze an elementary example illustrated by a part extracted from a sample activity graph (Fig. 3).

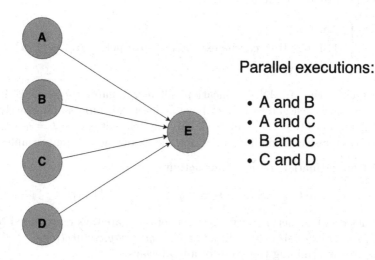

Fig. 3. Part of an activity graph

The figure contains also information about pairs of tasks that can be executed in parallel which are in the same time the elements of set PA. Based on this information, matrix AR for this structure was constructed (Formula 6).

$$AR = \begin{bmatrix} 0 & 1 & 1 & -1 \\ 1 & 0 & 1 & -1 \\ 1 & 1 & 0 & 1 \\ -1 & -1 & 1 & 0 \end{bmatrix} \tag{6}$$

The constraint satisfaction problem can be solved using MiniZinc Gecode solver [27] with the goal to satisfy the predefined constraints. Only first solution is sufficient for the purpose of structure identification. The generated results are the admissible values of decision variables (Formulae 7–10).

$$activity\_gateways = [\ 3\ \ 3\ \ 2\ \ 1] \tag{7}$$

$$gateway\_connections = \begin{bmatrix} 0 & 1 & -1 \\ -1 & 0 & 0 \\ 1 & 0 & 0 \end{bmatrix} \tag{8}$$

$$gateway\_types = [-1\ \ 1\ \ 1] \tag{9}$$

$$output\_gateway = 2 \tag{10}$$

If the required number of gateways is lower than maximum (e.g. 1) then the corresponding values in $gateway\_types$ vector are equal to zero. To illustrate such a case, let us reuse the example presented in Fig. 3, but omit the pairs of activities allowed to be executed in parallel. In such a situation, only one exclusive gateway would be needed to merge all the outgoing edges of the predecessors of v_0. Therefore, $gateway\_types$ vector would be equal to the one shown in Formula 11. In addition, all the elements in $activity\_gateway$ vector would be equal to 1 and so would be the value for $output\_gateway$. At the same time all the elements of $gateway\_connections$ matrix would be equal to zero.

$$gateway\_types' = [-1\ \ 0\ \ 0] \tag{11}$$

5   Generation of a Business Process Graph

The identified structure is used to build a business process graph which represents the complete BPMN model. In such case, the following algorithm should be used to include a merge gateway structure in an activity graph:

1. Create gateways whose number and types are determined by $gateway\_types$ vector.
2. Connect each vertex $v_1...v_n$ to gateways whose indices are present in $activity\_gateways$ vector.
3. Create connections between gateways using $gateway\_connections$ as incidence matrix.
4. Connect the output gateway with v_0.
5. Remove all edges leading to v_0 from its predecessors $v_1...v_n$.

The proposed algorithm was applied for the example presented in Sect. 4. A part of the BPMN model generated with use of the business process graph is shown in Fig. 4. In general, 3 new gateways as well as 7 sequence flows were created. At the same time, 4 sequence flows linking source activity directly to the destination one were removed.

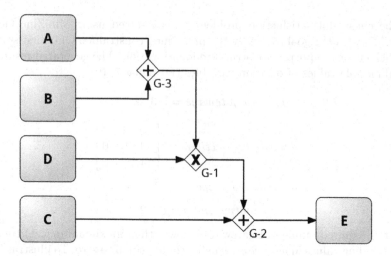

Fig. 4. Identified gateway structure in form of a BPMN diagram. Gateway names correspond to their indices in the generated data structures

6   Conclusions

In the paper, we presented a novel constraint-based model used to identify complex gateway structures. In contrast to the related approaches such as the discovery of logical propositions based on statistical data [30], our method focuses on the BPMN notation which makes it more specified for its purpose. Thus, it is easier to use this model as a part of a process composition algorithm or directly by a process designer. As future works, we would like to perform a comparative analysis between the proposed approach and related algorithms of causal structure discovery based on AI techniques. This would include rule identification using evolutionary algorithms [4] as well as graph structure discovery based on Bayesian Networks [16]. In addition, we plan to cover the whole scope of the business process in the constraint model in order to ensure the semantic correctness of the resulting BPMN diagram.

References

1. de Boer, M., Escher, C., Schutte, K.: Modelling temporal structures in video event retrieval using an AND-OR graph. In: The Ninth International Conferences on Advances in Multimedia, MMEDIA 2017. IARIA XPS Press: Sl (2017)
2. Dijkman, R.M., Dumas, M., Ouyang, C.: Semantics and analysis of business process models in BPMN. Inf. Softw. Technol. **50**(12), 1281–1294 (2008)
3. Favre, C., Völzer, H.: The difficulty of replacing an inclusive OR-join. In: Barros, A., Gal, A., Kindler, E. (eds.) BPM 2012. LNCS, vol. 7481, pp. 156–171. Springer, Heidelberg (2012). https://doi.org/10.1007/978-3-642-32885-5_12

4. Freitas, A.A.: A survey of evolutionary algorithms for data mining and knowledge discovery. In: Ghosh, A., Tsutsui, S. (eds.) Advances in Evolutionary Computing. Natural Computing Series, pp. 819–845. Springer, Heidelberg (2003). https://doi.org/10.1007/978-3-642-18965-4_33
5. Friedrich, F., Mendling, J., Puhlmann, F.: Process model generation from natural language text. In: Mouratidis, H., Rolland, C. (eds.) CAiSE 2011. LNCS, vol. 6741, pp. 482–496. Springer, Heidelberg (2011). https://doi.org/10.1007/978-3-642-21640-4_36
6. Hyttinen, A., Eberhardt, F., Järvisalo, M.: Constraint-based causal discovery: conflict resolution with answer set programming. In: UAI 2014, pp. 340–349 (2014)
7. BPM Academic Initiative: BPM Academic Initiative Model Collection. http://bpmai.org/BPMAcademicInitiative/
8. Kalenkova, A.A., van der Aalst, W.M., Lomazova, I.A., Rubin, V.A.: Process mining using BPMN: relating event logs and process models. Softw. Syst. Model. 16(4), 1019–1048 (2017)
9. Kalenkova, A.A., de Leoni, M., van der Aalst, W.M.: Discovering, analyzing and enhancing BPMN models using ProM? In: Business Process Management-12th International Conference, BPM, pp. 7–11 (2014)
10. Klimek, R.: A system for deduction-based formal verification of workflow-oriented software models. Int. J. Appl. Math. Comput. Sci. 24(4), 941–956 (2014)
11. Kluza, K., Honkisz, K.: From SBVR to BPMN and DMN models. Proposal of translation from rules to process and decision models. In: Rutkowski, L., Korytkowski, M., Scherer, R., Tadeusiewicz, R., Zadeh, L.A., Zurada, J.M. (eds.) ICAISC 2016. LNCS (LNAI), vol. 9693, pp. 453–462. Springer, Cham (2016). https://doi.org/10.1007/978-3-319-39384-1_39
12. Kluza, K., Nalepa, G.J.: A method for generation and design of business processes with business rules. Inf. Softw. Technol. 91, 123–141 (2017)
13. Kluza, K., Nalepa, G.J.: Formal model of business processes integrated with business rules. Inf. Syst. Front. 1–19 (2018). https://doi.org/10.1007/s10796-018-9826-y
14. Kluza, K., Nalepa, G.J., Ślażyński, M., Kutt, K., Kucharska, E., Kaczor, K., Łuszpaj, A.: Overview of selected business process semantization techniques. In: Pełech-Pilichowski, T., Mach-Król, M., Olszak, C.M. (eds.) Advances in Business ICT: New Ideas from Ongoing Research. SCI, vol. 658, pp. 45–64. Springer, Cham (2017). https://doi.org/10.1007/978-3-319-47208-9_4
15. Kluza, K., Wiśniewski, P.: Spreadsheet-based business process modeling. In: 2016 Federated Conference on Computer Science and Information Systems (FedCSIS), pp. 1355–1358. IEEE (2016)
16. Koivisto, M., Sood, K.: Exact Bayesian structure discovery in Bayesian networks. J. Mach. Learn. Res. 5(May), 549–573 (2004)
17. Kundu, D., Samanta, D.: A novel approach to generate test cases from UML activity diagrams. J. Object Technol. 8(3), 65–83 (2009)
18. Liang, Q.A., Su, S.Y.: And/or graph and search algorithm for discovering composite web services. Int. J. Web Serv. Res. 2(4), 48 (2005)
19. Ligęza, A.: An experiment in causal structure discovery. a constraint programming approach. In: Kryszkiewicz, M., Appice, A., Ślęzak, D., Rybinski, H., Skowron, A., Raś, Z.W. (eds.) ISMIS 2017. LNCS (LNAI), vol. 10352, pp. 261–268. Springer, Cham (2017). https://doi.org/10.1007/978-3-319-60438-1_26
20. Ligęza, A., Kluza, K., Potempa, T.: AI approach to formal analysis of BPMN models. Towards a logical model for BPMN diagrams. In: Proceedings of the Federated Conference on Computer Science and Information Systems - FedCSIS 2012, Wroclaw, Poland, 9–12 September 2012, pp. 931–934 (2012)

21. Maggi, F.M., Mooij, A.J., van der Aalst, W.M.: User-guided discovery of declarative process models. In: 2011 IEEE Symposium on Computational Intelligence and Data Mining (CIDM), pp. 192–199. IEEE (2011)
22. Mroczek, A., Ligeza, A.: A note on BPMN analysis. Towards a taxonomy of selected potential anomalies. In: 2014 Federated Conference on Computer Science and Information Systems (FedCSIS), pp. 1097–1102. IEEE (2014)
23. Nawrocki, J.R., Nedza, T., Ochodek, M., Olek, L.: Describing business processes with use cases. In: BIS 2006, pp. 13–27 (2006)
24. OMG: business process model and notation (BPMN): version 2.0 specification. Technical report formal/2011-01-03, Object Management Group, January 2011
25. Pearl, J.: Causality: models, reasoning, and inference. Econom. Theor. 19(675–685), 46 (2003)
26. Pesic, M., van der Aalst, W.M.P.: A declarative approach for flexible business processes management. In: Eder, J., Dustdar, S. (eds.) BPM 2006. LNCS, vol. 4103, pp. 169–180. Springer, Heidelberg (2006). https://doi.org/10.1007/11837862_18
27. Schulte, C., Stuckey, P.J.: Efficient constraint propagation engines. ACM Trans. Program. Lang. Syst. 31(1), 2:1–2:43 (2008). https://doi.org/10.1145/1452044.1452046
28. Mroczek, S.A., Wiśniewski, P., Ligęza, A.: Overview of verification tools for business process models. In: Ganzha, M., Maciaszek, L., Paprzycki, M. (eds.) Communication Papers of the 2017 Federated Conference on Computer Science and Information Systems. Annals of Computer Science and Information Systems, vol. 13, pp. 295–302. PTI, Mumbai (2017). https://doi.org/10.15439/2017F308
29. Trkman, M., Mendling, J., Krisper, M.: Using business process models to better understand the dependencies among user stories. Inf. Softw. Technol. 71, 58–76 (2016)
30. Tsukimoto, H.: The discovery of logical propositions in numermal data. In: AAAI-94 Workshop on Knowledge Discovery in Databases, pp. 205–216 (1994)
31. Vanderfeesten, I., Reijers, H.A., van der Aalst, W.M.P.: Product based workflow support: dynamic workflow execution. In: Bellahsène, Z., Léonard, M. (eds.) CAiSE 2008. LNCS, vol. 5074, pp. 571–574. Springer, Heidelberg (2008). https://doi.org/10.1007/978-3-540-69534-9_42
32. Weber, I.M.: Semantic Methods for Execution-level Business Process Modeling: Modeling Support Through Process Verification and Service Composition. LNBIP, vol. 40. Springer, Heidelberg (2009). https://doi.org/10.1007/978-3-642-05085-5
33. Wiśniewski, P.: Decomposition of business process models into reusable sub-diagrams. In: ITM Web of Conferences, vol. 15, p. 01002. EDP Sciences (2017)
34. Wiśniewski, P., Kluza, K., Ślażyński, M., Ligęza, A.: Constraint-based composition of business process models. In: Teniente, E., Weidlich, M. (eds.) BPM 2017. LNBIP, vol. 308, pp. 133–141. Springer, Cham (2018). https://doi.org/10.1007/978-3-319-74030-0_9
35. Wong, P.Y., Gibbons, J.: Formalisations and applications of BPMN. Sci. Comput. Program. 76(8), 633–650 (2011)
36. Zugal, S., Soffer, P., Haisjackl, C., Pinggera, J., Reichert, M., Weber, B.: Investigating expressiveness and understandability of hierarchy in declarative business process models. Softw. Syst. Model. 14(3), 1081–1103 (2015)

Developing a Fuzzy Knowledge Base and Filling It with Knowledge Extracted from Various Documents

Nadezhda Yarushkina(iD), Vadim Moshkin(iD), Aleksey Filippov(✉)(iD),
and Gleb Guskov(iD)

Ulyanovsk State Technical University, Street Severny Venets 32,
432027 Ulyanovsk, Russian Federation
{jng,v.moshkin,al.filippov,g.guskov}@ulstu.ru
http://www.ulstu.ru/

Abstract. The article describes the process of developing a fuzzy knowledge base. The content of the fuzzy knowledge base is the result of extracting knowledge from the set of documents by subject area. Set of documents consists of the wiki-resources, UML-diagrams, documents and source code of projects. Knowledge base based on the graph database Neo4j. An attempt to implement the mechanism of inference by the contents of a graph database was made. This mechanism is used to generate the screen forms of the user interface dynamically. The contexts allow representing the content of the fuzzy knowledge base in space and time. Each space context is assigned a linguistic label, for example, low, middle, high. This label determines the competence of the expert in the given subject area. Time contexts allow storing the history of the knowledge base content changes. It allows returning to a specific state of the contents of the knowledge base.

Keywords: Ontology · Fuzzy knowledge base · Context analysis
Subject area · Graph database

1 Introduction

People from all over the world operate with huge volumes of information both in everyday and professional activities. A large amount of information causes difficulties in decisions making within strict time constraints.

To solve this problem, many software tools are used to automate human activity. However, it is necessary to adapt them to the specifics of a particular subject area and its contexts for the effective operation of such tools [1,2,8,12, 20–22].

Thus, customized software for automation solves such tasks more efficiently, but customizing procedure requires considerable human and temporary resources.

© Springer International Publishing AG, part of Springer Nature 2018
L. Rutkowski et al. (Eds.): ICAISC 2018, LNAI 10842, pp. 799–810, 2018.
https://doi.org/10.1007/978-3-319-91262-2_70

In this article, an attempt is made to construct a fuzzy knowledge base as a core of the Athene platform [27]. The content of the fuzzy knowledge base is an applied ontology. The basic requirements for fuzzy knowledge base are:

- adaptation to the specifics of subject area based on contexts;
- reliability and speed of ontology storage;
- the presence of a mechanism of logical inference;
- availability of mechanisms for importing data from external information resources [13].

The contexts allow representing the content of the fuzzy knowledge base in space and time [7,10].

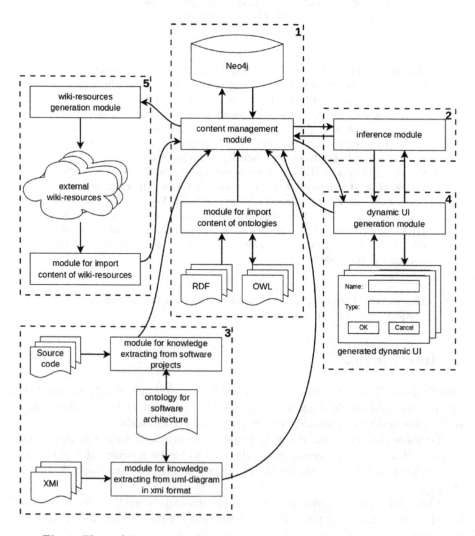

Fig. 1. The architecture of a fuzzy knowledge base of the Athene platform.

Space contexts allow taking into account the competence of the expert in the subject area. Each space context is assigned a linguistic label, for example, low, middle, high. This label determines the competence of the expert in the given subject area. The membership function is defined for each linguistic label. The resulting ontology is formed on the basis of the values of linguistic labels.

Time contexts allow storing the history of changes in the knowledge base content. It allows returning to a specific state of the contents of the knowledge base.

Figure 1 shows the system architecture which consists of the following sub-systems:

1. Ontology store:
   - Neo4j [17];
   - content management module;
   - ontology import/export module (RDF, OWL).
2. Inference subsystem:
   - inference module.
3. A subsystem for importing data from software source code and UML-diagram:
   - a module for importing data from software source code;
   - a module for importing data from UML-diagram in .XMI format;
4. A subsystem for interaction with users:
   - screen forms generation module.
5. A subsystem for importing data from external information resources:
   - a module for importing data from external wiki-resources;
   - a module for filling external wiki-resources.

2   The Organization of the Ontology Storage as a Fuzzy Knowledge Base

Ontology is a model of the subject area representation in the form of a semantic graph [11].

Ontology storage of knowledge base based on graph-oriented database management system (Graph DBMS) Neo4j. Neo4j is currently one of the most popular graph databases and it has the following advantages:

1. Native format for data storage.
2. One copy of Neo4j can work with graphs containing billions of nodes and relationships.
3. The presence of a graph-oriented query language Cypher.
4. Availability of transaction support [28].

Neo4j was chosen to store the subject area knowledges in the applied ontology form. It's possible because ontology can be presented as a graph. In this case, it is only necessary to limit the set of nodes and graph relations of ontologies in RDF or OWL format for successful transformation.

The context of an ontology is some state of ontology, obtained during building an ontology using different "points of view" or from many versions of an ontology.

Formally, the ontology can be represented by the following equation:

$$O = \langle T, C^{T_i}, I^{T_i}, P^{T_i}, S^{T_i}, F^{T_i}, R^{T_i} \rangle, i = \overline{1, t}, \tag{1}$$

where t is a number of the ontology contexts;

$T = \{T_1, T_2, \ldots, T_n\}$ is a set of ontology contexts;

$C^{T_i} = \{C_1^{T_i}, C_2^{T_i}, \ldots, C_n^{T_i}\}$ is a set of ontology classes within the i-th context;

$I^{T_i} = \{I_1^{T_i}, I_2^{T_i}, \ldots, I_n^{T_i}\}$ is a set of ontology objects within the i-th context;

$P^{T_i} = \{P_1^{T_i}, P_2^{T_i}, \ldots, P_n^{T_i}\}$ is a set of ontology classes properties within the i-th context;

$S^{T_i} = \{S_1^{T_i}, S_2^{T_i}, \ldots, S_n^{T_i}\}$ is a set of ontology objects states within the i-th context;

$F^{T_i} = \{F_1^{T_i}, F_2^{T_i}, \ldots, F_n^{T_i}\}$ is a set of the logical rules fixed in the ontology within the i-th context;

R^{T_i} is a set of ontology relations within the i-th context defined as:

$$R^{T_i} = \{R_C^{T_i}, R_I^{T_i}, R_P^{T_i}, R_S^{T_i}, R_F^{T_i}\}, \tag{2}$$

where $R_C^{T_i}$ is a set of relations defining a hierarchy of ontology classes within the i-th context;

$R_I^{T_i}$ is a set of relations defining the 'class-object' ontology tie within the i-th context;

$R_P^{T_i}$ is a set of relations defining the 'class-class property' ontology tie within the i-th context;

$R_S^{T_i}$ is a set of relations defining the 'object-object state' ontology tie within the i-th context;

$R_F^{T_i}$ is a set of relations generated on the basis of logical ontology rules in the context of i-th context.

Fuzzy knowledge base main principles are similar to object-oriented programming paradigm:

- ontology classes are concepts of the subject area;
- classes can have properties, the child-class inherits properties of the parent class;
- objects of ontology describe instances of the concepts of the subject ontology;
- specific values for the properties of objects inherited from the parent class are determined by the states;
- logical rules are used to implement the functions of inference by the content of a fuzzy knowledge base.

3   The Inference on the Contents of a Fuzzy Knowledge Base

The inference is the process of reasoning from the premises to the conclusion. Reasoners are used to implementing the function of inference. Reasoners form logical consequences on the basis of many statements, facts and axioms. The most popular at the moment reasoners are:

- Pellet [19];
- FaCT++;
- Hermit;
- Racer, etc [6].

These reasoners are actively used in the development of intelligent software. However, Neo4j does not assume the possibility of using any one of such reasoners. Thus, there is a need to develop a mechanism for inference based on the content of a fuzzy knowledge base [3,4]. Currently, the Semantic Web Rule Language (SWRL) is used to record logical rules [25]. These SWRL rules describe the conditions under which object a has "nephew-uncle" relation with object c. Formally the logical rule of the ontology of the fuzzy knowledge base is:

$$F^{T_i} = \langle A^{Tree}, A^{SWRL}, A^{Cypher} \rangle, \tag{3}$$

where T_i is i-th context of the ontology of the fuzzy knowledge base; A^{Tree} is a tree-like representation of a logical rule F^{T_i}; A^{SWRL} is an SWRL representation of a logical rule F^{T_i}; A^{Cypher} is a Cypher representation of a logical rule F^{T_i}.

The tree-view A^{Tree} of a logical rule F^{T_i} is:

$$A^{Tree} = \langle Ant, Cons \rangle, \tag{4}$$

where $Ant = Ant_1 \Theta Ant_2 \Theta \ldots Ant_n$ is the antecedent (condition) of the logical rule F^{T_i}; $\Theta \in \{AND, OR\}$ is a set of permissible logical operations between antecedent atoms; $Cons$ is a consequent (consequence) of a logical rule F^{T_i}.

Figure 2 shows an example of a tree-like representation of a logical rule. This rule describes the nephew-uncle relationship.

The tree-like logical rule (Fig. 2) is translated into the following SWRL:
hasParent(?a, ?b) ∧ *hasBrother(?b, ?c)* → *hasUncle(?a, ?c)*
hasChild(?b, ?a) ∧ *hasBrother(?b, ?c)* → *hasUncle(?a, ?c)*.
and the following Cypher view:
MATCH (a:Object)<-[:RANGE]
  -(:Statementname: "hasParent")
  -[:DOMAIN]->(b:Object)
MATCH (b:Object)<-[:RANGE]
  -(:Statementname: "hasBrother")
  -[:DOMAIN]->(c:Object)
MERGE (a)<-[:RANGE]
  -(:Statementname: "hasUncle")
  -[:DOMAIN]->(c)
MATCH (b:Object)<-[:RANGE]
  -(:Statementname: "hasChild")
  -[:DOMAIN]->(a:Object)
MATCH (c:Object)<-[:RANGE]

-(:Statementname: "hasSister")
-[:DOMAIN]->(b:Object)
MERGE (a)<-[:RANGE]
    -(:Statementname: "hasUncle")
    -[:DOMAIN]->(c).

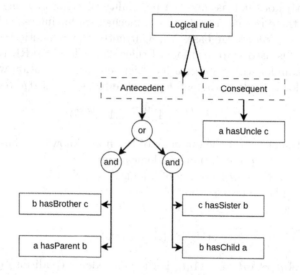

Fig. 2. Example of a tree-like representation of a logical rule.

Thus, the rules are translated into their tree-view when imported into the knowledge base of logical rules on SWRL language.

The presence of a tree-like representation of a logical rule allows forming both an SWRL-representation of a logical rule and a Cypher-representation based on it.

Relations of a special type are formed by using Cypher to represent the logical rule between entities of the ontology of the fuzzy knowledge base. These relations correspond to the antecedent atoms of the logical rule. Formed relationships provide the inference from the contents of the fuzzy knowledge base.

4   UML Meta-Model Based Ontology

A subsystem of extracting knowledge from UML-diagrams and source code based on specific ontology. T-box of this ontology corresponds to UML meta-model. UML class diagram contains all knowledge from the project. If class diagram for the project does not exist, representation for ontology could be obtained from project source code. Considered subsystem works with projects written in Java language [16].

As a structure for storing UML class diagrams was chosen an OWL ontology, because this format is the most expressive for representation of knowledge from complex subject areas [29]. The class diagram elements should be translated into ontology as concepts with considering to their semantics. The semantics of the whole diagram is being formed from the semantics of diagram elements and the semantics of their interaction. That is why the ontology was built on the basis of the UML meta-scheme, not a formal set of translated elements [5,9].

To solve the problem of intellectual analysis of project diagrams, included in the project documentation, it is necessary to have knowledge in the area of construction of formalized diagrams [14,26].

An ontology contains concepts that describe the most basic elements of the class diagram, but it can be expanded if necessary. When translating the meta-model of UML, the following notations were applied.

Formally, the ontology of project diagrams is represented as a set:

$$O^{prj} = \langle C^{prj}, R^{prj}, F^{prj}\rangle, \tag{5}$$

where $C^{prj} = \{c_1^{prj}, \ldots, c_i^{prj}\}$ – is a set of concepts that define main UML diagram elements such as: "Class", "Object", "Interface", "Relationship" and others;

R^{prj} – the set of connections between ontology concepts. These relationships allow us to correctly describe the rules of UML notation.
F^{prj} – is the set of interpretation functions defined on the relationships R^{prj}.

On Fig. 3 represented UML class diagram for a project that realize design pattern builder.

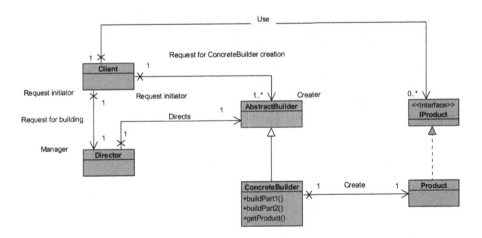

Fig. 3. A class diagram for a Builder design pattern.

Figure 4 shows the A-Box of ontology for representing a project that contains a Builder design pattern.

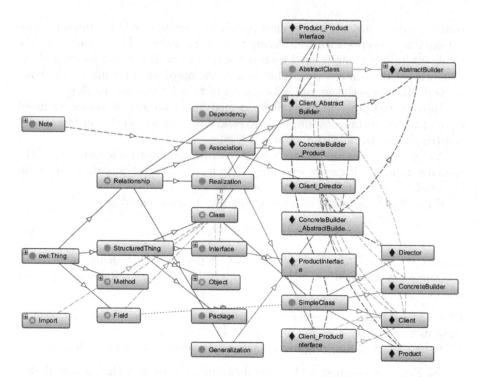

Fig. 4. Builder design pattern ontology presentation in Protege editor.

5   Building a Graphical User Interface Based on the Contents of a Fuzzy Knowledge Base

The dynamic graphical user interface (GUI) mechanism is used to simplify the work with knowledge base of untrained users and control of user input [15,18,23].

You need to map the fuzzy knowledge base ontology entities to the GUI elements to build a GUI based on the contents of the fuzzy knowledge base. Formally, the GUI model can be represented as follows:

$$UI = \langle L, C, I, P, S \rangle, \tag{6}$$

where $L = \{L_1, L_2, \ldots, L_n\}$ is a set of graphical GUI components (for example, ListBox, TextBox, ComboBox, etc.);
$C = \{C_1, C_2, \ldots, C_n\}$ is a set of ontology classes;
$I = \{I_1, I_2, \ldots, I_n\}$ is a set of ontology objects;
$P = \{P_1, P_2, \ldots, P_n\}$ is a set of properties of ontology classes;
$S = \{S_1, S_2, \ldots, S_n\}$ is a set of states of ontology objects of fuzzy knowledge base.

The following function is used to build a GUI based on fuzzy knowledge base:

$$F(O) = \{C^{T_i}, I^{T_i}, P^{T_i}, S^{T_i}, F^{T_i}, R^{T_i}\} \rightarrow \{L, C, I, P, S\},$$

where $\{C^{T_i}, I^{T_i}, P^{T_i}, S^{T_i}, F^{T_i}, R^{T_i}\}$ is a set of ontology entities of fuzzy knowledge base represented by expression 1 within the i-th context;

$\{L, C, I, P, S\}$ is a set of GUI entities of fuzzy KB represented by the expression 6.

Thus, the contents of the fuzzy KB are mapped to many GUI components. This makes it easier to work with KB for a user who does not have skills in ontological analysis and knowledge engineering. It also allows you to monitor the logical integrity of the user input, which leads to a reduction in the number of potential input errors.

6   Interaction of Fuzzy Knowledge Base with External Wiki-Resources

At present, wiki-technologies are used to organize corporate knowledge base [24]. It is necessary to solve the task of importing the content of such wiki-resources into a fuzzy knowledge base. Table 1 contains the result of mapping the fuzzy knowledge base ontology entities to the wiki resource entities.

Table 1. Correspondence between ontology elements and wiki elements

The ontology element of fuzzy knowledge base	The element of wiki-resources
Class	Category
Subclass	Subcategory
Object	Page
Class properties	The infobox elements (properties)
Object states	The infobox elements (values)
Relations	Hyperlinks

Thus, it becomes possible to import the content of external wiki resources for initial filling of knowledge base contents. There is also the possibility of the reverse process – generation of wiki-resources based on the contents of the fuzzy knowledge base.

7   Conclusion

Thus, the use of fuzzy knowledge base stored in the Graph DBMS in the decision support process presupposes the existence of a certain set of mechanisms:

- organization of inference on the content of the fuzzy knowledge base by translating SWRL-rules into Cypher-structures;
- building a graphical user interface based on the contents of fuzzy knowledge base;
- automated import of knowledge from internal and external wiki-resources.

These mechanisms allow you to automate the learning process of working with the system and simplify the work of specialists with a knowledge base.

The contexts of knowledge base allow representing the content of the fuzzy knowledge base in space and time.

Space contexts allow taking into account the competence of the expert in the subject area. Each space context is assigned a linguistic label, for example, low, middle, high. This label determines the competence of the expert in the given subject area. The membership function is defined for each linguistic label. The resulting ontology is formed on the basis of the values of linguistic labels.

Time contexts allows storing the history of the knowledge base content changes. It allows returning a specific state of the knowledge base content.

The application of a contextual approach to the storage of knowledge raises the using of subject ontologies effectiveness. This approach allows adapting the knowledge base to the subject area characteristics and requirements of specialists. A considered system provides them a convenient tool that is software dynamically changeable depending on the contents of the knowledge base.

Acknowledgments. The study was supported by the Ministry of Education and Science of the Russian Federation in the framework of the project No. 2.1182.2017/4.6. Development of methods and means for automation of production and technological preparation of aggregate-assembly aircraft production in the conditions of a multi-product production program.

References

1. Berant, J., Chou, A., Frostig, R., Liang, P.: Semantic parsing on freebase from question-answer pairs. In: Proceedings of the 2013 Conference on Empirical Methods in Natural Language Processing (EMNLP), pp. 1533–1544 (2013)
2. Bianchini, D., De Antonellis, V., Pernici, B., Plebani, P.: Ontology-based methodology for e-service discovery. Inf. Syst. **31**(4), 361–380 (2005)
3. Bobillo, F., Straccia, U.: FuzzyDL: an expressive fuzzy description logic reasoner. In: Proceedings of the 17th IEEE International Conference on Fuzzy Systems (FUZZ-IEEE 2008), pp. 923–930. IEEE Computer Society (2008)
4. Bobillo, F., Straccia, U.: Representing fuzzy ontologies in OWL 2. In: Proceedings of the 19th IEEE International Conference on Fuzzy Systems (FUZZ-IEEE 2010), pp. 2695–2700. IEEE Press (2010)
5. Carvalho, N.R., Almeida, J.J., Henriques, P.R., Pereira, M.J.V.: CONCLAVE: ontology-driven measurement of semantic relatedness between source code elements and problem domain concepts. In: Murgante, B., et al. (eds.) ICCSA 2014. LNCS, vol. 8584, pp. 116–131. Springer, Cham (2014). https://doi.org/10.1007/978-3-319-09153-2_9
6. Dentler, K., Cornet, R., ten Teije, A., de Keizer, N.: Comparison of reasoners for large ontologies in the OWL 2 EL profile. Semant. Web **2**, 71–87 (2011)
7. Falbo, R.A., Quirino, G.K., Nardi, J.C., Barcellos, M.P., Guizzardi, G., Guarino, N.: An ontology pattern language for service modeling. In: Proceedings of the 31st Annual ACM Symposium on Applied Computing, pp. 321–326 (2016)
8. Farid, D.M., Al-Mamun, M.A., Manderick, B., Nowe, A.: An adaptive rule-based classifier for mining big biological data. Expert Syst. Appl. **64**, 305–316 (2016)

9. Almeida Ferreira, D., Silva, A.: UML to OWL mapping overview an analysis of the translation process and supporting tools. In: 7th Conference of Portuguese Association of Information Systems, pp. 2536–2549 (2013)

10. Gao, M., Liu, C.: Extending OWL by fuzzy description logic. In: Proceedings of the 17th IEEE International Conference on Tools with Artificial Intelligence (ICTAI 2005), pp. 562–567. IEEE Computer Society (2005)

11. Guarino, N., Musen, M.A.: Ten years of applied ontology. Appl. Ontol. **10**(3–4), 169–170 (2015)

12. Guizzardi, G., Guarino, N., Almeida, J.P.A.: Ontological considerations about the representation of events and endurants in business models. In: La Rosa, M., Loos, P., Pastor, O. (eds.) BPM 2016. LNCS, vol. 9850, pp. 20–36. Springer, Cham (2016). https://doi.org/10.1007/978-3-319-45348-4_2

13. Gruber, T.: Ontology. http://tomgruber.org/writing/ontology-in-encyclopedia-of-dbs.pdf. Accessed 10 Jan 2018

14. Guskov, G., Namestnikov, A., Yarushkina, N.: Approach to the search for similar software projects based on the UML ontology. In: Abraham, A., Kovalev, S., Tarassov, V., Snasel, V., Vasileva, M., Sukhanov, A. (eds.) IITI 2017. AISC, vol. 680, pp. 3–10. Springer, Cham (2018). https://doi.org/10.1007/978-3-319-68324-9_1

15. Hattori, S., Takama, Y.: Recommender system employing personal-value-based user model. J. Adv. Comput. Intell. Intell. Inform. (JACIII) **18**(2), 157–165 (2014)

16. Koukias, A., Nadoveza, D., Kiritsis, D.: An ontology-based approach for modelling technical documentation towards ensuring asset optimisation. Int. J. Prod. Lifecycle Manag. **8**(1), 24–45 (2015)

17. Neo4j. https://neo4j.com/product. Accessed 10 Jan 2018

18. Ltifi, H., Kolski, C., Ayed, M.B., Alimi, A.M.: A human-centred design approach for developing dynamic decision support system based on knowledge discovery in databases. J. Decis. Syst. **22**, 69–96 (2013)

19. Pellet Framework. https://github.com/stardog-union/pellet. Accessed 10 Jan 2018

20. Rajpathak, D., Chougule, R., Bandyopadhyay, P.: A domain-specific decision support system for knowledge discovery using association and text mining. Knowl. Inf. Syst. **31**, 405–432 (2012)

21. Renu, R.S., Mocko, G., Koneru, A.: Use of big data and knowledge discovery to create data backbones for decision support systems. Procedia Comput. Sci. **20**, 446–453 (2013)

22. Rubiolo, M., Caliusco, M.L., Stegmayer, G., Coronel, M., Fabrizi, M.G.: Knowledge discovery through ontology matching: an approach based on an artificial neural network model. Inf. Sci. **194**, 107–119 (2012)

23. Ruy, F.B., Reginato, C.C., Santos, V.A., Falbo, R.A., Guizzardi, G.: Ontology engineering by combining ontology patterns. In: Johannesson, P., Lee, M.L., Liddle, S.W., Opdahl, A.L., López, Ó.P. (eds.) ER 2015. LNCS, vol. 9381, pp. 173–186. Springer, Cham (2015). https://doi.org/10.1007/978-3-319-25264-3_13

24. Suchanek, F.M., Kasneci, G., Weikum, G.: YAGO: a core of semantic knowledge unifying WordNet and Wikipedia. In: Proceedings of the 16th International Conference on World Wide Web, pp. 697–706 (2007)

25. SWRL: A Semantic Web Rule Language Combining OWL and RuleML. https://www.w3.org/Submission/SWRL. Accessed 20 Jan 2018

26. Wongthongtham, P., Pakdeetrakulwong, U., Marzooq, S.H.: Ontology annotation for software engineering project management in multisite distributed software development environments. In: Mahmood, Z. (ed.) Software Project Management for Distributed Computing, pp. 315–343. Springer, Heidelberg (2017). https://doi.org/10.1007/978-3-319-54325-3_13

27. Yarushkina, N., Filippov, A., Moshkin, V.: Development of the unified technological platform for constructing the domain knowledge base through the context analysis. In: Kravets, A., Shcherbakov, M., Kultsova, M., Groumpos, P. (eds.) CIT&DS 2017. CCIS, vol. 754, pp. 62–72. Springer, Heidelberg (2017). https://doi.org/10.1007/978-3-319-65551-2_5

28. Zarubin, A., Koval, A., Filippov, A., Moshkin, V.: Application of syntagmatic patterns to evaluate answers to open-ended questions. In: Kravets, A., Shcherbakov, M., Kultsova, M., Groumpos, P. (eds.) CIT&DS 2017, vol. 754, pp. 150–162. Springer, Heidelberg (2017). https://doi.org/10.1007/978-3-319-65551-2_11

29. Zedlitz, J., Jörke, J., Luttenberger, N.: From UML to OWL 2. In: Lukose, D., Ahmad, A.R., Suliman, A. (eds.) Proceedings of Knowledge Technology. CCIS, vol. 295, pp. 154–163. Springer, Heidelberg (2012). https://doi.org/10.1007/978-3-642-32826-8_16

Correction to: Analytical Realization of the EM Algorithm for Emission Positron Tomography

Robert Cierniak, Piotr Dobosz, Piotr Pluta, and Zbigniew Filutowicz

Correction to:
Chapter "Analytical Realization of the EM Algorithm for Emission Positron Tomography" in: L. Rutkowski et al. (Eds.): *Artificial Intelligence and Soft Computing*, **LNAI 10842,** **https://doi.org/10.1007/978-3-319-91262-2_12**

In the original version the name of the 4th author was incorrectly stated as "Piotr Filutowicz". It was corrected to "Zbigniew Filutowicz".

The updated version of this chapter can be found at
https://doi.org/10.1007/978-3-319-91262-2_12

© Springer International Publishing AG, part of Springer Nature 2018
L. Rutkowski et al. (Eds.): ICAISC 2018, LNAI 10842, p. C1, 2018.
https://doi.org/10.1007/978-3-319-91262-2_71

Author Index

Printed in the United States
By Bookmasters